Food Microbiology and Food Safety

Series Editor:
Michael P. Doyle

Food Microbiology and Food Safety Series

The Food Microbiology and Food Safety series is published in conjunction with the International Association for Food Protection, a non-profit association for food safety professionals. Dedicated to the life-long educational needs of its Members, IAFP provides an information network through its two scientific journals (Food Protection Trends and Journal of Food Protection), its educational Annual Meeting, international meetings and symposia, and interaction between food safety professionals.

For further volumes:
http://www.springer.com/series/7131

Fidel Toldrá • Leo M.L. Nollet
Editors

Proteomics in Foods

Principles and Applications

Springer

Editors
Fidel Toldrá
Department of Food Science
Instituto de Agroquimica y Tecnologia
de Alimentos (CSIC)
Valencia, Spain

Leo M.L. Nollet
Department of Applied Engineering Sciences
University College Ghent
Ghent, Belgium

ISBN 978-1-4614-5625-4 ISBN 978-1-4614-5626-1 (eBook)
DOI 10.1007/978-1-4614-5626-1
Springer New York Heidelberg Dordrecht London

Library of Congress Control Number: 2012954090

Printed on acid-free paper

Springer is part of Springer Science+Business Media (www.springer.com)

This book is dedicated to my father, Fidel Toldrá Graells, who suddenly passed away during the final preparation of this book. He was an enthusiastic and great supporter of my research activity. I will always remember him.

–Fidel Toldrá

Preface

Food proteomics constitute at this moment one of the most relevant and fast developing areas in food science. This is a very dynamic field, developing and changing very rapidly in recent years. This book provides readers with the recent advances and state of the art in food proteomics.

To achieve this goal, the book is divided into two approaches containing a total of 28 chapters. The first part deals with the principles of proteomics, especially focused to understand the proteome, the extraction and fractionation techniques for proteins and peptides, followed by separation techniques such as 2-D electrophoresis and chromatography and the mass spectrometry applications and search in protein databases. The second part is focused on the applications in foods. This part deals with quality issues related to post-mortem processes in animal foods and quality traits for a wide variety of foods such as meat, fish, dairy, eggs, wine, beer, cereals, fruits, and vegetables, but also for the identification of bioactive peptides and proteins which are very important from a nutritional point of view. Furthermore, consumers are now extremely susceptible towards food safety issues and proteomics can help in assuring different safety aspects including food authenticity, detection of animal species in the food, markers of pathogen microorganisms, or identification of prions. All these issues are considered in this book.

This book is written by 72 distinguished international contributors from 15 countries with solid experience and reputation, bringing together the existing knowledge with the current and future potential applications of proteomics in food science and technology. We wish to thank the production team at Springer and to express our gratitude to Susan Safren (Senior Editor) and Rita Beck (Assistant Editor) for their kind assistance in this book.

Valencia, Spain — Fidel Toldrá, Ph.D.
Ghent, Belgium — Leo M.L. Nollet, Ph.D.

Contents

Contributors

Francesco Addeo Istituto di Scienze dell'Alimentazione – Consiglio and Pasquale Ferranti Nazionale delle Ricerche (CNR), Avellino, Italy

Dipartimento di Scienza degli Alimenti, Università di Napoli "Federico II", Portici Napoli, Italy

Maria Filippa Addis Porto Conte Ricerche, Loc. Tramariglio, Alghero (SS), Italy

Lourdes Amigo Instituto de Investigación en Ciencias de la Alimentación (CIAL, CSIC-UAM, CEI UAM+CSIC), Madrid, Spain

Rudi Appels Centre for Comparative Genomics, Murdoch University, Western Australia, Australia

Keizo Arihara Department of Animal Science, Kitasato University, Towada-shi, Japan

M.-Concepción Aristoy Instituto de Agroquímica y Tecnología de Alimentos (CSIC), Paterna, Valencia, Spain

Emøke Bendixen Department of Molecular Biology and Genetics, Aarhus University, Aarhus, Denmark

Frank Békés FBFD PTY LTD, Beecroft, NSW, Australia

Matthew Bellgard Centre for Comparative Genomics, Murdoch University, Western Australia, Australia

Stephanie A. Booth Molecular Pathobiology, National Microbiology Laboratory, Public Health Agency of Canada, Winnipeg, MB, Canada

Department of Medical Microbiology, University of Manitoba, Winnipeg, MB, Canada

Peter M. Bramley School of Biological Sciences, Royal Holloway, University of London, Surrey, UK

Benito Cañas Complutense University of Madrid, Madrid, Spain

Catherine J. Carpenter Department of Chemistry and Biochemistry, University of California-Santa Barbara, Santa Barbara, CA, USA

Mónica Carrera Institute of Molecular Systems Biology (IMSB), ETH Zürich, Zürich, Switzerland

Christophe Chambon Plateforme d'exploration du métabolisme, des gènes aux métabolites, INRA Centre de Clermont Ferrand, Theix, France

Marie-Christine Champomier-Vergès INRA, UMR1319 Micalis, Jouy-en-Josas, France

AgroParisTech, UMR Micalis, Jouy-en-Josas, France

Thibault Chaze INRA, UR1213 Herbivores, Saint-Genès-Champanelle, France

Christophe Denoyelle Institut de l'Elevage, Paris, France

Aldo Di Luccia Department of SAFE, University of Foggia, Foggia, Italy

ConSDABI, Benevento, Italy

Elizabeth Escudero Instituto de Agroquímica y Tecnología de Alimentos (CSIC), Paterna (Valencia), Spain

Leo H. de Graaff Systems and Synthetic Biology, Wageningen University, Wageningen, The Netherlands

Michele Faccia Department of DIBCA, University of Bari, Bari, Italy

Pasquale Ferranti Istituto di Scienze dell'Alimentazione – Consiglio and Pasquale Ferranti Nazionale delle Ricerche (CNR), Avellino, Italy

Dipartimento di Scienza degli Alimenti, University of Naples "Federico II", Portici, Napoli, Italy

José M. Gallardo Marine Research Institute, CSIC, Vigo, Pontevedra, Spain

Dajana Gaso-Sokac Food Technology, J.J. Strossmayer University of Osijek, Osijek, Croatia

Department of Chemistry, J.J. Strossmayer University of Osijek, Osijek, Croatia

Jasminka Giacometti Department of Biotechnology, University of Rijeka, Rijeka, Croatia

Marzia Giribaldi National Research Council – Institute of Sciences of Food Production, Grugliasco, TO, Italy

José Ángel Gómez-Ruiz Instituto de Investigación en Ciencias de la Alimentación (CIAL, CSIC-UAM, CEI UAM+CSIC), Madrid, Spain

Blanca Hernández-Ledesma Instituto de Investigación en Ciencias de la Alimentación (CIAL, CSIC-UAM, CEI UAM+CSIC), Madrid, Spain

Jean-François Hocquette INRA, UR1213 Herbivores, Saint-Genès-Champanelle, France

Honggang Huang Department of Food Science, University of Copenhagen, Frederiksberg, Denmark

Rhiannon L.C.H. Huzarewich Viral Diseases, National Microbiology Laboratory, Public Health Agency of Canada, Winnipeg, MB, Canada

Takashi Iimure Bioresources Research and Development Department, Sapporo Breweries Ltd, Ota, Gunma, Japan

Caterina Incoronato ConSDABI, Benevento, Italy

Filomena Inglese ConSDABI, Benevento, Italy

Djuro Josić Department of Biotechnology, University of Rijeka, Rijeka, Croatia

Warren Alpert Medical School, Brown University, Providence, RI, USA

Angéla Juhász Agricultural Institute, Centre for Agricultural Research of the Hungarian Academy of Sciences, Martonvásár, Hungary

Laurent Journaux UNCEIA, MNE, Paris, France

Catherine Jurie INRA, UR1213 Herbivores, Saint-Genès-Champanelle, France

Setsuko Komatsu National Institute of Crop Science, Tsukuba, Japan

Spomenka Kovac Food Technology, J.J. Strossmayer University of Osijek, Osijek, Croatia

Department of Chemistry, J.J. Strossmayer University of Osijek, Osijek, Croatia

Marcel Kuntz Laboratoire de Physiologie Cellulaire Végétale (UMR5168), Université Joseph Fourier / CNRS / CEA / INRA, Grenoble, France

Carmela Lamacchia Department of SAFE, University of Foggia, Foggia, Italy

René Lametsch Department of Food Science, University of Copenhagen, Frederiksberg, Denmark

Sara Lamparelli Department of SAFE, University of Foggia, Foggia, Italy

Jacques Lepetit UR 370, Unité Qualité Produits Animaux, INRA, France

Iván López-Expósito Instituto de Investigación en Ciencias de la Alimentación, CIAL (CSIC-UAM), Madrid, Spain

Gianfranco Mamone Istituto di Scienze dell'Alimentazione (ISA) – CNR, Avellino, Italy

Jordi Mañes Departament de Medicina Preventiva i Salut Pública, Universitat de València, Burjassot, Spain

Iciar Martínez Department of Zoology and Cellular Biology, University of the Basque Country, Leioa, Spain

IKERBASQUE, Basque Foundation for Science, Bilbao, Spain

Daniel Martínez-Maqueda Instituto de Investigación en Ciencias de la Alimentación (CIAL, CSIC-UAM, CEI UAM+CSIC), Madrid, Spain

Gustavo Martos Instituto de Investigación en Ciencias de la Alimentación, CIAL (CSIC-UAM), Madrid, Spain

Donato Matassino ConSDABI, Benevento, Italy

Bruno Meunier INRA, UR1213 Herbivores, Saint-Genès-Champanelle, France

Beatriz Miralles Instituto de Investigación en Ciencias de la Alimentación (CIAL, CSIC-UAM, CEI UAM+CSIC), Madrid, Spain

Elena Molina Instituto de Investigación en Ciencias de la Alimentación, CIAL (CSIC-UAM), Madrid, Spain

Paula Moolhuijzen Centre for Comparative Genomics, Murdoch University, Western Australia, Australia

Leticia Mora Instituto de Agroquímica y Tecnología de Alimentos (CSIC), Paterna, Valencia, Spain

Chiara Nitride Dipartimento di Scienza degli Alimenti, University of Naples "Federico II", Portici I, Italy

Mohammad-Zaman Nouri Rice Research Institute of Iran-Deputy of Mazandaran, Amol, Iran

Mariaconsiglia Occidente ConSDABI, Benevento, Italy

Romina Pedreschi Food and Biobased Research, Wageningen, UR, The Netherlands

Brigitte Picard INRA, UR1213 Herbivores, Saint-Genès-Champanelle, France

Gianluca Picariello Istituto di Scienze dell'Alimentazione (ISA) – CNR, Avellino Italy

Carmen Piñeiro Institute for Marine Research (IIM), Spanish Council for Scientific Research (CSIC), Vigo, Spain

Gilles Renand INRA, UMR1313 Génétique Animale et Biologie Intégrative, Jouy-en-Josas, France

Agnès E. Ricroch Laboratoire Ecologie, Systématique et Evolution (UMR 8079), Université Paris-Sud / CNRS / AgroParisTech, Orsay, France

Sylvie Rousset Centre de Recherche en Nutrition Humaine Auvergne, INRA, France

Josep Rubert Departament de Medicina Preventiva i Salut Pública, Universitat de València, Burjassot, Spain

Françoise Rul INRA, UMR1319 Micalis, Jouy-en-Josas, France

AgroParisTech, UMR Micalis, Jouy-en-Josas, France

Peter J. Schaap Systems and Synthetic Biology, Wageningen University, Wageningen, The Netherlands

Philippe Simon Special Pathogens, National Microbiology Laboratory, Public Health Agency of Canada, Winnipeg, MB, Canada

Department of Medical Microbiology, University of Manitoba, Winnipeg, MB, Canada

Carla Soler Departament de Medicina Preventiva i Salut Pública, Universitat de València, Burjassot, Spain

Fidel Toldrá Instituto de Agroquímica y Tecnología de Alimentos (CSIC), Paterna, Valencia, Spain

Mark W. J. van Passel Systems and Synthetic Biology, Wageningen University, Wageningen, The Netherlands

Pingfang Yang Key Laboratory of Plant Germplasm Enhancement and Speciality Agriculture, Wuhan Botanical Garden, Chinese Academy of Science, Wuhan, China

Monique Zagorec INRA, UMR 1014 Secalim, 44307 Nantes, France

LUNAM Université, Oniris, Nantes, 44307, France

Qiang Zhang Mead Johnson Nutrition Pediatric Nutrition Institute, Evansville, IN, USA

Part I
Principles of Proteomics

Chapter 1
Understanding the Proteome

Emøke Bendixen

1.1 Introduction

The term "proteome" was suggested by Wilkins et al., in 1996, to meet the need for a new terminology that could reflect the growing interest in studying the entire protein complement expressed by a genome (Wilkins et al. 1996). This new term was presented in the shadows of the race for mapping the human genome, and may be seen as a reflection of a new post-genome way of understanding and approaching not just protein sciences but indeed all descriptive biological sciences. The post-genomic era presented new opportunities for developing better tools for investigating the molecular mechanisms of biological systems, and thereby led the way to discovery-based approaches that promised to give a more accurate picture of the impact of specific genes and proteins on health and disease. During the late nineties the biomedical research communities had adapted the idea of using the global "omics" technologies including genomics, proteomics, transcriptomics, and metabolomics, and these global approaches became preferred over the more reductionist approaches used in classic biochemistry and molecular biology. From the genome sequencing projects we learned that the human as well as most mammalian genomes include approximately 20,000 genes, meaning in principle that mammalian proteomes consist of 20,000 proteins. This is in striking contrast to the fact that proteome methods typically have the analytical power to display at best between 500 and 5,000 proteins in a single analytical comparison. Moreover, it also rapidly became clear that proteomes are far more complex than the genomes they reflect,

E. Bendixen (✉)
Department of Molecular Biology and Genetics, Aarhus University,
Aarhus, Denmark
e-mail: emoke.bendixen@agrsci.dk

F. Toldrá and L.M.L. Nollet (eds.), *Proteomics in Foods: Principles and Applications*,
Food Microbiology and Food Safety 2, DOI 10.1007/978-1-4614-5626-1_1,

mainly caused by post-translation modification (PTM) of proteins and by expression of alternative gene variants. This knowledge has over the past decades created awareness that specific and alternative methods for investigating parallel subsets of the proteome must be developed. The past decades have indeed seen the development of diverse and specific technologies and approaches that allow increasingly larger fractions of the proteome to be investigated, and although it is clear that we are still far from realizing the original aim to analyze "the complete protein output of a genome," the technological pace allows us to study increasingly larger subsets of proteomes, on increasingly accurate levels of detail and with rapidly increasing speed. Hence, proteomics have within the past three decades become a powerful approach to studying the role of proteins in all areas of biological sciences, including human medicine, animal sciences, veterinary medicine, and not the least within food sciences, which is the topic of this book. The aim of this introductory chapter is to present some basic concepts and principles of proteomics, and to provide an overview of the most common technologies and experimental approaches used to characterize the roles of proteins in food and nutritional sciences.

Chapter 2 reflects on the complexity of proteomes, and provides examples of how this relates to understanding food proteomes . Chapter 3 reviews the progress of technological developments, and introduces some of the most commonly used methods in food proteomics, and Chap. 4 provides a window to the future, discussing some aspects of food production where understanding proteomes is expected to be essential for production, quality, and safety of human food in the future.

1.2 How Complex Is the Proteome?

According to the very definition of the term, and according to common perception, the proteome is a reflection of the genome from which it is translated. However, a proteome is indeed a lot more complex than the genome it reflects, in molecular variation as well as regarding the diversity of technologies and methods needed for its complete characterization. Even before genome sequencing revealed to a great extent the complexity of genomes and proteomes, it was clear that the proteome is not just a parallel output of one protein per gene. On top of the complexity caused by the large diversity and variation in protein structures, added levels of complexity originate from the fact that although the genome of any individual is a constant feature, present in identical copies in every cell of an organism, and in principle remaining unchanged during its lifetime, the proteome is a dynamic entity. Unique subsets of proteomes are differentially expressed in cellular space and time and by different cells and tissue types. Cells particularly express the specific subset of proteins needed for their specific functions in any specific tissue type, and the quantities and patterns of expression fluctuate in order to meet the challenges imposed on cells and tissues within the organism. In this sense, any proteome analysis should be thought of as a snapshot of the proteome of a specific cell type from a given organism at a given time, and not as a diagnostic blueprint of the protein status of any specific condition or disease under investigation. Understanding the many factors that cause these levels of complexity and variation is a crucial part of characterizing and understanding proteomes. This chapter aims to introduce some important factors that contribute to the vast variation in forms, functions, and expression levels of proteins, and provide some examples of how these alternative protein forms influence the production and quality of food.

1.2.1 Post-Translational Modification (PTM)

PTM of proteins and peptides covers a wide range of covalent molecular changes imposed on the protein and peptide sequences. In particular, side chains of cysteine, asparagine, serine, threonine, tyrosine, and lysine can be modified by the addition of a wide range of molecules, most commonly including carbohydrates, lipids, phosphate, methyl, or nitrate groups. PTM is a major contributor to the diversity of protein structures. The fact that many proteins can harbor combinations of several types of modifications gives rise to a seemingly endless diversity of protein structures. Moreover, the modification of amino acid side chains is often associated with the introduction of altered functions of the protein, and certain types of modifications act as dynamic switches to modulate activation and deactivation of specific protein functions. Our comprehensive knowledge and annotation of the many different possible modification forms that proteins appear in is still far from complete, as is our understanding of the diversity of functional roles these specific modifications may play. An estimate of the diversity and frequency of PTMs was recently presented by Khoury et al. (2011). As of this writing, a subset of 431 different types of modifications were annotated in this database, and their relative abundances were estimated statistically, suggesting that glycosylation affects approximately 25% of all proteins and that phosphorylation may be an even more common event. A frequently updated count of PTM forms is available at http://selene.princeton.edu/PTMCuration.

The apparent importance of PTMs for controlling and modulation functions of proteins has resulted in much effort to develop specific methods that can be used for their further characterization. The following text provides a short introduction to the most prominent PTM forms, and presents a few examples as to how these variants may influence protein functions in food.

1.2.1.1 Glycosylation

Glycosylation is one of the most common forms of modification, affecting virtually all secreted proteins and membrane proteins whereas glycosylation of cytoplasmic and nuclear proteins has been observed, but to a much less frequent extent. Complex carbohydrates are assembled and appended by the concerted action of substrate-specific glycosyl transferases during synthesis and processing of proteins in RER and Golgi. Correct glycosylation is important for protein stability, cell–cell interactions, cell–protein interactions, and protein–protein interactions in live tissues (Spiro 2002) and is well known to play an important role in the solubility and structure of proteins (D'Ambrosio et al. 2008). It is therefore highly relevant for the production and processing of proteins for human consumption and particularly so for milk proteins. The vast abundance of milk proteins is indeed actively secreted by mammary gland cells, and therefore extensively glycosylated (Spiro 2002). Milk is a body fluid with a uniquely high concentration of proteins, designed for nourishing the newborn mammal, and the extensive glycosylation of milk proteins is likely to be important for supporting their solubility in this very dense body fluid. Caseins account for 80% of the entire load of milk proteins and are the main constituent of cheese. The glycosylation patterns of caseins play an important role in the stability and solubility of caseins during cheese production, and have therefore been studied in great detail in order to investigate their potentially functional roles in the production of cheese and yogurt (Robitaille et al. 1993; Kreuss et al. 2009). These studies are discussed in greater detail in Chaps. 16 and 17 which are dedicated to dairy production. Differential glycosylation of milk proteins may also influence allergenic properties of milk and dairy products (D'Alessandro and Zolla 2012) and influence the bioactive properties of proteins and peptides in milk (Panchaud et al. 2012).

Characterization of protein glycosylation is greatly complicated by the fact that glycan chains are widely heterogenic and glycosylation dramatically changes size, charge, and hydrophobicity of proteins, often causing unknown glycosylated forms to escape detection by standardized methods. Therefore enrichment steps combined with specific mass spectrometric methods are needed to characterize these biologically and functionally important protein isoforms from complex food matrices such as milk and cheese (Hernandez-Hernandez et al. 2010).

1.2.1.2 Phosphorylation

Phosphorylation of serine, threonine, and tyrosine side chains is also a very common form for modification. Phosphorylation of specific amino acid positions is tightly controlled by kinase and phosphatase proteins that mediate the addition (by kinases) and removal (by phosphatases) of phosphate groups. Phosphorylation provides a dynamic switch between activated and deactivated forms of many proteins, including enzymes in metabolism and in cellular signaling pathways in all eukaryotes (Larsen et al. 2006). Hence understanding the roles of specific phosphorylation events of specific proteins plays a crucial role in understanding proteomes of all crops and farm animal species important for the primary production of human foods. Phosphorylation of proteins can be challenging to analyze, partly because the modified sites are dynamically altered, and also because the addition of phosphate groups changes both charge and molecular mass of proteins and peptides, often causing phosphorylated proteins and peptides to escape analysis in LC-MS/MS-based methods. On the other hand, differentially phosphorylated isoforms of a protein may be readily distinguished using 2DE gels. Methods for capturing phosphorylated proteins and peptides through affinity purification have been developed (Carapito et al. 2009), (Larsen et al. 2005), and phospho-proteome mapping has been presented from a wide range of species and tissues including milk, meat, wheat, rice, and maize (Mamone et al. 2009).

Examples of the importance of phosphorylation in food proteins include their relevance to muscle growth, where phosphorylation is indeed a key regulator of metabolism, growth, and development and diversity of muscles, hence their characterization has been studied in order to understand their impact on growth and yield of muscles (Wang et al. 2012), and on the roles phosphorylation may play in post-mortem metabolism and meat quality (Lametsch et al. 2011; Huang et al. 2011) Also within milk and cheese production phosphorylation plays an important role, both for yield and quality of cheese and for the allergenic properties of milk and dairy products (Cases et al. 2011). Within crops, phosphorylation of proteins related to growth and stress in rice have been characterized (Mamone et al. 2009).

1.2.1.3 Controlled Degradation

Controlled degradation of proteins is a frequent biological event, and an important mechanism in controlling protein activity both in living organisms, and also during industrial processing of foods. Full-length proteins often support very different functions from those of its specific peptides, hence proteome methods sufficient for accurate characterization of exact length, sequence, and potential cleavage sites in proteins and peptides are key to understanding functional properties of proteins. Proteolysis is well known to play fundamental roles for developing optimal flavor, texture, and yield during storage and ripening of meat and cheese products (Sentandreu and Sentandreu 2011; Jardin et al. 2012; (Cunsolo et al. 2011) and for unleashing peptides with specific bioactive roles, primarily known from milk (Panchaud et al. 2012). The importance of analyzing the proteome at individual peptide levels is highlighted by the fact that many bioactive roles of proteins are activated only when specific peptides are released. Although much focus has already been given to studying bioactive peptides in milk, the exact mapping of specific cleavage sites and the extent to which these are processed within the complex protein matrices of meat and milk indeed remain challenging tasks (Panchaud et al. 2012). Important examples of the role of proteolysis for food quality include the production of cheese and yogurt, where the specific cleavage of K-casein initiates the formation of the cheese curd, hence properties interfering with control of cleavage sites greatly affects yield in cheese production (Leiber et al. 2005).

In meat the extent of post-mortem proteolysis of muscle proteins is believed to play an important role for development of meat tenderization, thus they may be used as quality markers during industrial processing of meat (Pomponio et al. 2008). This area is the topic of Chap. 10 where potential markers of tenderness in three beef breeds are discussed.

1.2.1.4 How PTM Interferes with Detection of Proteomes

PTM introduces changes in molecular weight as well as charge and hydrophobicity of proteins, and therefore interferes with most separation and detection methods. Hence PTM-mediated diversity is a common cause of considerable error in the interpretation of protein abundance data. Because modified proteins often escape detection by standardized methods, detection of PTM–protein isoforms typically requires specific adjustments of the purification and detection steps, or even new methods to be developed in order to characterize specific PTM variants of proteins, as reviewed by Carapito et al. (2009).

1.2.2 Homologous Proteins

Polymorphic genes and mutations, as well as alternative RNA processing, are also contributors to variation in the structure and function of proteins. All these events result in expression of protein isoforms that share long stretches of identical sequences, and therefore may not be possible to distinguish by common proteome methods, although these isoforms may indeed have very different functional properties. One important example includes the many parallel gene variants that code for gluten proteins in wheat (Muccilli et al. 2011). Because the different variants have very different impacts on baking quality, much effort has been made to distinguish and characterize the many protein isoforms in great detail by proteomic methods (Mamone et al. 2009). Also major muscle proteins such as troponin and myosin are differentially expressed in many variable but homologous isoforms to provide variation in form and function of different muscle types. The different muscle types have a great deal of variation in metabolic rates and growth phenotypes that eventually influences meat quality properties (Rehfeldt et al. 2011). Also the major proteins of milk, the caseins, appear in a variety of allelic and spliced isoforms (Ferranti et al. 1997), which may influence their properties related to cheese production and to allergenic properties of dairy products. Distinguishing homologous protein isoforms is a challenging task, because full sequence coverage is very rarely obtained in proteome studies. For the analyses of these protein families, it is necessary to cover the few unique sequences that distinguish the different isoforms, and this can in many cases only be achieved by targeted SRM (selected reaction monitoring) -based methods, (Calvo et al. 2011) which is a relatively new approach to mass spectral proteomics, and is discussed below.

1.2.3 Functional Protein Complexes

The ultimate aim of characterizing proteomes is to understand the functions of proteins in living organisms. However, the specific function of a protein often depends on its interaction with other proteins. Hence understanding protein function includes characterizing the functional consortiums (interactomes) in which proteins operate in vivo. This adds an extra level of complexity to proteome studies, as in principle a quantitative analysis of global proteome patterns cannot directly assign the role of any given specific protein just by estimating its abundance or differential expression, unless the stoichiometry of interacting proteins can be determined. Approaching this level of complexity requires isolation of the entire functional protein complexes, as seen in recent successful examples of characterizing proteomes of multimeric protein complexes (Przybylski et al. 2010).

1.3 Observing the Proteome: Developing Methods and Increasing Understanding of Proteomes

As discussed above, we are far from understanding the extent of variation of forms that proteins can appear in, and we still lack methods that can provide analyses of complete proteomes in a single analytical event, but the speed, accuracy, and level of detail with which proteomes are currently being characterized have taken a quantum leap over the past decades. For an excellent review of technological developments readers are referred to Malmstrom et al. (2007). Proteomics is a technology-driven research field in the sense that the interplay of developing better methods and achieving increased understanding of proteomes is the underlying dynamic duo that drives progress in this field of research. Achieving exact knowledge of the many forms and variations proteins may take is important for enhancing the methods needed for characterizing them. This chapter discusses the advances and progress in proteome research as they have unfolded over the past three decades. The most common technologies are introduced and discussed in the context of how these technological advances have influenced the shifting trends and approaches within proteome characterizations. Figure 1.1 summarizes this progression and illustrates how technological progress and scientific focus have worked in parallel to support our understanding of proteomes in ever-increasing detail.

1.3.1 Developing Proteomics Technology

The proteome of any biological sample is a complex matrix typically containing thousands of different protein molecules, each expressed in the adequate copy numbers needed to sustain their specific functions. Hence, characterizing proteomes includes analyzing each protein in its form and abundance within a complex protein matrix. As discussed in the previous section, any proteome appears in a near-endless variation, and proteomics is a technology-driven area of research where analyzing complete proteomes still remains to be realized. All proteome analyses rely, however,

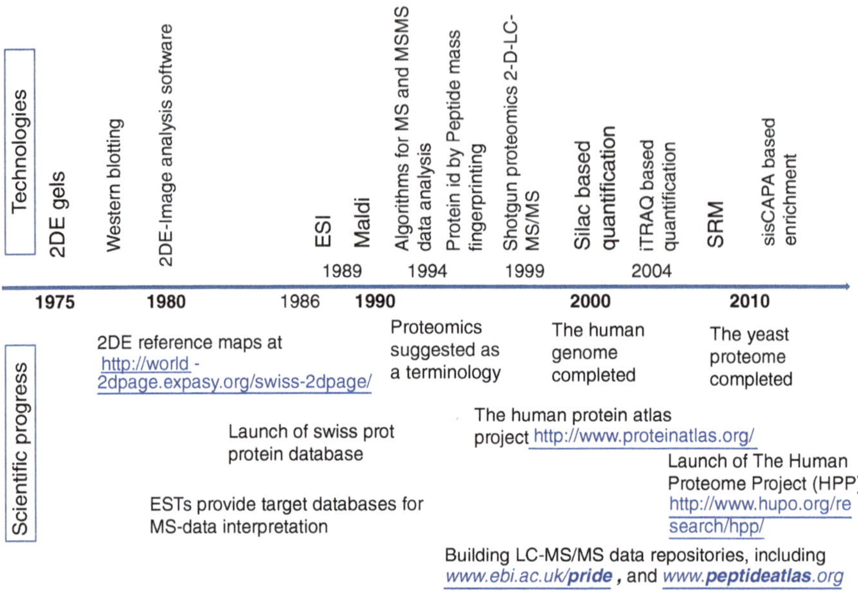

Fig. 1.1 *Timeline of technological advance and scientific progress in proteomics.* This timeline summarizes in parallel some key technological advances, listed above the line, and major scientific achievements closely associated with progress in proteomics listed under the line. Many of the technologies and methods that paved the way for proteome characterizations were already developed decades before the concept of proteomics was suggested in 1994. Abbreviations: *ESI* electrospray ionisation, *MALDI* Matrix Assisted Laser Desorption Ionization, *SRM* Selected Reaction Monitoring

on the basic steps of extraction, separation, molecular characterization, and quantification of each protein, hence integrating many different types and levels of data is needed in order to extract information from these individual but closely correlated analytical methods. Discussing these methods in detail is the aim of individual Chaps. 2, 3, 4, 5, 6, and 7 of this book. Here these methods are introduced with the aim of providing an overview of how the most important technological advances have influenced experimental approaches and scientific progress over the past three decades.

1.3.2 Two-Dimentional Electrophoresis (2DE): A First Approach to Proteome Mapping and Biomarker Discovery

Although biochemical methods have provided much knowledge of the structure and function of individual proteins, the invention of SDS-PAGE for separating proteins according to their molecular weight was indeed a major breakthrough in protein sciences, and offered for the first time a display method allowing many proteins to be analyzed by a single analytical event within complex biological samples (Laemmli 1970). It was, however, the introduction of 2DE that truly raised awareness that for understanding biological mechanisms, proteomes rather than proteins should be investigated, and that this could be achieved through display methods (Gorg et al. 2000). By adding a separation step based on isoelectric mobility prior to the SDS-PAGE separation, 2DE offered a display method that supported investigation of hundreds and even thousands of individual proteins in parallel from a single biological sample, with 2DE providing a separating matrix from which protein spots could be identified by western blotting, and later by mass spectrometry (Katayama et al. 2001; Jensen et al. 1998). 2DE technology was rapidly implemented in all areas of life sciences, first within human medicine, and with little delay to food and agricultural sciences (Lametsch et al. 2003). 2DE-based mapping of proteomes provided a first approach to discovery-based mapping of the concerted actions of proteins and pathways, and seemed promising for providing biomarkers that could prove useful in diagnostics as well as in understanding the functional roles of proteins in food and other biological samples. This first enthusiasm gave rise to many publications, regarding both mapping individual protein cells, tissues, and body fluids and to numerous biomarker studies, and indeed also served to reveal in more exact detail the complexity of proteomes. However, during two decades of 2DE-based proteomics it became increasingly clear that the extent of variation, and particularly the very broad range of protein copy numbers spanning from a few to several million copies per cell were incompatible with the relatively narrow analytical power of 2DE (Pedersen et al. 2003). It was clear that parallel methods with stronger analytical power were needed in order to map entire proteomes. By the turn of the century the mass spectrometry-based proteomics had become a new and promising alternative to the gel-based approaches (Haynes and Yates 2000), but 2DE has nevertheless remained a cornerstone in proteomics, partly because it still provides unmatched separation of full-length proteins and of proteins with post-translational modifications such as glycosylation and phosphorylation, and this classical method should be regarded as an important tool in the large toolbox needed for understanding entire proteomes, as well as for future proteome studies.

1.3.3 Mass Spectrometry-Based Proteomics: The Second Generation of Global Proteomics

Mass spectrometry (MS) became applicable for analyzing proteins and peptides with the invention of electrospray ionization (ESI) (Fenn 2003; Fenn et al. 1989) and matrix-assisted laser desorption ionization (MALDI) (Hillenkamp et al. 1991) in the late eighties. These are both soft ionization techniques that allowed molecules with the mass range of tryptic peptides and small proteins to be analyzed by mass spectrometry. For an excellent review of MS-based proteomics see Domon and Aebersold (2006). At the same time EST and genome sequencing projects started to provide access to well-covered protein sequence databases from a wide range of species, which greatly reduced the complexity of interpreting mass spectral data, as these could be directly aligned to global sequence databases (Nesvizhskii et al. 2003). MALDI–TOF methods provided a robust, fast, and affordable identification of proteins from 2DE gels by peptide mass fingerprinting (Jensen et al. 1998) but could not be interfaced easily to liquid chromatography (LC) separation. ESI on the other hand was easily coupled to LC-based separation systems, and in combination with tandem mass spectrometers became the cornerstone of a next generation of global proteome display technology that, just as with the 2DE-based approaches, aimed to provide discovery-based comparisons of complex proteomes, partly for investigating the protein diversity of biological samples, but also for discovering biomarkers for a wide range of applications within the medical and food sciences (Domon and Aebersold 2006). LC-MS/MS-based methods offered far more reproducible, informative, and sensitive analyses of complex proteomes than 2DE, and moreover provided peptide sequence information. Quantification could be made by multiplexed studies using differential heavy-isotope labeling or tagging of comparable samples, or by adding isotope-labeled standard peptides. This second generation of global and discovery-based proteomics renewed the enthusiasm for mapping entire proteomes, and indeed, the almost complete proteomes of small organisms such as yeast have been achieved (Nagaraj et al. 2012) and also a far more detailed mapping of the proteome parts-lists of biological samples was covered by LC-MS/MS. However, as with 2DE technology a decade earlier, it became clear that proteomes are indeed more complex and variable than what can be fully covered by any single global mapping technology known so far (Meyer et al. 2011). Nevertheless, data created through a decade of LC-MS/MS-based proteomics laid the foundation for developing targeted proteome methods. This technological approach can be seen as a reversal from the discovery-based to a more hypothesis-based proteome approach, and may be regarded as the next upcoming generation of global proteomics, and is discussed in the following.

1.3.4 *Targeted Proteomics: The Return to Hypothesis-Based Research*

As discussed above, three decades of proteome sciences have greatly improved our knowledge about the relative abundance and functional roles of proteins, in medicine as well as within food sciences, but have also provided the insight that global proteome approaches alone are not sufficient for understanding proteomes in their full complexity (Meyer et al. 2011). This insight has caused the move towards more targeted proteomics approaches, where specific subsets rather than entire proteomes are investigated using methods specifically adapted to that subset of proteins. Being hypothesis-based rather than discovery-based, targeted proteomics differ essentially from global proteomics in that they make use of a priori information about the proteins to be targeted, and this information to a wide extent is the result of the vast amount of data created from three decades of discovery-based proteomics (Pan et al. 2009). The currently favored method is termed selected reaction monitoring (SRM), and relies on using triple-quadrupole mass spectrometers in a highly selective mode, detecting only a few unique peptides from each specific targeted protein and only in selected time and mass windows during the instrument runtime (Calvo et al. 2011). SRM has proven successful in detecting specific proteins from complex samples over the entire dynamic range of protein abundance in yeast, down to as low as subfemtomolar levels, or corresponding to 50 copy numbers of a protein within a yeast cell (Picotti et al. 2009). SRM hereby offers a very promising approach for verifying potential biomarker candidates that have emerged from discovery-based studies (Huttenhain et al. 2009). SRM-based proteomics can furthermore be combined with affinity-based methods that allow pre-fractionating subproteomes to reduce the complexity of the biological sample considerably. This has proved particularly useful for characterization of PTM proteins (Chiou et al. 2010). Another promising approach is provided by antibody-based sorting (Whiteaker et al. 2010), and is particularly interesting in the light of recent progress in developing large-scale production of protein- and peptide-specific antibodies for all individual human proteins recognized so far. This effort is undertaken by the Human Protein Atlas project (http://www.proteinatlas.org/) and the Human Proteome Initiative (Uhlen et al. 2012; Paik et al. 2012). These approaches even seem to promise to mature into global-scale proteome mapping approaches, if and when thousands of proteins can be targeted in a display-like manner (Gillet et al. 2012).

1.4 The Importance of Understanding the Proteomes of Food

Understanding the biology, and also the proteomes of the many animal, plant, and bacterial species involved in the production of food is of key importance for human life.

On the other hand, understanding the structure and function of proteins present in the individual raw food materials and in the end products is essential for optimizing food-processing methods because proteins play important functional roles for texture, yield, flavor, and nutritional value of virtually all food products. In western societies, milk and meat are by far the most important sources of dietary proteins, hence farm animals have since prehistoric times been selectively bred for increasingly high yields of meat and milk. From the perspectives of food production, farm animals may be seen as species optimized to produce proteins, and the dairy and the meat-processing industries have indeed based their technologies on creating a variety of protein-based food products through directly modulating meat and milk proteomes at industrial scales. Moreover, protein-based foods are high-cost products, hence optimizing outcome and reducing waste during these protein-altering processes needed for turning milk into cheese and muscles into a variety of meat products has always been of great importance for success in these industries.

As such, it is not surprising that characterizing meat and milk proteomes has already gained so much attention (Bendixen et al. 2011). This book presents individual chapters dedicated to reviewing the many specific fields and applications of food proteomics. Here, the following sections serve to provide a window to the future by highlighting selected areas of food research where the understanding of proteomes is key to solving challenges in future food production.

1.4.1 Food Production

The key challenge of global food production is to supply enough food for a growing population, and under rapidly accelerating climate changes. The production of proteins of animal origin is limited due to large economic and environmental resources required for their production. Proteome studies have already been applied to investigate traits related to milk production, muscle growth, and feed conversion in most farm animals, but future studies that may help to further improve alternative production forms of meat without causing unacceptable harm to the environment and animal welfare are in great demand. Also improving yield and quality of crops is important, partly because of the more cost-efficient alternatives presented by plants as protein resources. More intensive production of both animal- and plant-based proteins can be achieved through understanding molecular mechanisms related to differences in metabolism of the different species, strains, and cultivars, hence the integration of proteome, genome, and metabolomics studies is widely expected to be a key to continuously enhance global food production. In particular, traits related to enhanced resistance to pathogens and to environmental factors including climate change and pollution are important to understand at the molecular level, and will greatly increase efficiency and quality in food production.

1.4.2 Understanding the Interplay of Food and Health

Foods that are important sources of bioactive components have gained much attention over the past decade because it is widely believed that foods may have preventive roles in human health (Senechal and Kussmann 2011). In some instances, even healing effects are anticipated (Thapa 2005). Nevertheless, our knowledge of the bioactive roles of foods is still very limited, and often scientific evidence as to their healing and health-improving properties remains to be validated. Bioactive proteins and peptides in milk have attracted particular attention, and have already been mapped in great detail by several research groups (Panchaud et al. 2012). Milk is evolutionary adapted food, rich in bioactive components and designed to nourish the neonate mammal. The importance of tightly controlled expression of the bioactive components in milk is clearly illustrated in the dramatic shifts in the proteomes of colostrum and milk over the timeline from birth to weaning (Hansson et al. 2011). The milk proteome is finely tuned to maintain the biological events of the specific developmental steps of early infancy. Mapping time-dependent fluctuations of bioactive components is an important field of research that still needs to be mapped in closer detail, and which will surely provide knowledge of growth, development, and nutritional needs of humans, as well as providing important knowledge for optimizing health in farm animals.

Plants are also rich in molecules with bioactive roles and of great importance for human health. Plant metabolites are relevant both for their medical as well as for their beneficial roles in human nutrition (Esfahani et al. 2011; Bach-Faig et al. 2011). Moreover, plant metabolites and proteins may have toxic and allergenic properties. Proteome studies are important for characterizing these properties in relation to human health, and are likely to play a key role in uncovering the molecular machinery needed for their biosynthesis. This knowledge is needed in order to cultivate plants with better and higher yields of preferred and beneficial metabolites while avoiding production of plants with inappropriate levels of toxic or allergenic properties. Finally, within nutrigenomics, the field of research aimed at understanding how the interplay of genes and nutrients acts on human health, proteomics will surely continue to be a key technology, providing knowledge of how structures, modifications, and genetic variants of proteins interact with human health and biology.

1.4.3 Food Safety

Proteomics are expected to play a pivotal role in a wide range of aspects related to food safety, as recently reviewed by D'Alessandro and Zolla (2012). These include characterizing food allergens, which very often are protein-based. Protein-based allergens may be eliminated or modulated by food-processing technologies if knowledge of their nature is available at molecular levels; for example, the structural differences between allergenic and nonallergenic isoforms are characterized. Also regarding adulteration of food, such as mixing cheaper products into more expensive ones as seen in mixing buffalo milk into cow milk and the cheaper chicken meat into beef and pork meat products can be detected by mass spectrometry-based methods (D'Alessandro and Zolla 2012). Other examples include documenting origin of foods, and contamination with food pathogens, which will likely be increasingly feasible with the development of more sophisticated and targeted mass spectrometry methods.

1.4.4 Food Quality and Food Technology

The importance of understanding proteomes in order to understand quality and processing of food is very clear, and is discussed throughout many of the separate chapters in this book. Here it should just be mentioned that although a vast abundance of meat, milk, cheese, and grain proteomes have already been studied in great details as to their quality traits (Mamone et al. 2009), it is clear that there is still even more knowledge that remains to be uncovered, and which can be applied to improve quality of food. The variables involved in controlling food quality are indeed many and complexly interacting, including biological factors such as genetics, age, and growth conditions, just to mention a few. On the other hand, industrial food processing is to a very large extent relying on processes that modify proteins, as when tenderizing, curing, fermenting, and heat-treating meat, and when precipitating milk proteins to produce cheese, and when fermenting grains for beer, wine, and baked goods. When considering the complexity of proteomes as discussed above, it is clear that future progress in understanding food proteomes will highly depend on the variety of technologies available, and on the dynamic tailoring of methods needed for investigating individual aims, biological samples, pathways, and selective subsets of food proteomes.

1.5 Concluding Remarks

Understanding production and quality of protein-based foods such as meat, milk, eggs, and beans greatly depends on our ability to characterize the biological processes in the animals and plants used for the production of these raw materials. One important aspect of understanding proteomics is to bear in mind that this relatively young field of research is still technology-driven in the sense that no single display method can fully cover the entire variation of proteomes. The sixteenth-century understanding of science as vocalized by Galileo Galilei, "to measure what is measurable and make measurable what cannot be measured," summarizes the current state of proteomics sciences very precisely.

The aim of all proteomics studies is to understand the roles of proteins and their concerted actions in controlling the intricate molecular systems of living organisms. As we have seen, all proteomes are far more complex than the genomes they reflect. Understanding proteomes of foods and food production primarily relies on the ability to choose, combine, and optimize the very large collection of tools and methods within proteomics, and to integrate these methods to achieve sufficient in-depth understanding of the roles of individual proteins for the production, quality, and safety of food.

The aim of this book is to present the many technological opportunities available, as well as the many different aspects and challenges within food proteomics.

Although it is clear that no single universal method can yet provide an in-depth understanding of the near-endless complexity of food proteomes, the available methods afford a rapidly increasing speed and accuracy by which these proteomes can be characterized.

References

Bach-Faig A, Berry EM, Lairon D, Reguant J, Trichopoulou A, Dernini S, Medina FX, Battino M, Belahsen R, Miranda G, Serra-Majem L (2011) Mediterranean diet pyramid today. Science and cultural updates. Public Health Nutr 14:2274–2284

Bendixen E, Danielsen M, Hollung K, Gianazza E, Miller I (2011) Farm animal proteomics – a review. J Proteomics 74:282–293

Calvo E, Camafeita E, Fernandez-Gutierrez B, Lopez JA (2011) Applying selected reaction monitoring to targeted proteomics. Expert Rev Proteomics 8:165–173

Carapito C, Klemm C, Aebersold R, Domon B (2009) Systematic LC-MS analysis of labile post-translational modifications in complex mixtures. J Proteome Res 8:2608–2614

Cases B, Garcia-Ara C, Boyano MT, Perez-Gordo M, Pedrosa M, Vivanco F, Quirce S, Pastor-Vargas C (2011) Phosphorylation reduces the allergenicity of cow casein in children with selective allergy to goat and sheep milk. J Investig Allergol Clin Immunol 21:398–400

Chiou SH, Huang CH, Lee IL, Wang YT, Liu NY, Tsay YG, Chen YJ (2010) Identification of in vivo phosphorylation sites of lens proteins from porcine eye lenses by a gel-free phosphoproteomics approach. Mol Vis 16:294–302

Cunsolo V, Muccilli V, Saletti R, Foti S (2011) Review: applications of mass spectrometry techniques in the investigation of milk proteome. Eur J Mass Spectrom (Chichester, Eng) 17:305–320

D'Alessandro A, Zolla L (2012) We are what we eat: food safety and proteomics. J Proteome Res 11:26–36

D'Ambrosio C, Arena S, Salzano AM, Renzone G, Ledda L, Scaloni A (2008) A proteomic characterization of water buffalo milk fractions describing PTM of major species and the identification of minor components involved in nutrient delivery and defense against pathogens. Proteomics 8:3657–3666

Domon B, Aebersold R (2006) Mass spectrometry and protein analysis. Science 312:212–217

Esfahani A, Wong JM, Truan J, Villa CR, Mirrahimi A, Srichaikul K, Kendall CW (2011) Health effects of mixed fruit and vegetable concentrates: a systematic review of the clinical interventions. J Am Coll Nutr 30:285–294

Fenn JB (2003) Electrospray wings for molecular elephants (Nobel lecture). Angew Chem Int Ed Engl 42:3871–3894

Fenn JB, Mann M, Meng CK, Wong SF, Whitehouse CM (1989) Electrospray ionization for mass spectrometry of large biomolecules. Science 246:64–71

Ferranti P, Addeo F, Malorni A, Chianese L, Leroux C, Martin P (1997) Differential splicing of pre-messenger RNA produces multiple forms of mature caprine alpha(s1)-casein. Eur J Biochem 249:1–7

GilletLC, Navarro P, Tate S, Roest H, Selevsek N, Reiter L, Bonner R, Aebersold R (2012) Targeted data extraction of the MS/MS spectra generated by data independent acquisition: a new concept for consistent and accurate proteome analysis. Mol Cell Proteomics 6:1–17

Gorg A, Obermaier C, Boguth G, Harder A, Scheibe B, Wildgruber R, Weiss W (2000) The current state of two-dimensional electrophoresis with immobilized pH gradients. Electrophoresis 21:1037–1053

Hansson J, Panchaud A, Favre L, Bosco N, Mansourian R, Benyacoub J, Blum S, Jensen ON, Kussmann M (2011) Time-resolved quantitative proteome analysis of in vivo intestinal development. Mol Cell Proteomics 10:M110

Haynes PA, Yates JR (2000) Proteome profiling – pitfalls and progress. Yeast 17:81–87

Hernandez-Hernandez O, Lebron-Aguilar R, Quintanilla-Lopez JE, Sanz ML, Moreno FJ (2010) Development of a new method using HILIC-tandem mass spectrometry for the characterization of O-sialoglycopeptides from proteolytically digested caseinomacropeptide. Proteomics 10:3699–3711

Hillenkamp F, Karas M, Beavis RC, Chait BT (1991) Matrix-assisted laser desorption/ionization mass spectrometry of biopolymers. Anal Chem 63:1193A–1203A

Huang H, Larsen MR, Karlsson AH, Pomponio L, Costa LN, Lametsch R (2011) Gel-based phosphoproteomics analysis of sarcoplasmic proteins in postmortem porcine muscle with pH decline rate and time differences. Proteomics 11:4063–4076

Huttenhain R, Malmstrom J, Picotti P, Aebersold R (2009) Perspectives of targeted mass spectrometry for protein biomarker verification. Curr Opin Chem Biol 13:518–525

Jardin J, Molle D, Piot M, Lortal S, Gagnaire V (2012) Quantitative proteomic analysis of bacterial enzymes released in cheese during ripening. Int J Food Microbiol 155:19–28

Jensen ON, Larsen MR, Roepstorff P (1998) Mass spectrometric identification and microcharacterization of proteins from electrophoretic gels: strategies and applications. Proteins 2:74–89

Katayama H, Nagasu T, Oda Y (2001) Improvement of in-gel digestion protocol for peptide mass fingerprinting by matrix-assisted laser desorption/ionization time-of-flight mass spectrometry. Rapid Commun Mass Spectrom 15:1416–1421

Khoury GA, Baliban RC, Floudas CA (2011) Proteome-wide post-translational modification statistics: frequency analysis and curation of the swiss-prot database. Sci Rep 1:srep00090

Kreuss M, Strixner T, Kulozik U (2009) The effect of glycosylation on the interfacial properties of bovine caseinomacropeptide. Food Hydrocoll 23:1818–1826

Laemmli UK (1970) Cleavage of structural proteins during the assembly of the head of bacteriophage T4. Nature 227:680–685

Lametsch R, Karlsson A, Rosenvold K, Andersen HJ, Roepstorff P, Bendixen E (2003) Postmortem proteome changes of porcine muscle related to tenderness. J Agric Food Chem 51:6992–6997

Lametsch R, Larsen MR, Essen-Gustavsson B, Jensen-Waern M, Lundstrom K, Lindahl G (2011) Postmortem changes in pork muscle protein phosphorylation in relation to the RN genotype. J Agric Food Chem 59:11608–11615

Larsen MR, Thingholm TE, Jensen ON, Roepstorff P, Jorgensen TJ (2005) Highly selective enrichment of phosphorylated peptides from peptide mixtures using titanium dioxide microcolumns. Mol Cell Proteomics 4:873–886

Larsen MR, Trelle MB, Thingholm TE, Jensen ON (2006) Analysis of posttranslational modifications of proteins by tandem mass spectrometry. Biotechniques 40:790–798

Leiber F, Nigg D, Kunz C, Scheeder MRL, Wettstein HR, Kreuzer M (2005) Protein composition, plasmin activity and cheesemaking properties of cows' milk produced at two altitudes from hay of lowland and high-alpine origins. J Dairy Res 72:65–74

Malmstrom J, Lee H, Aebersold R (2007) Advances in proteomic workflows for systems biology. Curr Opin Biotechnol 18:378–384

Mamone G, Picariello G, Caira S, Addeo F, Ferranti P (2009) Analysis of food proteins and peptides by mass spectrometry-based techniques. J Chromatogr A 1216:7130–7142

Meyer B, Papasotiriou DG, Karas M (2011) 100 % protein sequence coverage: a modern form of surrealism in proteomics. Amino Acids 41:291–310

Muccilli V, Lo BM, Cunsolo V, Saletti R, Gallo G, Foti S (2011) High molecular weight glutenin subunits in some durum wheat cultivars investigated by means of mass spectrometric techniques. J Agric Food Chem 59:12226–12237

Nagaraj N, Alexander KN, Cox J, Neuhauser N, Mayr K, Hoerning O, Vorm O, Mann M (2012) System-wide Perturbation Analysis with Nearly Complete Coverage of the Yeast Proteome by Single-shot Ultra HPLC Runs on a Bench Top Orbitrap. Mol Cell Proteomics 11:M111

Nesvizhskii AI, Keller A, Kolker E, Aebersold R (2003) A statistical model for identifying proteins by tandem mass spectrometry. Anal Chem 75:4646–4658

Paik YK, Jeong SK, Omenn GS, Uhlen M, Hanash S, Cho SY, Lee HJ, Na K, Choi EY, Yan F, Zhang F, Zhang Y, Snyder M, Cheng Y, Chen R, Marko-Varga G, Deutsch EW, Kim H, Kwon JY, Aebersold R, Bairoch A, Taylor AD, Kim KY, Lee EY, Hochstrasser D, Legrain P, Hancock

WS (2012) The chromosome-centric human proteome project for cataloging proteins encoded in the genome. Nat Biotechnol 30:221–223
Pan S, Aebersold R, Chen R, Rush J, Goodlett DR, McIntosh MW, Zhang J, Brentnall TA (2009) Mass spectrometry based targeted protein quantification: methods and applications. J Proteome Res 8:787–797
Panchaud A, Affolter M, Kussmann M (2012) Mass spectrometry for nutritional peptidomics: how to analyze food bioactives and their health effects. J Proteomics 12:3546–3559
Pedersen SK, Harry JL, Sebastian L, Baker J, Traini MD, McCarthy JT, Manoharan A, Wilkins MR, Gooley AA, Righetti PG, Packer NH, Williams KL, Herbert BR (2003) Unseen proteome: mining below the tip of the iceberg to find low abundance and membrane proteins. J Proteome Res 2:303–311
Picotti P, Bodenmiller B, Mueller LN, Domon B, Aebersold R (2009) Full dynamic range proteome analysis of S. cerevisiae by targeted proteomics. Cell 138:795–806
Pomponio L, Lametsch R, Karlsson AH, Costa LN, Grossi A, Ertbjerg P (2008) Evidence for postmortem m-calpain autolysis in porcine muscle. Meat Sci 80:761–764
Przybylski C, Junger MA, Aubertin J, Radvanyi F, Aebersold R, Pflieger D (2010) Quantitative analysis of protein complex constituents and their phosphorylation states on a LTQ-Orbitrap instrument. J Proteome Res 9:5118–5132
Rehfeldt C, Te Pas MF, Wimmers K, Brameld JM, Nissen PM, Berri C, Valente LM, Power DM, Picard B, Stickland NC, Oksbjerg N (2011) Advances in research on the prenatal development of skeletal muscle in animals in relation to the quality of muscle-based food. II – genetic factors related to animal performance and advances in methodology. Animal 5:718–730
Robitaille G, Ngkwaihang KF, Monardes HG (1993) Effect of kappa-casein glycosylation on cheese yielding capacity and coagulating properties of milk. Food Res Int 26:365–369
Senechal S, Kussmann M (2011) Nutriproteomics: technologies and applications for identification and quantification of biomarkers and ingredients. Proc Nutr Soc 70:351–364
Sentandreu MA, Sentandreu E (2011) Peptide biomarkers as a way to determine meat authenticity. Meat Sci 89:280–285
Spiro RG (2002) Protein glycosylation: nature, distribution, enzymatic formation, and disease implications of glycopeptide bonds. Glycobiology 12:43R–456R
Thapa BR (2005) Therapeutic potentials of bovine colostrums. Indian J Pediatr 72:849–852
Uhlen M, Oksvold P, Algenas C, Hamsten C, Fagerberg L, Klevebring D, Lundberg E, Odeberg J, Ponten F, Kondo T, Sivertsson A (2012) Antibody-based protein profiling of the human chromosome 21. Mol Cell Proteomics 11:M111
Wang L, Xiong Y, Zuo B, Lei M, Ren Z, Xu D (2012) Molecular and functional characterization of glycogen synthase in the porcine satellite cells under insulin treatment. Mol Cell Biochem 360:169–180
Whiteaker JR, Zhao L, Anderson L, Paulovich AG (2010) An automated and multiplexed method for high throughput peptide immunoaffinity enrichment and multiple reaction monitoring mass spectrometry-based quantification of protein biomarkers. Mol Cell Proteomics 9:184–196
Wilkins MR, Pasquali C, Appel RD, Ou K, Golaz O, Sanchez JC, Yan JX, Gooley AA, Hughes G, HumpherySmith I, Williams KL, Hochstrasser DF (1996) From proteins to proteomes: large scale protein identification by two-dimensional electrophoresis and amino acid analysis. Biotechnology 14:61–65

Chapter 2
Extraction/Fractionation Techniques for Proteins and Peptides and Protein Digestion

Daniel Martínez-Maqueda, Blanca Hernández-Ledesma, Lourdes Amigo, Beatriz Miralles, and José Ángel Gómez-Ruiz

2.1 Introduction

The use of proteomics approaches is a powerful tool in food science in terms of process optimization and monitoring, quality, traceability, safety, and nutritional assessment (Pedreschi et al. 2010). Proteins, together with peptides, are one of the major groups of food components, and they are found in many different organisms of both vegetal and animal origin. Peptides are also obtained during technological processes such as fermentation and storage of foods. Moreover, many experiments involve enzymatic hydrolysis of proteins from food resources such as milk, meat, fish, eggs, or plants to produce a variety of peptides (Minkiewicz et al. 2008).

The study of the food proteome at any specific time is extremely complex and diverse. The major limitations of proteome analysis are, in general, associated with the heterogeneity of proteins and peptides in terms of physicochemical properties and the vast differences in abundance. A typical proteomics workflow consists of (1) protein extraction, (2) protein or peptide separation and quantification, (3) protein identification, and (4) data analysis and interpretation (Carpentier et al. 2008). Sample preparation has a profound effect on the final outcome of protein and peptide separation and their subsequent analysis. These procedures need to be compatible with posterior analysis by two-dimensional electrophoresis (2DE) and/or liquid chromatography tandem mass spectrometry (LC-MS/MS). Therefore, sample preparation should include the steps needed to isolate and fractionate proteins and

D. Martínez-Maqueda • B. Hernández-Ledesma • L. Amigo
• B. Miralles • J.Á. Gómez-Ruiz (✉)
Instituto de Investigación en Ciencias de la Alimentación
(CIAL, CSIC-UAM, CEI UAM+CSIC),
Nicolás Cabrera 9, 28049 Madrid, Spain
e-mail: d.martinez@csic.es; b.hernandez@csic.es
lourdes.amigo@csic.es; beatriz.miralles@csic.es; joseangel.gomez.ruiz@csic.es

F. Toldrá and L.M.L. Nollet (eds.), *Proteomics in Foods: Principles and Applications*, Food Microbiology and Food Safety 2, DOI 10.1007/978-1-4614-5626-1_2,

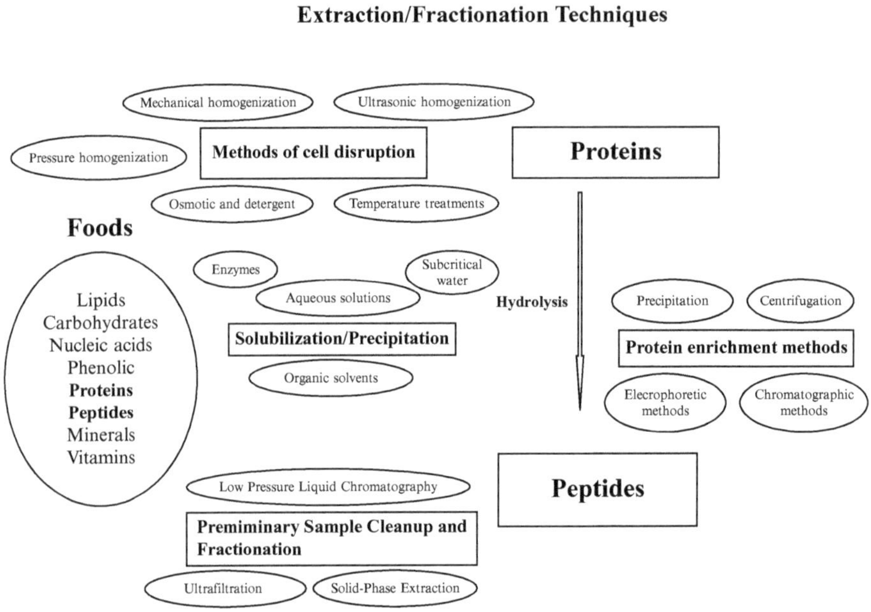

Fig. 2.1 Scheme illustrating integrated extraction and fractionation techniques for proteins and peptides employed on proteomics in foods

peptides ensuring an unbiased reliable map that gives an accurate representation of all proteins and peptides initially present in a particular food.

A wide variety of extraction and fractionation tools for proteins and peptides are available based on their physicochemical and structural characteristics such as solubility, hydrophobicity, molecular weight, isoelectric point (pI), and so on. Figure 2.1 shows an integrated view of extraction and fractionation techniques for proteins and peptides used in food proteomics studies. Generally, different technologies focused on cell disruption, solubilization/precipitation, and enrichment systems are needed to obtain the protein fraction of interest. Removal of interfering compounds (mainly lipids, nucleic acids, phenolic compounds, carbohydrates, proteolytic and oxidative enzymes, and pigments) is crucial. These procedures need to be optimized to minimize proteins´ modifications and proteolysis, as well as to be compatible with subsequent analysis.

This chapter describes the state-of-the-art of extraction and fractionation techniques for food proteins and peptides as a first step prior to proteome studies. The first part is dedicated to classical and novel extraction and fractionation techniques for food proteins, followed by a brief description on protein enzymatic digestion. The second part provides information about several extraction/fractionation techniques mainly used for food peptides.

2.2 Food Protein Extraction and Fractionation

2.2.1 *Cell Disruption Methods*

The preparation of any biological material as a sample for proteomic analysis requires homogenization. Plants are generally more problematic for protein extraction because tissues are rich in proteases and other interfering compounds (Wang et al. 2008a). Proteins are usually contained in protein bodies inside cell walls so cell disruption is required before they can be totally solubilized and extracted. The general procedure for sample preparation in this case strongly depends on the plant type, its fragment (leaf, fruit, seed, etc.), or even the stage of plant development. Generally, disruption of the cell wall and protein release is crucial for analytical success. Various chemical and physical techniques can be used to destroy the cell wall. These techniques can be grouped into five major categories: mechanical homogenization, ultrasound homogenization, pressure homogenization, temperature treatments, and osmotic and chemical lysis. A summary of these methods with their applications in different food matrices is shown in Table 2.1.

2.2.1.1 Mechanical Homogenization

Mechanical homogenization can be realized by at least two types of devices: so-called rotor–stator homogenizers and open blade mills. Rotor–stator homogenizers are one of the best homogenizing tools applied in laboratories. To homogenize dry samples using mechanical processing, open blade homogenizers, also called blenders, are used (Bodzon-Kulakowska et al. 2007). In the case of plant tissues, where cells are covered with strong cell walls, mechanical homogenization seems to be one of the best methods for disruption (Van Het Hof et al. 2000). Anderson and Guraya (2001) evaluated the use of colloid milling and homogenization to effect bran breakdown and extract rice protein. They demonstrated that the shearing actions of colloid milling and homogenization did not result in any significant denaturation of the proteins. Sometimes, a combination of mechanical homogenization with buffers is used. Examples of this are found in rice (Fukuda et al. 2003) and in olive tree seeds (Alche et al. 2006).

Wet-milling is a physicochemical separation of the components of grain, namely germ, bran, fiber, starch, and protein. Chemicals and enzymes can be added to the steeping water to facilitate the separation of grain components and increase starch recovery. Sulfur dioxide, sodium metabisulfite, sodium bisulfite, or sodium hydrogen sulfite, with variable effective concentrations, are typically added to solubilize the protein matrix enveloping the starch granules in the endosperm (De Mesa-Stonestreet et al. 2010).

Table 2.1 Examples of techniques used for food plant cell disruption

Type of disruption	Procedure	Food	Tissue	Reference
Mechanical homogenization	Colloid milling and homogenization	Rice	Bran	Anderson and Guraya 2001
	Centrifugal grinding and air dehulling	Pea, chickpea, lentil	Seed	Boye et al. 2010
	Cool mortar with lysis buffer	Rice	Embryo	Fukuda et al. 2003
	Cool mortar with Tris–HCl	Olive tree	Seed	Alche et al. 2006
	Wet-milling with sulphur dioxide	Sorghum	Seed	De Mesa-Stonestreet et al. 2010
Ultrasonic homogenization	Acoustic transducer	Soybean, rice	Root, sheath/hypocotyl, leaf	Toorchi et al. 2008
	Ultrasonic generator	Rice	Bran	Chittapalo and Noomhorm 2009
Pressure homogenization	High pressure homogenization	Peanut	Seed	Dong et al. 2011
		Rapeseed	Seed	Barbin et al. 2011
Temperature treatments	Mortar and pestle with liquid N_2	Tomato	Pollen	Sheoran et al. 2009
		Olive	Leaf	Wang et al. 2003
		Apple, banana	Mesocarp/exocarp	Song et al. 2006
		Peanut	Seed	Liang et al. 2006
		Maize	Endosperm	Méchin et al. 2007
		Potato	Tuber	Delaplace et al. 2006
	Pulverization in dry ice and grinding in liquid N_2	Grape	Berry cluster	Vincent et al. 2006
	Microwave, dry heating and parboiling	Rice	Bran	Khan et al. 2011
	Wet-milling with temperature	Sorghum	Seed	De Mesa-Stonestreet et al. 2010
Osmotic and chemical lysis	Hexadecyltrimethylammonium bromide (CTAB), sodium dodecyl sulphate (SDS) or isopropylalcohol (IPA)	Lactococcus lactis strains	–	Doolan and Wilkinson 2009

2.2.1.2 Ultrasonic Homogenization

In recent years, some researchers have used the ultrasonic method for protein pellet homogenization. The energy is produced by an acoustic transducer coupled with the pellet in the microtube, which is carried out manually in small batches. Toorchi et al. (2008) have attempted to use a high-performance, single-tube sample preparation device (Covaris) for the noncontact disruption and uniform preparation of three different plant tissues from soybean (root, hypocotyl, and leaf) and rice (root, leaf sheath, and leaf).

Many researchers have investigated the advantages of ultrasonic-assisted extraction compared with the conventional method. Chittapalo and Noomhorm (2009) reported that protein yield increased using ultrasound and that this process can enhance existing extraction processes and enable new commercial extraction opportunities.

2.2.1.3 Pressure Homogenization

The use of high-pressure homogenization (HPH) for the extraction of food proteins has been investigated. Higher pressures (40 and 80 MPa) produced approximately double protein extraction compared to atmospheric pressure. Dong et al. (2011) have suggested that HPH treatment could increase the susceptibility of peanut proteins to proteolytic enzymes such as alcalase. The increase may be related to the denaturation, unfolding, or dissociation of the proteins into monomers, allowing the accessibility of enzyme to the binding sites. HPH revealed no alteration of protein solubility when compared with the raw protein with pH adjusted in rapeseed protein concentrates (Barbin et al. 2011).

2.2.1.4 Temperature Treatments

Temperature treatments include the use of freeze–thaw and heat treatments. Freeze–thawing uses the effect of ice crystal formation in the tissue during the freezing process. Lysis of the cells or tissues is usually achieved by flash-freezing the cells in liquid nitrogen and homogenizing in a mortar with a pestle. Examples of this process are found in the analysis of leaves (Wang et al. 2003), fruits (Song et al. 2006), and seeds (Liang et al. 2006; Méchin et al. 2007). Vincent et al. (2006) developed a very efficient cell disruption method for grape berry clusters, which were pulverized frozen with dry ice using a stainless steel blender.

The use of heat is common in protein processing. Heating protein solutions usually improves their solubility, emulsifying, and foaming properties, but it makes protein extraction more difficult as reported in rice bran (Tang et al. 2002; Khan et al. 2011). Another approach is the application of heat during the wet-milling process. Steeping experiments have been done on temperature and holding time on sorghum grain (De Mesa-Stonestreet et al. 2010).

2.2.1.5 Osmotic and Chemical Lysis

Cell permeabilization or cell lysis can be performed by osmotic shock or chemical treatment. The use of osmotic shock implies cell suspension in a gently shaken hypertonic solution. Chemical treatment can include antibiotics, chelating agents, detergents, and solvents capable of disintegrating the cells. This procedure relies on the selective interaction of the chosen chemicals with components of the membrane and allows proteins to seep through the cell wall. The application of two or more procedures combined with the cell-wall disruption is also reported (Klimek-Ochab et al. 2011). These procedures are used in cell cultures of bacteria, yeast, or fungi. Doolan and Wilkinson (2009) have compared the effects of various chemicals on cell permeability in *Lactococcus lactis* strains with the aim of selectively releasing important intracellular ripening enzymes. Their findings permit a better understanding of methods affecting cell permeability and can allow development of food-grade technologies for protein released from cells.

2.2.2 *Protein Solubilization/Precipitation*

Protein solubilization is considered one of the key steps in proteomic sample preparation procedures. It is generally employed to separate proteins in the sample selectively from different substances that may interfere in the proteomic assay (Berkelman and Stenstedt 1998). The solubilization/precipitation process strongly affects the quality of the final results and thus determines the success of the entire experiment. Taking into account the immense variety of proteins and the huge number of interfering contaminants present in food-derived extracts, simultaneous solubilization of all proteins remains a great challenge.

Each food sample requires a specific protocol that needs to be optimized to minimize proteolysis and modification of proteins (Bodzon-Kulakowska et al. 2007). For animal tissues, which have higher protein yields, various protein solubilization buffers, including the use of chaotrophic agents, detergents, reducing agents, buffers, and ampholites are used (Pedreschi et al. 2010). The proper use of these additives avoids protein modifications, aggregation, or precipitation that may result in the occurrence of artifacts and the subsequent lowering of protein yield (Gorg et al. 2004).

2.2.2.1 Organic Solvents

The main organic solvents and additives used to extract proteins from food sources are shown in Table 2.2. Many studies performed in the last few years aimed to compare different protein solubilization methods suitable for proteomic analysis (Jiang et al. 2004; Natarajan et al. 2005; He and Wang 2008). The most common

Table 2.2 Examples of organic solvents and additives used to extract proteins from food sources

Solvent (s)	Food	Tissue	Reference
Acetic acid/urea/cetyltrimethylammonium bromide	Rice	Bran	Hamada 1997
Aqueous ethanol	Distiller's grain	Grain	Cookman and Glatz 2009
Aqueous isopropanol	Soybean	Seed	Natarajan et al. 2009
	Rapeseed	Seed	Barbin et al. 2011
Ethanol	*Saccharina japonica*	–	Kim et al. 2011
Glacial acetic acid	Sorghum	–	de Mesa-Stonestreet et al. 2010
Phenol	Tomato	Pollen grain	Sheoran et al. 2009
	Potato	Tuber	Delaplace et al. 2006
	Aloe vera	Leaf	He and Wang 2008
	Soybean	Seed	Natarajan et al. 2005
Phenol/ammonium acetate	Barley	Root	Hurkman and Tanaka 1986
	Avocado/tomato/orange/banana	Fruit	Saravanan and Rose 2004
	Banana	Leaf	Carpentier et al. 2007
	Grape	Fruit	Vincent et al. 2006
	Pear	Fruit	Pedreschi et al. 2007
	Apple/strawberry	Fruit	Zheng et al. 2007
Phenol/methanol-ammonium acetate	Coniferous	Seed	Zhen and Shi 2011
	Banana/apple/potato	Tissues	Carpentier et al. 2005
Sodium dodecyl sulphate/acetone	Coniferous	Seed	Zhen and Shi 2011
	Potato	Tuber	Delaplace et al. 2006
Sodium dodecyl sulphate/TCA/acetone	Apple/banana	Tissue	Song et al. 2006
TCA	Bean	Anther	Wu and Wang 1984
TCA/acetone	Citrus	Leaf	Maserti et al. 2007
	Soybean	Seed	Natarajan et al. 2006
	Soybean	Leaf	Xu et al. 2006
	Coniferous	Seed	Zhen and Shi 2011
	Tomato	Pollen grain	Sheoran et al. 2009
	Aloe vera	Leaf	He and Wang 2008
	Apple/banana	Tissues	Song et al. 2006
TCA/acetone/phenol	Olive	Leaf	Wang et al. 2003
	Bamboo/grape/lemon	Leaf	Wang et al. 2006
	Apple/orange/tomato	Fruit	Wang et al. 2006
Thiourea/urea	Soybean	Seed	Natarajan et al. 2005
	Apple/banana	Tissues	Song et al. 2006
Tris–HCl buffer	Tomato	Pollen grain	Sheoran et al. 2009

method used for the extraction of plant proteins is tricholoracetic acid (TCA)/acetone precipitation as proposed by Damerval et al. (1986). This method has been used to extract proteins from different tissues of cereals, legumes, and fruits. The extreme pH and negative charge of TCA and the addition of acetone realizes an immediate denaturation of the protein, along with precipitation, thereby instantly arresting the activity of proteolytic and other modifying enzymes. However, a disadvantage of TCA-precipitated proteins is that they are difficult to redissolve (Nandakumar et al. 2003). Sample solubility can be improved by using an appropriate mixture of chaotropic agents (urea or thiourea), and new efficient detergents (such as sodium dodecyl sulphate, SDS). In the last decade, the phenol extraction procedure has been widely used because of its high clean-up capacity. In contrast to its strong solvent action on proteins, phenol has little predisposition to dissolve polysaccharides and nucleic acids. However, phenol shows the disadvantages of being more time consuming than other sample precipitation procedures and of being toxic.

The alcoholic extraction process after dehulling and conventional deoiling has a high efficiency of protein recovery. Aqueous alcohols (ethanol, isopropyl alcohol, butanol) are widely used on a commercial scale to remove phenolics, oligosaccharides, or inhibitors from defatted meals and seeds (Moure et al. 2006). However, as a result of the extraction with these alcoholic solvents, protein structures can be coagulated and therefore show reduced functional properties. To avoid these problems and to obtain protein concentrates or isolates with good functionality and suitable as food ingredients, mechanical and thermal treatments are applied (Moure et al. 2006; Barbin et al. 2011). Recently, extractions with different organic solvents, such as n-hexane, 2-methyl pentane, diethyl ether, acetone, 2-propanol, and ethanol were compared regarding effectiveness, suitability, and protein solubility of the full-fat and defatted lupin (Bader et al. 2011).

2.2.2.2 Aqueous Solutions

In recent years, because of the growing environmental concerns over the use of organic solvents to extract oil/protein from oil-bearing food materials, aqueous extraction is gaining attention. Water is also operationally advantageous over alcohols because it is nonflammable and neither explosive nor toxic. Commercially, the production of protein concentrates (48–70% protein) or isolates (85–90% protein) consists of an aqueous solubilization of protein and carbohydrates at acid, neutral, or alkaline pH and the selective recovery of the solubilized protein, separation, and, optionally, washing and neutralization before drying. The protein extraction yield and properties are influenced by the type of extraction process and by different factors such as pH, salts concentration, the ionic strength of the medium, net charge, and electrostatic repulsions (Tan et al. 2011).

A number of acid and alkaline protein extraction protocols have been published from various plant and animal tissues. In the last decade, different studies have focused on evaluating the effect of extraction methods on the functional and rheological properties of proteins recovered from by-products of the meat and fish industry (Liang and Hultin 2003; Chaijan et al. 2006; Hrynets et al. 2010, 2011; Moayedi et al. 2010; Omana et al. 2010). In the case of plant proteins, the ideal extraction method is particularly challenging due to the metabolic and structural characteristics of plant tissues, including the cell wall matrix. The majority of alkaline extraction protocols are based on the so-called Osborne method (Osborne 1924), but each method is optimized according to the aim of the study and the type of vegetal protein source. Recent studies report the use of mainly sodium and calcium salts to extract proteins from different vegetal foods (Ghaly and Alkoaik 2010; Horax et al. 2010; Lestari et al. 2010; Karaca et al. 2011; Nadal et al. 2011). These extraction methods are simple because the agents required are easily available. However, as a result of the degradation at high pH conditions, the protein yield is generally low. Also, the protein quality can be altered by alkaline processing due to undesirable reactions involving racemization of amino acids, formation of toxic compounds such as lysinoalanine, reduction of digestibility, loss of essential amino acids, and decrease in nutritive value. Furthermore, the remaining alkali needs to be washed thoroughly from the final product, leading to generation of a large amount of wastewater (Sereewatthanawut et al. 2008). To optimize protein precipitation recovery different strategies have been developed. Use of additives, such as TCA or carboxymethylcellulose is generally accepted (Massoura et al. 1998). Extraction and further formation of protein micelles have also been proposed (Krause et al. 2002; Murray 2003; Ser et al. 2008; Green et al. 2010). This method has been demonstrated to reduce the concentration of problematic antinutritional or toxic factors, including the glucosinolates and their degradation products during canola protein extraction (Tan et al. 2011).

2.2.2.3 Aqueous Enzymatic Extraction

An alternative approach combining aqueous and enzymatic extraction is attracting attention. Studies using this extraction process are shown in Table 2.3. Enzymes can aid in the extraction of proteins in several ways. Carbohydrases, which can attack the cell wall components, may increase protein yield by liberating more protein from the matrix source (Ansharullah et al. 1997; Wang et al. 1999; Tang et al. 2002). A combination of cell wall-hydrolyzing enzymes (i.e., Viscozyme L) has been used to cleave linkages within the polysaccharide matrix effectively and hence, liberate more intracellular protein from oat bran (Guan and Yao 2008). In the last few years, different proteases, alone or in combination, have been used to partially hydrolyze proteins to peptides, increasing their solubility and making them more easily extractable. Recently, De Moura and co-workers (2011) developed a two-stage countercurrent aqueous enzymatic extraction process for soybean, significantly reducing the amount of water used. They achieved slightly higher oil and protein extraction yields than those from standard single-stage aqueous enzymatic extraction.

Aqueous enzymatic protein extraction has been defined as an environmentally friendly, safe, and cheap alternative to extract oil and protein simultaneously (Latif and Anwar 2009). Moreover, this process avoids serious damage to the proteins produced by the refining steps, improving their nutritional and functional properties (Domínguez et al. 1994; Moure et al. 2000). However, and although the enzymatic extraction process produces no toxic chemicals, it shows some disadvantages, such

Table 2.3 Aqueous-enzymatic extraction processes used to obtain protein isolates or concentrates from food sources

Enzyme	Food	Tissue/sample	Protein extracted (%)	Reference
Alcalase™	Rice	Bran	81.0	Hamada 2000
Alcalase 2.4 L	Rapeseed	Seed	66.7	Zhang et al. 2007
	Peanut	Seed	82.5	Wang et al. 2008b
	Peanut	Roasted seed	80.1	Zhang et al. 2011
Alcalase + Protamex (1:3)	Tea	Leave pulp	47.8	Shen et al. 2008
Alkaline protease	Rice	Broken rice	75.5	Hou et al. 2010
Flavourzyme	Rice	Bran	88.0	Hamada 2000
Glucoamylase	Lentil	Bean	–	Bildstein et al. 2008
Neutrase 1.5MG	Coconut	Meat	83.0	Sant'Anna et al. 2003
Olivex + Celluclast	*Guevina avellana*	Pressed cakes	85.8	Moure et al. 2002
Papain	Rice	Broken rice	46.3	Hou et al. 2010
Pectinase + Protease P	Rice	Bran	80.0	Tang et al. 2003
Phytase	Rice	Bran	80.0	Wang et al. 1999
Protex 6 L	Distiller's grain	Grain	90.0	Cookman and Glatz 2009
	Lupin	Seed	77.2	Jung 2009
	Soybean	Seed	84.6	Jung 2009
	Soybean	Flakes	96.0	De Moura et al. 2011
Protex 7 L	Sesame	Seed	87.1	Latif and Anwar 2011
	Moringa oleifera	Seed	75.4	Latif and Anwar 2009
Viscozyme L	Coconut	Meat	83.0	Sant'Anna et al. 2003
	Oat	Bran	56.2	Guan and Yao 2008
Viscozyme L + Cellulast 1.5 L	Rice	Bran	37.0	Tang et al. 2002
		Bran	53.0	Ansharullah et al. 1997
Xylanase	Rice	Bran	82.0	Wang et al. 1999
Xylanase + Phytase amylase	Rice	Bran	75.0	Wang et al. 1999

as the long time required and the high cost of enzymes that makes this strategy uneconomical. The use of immobilized enzyme in protein extraction may reduce the overall cost by allowing the reuse of enzymes.

2.2.2.4 Subcritical Water

Recent studies demonstrate the use of water at subcritical conditions as an environmentally friendly reaction medium to extract proteins from different food sources. Subcritical water is water that maintains its liquid state in the temperature range of 100–374°C under pressurized conditions. Its unique properties, such as a lower relative dielectric constant and a higher ion product than ambient water, make subcritical water a promising extraction solvent for various compounds, including proteins (Hata et al. 2008). Ho et al. (2007) used pressurized low-polarity water to extract proteins from defatted flaxseed meal. A number of studies have demonstrated the ability of water at subcritical conditions to extract proteins from rice bran and soybean meal with high protein yields and good functional properties (Watchararuji et al. 2008; Fabian and Ju 2011).

2.2.3 *Protein Enrichments Methods*

Once the protein fraction has been isolated from the rest of the constituents and the interference substances have been eliminated, there are still some other steps that are needed prior to the analysis of the sample by mass spectrometry (MS). Despite the last technological developments, no single analytical method exists covering the protein concentration range present in a specific sample. Sometimes the total protein content is very low or the objective is the determination of minor proteins with post-translational modifications (e.g., phosphorylation).

In many cases the methods described in this part are comparable (if not the same) to those previously seen in this chapter during protein extraction. However, this section is focused more on those steps to be applied once the protein fraction has been separated from other interfering components. As a matter of fact, the purpose of fractionation and enrichment methods is to obtain distinguishable fractions and increase the concentration of the proteins of interest.

2.2.3.1 Centrifugation

The use of centrifugation is one of the simplest methods used for isolation and enrichment/fractionation of proteins. Centrifugation can be used for different purposes. It can be a first step to separate different cell substructures where our proteins of interest are locally concentrated, for instance, mitochondria, membrane, or nucleus. This process involves multiple centrifugation steps and, as a result, the cellular homogenate is separated into different layers based on the molecular weight, size, and shape of each component. Afterwards, solubilization steps, as explained above, and enrichment and fractionation steps should be carried out to isolate the protein fraction from the selected layer prior to MS analysis.

Apart from its use separating crude mixtures of cell components, centrifugation is also commonly used to fractionate a protein mixture into different fractions. The separation takes place based on the coefficient of sedimentation of the proteins. This coefficient is usually expressed in *Svedberg units* (S), and the smaller the S value, the slower a molecule moves in a centrifugal field. Separation will depend on the mass, the shape, and the protein density. Numerous examples are found in the literature using the differential coefficient of sedimentation of the proteins to carry out fractionation (Sharma et al. 2010; Jiang et al. 2011). The efficiency of this fractionation step can be enhanced using gradient centrifugation, where the centrifuge tube is filled with a solution of sucrose, forming a density gradient.

2.2.3.2 Precipitation

It is recognized that among the different precipitants the most widespread is ammonium sulphate (Bodzon-Kulakowska et al. 2007). The addition of high amounts of this salt or other such as sodium chloride into a protein solution provokes an increase of protein interactions followed by protein aggregation and finally precipitation. This is known as a salting-out process and, as the salt concentration needed for protein precipitation varies from one protein to another, it allows selective protein separation. An alternative salting-out method using decreasing solutions of salt can also be used to enrich previously precipitated protein fractions. This salting-out approach has been used to separate the main storage soybean proteins, glycinin and β-conglycinin (Deak et al. 2006).

Another type of protein enrichment is immunoprecipitation, based upon the binding of the antigen to its specific antibody to form the antigen–antibody complex. In general it offers high recoveries of the proteins and it is widely used for food allergens (Pastorello and Trambaioli 2001).

2.2.3.3 Electrophoretic Methods

Electrophoresis separates mixtures of proteins based on charge, charge/mass ratio, size, or shape. This technique is mainly used as an analytical and preparative tool, especially one-dimensional separation, often employed as a pre-fractionating technique (Guttman et al. 2004; Jorgenson and Evans 2004). Often, laboratories dedicate one-dimensional gel electrophoresis (1DE) to evaluate the outcome of protein purification preceding the analysis by (2DE) (Chen et al. 2007).

Electrophoretic pre-fractionation methods include electrokinetic methodologies performed in free solution, essentially all relying on isoelectric focusing (IEF) steps. Purification using IEF is especially advantageous when protein activity must be maintained. Bioactivity is maintained because the proteins remain in solution in their native conformation. Based on the IEF principle, different instruments have been developed such as the Rotofor, a multicompartmental device with focusing chambers that allows the fractionation of volumes of sample (12–60 mL) containing micrograms to grams of protein (Hey et al. 2008). Another well-known device is the so-called "Off-Gel IEF" (Keidel et al. 2011). Upon application of an electric field perpendicularly to the liquid chamber, the current lines penetrate into the chamber and extract charged proteins from the solution into the IEF gel. In its multicompartment format, the protein fractions are separated by ranges of pI depending on their positioning over the IEF gel strip. Other instruments of interest are the Octopus, a continuous-flow device for isoelectric focusing in an upward flowing liquid curtain, and the Gradiflow, where different pI cuts are obtained by a multi-step passage through two compartments buffered at different pH values (Righetti et al. 2003).

Depending on the complexity of the samples, the separated fractions can be analyzed directly by MS or in some cases they may undergo a subsequent separation step in a second dimension, generally SDS-PAGE, to separate the proteins according to their molecular weight. In the first case, the possible presence of ampholytes may imply an extra step to remove them and avoid disturbance in MS.

2.2.3.4 Chromatographic Methods

Liquid chromatography (LC) techniques are the most commonly used in proteome pre-fractionation prior to in-depth analysis. The separation of the different proteins is achieved according to their charge, hydrophobicity, size, or specificity. In some cases, chromatographic methods can also be used to eliminate some interference substances (e.g., salts) coming from previous enrichment steps.

Among LC fractionation methods, ion-exchange chromatography (IEX) is probably the most used, with proteins being separated according to their pI. Acidic proteins are usually fractionated by anion-exchange chromatography whereas basic proteins are fractionated by cation-exchange chromatography. IEX has been often used to separate milk proteins as reported by Gómez-Ruiz et al. (2007a), who used cation-exchange chromatography to separate sheep milk caseins.

Reverse phase LC (RP-LC) separates proteins according to their hydrophobicity. Proteins are adsorbed on a stationary phase carrying hydrophobic groups, and are eluted with increasing concentration of an organic solvent, generally acetonitrile. RP-LC is widely used in proteomics in combination with IEX and MS analysis, usually in shotgun multidimensional strategies that are used as an alternative to 2-D-PAGE technology. A special case of chromatography based on hydrophobic interactions uses a high concentration of lyotropic salts (frequently ammonium sulphate) to expose the hydrophobic parts of proteins towards the hydrophobic patches of solid-phase sorbents. Desorption is promoted by using a decreasing concentration of the lyotropic salts.

Pre-fractionation with chromatographic methods is also used to investigate post-translational modifications such as glycosylation or phosphorylation, to cite perhaps the two most important. These modifications are mainly studied using affinity chromatography (AC). This chromatography utilizes highly specific biological interactions (i.e., antigen–antibody, receptor–ligand, enzyme–substrate/inhibitor, etc.). AC results are quite adequate for accessing low concentrated proteins in complex samples, in some cases through the depletion of high abundance proteins that remain bound to the column. Examples of AC are heparin chromatography, broadly used for studying microbial proteins, or lectin chromatography that is specially used for glycoproteins (Lee and Lee 2004; Azarkan et al. 2007). Immobilized metal affinity chromatography (IMAC) is used to enrich phosphoproteins. This chromatography is based on formation of coordinate bonds between basic groups on protein surface and metal ions. The major drawback is that little or no binding to Fe(III) or Ga(III) charged resins is observed at neutral pH, and using low-pH buffers may provoke protein denaturalization or precipitation in the column (Schmidt et al. 2007).

Some other techniques such as size-exclusion chromatography (SEC) separate proteins according to their molecular mass, as the second dimension of 2-D-PAGE. However, unlike 2-D-PAGE this chromatography can be used under nondenaturing conditions allowing the study of protein complexes. As an example, SEC has been used for the evaluation of the bread-making quality of hard spring wheat flours (Ohm et al. 2009). Additional methods based on the use of chip-based arrays are gaining importance recently, with surface-enhanced laser desorption/ionization (SELDI) one of their maximum exponents (Righetti et al. 2005). Unlike chromatography separation, here only retained proteins are eventually studied and the other proteins are removed by one or more washing steps. Subsequently, the use of a pulse of laser light provokes the desorption of the proteins of interest which are converted into gaseous ions and analyzed by MS, typically using time of flight (TOF) analyzers.

2.3 Protein Digestion

Once the proteins have been isolated from interfering compounds (other food components such as lipids, nucleic acids, phenolic compounds, or carbohydrates) they are usually analyzed by 1D or 2-D SDS-PAGE, depending of the complexity of the sample. Gel electrophoresis analysis is typically followed by protein digestion, a key procedure prior to the identification of proteins by MS. However, in some cases digestion is carried out without electrophoretic separation. For instance, direct digestion of a mixture of proteins is adequate when a broad survey of the identifiable protein components is desired or to minimize the loss of peptides by binding to the polyacrylamide when characterizing post-translational modifications (Kinter and Sherman 2005).

Different proteolytic agents are used for protein digestion, including enzymes such as trypsin, different endoproteases (Lys-C, Arg-C, Asp-N, Glu-C), or chymotrypsin, as well as chemical reagents such as hydroxylamine or cyanogens bromide. The specificity of the amide bond or bonds cleaved by these reagents allows the obtaining of specific peptides that facilitate the interpretation of their mass spectra and database search. Trypsin is certainly the most popular reagent because it shows many advantages compared to other enzymes and chemical reagents, in addition to its relatively low cost of production and high purity. This enzyme cleaves amide bonds at the C-terminal side of Lys and Arg residues except when these bonds are to Pro. Apart from this selectivity, Arg and Lys are common amino acids distributed through most proteomes such that tryptic cleavage yields peptides with an average length suitable for MS. Finally, trypsin cleavage yields peptides containing a strongly basic residue (Lys or Arg) at the C-terminal, a fact that facilitates the interpretation of collision-induced dissociation (CID) mass spectra (Couto et al. 2011).

When talking about protein digestion in proteomic studies we mainly think of two types of digestion: "in-gel" digestion and "in-solution" digestion. Most proteomic studies perform in-gel digestion of proteins previously separated by their charge and/or their molecular weight (1D or 2-D SDS-PAGE). Identification of proteins from polyacrylamide gels offers a number of important advantages compared to gel-free approaches, such as higher dynamic range of analysis of protein mixtures (ratio of lowest to highest abundance protein detectable) or removal of low molecular weight impurities before the MS analysis. In-gel protein digestion was first established by Rosenfeld et al. (1992). The typical steps of the method have remained the same since then, although small variations have been introduced to improve its performance. Destaining, reduction, and alkylation of Cys, enzymatic cleavage of proteins into peptides, and extraction of peptides from the gel are described as essential steps in obtaining high-quality mass spectra. Analysis by electrospray (ESI)-MS, less tolerant to salts, requires an additional desalting step which is optional for MALDI-MS (Granvogl et al. 2007)

Despite its widespread use, conventional tryptic digestion is very time consuming, with a typical digestion time in the range of several hours to half a day (Park and Russell 2000). This is a clear limitation to the production of high-throughputs in proteomic analysis. Therefore, in recent times many efforts have been focused on developing efficient and fast protein digestion methods. Several alternatives to the standard protocol have been proposed, many of them based on the use of electromagnetic waves (i.e., electromagnetic radiation), such as microwaves, infrared (IR) radiation, and ultraviolet (UV) light to accelerate protein digestion (Chen et al. 2011; Dycka et al. 2012). Among these strategies, the use of IR-assisted digestion seems to be the most promising approach due to its safety compared to the other electromagnetic waves. Other alternatives are the use of modified trypsin, for instance, by reductive methylation which decreases autolysis and shifts its optimal catalytic temperature to 50–60°C. This modified trypsin allows the reduction of digestion times from 16 h to 30 min without losing efficiency. For in-solution digestion, immobilized trypsin systems as part of a microchip bioreactor offer a very efficient alternative to conventional methods. As an example, the use of a fiber-based microchip bioreactor provides on-chip digestion in less than 5 s with similar tryptic digests to those obtained by the conventional in-solution tryptic digestion (Fan and Chen 2007).

During each step along the protein digestion experiments extreme care is necessary to avoid contamination that can compromise MS analyses and the outcome of the study. In some cases the contamination refers to low molecular weight compounds (from either the polyacrylamide gel or the subsequent digestion of specific bands) that will not provoke erroneous protein identification but will complicate peptide detection due to higher noise levels. The use of high-purity reagents throughout the experiment, especially for gel-electrophoresis and digestion can significantly minimize this type of contamination. On other occasions, contamination with keratins, proteins derived from skin and hair, is the main problem. In this case wearing gloves and an adequate handling of the laboratory consumables (e.g., pipette tips) and reagents will limit this contamination.

2.4 Food Peptide Extraction and Fractionation

Generally, food peptide content is not as abundant as would be desirable. In addition to this, the presence of nonpeptidic constituents (i.e., lipids, sugars) may also interfere in peptide analysis. Therefore, in practice it is difficult to analyze food peptides with good accuracy without performing a sample preparation step. This sample preparation can comprise diverse procedures for isolation, purification, and preconcentration of the analyte, more than one step being required in many cases (Poliwoda and Wieczorek 2009).

RP-LC and capillary electrophoresis (CE), are the basic analytical methods used for chemometrical analysis of food peptidome (Minkiewicz et al. 2008). In relation to CE and capillary electrochromatography (CEC), restrictions come from the small sample volume applied (nano- to picolitre) that necessitates the application of pre-concentration and pre-separation steps in samples with low peptide concentration or complex mixtures (Kasicka 2012).

In general, food samples are first subjected to a preliminary sample cleanup step to remove interfering substances and then, different fractionation steps are applied, as has been widely revised (González de Llano et al. 2004; Asensio-Ramos et al. 2009; Hernández-Ledesma et al. 2012). Several options that are summarized in Table 2.4 may be taken. Direct peptide analysis on food samples without any preparation treatment is not often reported in the literature (Cheison et al. 2010).

Peptide derivatization may be necessary in some analyses for better detection (Wang et al. 2011). Most derivatizations are developed with fluorescent labels to become detectable with fluorescence detection whose limit of detections (LODs) is about two to three orders of magnitude lower than common UV-absorption detections (Kasicka 2012). An example may be found in the determination of glutathione in must and white wine during alcoholic fermentation (Lavigne et al. 2007).

2.4.1 *Extraction and Preliminary Sample Cleanup*

Hydrophilic peptides are generally extracted with homogenization in water or in solutions of organic acids whereas organic solvents are used to obtain highly hydrophobic peptides. Homogenization in a mixture of organic solvents (chloroform/methanol) can be used for peptide extraction as well as for the removal of sample interferences after producing a biphasic system. By using this method, Kostyra et al. (2003) studied the opioid activity of cheese and fermented milk samples, On the other hand, homogenization in water has been widely applied on cheese, fish, meat, and cereals samples as shown in Table 2.4. Typically, the ratio of water to cheese used was 2:1 in the homogenization process, followed by an incubation step of an hour at 60°C (Gómez-Ruiz et al. 2002; Meyer et al. 2009).

Peptide extraction is usually followed by a preliminary sample cleanup for removal of other food components (i.e., proteins, lipids). Deproteinization, the most important preliminary cleanup procedure in peptide analysis, is carried out by precipitation of protein using several agents. Deproteinization could also act as a fractionation procedure for peptides because their solubility depends on the precipitant agent and its proportion (Cheng et al. 2010a,b). After precipitation, centrifugation and filtration methods are used to separate proteins from soluble peptides. In addition, the application of heat treatments or ultracentrifugation steps at high speed to eliminate the proteins has been reported (Gómez-Ruiz et al. 2007b; Ho et al. 2010).

The selectivity of precipitation directly depends on the type of precipitating agent applied. In addition to the use of organic solvents such as ethanol, methanol, or acetone, solutions containing acids such as TCA or trifluoroacetic acid (TFA) are classical protein precipitants (Juan-García et al. 2009; Escudero et al. 2010). Salting-out precipitation, based on polarity, with high concentrations of salts or precipitation by adjusting the pH to the pI of protein (Contreras et al. 2010; Pihlanto et al. 2010) are other options. A representative example is found in the isolation and identification of an angiotensin I-converting enzyme (ACE) inhibitory peptide from whole buckwheat seeds after consecutive cleanup steps of diethyl ether extraction in order to remove most of the fat content and deproteinization by adjusting the pH to the pI of buckwheat protein (Ma et al. 2006).

In some cases, the application of homogenization and/or deproteinization is enough to proceed with peptide analysis (Contreras et al. 2010). Unfortunately, most of the samples need additional steps to achieve suitable peptide isolation and concentration levels before the analysis.

2.4.2 *Fractionation*

2.4.2.1 Ultrafiltration

Ultrafiltration is mainly useful for fractionating peptides as well as the removal of proteins and other macromolecules based on their molecular size. Dedicated

Table 2.4 Examples of food sample preparation for peptidomic analysis

Food matrix	Extraction and clean-up	Fractionation techniques	References
Dairy products			
Cheese, milk, yoghurt and infant formula	Homogenization, centrifugation and deproteinization by pH adjustment	Ultrafiltration	De Noni and Cattaneo 2010
Dry-off cows milk	Centrifugation and ultracentrifugation	Ultrafiltration	Ho et al. 2010
Fermented milk	Centrifugation	SPE	Hernández-Ledesma et al. 2005
Manchego cheese	Homogenization, centrifugation, filtration and ultracentrifugation	Ultrafiltration and SEC	Taborda et al. 2007; Gómez-Ruiz et al. 2007b
Whey protein concentrate hydrolysate	Centrifugation and filtration	Ultrafiltration	Contreras et al. 2011
Fish and meat			
Rainbow trout muscle	Homogenization and centrifugation	Ultrafiltration and SPE Derivatization of anserine	Bauchart et al. 2007
Cuttlefish protein hydrolysate	Centrifugation	SEC	Balti et al. 2010
Loach protein hydrolysate	Centrifugation	Ultrafiltration, IEX and SEC	You et al. 2010
Dry-cured ham	Homogenization, centrifugation, filtration and deproteinization by ethanol addition	SEC	Mora et al. 2010, 2011
Pork meat digest	Centrifugation and deproteinization by ethanol addition.	–	Escudero et al. 2010
Eggs			
Egg white protein hydrolysate	Centrifugation	SEC and IEX	Liu et al. 2010
Egg white protein hydrolysate	Centrifugation	Ultrafiltration	Miguel et al. 2004
Drinks			
Beer	–	SPE and SEC	Picariello et al. 2011
Must and wines	Centrifugation (not bottled wines)	Derivatization of SH functions	Lavigne et al. 2007
White and red wines	Centrifugation	Ultrafiltration and SEC	Pozo-Bayón et al. 2007

Vegetable foods			
Corn zein hydrolysate	Pigments extraction and centrifugation	Ultrafiltration	Tang et al. 2010
Fermented soybean extract	Filtration and dialysis	Ultrafiltration and IEX	Rho et al. 2009
Soy protein hydrolysate product	pH adjustment and centrifugation	–	Johns et al. 2011
Pea protein hydrolysate	Centrifugation	Ultrafiltration and SPE	Li and Aluko 2010
Potato protein hydrolysate	Salting-out precipitation and centrifugation	SEC	Cheng et al. 2010a, b

IEX ion exchange chromatography, *SEC* size exclusion chromatography, *SPE* solid phase extraction membranes are mostly made of polysulfone or cellulose derivatives. Cellulose membranes have excellent hydrophilicity, which is very important in minimizing fouling, but they possess low chemical resistance and poor mechanical strength. However, polysulfone membranes provide high rigidity but foul earlier because of their hydrophobicity (Doyen et al. 2011). Commercially, membranes offer a wide range of cutoffs (500–100 kDa) and different formats including centrifugal units or cassettes for peristaltic lab systems. Fractionation of peptides has been achieved in food samples by applying ultrafiltration with more than one cutoff membrane. As an example, Samaranayaka et al. (2010) searched the presence of antioxidant and ACE inhibitory peptides in a hake protein hydrolyzate using an ultrafiltration unit with different molecular mass cutoff membranes (10, 3, and 1 kDa). In summary, ultrafiltration presents some advantages as the sample is not diluted or organic solvents are not required. Therefore, in some cases after the ultrafiltration step no additional fractionation processes are applied before analysis, such as in cheese (Bütikofer et al. 2008) or champagne wine samples (Person et al. 2004). Nevertheless, samples often need further pre-treatment procedures that mainly improve the analyte concentration.

A recent technology named electrodialysis with ultrafiltration membranes (EDUF) has been developed to fractionate peptides from complex mixtures on the basis of their electrical charge, size, or molecular weight. A conventional electrodialysis is used but some ion exchange membranes are replaced by ultrafiltration ones. This equipment has been employed for the concentration and selective separation of bioactive peptides from an alfalfa white protein hydrolyzate (Firdaous et al. 2009). A successful use of these membranes has also been reported, isolating an antihypertensive peptide from a tryptic hydrolyzate of β-lactoglobulin (Poulin et al. 2007).

2.4.2.2 Low-Pressure Liquid Chromatography

Low-pressure size exclusion chromatography (SEC) fractionates peptides on the basis of their molecular size. This technique separates analytes through a bed of porous beads where they can either enter or be excluded from the internal space of the beads based on their size. Elution occurs from the largest to the smallest analyte over time (Ly and Wasinger 2011). Several resins with different pore sizes are commercially available. Cross-linked dextran (Sephadex) resins are mostly used but polyacrylamide (BioGel P) or divinylbenzene polymers are also available as stationary phases (Poliwoda and Wieczorek 2009). Depending on the resin composition, peptides are eluted with water, organic acids, ammonia, or ammonium salts, even as alcoholic solutions that reduce potential hydrophobic interactions. For instance, Mora et al. (2011) applied SEC to fractionate peptides released during dry-cured ham processing in a Sephadex G25 column under isocratic conditions in 0.01 N HCl. Other uses have been reported, for instance, to identify ACE inhibitory peptides in white and red wines (Pozo-Bayón et al. 2007).

Low-pressure ion exchange chromatography (IEX) constitutes another technique for peptide fractionation in food analysis. In this case, peptides are fractionated according to their net surface charge/polarity. Porous or nonporous matrices with hydrophilic materials such as cellulose, cross-linked dextrans, polystyrene polymers (Dowex resins), or Bio-Rex membranes are very useful as anion or cation exchange stationary phases. These matrices are substituted with functional groups that determine the charge of the medium (e.g., quaternary ammonium, diethylaminoethyl, sulfopropyl, carboxymethyl, etc.). The ionic strength increases as the elution method can carry a large amount of salts in the elution buffer that makes samples incompatible with techniques such as MS (Ly and Wasinger 2011).

Off-line combination of IEX and SEC has been reported in some food peptide analyses. A representative example is found in the work of Liu et al. (2010), who fractionated an egg white protein hydrolyzate by SEC with Sephadex G-25 resin followed by IEX (Sephadex C-25 column) of those fractions with the highest ACE inhibitory activity. A similar fractionation strategy has also been used in the study of antioxidant peptides in a fish protein hydrolyzate (You et al. 2010), and for the evaluation of the peptide contribution to the umami taste of soy sauces (Lioe et al. 2006).

2.4.2.3 Solid-Phase Extraction

Solid-phase extraction (SPE) is based on the same principle of affinity-based separation as liquid chromatography. SPE enables retention and elution of analytes from complex mixtures, removal of interfering compounds, and sample concentration. SPE is available in normal phase, reverse-phase, and ion exchange modes, reversed-phase being one of the most used formats (Kole et al. 2011). Based on the wide range of physicochemical properties of the analytes, several commercial sorbents (e.g., C_{18}, C_8, C_2, phenyl, cyanopropyl, and ion exchange bonded materials, among others) are supplied to improve the versatility of SPE. Regarding this, in the peptidomic characterization of beer, Picariello et al., (2011) applied the samples directly onto the C_{18} pre-packed cartridges and eluted with acetonitrile/TFA to RP-LC. In other examples, Hernández-Ledesma et al. (2005) treated the water-soluble extract of fermented milk with a Sep-Pak C_{18} cartridge and acetonitrile elution, and a similar extraction step was used by Muguruma et al. (2009) to desalt SEC-eluted fractions from porcine myosin B.

Based on similar principles of SPE techniques, innovative size-reduced devices have recently appeared for concentration, purification, and desalting of peptides prior to analysis by MS. These devices support a membrane or microcolumn that can be of diverse nature (polar, nonpolar, and ion exchange) and feature an optimized procedure for sample preparation. For instance, in the study of trout peptidome changes during storage, Bauchart et al. (2007) used a C_{18} membrane device prior to MALDI-TOF analysis with the aim of removing the perchloric acid used in the previous extraction. Another case is found in the study of beer peptidome in which residual interfering sugars are removed by applying the sample on C_{18} Zip-Tip microcolumns and peptides eluted with acetonitrile/TFA (Picariello et al. 2011).

2.5 Future Perspectives

Current efforts are mainly focused on the improvement and development of automated systems as today sample preparation implies several labor-intensive and time-consuming handling steps. Despite the generalized use of 2-D electrophoresis, this technology has limitations mainly when dealing with proteins at varying expression levels. An alternative could be the use of automated pre-fractionation methods such as electrokinetic methodologies performed in free solution combined with one-dimensional PAGE and capillary LC-MS/MS. Recent technologies such as SELDI also imply minimal requirements for purification and separation of proteins prior to their analysis by MS. For the investigation of post-translationally modified proteins the future approach seems to be the combined use of affinity-based enrichment and extraction methods and multidimensional separation technologies prior to MS analysis. A persisting challenge is still the development of appropriate enrichment/fractionation techniques to facilitate MS analysis of membrane proteins.

Acknowledgments The authors thank the Spanish Ministry of Economy and Competitiveness for the financial support from Projects AGL2011-24643, FUN-C-Food CSD2007-0063, and Project P2009/AGR-1469. D. Martínez Maqueda and B. Hernández-Ledesma thank Spanish National Research Council (CSIC) for their JAE-Pre and Ramón y Cajal Contracts, respectively.

References

Alche JD, Jiménez-López JC, Wang W, Castro-López AJ, Rodriguez-García MI (2006) Biochemical characterization and cellular localization of 11S type storage proteins in olive (*Olea europaea L.*) seeds. J Agric Food Chem 54:5562–5570

Anderson AK, Guraya HS (2001) Extractability of protein in physically processed rice bran. J Am Oil Chem Soc 78:969–973

Ansharullah A, Hourigan JA, Chesterman CF (1997) Application of carbohydrases in extracting protein from rice bran. J Sci Food Agric 74:141–146

Asensio-Ramos M, Hernández-Borges J, Rocco A, Fanali S (2009) Food analysis: a continuous challenge for miniaturized separation techniques. J Sep Sci 32:3764–3800

Azarkan M, Huet J, Baeyens-Volant D, Looze Y, Vandenbussche G (2007) Affinity chromatography: a useful tool in proteomics studies. J Chromatogr B 849:81–90

Bader S, Oviedo JP, Pickardt C, Eisner P (2011) Influence of different organic solvents on the functional and sensory properties of lupin (*Lupinus angustifolius* L.) proteins. LWT Food Sci Technol 44:1396–1404

Balti R, Nedjar-Arroume N, Adjé EY, Guillochon D, Nasri M (2010) Analysis of novel angiotensin I-converting enzyme inhibitory peptides from enzymatic hydrolysates of cuttlefish (*Sepia officinalis*) muscle proteins. J Agric Food Chem 58:3840–3846

Barbin DF, Natsch A, Müller RK (2011) Improvement of functional properties of rapeseed protein concentrates produced via alcoholic processes by thermal and mechanical treatments. J Food Process Preserve 35:369–375

Bauchart C, Chambon C, Mirand PP, Savary-Auzeloux I, Rémond D, Morzel M (2007) Peptides in rainbow trout *(Oncorhynchus mykiss)* muscle subjected to ice storage and cooking. Food Chem 100:1566–1572

Berkelman T, Stenstedt T (1998) 2-D electrophoresis using immobilized pH gradients. Principles and methods. Amersham Biosciences, San Francisco, CA, USA

Bildstein M, Lohmann M, Hennigs C, Krause A, Hilz H (2008) An enzyme-based extraction process for the purification and enrichment of vegetable proteins to be applied in bakery products. Eur Food Res Technol 228:177–186

Bodzon-Kulakowska A, Bierczynska-Krzysik A, Dylag T, Drabik A, Suder P, Noga M, Jarzebinska J, Silberring J (2007) Methods for samples preparation in proteomic research. J Chromatogr B 849:1–31

Boye JI, Aksay S, Roufik S, Ribéreau S, Mondor M, Farnworth E, Rajamohamed SH (2010) Comparison of the functional properties of pea, chickpea and lentil protein concentrates processed using ultrafiltration and isoelectric precipitation techniques. Food Res Int 43:537–546

Bütikofer U, Meyer J, Sieber R, Walther B, Wechsler D (2008) Occurrence of the angiotensin-converting enzyme inhibiting tripeptides Val-Pro-Pro and Ile-Pro-Pro in different cheese varieties of Swiss origin. J Dairy Sci 91:29–38

Carpentier SC, Kris Laukens EW, Deckers P, Swennen R, Panis B (2005) Preparation of protein extracts from recalcitrant plant tissues: an evaluation of different methods for two-dimensional gel electrophoresis analysis. Proteomics 5:2497–2507

Carpentier SC, Dens K, Van den houwe I, Swennen R, Panis B (2007) Lyophilization, a practical way to store and transport tissues prior to protein extraction for 2DE analysis? Proteomics 7:S64–S69

Carpentier SC, Panis B, Vertommen A, Swennen R, Sergeant K, Renaut J, Laukens K, Witters E, Samyn B, Devresee B (2008) Proteome analysis of non-model plants: a challenging but powerful approach. Mass Spectrom Rev 27:354–377

Chaijan M, Benjakul S, Visessanguan W, Faustman C (2006) Physicochemical properties, gel forming ability and myoglobin content of sardine (*Sardinalla gibbosa*) and mackerel (*Rastrelliger kanagurta*) surimi produced by conventional method and alkaline solubilisation process. Eur Food Res Technol 222:58–63

Cheison SC, Schmitt M, Leeb E, Letzel T, Kulozik U (2010) Influence of temperature and degree of hydrolysis on the peptide composition of trypsin hydrolysates of β-lactoglobulin: analysis by LC–ESI-TOF/MS. Food Chem 121:457–467

Chen B, Dong JQ, Chen YJ, Wang JM, Tian J, Wang CB, Zou SQ (2007) Two-dimensional electrophoresis for comparative proteomic analysis of human bile. Hepatob Pancreat Dis Int 6:402–406

Chen Q, Liu T, Chen G (2011) Highly efficient proteolysis accelerated by electromagnetic waves for peptide mapping. Curr Genomics 12:380–390

Cheng Y, Chen J, Xiong YL (2010a) Chromatographic separation and tandem MS identification of active peptides in potato protein hydrolysate that inhibit autoxidation of sybean oil-in-water emulsions. J Agric Food Chem 58:8825–8832

Cheng Y, Xiong Y, Chen J (2010b) Fractionation, separation, and identification of antioxidative peptides in potato protein hydrolysate that enhance oxidative stability of soybean oil emulsions. J Food Sci 75:760–765

Chittapalo T, Noomhorm A (2009) Ultrasonic assisted alkali extraction of protein from defatted rice bran and properties of the protein concentrates. Int J Food Sci Technol 44:1843–1849

Contreras MM, Gómez-Sala B, Martín-Alvarez PJ, Amigo L, Ramos M, Recio I (2010) Monitoring the large-scale production of the antihypertensive peptides RYLGY and AYFYPEL by HPLC-MS. Anal Bioanal Chem 397:2825–2832

Contreras MM, Hernández-Ledesma B, Amigo L, Martín-Alvarez PJ, Recio I (2011) Production of antioxidant hydrolyzates from a whey protein concentrate with thermolysin: optimization by response surface methodology. LWT Food Sci Technol 44:9–15
Cookman DJ, Glatz CE (2009) Extraction of protein from distiller's grain. Bioresour Technol 100:2012–2017
Couto N, Barber J, Gaskell SJ (2011) Matrix-assisted laser desorption/ionisation mass spectrometric response factors of peptides generated using different proteolytic enzymes. J Mass Spectrom 46:1233–1240
Damerval C, Devienne D, Zyvy M, Thiellement H (1986) Technical improvements in two-dimensional electrophoresis increase the level of genetic-variation detected in wheat-seedling proteins. Electrophoresis 7:52–54
De Mesa-Stonestreet NJ, Alavi S, Bean SR (2010) Sorghum proteins: the concentration, isolation, modification, and food applications of kafirins. J Food Sci 75:R90–R104
De Moura JMLN, Campbell K, de Almeida NM, Glatz CE, Johnson LA (2011) Protein extraction and membrane recovery in enzyme-assisted aqueous extraction processing of soybeans. J Am Oil Chem Soc 88:877–889
De Noni I, Cattaneo S (2010) Occurrence of β-casomorphins 5 and 7 in commercial dairy products and in their digests following in vitro simulated gastro-intestinal digestion. Food Chem 119:560–566
Deak NA, Murphy PA, Johnson LA (2006) Effects of NaCl concentration on salting-in and dilution during salting-out on soy protein fractionation. J Food Sci 71:247–254
Delaplace P, van der Wal F, Dierick JF, Cordewener JHG, Fauconnier ML, du Jardin P, America AHP (2006) Potato tuber proteomics: comparison of two complementary extraction methods designed for 2-DE of acidic proteins. Proteomics 6:6494–6497
Domínguez H, Núñez MJ, Lema JM (1994) Enzymatic pretreatment to enhance oil extraction from fruits and oilseeds – a review. Food Chem 49:271–286
Dong X, Zhao M, Shi J, Yang B, Li J, Luo D, Jiang G, Jiang Y (2011) Effects of combined high-pressure homogenization and enzymatic treatment on extraction yield, hydrolysis and function properties of peanut proteins. Innov Food Sci Emerg Technol 12:478–483
Doolan IA, Wilkinson MG (2009) Comparison of the effects of various attenuation methods on cell permeability and accessibility of intracellular enzymes in *Lactococcus lactis* strains. Int Dairy J 19:215–221
Doyen A, Beaulieu L, Saucier L, Pouliot Y, Bazinet L (2011) Impact of ultrafiltration membrane material on peptide separation from a snow crab byproduct hydrolysate by electrodialysis with ultrafiltration membranes. J Agric Food Chem 59:1784–1792
Dycka F, Bobal P, Mazanec K, Bobalova J (2012) Rapid and efficient protein enzymatic digestion: an experimental comparison. Electrophoresis 33:288–295
Escudero E, Sentandreu MA, Toldrá F (2010) Characterization of peptides released by in vitro digestion of pork meat. J Agric Food Chem 58:5160–5165
Fabian C, Ju YH (2011) A review on rice bran protein: its properties and extraction methods. Crit Rev Food Sci Nutr 51:816–827
Fan H, Chen G (2007) Fiber-packed channel bioreactor for microfluidic protein digestion. Proteomics 7:3445–3449
Firdaous L, Dhulster P, Amiot J, Gaudreau A, Lecouturier D, Kapel R (2009) Concentration and selective separation of bioactive peptides from an alfalfa white protein hydrolysate by electrodialysis with ultrafiltration membranes. J Membr Sci 329:60–67
Fukuda M, Islam N, Woo SH, Yamagishi A, Takaoka M, Hirano H (2003) Assessing matrix assisted laser desorption/ionization-time of flight-mass spectrometry as a means of rapid embryo protein identification in rice. Electrophoresis 24:1319–1329
Ghaly AE, Alkoaik FN (2010) Extraction of protein from common plant leaves for use as human food. Am J Appl Sci 7:323–334
Gómez-Ruiz JA, Ramos M, Recio I (2002) Angiotensin-converting enzyme-inhibitory peptides in Manchego cheese manufactured with different starter cultures. Int Dairy J 12:697–706

Gómez-Ruiz JA, Ramos M, Recio I (2007a) Identification of novel angiotensin-converting enzyme-inhibitory peptides from ovine milk proteins by CE-MS and chromatographic techniques. Electrophoresis 28:4202–4211
Gómez-Ruiz JA, Taborda G, Amigo L, Ramos M, Molina E (2007b) Sensory and mass spectrometric analysis of the peptidic fraction lower than one thousand daltons in Manchego cheese. J Dairy Sci 90:4966–4973
González de Llano D, Herráiz T, Polo MC (2004) Peptides. In: Nollet ML (ed) Handbook of food analysis, vol. 1. Physical characterization and nutrient analysis, vol 1. Marcel Dekker, New York, pp 125–166
Gorg A, Weiss W, Dunn MJ (2004) Current two-dimensional electrophoresis technology for proteomics. Proteomics 4:3665–3685
Granvogl B, Plösche M, Eichacker LA (2007) Sample preparation by in-gel digestion for mass spectrometry-based proteomics. Anal Bioanal Chem 389:991–1002
Green BE, Martens RW, Tergesen JF, Milanova R (2010) Production of flax protein isolate. US Patent 7,820,226 B2
Guan X, Yao HY (2008) Optimization of viscozyme L-assisted extraction of oat bran protein using response surface methodology. Food Chem 106:345–351
Guttman A, Varoglu M, Khandurina J (2004) Multidimensional separations in the pharmaceutical arena. Drug Discov Today 9:136–144
Hamada JS (1997) Characterization of protein fractions of rice bran to devise effective methods of protein solubilization. Cereal Chem 74:662–668
Hamada JS (2000) Characterization and functional properties of rice bran proteins modified by commercial exoproteases and endoproteases. J Food Sci 65:305–310
Hata S, Wiboonsirikul J, Maedab A, Kimura Y, Adachi S (2008) Extraction of defatted rice bran by subcritical water treatment. Biochem Eng J 40:44–53
He C-F, Wang Y-M (2008) Protein extraction from leaves of *Aloe vera* L., a succulent and recalcitrant plant, for proteomic analysis. Plant Mol Biol Rep 26:292–300
Hernández-Ledesma B, Miralles B, Amigo L, Ramos M, Recio I (2005) Identification of antioxidant and ACE-inhibitory peptides in fermented milk. J Agric Food Chem 85:1041–1048
Hernández-Ledesma B, Martínez-Maqueda D, Miralles B, Amigo L, Gómez-Ruiz JA (2012) Peptides. In: Nollet ML, Toldrá F (eds) Food analysis by HPLC. Marcel Dekker, New York, in press
Hey J, Posch A, Cohen A, Liu N, Harbers A (2008) Fractionation of complex protein mixtures by liquid-phase isoelectric focusing. Methods Mol Biol 424:225–239
Ho CHL, Cacace JE, Mazza G (2007) Extraction of lignans, proteins and carbohydrates from flaxseed meal with pressurized low polarity water. LWT Food Sci Technol 40:1637–1647
Ho CH, Chang CJ, Liu WB, Peh HC, Chen SE, Chen HY, Ho TH, Chen MT, Nagahata H (2010) In situ generation of milk protein-derived peptides in drying-off cows. J Dairy Res 77:487–497
Horax R, Hettiarachchy N, Over K, Chen P, Gbur E (2010) Extraction, fractionation and characterization of bitter melon seed proteins. J Agric Food Chem 58:1892–1897
Hou L, Zhu Y, Li Q (2010) Characterization and preparation of broken rice proteins modified by proteases. Food Technol Biotechnol 48:50–55
Hrynets Y, Dileep AO, Xu Y, Betti M (2010) Effect of acid- and alkaline-aided extractions on functional and rheological properties of proteins recovered from mechanically separated turkey meat (MSTM). J Food Sci 75:E477–E486
Hrynets Y, Dileep AO, Xu Y, Betti M (2011) Comparative study on the effect of acid- and alkaline-aided extractions on mechanically separated turkey meat (MSTM): chemical characteristics of recovered proteins. Process Biochem 46:335–346
Hurkman WJ, Tanaka CK (1986) Solubilization of plant membrane-proteins for analysis by two-dimensional gel-electrophoresis. Plant Physiol 81:802–806
Jiang L, Hea L, Fountoulakis M (2004) Comparison of protein precipitation methods for sample preparation prior to proteomic analysis. J Chromatogr A 1023:317–320

Jiang J, Xiong YL, Chen J (2011) Role of β-conglycinin and glycinin subunits in the pH-shifting-induced structural and physicochemical changes of soy protein isolate. J Food Sci 76:293–302
Johns PW, Jacobs WA, Phillips RR, McKenna RJ, O'Kane KA, McEwen JW (2011) Characterisation of peptide molecular mass distribution in commercial hydrolysates and hydrolysate-based nutritional products. Food Chem 125:1041–1050
Jorgenson JW, Evans CR (2004) Multidimensional LC-LC and LC-CE for high-resolution separations of biological molecules. Anal Bioanal Chem 378:1952–1961
Juan-García A, Font G, Juan C, Picó Y (2009) Nanoelectrospray with ion-trap mass spectrometry for the determination of beta-casomorphins in derived milk products. Talanta 80:294–306
Jung S (2009) Aqueous extraction of oil and protein from soybean and lupin: a comparative study. J Food Process Preserv 33:547–559
Karaca AC, Low N, Nickerson M (2011) Emulsifying properties of chickpea, faba bean, lentil and pea proteins produced by isoelectric precipitation and salt extraction. Food Res Int 44:2742–2750
Kašička V (2012) Recent developments in CE and CEC of peptides (2009–2011). Electrophoresis 33:48–73
Keidel EM, Dosch D, Brunner A, Kellermann J, Lottspeich F (2011) Evaluation of protein loading techniques and improved separation in OFFGEL isoelectric focusing. Electrophoresis 32:1659–1666
Khan SH, Butt MS, Sharif MK, Sameen A, Mumtaz S, Sultan MT (2011) Functional properties of protein isolates extracted from stabilized rice bran by microwave, dry heat, and parboiling. J Agric Food Chem 59:2416–2420
Kim E-Y, Kim D-G, Kim Y-R, Hwang H-J, Nam T-J, Kong I-S (2011) An improved method of protein isolation and proteome analysis with *Saccharina japonica* (Laminariales) incubated under different pH conditions. J Appl Phycol 23:123–130
Kinter M, Sherman NE (2005) The preparation of protein digests for mass spectrometric sequencing experiments, in protein sequencing and identification using tandem Mass Spectrometry. Wiley, Hoboken
Klimek-Ochab M, Brzezinska-Rodak M, Zymanczyk-Duda E, Lejczak B, Kafarski P (2011) Comparative study of fungal cell disruption-scope and limitations of the methods. Folia Microbiol 56:469–475
Kole PL, Venkatesh G, Kotecha J, Sheshala R (2011) Recent advances in sample preparation techniques for effective bioanalytical methods. Biomed Chromatogr 25:199–217
Kostyra E, Kostyra H, Krawczuk S, Jarmotowska B, Sienkiewicz-Szatapka E, Grzegorzewskic W (2003) Procedure for the evaluation of the opioid activity of milk products. Pol J Food Nutr Sci 12:39–42
Krause JP, Schultz M, Dudek S (2002) Effect of extraction conditions on composition, surface activity and rheological properties of protein isolates from flaxseed (*Linum usitativissimum* L). J Sci Food Agric 82:970–976
Latif S, Anwar F (2009) Effect of aqueous enzymatic processes on sunflower oil quality. J Am Oil Chem Soc 86:393–400
Latif S, Anwar F (2011) Aqueous enzymatic sesame oil and protein extraction. Food Chem 125:679–684
Lavigne V, Pons A, Dubourdieu D (2007) Assay of glutathione in must and wines using capillary electrophoresis and laser-induced fluorescence detection. Changes in concentration in dry white wines during alcoholic fermentation and aging. J Chromatogr A 1139:130–135
Lee WC, Lee KH (2004) Applications of affinity chromatography in proteomics. Anal Biochem 324:1–10
Lestari D, Mulderb W, Sandersa J (2010) Improving *Jatropha curcas* seed protein recovery by using counter current multistage extraction. Biochem Eng J 50:16–23
Li H, Aluko RE (2010) Identification and inhibitory properties of multifunctional peptides from pea protein hydrolysate. J Agric Food Chem 58:493–501
Liang Y, Hultin HO (2003) Functional protein isolates from mechanically deboned turkey by alkaline solubilization with isoelectric precipitation. J Muscle Food 14:195–205

Liang XQ, Luo M, Holbrook CC, Guo BZ (2006) Storage protein profiles in Spanish and runner market type peanuts and potential markers. BMC Plant Biol 6:24–32
Lioe HN, Takara K, Yasuda M (2006) Evaluation of peptide contribution to the intense umami taste of Japanese soy sauces. J Food Sci 71:277–283
Liu J, Yu Z, Zhao W, Lin S, Wang E, Zhang Y, Hao H, Wang Z, Chen F (2010) Isolation and identification of angiotensin-converting enzyme inhibitory peptides from egg white protein hydrolysates. Food Chem 122:1159–1163
Ly L, Wasinger VC (2011) Protein and peptide fractionation, enrichment and depletion: tools for the complex proteome. Proteomics 11:513–534
Ma M, In YB, Hyeon GL, Yang C (2006) Purification and identification of angiotensin I-converting enzyme inhibitory peptide from buckwheat (*Fagopyrum esculentum moench*). Food Chem 96:36–42
Maserti BE, Della Croce CM, Luro F, Morillon R, Cini M, Caltavuturo L (2007) A general method for the extraction of citrus leaf proteins and separation by 2-D electrophoresis: a follow up. J Chromatogr B 849:351–356
Massoura E, Vereijken JM, Kolster P, Derksen JTP (1998) Proteins from *Crambe abyssinica* oil seed. I. Isolation procedure. J Am Oil Chem Soc 75:323–327
Méchin V, Thévenot C, Le Guilloux M, Prioul JL, Damerval C (2007) Developmental analysis of maize endosperm proteome suggests a pivotal role for pyruvate orthophosphate dikinase. Plant Physiol 143:1213–1219
Meyer J, Bütikofer U, Walther B, Wechsler D, Sieber R (2009) Hot topic: changes in angiotensin-converting enzyme inhibition and concentrations of the tripeptides Val-Pro-Pro and Ile-Pro-Pro during ripening of different Swiss cheese varieties. J Dairy Sci 92:826–836
Miguel M, Recio I, Gómez-Ruiz JA, Ramos M, López-Fandiño R (2004) Angiotensin I-converting enzyme inhibitory activity of peptides derived from egg white proteins by enzymatic hydrolysis. J Food Prot 67:1914–1920
Minkiewicz P, Dziuba J, Darewicz M, Iwaniak A, Dziuba M, Nalecz D (2008) Food peptidomics. Food Technol Biotechnol 46:1–10
Moayedi V, Omana DA, Chan J, Xu Y, Betti M (2010) Alkali aided protein extraction from chicken dark meat: composition and stability to lipid oxidation of recovered proteins. Poult Sci 89:766–775
Mora L, Sentandreu MA, Toldrá F (2010) Identification of small troponin T peptides generated in dry-cured ham. Food Chem 123:691–697
Mora L, Valero ML, Sánchez del Pino MM, Sentandreu MA, Toldrá F (2011) Small peptides released from muscle glycolytic enzymes during dry-cured ham processing. J Proteomics 74:442–450
Moure A, Franco D, Sineiro J, Domínguez H, Núñez MJ, Lema JM (2000) Evaluation of extracts from *Gevuina avellana* hulls as antioxidants. J Agric Food Chem 48:3890–3897
Moure A, Domínguez H, Zuñiga ME, Soto C, Chamy R (2002) Characterisation of protein concentrates from pressed cakes of *Guevina avellana* (Chilean hazelnut). Food Chem 78:179–186
Moure A, Sineiro J, Domínguez H, Parajó JC (2006) Functionality of oilseed protein products: a review. Food Res Int 39:945–963
Muguruma M, Ahhmed AM, Katayama K, Kawahara S, Maruyama M, Nakamura T (2009) Identification of pro-drug type ACE inhibitory peptide sourced from porcine myosin B: evaluation of its antihypertensive effects in vivo. Food Chem 114:516–522
Murray ED (2003) Canola protein isolate functionality I. US Patent 2003/0021884 A1
Nadal P, Canela N, Katakis I, O'Sullivan CK (2011) Extraction, isolation, and characterization of globulin proteins from *Lupinus albus*. J Agric Food Chem 59:2752–2758
Nandakumar MP, Shen J, Raman B, Marten MR (2003) Solubilization of trichloroacetic acid (TCA) precipitated microbial proteins via NaOH for two-dimensional electrophoresis. J Proteome Res 2:89–93
Natarajan SS, Xu C, Caperna TJ, Garrett WM (2005) Comparison of protein solubilization methods suitable for proteomic analysis of soybean seed proteins. Anal Biochem 342:214–220

Natarajan SS, Xu C, Bae H, Caperna TJ, Garrett WM (2006) Characterization of storage proteins in wild (*Glycine soja*) and cultivated (*Glycine max*) soybean seeds using proteomics analysis. J Agric Food Chem 54:3114–3120

Natarajan SS, Krishnan HB, Lakshman S, Garrett GM (2009) An efficient extraction method to enhance analysis of low abundant proteins from soybean seed. Anal Biochem 394:259–268

Ohm J-B, Hareland G, Simsek S, Seabourn B (2009) Size-exclusion HPLC of protein using a narrow-bore column for evaluation of breadmaking quality of hard spring wheat flours. Cereal Chem 86:463–469

Omana DA, Moayedi V, Xu Y, Betti M (2010) Alkali-aided protein extraction from chicken dark meat: textural properties and color characteristics of recovered proteins. Poult Sci 89:1056–1064

Osborne TB (1924) The vegetable protein. Longmans Green, New York

Park ZY, Russell DH (2000) Thermal denaturation: a useful technique in peptide mass mapping. Anal Chem 72:2667–2670

Pastorello EA, Trambaioli C (2001) Isolation of food allergens. J Chromatogr B 756:71–84

Pedreschi R, Vanstreels E, Carpentier S, Hertog M, Lammertyn J, Robben J, Noben JP, Swennen R, Vanderleyden J, Nicola B (2007) Proteomic analysis of core breakdown disorder in conference pears (*Pyrus communis* L.). Proteomics 7:2083–2099

Pedreschi R, Hertogi M, Lilley KS, Nicolaïl B (2010) Proteomics for the food industry: opportunities and challenges. Crit Rev Food Sci Nutr 50:680–692

Person M, Sevestre A, Chaimbault P, Perrot L, Duchiron F, Elfakir C (2004) Characterization of low-molecular weight peptides in champagne wine by liquid chromatography/tandem mass spectrometry. Anal Chim Acta 520:149–158

Picariello G, Bonomi F, Iametti S, Rasmussen P, Pepe C, Lilla S, Ferranti P (2011) Proteomic and peptidomic characterisation of beer: immunological and technological implications. Food Chem 124:1718–1726

Pihlanto A, Virtanen T, Korhonen H (2010) Angiotensin I converting enzyme (ACE) inhibitory activity and antihypertensive effect of fermented milk. Int Dairy J 20:3–10

Poliwoda A, Wieczorek PP (2009) Sample pretreatment techniques for oligopeptide analysis from natural sources. Anal Bioanal Chem 393:885–897

Poulin J, Amiot J, Bazinet L (2007) Improved peptide fractionation by electrodialysis with ultrafiltration membrane: influence of ultrafiltration membrane stacking and electrical field strength. J Membr Sci 299:83–90

Pozo-Bayón M, Alcaide J, Polo MC, Pueyo E (2007) Angiotensin I-converting enzyme inhibitory compounds in white and red wines. Food Chem 100:43–47

Rho SJ, Lee J, Chung YI, Kim Y, Lee HG (2009) Purification and identification of an angiotensin I-converting enzyme inhibitory peptide from fermented soybean extract. Process Biochem 44:490–493

Righetti PG, Castagna A, Herbert B, Reymond F, Rossier JS (2003) Prefractionation techniques in proteome analysis. Proteomics 3:1397–1407

Righetti PG, Castagna A, Antonioli P, Boschetti E (2005) Prefractionation techniques in proteome analysis: the mining tools of the third millennium. Electrophoresis 26:297–319

Rosenfeld J, Capdevielle J, Guillemot JC, Ferrara P (1992) In-gel digestion of proteins for internal sequence analysis after one- or two-dimensional gel electrophoresis. Anal Biochem 203:173–179

Samaranayaka AGP, Kitts DD, Li-Chan ECY (2010) Antioxidative and angiotensin-I-converting enzyme inhibitory potential of a pacific hake (*Merluccius productus*) fish protein hydrolysate subjected to simulated gastrointestinal digestion and Caco-2 cell permeation. J Agric Food Chem 58:1535–1542

Sant'Anna BPM, Freitas SP, Coelho MAZ (2003) Enzymatic aqueous technology for simultaneous coconut protein and oil extraction. Grasas y Aceites 54:77–80

Saravanan RS, Rose JKC (2004) A critical evaluation of sample extraction techniques for enhanced proteomic analysis of recalcitrant plant tissues. Proteomics 4:2522–2532

Schmidt SR, Schweikart F, Andersson ME (2007) Current methods for phosphoprotein isolation and enrichment. J Chromatogr B 849:154–162
Ser WY, Arntfield SD, Hydamaka AW, Slominski BA (2008) Use of diabetic test kits to assess the recovery of glucosinolates during isolation of canola protein. LWT Food Sci Technol 41:934–941
Sereewatthanawut I, Prapintip S, Watchiraruji K, Goto M, Sasaki M, Shotipruk A (2008) Extraction of protein and amino acids from deoiled rice bran by subcritical water hydrolysis. Bioresour Technol 99:555–561
Sharma GM, Mundoma C, Seavy M, Roux KH, Sathe SK (2010) Purification and biochemical characterization of Brazil nut (*Bertholletia excelsa L.*) seed storage proteins. J Agric Food Chem 58:5714–5723
Shen L, Wang X, Wang Z, Wua Y, Chen J (2008) Studies on tea protein extraction using alkaline and enzyme methods. Food Chem 107:929–938
Sheoran IS, Ross ARS, Olson DJH, Sawhney VK (2009) Compatibility of plant protein extraction methods with mass spectrometry for proteome analysis. Plant Sci 176:99–104
Song J, Braun G, Bevis E, Doncaster K (2006) A simple protocol for protein extraction of recalcitrant fruit tissues suitable for 2-DE and MS analysis. Electrophoresis 27:3144–3151
Taborda G, Gómez-Ruiz JA, Martínez-Castro I, Amigo L, Ramos M, Molina E (2007) Taste and flavor of artisan and industrial Manchego cheese as influenced by the water-soluble extract compounds. Eur Food Res Technol 227:323–330
Tan SH, Mailer RJ, Blanchard CL, Agboola SO (2011) Canola proteins for human consumption: extraction, profile, and functional properties. J Food Sci 76:R16–R28
Tang S, Hettiarachchy NS, Shellhammer TH (2002) Protein extraction from heat-stabilized defatted rice bran. 1. Physical processing and enzyme treatments. J Agric Food Chem 50:7444–7448
Tang S, Hettiarachchy NS, Horax R, Eswaranandam S (2003) Physicochemical properties and functionality of rice bran protein hydrolyzate prepared from heat-stabilized defatted rice bran with the aid of enzymes. J Food Sci 68:152–157
Tang X, He Z, Dai Y, Xiong YL, Xie M, Chen J (2010) Peptide fractionation and free radical scavenging activity of zein hydrolysate. J Agric Food Chem 58:587–593
Toorchi M, Nouri MZ, Tsumura M, Komatsu S (2008) Acoustic technology for high-performance disruption and extraction of plant proteins. J Proteome Res 7:3035–3041
Van Het Hof KH, De Boer BCJ, Tijburg LBM, Lucius BRHM, Zijp I, West CE, Hautvast JGAJ, Weststrate JA (2000) Carotenoid bioavailability in humans from tomatoes processed in different ways determined from the carotenoid response in the triglyceride- rich lipoprotein fraction of plasma after a single consumption and in plasma after four days of consumption. J Nutr 130:1189–1196
Vincent D, Wheatley MD, Cramer GR (2006) Optimization of protein extraction and solubilization for mature grape berry clusters. Electrophoresis 27:1853–1865
Wang M, Hettiarachchy NS, Qi M, Burks W, Siebenmorgen T (1999) Preparation and functional properties of rice bran protein isolate. J Agric Food Chem 47:411–416
Wang W, Scali M, Vignani R, Spadafora A, Sensi E, Mazzuca S, Cresti M (2003) Protein extraction for two-dimensional electrophoresis from olive leaf, a plant tissue containing high levels of interfering compounds. Electrophoresis 24:2369–2375
Wang W, Vignani R, Scali M, Cresti M (2006) A universal and rapid protocol for protein extraction from recalcitrant plant tissues for proteomic analysis. Electrophoresis 27:2782–2786
Wang W, Tai F, Chen S (2008a) Optimizing protein extraction from plant tissues for enhanced proteomics analysis. J Sep Sci 31:2032–2039
Wang Y, Wang Z, Cheng S, Han F (2008b) Aqueous enzymatic extraction of oil and protein hydrolysates from peanut. Food Sci Technol Res 14:533–540
Wang Z, Wang W, Wang W, Xu L, Chen G, Fu F (2011) Separation and determination of β-casomorphins by using glass microfluidic chip electrophoresis together with laser-induced fluorescence detection. J Sep Sci 34:196–201

Watchararuji K, Goto M, Sasaki M, Shotipruk A (2008) Value-added subcritical water hydrolysate from rice bran and soybean meal. Bioresour Technol 99:6207–6213
Wu FS, Wang MY (1984) Extraction of proteins for sodium dodecyl-sulfate polyacrylamide-gel electrophoresis from protease-rich plant-tissues. Anal Biochem 139:100–103
Xu C, Garrett WM, Sullivan J, Caperna TJ, Natarajan S (2006) Separation and identification of soybean leaf proteins by two-dimensional gel electrophoresis and mass spectrometry. Phytochemistry 67:2431–2440
You L, Zhao M, Regenstein JM, Ren J (2010) Purification and identification of antioxidative peptides from loach (*Misgurnus anguillicaudatus*) protein hydrolysate by consecutive chromatography and electrospray ionization-mass spectrometry. Food Res Int 43:1167–1173
Zhang SB, Wang Z, Xu SY (2007) Downstream processes for aqueous enzymatic extraction of rapessed oil and protein hydrolysates. Am Oil Chem Soc 84:693–700
Zhang SB, Lu QY, Yang H, Li Y, Wang S (2011) Aqueous enzymatic extraction of oil and protein hydrolysates from roasted peanut seeds. J Am Oil Chem Soc 88:727–732
Zhen Y, Shi J (2011) Evaluation of sample extraction methods for proteomic analysis of coniferous seeds. Acta Physiol Plant 33:1623–1630
Zheng Q, Song J, Doncaster K, Rowland E, Byers DM (2007) Qualitative and quantitative evaluation of protein extraction protocols for apple and strawberry fruit suitable for two-dimensional electrophoresis and mass spectrometry analysis. J Agric Food Chem 55:1663–1673

Chapter 3
Primary Separation: 2-D Electrophoresis

Romina Pedreschi

3.1 Introduction

A typical workflow (Fig. 3.1) in proteomics involves: (1) protein extraction, (2) protein/peptide separation and quantification, (3) protein identification/quantification, and (4) data analysis and interpretation. Protein/peptide separation can be either gel-based or gel-free. They differ in the way proteins are isolated, separated, and detected. A gel-based approach relies on 2DE for protein separation and quantification whereas a gel-free approach relies on LC for separation of either entire proteins or peptides. This chapter focuses on protein separation and quantification using two-dimensional electrophoresis (2DE).

3.2 Standard 2DE

Two-dimensional electrophoresis dates back to 1975 when it was introduced for the first time by O'Farrell (O'Farrell 1975). Standard 2DE is carried out under denaturing conditions. This is accomplished by the use of disulphide-reducing (e.g., dithiothreitol, mercaptoethanol), chaotropic (urea, thiourea), and thiol alkylating additives (e.g., iodoacetamide). It involves separation of proteins in two dimensions using different protein properties. The complete workflow involves several steps: first-dimension separation or isoelectric focusing of proteins (IEF), second-dimension

R. Pedreschi (✉)
Food and Biobased Research,
P.O Box 17, Wageningen, UR 6700 AA, The Netherlands
e-mail: romina.pedreschiplasencia@wur.nl; r_pedreschi@yahoo.com

F. Toldrá and L.M.L. Nollet (eds.), *Proteomics in Foods: Principles and Applications*, Food Microbiology and Food Safety 2, DOI 10.1007/978-1-4614-5626-1_3,

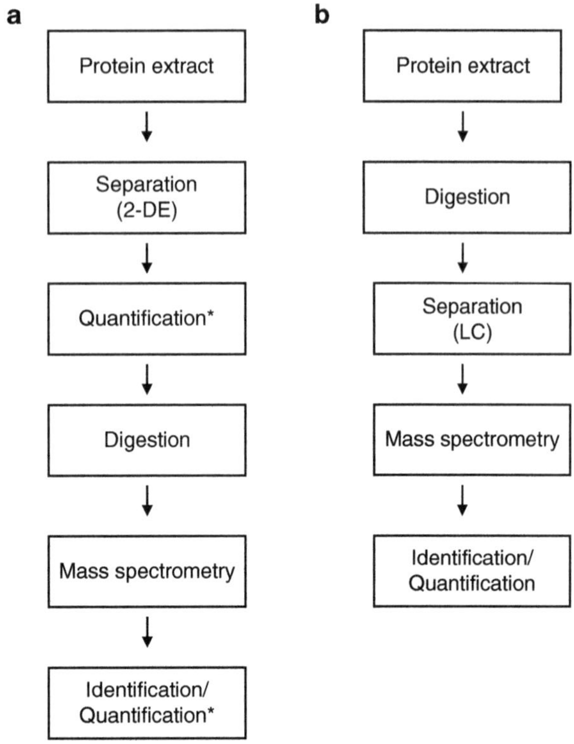

Fig. 3.1 Gel-based (**a**) versus gel-free (**b**) typical protein analysis workflows. In a gel-based approach, proteins are separated using 2-DE. After image analysis proteins can be relatively quantified and then selected proteins are further digested for identification via mass spectrometry. In a gel-free approach, protein digests (peptides) are separated via LC and further identified via mass spectrometry analysis. Label free quantification or absolute quantification using stable isotope peptides is feasible

separation or Dalt separation of proteins, protein staining and visualization, image analysis, data pre-processing, and statistical analysis to account for differentially expressed proteins that will be further sent for mass spectrometry identification.

The first-dimension separation or IEF consists of using the differences in the electric charge of proteins, taking advantage of their amphoteric character. Depending on the pH of the environment, proteins change their charge. Separation is accomplished by applying an electric current to an IPG (immobilized pH gradient) strip. Thus, if a protein is in a pH region below its isoelectric point (pI), it will be positively charged and will migrate towards the cathode. When proteins reach their pI, they will stop migrating and be "focused." Thanks to the introduction in the middle 1990s of IPG gels with a plastic backing, so-called strips instead of tubes of carrier ampholites, superior reproducibility, resolution, and loading capacity are possible (Penque 2009). The commercial availability of precast IPG gels from narrow to broad IPG ranges allows a wide range of choices. For extensive and detailed information on IEF and IPG technology, the reader is referred to Righetti and Bossi (1997) and Görg et al. (1999, 2000).

In the second dimension, the IPG strip with the separated proteins via IEF is used as the starting point for second-dimension separation. Molecular weight-based protein separation is carried out in sodium dodecyl sulphate (SDS) polyacrylamide

(PAGE) gels better known as SDS-PAGE. Before this second separation is performed, proteins on the IPG strip are equilibrated in an SDS excess in order to eliminate the intrinsic charges of proteins. Thus, secondary and tertiary structures of proteins are eliminated and after reduction of disulphide bridges between cysteins, the electrophoretic mobility of proteins is solely dependent on molecular weight. Given the different particle sizes of the separating acrylamide gel, smaller proteins will move faster than larger molecules. A two-dimensional gel containing hundreds to thousands of proteins can be resolved on a single run (Fig. 3.2).

Once the separation is accomplished, proteins need to be visualized. A wide range of visualisation strategies exists for quantitative analysis of the separated proteins. The most commonly used are colloidal coomassie blue (CCB) (Neuhoff et al. 1988), silver staining (Blum et al. 1987), radiolabeling (Patton 2002), and fluorescence (Chevalier et al. 2004). For more extensive and detailed information about the different staining strategies, the reader is referred to Miller et al. (2006). Before a staining strategy is chosen, the different limitations and advantages of each should be evaluated. The different staining strategies (e.g., colloidal coomassie blue vs. silver) differ in achievable detection limits (8–10 ng for CCB vs. 0.1–1 ng for silver), dynamic range (10^3 for CCB vs. 10^2 for silver), and reproducibility. Classical 2DE comparative analysis presents the limitation of high gel to gel variation which renders quantitative or differential analysis difficult because technical and biological variability can be confused with true treatments. Differential gel electrophoresis technology known as DIGE was developed (Űnlü et al. 1997) to circumvent the inherent variability of classical 2DE, thus increasing confidence in terms of detection and quantification of differences in protein abundance and reducing the number of gels needed in an experiment (Alban et al. 2003; Timms and Cramer 2008; Minden et al. 2009).

3.2.1 Difference Gel Electrophoresis (DIGE)

Differential in gel electrophoresis combined with 2DE is well suited to separate and account for differential protein analysis. DIGE was introduced to overcome the poor reproducibility of 2DE. Compared to classical staining (e.g., silver, CBB), proteins present in a sample are labeled with spectrally resolvable fluorescent dyes: NHS-Cy2 (MW: 550.6 Da, excitation: 480 nm, emission: 530 nm), NHS-Cy3 (MW: 582.8 Da, excitation: 540 nm, emission: 590 nm), and NHS-Cy5 (MW: 580.7 Da, excitation: 620 nm, emission: 680 nm) prior to 2DE separation. The sensitivity of DIGE is extremely high: <1 *f*mol of protein and a dynamic range of 10^5 (Lilley et al. 2002; Viswanathan et al. 2006).

The use of labels that have different wavelengths for detection allows the co-separation of two or more differentially labeled samples on the same gel (Timms and Cramer 2008). These reagents react with primary amine groups (N-terminal α-amino and lysine ε-amino groups) in the protein sample through nucleophilic substitution. This is known as "minimal labeling" because two different species per

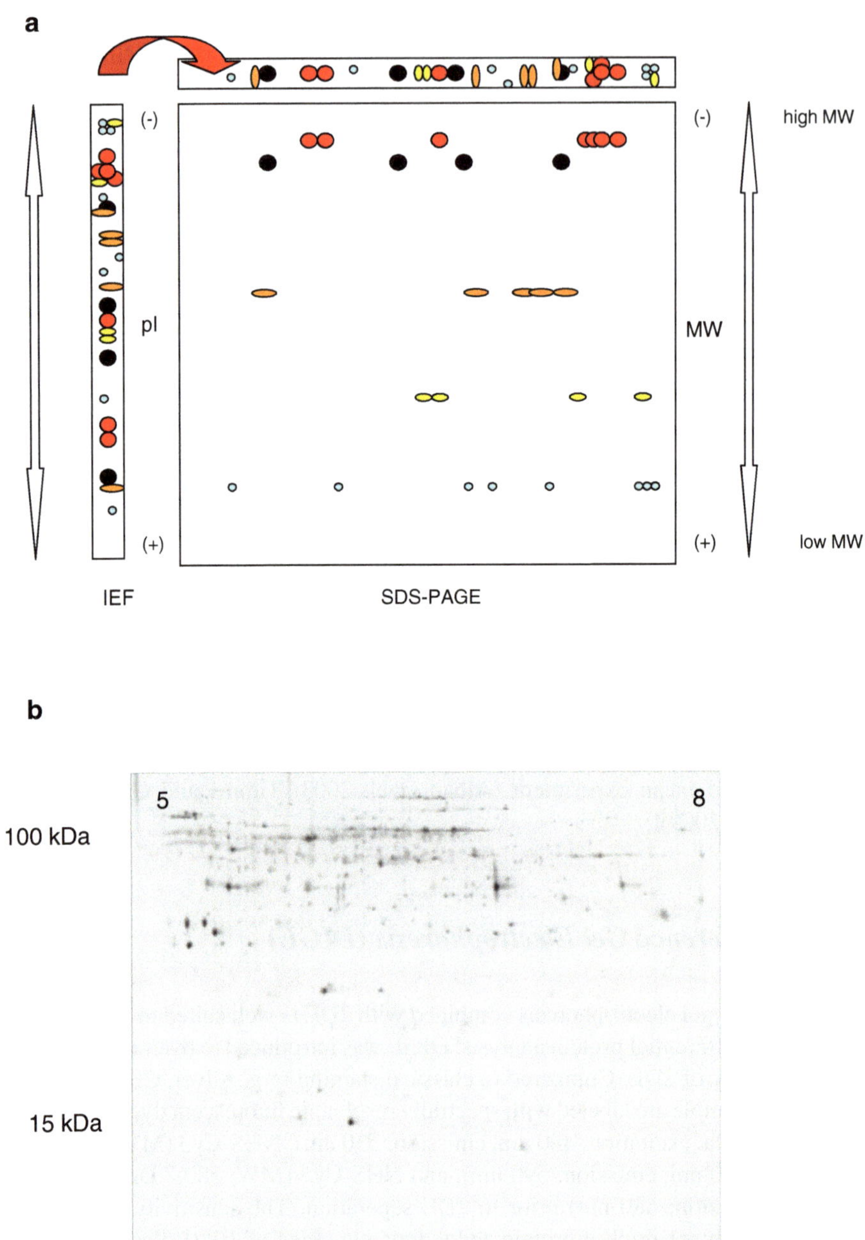

Fig. 3.2 (**a**) Schematic representation of a two-dimensional electrophoresis separation. Proteins are separated based on isoelectric point (pI) and molecular weight (MW). (**b**) Two-dimensional electrophoresis (2-DE). Proteins are separated based on isoelectric point (pI) on the horizontal direction and based on molecular weight (MW) on the vertical direction. A representative gel for pear tissue stored under optimal conditions. Proteins were separated via 2-DE (24 cm gel, pI 5–8, 12.5% acrylamide) and the gel was visualized with silver staining. A total of 35 μg protein was loaded on the gel

protein are created representing the labeled ones 1–2% and the unlabeled ones 98–99% (Hrebicek et al. 2007). Protein samples are labeled in denaturing 2DE lysis buffer at an optimal pH 8.0–8.5 in the absence of primary amines (e.g., buffers and carrier ampholites) and other reducing reagents that show reactivity (Timms and Cramer 2008). The labeling reaction is stopped with an excess of primary amine (free lysine) and then samples are reduced. These dyes are pKa matched with the ε-amino group of lysine to avoid a pI shift upon derivatization (Hrebicek et al. 2007). Depending on the dye, the molecular weight increase of proteins is between 434 and 464. The dye effect is supposed to have little impact on most proteins, except for proteins of low molecular mass (Hrebicek et al. 2007).

Equal protein amounts of differentially labeled samples with NHS-Cy3 and NHS-Cy5 are mixed and 2DE run. A third sample or "pooled internal standard" labeled with NHS-Cy2 consisting of an equal mixture of all test samples to ensure complete representation of all protein species can be introduced. The main advantage of multiplexing samples is that the pooled internal standard is used to normalize protein abundance across multiple gels. Thus, each gel will contain an image with a highly similar spot pattern so improved confidence will be attained for inter-gel spot matching and quantification. The complete DIGE process can be summarized as: labeling reaction, 2DE, acquisition of separate images for each sample, and software-aided analysis of the images for assessment of differences in expression (Fig. 3.3).

"Saturation" labeling is also possible with DIGE; it includes cysteine-reactive labels (Cy3 and Cy5 maleimide derivatives). It can be an interesting option when the protein amount is scarce. Fewer cysteine residues than lysine residues exist in proteins, so theoretically a higher stoichiometric labeling of protein cysteines can be obtained without any compromise in sample solubility (Timms and Cramer 2008).

3.2.2 Image Analysis

After classical staining or DIGE 2DE, the different gels corresponding to the different treatments need to be submitted to image analysis to account for differences in protein expression. The performance of image and data analysis to assess differences in expression is dependent on the quality of the raw data which will have a huge impact on the final conclusions. Thus, it is essential that experimentally caused artifacts are reduced or eliminated and that the scanning devices are configured in an optimal manner (Berth et al. 2007). Image and data analysis involves: image acquisition, image warping, spot detection and editing, quantification and normalization, application of a consensus spot pattern (optional), spot matching, and statistical analysis. There are several commercially available software packages for 2DE and DIGE image analysis such as: PDQuest, Decyder 2-D, ImageMaster Platinum, and Progenesis SameSpots among others (Berth et al. 2007). For detailed information, the reader is referred to Berth et al. (2007).

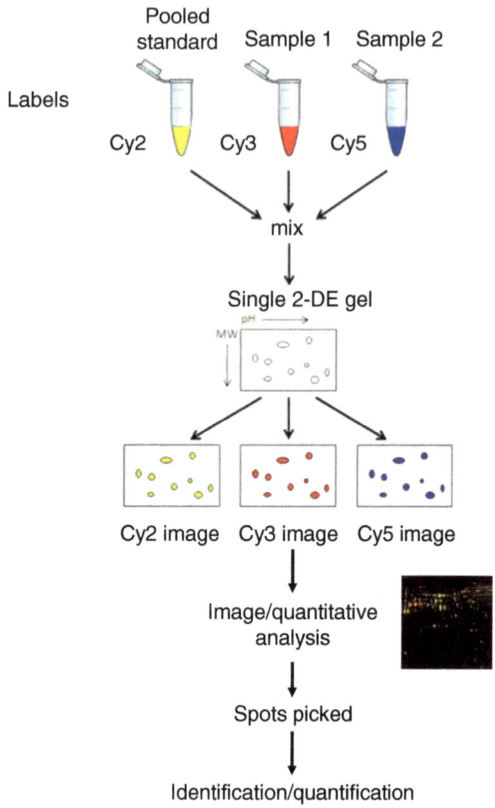

Fig. 3.3 Difference in gel electrophoresis (DIGE) 2-DE workflow. Up to three samples can be run on a single run after labelling with NHS-Cy dyes. In a 3 dye approach, samples are labelled with Cy3 and Cy5 and a pooled standard (mix of equal amounts of each sample analyzed) is labelled with Cy2

The first step involves the acquisition of the gel images with scanners coupled to charged couple devices (CCD), or camera-based or laser imaging devices. During this step, quantitative information from the gel will be transformed into computer-readable data. The next step is the removal of variations in spot positions known as "warping." There are two main ways commercial software performs the work of analyzing gel images to match the same spots on different gels: (1) one involves image warping and the creation of a fusion gel before spot detection. The fusion gel contains all spots from all images. Thus, the boundaries of all spots from the fusion gel are transferred back to the other images, so no missing values are generated (Albrecht et al. 2010). However, there are technical issues such as gel distortions, incorrect merging/splitting, and others that need to be corrected; (2) the second approach consists of detecting spots on every single gel before spot matching across the different gels. The image containing the maximum number of spots is usually chosen as the master gel and the different other images are matched to it. Differences in spot positions across different gels will be corrected as it is extremely important for accurate spot matching and further construction of expression profiles. The differences observed in the position of proteins on 2DE from separation to separation are due to variations in the pH value of the running buffer, incomplete polymerization of the gel matrix, or interference of high abundant proteins with the pH gradient in the IPG gel, among others (Berth et al. 2007). This procedure generates missing data which are discussed further. These variations can be circumvented up to a certain point by multiplexing samples (DIGE) because multiple samples are run on the same gel, but biological replicates are still run per treatment in a randomized form, in practice meaning that multiple gels are run.

The next step involves spot detection and normalization of spot quantities. Normalization deals with the mitigation of systematic differences between images due to protein loading, imaging exposure time, and dye/staining efficiency. During this step raw spot volumes are related to spot volumes on the same (vertical) and other gels (horizontal). The DIGE setup uses a horizontal normalization. The Cy2 channel is loaded with the common internal standard (composed of equal amounts of all the samples used in the experiment). Spot intensities in the other channels (Cy3 and Cy5) are normalized by dividing them with the corresponding one on the Cy2 channel on the same gel. The last step before comparison of expression profiles is spot matching or the comparison of spot intensities over the complete experiment (each spot needs to be mapped to corresponding spots in the other gels). This process is not trivial due to differences in migration in spot patterns across different gels, spots being in the threshold of detection, badly resolved spot clusters. Whatever the reason, problems related to spot matching will have a huge impact on the subsequent statistical analysis because they will generate missing values or wrong values that need to be manually corrected and treated as missing values. How missing values are tackled and handled will have a significant impact on the statistical analysis.

3.2.3 *Data Analysis*

Gel-based proteomics that rely on 2DE generate huge amounts of data which if improperly handled can bias conclusions. Thus, a good experimental design is the starting point to limit systematic errors, improve performance and precision of any statistical test used, and reduce or limit the number of false positive calls.

A good experimental design demands careful assessment of the sources of variation which can be classified either as technical or biological. Technical replicates are repeated measures from the same biological sample whereas biological replicates are different replicate samples from the same treatment group (Karp et al. 2005; Hollywood et al. 2006). It is very important to be aware of the differences in the sources of variation, because the type of replicate will limit the statistical test that can be used and the conclusions that can be drawn.

Systematic errors due to unknown sources of variation can be prevented by randomization, facilitating the comparison of treatments and precision in the estimation of the results (Chich et al. 2007).

3.2.3.1 Data Pre-processing Steps

The statistical analysis of 2DE experiments demands some pre-processing steps beforehand. The normalization of the spot volumes to remove effects related to differences in loading and staining is one step as previously discussed in the image analysis section. Many statistical tests are based on the normal distribution assumption. It might be necessary to apply some type of data transformation (e.g., logarithmic, sinh transformation gives the hyperbolic sin) to stabilize the variance (e.g., highly abundant proteins have larger variance than low abundant proteins) and thus fulfill the second assumption of "homoscedasticity" of parametric statistical methods (Urfer et al. 2006; Jung et al. 2006).

Missing values, typical of 2DE-generated data should be handled with extreme care (Krogh et al. 2007; Pedreschi et al. 2008). Particular interest should be given to the potential causes of missing data, because how they are handled will have a huge impact on the making of further conclusions. The main causes of missing value occurrence are (1) spots below a threshold of detection limit, (2) mismatches caused by distortions in protein patterns, (3) spot absence due to bad transfer from the first to the second dimension, or (4) truly absent spots from the samples (Pedreschi et al. 2008). 2DE data can present up to 50% missing data (Krogh et al. 2007). How gel images are analyzed by commercial software (refer to Sect. 4.1.2) will determine the presence of missing data. Regardless of the approach used: warping and generation of a fuse gel (supposedly no missing data) or spot detection in all gels and matching of gels to the master, pros and cons are encountered (Karp et al. 2008; Silva et al. 2010). Several methods are proposed to deal with and/or impute missing data: row average method, *k*-nearest neighbor (KNN), singular value decomposition (SVD), impute algorithm (Troyanskaya et al. 2001), Bayesian principal component analysis (BPCA), missing value estimation method (Oba et al. 2003; Pedreschi et al. 2008), and maximum likelihood algorithm (Krogh et al. 2007), among others. The performance of the methods is dependent on the structure of the data. The missing data problem is being acknowledged and several approaches to deal with it have been proposed (Pedreschi et al. 2008; Rosenberg et al. 2010; Miecznikowski et al. 2010; Li et al. 2011).

3.2.3.2 Univariate Data Analysis: Hypothesis-Driven Methods

Probably the most common scenario in 2DE studies is the comparison of two treatment groups (e.g., diseased condition) versus a control group (e.g., healthy condition). 2DE experiments will generate hundreds or thousands of proteins to be compared across different treatments. Univariate statistical tests will compare protein by protein looking for significant differences in expression between the treatment and control group. This is an hypothesis-driven scenario in which the null hypothesis will test for no significant differences. A significance level α is chosen and if the p-value of the test is lower than the α-value, then the null hypothesis is rejected.

Parametric univariate tests such as the *t*-test (to compare two treatments) and analysis of variance, ANOVA (to compare more than two treatments), and post hoc tests (e.g., Duncan, Tukey, etc.) are powerful but require that the normal distribution assumption be fulfilled as well as the homoscedasticity and independent sampling assumptions. The main problem with univariate statistical tests for 2DE data is the multiple testing scenario which can generate a large number of false positives. Thus, it is important to control the number of false positives but still be able to detect true differences or the so-called "power." The power of the test needs to be chosen in order to determine an appropriate sample size. The power is also dependent on the variance of the data, significance level α, the magnitude of the expression change to be detected, and the type of test procedure (Urfer et al. 2006). To anticipate and accept a certain amount of false positives, the false discovery rate approach (FDR) seems a reasonable procedure (Karp et al. 2007). In FDR, *q*- and *p*-values are in play. The *q*-value is the adjusted *p*-value for each test. Thus, the number of false positives is controlled in those tests that were significant. It is less conservative than the Bonferroni approach and has greater ability to find truly significant differences (Karp et al. 2007; Berth et al. 2007).

Nonparametric statistical tests are needed when the normality assumption is not fulfilled. They do not assume any distribution of the data but have less power than parametric tests. However, other assumptions still hold (e.g., independent sampling, continuous data). The Kolmogorov–Smirnov test is equivalent to the *t*-test in the case of comparing two treatments and the Kruskal–Wallis test is the equivalent to ANOVA when comparing more than two treatments. Technical replicates should be clearly differentiated from biological replicates in order to apply any statistical test correctly. In the case where technical and biological replicates are available, a nested ANOVA is able to assess that the variance due to treatments is greater than the variance related to technical aspects (Karp et al. 2007).

3.2.3.3 Multivariate Data Analysis: Hypothesis-Independent Methods

Hypothesis-independent methods allow the analysis of high-dimensional data such as 2DE data (Berth et al. 2007). Meaningful and useful information such as similarities and correlations can be extracted by exploiting the high-dimensional character of 2DE data. Multivariate statistical techniques will treat the variables (proteins) as a group of correlated variables rather than focusing on one protein at a time as univariate statistics does. It should not be forgotten that proteins are part of the same pathways and/or are interconnected with other pathways. The strong correlation that exists among the different variables can be used and extracted by utilizing multivariate statistical tools such as hierarchical clustering, principal component analysis (PCA), or partial least squares discriminant analysis (PLS-DA) (Karp et al. 2005; Pedreschi et al. 2008, 2010 and others).

Hierarchical clustering allows the grouping of expression profiles (gels/proteins) by similarity. It can be used for outlier detection and to identify structures in the data (e.g., cluster composition allows identification of the structure of the experiment so replicates from the same sample/treatment should end up in the same cluster) (Beth et al. 2007).

Principal component analysis is an unsupervised technique also referred to as explorative data analysis (Carpentier et al. 2008). New variables (principal components) that are linear combinations of the original variables are obtained and thus the essential data patterns of the original data are captured in a reduced form. PCA is recommended as a useful technique to examine datasets with multiple collinearity (e.g., proteins that exert functions in concert with other proteins) and to get insight into patterns and trends (Karp et al. 2005). PCA will display score plots showing the distribution of the samples (gels) allowing, for example, the identification in a first instance of outliers through the Hotelling T^2 ellipse. In addition, loading plots will show the distribution of the different variables (proteins). The further the variable from the origin, the more influential is the variable in explaining relationships in the dataset. The superimposition of both the score and loading plots will yield a biplot in which discrimination between treatments/samples and relevant proteins can be assessed directly.

Partial least squares discriminant analysis (PLS-DA) is a supervised bilinear regression model that creates prediction models of one or several responses from a set of factors (Wold et al. 1987). In PLS-DA latent variables (LV) are constructed in such a way that a maximum separation is obtained among them. In addition to PCA, PLS-DA is useful for correlating variables in a dataset with membership (Karp et al. 2005) and selecting important variables involved in class discrimination. Score and loading plots are obtained and can be interpreted the same way as in PCA. Other plots such as variable importance (VIP), model coefficients, residuals, distances to model plots, and validation plots exist (Danvind 2002).

It might be technically and economically infeasible to try to identify (e.g., through mass spectrometry) all proteins from a 2DE experiment even though it might be biologically relevant not only to look for differentially expressed proteins but also for the ones that do not change. In addition, treatment distinction might be due to subtle changes in protein expression. The variable importance procedure has been suggested as a practical approach to identify the variables that are more important in class distinction (Karp et al. 2005) and it seems reasonable to use it when the number of proteins to be identified is dependent on budget limitations, to get a general snapshot, and so on (Pedreschi et al. 2007). In this procedure, the VIP coefficient of a protein is calculated as a weighted sum of the squared correlations between the PLS-DA components and the original variables. The weights correspond to the percentage variation explained by the PLS-DA component in the model. The number of terms in the sum is dependent on the number of PLS-DA components found to be necessary to allow class discrimination (Fig. 3.4). Exclusion of many variables from the model should receive special attention, because important explanatory information can be lost. This approach has been used with 2DE data and combined with visualization tools (Karp et al. 2005; Pedreschi et al. 2007) (Fig. 3.5). For more details about PLS and VIP, the reader is referred to Norden et al. (2005).

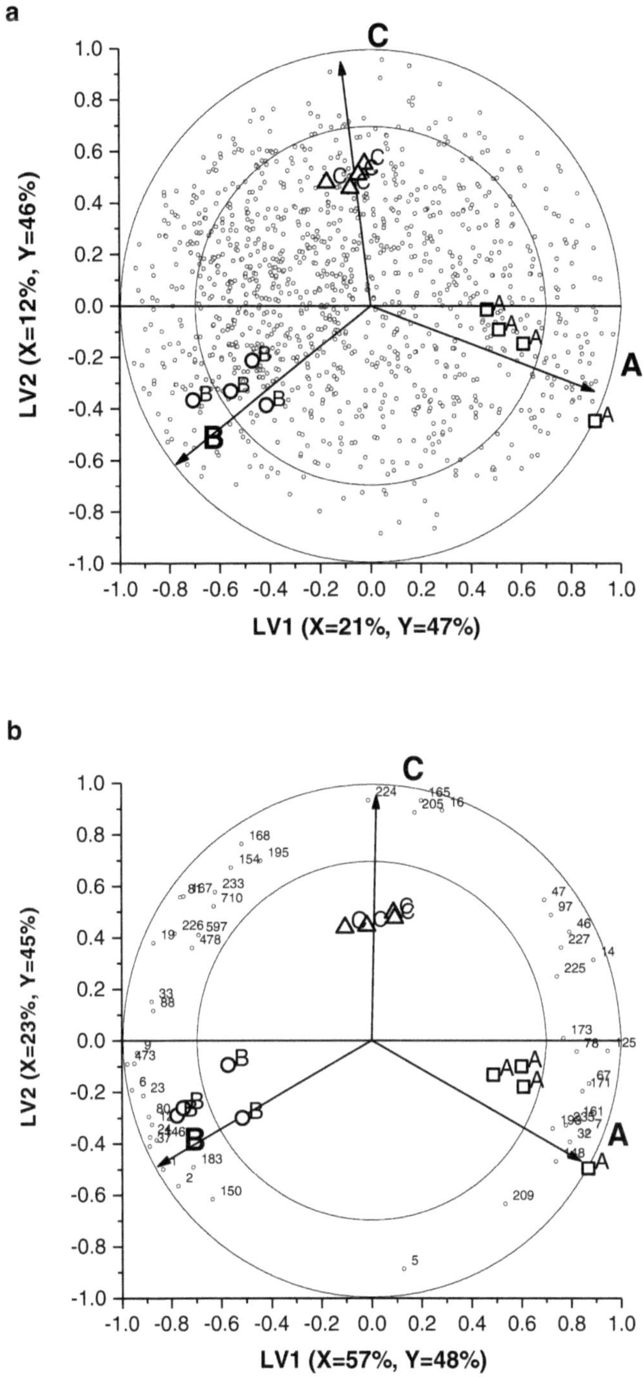

Fig. 3.4 Partial least square discriminant analysis (PLS-DA) of samples representing three different treatments (A, B, C). Discrimination among the treatments can be observed. The *small dots* represent all the proteins separated on 2-DE gels (variables). (**a**) PLS-DA based on all variables (proteins). (**b**) After the VIP procedure was applied. The most representative proteins of each treatment are represented (Reproduced from [Pedreschi et al., 2010] with permission from [Copyright © 2010, Taylor & Francis])

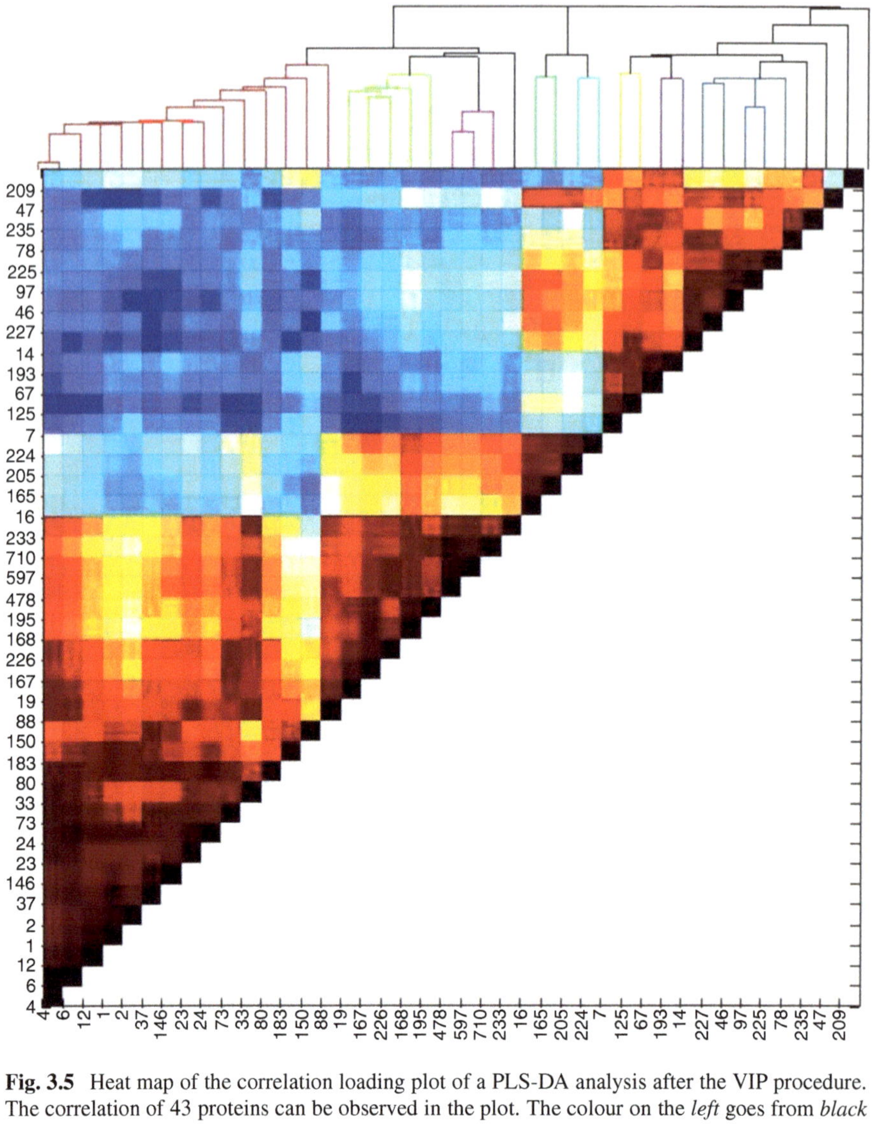

Fig. 3.5 Heat map of the correlation loading plot of a PLS-DA analysis after the VIP procedure. The correlation of 43 proteins can be observed in the plot. The colour on the *left* goes from *black* '1' indicative of positive correlation to *blue* '-1' indicative of negative correlation. LC-MS/MS identification of the different correlated proteins showed that many belong to similar pathways or interconnected pathways (Reproduced from [Pedreschi et al. 2009] with permission from [Copyright © 2009, American Chemical Society])

3.2.4 The Pros and Cons of 2DE

2DE turns difficult for very acidic or very hydrophobic proteins. However, it is possible to deal with these proteins with special characteristics up to a certain extent by introducing variants in gel media composition or by using alternative surfactants and solvents (Miller et al. 2010). Extremely basic proteins are problematic during 2DE because of hydrolysis of acrylamide at pH >10 (Carpentier et al. 2008). Modifications in gel composition have been introduced in order to obtain reproducibly and commercially available basic IPG strips (Görg et al. 1997). Membrane proteins, for example, are problematic for 2DE but they are key players in cell biology (e.g., signal transduction, stress sensors, cell–cell communication, cellular and organellar trafficking and transport, and formation of electron transfer chains in the mitochondria and plastids; Carpentier et al. 2008). They can represent up to 30% of the total proteome (Santoni et al. 2000). This category of proteins is very important in food applications such as those involving temperature reduction (Gómez-Galindo et al. 2007; Pedreschi et al. 2010). Alternative techniques to 2DE that involve use of different ionic detergents and acrylamide concentrations are recommended (Buxbaum 2003; Rais et al. 2004) to deal with these problematic proteins.

Proteins with high (>150 kDa) or low (<10 kDa) molecular mass cannot be resolved with 2DE. High molecular weight proteins are poorly transferred from the first to second dimension whereas the classical Laemmli SDS-PAGE system lacks resolving power below 10 kDa (Carpentier et al. 2008). For high molecular weight proteins, different strategies to optimize separation mainly focused on specific matrices are discussed and detailed by Miller et al. (2010). The incorporation of DIGE in 2DE analysis might introduce some shifts in molecular weight especially in the low molecular range. During minimal labeling, the additional mass of the dye will cause labeled proteins (1–2%) to migrate more slowly than the unlabeled part. This difference in migration might be an issue after image analysis of differentially expressed proteins sent for identification. Thus, it is recommended to post-stain the DIGE gels and align the fluorescent and post-stained images to assure accuracy in spot picking (Timms and Cramer 2008). Protein-specific dye bias during DIGE labeling for certain proteins complicates quantification. This issue cannot be corrected by global normalization. Thus, there is need for this type of protein either to apply alternative normalization strategies or validate the differentially expressed proteins by more specific methods. Mass spectrometry analysis unless based on high resolution mass spectrometry faces a similar problem (Timms and Cramer 2008). The assumption that in spots where co-migration of proteins (confirmed through mass spectrometry tools) is observed the most abundant protein species in that spot is responsible for the difference in spot abundance might be and is often wrong. When this problem is encountered, validation is needed. Approaches that can be used to counteract this problem are the use of narrow-range IPG strips, introduction of sample fractionation methods, different sample preparation methods, and modifications during 2DE (Hunsucker and Duncan 2006).

Despite all the limitations listed above, 2DE seems to be the most logical choice for the separation of entire proteins from nonmodel organisms such as would be the case for many food proteomes. 2DE has the advantage in contrast to other proteomics techniques that base the quantitative protein analysis on peptide and not at the protein level, losing important information on isoform specific expression and post-translational modifications (Timms and Cramer 2008; Carpentier et al. 2008). Modifications introduced, for example, during food processing might be potentially noticed by 2DE (Pischetsrieder and Bauerlein 2009; Pedreschi et al. 2010). 2DE is difficult to automate and operator skills have a great influence on control of technical variation.

Standard 2DE as mentioned before is not suitable for the study of hydrophobic (e.g., membrane proteins) (Vertommen et al. 2011; Kota and Goshe 2011) or native proteins, and protein complexes (e.g., electrophoresis needs to be run under nondenaturing conditions). For a complete overview of alternative approaches to 2DE, the reader is referred to Miller et al. (2010). Variants to standard 2DE protocols can be divided into (1) run under native conditions during one or both electrophoretic runs and (2) run under nonreducing conditions during one or both electrophoretic runs.

3.3 2DE Under Native Conditions

IEF followed by gradient gel electrophoresis (GGE) is suitable, for example, to resolve proteins based on native pI and native molecular weight. In this case, the separation is done in the first and native in the second dimension. Thus, the native structure is preserved and it finds applications, for example, in the study of protein–protein interactions (Miller et al. 2010). It is possible instead of IEF in the first dimension, to perform separation based on charge density (agarose electrophoresis).

For the analysis of protein complexes in vitro, running native in the first dimension and denaturing in the second dimension is being used. It is based on the assumption that proteins that form complexes with specific ligands migrate at a different pI than their unbound counterparts (Miller et al. 2010). Native one-dimensional electrophoresis makes use of nonionic detergents both in the sample as well as in the gel phase. Both unfolding and dissociation of interacting proteins is prevented. The two protocols more extensively mentioned are clear native electrophoresis (CNE) and blue native electrophoresis (BNE).

3.4 2DE Under Nonreducing Conditions

Nonreducing conditions in both dimensions find useful applications with samples that generate spot aggregation in gels (e.g., biological fluids). By using nonreducing conditions, image analysis, quantification, spot analysis, and MS analysis are favored (Miller et al. 2010). Nonreducing conditions in the first dimension and reducing conditions in the second dimension (known as diagonal electrophoresis) are useful. For more specific details on the available protocols, the reader is referred to Miller et al. (2010).

References

Alban A, Olu S, Bjorkesten L, Andersson C, Sloge E, Lewis S, Currie I (2003) A novel experimental design for comparative two-dimensional gel analysis: two-dimensional gel electrophoresis incorporating a pooled internal standard. Proteomics 3:36–44

Albrecht D, Kniemeyer O, Brakhage A, Guthke R (2010) Missing values in gel based proteomics. Proteomics 10:1202–1211

Berth M, Moser FM, Kolbe M, Berhardt J (2007) The state of the art in the analysis of two-dimensional gel electrophoresis images. Appl Microbiol Biot 76:1223–1243

Blum H, Beier H, Gross H (1987) Improved silver staining of plant-proteins, RNA and DNA in polyacrylamide gels. Electrophoresis 8:93–99

Buxbaum E (2003) Cationic electrophoresis and electrotransfer of membrane glycoproteins. Anal Biochem 314:70–76

Carpentier SC, Panis B, Vertommen A, Swennen R, Sergeant K, Renaut J, Laukens K, Witters E, Samyn B, Devreese B (2008) Proteome analysis of non-model plants: a challenging but powerful approach. Mass Spectrom Rev 27:354–377

Chevalier F, Rofidal V, Vanova P, Bergoin A, Rossignol M (2004) Proteomic capacity of recent fluorescent dyes for protein staining. Phytochemistry 65:1499–1506

Chich JF, David O, Villers F, Schaeffer B, Lutomski D, Huet S (2007) Statistics for proteomics: experimental design and 2-DE differential analysis. J Chromatogr B 849:261–272

Danvind J (2002) PLS prediction as a tool for modelling wood properties. Holz als Rob-und Werstoff 60:130–140

Gómez-Galindo F, Sjöholm I, Rasmussion A, Widell S, Kaack K (2007) Plant stress physiology: opportunities and challenges for the food industry. Crit Rev Food Sci Nutr 46:749–763

Görg A, Obermaier C, Boguth G, Csordas A, Diaz J, Madjar J (1997) Very alkaline immobilized pH gradients for two dimensional electrophoresis of ribosomal and nuclear proteins. Electrophoresis 18:328–337

Görg A, Obermaier C, Boguth G, Weiss W (1999) Recent developments in two-dimensional gel electrophoresis with immobilized pH gradients: wide pH gradients up to 12, longer separation distances and simplified procedures. Electrophoresis 20:712–720

Görg A, Obermaier C, Boguth G, Harder A, Scheibe B, Wildgruber R, Weiss W (2000) The current state of two dimensional electrophoresis with immobilized pH gradients. Electrophoresis 21:1037–1053

Hollywood K, Brison D, Goodacre R (2006) Metabolomics: current technologies and future trends. Proteomics 6:4716–4723

Hrebicek T, Dürrschmid K, Auer N, Bayer K, Rizzi A (2007) Effect of CyDye minimun labeling in differential gel electrophoresis on the reliability of protein identification. Electrophoresis 28:1161–1169

Hunsucker SW, Duncan MW (2006) Is protein overlap in two-dimensional gels a serious practical problem? Proteomics 6:1374–1375

Jung K, Gannoun A, Sitek B, Apostolov O, Schramm A, Meyer H, Stuhler K, Urfer W (2006) Statistical evaluation of methods for the analysis of dynamic protein expression data from a tumor study. REVSTAT-Stat J 4:67–80

Karp N, Spencer M, Lindsay H, O'Dell K, Lilley K (2005) Impact of replicate types on proteomic expression analysis. J Proteome Res 4:1867–1871

Karp N, McCormick P, Russell M, Lilley K (2007) Experimental and statistical considerations to avoid false conclusions in proteomic studies using differential in gel electrophoresis. Mol Cell Proteomics 6:1354–1364

Karp N, Feret R, Rubtsov D, Lilley K (2008) Comparison of DIGE and post-stained gel electrophoresis with both traditional and SameSpots analysis for quantitative proteomics. Proteomics 8:948–960

Kota U, Goshe M (2011) Advances in qualitative and quantitative plant membrane proteomics. Phytochemistry 72:1040–1060

Krogh M, Fernandez C, Teilum M, Bengtssin S, James P (2007) A probabilistic treatment of the missing spot problem in 2-D gel electrophoresis experiments. J Proteome Res 6:3335–3343
Li F, Nie L, Wu G, Qiao J, Zhang W (2011) Prediction and characterization of missing proteomic data in *Desulfovibrio vulgaris*. Comp Funct Genom. doi:10.1155/2011/780973
Lilley KS, Razzaq A, Dupree P (2002) Two-dimensional gel electrophoresis: recent advances in sample preparation, detection and quantification. Curr Opin Chem Biol 6:46–50
Miecznikowski J, Damodaran S, Sellers K, Rabin R (2010) A comparison of imputation procedures and statistical tests for the analysis of two-dimensional electrophoresis data. Proteome Sci 8:66
Miller I, Crawford J, Gianazza E (2006) Protein stains for proteomic applications: which, when and why? Proteomics 6:5385–5408
Miller I, Eberini I, Gianazza E (2010) Other than IPG-DALT: 2-DE variants. Proteomics 10:586–610
Minden J, Dowd SR, Meyer H, Stühler K (2009) Difference gel electrophoresis. Electrophoresis 30:S156–S161
Neuhoff V, Arold N, Taube D, Ehrhardt W (1988) Improved staining of proteins in polyacrylamide gels including isoelectric focusing gels with clear background at nanogram sensitivity using Coomassie Brilliant Blue G-250 and R-250. Electrophoresis 9:255–262
Norden B, Broberg P, Lindberg C, Plymoth A (2005) Analysis and understanding of high dimensionality data by means of multivariate data analysis. Chem Biodivers 2:1487–1494
Oba S, Sato M, Takemasa I, Monden M, Matsubara K, Ishii S (2003) A Bayesian missing value estimation method for gene expression profile data. Bioinformatics 19:2088–2096
O'Farrell PH (1975) High resolution two-dimensional electrophoresis of proteins. J Biol Chem 250(10):4007–4021
Patton WF (2002) Detection technologies in proteome analysis. J Chromatogr B 771:3–31
Pedreschi R, Vanstreels E, Carpentier S, Hertog M, Lammertyn J, Noben J, Swennen R, Vanderleyden J, Nicolaï B (2007) Proteomic analysis of core breakdown disorder in conference pears (*Pyrus communis* L.). Proteomics 7:2083–2099
Pedreschi R, Hertog M, Carpentier SC, Lammertyn J, Noben J, Panis B, Swennen R, Nicolaï B (2008) Treatment of missing values for multivariate statistical analysis of gel-based proteomics data. Proteomics 8:1371–1383
Pedreschi R, Hertog M, Robben J, Lilley K, Karp N, Baggerman G, Vanderleyden J, Nicolai B (2009) Gel-based proteomics approach to the study of metabolic changes in pear tissue during storage. J Agric Food Chem 57:6997–7004
Pedreschi R, Hertog M, Lilley KS, Nicolaï B (2010) Proteomics for the food industry: opportunities and challenges. Crit Rev Food Sci Nutr 50:680–692
Penque D (2009) Two dimensional gel electrophoresis and mass spectrometry for biomarker discovery. Proteomics Clin Appl 3:155–172
Pischetsrieder M, Bauerlein R (2009) Proteome research in food science. Chem Soc Rev 38:2600–2608
Rais I, Karas M, Schagger H (2004) Two-dimensional electrophoresis for the isolation of integral membrane proteins and mass spectrometry identification. Proteomics 4:2567–2571
Righetti P, Bossi A (1997) Isoelectric focusing in immobilized pH gradients: an update. J Chromatogr B 699:77–89
Rosenberg L, Franzén B, Auer G, Lechtio J, Forshed J (2010) Multivariate meta-analysis of proteomics data from human prostate and colon cancers. BMC Bioinformatics 11:468
Santoni V, Molloy M, Rabilloud T (2000) Membrane proteins and proteomics: un amour impossibile. Electrophoresis 21:1054–1070
Silva E, O''Gorman M, Becker S, Aver G, Eklund A, Grunewald J, Wheelock A (2010) In the eye of the beholder: does the master see the SameSpots as the novice? J Proteome Res 9:1522–1532
Timms JF, Cramer R (2008) Difference gel electrophoresis. Proteomics 8:4886–4897
Troyanskaya O, Cantor M, Sherlock G, Brown P, Hastie T, Tibshirani R, Botstein D, Altman R (2001) Missing value estimation methods for DNA microarrays. Bioinformatics 17:520–525

Űnlü M, Morgan M, Minden J (1997) Difference gel electrophoresis: a single gel method for detecting changes in protein extracts. Electrophoresis 18:2071–2077

Urfer W, Grzegorczyk M, Jung K (2006) Statistics for proteomics: a review of tools for analyzing experimental data. Proteomics Pract Appl 1:48–55

Vertommen A, Panis B, Swennen R, Carpentier S (2011) Challenges and solutions for the identification of membrane proteins in non-model plants. J Proteomics 74:1165–1181

Viswanathan S, Unlu M, Minden JS (2006) Two dimensional difference gel electrophoresis. Nat Prot 1:1351–1358

Wold S, Esbensen K, Geladi P (1987) Principal component analysis. Chemometr Intell Lab 2(1–3):37–52

Chapter 4
Primary Separation: Chromatography

M.-Concepción Aristoy, Leticia Mora, Elizabeth Escudero, and Fidel Toldrá

4.1 Introduction

Once the peptides are extracted from the food sample, or the peptides are obtained after enzymatic digestion (described in Chap. 2), further fractionation and/or concentration of the peptides must be achieved for their subsequent mass spectrometry (MS) analysis. Foods and especially food hydrolyzates, may contain a large amount of peptides that require primary separation prior to MS identification. On the other hand, some target peptides might be present in very low concentration, requiring a pre-concentration of the sample before MS. So, the final quality of the analysis will be conditioned by a good pre-fractionation and/or concentration of the peptide extract.

Physical and physicochemical properties of the peptides are considered in their fractionation. The molecular size (using ultrafiltration membranes and/or size exclusion chromatography) is the most used characteristic as a first approach. In some cases, molecular size is combined with charge in electrodialysis with ultrafiltration membranes (Doyen et al. 2011) or 2-D- or 1D-electrophoresis. Electrophoresis as the primary peptide separation is dealt with in chapter 3.

The peptide purification process should be as simple as possible and may contain a minimum number of steps. A combination of at least two complementary methods operating via different chromatographic principles, such as ion exchange chromatography and/or gel permeation chromatography and reversed phase or hydrophilic interaction chromatography result in powerful purification processes. This chapter is focused on the chromatographic methods used in the primary separation of peptides.

M.-C. Aristoy • L. Mora • E. Escudero • F. Toldrá (✉)
Instituto de Agroquímica y Tecnología de Alimentos (CSIC),
Avenida Agustín Escardino 7, 46980 Paterna, Valencia, Spain
e-mail: ftoldra@iata.csic.es

F. Toldrá and L.M.L. Nollet (eds.), *Proteomics in Foods: Principles and Applications*, Food Microbiology and Food Safety 2, DOI 10.1007/978-1-4614-5626-1_4,

4.2 First Fractionation

Whether the samples consist of food (solid or liquid) or food hydrolyzates, high amounts of other compounds such as salts or even proteins coexist with the peptides and should be removed to avoid ionization suppression in subsequent steps of mass spectrometry. Adequate methods are described in the literature and include (1) ultrafiltration with molecular weight cutoff (MWCO) membranes; (2) protein precipitation with acids, organic solvents, concentrated saline solutions, or heat; and (3) chromatographic methods such as ion exchange chromatography (IEC) and/or gel filtration chromatography (GFC).

There are many types of ultrafiltration MWCO membranes with controlled pore size disposable for the deproteinization of biological samples. Adequate membranes are those that do not interact with peptides reducing the recoveries. Materials such as cellulose, esters, or polyethersulfone, are the most used. Among them, polyethersulfone membranes exhibit no hydrophobic or hydrophilic interactions and are usually preferred for their low fouling characteristics, exceptional flux, and broad pH range compatibility. These membranes can contain trace amounts of additives (glycerin, sodium azide, etc.) that could interfere with analysis and that must be removed by rinsing with buffer solution or deionized water through the concentrator. Also the pretreatment of the device overnight with a passivation solution, such as 5% SDS, Tween® 20, or Triton® X in distilled water would optimize the peptides' recovery.

This method has also been used to obtain a solution rich in peptides under a certain molecular size (Ghassem et al. 2011; Molina et al. 1999; Sommerer et al. 1998; Moreno Arribas et al. 1998). It is also often used as a filtration technique in a series with filters of decreasing pore size (some among 10, 5, 3, 1 KDa) to obtain fractions with different molecular size peptides (Ko et al. 2012; Cho et al. 2004; Lioe et al. 2004; Park et al. 2001; Kaiser et al. 1992).

Other methods of deproteinization by using concentrated saline solutions (Cheng et al. 2010a), the adjustment of pH to 5.4 (Matoba et al. 1970), strong acids such as 5–10% trichloroacetic acid (TCA), perchloric acid (PCA) (Stoeva et al. 2000; Cambero et al. 1992), or organic solvents such as methanol or ethanol (Mora et al. 2010; Flores et al. 1997; Combes et al. 2001; Moreno Arribas et al. 1998), allow an easy handling of larger sample volume with less expense. As with ultrafiltration, these protein denaturing agents have been used to fractionate peptide extracts only by finely controlling their concentration as in cheese extracts with 2% and 12% TCA or 70% ethanol (Rohm et al. 1996; Kuchroo and Fox 1982).

Enzymatic hydrolyzates are often submitted to heat treatment in order to stop the hydrolytic reaction by enzyme denaturation. This denatured enzyme is afterwards removed from the sample by centrifugation (Ko et al. 2012). Ammonium sulfate precipitation was used by Cheng et al. (2010a) to fractionate potato protein hydrolyzate, when searching for antioxidative peptides.

Many investigations have used semi-preparative low-pressure chromatographic methods based on ion exchange (IEC) or size exclusion chromatography

(SEC) as a primary isolation or fractionation of the peptidic material. Also solid-phase extraction has been used as a pre-fractionation method of peptide extracts. These three methodologies, described below, can be used alone or combined in series.

4.2.1 Ion-Exchange Chromatography (IEC)

Ion-exchange chromatography uses the differential charge of peptides with respect to other sample compounds to isolate them. Some researchers prefer to use the IEC as a primary method for peptide fraction extraction (Suetsuna and Chen 2002; Park et al. 2001; Suetsuna et al. 2000; Lee et al. 1999) but others use IEC for selected fractions after GFC (Kim et al. 2001) or solid-phase extraction methods. Ion-exchange chromatography in HPLC columns has also been used when better resolution is required and often combined with other methods as shown in Fig. 4.1 (Kim et al. 2001; Flores et al. 2000; Sommerer et al. 1998).

Ion-exchange resins have a high loading capacity and therefore they have been used in the initial stages of peptide analysis and purification. This chromatography, when used in these early steps, is mainly performed under low pressure. Separation is achieved through electrostatic interaction between charged amino acid side chains and the surface charge of the ion-exchange resin. Two types of interactions give name to two types of IEC: in anion exchange, amine groups on the resin interact with aspartic or glutamic acid side chains; in cation exchange, acid groups on the surface of the resin interact with histidine, lysine, and arginine.

The types of matrices used as support for stationary phases in ion chromatography are shown in Table 4.1. Functional groups are attached to these matrices (except in the hydrous oxides group that is used without modification) constituting the stationary phases. The most common functional groups are based on sulfonic acids (S, SP) and carboxymethyl- (CM) for strong and weak cation exchange, respectively, or in quaternary amine (Q, QAE) and tertiary amine (diethylaminoethyl-, DEAE) for strong and weak anion exchange, respectively. Ion-exchanger functional groups can function as such only when they are ionized, therefore they are classified into strong acid or base and weak acid or base types accordingly. The strong acidic or basic functional groups are ionized over a wide pH range, in contrast to the weak acidic or basic functional groups, which are ionized over a limited pH range.

Most of the IE separations of peptides perform on strong anion-exchanger (SAX) or strong cation exchangers (SCX). The normal procedure is to adjust the pH of the sample so that the peptides are charged and thus retained in the column, and other noncharged contaminants are eluted. Afterwards, peptides are eluted by changing the pH or increasing the ionic strength by means of the eluant. In general, columns don't need to be very long because the resin only has to retain and release the peptides when the composition (ionic strength) of the eluant is adequate.

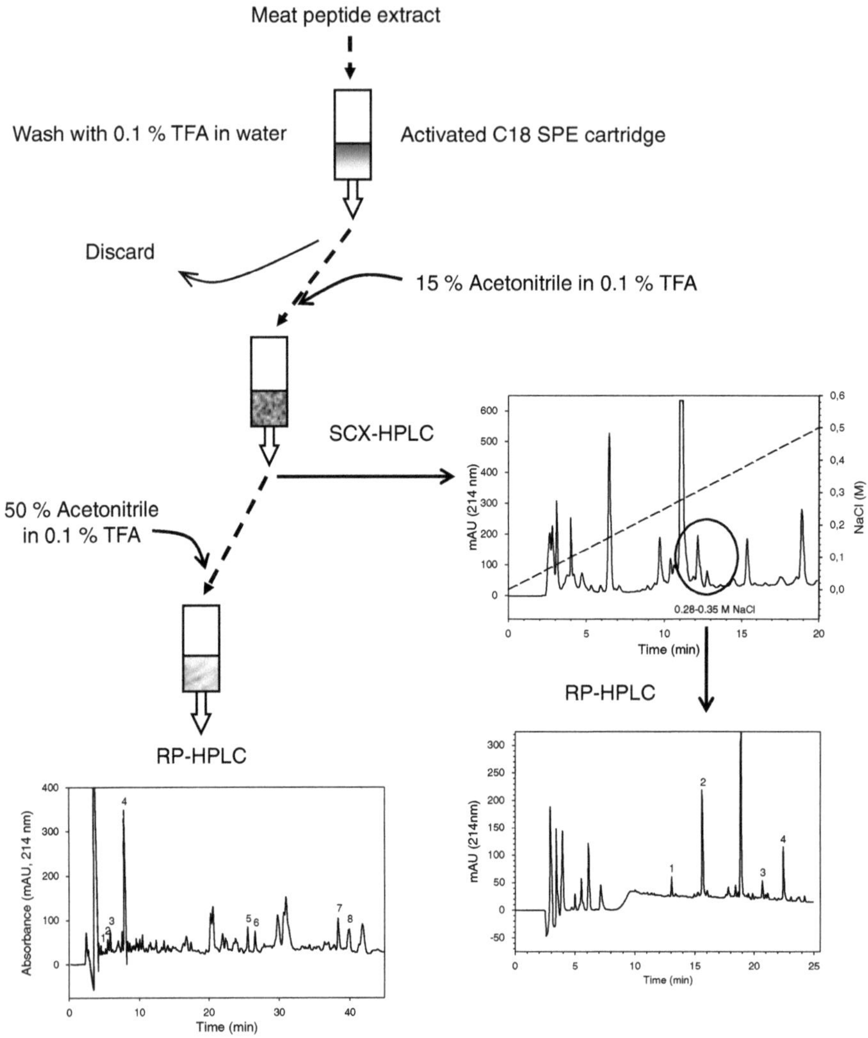

Fig. 4.1 Scheme for the purification of pork meat peptides used as quality markers (Extracted from Flores et al. (2000) and Moya et al. (2001) with permission from Elsevier. For abbreviations see text)

4.2.2 Molecular Exclusion Chromatography

Molecular exclusion or size exclusion chromatography refers to a chromatographic separation based on size. SEC, which is also known as gel filtration chromatography (GFC), is widely used as an initial step for food peptide analysis and is often performed under low pressure. This technique fractionates peptides according to their molecular size and this is a good and intuitive starting point before further resolutive

Table 4.1 Some types of stationary phase matrixes for ionic exchange chromatography (IEC) and gel filtration chromatography (GFC) peptide analysis

	IEC		GFC	
Chemically based on	Cross-linked structure	Commercial brand	Cross-linked structure	Commercial brand
Polysaccharide	Dextran	Sephadex™SP	Dextran	Shephadex™G
	Agarose	Sepharose™	Dextran with agarose	Superdex™
	Cellulose	Sephacel™		
Synthetic organic polymers	Polystyrene-divinyl benzene (PS-DVB)	Dowex™, Vydac™	Vinyl polymers with hydroxyl groups	Toyopearl™ HW
			Polyacrylamide	BioGel™ P
Mixed of above			Dextran with acrylamide	Sephacryl™
Hydrous oxides	Aluminosilicates, alumina, silica or zirconia	Nucleogen™ (silica)		

techniques are used. This separation technique is performed under atmospheric pressure or in HPLC, depending on the stationary phase characteristics (particle size and thus, pressure resistance).

Stationary phases used to be hydrophilic cross-linked structures and the most used are shown in Table 4.1. Each stationary phase is available with a variety of different ranges of pore size in beds, allowing the separation of peptides of different size.

Manufacturers provide tables for selecting the most appropriate gel for a specific application, in which the fractionation range is shown. The upper limit is the exclusion limit, which coincides with the column void volume. Operating conditions and gel selection depend on the application and the desired resolution. The use of CFG can have two different objectives of fractionation or group separation, this last including desalting and buffer exchange. In desalting, the molecule of interest is eluted in the void volume and smaller molecules are retained in the gel pores. To obtain the desired separation, the gel should have an exclusion limit smaller than the molecule of interest. In fractionation, molecules of different molecular weight are separated within the gel matrix. With this method, the molecules of interest should fall within the fractionation range of the gel. A gel with a smaller range of pore sizes will give a higher resolution; a gel with a wider range will give lower resolution, but will permit fractionation of a wider range of sizes. The particle size of the gel beads (the mesh size classified as superfine, fine, medium, coarse) also affects resolution; smaller beads permit higher resolution, but a lower flowrate through the column (and hence a slower separation). Other factors to take into account for the selection of the more appropriate stationary phase are the range of pH, the compatibility with salts or denaturing reagents, and the mechanical strength among others.

The stationary phase is supplied either swollen (ready to use) or as a dry powder that has to be swollen by following the manufacturer's instructions. Prepared media will preferably be packed in a glass column. The resolution in gel filtration increases as the square root of column length but limitations due to pressure limits given by the stationary phase manufacturers should be taking into account. Twenty-five cm for a Bio-Gel P2 (Ghassem et al. 2011) or 140 cm for Sephadex G-25 (Chen et al. 2003) columns are two extreme examples of length. Column diameter of 1.5 cm for Bio-Gel P2 and 2.6 cm for Sephadex G-25 are good options for the majority of applications. In the case of Sephadex, narrower columns may give slight wall effects, whereas wider columns imply excessive dilution.

GF columns for HPLC (TSK gel 2000 from Tosoh, Shodex Protein KW-802.5 from Waters, Superdex peptide from Amersham Bioscience, or Bio-gel P-2 from Bio-Rad, etc.) have also been used to fractionate peptide extracts (Yuk et al. 2000) but their main use is for peptide molecular mass determination (Gu et al. 2011; Ghassem et al. 2011; Cheng et al. 2010b, Rodríguez-Nuñez et al. 1995).

The selected eluant should provide adequate stability to the solutes. The ionic strength should be at least 20 mM to eliminate the effect of small amounts of negatively charged groups on the gel. The use of organic solvents, additives such as salts or denaturing reagents, and pH must be tested for adequate compatibility with the gel. Some usual eluants used for peptide separation are water, 0.05–0.1% ammonium

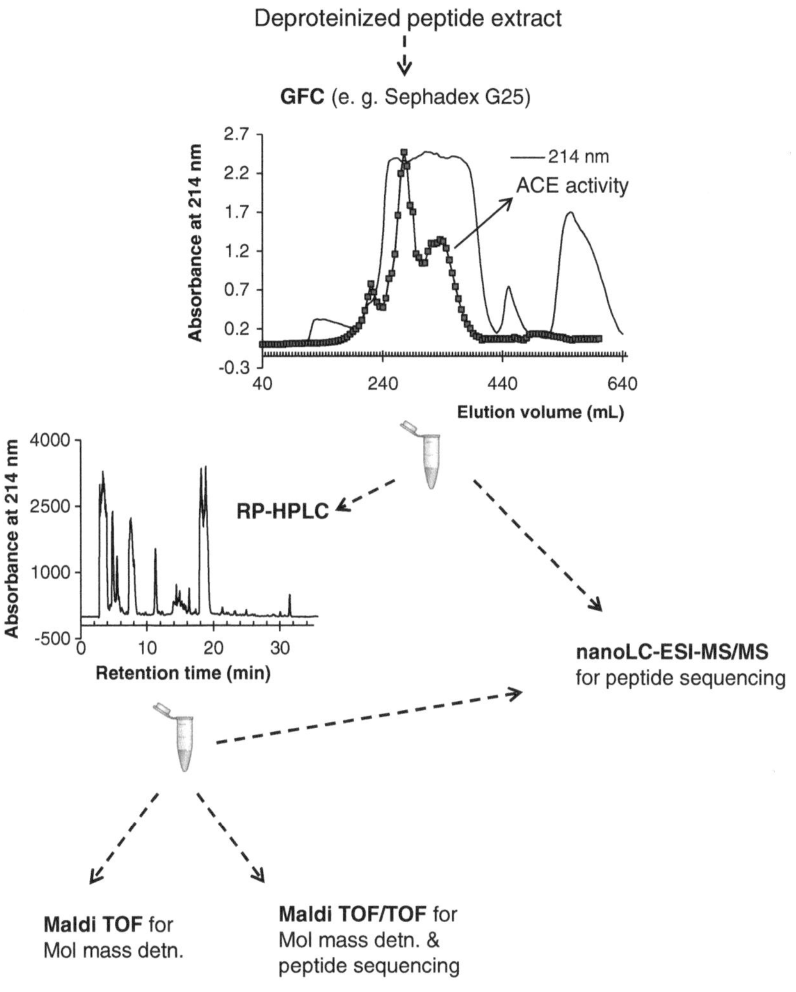

Fig. 4.2 Scheme for the analysis of food peptides with inhibitory action against the angiotensin I-converting (ACE) enzyme. For abbreviations see text

bicarbonate, diluted ammonia, diluted (0.01 N) hydrochloric acid, 0.1% trifluoroacetic acid, 5% formic acid, 30% acetic acid, pyridine-acetic acid buffers, sodium phosphate buffer, or Tris-K buffer.

Fractions from GFC are collected and may be submitted to either sensory tests (Toelstede and Hofmann 2008; Salles et al. 1995) or bioactivity (Mora et al. 2010; Suetsuna et al. 2000) evaluation. Selected fractions from these assays are usually submitted to further peptide fractionation prior to MS analysis although, in some cases, lyophilized fractions from GFC have been analyzed directly by MS using a tandem liquid chromatography (LC)-MS system (Schindler et al. 2011). These procedures are schematized in Fig. 4.2.

4.2.3 Solid-Phase Extraction (SPE)

Solid-phase extraction (SPE), thanks to the rapid development of disposable cartridges, has been successfully applied to the isolation and fractionation of peptides. This technique is versatile, permitting orthogonal strategies of fractionation by combining two or three methods of chromatographic separation: polar interaction (normal-phase), hydrophobic interaction (reversed-phase), and ion exchange (SCX and SAX). Some examples of these procedures have been used to fractionate peptides from meat (Moya et al. 2001; Flores et al. 2000), milk (Catalá-Clariana et al. 2010; Voirin et al. 1991), cheese (Toelstede and Hofmann 2008; Singh et al. 1994), and vegetables (Huan and Aluko 2010) among others. One example of recent combined use of SPE for the characterization strategy of an antihypertensive peptide from mushroom, a reversed phase (C_{18})- and SCX solid-phase cartridges were successively used to purify active fractions obtained from a gel-filtration column (Jang et al. 2011).

An example of SPE used for the study of peptides from pork meat in search of biochemical markers of quality (Moya et al. 2001; Flores et al. 2000) is shown in Fig. 4.1. Meat extract was fractionated using a C_{18} SPE cartridge. Mineral salts, very polar substances including amino acids, some peptides, and proteins were eluted in the washings and peptides of interest were retained. A first peptide fraction was collected by elution using 15% acetonitrile and a second fraction containing the most hydrophobic peptides was eluted from the cartridge using 50% acetonitrile in 0.1% TFA. Both fractions were submitted to further purification as shown in the figure and discussed below.

4.3 Second Fractionation

After a first peptide fractionation, it is very often recommended to proceed to a precise separation of the peptides of interest, prior to the analysis by MS.

This stage is mainly achieved by HPLC methods, with reversed phase (RP) the most used. Lately, hydrophilic interaction chromatography (HILIC) has also shown satisfactory results when connected directly to MS analyzers for very short, polar peptide analysis. Also, both methods are useful for desalting the samples because nonvolatile salts used in the previous preparative chromatographies (IEC and GFC) are not compatible with MS.

4.3.1 RP- HPLC

This method is the most used to resolve peptide mixtures based on differences in polarity, prior to analysis by mass spectrometry. The most common method is based on octadecyl-hydrophobic chains (C_{18}) as a stationary phase and an acidic mobile phase mixed with organic solvent, preferably acetonitrile for its spectral characteristics and low viscosity.

The use of trifluoroacetic acid (TFA) as acidic modifier in the mobile phase has been generally adopted for RP-HPLC peptide analysis because it acts as an ion pair improving peptide resolution (Cliffe et al. 1989) and its volatility allows for evaporation prior to MS analysis. However, TFA should be avoided if the RP-HPLC column is connected directly to the MS detector using electrospray ionization (ESI) because TFA inhibits the signal (Eshraghi and Chowdhury 1993). In such cases, mainly formic or acetic acids should be used (Ghassem et al. 2011). As an exception, Gartenmann and Kochhar (1999) found that if the peptides were precolumn derivatized with 9-fluorenysmethyloxycarbonyl fluoride (Fmoc), an improvement was obtained in the ionization even in the presence of TFA, and better benefit achieved (even with the peptides derivatized) from the chromatographic separation in relation to that obtained using formic acid.

In contrast, TFA works very well with matrix-assisted laser desorption/ionization (MALDI) and its use at 0.1% is recommended in this MS equipment, where heptafluorobutyric (HFBA), acetic, and formic acids do not work properly (Mora et al. 2010; Escudero et al. 2010a).

The use of commercially available C18 stationary phases assembled in tips is a very frequently used option before MS analysis when ionization sources are both ESI and MALDI. These tips permit the elimination of possible contaminants or ionization inhibitors prior to MS analysis as well as concentrate the sample. The C18 filling of the tip acts as a RP column and their use is very simple and reliable, similar to SPE cartridges.

Figures 4.1 and 4.2 show how RP-HPLC is used as the last purification step prior to MS analysis. In Fig. 4.1 the fraction eluted from a C_{18} SPE cartridge using 15% acetonitrile (from Flores et al. 2000) was submitted to an intermediate fractionation step using a strong cation exchange column (Spherisorb S10SCX 10×250 mm, Waters Corp., MA, USA) and a gradient of NaCl in 0.1 N HCl and 20% acetonitrile. The fraction eluted between a NaCl concentration of 0.28–0.35 M was collected, concentrated, and reinjected in a RP-HPLC column (Symetry C18, 4.6×250 mm, from Waters) for a better peptide isolation, using an acetonitrile gradient in 0.1% TFA. Meanwhile the fraction eluted from the cartridge with 50% acetonitrile (from Moya et al. 2001) was readily separated in the RP-HPLC column and also using an adequate acetonitrile gradient in 0.1% TFA. Fractions (numbered peaks in the figure) are thus ready for further analysis by MS, if required. Figure 4.2 shows how after a GFC (Sephadex G25, 60×2.5 cm) as the prefractionation method for a deproteinized meat extract, RP-HPLC was used as the last purification step prior to MS analysis. In this example, fractions from GFC were tested for inhibitory action against ACE activity and those of interest were taken to dryness after being submitted to the second step purification by RP-HPLC. Fractions were again tested for activity and those of interest were lyophilized and submitted to MS analysis.

The development of monolithic stationary phases, and especially those based on poly(styrene-divinylbenzene) (PS-DVB) has constituted a novelty that is worthwhile exploring. These columns (mainly capillary columns) work in reversed-phase mode and are suitable for robust and high-resolution separations of peptides at elevated temperatures using TFA, HFBA, and formic acid as mobile phase additives but also they are stable at elevated pH using basic volatile buffers such as triethylamine, tetrabutylamine, or decyltrimethylamine and obtaining a different selectivity to that working at acidic pH (Melchior et al. 2010; Toll et al. 2005).

4.3.2 Hydrophilic Interaction Chromatography (HILIC)

HILIC is a special case of normal-phase chromatography because the stationary phase is more polar than the mobile phase. It is used primarily for separation of very polar compounds with a negative log P (1-octanol/water partition coefficient) factor. Some small polar peptides satisfy this condition and consequently could be correctly separated by HILIC (Mora et al. 2007). The HILIC method has solved the problems related to the analysis of very polar, even ionic, compounds with comparable efficiency to that of RP-HPLC, without the need of either derivatization or adding nonvolatile ion pairs or other additives to improve the separation.

HILIC employs semi-aqueous mobile phases in which peptides are soluble, and thus it is more convenient than normal-phase chromatography for this analysis. Typical eluents for HILIC consist of 40–97% acetonitrile in water or in a volatile buffer (below 20 mM ammonium salts of acetate or formate) or organic acid (acetic or formic acids), making this technique compatible with subsequent MS analysis.

Columns for HILIC are hydrophilic, made of either bare silica, silica, or polymers with attached charged (in a range of pH) molecules of a different nature (hydroxyl, amine, or zwitterionic molecule). The presence of water in the mobile phase is necessary because, in some way, water takes part of the stationary phase creating a water-enriched liquid layer. The retention process in HILIC is the result of two combined forces: hydrophilic partitioning and electrostatic interaction with either positive or negative charges. The retention is controlled by the constituents of mobile phase. Thus, the organic/water ratio controls the hydrophilic partitioning and the acid or saline buffer controls the electrostatic interaction.

HILIC separations can be easily combined with several detection techniques such as ultraviolet absorbance, fluorescence, light scattering, refractive index, and MS. HILIC methodology used with electrospray ionization and coupled with MS detection results in higher sensitivity compared to RP-HPLC, probably due to the high content of organic solvent in the mobile phase, which simplifies the drop formation during the spray process (Mora et al. 2011). Indeed, two types of MS techniques, using atmospheric pressure ionization (APCI) and electrospray ionization (ESI), can handle a wide range of flowrates and low polarity to ionic compounds, which make them appreciable for the analysis of low-molecular-weight compounds in biological samples, as is the case of some peptides. Some researchers have used these combined techniques in the study of small peptides (di- or tri-peptides) related to taste (Schindler et al. 2011; Dunkel and Hofmann 2009; Toelstede and Hofmann 2008; Andersen et al. 2008), or antioxidants (Broncano et al. 2012) activity.

As mentioned above, RP-HPLC or HILIC are used for the last separation of peptides prior to MS detection and/or peptide sequencing. Two working procedures are the most usual: to collect fractions from the chromatography column, choose those that are of interest (bioactivity, taste, quality markers, etc.), lyophilize them, resuspend, and analyze with the MS (Ko et al. 2012; Mora et al. 2010; Escudero et al. 2010a), or directly connect both pieces of equipment. In the case of direct connection, a split for reducing the incoming effluents to the MS-ionization source can be necessary when using chromatographic columns of standard diameters (Toelstede and Hofmann 2008). The use or micro- or better nano-LC columns avoids the necessity of splitting (Ghassem et al. 2011; Schindler et al. 2011; Escudero et al. 2010a, b) having a double advantage: it has permitted the direct connection between the exit of the column and the MS entrance for a more efficient analysis (even high-throughput screenings) and secondly, the analyses are more environmentally friendly.

The rapid development experienced in recent years by "omics" techniques, such as proteomics or peptidomics (Toelstede and Hofmann 2008), has driven the development of chromatographic columns capable of very rapidly resolving complex peptide mixtures in low sample loads with a low amount of solvents (mobile phase), while maintaining good sensitivity and reliability. The column modifications leading to these advances include the reduction of the column dimensions with regard to diameter, length, and particle size.

References

Andersen LT, Schlichtherle-Cerny H, Ardo Y (2008) Hydrophilic di- and tripeptides are not a precondition for savoury flavour in mature Cheddar cheese. Dairy Sci Technol 88:467–475

Broncano JM, Otte J, Petron MJ, Parra V, Timon ML (2012) Isolation and identification of low molecular weight antioxidant compounds from fermented "chorizo" sausages. Meat Sci 90:494–501

Cambero MI, Seuss I, Honikel KO (1992) Flavor compounds of beef broth as affected by cooking temperature. J Food Sci 57:1285–1290

Catalá-Clariana S, Benavente F, Giménez E, Barbosa J, Sanz-Nebot V (2010) Identification of bioactive peptides in hypoallergenic infant milk formulas by capillary electrophoresis-mass spectrometry. Anal Chim Acta 683:119–125

Chen J-R, Okada T, Muramoto K, Suetsuna K, Yang S-C (2003) Identification of angiotensin I-converting enzyme inhibitory peptides derived from the peptic digest of soybean protein. J Food Biochem 26:543–554

Cheng Y, Chen J, Xiong YL (2010a) Chromatographic separation and tandem MS identification of active peptides in potato protein hydrolysate that inhibit autoxidation of soybean oil-in-water emulsions. J Agric Food Chem 58:8825–8832

Cheng Y, Xiong YLL, Chen J (2010b) Fractionation, separation, and identification of antioxidative peptides in potato protein hydrolysate that enhance oxidative stability of soybean oil emulsions. J Food Sci 75:C760–C765

Cho MJ, Unklesbay N, Hsieh F-H, Clarke AD (2004) Hydrophobicity of bitter peptides from soy protein hydrolysates. J Agric Food Chem 52:5895–5901

Cliffe AJ, Revell D, Law BA (1989) A method for the reverse phase high performance liquid chromatographjy of peptides from Cheddar cheese. Food Chem 34:147–160

Combes C, Paterson E, Amadò R (2001) Isolation and identification of low-molecular weight peptides from emmental cheese. J Food Sci 67:553–559

Doyen A, Beaulieu L, Saucier L, Pouliot Y, Bazinet L (2011) Impact of ultrafiltration membrane material on peptide separation from a snow crab byproduct hydrolysate by electrodialysis with ultrafiltration membranes. J Agric Food Chem 59:1784–1792

Dunkel A, Hofmann T (2009) Sensory-directed identification of beta-alanyl dipeptides as contributors to the thick-sour and white-meaty orosensation induced by chicken broth. J Agric Food Chem 59:9867–9877

Escudero E, Sentandreu MA, Arihara K, Toldrá F (2010a) Angiotensin I-converting enzyme inhibitory peptides generated from in vitro gastrointestinal digestion of pork meat. J Agric Food Chem 58:2895–2901

Escudero E, Sentandreu MA, Toldra F (2010b) Characterization of peptides released by in vitro digestion of pork meat. J Agric Food Chem 58:5160–5165

Eshraghi J, Chowdhury SK (1993) Factor affecting elestrospray ionization of effluents containing trifluoroacetic acid for high performance liquid chromatography-mass spectrometry. Anal Chem 65:3528–3533

Flores M, Aristoy MC, Spanier AM, Toldrá F (1997) Non-volatile components effects on quality of "Serrano" dry-cured ham as related to processing time. Food Sci 62:1235–1239

Flores M, Moya VJ, Aristoy MC, Toldrá F (2000) Nitrogen compounds as potential biochemical markers of pork meat quality. Food Chem 69:371–377

Gartenmann K, Kochhar S (1999) Short-chain peptide analysis by high performance liquid chromatography coupled to elestrospray ionization mass spectrometer after derivatization with 9-fluorenylmethyl chloroformate. J Agric Food Chem 47:5068–5071

Ghassem M, Arihara K, Babji AS, Daid M, Ibrahim S (2011) Purification and identification of ACE inhibitory peptides from Haruan (Chana striatus) myofibrillar protein hydrolysate using HPLC-ESI-TOF MS/MS. Food Chem 129:1771–1777

Gu R-Z, Li C-Y, Liu W-Y, Yi W-X, Cai M-Y (2011) Angiotensin I-converting enzyme inhibitory activity of low-molecular-weight peptides from Atlantic salmon (*Salmo salar* L.) skin. Food Res Int 44:1536–1540

Huan L, Aluko RE (2010) Identification and inhibitory properties of multifunctional peptides from pea protein hydrolysate. J Agric Food Chem 58:11471–11476

Jang J-H, Jeong S-C, Kim J-H, Lee Y-H, Ju Y-C, Lee J-S (2011) Characterisation of a new antihypertensive angiotensin I-converting enzyme inhibitory peptide from *Pleurotus cornucopiae*. Food Chem 127:412–418

Kaiser KP, Belitz HD, Fritsch RJ (1992) Monitoring Cheddar cheese ripening by chemical indices of proteolysis. II. Peptide mapping of casein fragments by reverse-phase high-performance liquid chromatography. Z Lebensm Unters Forsch 195:8–14

Kim S-K, Byun H-G, Park P-J, Shahidi F (2001) Angiotensin I converting enzyme inhibitory peptides purified from bovine skin gelatin hydrolysate. J Agric Food Chem 49:2992–2997

Ko S-C, Lee J-K, Byun H-G, Lee S-C, Jeon Y-J (2012) Purification and characterization of angiotensin I-converting enzyme inhibitory peptide from enzymatic hydrolysates of *Styela clava* flesh tissue. Proc Biochem 47:34–40

Kuchroo CN, Fox PF (1982) Fractionation of the water-soluble-nitrogen from Cheddar cheese: chemical methods. Milchwissenschaft 37:651–653

Lee JR, Kwon DY, Shin H-K, Yang C-B (1999) Purification and identification of angiotensin-I converting enzyme inhibitory peptide from kidney bean protein hydrolyzate. Food Sci Biotechnol 8:172–178

Lioe HN, Apriyantono A, Takara K, Wada K, Naoki H, Yasuda M (2004) Low molecular weight compounds responsible for savory taste of Indonesian soy sauce. M J Agric Food Chem 52:5950–5956

Matoba T, Hayashi R, Hata T (1970) Isolation of bitter peptides from tryptic hydrolysate of casein and their chemical structure. Agric Biol Chem 34:1235–1243

Melchior K, Tholey A, Heisel S, Keller A, Lenhof HP, Meese E, Huber CG (2010) Protein-versus peptide fractionation in the first dimension of two-dimensional high-performance

liquid chromatography-matrix-assisted laser desorption/ionization tandem mass spectrometry for qualitative proteome analysis of tissue samples. J Chromatogr A 1217:6159–6168
Molina E, Ramos M, Alonso L, Lopez-Fandiño R (1999) Contribution of low molecular weight water soluble compounds to the taste of cheeses made of cows', ewes' and goats' milk. Int Dairy J 9:613–621
Mora L, Sentandreu MA, Toldrá F (2007) Hydrophilic chromatographic determination of carnosine, anserine, balenine, creatine, and creatinine. J Agric Food Chem 55:4664–4669
Mora L, Sentandreu MA, Toldrá F (2010) Identification of small troponin T peptides generated in dry-cured ham. Food Chem 123:691–697
Mora L, Hernández-Cázares AS, Aristoy MC, Toldra F, Reig M (2011) Hydrophilic interaction chromatography (HILIC) in the analysis of relevant quality and safety biochemical compounds in meat, poultry and processed meats. Food Anal Meth 4:121–129
Moreno Arribas MV, Bartolomé B, Pueyo E, Polo MC (1998) Isolation and characterization of individual peptides from wine. J Agric Food Chem 46:3422–3425
Moya VJ, Flores M, Aristoy MC, Toldrá F (2001) Evolution of hydrophobic polypeptides during the ageing of exudative and non-exudative pork meat. Meat Sci 57:395–401
Park P-J, Jung W-K, Nam K-S, Shahidi F, Kim S-K (2001) Purification and characterization of antioxidative peptides from protein hydrolysate of lecithin-free egg yolk. J Am Oil Chem Soc 78:651–656
Rodríguez-Núñez E, Aristoy M-C, Toldrá F (1995) Peptide generation in the processing of dry-cured ham. Food Chem 53:187–190
Rohm H, Jaros D, Rockenbauer C, Riedler-Hellrigl M, Uniacke-Lowe T, Fox PF (1996) Comparison of ethanol and trichloroacetic acid fractionation for measurement of proteolysis in emmental cheese. Int Dairy J 6:1069–1077
Salles C, Septier C, Roudot-Algaron F, Guillot A, Etievant PX (1995) Sensory and chemical analysis of fractions obtained by gel permeation of water-soluble Comte cheese extracts. J Agric Food Chem 43:1659–1668
Schindler A, Dunkel A, Staehler F, Backes M, Ley J, Meyerhof W, Hofmann T (2011) Discovery of salt taste enhancing arginyl dipeptides in protein digests and fermented fish sauces by means of a sensomics approach. J Agric Food Chem 59:12578–12588
Singh TK, Fox PF, Hojrup P, Healy A (1994) A scheme for the fractionation of cheese nitrogen and identification of principal peptides. Int Dairy J 4:111–122
Sommerer N, Septier C, Salles C, le Quere JL (1998) A liquid chromatography purification method to isolate small peptides from goat cheese for their mass spectrometry analysis. Sci Aliment 18:537–551
Stoeva S, Byrne CE, Mullen AM, Troy DJ, Voelter W (2000) Isolation and identification of proteolytic fragments form TCA soluble extracts of bovine M. longissimus dorsi. Food Chem 69:365–370
Suetsuna K, Chen JR (2002) Isolation and characterization of peptides with antioxidant activity derived from wheat gluten. Food Sci Technol Res 8:227–230
Suetsuna K, Ukeda H, Ochi H (2000) Isolation and characterization of free radical scavenging activities peptides derived from casein. J Nutr Biochem 11:128–131
Toelstede S, Hofmann T (2008) Sensomics mapping and identification of the key bitter metabolites in Gouda cheese. J Agric Food Chem 56:2795–2804
Toll H, Oberacher H, Swart R, Huber CG (2005) Separation, detection, and identification of peptides by ion-pair reversed-phase high-performance liquid chromatography-electrospray ionization mass spectrometry at high and low pH. J Chromatogr A 1079:274–286
Voirin A, Letavernier JF, Sebille B (1991) Improvement of extraction and concentration of milk peptides with solid-phase cartridges for analysis by high-performance liquid-chromatography. J Chromatogr 553:155–164
Yuk J-S, Lim Y-H, Cho H-Y (2000) Purification and characterization of angiotensin I-converting enzyme inhibitors from Sinapisalba L. J Food Sci Nutr 5:75–80

Chapter 5
Mass Spectrometry Applications

Carla Soler, Josep Rubert, and Jordi Mañes

5.1 Proteomics and Mass Spectrometry

Over the last decades, food safety and quality control in all production processes have attracted a great deal of attention. Food safety itself might be implicit in the broader concept of food security, defined as a situation in which all people at all times have physical and economic access to sufficient, safe, and nutritious food to meet their dietary needs and live an active healthy life (World Food 1996).

In recent years, food safety is an increasingly broad concept that encompasses mainly three main areas: (1) food quality (food composition), (2) traceability (food origin), and (3) food safety per se (absence of allergens, pathogens, or other contaminants) (D'Alessandro and Zolla 2012). Early attempts to individuate quality standards through biomolecules mainly relied on rough biochemical parameters, such as product composition (e.g., through labeling of percentages of lipids, carbohydrates, proteins, and vitamins). Recent advancements in the fields of biochemistry and molecular biology allowed us to look for actual biomarkers to be potentially exploited as indicators of product quality and traceability (Cairns 2011). In this frame, proteomics have already contributed to accumulating a relevant body of knowledge that might result not only in improving food safety through enhancing selection of food quality and traceability markers but also in providing the end-user consumer with a unique tool to make a fully aware alimentary choice (D'Alessandro and Zolla 2012)

C. Soler (✉) • J. Rubert • J. Mañes
Departament de Medicina Preventiva i Salut Pública, Universitat de València,
Av. Vicent Andrès Estellès, Burjassot 46100, Spain
e-mail: carla.soler@uv.es

F. Toldrá and L.M.L. Nollet (eds.), *Proteomics in Foods: Principles and Applications*,
Food Microbiology and Food Safety 2, DOI 10.1007/978-1-4614-5626-1_5,

The history of proteomics dates back to the discovery of two-dimensional gels in the 1970s, which provided the first feasible way of displaying hundreds or thousands of proteins on a single gel (Klose 1975; O'Farrell 1975). Identification of the spots separated on these gels remained laborious and was limited to the most abundant proteins until the 1990s, when biological mass spectrometry had developed into a sufficiently sensitive and robust technique.

Mass spectrometry was restricted for a long time to small and thermostable compounds because of the lack of effective techniques to softly ionize and transfer the ionized molecules from the condensed phase into the gas phase without excessive fragmentation. The development in the late 1980s of two techniques for the routine and general formation of molecular ions of intact biomolecules (electrospray ionization, ESI) (Fenn et al. 1989) and matrix-assisted laser desorption/ionization (MALDI) (Karas and Hillenkamp 1988) dramatically changed this situation and made polypeptides accessible to mass spectrometry analysis.

In recent years, mass spectrometry has evolved into an indispensable tool for proteomics research. Coupled with an isolation technique to purify a specific protein complex of interest, MS can rapidly and reliably identify the components of complexes. For MS experiments, consideration should be given to the type of instrumentation, fragmentation method, and analysis strategy best suitable to an individual sample (Aerbersold and Goodlett 2001).

The strategy followed to characterize protein complex composition using MS has been well schematized by Gingras et al. (2005): a protein complex to be analyzed is first purified using an appropriate approach. The purified protein may then be separated using techniques such as SDS-PAGE, isoelectric focusing, or various two-dimensional separation methods. Individual proteins may then be visualized by staining, and recovered for analysis. Isolated protein can be proteolytically digested (most often using trypsin) to generate a mixture of peptides that can be identified by MS. The proteins present in the sample are deduced by recombining the identified peptides, using suitable informatics tools.

Currently, the uses of MS in proteomics are in three major areas. MS is the preferred technique in the field of biotechnology for characterization and quality control of recombinant proteins and other macromolecules. It is also commonly used for protein identification, either in classical biochemical projects or in large-scale proteomics ones. Finally, since MS measures the molecular weight of a protein, it is the method of choice for the detection and characterization of post-translational modifications and potentially can identify any covalent modification that alters the mass protein (Mann et al. 2001).

Different strategies can be followed to carry out a proteomic study including "top-down proteomics" (MS is carried out in the form of whole-protein analysis) and "bottom-up proteomics" (protein identification is carried out by the analysis of enzymatically or chemically produced peptides) (Han et al. 2008). In the first one, MS analysis can provide accurate molecular mass measurement of intact proteins as large as 100 kDa or more. Highly accurate protein mass measurements generally are of limited utility, because they often are not sufficiently sensitive and because net mass often is insufficient for unambiguous protein identification.

In the second one, MS analysis can provide accurate mass measurements of peptides from proteolytic digests. In contrast to whole protein mass measurements, peptide mass measurements can be done with higher sensitivity and mass accuracy (Liebler and Yates 2002).

There are four basic types of mass analyzer currently used in proteomic research; ion trap, time-of-flight, quadrupole, and Fourier transform analyzers. They are very different in design and performance, each with its own strength and weakness. These analyzers can be standalone or in some cases put together in tandem to take advantage of the strengths of each (Aebersold and Mann 2003).

The aim of this chapter is to review the major types of MS instruments, both currently in use and under development, and to discuss strategies for the analysis of intact proteins or for the analysis of peptides obtained after enzymatic degradation of the protein. Finally, application of mass spectrometry to three areas related to food safety such as the detection of micro-organisms, the determination of modifications induced by food processing, and the identification of protein components responsible for food allergy and intolerance are summarized.

However, it is important to keep in mind that proteomics still remain a multifaceted, rapidly developing, and open-ended endeavor. Although they have enjoyed tremendous recent success, proteomics still face significant technical challenges; each breakthrough that either allows a new type of measurement or improves the quality of data made by traditional types of measurements expands the range of potential application of MS to molecular and cellular biology. Indeed, this field is already too extensive for a comprehensive single review; thus it has been impossible to include all the achievements in MS-based proteomics.

5.2 Ionization Techniques

Ionization techniques are critical to convert molecules into ions that can then be manipulated within electric or magnetic fields. The challenge with biological molecules such as peptides and proteins is to convert polar molecules into gas-phase ions without degradation or fragmentation (Yates 2004). The development of ESI and MALDI solved this difficult problem of generating ions from large nonvolatile analytes such as protein and peptides without significant analyte fragmentation.

ESI gained immediate popularity because of the ease with which it could be interfaced with popular chromatographic and electrophoretic liquid-phase separation techniques and quickly supplanted fast atom bombardment as the ionization method of choice for protein and peptide samples dissolved in a liquid phase. This technique creates ions by spraying an electrically generated fine mist of ions into the inlet of a mass spectrometer at atmospheric pressure (Fenn et al. 1989). Steady advances in the application of ESI to the analysis of peptides and proteins have been made. The most notable improvements have come from reduction in the flowrate of the liquid used to create the electrospray and thus more efficiently create ions (Emmet and Caprioli 1994; Smith et al. 1990). Furthermore, due to the propensity of ESI to produce multiple charged analytes, mass analyzers with limited *m/z* range could be used to detect analytes with masses exceeding the nominal *m/z* range of instruments (Aerbersold and Goodlett 2001).

For different but no less compelling reasons, MALDI also rapidly gained popularity. In this technique, a pulsed laser is used to deposit energy into the matrix, causing rapid thermal heating of the molecules to desorb molecules and ions into the gas phase. Because a pulsed laser is used, this ionization technique produces ions in packets rather than a continuous beam and thus requires a mass spectrometer that can either measure a complete mass spectrum without scanning a mass range or trap of all the ions for a subsequent mass analysis (Yates 2004).

MALDI mass spectra are simple to interpret due to the propensity of the method to generate predominantly singly charged ions. The method is relatively resistant to interference with matrices commonly used in protein chemistry (Aerbersold and Goodlett 2001). However, preparation of reliable sample spots onto the MALDI target demands a homogeneously applied peptide sample at the region upon which the laser shots will take place. Several factors affect the efficacious preparation of a MALDI target. These factors include the accurate application of matrix (pipette tip must not excessively touch the steel target but yet apply the full liquid volume), sample, and recrystallization solutions onto the steel plate spots, and the use of freshly prepared matrix and recrystallization solutions (Garbis et al. 2005).

5.3 Mass Spectrometers for Protein and Peptide Analysis

MS has played a pivotal role in proteomics research, with mass analyzers being central to the technology, and has proved to be a powerful analytical technique for both protein and peptide analysis (Chen 2008). Tandem MS (MS/MS) consists of either the fragmentation of a selected precursor peptide ion to generate specific fragment ions for sequence elucidation (data-dependent acquisition), or uncoupled acquisitions of intact and fragment masses with retrospective reconstitution of the parent–daughter ion context (data-independent acquisition) (Sénéchal and Kussmann 2011). By using tandem MS, a complete sequence of the peptides of interest is often achieved. To identify these peptides, spectra are scanned against protein-sequence databases using search algorithms (Lane 2005).

The most popular analyzers in proteomics are ion traps, triple quadrupoles, time-of-flight tubes, orbitrap, and Fourier transform ion cyclotron resonance, with their specific advantages: high sensitivity and multiple-stage fragmentation for ion traps, high selectivity for triple-quadrupoles, high sensitivity and speed for time-of-flight.

Figure 5.1 depicts the principal ionization sources and principal mass spectrometers used in proteome research.

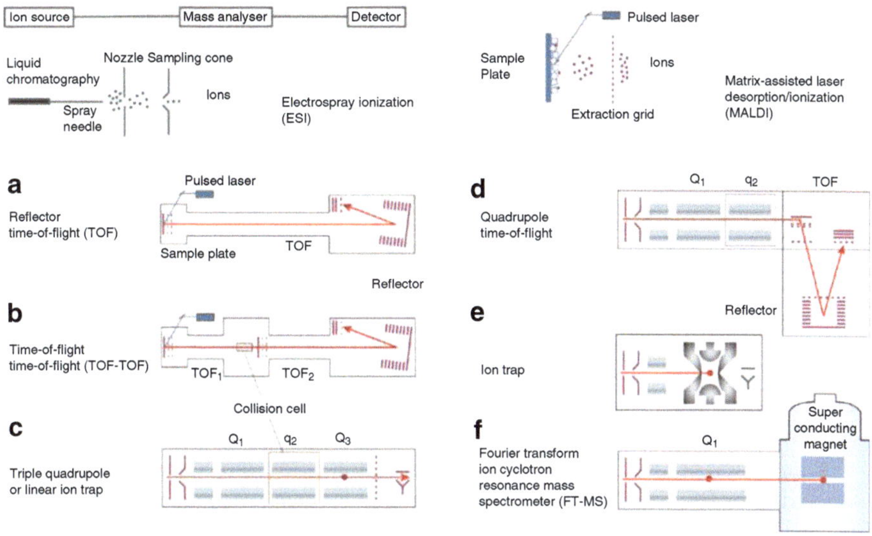

Fig. 5.1 The *left* and *right upper* panels depict the ionization and sample introduction process in electrospray ionization (ESI) and matrix-assisted laser desorption/ionization (MALDI). The different instrumental configurations (**a**–**f**) are shown with their typical ion source (Reproduced from Aerbersold and Mann (2003) with permission from Nature Publishing Group)

5.3.1 Time-of-Flight (TOF) Mass Analyzer

In theory the time-of-flight mass spectrometer is the simplest mass analyzer. Mass-to-charge (*m/z*) ratios are determined from the flight times of analyte ions traveling through a field-free region or a flight tube. At a constant accelerating voltage, the flight time for an ion is proportional to the square of the *m/z* ratio. The lighter ions (with lower *m/z* values) have a higher velocity and are recorded on a detector earlier than the heavier ones (higher *m/z* values), thus producing the TOF spectrum as a function of time (Shen and Noon 2004). The biggest drawback to TOF analyzers is their inability to perform true MS/MS. However, in 1993 the tandem time-of-flight (TOF/TOF) mass spectrometer was introduced (Cornish and Cotter 1993a, b)with two reflectron mass analyzers separated by a mass-selector gate and collision region into which a collision gas was pulsed once each mass recording cycle. These TOF/TOF instruments incorporate a collision cell between two TOF sections. Ions of one *m/z* ratio are selected in the first TOF section, fragmented in the collision cell, and the fragments are separated in the second TOF section (Guerrera and Kleiner 2005)

The scanning mass analyzers such as TOF are usually interfaced with MALDI to perform pulsed analysis. In fact MALDI-TOF or MALDI-TOF/TOF are relatively simple and robust to operate, have good mass accuracy, and high resolution and sensitivity. They are widely used in proteomics to identify proteins from simple mixtures by a process called peptide mass fingerprinting. In this approach, proteins of interest are digested with a sequence-specific enzyme such as trypsin and the generated peptides are analyzed by MS. The determined masses, usually in the range of 800–3,000 *m/z*, are then compared against a database comprising peptide masses from a virtual digest of all proteins from a given organism with the same sequence-specific protease (Cottrell 1994).

5.3.1.1 Drawbacks

The major drawbacks of this method are that (1) hydrophobic proteins such as membrane proteins, very basic proteins, and low abundant proteins are not well represented on the gels and (2) the procedure has low throughput and is difficult to standardize and automate.

A limitation of MALDI-TOF is the identification of low molecular mass proteins, which deliver few peptides, and the identification is often based on a low number of matches (Fountoulakis et al. 1998). In that case, it should be confirmed that the matching peptides are simultaneously the major peptides, and in general, identification of small proteins by MALDI-TOF-MS is not efficient, and the application of MS/MS technologies such as TOF-TOF, Qq-TOF, or LC–MS may be more advantageous.

Moreover, the MALDI-TOF-based protein identification approach cannot identify multiple components of a mixture. In most cases, in 2-D gels, the major component of a protein mixture is identified by MALDI-TOF from one spot. A spot often contains more than one protein and two or more additional proteins can be identified from the same spot as well, but they are usually highly homologous, and the software cannot distinguish between the identification hits. If experimentally derived peptide masses for a given mass accuracy tolerance correspond to more than one protein, the resulting search score becomes low (Westermeier and Marouga 2005). This problem is further aggravated when the number of experimentally derived peptides is small and the intensity of the peptide signal is weak. Additional proteins can be identified from spots by tandem mass spectrometry if a sufficient number of peptides have been analyzed in the MS/MS mode.

In most analyses, only a subset of the predicted masses is actually observed. Such a limitation is the result of multiple contributing factors including loss of peptides during the sample handling process, as described before, ion suppression effects due to the presence of trifluoroacetic acid, surfactant residues and salts, the selective ionization of certain peptide species over others depending on their amino acid sequence, which typically exhibit poor ionization potential, and small peptide masses that are beyond the optimum mass range of the MALDI-TOF instrument.

5.3.2 Ion Trap

In the IT analyzers ions are trapped and can therefore be accumulated over time in a physical device. These ions are then subjected to an additional electric field, which ejects one ion species after another from the trap, and are detected, to produce a mass spectrum. The IT technology is characterized by MS/MS capabilities with unmatched sensibility and fast data acquisition. Used in conjunction with data-dependent acquisition, IT technology allows high-throughput (Domon and Aaebersold 2006).

Many proteomic studies use linear ion trap mass spectrometers to analyze complex peptide mixtures derived from proteolytic digestion of protein samples via online microcapillary liquid chromatography (μLC) and tandem mass spectrometry (MS/MS). Compared to earlier ion trap instruments, the trapping of ions in an axial fashion in the linear ion trap provides significant improvements, increasing ion capacity, scan rate, and improved detection and trapping efficiency. Consequently, these instruments identify proteins by MS/MS from complex biological mixtures with unsurpassed sensitivity, which, combined with their low maintenance, ease of use, and relatively low cost, have made linear ion traps the instrument of choice for a wide variety of proteomic studies. (Griffin et al. 2007).

5.3.2.1 Drawbacks

When only low-resolution spectrometers and low-energy fragmentation techniques are available the MS/MS-based identification of nontryptic peptides becomes a very hard task. In this case even more refined programs developed for peptide identification, such as those also exploiting relative abundances produced at different levels of the procedure (Eng et al. 1994; Narasimhan et al. 2005), may not be able to provide a unique reliable result.

Another limitation of linear ion traps, commonly known as the "one-third rule," arises from decreased stability of fragment ions with *m/z* values less than 30% of the *m/z* for the precursor peptide selected for fragmentation by collision-induced dissociation (CID) during automated MS/MS analysis. For example, fragment ions from an ion at *m/z* 1,500 will not be detected below *m/z* 500, a significant limitation for the de novo sequencing of peptides.

5.3.3 Triple Quadrupole

ESI-MS with quadrupole instruments has been demonstrated to be a powerful tool for mass determination of food proteins and for detection of protein modifications induced by processing (Careri et al. 2002). The quadrupole is a mass filter consisting of four rods to which an oscillating electric field is applied and which lets only a certain mass pass through. The triple quadrupole is composed of two of these quadrupoles; these are separated by a somewhat different quadrupole which serves as a collision cell leading to peptide ion fragmentation (Mann et al. 2001).

One of the first people to exploit the triple quadrupole for protein sequencing by tandem mass spectrometry was Don Hunt (Hunt et al. 1986). Hunt championed the use of low-resolution parent ion selection and daughter ion detection to maximize sensitivity, and chemical modification (peptide methylation and acetylation) to aid the determination of ion series for interpretation.

Triple quadrupoles were the original instruments used for tandem MS in proteomics studies. The accuracy of quadrupole mass analyzers allows selection of specific peptide ions and analysis of fragment ions from MS-MS to within at least ±0.5 amu of their true *m/z* values. This degree of mass accuracy is sufficient to allow direct interpretation of amino acid sequences from peptide MS-MS data obtained with this mass spectrometer (Liebler and Yates 2002).

5.3.3.1 Drawbacks

To apply a "top-down" approach, selected reaction monitoring (SRM) using triple-quadrupole mass spectrometers has been developed (Ji et al. 2003); in this mode, the "parent" ion and its principal obtained fragments are selected. Although SRM detection provides excellent selectivity and duty cycle, it also has some inherent disadvantages in fragmentation-dependent MS detection. As the molecular weight of a protein increases, the fragmentation efficiency of the protein through the SRM process often decreases. Even when the protein is fragmented, the charges are more evenly distributed into many fragments rather than concentrated onto a few major fragments as with small molecules. Furthermore, the electrospray ion source commonly used on triple-quadrupole MS tends to ionize a protein into different charge states. SRM can only detect the one particular fragment transition from a precursor ion at a specific mass-to-charge ratio, whereas the majority of the ions in all the other charge states cannot be utilized at the same moment (Ruan et al. 2011).

5.3.4 Orbitrap

As its name suggests, orbitrap is an ion trap. But it is not a conventional ion trap; there is neither RF (radio frequency) nor a magnet to hold ions inside. Instead, moving ions are trapped in an electrostatic field (Makarov 2000). The electrostatic attraction towards the central electrode is compensated by a centrifugal force that arises from the initial tangential velocity of ions, very much like a satellite in orbit. The electrostatic field that ions experience inside the orbitrap forces them to move in complex spiral patterns. The axial component of these oscillations is independent of initial energy, angles, and positions, and can be detected as an image current on the two halves of an electrode encapsulating the orbitrap. A Fourier transform is employed to obtain oscillation frequencies for ions with different masses, resulting in an accurate reading of their *m/z* (Scigelova and Makarov 2006).

The orbitrap is a high mass accuracy detector. High and accurate mass measurement has led to new proteomic methodologies; the better characterization of the intact protein state and on the other hand the use of accurate mass tags for digested proteins minimizes the need for time-consuming MS/MS-based peptide identification and can, in principle, improve throughput. Protein quantification is also improved by accurate mass analysis (Marshall and Hendrickson 2008). In fact, the literature supports claims of routine mass measurement accuracies of less than 2 ppm for the analyses of complex peptide mixtures (Yates et al. 2006) and even intact proteins (Macek et al. 2006).

5.3.5 *New Types of Mass Spectrometers: The Hybrid Instruments*

The general goal in the design of a hybrid instrument is to combine different performance characteristics offered by various types of analyzers into one mass spectrometer. These performance characteristics may include mass resolving power, the ion kinetic energy for collision-induced dissociation, and speed of analysis (Glish and Burinsky 2008). Table 5.1 summarizes the performance parameters for the various analyzers.

For example, as commented before, the biggest drawback to the TOF analyzers is their inability to perform true MS-MS, however, developed TOF-TOF instruments have overcome this problem. These TOF-TOF instruments incorporate a collision cell between two TOF sections; ions of one *m/z* ratio are selected in the first TOF section, fragmented in the collision cell, and the fragments are separated in the second TOF section (Guerrera and Kleiner 2005).

The combination of triple-quadrupole and ion-trap mass spectrometer functionality inherent in the QTRAP™ instruments means that highly selective scan modes for analyte identification can be combined with very sensitive scans for structural identification on a chromatographic timescale (Hager 2004).

The unique scanning capabilities of a hybrid linear ion trap (Q TRAP) mass spectrometer are described with an emphasis on proteomics applications. The combination of the very selective triple-quadrupole-based tandem mass spectrometry scans with the very sensitive ion trap product ion scans allows rapid identification of peptides at low concentrations derived from post-translationally modified proteins on chromatographic timescales. The Q TRAP instrument also offers the opportunity to conduct a variety of ion processing steps prior to performing a mass scan. For example, the enhancement of the multiple-charge ion contents of the ion trap can be performed resulting in a survey mass spectrum dominated by double- and triple-charge peptides. This facilitates the identification of relevant biological species in both separated and unseparated peptide mixtures for further MS/MS experiments (Le Blanc et al. 2003).

In addition, highly resolved and accurate orbitrap measurements for the parent ions are complemented with tandem mass spectra detected in a linear ion trap with high speed and sensitivity. LTQ orbitrap hybrid instruments offer the additional

Table 5.1 Comparison of mass analyzers with respect to key performance attributes (Reproduced from Glish and Burinsky (2008) with permission from Elsevier)

Analyzer type	Kinetic energy	Resolving power		Mass accuracy		Measurement time[a] (s)
		MS[1]	MS[2]	MS[1]	MS[2]	
Magnetic (B) sector	High	Medium	Medium	ppt	ppt	10^{-5}
Electric (E) sector	High	na[b]	Low	na[b]	ppt	10^{-5}
Electric and magnetic sector[c]	High	High	Medium	ppm	ppt	10^{-5}
Time-of-flight (TOF)	High	Low	Medium[d]	ppm	ppm	10^{-5}
Quadrupole mass filter (Q)	Low	Low	Low	ppt	ppt	10^{-4}
Quadrupole ion trap (QIT)	Low	Low	Low	ppt	ppt	10^{-2}
FTICR	Low	High	High	ppm	ppm	10^{-1}
Orbitrap	High	na[b]	High	na[b]	ppm	10^{-1}

[a]The time, after activation, for the ions to react prior to the second stage of analysis

[b]na: not applicable; this analyzer has not been used for the first stage of an MS/MS experiment. There is currently no mode of operation in which a parent ion can be mass selected for a subsequent stage of analysis

[c]When electric and magnetic sectors are used as a double focusing mass spectrometer for one stage of analysis, either prior to (MS-1) the reaction, or subsequent to (MS-2) the reaction

[d]The resolving power is medium if the TOF ion axis is orthogonal to the first stage of MS; if the axes are coaxial, the resolving power is low

option of transferring the product ions (created in the linear trap) to the orbitrap mass analyzer for analysis with high resolution and high mass accuracy. An orbitrap mass analyzer is the most recent addition to the set of tools that can be applied to identification, characterization, and quantitation of components in biological systems. With its ability to deliver low-ppm mass accuracy and extremely high resolution, all within a timescale compatible with nano-LC separations, the orbitrap has become an instrument of choice for many proteomics applications since its commercial introduction in 2005 (Scigelova and Makarov 2006).

More recently the Q-Exactive instrument has appeared, which features high ion currents because of an S-lens, and fast high-energy collision-induced dissociation peptide fragmentation because of parallel filling and detection modes. In this mass analyzer the orbitrap is coupled with a quadrupole and the image current from the detector is processed by an "enhanced Fourier transformation" algorithm, doubling mass spectrometric resolution. This is demonstrated in a multiplexed single ion monitoring mode, in which the quadrupole rapidly switches among different narrow mass ranges that are analyzed in a single composite MS spectrum. Similarly, the quadrupole allows fragmentation of different precursor masses in rapid succession, followed by joint analysis of the higher energy collisional dissociation fragment ions in the orbitrap analyzer.

More than 2,500 proteins can be identified in standard 90-min gradients of tryptic digests of mammalian cell lysate, a significant improvement over previous orbitrap mass spectrometers. Furthermore, the quadrupole orbitrap analyzer combination enables multiplexed operation at the MS and tandem MS levels. The high performance together with the ability to perform complex multiplexed scan modes makes the Q Exactive an exciting new instrument for proteomics (Michalski et al. 2011).

5.4 Mass Spectrometry for Proteomics: Major Application in Food Safety and Quality

The development of molecular medicine is associated with the development of proteomics. In fact, proteomic analyses of early-stage cancers have provided new insights into the changes that occur in the early phases of tumorigenesis and represent a new resource of candidate biomarkers for early-stage disease. Studies that profile proteomic patterns in body fluids also present new opportunities for the development of novel, highly sensitive diagnostic tools for the early detection of cancer (Wulfkuhle et al. 2003).

This is the most relevant and well-known application of proteomics. However, the tremendous improvement in proteomics sensitivity and specificity due to the progress of mass spectrometric methodologies has reached the scientifically and economically important food management chain. The search for markers of authenticity, quality, and safety of food, as well as the discovery of signature peptides in allergonomics and fraud detection has flourished in the last 10 years (Aiello et al. 2011).

5.4.1 Detection of Food Allergens

Food allergies are an increasingly emergent issue in food science and technology due to the marked increase in these allergies recorded among the population. Allergenic food proteins are numerous and heterogeneous, with a genetic polymorphism resulting in several variants for each protein (Picariello et al. 2011; Monaci and Visconti 2009). Up to now, the bottom-up and middle-down (using limited digestion to produce larger peptides) strategies are the most widely used methods for the detection of allergens in food.

Mass spectrometry has allowed the identification of either genes for allergic diseases or allergenic proteins. In this way, MALDI-TOF-MS has been used to quantify gluten gliadins in both processed and unprocessed foods (Camafeita et al. 1997a, b); the procedure is rapid and sensitive with good correlation with data from an immunological assay method. MALDI-TOF-MS can be used as a rapid screening technique for (a) the presence of gliadins in foods by monitoring the occurrence of the protonated gliadin mass pattern in the mass range from 25,000 to 40,000 and (b) the presence of other toxic gluten cereal prolamins fractions, such as barley hordeins, rye secalins, and oat avenins (Camafeita and Mendez 1998) and (Camafeita et al. 1998).

Although MALDI-TOF-MS (Ferranti et al. 2007) and LC-QTOF-MS (Monaci et al. 2010a, b) are widely used instruments for this purpose, other mass analyzers have been applied. For example, Heick et al. (2011) have developed a method based on liquid chromatography–triple-quadrupole mass spectrometry in selected reaction monitoring (SRM) mode for the simultaneous detection of seven allergens in milk, egg, soy, hazelnut, peanut, walnut, and almond. The method was based on extraction of the allergens from the food matrix, followed by an enzymatic digestion with trypsin. The MS method was used to detect all seven allergens from an incurred bread matrix.

In light of the bibliography, proteomics MS-based methods for allergenic protein identification, characterization, and quantification deserve a great deal of attention. New MS analyzers such as orbitrap instruments are being focused on analysis of intact protein or the use of accurate peptide mass for high-throughput analysis (Monaci et al. 2011).

5.4.2 Detection of Pathogens

The first attempts to identify microorganisms using mass spectrometry were performed as early as 1975 (Anhalt and Fenselau 1975). However, these experiments suffered from irreproducible results due to the variabilities caused by growth conditions and media. Only with the discovery of MALDI-TOF MS in the 1980s did the analysis of relatively large biomolecules, including larger ribosomal proteins, become possible (Hillenkamp and Karas 1990). The latter are less influenced by culture conditions allowing MALDI-TOF MS to be consistently used to differentiate bacterial species. In recent years, MALDI-TOF MS has been implemented in routine laboratories and utilized as a completely new approach for the identification of bacteria and yeast (Wieser et al. 2012).

Regarding other mass analyzers, Sospedra et al. (2012) have developed a simple, rapid, economic, and effective procedure for the simultaneous analysis of two staphylococcal enterotoxins using a triple-quadrupole instrument. The authors concluded that this detector has many advantages over traditional techniques such as avoiding the necessity of preliminary steps to isolate the bacteria from food, possibility of quantification, and lower limits of detection; moreover, under the proposed conditions, SEA and SEB (enterotoxins A and B) could be directly detected from milk and other food samples.

Melanson et al. (2006) have reported the first application of the MALDI-QqQ to validate an important virulence biomarker of *Candida albicans* employing an amine-specific isotopic labeling approach; the technique was validated using standard peptides differentially labeled at known concentrations.

5.4.3 Food Processing

Proteins are largely responsible for the characteristics of many food products during the manufacturing process: physicochemical properties, such as viscosity, thermal conductivity, and vapor pressure, but also nutritional and sensory properties depend on their protein composition and content. This manufacturing process induces significant changes including chemical modification, heat-shock protein responses, and proteolytic degradation (Gašo-Sokač et al. 2010).

As an example, many different thermal treatments are used in milk products, typically for the improvement of shelf life, which can, however, potentially affect the nutritional and functional qualities of milk proteins. Structurally sensitive techniques must rely on the detection of intact proteins by MS; for example, MALDI-ToF-MS has been applied to unfractionated milk for the characterization of milk-processing history with some success (Holland et al. 2011). In this work 2DE coupled with MALDI-TOF MS was used to examine the effects of storage at elevated temperature on UHT milk.

Zhang et al. (2012) developed a reliable LC-triple quadrupole method based on tryptic product peptides for determination of total α-lactalbumin in infant formulas and whey protein concentrates. The quantitation of the α-lactalbumin content was carried out by digesting the infant formula or whey protein concentrates followed by MRM-based LC–MS/MS analysis at the peptide level. An additional advantage of the current method was a measure of the whole bovine α-lactalbumin including the native form and thermally induced denatured form. Finally, it was successfully applied to determine bovine α-lactalbumin in 21 different infant formulas and eight whey protein concentrates. Because the established method could determine native and thermal denatured bovine α-lactalbumin, as well as soluble and therefore functional bovine α-lactalbumin, its application promoted the development of nutrient investigation and quality control of infant formulas and other dairy products containing α-lactalbumin.

Protein composition of other foods such as meat and meat products are more complex, and the change of physicochemical properties during processing depends on more than one highly abundant protein (Lametsch et al. 2002). When looking for meat quality indicators, researchers often wonder which compounds can best be used to represent them. Inasmuch as proteins make up a considerable part of the meat basic composition and changes are associated with its quality, it is not surprising that these compounds are frequently looked to as a source of possible indicators/markers of meat quality (Bendixen 2005).

Bernevic et al. (2011) reported the identification and structural characterization of post-mortem degradation and oxidation of myofibrillar proteins using high-resolution mass spectrometric proteomics. The combination of 2D-PAGE and FTICR-MS was found to be a powerful approach for identification of muscle protein degradation products, providing identification of several truncation forms of creatine kinase and troponin T.

5.4.4 Food Adulteration

Proteomics approaches have also been applied to assess other important points included in food quality, because protein profiling can give useful information on food origin or adulteration. Protein profiling with MS can search for biomarkers that permit the characterization of food samples according to their origin (Herrero et al. 2012).

In this sense, Di Girolamo et al. (2012) analyzed honeys of different origin (from chestnut, acacia, sunflower, eucalyptus, and orange) for their proteome content, in order to see if any plant proteins present would allow the proteo-typing of these different varieties. All bands visible in the SDS-PAGE profile of each type of honey were eluted, digested, and identified by mass spectrometry in a hybrid LTQ-orbitrap instrument. It turned out that all proteins identified were not of plant origin but belonged to the *Apis mellifera* proteome. Among the total proteins identified (eight, but only seven as basic constituents of all types of honey) five belonged to the family of major royal jelly proteins, and were also the most abundant ones in any type of honey. It thus appears that honey has a proteome resembling the royal jelly proteome (but with considerably fewer species), except that its protein concentration is lower by three to four orders of magnitude as compared to royal jelly. Attempts at identifying additional plant (pollen, nectar) proteins via peptidome analysis were unsuccessful.

The study of proteins can be also useful to detect adulterations in food products; the aim of these applications usually is the detection of proteins that are not a part of the proteome of the studied product. An example of this approach was the study of Angeletti et al. that examined the capabilities of MALDI-TOF-MS for the characterization of water-buffalo milk and mozzarella cheese to detect possible fraudulence in mozzarella cheese production (Angeletti et al. 1998).

Another example was the development of an untargeted LC-QTOF-MS method for protein analysis in skimmed milk powder (Cordawener et al. 2009). The significantly lower price of other vegetable-protein preparations induces the partial adulteration of the dairy product with soy or pea proteins. It was demonstrated that this procedure detected the presence of adulterations in the milk powder based on differential peptide profiling.

5.5 Mass Spectrometry-Based Proteomics Future Perspectives

Nutrition is still an expanding field for proteomics compared to well-established clinical and medical applications. The success of proteomics in nutrition will depend on multiple factors; one of them concerns the analytical strategies.

In proteomics, MS as a standalone technique or combined with LC has become widespread. However, there is an evident need to develop improved technologies to become a reality for the routine analysis for proteome research, including improvements in the resolution of peptides to provide increased protein coverage. In any case, MS will continue being essential for the systematic investigation in proteomics; in this sense, conventional mass spectrometers as quadrupoles or ion traps are replaced by the more sophisticated and compact mass spectrometers, most of them hybrid instruments in a combination of two or more analyzers.

New applications of proteomics technologies are expected in food safety and quality; as an example, although MS-based proteomics have proven to be a very useful tool for the identification, characterization, and detection of food allergens, there are still some issues that have not been successfully resolved, such as the development of MS-based methods for the simultaneous determination of multiple food allergens in food products and commodities.

References

Aebersold R, Mann M (2003) Mass spectrometry-based proteomics. Nature 422:198–207

Aerbersold R, Goodlett DR (2001) Mass spectrometry in proteomics. Chem Rev 101:269–295

Aiello D, De Luca D, Gionfriddo E, Naccarato A, Napoli A, Romano E, Russo A, Sindona G, Tagarelli A (2011) Multistage mass spectrometry in quality, safety and origin of foods. Eur J Mass Spectrom 17:1–31

Angeletti R, Gioacchini AM, Seraglia R, Piro R, Traldi P (1998) The potential of matrix-assisted laser desorption/ionization mass spectrometry in the quality control of water buffalo mozzarella cheese. J Mass Spectrom 33:525–531

Anhalt JP, Fenselau C (1975) Identification of bacteria using mass spectrometry. Anal Chem 47:219–225

Bendixen E (2005) The use of proteomics in meat science. Meat Sci 71:138149

Bernevic B, Petre BA, Galetskiy D, Werner C, Wicke M, Schellander K, Przybylski M (2011) Degradation and oxidation postmortem of myofibrillar proteins in porcine skeleton muscle revealed by high resolution mass spectrometric proteome analysis. Int J Mass Spectrom 305:217–227

Cairns DA (2011) Statistical issues in quality control of proteomic analyses, good experimental design and planning. Proteomics 11:1037–1048
Camafeita E, Mendez E (1998) Screening of gluten avenins in foods by matrix-assisted laser desorption/ionization time-of-flight mass spectrometry. J Mass Spectrom 33:1023–1028
Camafeita E, Alfonso P, Acevedo B, Mendez E (1997a) Sample preparation optimization for the analysis of gliadins in food by matrix-assisted laser desorption/ionization time-of-flight mass spectrometry. J Mass Spectrom 32:444–449
Camafeita E, Alfonso P, Mothes T, Mendez E (1997b) Matrix-assisted laser desorption/ionization time-of-flight mass spectrometric micro-analysis: the first non-immunological alternative attempt to quantify gluten gliadins in food samples. J Mass Spectrom 32:940–947
Camafeita E, Solis J, Alfonso P, Lopez JA, Sorell L, Mendez E (1998) Selective identification by matrix-assisted laser desorption/ionization time-of-flight mass spectrometry of different types of gluten in foods made with cereal mixtures. J Chromatogr A 823:299–306
Careri M, Bianchi F, Corradini C (2002) Recent advances in the application of mass spectrometry in food-related analysis. J Chromatogr A 970:3–64
Chen CH (2008) Review of a current role of mass spectrometry for proteome research. Anal Chim Acta 624:16–36
Cordawener JHG, Luykx DMAM, Frankhuizen R, Bremer MGEG, Hooijerink H, America AHP (2009) Untargeted LC-Q-TOF mass spectrometry method for the detection of adulterations in skimmed-milk powder. J Sep Sci 32:1216–1223
Cornish TJ, Cotter RJ (1993a) Tandem time-of-flight mass spectrometer. Anal Chem 65:1043–1047
Cornish TJ, Cotter RJ (1993b) Collision-induced dissociation in a tandem time-of-flight mass spectrometer with two single-stage reflectrons. Org Mass Spectrom 28:1129–1134
Cottrell JS (1994) Protein identification by peptide mass fingerprinting. Pept Res 7:115.124
D'Alessandro A, Zolla L (2012) We are what we eat: food safety and proteomics. J Proteome Res 11:26–36
Di Girolamo F, D'Amato A, Righetti PG (2012) Assessment of the floral origin of honey via proteomic tools. J Proteomics 75:3688–3693
Domon B, Aaebersold R (2006) Mass spectrometry and protein analysis. Science 312:212–217
Emmet MR, Caprioli RM (1994) Microelectrospray mass spectrometry: ultra high-sensitivity analysis of peptides and proteins. J Am Soc Mass Spectrom 5:605–613
Eng JK, McCormack AL, Yates JR (1994) An approach to correlate tandem mass spectral data of peptides with amino acid sequences in a protein database. J Am Soc Mass Spectrom 5:976–989
Fenn JB, Mann M, Meng CK, Wong SF, Whitehouse CM (1989) Electrospray ionization for mass spectrometry of large biomolecules. Science 246:64–71
Ferranti P, Mamone G, Picariello G, Addeo F (2007) Mass spectrometry analysis of gliadins in celiac disease. J Mass Spectrom 42:1531–1548
Fountoulakis M, Juranville JF, Roeder D, Evers S, Berndt P, Langen H (1998) Reference map of the low molecular mass proteins of Haemophilus influenzae. Electrophoresis 19:1819–1827
Garbis S, Lubec G, Fountoulakis M (2005) Limitations of current proteomics technologies. J Chromatogr A 1077:1–18
Gašo-Sokač D, Kovač S, Josić D (2010) Application of proteomics in food technology and food biotechnology: process development, quality control and product safety. Food Technol Biotechnol 48:284–295
Gingras AC, Aebersold R, Raught B (2005) Advances in protein complex analysis using mass spectrometry. J Physiol 563:11–21
Glish GL, Burinsky DJ (2008) Hybrid mass spectrometers for tandem mass spectrometry. J Am Soc Mass Spectrom 19:161–172
Griffin TJ, Xie H, Bandhakavi S, Popko J, Mohan A, Carlis JV, Higgins L (2007) iTRAQ reagent-based quantitative proteomic analysis on a linear ion trap mass spectrometer. J Proteome Res 6:4200–4209
Guerrera IC, Kleiner O (2005) Application of mass spectrometry in proteomics. Biosci Rep 25:71–93

Hager JW (2004) QTRAP™ mass spectrometer technology for proteomics applications. Drug Discov Today Targets 3:31–36

Han X, Aslanian A, Yates JR (2008) Mass spectrometry for proteomics. Curr Opin Chem Biol 12:483–490

Heick J, Fischer M, Pöpping B (2011) First screening method for the simultaneous detection of seven allergens by liquid chromatography mass spectrometry. J Chromatogr A 1218:938–943

Herrero M, Simó C, Garcia-Cañas V, Ibáñez E, Cifuentes A (2012) Foodomics: MS-based strategies in modern food science and nutrition. Mass Spectrom Rev 31:49–69

Hillenkamp F, Karas M (1990) Mass spectrometry of peptides and proteins by matrix-assisted ultraviolet laser desorption/ionization. Methods Enzymol 193:280–295

Holland JW, Gupta R, Deeth HC, Alewood PF (2011) Proteomic analysis of temperature-dependent changes in stored UHT milk. J Agr Food Chem 59:1837–1846

Hunt DF, Yates JR, Shabanowitz J, Winston S, Hauer CR (1986) Protein sequencing by tandem mass spectrometry. Proc Natn Acad Sci U S A 83:6233–6237

Ji QC, Rodila R, Gage EM, El-Shourbagy TA (2003) A strategy of plasma protein quantitation by selective reaction monitoring of an intact protein. Anal Chem 75:7008–7014

Karas M, Hillenkamp F (1988) Laser desorption ionization of proteins with molecular mass exceeding 10,000 Daltons. Anal Chem 60:2299–2301

Klose J (1975) Protein mapping by combined isoelectric focusing and electrophoresis of mouse tissues. A novel approach to testing for induced point mutations in mammals. Humangenetik 26:231–243

Lametsch R, Roepstorff P, Bendixen E (2002) Identification of protein degradation during post-mortem storage of pig meat. J Agr Food Chem 50:5508–5512

Lane CS (2005) Mass spectrometry-based proteomics in the life sciences. Cell Mol Life Sci 62:848–869

Le Blanc JC, Hager JW, Ilisiu AM, Hunter C, Zhong F, Chu I (2003) Unique scanning capabilities of a new hybrid linear ion trap mass spectrometer (QTRAP) used for high sensitivity proteomics applications. Proteomics 3:859–869

Liebler DC, Yates JR (2002) Introduction to proteomics. Tools for the new biology. Humana Press, Totowa

Macek B, Waanders L, Olsen JV, Mann M (2006) Top-down protein sequencing and MS3 on a hybrid linear quadrupole ion trap-orbitrap mass spectrometer. Mol Cell Proteom 5:949–958

Makarov A (2000) Electrostatic axially harmonic orbital trapping: a high-performance technique of mass analysis. Anal Chem 72:1156–1162

Mann M, Hendrickson RC, Pandey A (2001) Analysis of proteins and proteomes by mass spectrometry. Annu Rev Biochem 70:437–473

Marshall AG, Hendrickson CL (2008) High-resolution mass spectrometers. Annu Rev Anal Chem 1:579–599

Melanson JE, Chisholm KA, Pinto DM (2006) Targeted comparative proteomics by liquid chromatography/matrix-assisted laser desorption/ionization triple-quadrupole mass spectrometry. Rapid Comm Mass Spectrom 20:904–910

Michalski A, Damoc E, Hauschild JP, Lange O, Wieghaus A, Makarov A, Nagaraj N, Cox J, Mann M, Horning S (2011) Mass spectrometry-based proteomics using Q Exactive, a high-performance benchtop quadrupole Orbitrap mass spectrometer. Mol Cell Proteom 10: M111.011015-(1–15)

Monaci L, Visconti A (2009) Mass spectrometry-based proteomics methods for analysis of food allergens. TrACs Trends Anal Chem 28:581–591

Monaci L, Losito I, Palmisano F, Visconti A (2010a) Identification of allergenic milk proteins markers in fined white wines by capillary liquid chromatography-electrospray ionization-tandem mass spectrometry. J Chromatogr A 1217:4300–4305

Monaci L, Nørgaard JV, van Hengel AJ (2010b) Feasibility of a capillary LC/ESI-Q-TOF MS method for the detection of milk allergens in an incurred model food matrix. Anal Methods 2:967–972

Monaci L, Losito I, Palmisano F, Godula M, Visconti A (2011) Towards the quantification of residual milk allergens in caseinate-fined white wines using HPLC coupled with single-stage Orbitrap mass spectrometry. Food Add Contam A 28:1304–1314
Narasimhan C, Tabb DL, VerBerkmoes NC, Thompson MR, Hettich RL, Uberbacher EC (2005) MASPIC: intensity-based tandem mass spectrometry scoring scheme that improves peptide identification at high confidence. Anal Chem 77:7581–7593
O'Farrell PH (1975) High resolution two-dimensional electrophoresis of proteins. J Biol Chem 250:4007–4021
Picariello G, Mamone G, Addeo F, Ferranti P (2011) The frontiers of mass spectrometry-based techniques in food allergenomics. J Chromatogr A 1218:7386–7398
Ruan Q, Ji QC, Arnold ME, Griffith Humphreys W, Zhu M (2011) Strategy and its Implications of protein bioanalysis utilizing high-resolution mass spectrometric detection of intact protein. Anal Chem 83:8937–8944
Scigelova M, Makarov A (2006) Orbitrap mass analyzer -overview and applications in proteomics. Prac Proteom 1–2:16–21
Sénéchal S, Kussmann M (2011) Nutriproteomics: technologies and applications for identification and quantification of biomarkers and ingredients. P Nutr Soc 70:351–354
Shen TL, Noon KR (2004) Liquid cromatography-mass spectrometry and tandem mass spectrometry of peptides and proteins. In: Aguilar MI (ed) HPLC of peptides and proteins. Methods and protocols. Humana Press, Totowa, pp 111–139
Smith RD, Loo JA, Edmonds CG, Barinaga CJ, Udseth HR (1990) Sensitivity considerations for large molecule detection by capillary electrophoresis-electrospray ionization mass spectrometry. J Chromatogr 516:157–165
Sospedra I, Soler C, Mañes J, Soriano JM (2012) Rapid whole protein quantitation of staphylococcal enterotoxins A and B by liquid chromatography/mass spectrometry. J Chromatogr A 1238:54–59
Westermeier R, Marouga R (2005) Protein detection methods in proteomics research. Bioscience Rep 25:19–32
Wieser A, Schneider L, Jung J, Schubert S (2012) MALDI-TOF MS in microbiological diagnostics—identification of microorganisms and beyond (mini review). Appl Microbiol Biotechnol 93:965–974
World Food Summit (1996) http,//www.fao.org/wfs/index_en.htm
Wulfkuhle JD, Liotta LA, Petricoin EF (2003) Early detection: proteomic applications for the early detection of cancer. Nat Rev Cancer 3:267–275
Yates JR III (2004) Mass spectral analysis in proteomics. Annu Rev Biophys Biomol Struct 33:297–316
Yates JR III, Cociorva D, Liao J, Zabrouskov V (2006) Performance of a linear ion trap-orbitrap hybrid for peptide analysis. Anal Chem 78:493–500
Zhang J, Lai S, Zhang Y, Huang B, Li D, Ren Y (2012) Multiple reaction monitoring-based determination of bovine α-lactalbumin in infant formulas and whey protein concentrates by ultra-high performance liquid chromatography–tandem mass spectrometry using tryptic signature peptides and synthetic peptide standards. Anal Chimica Acta 727:47–53

Part II
Food Applications of Proteomics

Chapter 6
Challenges and Applications of Proteomics for Analysis of Changes in Early Postmortem Meat

Honggang Huang and René Lametsch

6.1 Introduction

Skeletal muscle is the most abundant tissue in animals; on average, 30–40% of the live weight or 40–60% of the carcass of domestic animals consists of muscular tissue (Lawrie and Ledward 2006). In living animals, contractile fibers of skeletal muscle tissues constitute the cellular units that support coordinated excitation–contraction–relaxation cycles for movements and postural control (Gordon et al. 2000). Nutritionally, muscle is a great source of essential amino acids, and to a lesser extent, of other important nutrients such as minerals, vitamins, and fatty acids, so that muscle could provide the majority of the nutrients required for human health. From the food science perspective, due to the high abundance and nutritional value, skeletal muscle is one of the most important food resources for consumers.

Variation in meat quality traits is a well-known problem in the meat industry. Although extensively studied, the underlying mechanisms contributing to different meat quality traits are not fully understood which is essential in order to improve meat quality. Hence, new tools and strategies must be applied in meat research. Proteomics is one of these tools and has been shown to be a powerful tool to investigate post-mortem protein changes.

In proteomics it is possible to separate and quantify hundreds of proteins in one analysis. A major advantage of proteomics is that it is possible to carry out a detailed characterization of the single proteins and identify cleavage sites and protein modifications such as protein oxidation and phosphorylation. The use of proteomics in meat research has a large potential and will lead to a much more detailed characterization of the post-mortem protein changes. Proteomics can

H. Huang • R. Lametsch (✉)
Department of Food Science, University of Copenhagen,
Frederiksberg DK-1958, Denmark
e-mail: rla@life.ku.dk

F. Toldrá and L.M.L. Nollet (eds.), *Proteomics in Foods: Principles and Applications*,
Food Microbiology and Food Safety 2, DOI 10.1007/978-1-4614-5626-1_6,

also be applied in other aspects of meat research and become a valuable tool in the investigation of protein changes during meat processing, effects of meat packaging, animal growth, and genetic variation. Hence, proteomics can provide valuable information on the mechanisms influencing the different quality traits to gain a better understanding of these mechanisms. This information can be used to optimize meat production and improve meat quality. This chapter focuses on the challenges regarding muscle proteomic study, the technologies used in proteomics, and the recent results obtained with proteomics in meat research.

6.2 The Challenges Regarding Proteomic Analysis of Skeletal Muscle

Skeletal muscle proteomics aims at the global identification, characterization of the entire protein complement in muscle under normal or abnormal development, or different treatments. Over the past decade, the rapid developments in mass spectrometry and bioinformatics tools have driven the remarkable progress of proteomic science. However, proteomics study in skeletal muscle is still a very challenging task because of the wide-ranging biochemical heterogeneity of the muscle proteins.

Muscle proteome is notorious for its high dynamicity and diversity (Ohlendieck 2010). The amount of protein expression varies greatly in muscle. Myosin heavy chain, actin, titin, and nebulin account for the majority of muscle proteins whereas the other proteins are only present in relatively low amounts, in a large dynamic range on the order of 10^6 magnitude. The wide and dynamic expression range of proteins makes it impossible to separate and detect all protein species with current techniques because the extremely intensive signals from high abundant proteins often overlap or cover the signals from other low abundant proteins during separation and MS identification. In addition, skeletal muscle is heterogeneous in composition. Aside from the multiple fiber types, such as slow oxidative (type I), moderately fast oxidative glycolytic (type IIa), and fast glycolytic (type IIb) fibers, muscle also contains connective tissues, capillaries, and nerve cells.

The complex tasks of muscles are performed by numerous muscle proteins with specialized functions, structures, and interactions. A considerable amount of muscle proteins are integral membrane proteins and high molecular mass complexes. It is a notable feature of contractile fibers that some of the largest protein species are highly expressed in skeletal muscle, such as nebulin of 600–800 kDa and titin with a molecular mass exceeding 1,200 kDa (Ottenheijm and Granzier 2010a, b). Supramolecular membrane assemblies are also highly present in muscle, such as the abundant actomyosin machinery with its regulatory troponin–tropomyosin system, the ryanodine receptor Ca^{2+} release channel of the triad junctions, the dystrophin–glycoprotein complex of the sarcolemma, and the respiratory chain of muscle mitochondria (Ohlendieck 2011). The large number of membrane-associated proteins, the high molecular mass of many muscle components, and extensive post-translational modifications in various muscle proteins and their assembly in highly complex supramolecular structures make it extremely difficult to carry out conventional biochemical studies of muscle. All these analytical facts have to be considered when one performs a skeletal muscle proteomic study. Standard proteomic methods such as gel electrophoresis and liquid chromatography are not capable of separating all muscle proteins in one run. Some sample fractionation and subcellular proteomic strategies can be employed to analyze membrane proteins and proteins from subcellular organelles (Ohlendieck 2011).

6.3 Approaches for Proteomic Study of Muscle Food

For muscle food, mass spectrometry-based large-scale proteomic studies have already provided a plethora of new information by comparison of different meat quality, processing condition, and genetic backgrounds in most farm animals (Bendixen 2005; Bendixen et al. 2011; Hollung et al. 2007). Methodologically, proteomics routinely employ diverse separation techniques including gel-based approaches (one and two-dimensional gel electrophoresis) and gel-free approaches (different chromatographic methods), in combination with advanced mass spectrometric methods for the identification of peptides and proteins of interest (Baggerman et al. 2005; Falk et al. 2007). The proteomic data are interpreted by diverse bioinformatics tools and can be verified by using immunoblotting, activity assays, or other methods.

6.3.1 Gel-Based Proteomic Approaches

Gel-based proteomic (one-dimensional gel electrophoresis (1DE), two-dimensional polyacrylamide gel electrophoresisis (2DE)) comprise the most classic and versatile methods for global protein separation and quantification (Baggerman et al. 2005). They can screen the protein expression at large scale with lower cost as compared with gel-free proteomics. Despite the apparent advantages of 2DE for separation of complex protein mixtures, the technique suffers from a number of major drawbacks. The proteins identified from 2-D gel are mainly high abundant proteins; the low abundant proteins are not easily detected by gel-based approaches. Some prefractionation methods can be used to overcome this drawback to some extent. Proteins with high (>150 kDa) and low (<10 kDa) molecular weight and proteins with extreme isoelectric points, in particular basic proteins, are usually not detected in standard 2DE. Another drawback concerns the hydrophobic proteins that are not soluble in the buffers used for sample loading or precipitate during the electrophoresis process (Lescuyer et al. 2004). These limitations of the 2DE make it very difficult to investigate many of the structural proteins in muscle as discussed before. 2DE also has some technical limitations. Briefly, the process is time consuming, labor intensive, and requires significant technical expertise to generate reproducible gels (Rabilloud 2002).

6.3.2 *Gel-Free Proteomic Approaches*

The limitations of gel-based approaches have promoted the development of alternative gel-free proteomic approaches. In these approaches, complex protein mixtures are digested in solution. The resulting peptide mixture is fractionated by one or several steps of chromatographic methods and analyzed in a data-dependent manner by LC-MS/MS for separation, quantification, and identification of the proteins (Aebersold and Mann 2003; Yates 2004). However, no single chromatographic method is capable of separating the complex proteome of a cell or tissue. Hence, it is necessary to use a combination of different separation steps such as ion exchange, size exclusion, and reverse-phase chromatography. The gel-free methods have the advantage over 2DE in that they allow examination of high or low abundance proteins in the same analysis and are unbiased with respect to molecular weight, isoelectric point, and hydrophobicity of the proteins. Furthermore, all steps may be automated for high-throughput analysis.

The development in LC-MS/MS methods in proteomics is enormous and many new and smarter methods are constantly being introduced. However, the methods all have some common limitations. For instance, the use of LC-MS/MS in the high-throughput gel-free approaches results in an enormous dataflow and thousands of spectra are generated during the MS analysis; a very strong bioinformatics platform is required to analyze the raw data. Datasets from different LC-MS/MS runs can be combined, but there is a high likelihood that different sets of peptides will be identified in each experiment, so how to interpret the preliminary results is also a big challenge. Another limitation is that only a few samples can be analyzed in each LC-MS/MS run; this makes comparison of multiple samples difficult to undertake. Lastly, in order to get satisfactory results, the requirement for advanced equipment and experienced personal is also very high and expensive.

6.4 Application of Proteomics in Early Post-Mortem Meat

The quality of raw pig meat is influenced by changes in the muscle/meat proteome caused by different factors such as animal growth, age, rate of glycolysis, and post-mortem protein degradation (Hopkins and Thompson 2002; Rosenvold and Andersen 2003). Meat scientists have performed a substantial amount of research on their factors, which has led to considerable quality and compositional improvements. However, the underlying biochemical and physicochemical mechanisms behind the

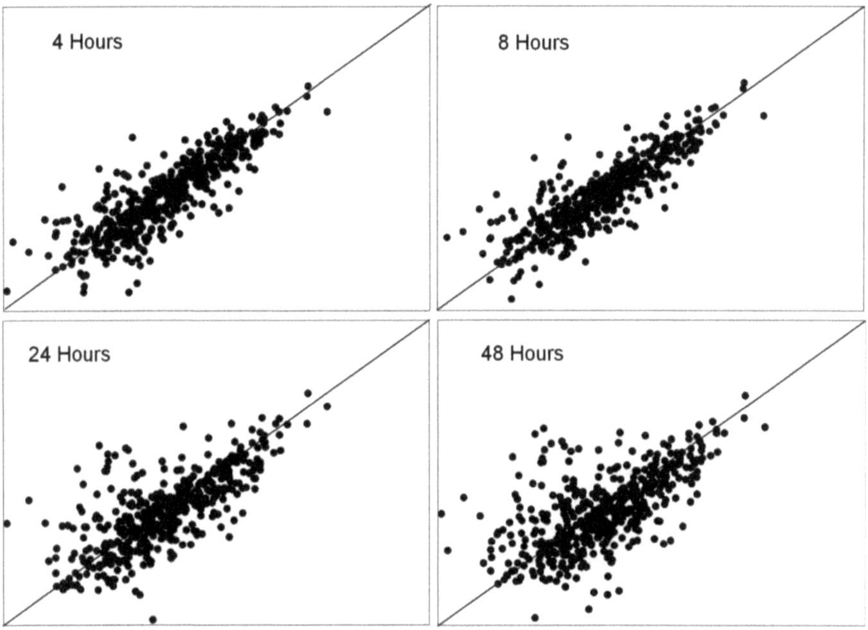

Fig. 6.1 Postmortem protein changes. The spot intensity at 4, 8, 24 and 48 h postmortem of more than 600 individual spots are plotted against the spot intensity at slaughter. If the spot intensity does not change postmortem it is plotted on the diagonal. If the spot intensity increases it is plotted above the diagonal and vice versa

influence of these factors on meat are to some extent still not fully understood. The recent application of proteomics in the field of meat science has provided some interesting and promising results. Early post-mortem protein changes in muscle have been investigated with proteomics in several studies (D'Alessandro et al. 2011; Jia et al. 2007; Lametsch et al. 2003; Promeyrat et al. 2011). The studies revealed that a large part of the proteome changes post mortem. The post-mortem changes in pig LM during the first 48 h after slaughter were monitored with 2DE and more than 600 protein spots were separated and quantified; more than 100 proteins were found to change post mortem as illustrated in Fig. 6.1 (Lametsch and Bendixen 2001). The mechanism behind these post-mortem changes are to some extent still unclear. However, the main cause of post-mortem protein changes is probably protein degradation, as many of the identified changes are protein fragments that increase in spot intensity post mortem (Lametsch et al. 2002). It has been suggested that protein stability is changed post mortem leading to precipitation and aggregation (Bjarnadottir et al. 2010). But changes in protein modification such as phosphorylation or oxidation that change the isoelectric point of the proteins or release from protein complexes have also been found to contribute to post-mortem protein changes (Lametsch et al. 2011; Lund et al. 2007). Even protein expression may to some extent cause some changes post mortem, however, it is unlikely that protein expression causes major changes once the muscle has entered the state of rigor mortis because protein expression is an energy-requiring process and the energy is nearly depleted after reaching rigor mortis (Henckel et al. 2002).

6.5 Future Perspectives

The proteomics studies made on the post-mortem changes in meat clearly illustrate their large potential in meat research. Proteomics have especially proven to be a powerful tool to investigate post-mortem protein degradation in meat because they provide valuable information about the complex mechanisms behind post-mortem proteolysis. Proteomics can provide information on the cleavage sites and post-mortem degradation pattern of the proteins. Furthermore the resulting protein fragments may be used as biomarkers to measure the activity of specific proteolytic enzymes. Such biomarkers could be applied in breeding or to optimize animal slaughter and meat processing to gain more tender meat. Proteomics can also be applied to study post-mortem metabolism to provide further knowledge of undesirable meat characteristics such as PSE. Proteomics will also be an effective research tool to investigate the relation between meat quality and post-mortem protein modifications such as protein oxidation. However, it is important to emphasize that the proteome of a muscle cell or a meat product is extremely complex and that none of the proteomic methods is able to profile a complete proteome. Moreover, all methods in proteomics are very complicated and time consuming and only a limited number of samples can be analyzed in each experiment. Hence, it is often necessary to perform several different proteomics studies to reveal a specific mechanism.

References

Aebersold R, Mann M (2003) Mass spectrometry-based proteomics. Nature 422:198–207

Baggerman G, Vierstraete E, De Loof A, Schoofs L (2005) Gel-based versus gel-free proteomics: a review. Comb Chem High Throughput Screen 8:669–677

Bendixen E (2005) The use of proteomics in meat science. Meat Sci 71:138–149

Bendixen E, Danielsen M, Hollung K, Gianazza E, Miller I (2011) Farm animal proteomics – a review. J Proteomics 74:282–293

Bjarnadottir SG, Hollung K, Faergestad EM, Veiseth-Kent E (2010) Proteome changes in bovine longissimus thoracis muscle during the first 48 h postmortem: shifts in energy status and myofibrillar stability. J Agric Food Chem 58:7408–7414

D'Alessandro A, Marrocco C, Zolla V, D'Andrea M, Zolla L (2011) Meat quality of the longissimus lumborum muscle of Casertana and large white pigs: metabolomics and proteomics intertwined. J Proteomics 75:610–627

Falk R, Ramstrom M, Stahl S, Hober S (2007) Approaches for systematic proteome exploration. Biomol Eng 24:155–168

Gordon AM, Homsher E, Regnier M (2000) Regulation of contraction in striated muscle. Physiol Rev 80:853–924

Henckel P, Karlsson A, Jensen MT, Oksbjerg N, Petersen JS (2002) Metabolic conditions in porcine longissimus muscle immediately pre-slaughter and its influence on peri- and post mortem energy metabolism. Meat Sci 62:145–155

Hollung K, Veiseth E, Jia XH, Faergestad EM, Hildrum KI (2007) Application of proteomics to understand the molecular mechanisms behind meat quality. Meat Sci 77:97–104

Hopkins DL, Thompson JM (2002) Factors contributing to proteolysis and disruption of myofibrillar proteins and the impact on tenderisation in beef and sheep meat. Aus J Agric Res 53:149–166

Jia XH, Ekman M, Grove H, Faergestad EM, Aass L, Hildrum KI, Hollung K (2007) Proteome changes in bovine longissimus thoracis muscle during the early postmortem storage period. J Proteome Res 6:2720–2731

Lametsch R, Bendixen E (2001) Proteome analysis applied to meat science: characterizing post mortem changes in porcine muscle. J Agric Food Chem 49:4531–4537

Lametsch R, Roepstorff P, Bendixen E (2002) Identification of protein degradation during post-mortem storage of pig meat. J Agric Food Chem 50:5508–5512

Lametsch R, Karlsson A, Rosenvold K, Andersen HJ, Roepstorff P, Bendixen E (2003) Postmortem proteome changes of porcine muscle related to tenderness. J Agric Food Chem 51:6992–6997

Lametsch R, Larsen MR, Essen-Gustavsson B, Jensen-Waern M, Lundstrom K, Lindahl G (2011) Postmortem changes in pork muscle protein phosphorylation in relation to the RN genotype. J Agric Food Chem 59:11608–11615

Lawrie RA, Ledward DA (2006) Lawrie's meat science, 7th edn. Woodhead, Cambridge

Lescuyer P, Hochstrasser DF, Sanchez JC (2004) Comprehensive proteome analysis by chromatographic protein prefractionation. Electrophoresis 25:1125–1135

Lund MN, Lametsch R, Hviid MS, Jensen ON, Skibsted LH (2007) High-oxygen packaging atmosphere influences protein oxidation and tenderness of porcine longissimus dorsi during chill storage. Meat Sci 77:295–303

Ohlendieck K (2010) Proteomics of skeletal muscle glycolysis. Biochim Biophys Acta 1804:2089–2101

Ohlendieck K (2011) Skeletal muscle proteomics: current approaches, technical challenges and emerging techniques. Skelet Muscle 1:6

Ottenheijm CAC, Granzier H (2010a) New insights into the structural roles of nebulin in skeletal muscle. J Biomed Biotech 968139:1–6

Ottenheijm CAC, Granzier H (2010b) Role of titin in skeletal muscle function and disease. Muscle Biophys Mol Cells 682:105–122

Promeyrat A, Sayd T, Laville E, Chambon C, Lebret B, Gatellier P (2011) Early post-mortem sarcoplasmic proteome of porcine muscle related to protein oxidation. Food Chem 127:1097–1104

Rabilloud T (2002) Two-dimensional gel electrophoresis in proteomics: old, old fashioned, but it still climbs up the mountains. Proteomics 2:3–10

Rosenvold K, Andersen HJ (2003) Factors of significance, for pork quality – a review. Meat Sci 64:219–237

Yates JR (2004) Mass spectral analysis in proteomics. Annu Rev Biophys Biomol Struct 33:297–316

Chapter 7
Application of Proteomics for Analysis of Protein Modifications in Postmortem Meat

Honggang Huang and René Lametsch

7.1 Introduction

During the post-mortem (PM) conversion of muscle into meat, a significant series of events occurs in response to the stoppage of the respiratory system and blood circulation; the metabolism in the muscle is known to be changed substantially (Scheffler and Gerrard 2007). The rate and extent of post-mortem metabolic processes could greatly influence many important meat quality properties such as tenderness, water-holding capacity, and color. Underlying these processes, muscle proteins are the fundamental targets and factors contributing to the above-mentioned changes and properties. In PM muscle, proteins are subjected to a series of protein modifications, such as reversible phosphorylation, oxidation, degradation, and denaturation, and all these modifications are critical for the formation of different meat quality traits. Protein modifications could fundamentally affect the biological and chemical properties of proteins, and thereby they can be involved in all aspects of meat quality development.

In recent years, protein modifications have been studied by several research groups that have looked at the potential effects on muscle foods. They revealed that muscle protein modifications could significantly affect meat quality development. Oxidation of proteins in processed meat products leads to poor meat quality and lesser nutritional value (Lund et al. 2007; Morzel et al. 2006; Rowe et al. 2004). Protein phosphorylation in PM meat was supposed to regulate the activity of enzymes and rigor mortis development, consequently affecting pH decline rate and meat quality development (Huang et al. 2011; Lametsch et al. 2011; Muroya et al.

H. Huang • R. Lametsch (✉)
Department of Food Science, University of Copenhagen,
Frederiksberg DK-1958, Denmark
e-mail: rla@life.ku.dk

F. Toldrá and L.M.L. Nollet (eds.), *Proteomics in Foods: Principles and Applications*, Food Microbiology and Food Safety 2, DOI 10.1007/978-1-4614-5626-1_7,

2007). It is generally accepted that degradation and denaturation of proteins in PM meat are responsible for the tenderness development of meat (Koohmaraie 1996). Oxidation, phosphorylation, and degradation are the three protein modifications mostly studied in muscle foods, and proteomics are the key tool for analysis of these muscle protein modifications.

Proteomics are the systematically unbiased study of entire protein complements expressed by the genome of an organism (proteome) for the high-throughput discovery of protein alterations. Mass spectrometry (MS) -based proteomics are dedicated to the global analysis of protein composition, modifications (PTMs), and the dynamic range of expression levels (Aebersold and Mann 2003; Yates et al. 2009). In the meat science area, proteomics were confirmed to be a powerful tool for studies with regard to different qualities or processing parameters, as described in several reviews (Bendixen 2005; Bendixen et al. 2011; Hollung et al. 2007). Proteomics can provide valuable information for a better understanding of the mechanisms influencing the different quality traits. This information can be used to optimize meat production and improve meat quality. Advanced MS-based proteomics are also a central tool for detection, site mapping, and quantification of modifications on proteins (Witze et al. 2007). Both gel-based and gel-free proteomic approaches were developed for the analysis of different protein modifications in biological samples, and these approaches were also employed for studying the protein modifications in muscle food, mainly focusing on protein oxidation, phosphorylation, and degradation.

This chapter outlines current achievements in the study of protein modifications in muscle food using proteomic approaches. First we describe the general knowledge of protein modifications and then the development of gel-based and gel-free proteomic approaches for the characterization of such modifications. Finally, we elucidate the effects of protein modifications on muscle foods and devote our main attention to the application of proteomic approaches for the analysis of these modifications.

7.2 Protein Modifications

Most proteins are subjected to some form of protein modifications that can occur on protein both in vivo as post-translational modifications and in response to different environmental factors. Protein modifications increase the diversity of proteins because many types of modifications are usually covalently present at different amino acid residues of the protein. The heterogeneous modifications at distinct amino acid residues lead to further complexity at the protein level (Jensen 2006). Protein modifications can be divided into two categories according to their modification forms: covalent modification of a nucleophilic amino acid side chain by an electrophilic fragment of a cosubstrate, and cleavage of a protein backbone at a specific peptide bond (Walsh et al. 2005).

The most widely studied protein modifications in biological science include protein phosphorylation, acylation, glycosylation, ubiquitination, acetylation, oxidation, and so on. One protein is often subjected to several modifications at a time. However, PTMs are usually present at substoichiometric levels, because a protein modification at a given site is often present in a small fraction of the protein. PTMs can affect the activity, structure, location, and lifetime of proteins, thereby playing essential roles in most cellular processes such as the maintenance of protein structure and integrity, regulation of metabolism and defense processes, cell signaling pathways, and protein spatial–temporal distribution (Jensen 2006).

Protein modifications can affect many food properties such as shelf-life, nutritional value, digestibility, health benefits, and consumer appeal (Kerwin and Remmele 2007). For instance, thermal treatment is widely used in processing and manufacturing steps for many foods and ingredients, including dairy, meat, and cereal products. However, thermal treatment can lead to protein oxidation, the Maillard reaction, protein aggregates, and cross-linking modifications that have been implicated in quality deterioration, nutritional damage, and adverse health effects (Promeyrat et al. 2010; Silvestre et al. 2006).

7.3 Proteomic Approaches for Detection of Protein Modifications

Compared to the total proteins, the proteins with site-specific modifications are only present at substoichiometric levels, and it is therefore a great analytical challenge in proteomics to analyze the specific modifications. In the past two decades, many proteomic strategies were developed for the analysis of protein modification in complex samples ranging from the traditional gel-based Western-blot to the high-throughput advanced MS-based proteomic strategies. Typically, the identification and characterization of protein modification are achieved by combining protein-extraction methods, affinity enrichment, and chromatographic and/or electrophoretic separation with peptide-mass determination and amino acid sequencing by high-performance MS (Jensen 2006; Witze et al. 2007).

7.3.1 Gel-Based Proteomic Approaches

Western-blot analysis is a widespread and traditional method for analyzing protein modifications. Western-blot can be specific and relatively quantitative, and specific antibodies can be utilized to study the modified forms of proteins, such as protein cleavage and post-translational modifications. However, Western-blot analysis relies heavily on prior knowledge of the type and position of specific modifications in proteins and is limited by the availability and specificity of antibodies.

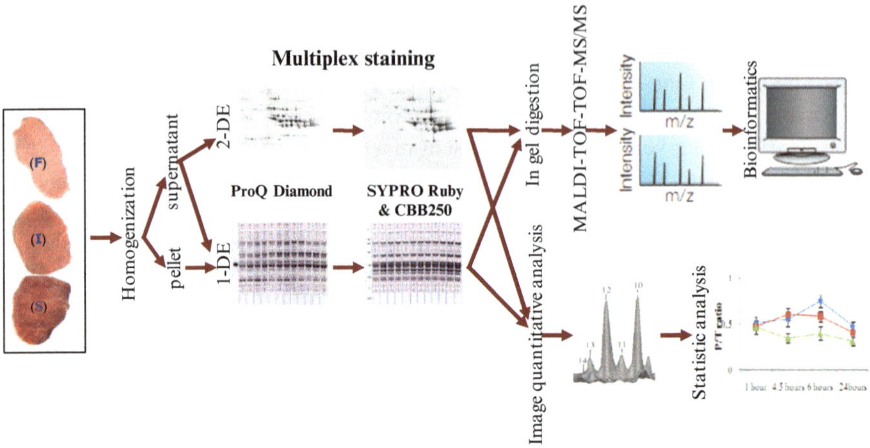

Fig. 7.1 The scheme for gel-based phosphoproteomic analysis of postmortem porcine muscle

One- and two-dimensional gel electrophoresis (1DE and 2DE) -based proteomic strategies are widely used for the detection of variation in protein modification, because such modifications can result in the shift of the isoelectric point or molecular weight of protein. Protein phosphorylation could affect the isoelectric point, as neutral hydroxyl groups on serines, threonines, or tyrosines are replaced with negatively charged phosphates, and cause the protein to shift towards the acid end of the 2-D gel creating the "beads-on-a-string" look for multiple phosphates. If the proteins are modified by degradation and/or glycolyzation, the molecular weight of the resulting fragments will change and migrate differently in the second dimension. Therefore, the proteins with modifications would show a different mobility pattern on a 2-D gel, and different isoforms of the same protein can be visible as different spots on the 2DE gel.

Different types of nonradioactive fluorescence staining dyes have been developed for the in-gel detection of proteins with specific modifications. The fluorescent dyes have a good dynamic range of quantification and good compatibility with subsequent protein characterization and identification (Patton 2002). Pro-Q Diamond phosphoprotein stain (Molecular Probes) was developed for the detection of phosphoserine-, phosphothreonine-, and phosphotyrosine-containing proteins on 1D and 2-D gels, electroblots, and protein microarrays (Steinberg et al. 2003). Recently, it was shown that a combination of one- and two-dimensional gel electrophoresis stained with Pro-Q Diamond could be used to detect changes in protein phosphorylation in PM porcine meat samples (Huang et al. 2011, 2012). This strategy can be applied for semi-quantitative analysis of protein phosphorylation changes in multiple samples (Fig. 7.1). Protein oxidation can be analyzed by detection of carbonyl derivatives which involves a pre-derivatization of the carbonyl groups with dinitrophenylhydrazine (DNPH) prior to electrophoresis, followed by immunoblotting with an anti-DNP antibody (Nakamura and Goto 1996). Many protein oxidation assay kits are commercially available, such as the OxyBlot Protein Oxidation Detection Kit. After modification-specific staining and imaging, the same gel will be subjected to total-protein stains, such as SYPRO Ruby Fluorescent dye; the images from both modification staining and total protein staining can be obtained with gel scanners with multiple laser excitation sources and analyzed with proteomic image software. Changes at the protein modification level can be semi-quantified by comparing the intensity of the protein in the modification staining image and its intensity in the total protein image.

After gel separation, the proteins will usually be cut off from the gel and subjected to ingel digestion and MS identification. From the MS data it is possible to identify the protein modification sites and type (Jensen 2004). It is also possible to predict the cleavage site using tandem MS if the protein is degraded (Lametsch et al. 2002; Larsen et al. 2001).

7.3.2 Gel-Free Proteomic Approaches

Compared to gel-based approaches, gel-free liquid chromatography coupled with tandem mass spectrometry (LC-MS/MS) approaches seems to be much more powerful for high-throughput modification-specific proteomics, inasmuch as many specific techniques for modified proteins or peptides can be incorporated prior to LC-MS/MS analysis, and it is also possible for quantitative study by introducing different isotopic labelings into the protein or peptides. For noncomplicated samples, it is possible to detect the protein modification with shotgun sequencing. The protein samples are treated with multiple proteases to generate complementary and redundant sets of overlapping peptides, and then subjected to shotgun sequencing by LC-MS/MS; finally extensive data analysis is needed. This strategy is suitable for subproteomes (Wu et al. 2003), simple protein mixtures, and protein complexes. However, for most of the complicated biological samples, it is difficult to detect and characterize protein modification using common global proteomics strategies. A key to protein modification-specific proteomics approaches is to enrich and purify protein or peptide species with specific modification prior to characterization by MS/MS.

Different enrichment methods for specific modification in combination with LC/MS/MS approaches are the most popular strategy for high-throughput protein modification-specific proteomics. The enrichment of modified proteins or peptides is often achieved by using affinity chromatographic techniques. Immunoprecipitation with modification-specific antibodies is highly helpful for enriching the corresponding modified proteins. Commercially available phosphotyrosine (pTyr)-specific antibodies are widely used for cell-signaling study (Rush et al. 2005). Phosphopeptides can also be enriched by immobilized metal affinity chromatography (IMAC) (Ficarro et al. 2002; Posewitz and Tempst 1999), or by titanium dioxide (TiO_2) columns (Larsen et al. 2005; Thingholm et al. 2006) prior to MS/MS analysis. Strong cation exchange and anion exchange chromatography can be used to reduce peptide complexity (Beausoleil et al. 2004; Nuhse et al. 2003). Glycoproteins can be enriched by using sugar-specific antibody lectins (Gabius et al. 2002; Yang and Hancock 2004) and TiO_2 columns (Larsen et al. 2007). Glycosidase D/H treatment and MS/MS facilitated protein identification and assignment of glycosylation sites (Hagglund et al. 2004). Acetylated peptides can be enriched by using resin-coupled antibodies to acetyllysine (Kim et al. 2006). Chemistry-based methods for modification-specific covalent capture of proteins and/or peptides are another choice for characterization of protein modification. These methods are often based on β-elimination/Michael addition chemistry, phosphoramidate chemistry, or other PTM-specific chemistries. The targeted PTMs can be converted to stable and manageable status and then identified by MS/MS sequencing (Oda et al. 2001; Wells et al. 2002).

Most modifications can result in a mass increment or a mass deficit relative to the nascent unmodified protein, as summarized by Jensen (Jensen 2006). For example, the phosphorylation of a Tyr residue can increase its mass from 163 Da to 243 Da by the addition of an HPO_3 group (80 Da). After enrichment and pre-fractionation, the modification site can be assigned through the observation of a discrete mass change of the peptide in MS or of the residue in MS/MS (Larsen et al. 2006). The data acquisition software can then be programmed to monitor and sequence all of the pre-defined candidate modified peptides. Huge amounts of data are normally generated from LC-MS/MS; how to extract the useful information and interpret the data is also a great challenge. The integration of computational tools into modification-specific proteomics studies is a prerequisite for the interpretation of large-scale datasets into meaningful biological information; the explosive accumulation of large-scale data and development of a PTM database will strongly accelerate this step (Jensen 2006; Witze et al. 2007).

7.4 Impacts of Protein Modifications on Muscle Food Quality

During the conversion of muscle into meat, the occurrence of distinct protein modifications is also involved in the whole process of meat quality development. Protein modifications in meat are influenced by conditions prior to and after animal slaughter. Prior to slaughter, physiological reactions to stress may influence the rate and extent of PM pH decline and alter the protein modification pattern, thereby affecting meat quality traits (Ferguson and Warner 2008). Shortly after slaughter, drastic metabolic changes can result in the activation of different protein modification mechanisms (mainly PTMs) to regulate the activity of enzymes and structural proteins in response to the shortage of ATP and the development of rigor mortis. For example, the phosphorylation levels of many muscle proteins were known to change significantly during PM 1–24 h, and the reversible phosphorylation can indirectly affect the pH decline rate and the development of rigor mortis (Huang et al. 2011, 2012; Lametsch et al. 2011). Protein oxidation occurs during meat processing and has numerous negative effects on meat quality and it is therefore important to control this oxidation (Lund et al. 2011). The Maillard reaction occurring during the cooking of meat is important for the development of meat flavor and color.

7.5 Proteomic Analysis of Protein Modifications in Muscle Food

In muscle, intensive studies were performed using proteomic approaches for studying protein modifications of muscle protein, such as oxidation (Feng et al. 2008), phosphorylation (Hojlund et al. 2009), nitration (Kanski et al. 2005), and glycosylation (Martin-Rendon and Blake 2003). These modifications were found to play essential roles in metabolism regulation and muscle contraction. Whereas until recently protein modifications were largely unexplored in muscle food systems, more attention was drawn to characterizing protein modifications in meat during intensive exploration of the mechanisms of regulating meat quality development and storage. Specifically, researchers focus much more on oxidation and phosphorylation. Meanwhile, protein degradation has also been intensively studied in the meat science area, as it plays a key role in regulating the development of meat tenderness.

7.5.1 Protein Phosphorylation

Reversible protein phosphorylation is one of the most widespread regulatory mechanisms in nature. Phosphorylation and dephosphorylation of proteins regulate critical biological processes including metabolism, signaling transduction, proliferation, and differentiation (Graves and Krebs 1999; Hunter 2000). In recent years, with the development of phosphoproteomic methods, protein phosphorylation has been comprehensively studied in various muscle samples (Gannon et al. 2008; Hojlund et al. 2009; Hou et al. 2010). Many sarcoplasmic and myofibrillar proteins were identified to be phosphorylated, and the phosphorylation could affect the metabolism and contraction of muscle. In PM muscle, enzymes catalyzing the glycolysis reactions affect the rate and extent of pH decline (Scheffler and Gerrard 2007). Most of the glycolytic enzymes are phosphoproteins; several studies that focused on individual or few glycolytic enzymes indicated that protein phosphorylation plays critical roles in several key steps of glycolysis in PM muscle. Phosphorylase kinase can phosphorylate glycogen phosphorylase b on serine 14, change the structure, and transform it into the active form (Johnson 1992; Sprang et al. 1988). Phosphorylation of pyruvate kinase could result in an additional, more acid stable enzyme isoform and maintain high activity in PSE meat (Schwagele et al. 1996). The phosphorylation status of AMP-activated protein kinase (AMPK) could indirectly influence the glycolysis and pH decline in PM muscle (Shen and Du 2005). It has also been reported that myosin regulatory light chain 2 (MyLC2) became doubly phosphorylated during rigor formation in bovine longissimus (Muroya et al. 2007). Studying the dynamic changes of protein phosphorylation in PM meat can lead to the identification of candidate regulatory proteins from a new perspective.

A combination of one- and two-dimensional gel electrophoresis (1DE and 2DE) coupled with Pro-Q Diamond-SYPRO Ruby staining and tandem MS strategy was employed to semi-quantitatively analyze the protein phosphorylation changes of sarcoplasmic proteins (Fig. 7.1) (Huang et al. 2011, 2012). The staining method showed high specificity for both phosphoprotein and total protein in 1DE and 2DE (Fig. 7.2). The results revealed that the fast pH decline group had the highest global phosphorylation level at PM 1 h, but lowest at 24 h, whereas the slow pH decline group showed the reverse case. The phosphorylation levels of many proteins were significantly affected by the synergy effects of pH and time ($p < 0.05$). Most of the phosphoproteins were identified as glycometabolism-related enzymes. The phosphorylation of pyruvate kinase and triosephosphate isomerase-1 were shown to be related to the PM muscle pH decline rate. It was suggested that the phosphorylation level might indicate the activity of corresponding proteins in PM muscle. A similar study was also performed for the myofibrillar proteins, MyLC2, troponin T, and tropomyosin, and several other structural proteins were identified as highly phosphorylated and changed with PM time. Interestingly, unlike the sarcoplasmic proteins, the phosphorylation pattern of myofibrillar proteins in PM muscle is mainly changed with PM time, but only to a minor extent influenced by the rate of pH decline, suggesting the phosphorylation of myofibrillar proteins may be related to the meat rigor mortis and quality development (Huang et al. 2012).

PM changes in porcine muscle protein phosphorylation in relation to the RN^- and normal genotypes have also been investigated (Lametsch et al. 2011). Glycogen phosphorylase, phosphofructokinase, and pyruvate kinase were found in protein bands affected by the RN^- genotype, the protein phosphorylation level of the muscle proteins could be interpreted as a global metabolic fingerprint containing information about the activity status of the enzymes in the PM metabolism. Another related study used this approach to analyze the response of sarcoplasmic proteins in PM bovine longissimus muscle to electrical stimulation and its effects on meat tenderization; the proteomic analysis showed that ES resulted in lower ($p < 0.05$) phosphorylation levels of creatine kinase M chain, fructose bisphosphate aldolase C-A, β-enolase, and pyruvate kinase at PM 3 h (Li et al. 2011).

7.5.2 *Protein Oxidation*

Protein oxidation in muscle foods results in the loss of meat product quality, such as reduced water-holding capacity (WHC), tenderness, and juiciness (Lund et al. 2007; Rowe et al. 2004). Oxidation of proteins may affect the susceptibility of protein substrates to proteolytic enzymes and result in low digestibility and poor nutritional value (Morzel et al. 2006). Oxidation of numerous amino acids leads to the

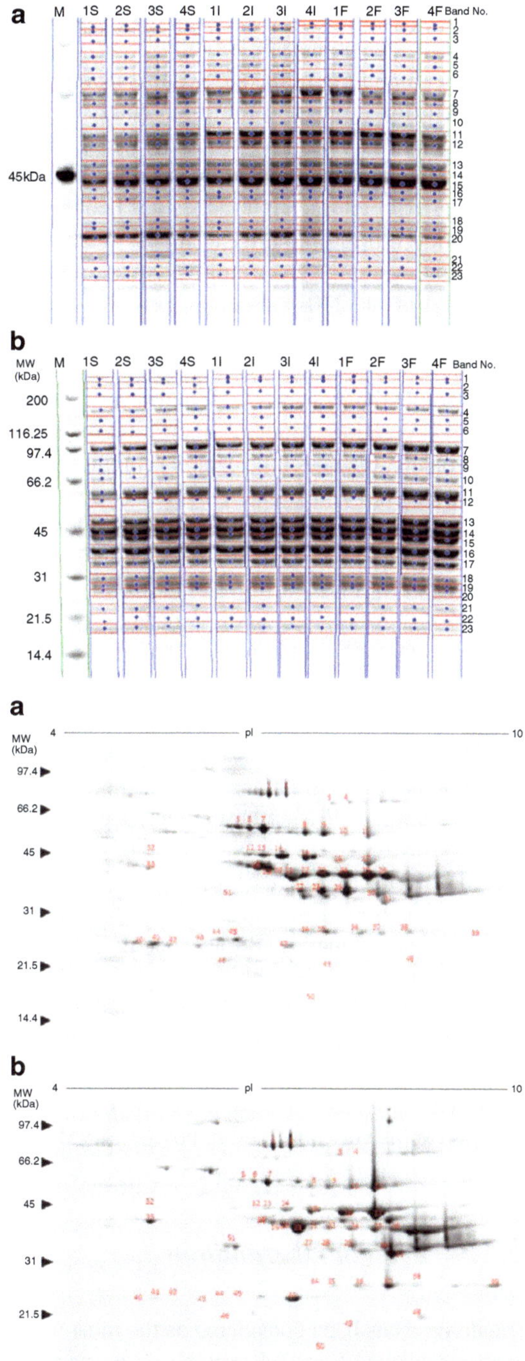

Fig. 7.2 1DE (*up*) and 2 DE (*down*) images of sarcoplasmic proteins. (**a**) ProQ Diamond staining for phosphoprotein. (**b**) SYPRO Ruby staining for total protein (Adapted from Huang et al. 2011)

formation of carbonyl groups and other derivatives, and causes a depletion of essential amino acids in muscle foods.

Direct oxidation of the side chains from lysine, threonine, arginine, and proline has been highlighted as the main route for protein carbonylation and the main mechanism that has been proved to yield carbonyls from meat proteins (Estevez 2011). The 2,4- dinitrophenylhydrazine (DNPH)-method (Oliver et al. 1987) combined with proteomic approaches has been widely used for evaluating protein carbonylation in multiple muscle foods. A combined immunologic and 1D and 2-D proteomic approach was employed to address protein oxidation in chicken muscles (Stagsted et al. 2004). Specific proteins containing carbonyls and/or 3-nitrotyrosine (3-NT) were detected by the DNPH method and antibody against 3-NT. It was found that enolase was the predominant carbonyl-reactive species among the water-soluble muscle proteins, and several other proteins (actin, heat shock protein 70, and creatine kinase) also contained carbonyls and/or 3-nitrotyrosine. Additionally, the effect of feed on protein oxidation was checked as well. A similar study used DNPH and anti-DNP antibody to detect oxidized myofibrillar proteins from PM porcine muscle on 2-D gel and identified about 70 oxidatively modified proteins (Bernevic et al. 2011). Another study employed 2DE combined with DNPH and fluorescent thiosemicarbazide (FTSC) staining and MS/MS to detect the protein carbonyls from bonito muscle during storage; the oxidation status of enolase, aldolase, and L-lactate dehydrogenase A chain (LDH-A) were identified as changing during storage (Kinoshita et al. 2007). Significant correlations ($p < 0.05$) were observed between the level of carbonyl groups and the intensities of 52 proteins on the 2-D electrophoresis of PM porcine sarcoplasmic proteins (Promeyrat et al. 2011).

In addition to the gel-based approaches, some recent studies started to perform LC-MS/MS-based proteomics to analyze oxidation in meat. Researchers used LC-MS/MS to characterize the interesting protein spots cut from 2-D gel, which led to the identification of several unique oxidation sites on creatine kinase, actin, and triosephosphate isomerise (Bernevic et al. 2011). Myoglobin (Mb) redox status affects meat color and is destabilized by lipid oxidation products. Researchers utilized LC-ESI-MS/MS and other proteomic methods to investigate oxymyoglobin (OxyMb) oxidation in Mb from pork and beef; they identified several adducted histidine (HIS) residues (more in beef), and revealed that HNE-induced HIS residues adduction occurred in a species-specific preferential manner, Preferential HNE adduction at HIS 93 was exclusively observed in bovine OxyMb (Suman et al. 2007).

7.5.3 *Protein Degradation*

Tenderness has been considered as the most important quality attribute of meat; it is well established that PM proteolysis is mainly responsible for meat tenderization (Koohmaraie and Geesink 2006). Proteomics has been widely used to investigate PM protein degradation and has provided global insight into the process of proteolysis.

Using the classic 2-D method, many proteins were found to be fragmented as a consequence of protein degradation in PM pork; identified proteins included structural proteins (such as actin, myosin heavy chain, troponin T, etc.) and sarcoplasmic proteins (such as creatine kinase, carbonate dehydratase, triosephosphate isomerase, pyruvate kinase, etc.) (Hwang et al. 2005; Lametsch et al. 2002). Moreover, the actin and the myosin heavy chain fragments have been found to correlate significantly with meat tenderness (Hwang et al. 2005; Lametsch et al. 2003), indicating that the PM degradation of structural proteins contributes to meat tenderization. It is speculated that myosin heavy chain degradation leads to disruption of the myosin–actin interaction, and has an effect on the integrity of the thin flament, which results in meat tenderization. Recently, the combination of 2DE and FTICR-MS was utilized for the identification of muscle protein degradation products in pork (Bernevic et al. 2011), and several truncation forms of creatine kinase and troponin T were identified and confirmed the former results. PM proteolysis in muscle has also been analyzed using proteomics in beef. In *M. longissimus dorsi* (LD) and *M. semitendinosis* (ST) muscle from 0 to 24 h, the intact form of cofilin, lactoylglutathione lyase, and substrate protein of mitochondrial ATP-dependent proteinase SP-22, HSP27, and HSP20 were detected as decreasing (Jia et al. 2006). It was also found that the abundance of fragments of actin, creatine kinase, HSP27, and crystallin increased, whereas the amount of intact molecules decreased over 14 days of cold storage (Morzel et al. 2008). The fragmentation of creatine kinase and other proteins was also observed in a proteomic study of fish (carp) during cold acclimation (McLean et al. 2007).

7.6 Conclusion

Even though it is still a very challenging task to characterize the protein modifications in complex food and biological systems, over the past 10 years development in the areas of fluorescent detection, affinity-enrichment techniques, and quantitative protein and advanced MS greatly accelerate the application of modification-specific proteomics (mainly PTMs) to reveal the molecular features and functions of proteins in life science. The global and systematic characterization and interpretation of protein modifications can be performed through the integration of computational tools into large-scale high-throughput modification-specific proteomics studies, and will offer meaningful information of biological events. Compared to the life science area, protein modifications in food science are rarely explored. The application of proteomic approaches to analyze food protein modification is predicted to become increasingly important in the area of general food science, quality assurance, and product differentiation. The knowledge and approaches of modification-specific proteomics developed in the life sciences area can be referenced and will contribute to the knowledge development of protein modifications in food sciences.

The proteins in PM muscle are subjected to both post-translational modifications and extracellular modifications. Importantly, these protein modifications mainly determine the biochemical and physical properties of meat proteins and thereby affect final meat quality. Until now, protein modification studies in meat science are mainly limited to phosphorylation, oxidation, and degradation, and to some extent, knowledge of these modifications in meat is still at the preliminary level. We can expect that high-throughput modification-specific proteomics will be employed to systematically analyze the qualitative and quantitative differences of these protein modifications in meat concerning the development, genetic background, processing, and storage, which will greatly contribute to our understanding of the mechanisms underlying meat quality difference. In addition, it is also of great interest to explore the potential roles of other important protein modifications in meat systems using modification-specific proteomics.

References

Aebersold R, Mann M (2003) Mass spectrometry-based proteomics. Nature 422(6928):198–207

Beausoleil SA, Jedrychowski M, Schwartz D, Elias JE, Villen J, Li JX, Cohn MA, Cantley LC, Gygi SP (2004) Large-scale characterization of HeLa cell nuclear phosphoproteins. Proc Natl Acad Sci USA 101(33):12130–12135

Bendixen E (2005) The use of proteomics in meat science. Meat Sci 71(1):138–149

Bendixen E, Danielsen M, Hollung K, Gianazza E, Miller I (2011) Farm animal proteomics – a review. J Proteomics 74(3):282–293

Bernevic B, Petre BA, Galetskiy D, Werner C, Wicke M, Schellander K, Przybylski M (2011) Degradation and oxidation postmortem of myofibrillar proteins in porcine skeleton muscle revealed by high resolution mass spectrometric proteome analysis. Int J Mass Spectrom 305(2–3):217–227

Estevez M (2011) Protein carbonyls in meat systems: a review. Meat Sci 89(3):259–279

Feng J, Xie HW, Meany DL, Thompson LV, Arriaga EA, Griffin TJ (2008) Quantitative proteomic profiling of muscle type-dependent and age-dependent protein carbonylation in rat skeletal muscle mitochondria. J Gerontol Ser A Biol 63(11):1137–1152

Ferguson DM, Warner RD (2008) Have we underestimated the impact of pre-slaughter stress on meat quality in ruminants? Meat Sci 80(1):12–19

Ficarro SB, McCleland ML, Stukenberg PT, Burke DJ, Ross MM, Shabanowitz J, Hunt DF, White FM (2002) Phosphoproteome analysis by mass spectrometry and its application to Saccharomyces cerevisiae. Nat Biotechnol 20(3):301–305

Gabius HJ, Andre S, Kaltner H, Siebert HC (2002) The sugar code: functional lectinomics. BBA Gen Subjects 1572(2–3):165–177

Gannon J, Staunton L, O'Connell K, Doran P, Ohlendieck K (2008) Phosphoproteomic analysis of aged skeletal muscle. Int J Mol Med 22(1):33–42

Graves JD, Krebs EG (1999) Protein phosphorylation and signal transduction. Pharmacol Ther 82(2–3):111–121

Hagglund P, Bunkenborg J, Elortza F, Jensen ON, Roepstorff P (2004) A new strategy for identification of N-glycosylated proteins and unambiguous assignment of their glycosylation sites using HILIC enrichment and partial deglycosylation. J Proteome Res 3(3):556–566

Hojlund K, Bowen BP, Hwang H, Flynn CR, Madireddy L, Geetha T, Langlais P, Meyer C, Mandarino LJ, Yi ZP (2009) In vivo phosphoproteome of human skeletal muscle revealed by phosphopeptide enrichment and HPLC-ESI-MS/MS. J Proteome Res 8(11):4954–4965

Hollung K, Veiseth E, Jia XH, Faergestad EM, Hildrum KI (2007) Application of proteomics to understand the molecular mechanisms behind meat quality. Meat Sci 77(1):97–104

Hou JJ, Cui ZY, Xie ZS, Xue P, Wu P, Chen XL, Li J, Cai TX, Yang FQ (2010) Phosphoproteome analysis of rat L6 myotubes using reversed-phase C18 prefractionation and titanium dioxide enrichment. J Proteome Res 9(2):777–788

Huang HG, Larsen MR, Karlsson AH, Pomponio L, Costa LN, Lametsch R (2011) Gel-based phosphoproteomics analysis of sarcoplasmic proteins in postmortem porcine muscle with pH decline rate and time differences. Proteomics 11(20):4063–4076

Huang HG, Larsen MR, Lametsch R (2012) Changes in phosphorylation of myofibrillar proteins during postmortem development of porcine muscle. Food Chem. doi: http://dx.doi.org/10.1016/j.foodchem.2012.03.132

Hunter T (2000) Signaling – 2000 and beyond. Cell 100(1):113–127

Hwang IH, Park BY, Kim JH, Cho SH, Lee JM (2005) Assessment of postmortem proteolysis by gel-based proteome analysis and its relationship to meat quality traits in pig longissimus. Meat Sci 69(1):79–91

Jensen ON (2004) Modification-specific proteomics: characterization of post-translational modifications by mass spectrometry. Curr Opin Chem Biol 8(1):33–41

Jensen ON (2006) Interpreting the protein language using proteomics. Nat Rev Mol Cell Biol 7(6):391–403

Jia XH, Hollung K, Therkildsen M, Hildrum KI, Bendixen E (2006) Proteome analysis of early post-mortem changes in two bovine muscle types: M-longissimus dorsi and M-semitendinosis. Proteomics 6(3):936–944

Johnson LN (1992) Glycogen-phosphorylase – control by phosphorylation and allosteric effectors. FASEB J 6(6):2274–2282

Kanski J, Hong SJ, Schoneich C (2005) Proteomic analysis of protein nitration in aging skeletal muscle and identification of nitrotyrosine-containing sequences in vivo by nanoelectrospray ionization tandem mass spectrometry. J Biol Chem 280(25):24261–24266

Kerwin BA, Remmele RL (2007) Protect from light: photodegradation and protein biologics. J Pharm Sci 96(6):1468–1479

Kim SC, Sprung R, Chen Y, Xu YD, Ball H, Pei JM, Cheng TL, Kho Y, Xiao H, Xiao L, Grishin NV, White M, Yang XJ, Zhao YM (2006) Substrate and functional diversity of lysine acetylation revealed by a proteomics survey. Mol Cell 23(4):607–618

Kinoshita Y, Sato T, Naitou H, Ohashi N, Kumazawa S (2007) Proteomic studies on protein oxidation in bonito (Katsuwonus pelamis) muscle. Food Sci Technol Res 13(2):133–138

Koohmaraie M (1996) Biochemical factors regulating the toughening and tenderization processes of meat. Meat Sci 43:S193–S201

Koohmaraie M, Geesink GH (2006) Contribution of postmortem muscle biochemistry to the delivery of consistent meat quality with particular focus on the calpain system. Meat Sci 74(1):34–43

Lametsch R, Roepstorff P, Bendixen E (2002) Identification of protein degradation during post-mortem storage of pig meat. J Agric Food Chem 50(20):5508–5512

Lametsch R, Karlsson A, Rosenvold K, Andersen HJ, Roepstorff P, Bendixen E (2003) Postmortem proteome changes of porcine muscle related to tenderness. J Agric Food Chem 51(24):6992–6997

Lametsch R, Larsen MR, Essen-Gustavsson B, Jensen-Waern M, Lundstrom K, Lindahl G (2011) Postmortem changes in pork muscle protein phosphorylation in relation to the RN genotype. J Agric Food Chem 59(21):11608–11615

Larsen MR, Larsen PM, Fey SJ, Roepstorff P (2001) Characterization of differently processed forms of enolase 2 from Saccharomyces cerevisiae by two-dimensional gel electrophoresis and mass spectrometry. Electrophoresis 22(3):566–575

Larsen MR, Thingholm TE, Jensen ON, Roepstorff P, Jorgensen TJD (2005) Highly selective enrichment of phosphorylated peptides from peptide mixtures using titanium dioxide micro-columns. Mol Cell Proteomics 4(7):873–886

Larsen MR, Trelle MB, Thingholm TE, Jensen ON (2006) Analysis of posttranslational modifications of proteins by tandem mass spectrometry. Biotechniques 40(6):790–798

Larsen MR, Jensen SS, Jakobsen LA, Heegaard NHH (2007) Exploring the sialiome using titanium dioxide chromatography and mass spectrometry. Mol Cell Proteomics 6(10):1778–1787

Li CB, Li J, Zhou GH, Lametsch R, Ertbjerg P, Bruggemann DA, Huang HG, Karlsson AH, Hviid M, Lundstrom K (2011) Electrical stimulation affects metabolic enzyme phosphorylation, protease activation and meat tenderization in beef. J Anim Sci 90(5):1638–1649

Lund MN, Lametsch R, Hviid MS, Jensen ON, Skibsted LH (2007) High-oxygen packaging atmosphere influences protein oxidation and tenderness of porcine longissimus dorsi during chill storage. Meat Sci 77(3):295–303

Lund MN, Heinonen M, Baron CP, Estevez M (2011) Protein oxidation in muscle foods: a review. Mol Nutr Food Res 55(1):83–95

Martin-Rendon E, Blake DJ (2003) Protein glycosylation in disease: new insights into the congenital muscular dystrophies. Trends Pharmacol Sci 24(4):178–183

McLean L, Young IS, Doherty MK, Robertson DHL, Cossins AR, Gracey AY, Beynon RJ, Whitfield PD (2007) Global cooling: cold acclimation and the expression of soluble proteins in carp skeletal muscle. Proteomics 7(15):2667–2681

Morzel M, Gatellier P, Sayd T, Renerre M, Laville E (2006) Chemical oxidation decreases proteolytic susceptibility of skeletal muscle myofibrillar proteins. Meat Sci 73(3):536–543

Morzel M, Terlouw C, Chambon C, Micol D, Picard B (2008) Muscle proteome and meat eating qualities of Longissimus thoracis of "Blonde d'Aquitaine" young bulls: a central role of HSP27 isoforms. Meat Sci 78(3):297–304

Muroya S, Ohnishi-Kameyama M, Oe M, Nakajima I, Shibata M, Chikuni K (2007) Double phosphorylation of the myosin regulatory light chain during rigor mortis of bovine longissimus muscle. J Agric Food Chem 55(10):3998–4004

Nakamura A, Goto S (1996) Analysis of protein carbonyls with 2,4-dinitrophenyl hydrazine and its antibodies by immunoblot in two-dimensional gel electrophoresis. J Biochem 119(4):768–774

Nuhse TS, Stensballe A, Jensen ON, Peck SC (2003) Large-scale analysis of in vivo phosphorylated membrane proteins by immobilized metal ion affinity chromatography and mass spectrometry. Mol Cell Proteomics 2(11):1234–1243

Oda Y, Nagasu T, Chait BT (2001) Enrichment analysis of phosphorylated proteins as a tool for probing the phosphoproteome. Nat Biotechnol 19(4):379–382

Oliver CN, Ahn BW, Moerman EJ, Goldstein S, Stadtman ER (1987) Age-related-changes in oxidized proteins. J Biol Chem 262(12):5488–5491

Patton WF (2002) Detection technologies in proteome analysis. J Chromatogr B Analyt Technol Biomed Life Sci 771(1–2):3–31

Posewitz MC, Tempst P (1999) Immobilized gallium(III) affinity chromatography of phosphopeptides. Anal Chem 71(14):2883–2892

Promeyrat A, Gatellier P, Lebret B, Kajak-Siemaszko K, Aubry L, Sante-Lhoutellier V (2010) Evaluation of protein aggregation in cooked meat. Food Chem 121(2):412–417

Promeyrat A, Sayd T, Laville E, Chambon C, Lebret B, Gatellier P (2011) Early post-mortem sarcoplasmic proteome of porcine muscle related to protein oxidation. Food Chem 127(3):1097–1104

Rowe LJ, Maddock KR, Lonergan SM, Huff-Lonergan E (2004) Influence of early postmortem protein oxidation on beef quality. J Anim Sci 82(3):785–793

Rush J, Moritz A, Lee KA, Guo A, Goss VL, Spek EJ, Zhang H, Zha XM, Polakiewicz RD, Comb MJ (2005) Immunoaffinity profiling of tyrosine phosphorylation in cancer cells. Nat Biotechnol 23(1):94–101

Scheffler TL, Gerrard DE (2007) Mechanisms controlling pork quality development: the biochemistry controlling postmortem energy metabolism. Meat Sci 77(1):7–16

Schwagele F, Haschke C, Honikel KO, Krauss G (1996) Enzymological investigations on the causes for the PSE-syndrome, 1. Comparative studies on pyruvate kinase from PSE- and normal pig muscles. Meat Sci 44(1–2):27–40

Shen QW, Du M (2005) Role of AMP-activated protein kinase in the glycolysis of postmortem muscle. J Sci Food Agric 85(14):2401–2406

Silvestre D, Ferrer E, Gaya J, Jareno E, Miranda M, Muriach M, Romero FJ (2006) Available lysine content in human milk: stability during manipulation prior to ingestion. Biofactors 26(1):71–79

Sprang SR, Acharya KR, Goldsmith EJ, Stuart DI, Varvill K, Fletterick RJ, Madsen NB, Johnson LN (1988) Structural-changes in glycogen-phosphorylase induced by phosphorylation. Nature 336(6196):215–221

Stagsted J, Bendixen E, Andersen HJ (2004) Identification of specific oxidatively modified proteins in chicken muscles using a combined immunologic and proteomic approach. J Agric Food Chem 52(12):3967–3974

Steinberg TH, Agnew BJ, Gee KR, Leung WY, Goodman T, Schulenberg B, Hendrickson J, Beechem JM, Haugland RP, Patton WF (2003) Global quantitative phosphoprotein analysis using multiplexed proteomics technology. Proteomics 3(7):1128–1144

Suman SP, Faustman C, Stamer SL, Liebler DC (2007) Proteomics of lipid oxidation-induced oxidation of porcine and bovine oxymyoglobins. Proteomics 7(4):628–640

Thingholm TE, Jorgensen TJD, Jensen ON, Larsen MR (2006) Highly selective enrichment of phosphorylated peptides using titanium dioxide. Nat Protoc 1(4):1929–1935

Walsh CT, Garneau-Tsodikova S, Gatto GJ (2005) Protein posttranslational modifications: the chemistry of proteome diversifications. Angew Chem Int Edit 44(45):7342–7372

Wells L, Vosseller K, Cole RN, Cronshaw JM, Matunis MJ, Hart GW (2002) Mapping sites of O-GlcNAc modification using affinity tags for serine and threonine post-translational modifications. Mol Cell Proteomics 1(10):791–804

Witze ES, Old WM, Resing KA, Ahn NG (2007) Mapping protein post-translational modifications with mass spectrometry. Nat Methods 4(10):798–806

Wu CC, MacCoss MJ, Howell KE, Yates JR (2003) A method for the comprehensive proteomic analysis of membrane proteins. Nat Biotechnol 21(5):532–538

Yang ZP, Hancock WS (2004) Approach to the comprehensive analysis of glycoproteins isolated from human serum using a multi-lectin affinity column. J Chromatogr A 1053(1–2):79–88

Yates JR, Ruse CI, Nakorchevsky A (2009) Proteomics by mass spectrometry: approaches, advances, and applications. Annu Rev Biomed Eng 11:49–79

Chapter 8
Biological Markers for Meat Tenderness of the Three Main French Beef Breeds Using 2-DE and MS Approach

Thibault Chaze, Jean-François Hocquette, Bruno Meunier, Gilles Renand, Catherine Jurie, Christophe Chambon, Laurent Journaux, Sylvie Rousset, Christophe Denoyelle, Jacques Lepetit, and Brigitte Picard

8.1 Introduction

Managing beef quality and offering a high and consistent sensorial quality to consumers is a challenge in the meat industry. Consumers are looking for tender meat as a first quality among flavor, juiciness, and color, and this quality is one of most important social and economic challenges for beef producers and retailers (Hocquette et al. 2005). In order, first, to understand how meat quality traits, particularly tenderness, are biologically characterized and, second, to guarantee meat quality as desired by consumers, scientific programs have been conducted for many years. These programs looking for meat quality traits have focused on different breeds (Dransfield et al. 2003) and different muscles (Bernard et al. 2007), and have used different approaches ranging from

T. Chaze • J. Hocquette • B. Meunier • C. Jurie • B. Picard (✉)
INRA, UR1213 Herbivores,
Saint-Genès-Champanelle 63122, France
e-mail: brigitte.picard@clermont.inra.fr

G. Renand
INRA, UMR1313 Génétique Animale et Biologie Intégrative,
Jouy-en-Josas 78352, France

C. Chambon
Plateforme d'exploration du métabolisme, des gènes aux métabolites,
INRA Centre de Clermont Ferrand, Theix, France

L. Journaux
UNCEIA, MNE, 149 rue de Bercy 75595 Cedex 12, Paris, France

S. Rousset
Centre de Recherche en Nutrition Humaine Auvergne, INRA, UMR 1019, INRA, France

C. Denoyelle
Institut de l'Elevage, 149 rue de Bercy, Saint Genès Champanelle, 63122, France

J. Lepetit
UR 370, Unité Qualité Produits Animaux, INRA, Saint Genès Champanelle, 63122, France

F. Toldrá and L.M.L. Nollet (eds.), *Proteomics in Foods: Principles and Applications*, Food Microbiology and Food Safety 2, DOI 10.1007/978-1-4614-5626-1_8,

histologically based experiments, enzymatic activities, and SNP screening, to genomewide techniques (Andersson and Georges 2004; Hocquette et al. 2007b). Concerning meat tenderness and despite these efforts, only a small part of meat tenderness can be explained by biochemical muscle traits (Renand et al. 2001). Thus the need remains to find relevant biological markers of meat quality traits for French breeds to complement the genetic markers already found in meat science worldwide (Hocquette et al. 2007b).

High-throughput "omics" sciences are the golden tools for the discovery of new indicators of beef quality. More particularly, proteomics have now become a mature science which has already led to new insights in animal and meat science (Mullen et al. 2006). One of the main limitations of using omics experiments in meat science is that meat quality is often classified according to its appearance (color, texture, taste) (Bendixen 2005), whereas meat quality is known to be characterized by a larger number of parameters and a larger number of biological interactions and complex environmental factors (Hocquette et al. 2007a).

In this context of continuing investigations into meat quality, the Qualvigène program (Allais et al. 2010), a French genomic program based on a thorough analysis of more than 3,300 young bulls from the three main French beef breeds, aims (1) to analyze the genetic variability of cattle, (2) to look for functional and genetic markers of meat quality traits, and (3) to confirm published and patented markers of meat quality.

Results from proteomic analyses are presented here, based on the Qualvigène system, leading to the discovery of original biological markers for three main French beef breeds. This work is defined by three main originalities. The first one lies in the thorough characterization of animals from a large cohort of more than 3,300 individuals. All animals in this cohort were characterized according to 30 different parameters representative of beef traits, muscle characteristics, and meat qualities. The second originality of our work is that the choice of animals was made thanks to the creation of a synthetic index of meat tenderness combining (1) sensorial analysis, reported by a trained panelist, and (2) the mechanical shear force score (Warner–Bratzler shear force). Animals were ranked according to this synthetic index and grouped in tender or tough meat-producing animal groups for each breed. The third and last originality lies in the choice of animal defining the groups and comparing in tenderness. Thus, for the first time, the choice of animal and muscle sample was made according to tenderness of the meat and with no significant differences for the other parameters measured (i.e., selection on meat tenderness *ceteris paribus*).

This proteomic comparison leads to a list of proteins differentially abundant between tough and tender meat. This work also leads to new insights concerning the contribution of the HSP (heat shock protein) family to meat tenderness by the study of multi-isoforms and post-translational modifications of chaperone protein HSP27.

8.2 Methodology for Sample Choice and Preparation

This work is part of a large research program (Qualvigène program, French national agency number ANR-06-GANI-001) based on the observation and deep analysis of 3,349 young beef bulls, the progeny of 114 sires (Allais et al. 2010). A total of 981 young Blond d'Aquitaine breed bulls, 1,114 young Charolais breed bulls, and 1,254 young Limousin breed bulls were slaughtered in different industrial slaughterhouses (depending on the regions in France where reared) using strictly the same protocol. Young Charolais bulls were slaughtered at constant weight (730 kg) whereas young Blond d'Aquitaine and Limousin bulls were slaughtered at constant age (14 months and 16 months, respectively).

Each animal was genotyped for two mutations (Q204× and nt821) in the *myostatin* gene on chromosome 2. A consistent phenotypic database was built using characteristics from the slaughtered animals. In fact every animal was characterized by 30 different parameters representative of beef traits (carcass composition and conformation, growth potential, internal fat deposition), muscle characteristics (fiber section area, soluble and insoluble collagen content, intramuscular lipid content, pH), and meat qualities (sensory analysis of tenderness by a trained panelist, specifically trained for each breed, and mechanical tenderness using Warner–Bratzler shear force).

In order to select two samples of 10 young bulls in each breed with the most extreme tenderness attribute *ceteris paribus* the following selection process was implemented in each breed: (1) young bulls with excessive pH value were discarded (two of them with pH value of 6.15 and 6.4, i.e., above 5 standard deviations); (2) a logarithm transformation of the shear force measurements was computed in order to obtain a more normal distribution; (3) tenderness scores and transformed shear force values on the one hand, and intramuscular and insoluble collagen contents and muscle fiber section area were analyzed in a linear model and adjusted to the effect of contemporary slaughter groups; (4) a tenderness synthetic index was computed that combined the standardized normal values of the tenderness scores and of the logarithm of shear force measurements with weights +1 and −1, respectively; (5) bulls with a mutated copy in the *myostatin* gene or with one of the muscle characteristics exceeding 0.89 residual standard deviation from the breed average were discarded; (6) bulls were ranked according to their tenderness synthetic index and the 10 bulls ranking first and last were selected.

For each breed, the *Longissimus thoracis* (LT) muscle was excised 24 h after slaughter. A portion of this muscle was directly frozen in liquid nitrogen to perform biochemical measurements and proteomic analysis. The other portion of the LT was aged for 14 days at 4 °C and kept for panelist tasting and sensorial analysis as reported previously (Dransfield et al. 2003).

8.3 Meat Tenderness Using 2DE and MS Approach

Figure 8.1 displays the population of young bulls according to the two components of the synthetic index: Warner–Bratzler shear force (logarithm) in the abscissa and the panelist's tenderness score in the ordinate, both components adjusted to the contemporary group means and standardized. The candidate animals for proteomics analysis (i.e., the 20 selected young bulls presenting extreme scores in the synthetic index) are represented either by a triangle (young bull with tender meat) or square (young bull with tough meat).

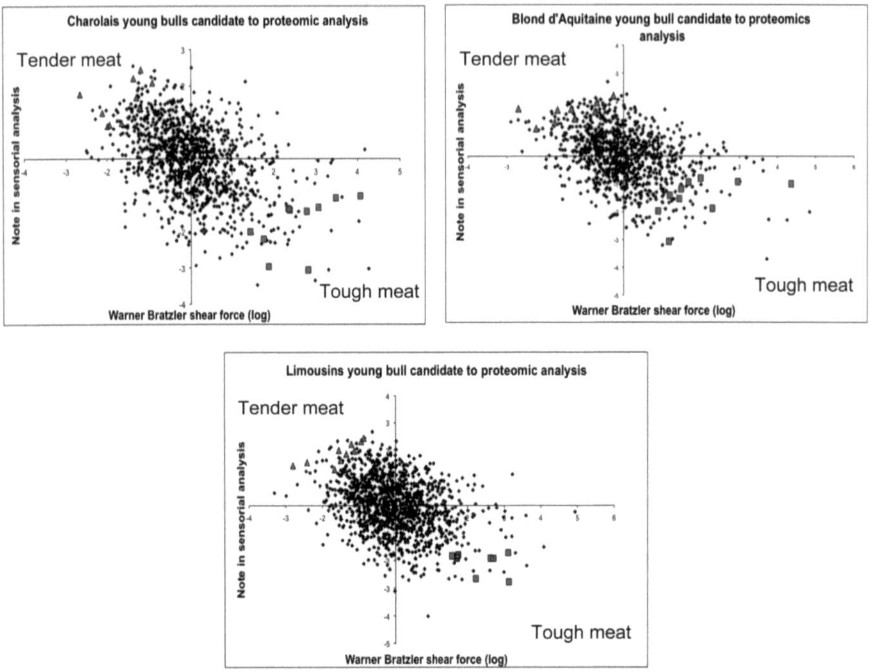

Fig. 8.1 Assessment in the difference of animal groups, compared in the proteomic analysis. Meat tenderness of each young bull was appreciated using a combination of sensorial analysis and mechanical analysis. Animals presenting significant differences in meat tenderness *ceteris paribus* are indicated either with a *triangle* (tender meat) or a *square* (tough meat)

The standard deviations in the whole population and the contrasts between the selected extreme samples are reported in Table 8.1. Large differences were generated between extreme samples for tenderness score and shear force: 3.1 to 3.9 standard deviations, and for the synthetic index: 3.8 to 4.7 standard deviations, and these extreme samples did not differ for intramuscular lipid and collagen content, and muscle fiber cross-section area.

8.3.1 General Proteomics Observations

For the purpose of evaluating the protein profile of muscle using 2-D gels, muscle proteins were extracted, as described in Bouley et al. (2004). For each breed, each tenderness group was represented by 10 different 2DE of the 10 different animals. Proteins were separated using 2DE (first-dimension IEF range pH 4 to 7, second 12% acrylamide SDS-PAGE) and visualized using colloidal Coomassie blue staining. A one-way ANOVA of computed gels was processed under SameSpots to highlight spots differentially expressed between groups presenting meats differing in their tenderness, with $p < 0.05$.

Table 8.1 Variability and selection differential in each breed

	Charolais n = 1,114		Limousin n = 1,254		Blonde d'Aquitaine n = 981	
Trait	RSD*	Diff/RSD	RSD*	Diff/RSD	RSD*	Diff/RSD
Synt. index	1.71	+4.1	1.65	+4.7	1.68	+3.8
Tend. score	7.5	+3.2	7.3	+3.9	10.8	+3.1
Shear force (ln)	0.17	−3.8	0.16	−3.8	0.22	−3.3
Intramuscular lipids	0.81	+0.4	0.48	+0.3	0.36	+0.1
Insoluble collagen	0.037	+0.3	0.037	+0.1	0.049	+0.1
Muscle fibre section area	746	−0.1	622	+0.1	541	−.05

*RSD** residual standard deviation
Diff/RSD: tender – tough groups contrast, divided by the residual standard deviation

Identification of proteins was performed using a MALDI-TOF peptide mass fingerprint (PMF) strategy of trypsin-digested spots. PMF were compared to the Bos taurus Swiss Prot 51.6 database (October 2008, 257964 seq) or to another mammalian using MASCOT 2.1 software. PMF of unidentified proteins (using the bovine database) were compared to the nrMammalia database (06/2008, 819370seq).

Using this 2-D MS strategy, ANOVA revealed 15% of spots with a differential abundance between the two tenderness groups in BA, 8% in the Li and 10% in the Ch breed. Table 8.1 presents a synthetic view of all proteins differentially expressed between the meat tenderness groups and identified by mass spectrometry in this experiment for the three breeds. Only two proteins were found in common for the three breeds (HSP27 and α skeletal actin). HSP27 proteins were detected in many different isoforms (9 for Li, 2 for CH, and 6 for BA) which appear to be differentially expressed among tenderness groups. Limousin and Blond d'Aquitaine shared four proteins in common such as α crystalline B chain and HSP20. Six proteins were differentially expressed in two breeds. Among these six proteins, Troponin T (TnT) fast skeletal muscle was detected in Ch and BA breeds in many isoforms (respectively, 3 and 6), only in tough meat. The differences between tenderness groups revealed by comparative gel-based proteomics were validated by Western blot when antibodies were available as described in Chaze et al. (2008) (Table 8.2, Fig. 8.2).

8.3.2 *Analysis of TNNT3 and HSP27 Isoforms*

In order to understand variations in protein abundance between the two observed conditions, immunodetections by western blot of a targeted zone of 2DE gels were done using available anti-HSP27 and anti-TNNT3 antibodies using infrared detection (Odyssey Infrared Imaging System fluorescent scanner (LI-COR Bioscience)) as described in Guillemin et al. (2009). In each analysis, a general canvas was created and the same number of isoforms was kept between breeds to make the observation easier.

The number of troponin T spots was defined according to the data of Bouley et al. (2005), which allowed the characterization of troponin T fast composition in exon 16 or 17

Table 8.2 Exhaustive list of protein resulting from the identification of spot differentially expressed between groups of tenderness in each breed. This table aims to show the overall result of biomarker discovery for each breed and its relation to the others breeds. The table reports the differences noticed in each breed concerning the discovery of a potential biological marker and its discovery in the other breeds or not. Furthermore, this table indicates the number of different protein isoforms declared differentially expressed between conditions. For example, Li breed shows one isoform of HSPB6 over expressed in tough meat, 2 over expressed in tender one. This protein is not differentially expressed in Ch breed and BA breed display one isoform over expressed in tough meat

		Breed					
		Li		Ch		BA	
Identified protein	WB validation	Over-exp tough	Over-exp tend	Over-exp tough	Over-exp tend	Over-exp tough	Over-exp tend
Proteins differentially expressed in the three breeds							
Heat shock protein beta-1 – Bos taurus (Bovine)	Ok	3	6	1	1	3	3
Actin, alpha skeletal muscle – Bos taurus (Bovine)	Ok	–	1	–	1	–	1
Proteins differentially expressed in two of the three breeds							
Heat shock protein beta-6 – Bos taurus (Bovine)	Ok	1	2	–	–	1	–
Alpha crystallin B chain – Bos taurus (Bovine)	Ok	–	1	–	–	1	–
Beta-enolase – Bos taurus (Bovine)	Ok	1	–	–	–	1	-
F-actin capping protein subunit beta (CapZ beta) – Homo sapiens (Human)	Ok	1	–	–	–	–	1
Protein DJ-1 – Bos taurus (Bovine)	Ok	1	–	–	2	–	-
Troponin T, fast skeletal muscle – Oryctolagus cuniculus (Rabbit) TNNT3	Ok	–	–	3	–	6	–
Proteins differentially expressed only in the Limousin breed							
Alpha-enolase – Mus musculus (Mouse)	Ok	–	1	–	–	–	–
26S protease regulatory subunit 7 – Bos taurus (Bovine)		–	1	–	–	–	–
Rab GDP dissociation inhibitor beta – Bos taurus (Bovine)		–	1	–	–	–	–
capping protein (actin filament) muscle Z-line, alpha 2 [Bos taurus]		–	1	–	–	–	–
Stress-induced-phosphoprotein 1 – Bos taurus (Bovine)	Ok	1	–	–	–	–	–
Geranylgeranyl transferase type-2 subunit alpha – Bos taurus (Bovine)		1	–	–	–	–	–
Glyceraldehyde-3-phosphate dehydrogenase (GAPDH)	Ok	1	–	–	–	–	–
Proteins differentially expressed only in the Charolais breed							
Myosin regulatory light chain 2, skeletal muscle isoform – Bos taurus (Bovine)	Ok	–	–	1	1	–	–

Phosphoglucomutase-1 (EC 5.4.2.2) (Glucose phosphomutase 1) (PGM 1) – Mus musculus (Mouse)	Ok	–	–	1	–	–	–
Proteasome subunit beta type 2 (EC 3.4.25.1) – Bos taurus (Bovine)		–	–	–	1	–	–
Glutathione S-transferase P – Bos taurus (Bovine)		–	–	–	1	–	–
Proteins differentially expressed only in the Blond d'Aquitaine breed							
Troponin T, slow skeletal muscle (TnTs) (Slow skeletal muscle troponin T) (sTnT) – Bos taurus (Bovine)	Ok	–	–	–	–	1	–
WD repeat-containing protein 1 – Bos taurus (Bovine)		–	–	–	–	1	–
Myosin light chain 3, skeletal muscle isoform (A2 catalytic) (Alkali myosin light chain 3) (MLC3F) – Oryctolagus cuniculus (Rabbit)		–	–	–	–	–	1
Creatine kinase M-type – Bos taurus (Bovine)		–	–	–	–	–	2
Myosin regulatory light chain 2, ventricular/cardiac muscle isoform (MLC-2) (MLC-2v) – Rattus norvegicus (Rat)	Ok	–	–	–	–	–	1
NADH-ubiquinone oxidoreductase 75 kDa subunit, mitochondrial precursor (EC 1.6.5.3) (EC 1.6.99.3) (Complex I-75kD) (CI-75kD) – Bos taurus (Bovine)		–	–	–	–	–	1
14-3-3 protein epsilon (14-3-3E) – Bos taurus (Bovine)	Ok	–	–	–	–	–	1

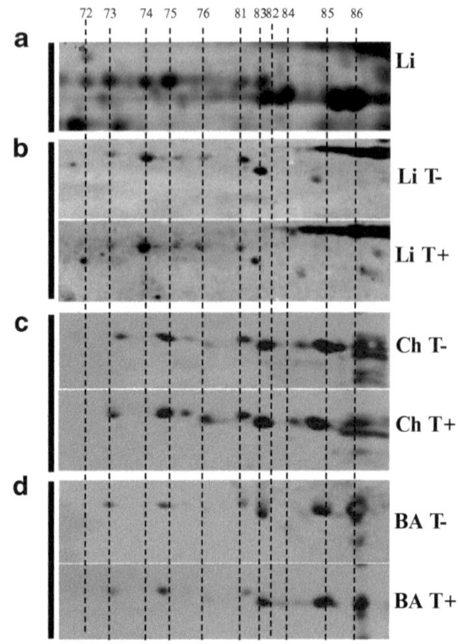

Fig. 8.2 Analysis of the different Troponin T fast spots pattern. The spots were numbered (at the *top* of the figure) according to the article of Bouley et al. 2005. (**a**) Coomassie blue staining in Limousine LT muscle. (**b**) Immunodetection of spot by western-blot using an anti TnTf antibody in Limousin (Li T+ for tender meat, Li T– for tough meat). (**c**) Immunodetection of spot by western-blot using an anti TnTf antibody in Charolais (Ch T+ for tender meat, Ch T– for tough meat). (**d**) Immunodetection of spot by western-blot using an anti TnTf antibody in Blond d'Aquitaine (BA T+ for tender meat, BA T– for tough meat)

(Fig. 8.2). The spots revealed with the antibody were numbered and classified in exon 16 and 17 according to Bouley et al. (2005). The patterns observed in the three breeds were very close (Fig. 8.2). The Western blot pattern of TNNT3 gave access to 6 spots in Blond d'Aquitaine, 9 spots in Charolais, and 8 spots in Limousin. Only proteins 85 and 86 were not found in the Limousin breed Western blot (Fig. 8.2).

The analysis of Troponin T pattern variation was assessed using HCA (Fig. 8.3). This demonstrated a lower abundance of all of these spots in the tender groups of the BA and Ch breeds, not observed in Li. However, the classification in exons 16 and 17 did not allow discrimination between tenderness groups in any breed.

Concerning HSP27 analysis, Western blot detected a distinct, but stackable, pattern between the three breeds and the tenderness groups (Fig. 8.4). After revelation with the HSP27 specific antibody we were able to reveal a total of 12 different isoforms (Fig. 8.4). The maximum number of spots was observed in Li. In this breed we observed that spots 10, 11, 12 were specific to the tender group. We also found these spots in the BA breed tender group. On the contrary, in the Ch breed these spots were not present in the tender group which was characterized by a higher level of spots 1, 2, 4. Moreover, spot 3 was only detected in the Limousin breed, whereas on the contrary, spot 5 was present in all breeds. These data illustrate differences between the two tenderness groups for HSP27 spots. However, these differences were not the same in each breed.

Hierarchical clustering analysis (HCA) was performed to highlight relationships between protein isoform abundances. HCA was conducted according to the Pearson distance and Ward's aggregation method (Meunier et al. 2007) using the

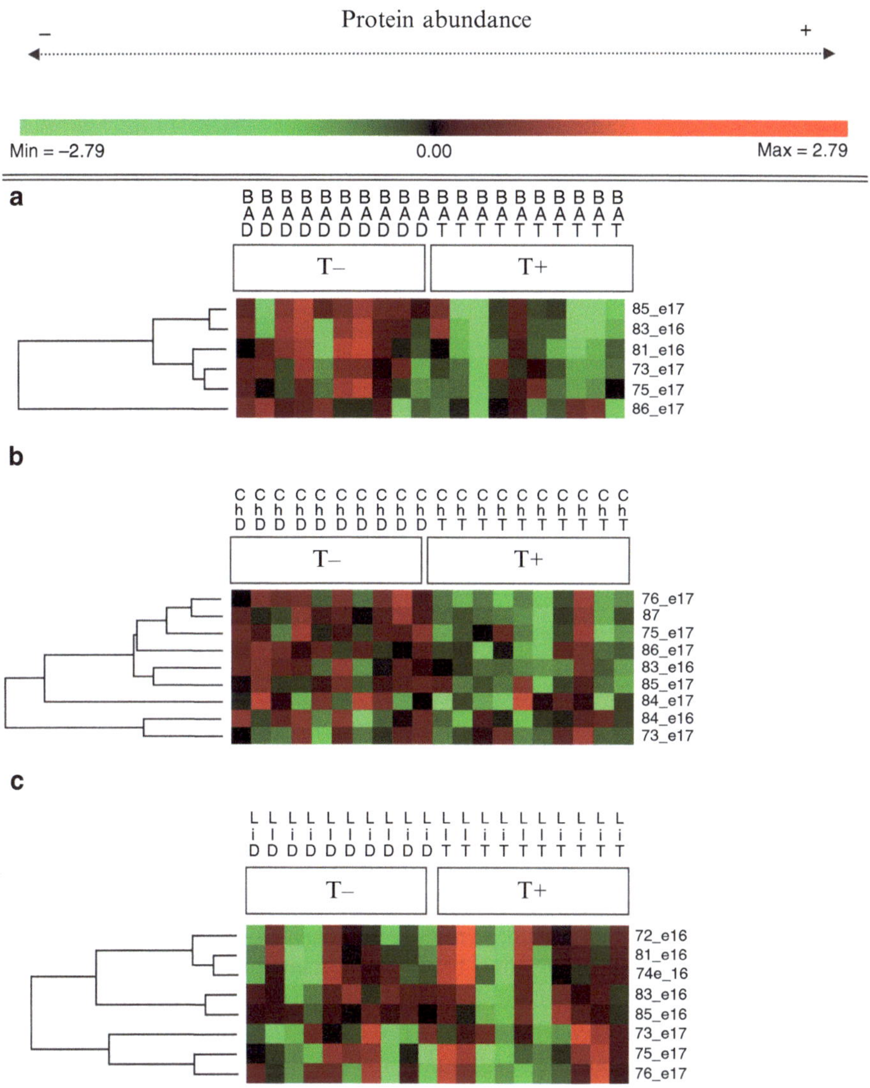

Fig. 8.3 One way hierarchical clustering of immunodetected TnT spots template and the two groups of tenderness (high: T+; low: T–) in (**a**) Blond d'Aquitaine (**b**) Charolais (**c**) Limousin breeds. The spot numbers on the *right* corresponds to those of cartography (Fig. 8.2e). The dendrogram on the *left* indicates the profile similarities. The heat map represents the logged-ratio protein abundance: from *green* (under expressed) to *red* (over expressed)

freeware PermuMatrix version 1.8.1 (Caraux and Pinloche 2005). The use of hierarchical clustering classification analysis of these 12 isoforms showed a good discrimination of the two tenderness groups in the BA breed (Fig. 8.5). In this breed spots 6, 8, 9, 11, 12, 13 were less abundant in the tender group, with a good level of significance for spots 8 ($p<0.001$) and 9 ($p<0.05$). Spots 1 and 7 were globally

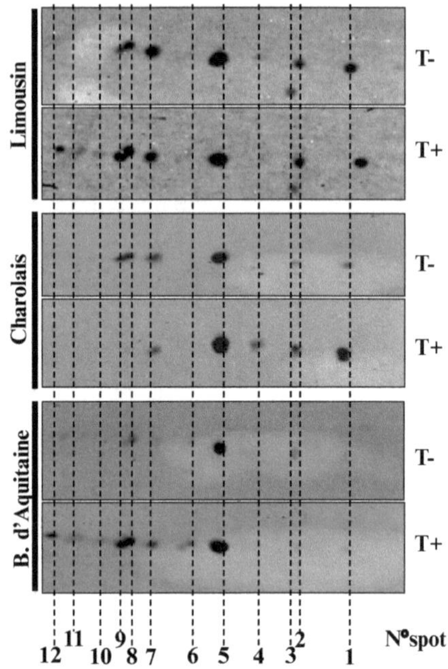

Fig. 8.4 Immunodetection of HSP27 spots by western-blotting in the three breeds for the two groups of meat tenderness (T+ tender meat, T– tough meat)

more abundant in the tender group with a *p*-value in *t*-test less than 0.01. Spots 2 and 5 were less discriminating.

HCA of HSP27 immunodetected spots from the Limousin breed also displayed some interesting data. In this case, spots 1 ($p<0.01$), 2 and 5 ($p<0.05$) were more abundant in tough meat, whereas spots 8 ($p<0.05$) and 9 were more abundant in tender meat. Limousin and Blond d'Aquitaine breeds seem to display opposite data concerning the implication of HSP27 in meat tenderness.

Unfortunately this discrimination between the tenderness groups with HSP27 spots was not observed for the Charolais breed, with no statistical significance in the difference between meat tenderness values (Fig. 8.5).

The detection of phosphorylated isoforms (protocols described by the manufacturer and by the work of Agrawal and Thelen (2005) were strictly followed) using Pro Q Diamond staining (Molecular Probes, Eugene, OR, USA) demonstrated that spots number 2, 5, 8, and 9 were phosphorylated (Fig. 8.6).

8.3.3 Proteomic Network Analysis

In order to have a global visualization of proteomic data concerning each breed in this context of meat tenderness, a network analysis was performed using Pathway Studio 6. The complexity of each network is related to the number of proteins found

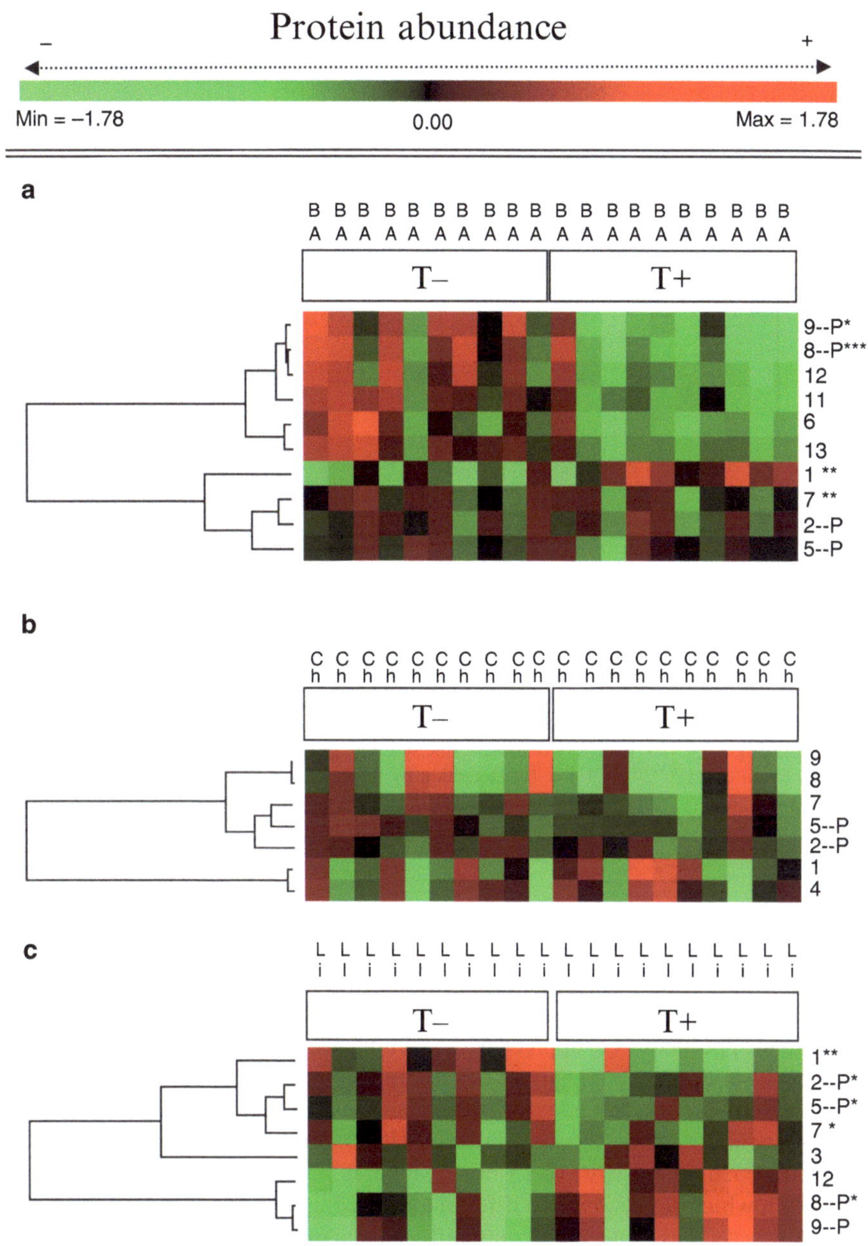

Fig. 8.5 One way hierarchical clustering of immunodetected HSP27 spots template and the two groups of tenderness (high: T+; low: T–) in (**a**) Blond'Aquitaine, (**b**) Charolais, (**c**) Limousin breeds. The spot numbers on the *right* corresponds to those of Fig. 8.4. The dendrogram on the *left* indicates the profile similarities. The heat map represents the logged-ratio protein abundance: from *green* (under expressed) to *red* (over expressed)

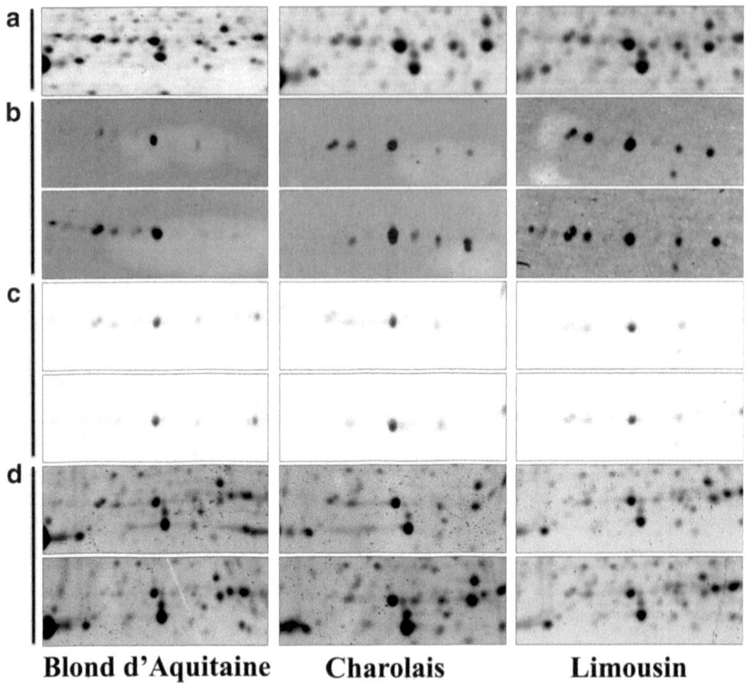

Fig. 8.6 Detection of HSP27 spot by western-blotting in the three breeds for the two groups of tenderness (high: T+, low: T−) (**a**) Coomassie blue staining, general pattern. (**b**) Immuno-detection of spots by an anti HSP27 antibody. (**c**) Detection by fluorescence of phosphorylated proteins in the same gel area. (**d**) Sypro Ruby control (associated with phosphoprotein detection) of protein detected in the gel

differentially expressed in each breed between the meat tenderness groups. In each breed, the network confirms the central role of HSP27 (Fig. 8.7). In the Blond d'Aquitaine and Limousin breeds, apoptosis is the other main node confirming the hypothesis proposed by Morzel et al. (2008). This graphic kind of representation offers new opportunities to appreciate the complexity of meat tenderness.

A gene set enrichment analysis of proteins identified in each breed and independently of the meat tenderness was also performed using Pathway Studio (v.6, Ariadne genomics, Rockville, MD, USA). Table 8.3 displays the GSEA result for each breed. BA and Li breeds display 23 and 22 different GO terms unlike Ch which shows only 10 different terms. Eight terms were only found in the BA breed. They concerned contractile-related proteins such as troponin, tropomyosin, and actin terms. Nine terms were only found in the Li breed and they were more concerned with enzymatic activity (phosphopyruvate hydratase, glycolysis) and cell localization (Golgi apparatus, nucleus). Interestingly, the Ch breed was always characterized by terms shared with another breed, and no term was found to be specific to this breed.

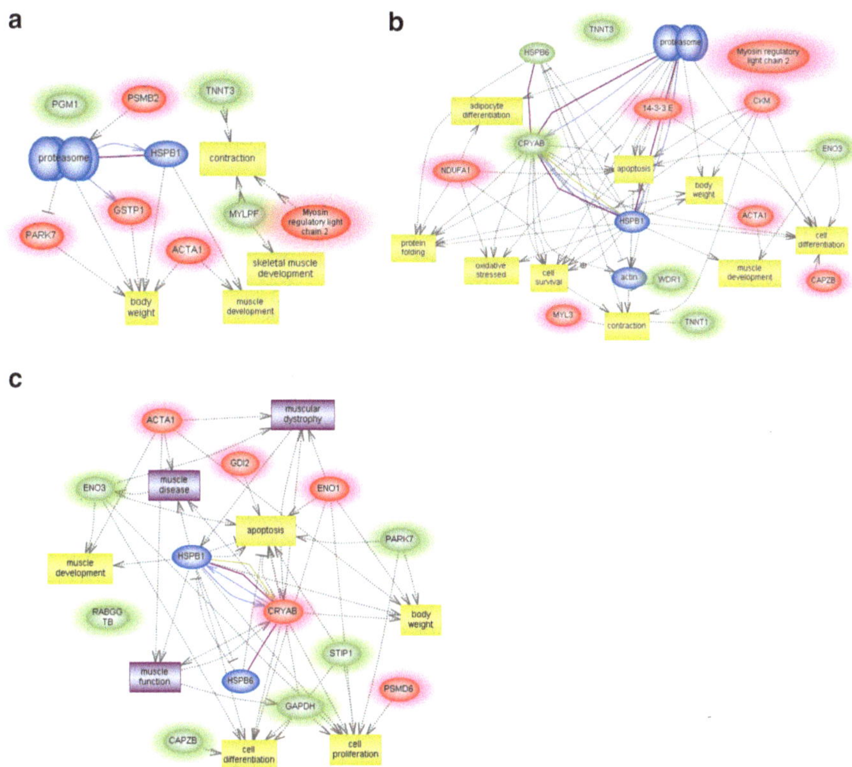

Fig. 8.7 Biological network of proteins declared involved in meat tenderness. *Colour code* refers to the issue of statistical test comparing differences in meat tenderness. *Red circled* proteins are over expressed in tender meat while *green circled* proteins are over expressed in tough meat. *Blue circle* proteins are protein present in multi-isoforms which are over expressed in one or another group of tenderness. *Yellow squares* are cellular process. Networks were made using Pathway Studio 6 (Ariadne genomics, Rockville MD). (**a**) Charolais network. (**b**) Blond d'Aquitaine network. (**c**) Limousin network

8.4 Discussion

This study presents great originality as it is the first time that animals to be compared were chosen on the basis of a synthetic index reflecting a phenotype (here tenderness). In the previously published studies the tenderness groups were constituted on the basis of mechanical analysis (Warner–Bratzler test) (Laville et al. 2009; Zapata et al. 2009) or on the basis of sensory analysis (Bouley et al. 2005; Bernard et al. 2007; Morzel et al. 2008). Here, moreover, the samples were chosen without differences in other characteristics involved in tenderness such as collagen and lipids content, or fiber cross-section areas.

Table 8.3 GSEA (Gene Set Enrichment Analysis) of each breed concerning the GO (Gene Ontology) term statistically found. Data refers to number of protein matching the GO term and the p-value associated, all computed using Pathway Studio 6. Overlap indicates the number of proteins including in the GO term from our analysis. P-value corresponds to the statistical signification of the GSEA. *NA* Not available, meaning that any protein of that breed refers to the GO term associated

	Blonde d'Aquitaine		Charolais		Limousin	
GO term name	Overlap	*p*-value	Overlap	*p*-value	Overlap	*p*-value
Actin binding	3	6,854E-04	0	NA	0	NA
Actin cytoskeleton	2	2,928E-03	0	NA	0	NA
Anti-apoptosis	2	3,974E-03	2	1,462E-03	2	3,974E-03
Cell motility	2	2,561E-03	0	NA	2	2,561E-03
Cell surface	0	NA	0	NA	2	3,376E-03
Contractile fiber	2	2,583E-05	0	NA	2	2,154E-05
Cytoplasm	8	4,406E-04	6	9,924E-04	10	6,297E-07
Cytoskeleton	5	5,934E-06	3	7,456E-04	3	2,079E-03
Cytoskeleton organization and biogenesis	2	3,446E-03	2	1,265E-03	0	NA
Cytosol	0	NA	2	1,834E-02	4	1,500E-04
Glucose metabolic process	0	NA	0	NA	2	4,351E-04
Glycolysis	0	NA	0	NA	3	4,447E-06
Golgi apparatus	0	NA	0	NA	3	5,785E-03
Insoluble fraction	2	2,415E-04	0	NA	2	2,015E-04
Lyase activity	0	NA	0	NA	2	2,240E-03
Motor activity	2	2,399E-03	2	8,775E-04	0	NA
Muscle contraction	4	1,603E-07	3	3,311E-06	2	1,173E-03
Muscle development	2	1,151E-03	0	NA	2	1,151E-03
Nucleus	0	NA	0	NA	6	2,685E-02
Phosphopyruvate hydratase activity	0	NA	0	NA	2	2,360E-06
Phosphopyruvate hydratase complex	0	NA	0	NA	2	2,056E-06
Proteasome complex	0	NA	0	NA	2	1,078E-04
Protein binding	7	3,535E-02	5	3,673E-02	8	8,968E-03
Regulation of muscle contraction	3	5,608E-07	0	NA	0	NA
Regulation of striated muscle contraction	2	1,400E-05	0	NA	0	NA
Response to stress	3	1,190E-04	2	1,580E-03	5	3,046E-08
Skeletal muscle contraction	2	5,948E-06	0	NA	0	NA
Soluble fraction	2	1,765E-02	0	NA	3	7,704E-04
Structural constituent of cytoskeleton	3	4,728E-05	2	8,538E-04	0	NA
Tropomyosin binding	2	1,836E-05	0	NA	0	NA
Troponin complex	2	6,903E-06	0	NA	0	NA
Troponin T binding	2	2,360E-06	0	NA	0	NA
Z disc	3	1,710E-06	0	NA	3	1,284E-06

8.4.1 The Important Role of HSP

In agreement with the role of apoptosis in tenderness (Ouali et al. 2006), several proteins of the heat shock protein family were found to be related to tenderness in numerous studies (Picard et al. 2010). For example, Bernard et al. (2007) revealed a lower expression of DNAJA1 in term of mRNA in Charolais young bulls with tender meat (Patent application number 20090311689). The DNAJA1 gene encodes Hsp40, a chaperone involved in protein import into mitochondria. In our study we did not reveal any differences in HSP 40 abundance, however, we found differences in the abundance of other proteins of the same family involved in similar functions: HSP20, HSP27, and α crystallin B. These three proteins belong to the same family called small HSP (Golenhofen et al. 2004). In our results, one spot of HSP20 was less abundant in the more tender meat as observed for HSP40 at the mRNA level (Bernard et al. 2007). Bouley et al. (2003) also found lower levels of HSP20 and HSP27 in the more tender groups of ST muscle from young Charolais and Salers bulls. The mechanisms underlying the negative relationships between certain HSP levels and tenderness could be related to the antiapoptotic action of HSP (Solary et al. 1998; Creagh et al. 2000; Arrigo 2007). However, these data are a matter of discussion in other experiments such as in Morzel et al. (2008) on LT muscle from BA young bulls. In that study tenderness scores were positively correlated with levels of HSP27 in the early post-mortem muscle. These authors explained their results by the fact that the tenderizing process relies not only on proteolytic but also apoptotic processes and consequently increased HSP levels may reduce protection of the acto–myosin complex exercised in muscle cells. The controversy concerning results of the relation between HSP 27 and tenderness found in the literature could be explained by the different spots identified in our study. We found several spots of this protein (some with higher abundance in tender meat and other with lower abundance in less tender meat). We observed more differences in HSP27 spots between tenderness groups in BA and Li breeds and less in Ch (Table 8.1). The fact that many different HSP27 and troponin T fast spots were found with different abundances for a particular breed constitutes the most obvious advantage of proteomics compared to other functional genomic approaches.

8.4.2 Other Proteins Related to Tenderness

Protein DJ1 showed some differences according to the breed. In the Ch breed it was more abundant in the tender group and the opposite was observed in Li with no differences in the BA breed. This result was confirmed by Western blot. A relation between DJ1 and tenderness was also observed in young Charolais bulls in a previous experiment (Hocquette et al. 2007b). This protein is involved in cellular defense against the effects of oxidative stress by protecting proteins from aggregation (Junn et al. 2005). It also reduces apoptosis, and as for HSP 27 various hypotheses could be proposed.

Glutathione S transferase was also more abundant in the tender meat group of Ch only. This cytosolic protein plays an important role in cell detoxification.

It utilizes glutathione in reactions contributing to the transformation of a wide range of compounds, including carcinogens, therapeutic drugs, and products of oxidative stress. As with protein DJ1, this protein is involved in oxidative stress. Consequently the higher abundance of these two proteins in tender Ch meat is coherent. Other proteins involved in oxidative stress have already been described as good markers of tenderness. For example, Jia et al. (2009) suggested that protein PRDX6 is a potential marker for meat tenderness in *Longissimus thoracis* muscle. These authors also showed a putative role of DJ1 that is coherent with our results.

The protein 14-3-3 epsilon for which a role in meat tenderness had never been described in the literature was found with a higher abundance in the tender group in the BA breed only. This protein belongs to a large family functioning as a dimer in diverse signaling pathways. It is involved in various functions: differentiation, structural organization of intermediate filaments, and cell survival. It was revealed in several neuronal diseases such as Alzheimer's, Parkinson's and Creutzfeldt–Jacob diseases as protein DJ1 and glutathione S transferase. Han et al. (2010) suggested that the monomer of 14-3-3 epsilon is sufficient to protect cells from apoptosis. Consequently, its role in tenderness could be explained by its implication in apoptosis.

Two proteins associated with proteasome activity were revealed: β type 2 subunit of proteasome in Ch and regulatory subunit 7 of proteasome 26S in BA. As proteolytic systems are involved in meat aging, the role of these proteins in tenderness is coherent.

Two new proteins never identified previously were revealed in the Li breed: Geranyl transferase α type 2 subunit was more abundant in the less tender group and Rab GDP dissociation inhibitor factor β was more abundant in the more tender group. These proteins are involved in cellular transport and play an important role in calcium release from sarcoplasmic reticulum. They could have a role in calcic flux post mortem. This is in accordance with the finding that several proteins involved in calcium metabolism have also been identified as positive markers of beef tenderness. For example, the abundance of parvalbumin is greatly increased in tender muscles in Charolais and Limousin cattle (Bouley et al. 2004). This protein has a strong affinity for calcium ions. It participates in calcium cycling between cytoplasm and sarcoplasmic reticulum (Berchtold et al. 2000) especially in fast fibers.

Another protein never revealed in our previous experiments is a stress-induced phosphoprotein, STIP1. This protein regulates the association of chaperones HSP70-1B and HSP90. STIP1 stimulates the ATPase activity of HSP70 and inhibits the ATPase activity of HSP90, suggesting that it regulates both the conformations and ATPase cycles of these chaperones (Song and Masison 2005). Interestingly, HSP70 has been identified as a potential marker of tenderness in other experiments (Jia et al. 2009 and Picard et al. 2010). We can note that in our study STIP1 was over-abundant in the less tender meat only in the LI breed. As the Limousin breed has the reputation of being more aggressive and more sensitive to stress than the other two, this could explain why we found a relationship between this protein and tenderness only in this breed.

8.4.3 Some Similarities Among the Three Breeds

Even if we observed different proteins with differential expression between the two tenderness groups for each breed (Table 8.1), we found the same biological functions for all three breeds. For example, the discussion above demonstrates the implication of proteins involved in apoptosis and oxidative stress in all breeds. Moreover, we can propose a general trend: in each breed, proteins corresponding to slow oxidative characteristics were more abundant in the group with higher tenderness: α enolase in Li, an isoform of troponin T slow, isoforms of slow MyLC (MyLC2), M creatine kinase and the mitochondrial protein NADH ubiquinnone oxydoreductase in BA. On the contrary, proteins corresponding to fast glycolytic characteristics were more abundant in the less tender group: PGM in the Ch breed, GAPDH in Li, fast isoforms of TnT in Ch and BA, and β enolase in Li and BA. This was confirmed by checking out different spots of fast TnT, allowing a good discrimination between tenderness groups in Ch and BA breeds but not in Li. Bouley et al. (2005) demonstrated that the ratio of TnT exon 16/exon 17 was a good marker of muscle mass (with a higher ratio in high-muscled animals). From these data we assumed that this ratio could also be related to meat tenderness. However, our hierarchical analysis results allow the conclusion of the absence of any relation.

Our results indicate that there is a general relationship between contractile and metabolic properties and meat tenderness, with a positive relation between tenderness and slow oxidative properties, and a negative relation with fast glycolytic properties and tenderness. This is particularly well illustrated in Fig. 8.7 which shows the network of proteins involved in meat tenderness and the general biological processes in which they are involved. Most particularly, control of body weight is always controlled by a majority of over-expressed proteins in the three breeds.

Various proteins appear to be more related to tenderness in a breed-dependent manner. This general observation is coherent with previous proteomic results. For example, Morzel et al. (2008) by comparative proteomic analysis on LT muscle of BA young bulls found that the oxidative enzyme succinate dehydrogenase (SDH) was the best predictor of beef tenderness. In our case we did not reveal this enzyme but did find other proteins from the same functionality. Laville et al. (2009) also demonstrated that tender meat from Charolais young bulls showed a higher abundance of mitochondrial membrane fragments in the early post-mortem stages, suggesting increased degradation of the mitochondrial membrane in slow oxidative muscles. This could be explained by a faster or earlier onset of post-mortem apoptosis according to the hypothesis of Ouali et al. (2006). Mitochondria play a central role in the apoptotic process and a large set of pro-apoptotic proteins are translocated from these organelles to the cytoplasm and/or to the nucleus (cytochrome c, Smac/Diablo, apoptosis-inducing factor or AIP, endonuclease G, etc.) according to their biological function (Suen et al. 2008; Taylor et al. 2008).

Interestingly in our experiments the alpha actin protein was more abundant in the more tender group in all three breeds. This constitutes a good candidate as a biological marker of beef tenderness. This difference was confirmed by Western blot with a specific antibody. This protein has been identified as differentially expressed between extreme tenderness groups in Charolais (Hocquette et al. 2007b) and

Blonde d'Aquitaine young bulls (Morzel et al. 2008). In these experiments another structural protein CapZβ was also observed with a higher abundance in the tender group. This result was also observed in this study, but only in the Li breed. These structural proteins are destructured during post-mortem aging by proteolytic systems. Ouali et al. (2006) suggest that α actin constitutes a good marker of post-mortem apoptosis. Further analysis will be conducted on different muscles and breeds to confirm its relationship with tenderness in different contexts.

8.5 Conclusion

This study shows important new results concerning meat quality as it is constructed on the basis of groups with high standard deviation on an index of tenderness combining both mechanical and sensory tenderness *ceteris paribus* for other muscle characteristics involved in meat tenderness (fibers, collagen, and lipids). The comparative proteomic analysis allows confirmation of an important role of proteins from contractile and metabolic properties, structural proteins, and from the HSP family observed in previous studies at the mRNA or protein level. For example, proteins such as alpha actin, PGM, enolase, alpha crystalline B, and HSP20 appeared to be good potential markers of tenderness. This study also revealed new proteins whose role in tenderness has never been described, such as 14-3-3, STIP1 proteins. The results demonstrated differences among the three breeds for the proteins best able to explain meat tenderness. For example, HSP27 isoforms constituted good markers of tenderness only in the Blond d'Aquitaine breed. All the data obtained constitute new insights for beef tenderness. However, further analysis must be scheduled on the principal proteins in order to validate their relation to tenderness in different muscle and animal types. As meat science now has the opportunity to embark on system biology and modeling of meat characteristics, other studies using deep proteomics investigation will have to be employed. In the end, these results will be used to produce a phenotyping tool such as an antibody micro-array grouping all the antibodies specific to the proteins validated as good biological markers of beef tenderness.

Acknowledgments The authors thank the national program ANR AGENAE and APIS-GENE for the funding of the Qualvigène B program (ANR AGENAE/APIS-GENE: contract 23000160). They are grateful to the national consortium of Qualvigène (ANR AGENAE/APIS-GENE: contract 01-2005-QualviGenA-02) for providing muscle samples and data on tenderness (mechanical and sensorial) and on muscle characteristics for the constitution of extreme groups of 10 animals from all 3,349 animals of the Qualvigène project.

References

Agrawal GK, Thelen JJ (2005) Development of a simplified, economical polyacrylamide gel staining protocol for phosphoproteins. Proteomics 5:4684–4688

Allais S, Leveziel H, Payet-Duprat N, Hocquette JF, Lepetit J, Rousset S, Denoyelle C, Bernard-Capel C, Journaux L, Bonnot A, Renand G (2010) The two mutations, Q204X and nt821, of the

myostatin gene affect carcass and meat quality in young heterozygous bulls of French beef breeds. J Anim Sci 88:446–454
Andersson L, Georges M (2004) Domestic-animal genomics: deciphering the genetics of complex traits. Nat Rev Genet 5:202–212
Arrigo AP (2007) The cellular "networking" of mammalian Hsp27 and its functions in the control of protein folding, redox state and apoptosis. Molecular aspects of the stress response: chaperones, membranes and networks. Adv Exp Med Biol 594:14–26
Bendixen E (2005) The use of proteomics in meat science. Meat Sci 71:138–149
Berchtold MW, Brinkmeier H, Muntener M (2000) Calcium ion in skeletal muscle: its crucial role for muscle function, plasticity, and disease. Physiol Rev 80:1215–1265
Bernard C, Cassar-Malek I, Le Cunff M, Dubroeucq H, Renand G, Hocquette JF (2007) New indicators of beef sensory quality revealed by expression of specific genes. J Agric Food Chem 55:5229–5237
Bouley J, Chambon C, Picard B (2003) Proteome analysis applied to the study of muscle development and sensorial qualities of bovine meat. Sci Aliment 23:75–78
Bouley J, Chambon C, Picard B (2004) Mapping of bovine skeletal muscle proteins using two-dimensional gel electrophoresis and mass spectrometry. Proteomics 4:1811–1824
Bouley J, Meunier B, Chambon C, De Smet S, Hocquette JF, Picard B (2005) Proteomic analysis of bovine skeletal muscle hypertrophy. Proteomics 5:490–500
Caraux G, Pinloche S (2005) PermutMatrix: a graphical environment to arrange gene expression profiles in optimal linear order. Bioinformatics 21:1280–1281
Chaze T, Meunier B, Chambon C, Jurie C, Picard B (2008) In vivo proteome dynamics during early bovine myogenesis. Proteomics 8:4236–4248
Creagh EM, Sheehan D, Cotter TG (2000) Heat shock proteins – modulators of apoptosis in tumour cells. Leukemia 14:1161–1173
Dransfield E, Martin JF, Bauchart D, Abouelkaram S, Lepetit J, Culioli J, Jurie C, Picard B (2003) Meat quality and composition of three muscles from French cull cows and young bulls. Anim Sci 76:387–399
Golenhofen N, Perng MD, Quinlan RA, Drenckhahn D (2004) Comparison of the small heat shock proteins alpha B-crystallin, MKBP, HSP25, HSP20, and cvHSP in heart and skeletal muscle. Histochem Cell Biol 122:415–425
Guillemin N, Meunier B, Jurie C, Cassar-Malek I, Hocquette JF, Leveziel H, Picard B (2009) Validation of a dot-blot quantitative technique for large scale analysis of beef tenderness biomarkers. J Physiol Pharmacol 60:91–97
Han DD, Ye GM, Liu TT, Chen C, Yang XM, Wan B, Pan YW, Yu L (2010) Functional identification of a novel 14-3-3 epsilon splicing variant suggests dimerization is not necessary for 14-3-3 epsilon to inhibit UV-induced apoptosis. Biochem Biophys Res Commun 396:401–406
Hocquette JF, Richardson RI, Prache S, Medale F, Duffy G, Scollan ND (2005) The future trends for research on quality and safety of animal products. Ital J Anim Sci 4:49–72
Hocquette JF, Jurie C, Picard B, Alberti P, Panea B, Christensen M, Failla S, Gigli S, Leveziel H, Olleta JL, Sanudo C, Ertbjerg P, Nute GR, Williams JL (2007a) Metabolic and contractile characteristics of Longissimus thoracis muscle of young bulls from 15 European breeds in relationship with body composition. In: OrtiguesMarty I (ed) Energy and protein metabolism and nutrition. Wageningen Academic, Wageningen, pp 111–112
Hocquette JF, Lehnert S, Barendse W, Cassar-Malek I, Picard B (2007b) Recent advances in cattle functional genomics and their application to beef quality. Animal 1:159–173
Jia X, Veiseth-Kent E, Grove H, Kuziora P, Aass L, Hildrum KI, Hollung K (2009) Peroxiredoxin-6-A potential protein marker for meat tenderness in bovine longissimus thoracis muscle. J Anim Sci 87:2391–2399
Junn E, Taniguchi H, Jeong BS, Zhao X, Ichijo H, Mouradian MM (2005) Interaction of DJ-1 with Daxx inhibits apoptosis signal-regulating kinase 1 activity and cell death. Proc Natl Acad Sci USA 102:9691–9696
Laville E, Sayd T, Morzel M, Blinet S, Chambon C, Lepetit J, Renand G, Hocquette JF (2009) Proteome changes during meat aging in tough and tender beef suggest the importance of apop-

tosis and protein solubility for beef aging and tenderization. J Agric Food Chem 57:10755–10764
Meunier B, Dumas E, Piec I, Bechet D, Hebraud M, Hocquette JF (2007) Assessment of hierarchical clustering methodologies for proteomic data mining. J Proteome Res 6:358–366
Morzel M, Terlouw C, Chambon C, Micol D, Picard B (2008) Muscle proteome and meat eating qualities of Longissimus thoracis of "Blonde d'Aquitaine" young bulls: A central role of HSP27 isoforms. Meat Sci 78:297–304
Mullen AM, Stapleton PC, Corcoran D, Hamill RM, White A (2006) Understanding meat quality through the application of genomic and proteomic approaches. Meat Sci 74:3–16
Ouali A, Herrera-Mendez CH, Coulis G, Becila S, Boudjellal A, Aubry L, Sentandreu MA (2006) Revisiting the conversion of muscle into meat and the underlying mechanisms. Meat Sci 74:44–58
Picard B, Berri C, Lefaucheur L, Molette C, Sayd T, Terlouw C (2010) Skeletal muscle proteomics in livestock production. Brief Funct Genomics 9:259–278
Renand G, Picard B, Touraille C, Berge P, Lepetit J (2001) Relationships between muscle characteristics and meat quality traits of young Charolais bulls. Meat Sci 59:49–60
Solary E, Eymin B, Droin N, Haugg M (1998) Proteases, proteolysis, and apoptosis. Cell Biol Toxicol 14:121–132
Song YT, Masison DC (2005) Independent regulation of Hsp70 and Hsp90 chaperones by Hsp70/Hsp90-organizing protein Sti1 (Hop1). J Biol Chem 280:34178–34185
Suen DF, Norris KL, Youle RJ (2008) Mitochondrial dynamics and apoptosis. Genes Devel 22:1577–1590
Taylor RC, Cullen SP, Martin SJ (2008) Apoptosis: controlled demolition at the cellular level. Nat Rev Mol Cell Biol 9:231–241
Zapata I, Zerby HN, Wick M (2009) Functional proteomic analysis predicts beef tenderness and the tenderness differential. J Agric Food Chem 57:4956–4963

Chapter 9
Dry-Cured Ham

Leticia Mora and Fidel Toldrá

9.1 Introduction

Dry-cured ham has been consumed for centuries in many countries and is a typical high-quality product with characteristic texture and flavor. The processing of dry-cured ham is very long, lasting several months or even a few years, including the stages of salting, post-salting, and ripening/drying. The water activity of the ham is progressively reduced by the use of sodium chloride and drying. Many biochemical reactions take place during the process and are responsible for its final characteristic texture and flavor (Toldrá 1998, 2004).

One of the most important biochemical phenomena is the intense proteolysis that takes place as a result of the action of endogenous muscle peptidases. The main enzymes responsible for the degradation of muscle proteins are endopeptidases and certain groups of exopeptidases that have been widely reviewed (Toldrá 2002). Endopeptidases, mainly cathepsins and calpains, are able to degrade a large portion of the muscle protein structure by cleaving myofibrillar proteins, affecting the texture, and also giving rise to large polypeptides that are further degraded by exopeptidases into small peptides and free amino acids (Toldrá and Flores 1998; Toldrá 2006).

This chapter describes the latest advances in the application of proteomic tools to dry-cured ham in order to better understand the intense proteolysis phenomena that occur during its processing.

L. Mora • F. Toldrá (✉)
Department of Food Science, Instituto de Agroquímica y
Tecnología de Alimentos (CSIC), Avenue Agustín Escardino 7,
Paterna, Valencia, 46980, Spain
e-mail: ftoldra@iata.csic.es

F. Toldrá and L.M.L. Nollet (eds.), *Proteomics in Foods: Principles and Applications*, Food Microbiology and Food Safety 2, DOI 10.1007/978-1-4614-5626-1_9,

9.2 Muscle Proteins

Myofibrillar and sarcoplasmic proteins, together with the stroma proteins such as collagen, constitute the main groups of proteins in skeletal muscle. Sarcoplasmic proteins are water soluble or soluble in diluted salt solutions, whereas myofibrillar proteins are soluble in highly concentrated salt solutions. The structure of the skeletal muscle is mainly constituted of muscle fibers, which are essential for the generation and transmission of the contractile force. In fact, the muscle fiber is the basic machinery necessary for muscle contraction (Yamada et al. 2004). Each muscle fiber contains hundreds of myofibrils, which are long bundled tubes of cytoskeleton, made of long chains of a repeating unit called the sarcomere, the basic functional unit of the muscle fiber. These units are composed of actin and myosin proteins that act as thin and thick filaments, respectively, and are responsible for skeletal muscle's striated appearance (Lawrie and Ledward 2006). Other relevant myofibrillar proteins in skeletal muscle are titin, nebulin, actinins (α and β forms), tropomyosin, and troponins C, I, and T (MacIntosh et al. 1996).

Sarcoplasmic proteins are those proteins found in the sarcoplasm, or the fluid surrounding the myofibrils. Myoglobin is considered the most important protein of sarcoplasm as it is the most abundant water-soluble protein and gives the typical red color to the sarcoplasm (Toldrá 2006). However, the sarcoplasmic fraction also contains lysosomal enzymes including cathepsins, which participate in hydrolytic degradation of waste material, and nucleoproteins, that regulate protein synthesis and deposition (Lawrie and Ledward 2006), as well as oxidative enzymes, including cytochromes, flavin nucleotides, various heme pigments, mitochondrial oxidative enzymes, and glycolytic enzymes, which control both aerobic and anaerobic glycolysis (Toldrá 2006).

9.3 Muscle Enzymes and Their Role in Ham Processing

Cathepsins and calpains have been the object of research in the last decades regarding post-mortem proteolysis. Thus, calpains have been considered the major contributors to meat tenderization (Lawrie and Ledward 2006) even though in recent years it has been also attributed to the action of other muscle proteolytic enzymes such as caspases and the proteasome (Ouali et al. 2006; Kemp et al. 2010). On the other hand, cathepsins B (EC 3.4.22.1), H (EC 3.4.22.16), and L (EC 3.4.22.15) have been reported to be stable during dry-cured ham processing, showing activity even after 15 months of processing (Toldrá et al. 1993a; Toldrá 1998), whereas cathepsin D activity disappears after 6–10 months of processing (Toldrá et al. 1993a; Rico et al. 1991). Calpains (EC 3.4.22.17) can participate in the post-mortem muscle proteolysis only during the first weeks of the curing process, especially due to their relatively poor stability (Toldrá and Flores 1998). All these endopeptidases contribute to the release of numerous polypeptides and peptides during the processing of dry-cured ham.

Exopeptidases degrade the large polypeptides and peptides generated by endopeptidases, giving rise to small peptides and free amino acids that, together with volatile compounds, are responsible for the characteristic flavor of dry-cured ham (Toldrá et al. 1997). Dipeptidyl peptidases (DPP I, II, III, and IV) are exopeptidases able to release different dipeptides from the N-terminal site of polypeptides. The action and stability of these enzymes during the processing of Spanish dry-cured ham are variable but DPP II and DPP IV contribute even though at slow rates during the process (Sentandreu and Toldrá 2001).

On the other hand, a remarkable increase in the concentrations of free amino acids has been reported during dry-cured ham processing. Glutamic and aspartic acids, alanine, leucine, lysine, valine, and threonine appear to be some of the amino acids experiencing larger increases (Toldrá et al. 2000). This accumulation of free amino acids, which is relevant for the development of the characteristic dry-cured flavor, has been attributed to muscle aminopeptidases, responsible for the release of amino acids from the N-terminus of peptides and proteins. Aminopeptidase activity has been detected in meat products even after more than 12 months of processing, suggesting that these enzymes are involved in the later stages of protein degradation. Many factors, such as curing agent or the presence of other peptides, can modulate the activity of these enzymes (Toldrá et al. 1993b; Gianelli et al. 2000).

Carboxypeptidases and peptidyl dipeptidases constitute other groups of exopeptidases responsible for the hydrolysis of amino acids and dipeptides, respectively, from the C-terminal side of the protein fragments (Bodwell and Meyer 1981) although their action and contribution to the process remain quite unknown.

9.4 Proteolysis of Myofibrillar and Sarcoplasmic Proteins During the Processing of Dry-Cured Ham

Techniques such as sodium dodecyl sulfate polyacrylamide gel electrophoresis (SDS-PAGE), two-dimensional gel electrophoresis (2-DGE), free solution conjugate electrophoresis (FSCE), size-exclusion chromatography (SEC) also known as gel filtration chromatography (GFC), and reversed-phase high-performance liquid chromatography (RP-HPLC) have been used during the last decades to study proteolysis and to separate and identify the proteins from complex mixtures.

In this sense, SDS-PAGE has been the method of choice in the last decades for the separation and quantitation of protein changes in meat and processed meats. This method is simple, reproducible, and reliable, and also easier to interpret in comparison to 2-D separations as they are mainly based on separation according to molecular mass, and do not require expensive material and instrumentation. Many authors have reported the structural alteration and progressive disappearance of muscle proteins as well as the generation of polypeptides during tenderization and processing of dry-cured ham. Muscle cathepsins play a relevant role in protein breakdown during dry-curing (Toldrá and Etherington 1988). In this way, a progressive disappearance of myofibrillar proteins including myosin heavy chain, myosin light chains 1 and 2 (MLC 1 and MLC 2), and troponins C and I, as well as the appearance of numerous smaller fragments in the 50–100 and 20–45 kDa regions have been reported (Toldrá et al. 1993b; Buscailhon et al. 1994). The generation of polypeptides (Rodríguez-Núñez et al. 1995) and small peptides of 2–4 amino acid residues (Sentandreu and Toldrá 2007a) was also reported. More recently, an intense proteolysis of actin, tropomyosin, and myosin light chains extracted in the myofibrillar fraction of dry-cured hams with different ripening times was also reported (Di Luccia et al. 2005). These authors concluded that after 12 months of ripening, most myofibrillar proteins were completely hydrolyzed.

The hydrolysis of sarcoplasmic proteins during dry-curing has also been widely described. Myoglobin protein has been described as degraded during the post-mortem storage of pig meat (Lametch et al. 2006) as well as during the processing of dry-cured ham. In this respect, electrophoretic bands for the sarcoplasmic proteins at different stages of the dry-curing process of Iberian ham were reported up to 17.5 months of curing, showing that the levels of myoglobin protein were undetectable after this period of time (Córdoba et al. 1994). A similar study was done on Bayonne hams cured for 8 months (Monin et al. 1997) and Serrano hams cured for 15 months (Toldrá et al. 1993a), and also a gradual decrease in the intensity of the electrophoretic band corresponding to myoglobin was detected until the end of curing. These results agree with those reported by other authors (Soriano-Pérez et al. 2003) who identified a 16 kDa electrophoretic band obtained after SDS-PAGE separation as myoglobin. Other authors (Picariello et al. 2006) have also described the partial hydrolysis of myoglobin protein in dry-cured ham after 12 months and after the analysis of the bands using two-dimensional gel electrophoresis. The same authors also reported the total disappearance of phosphoglycerate kinase (PGK), lactate dehydrogenase (LDH), glyceraldehyde 3-phosphate dehydrogenase (GAPDH), glycogen phosphorylase (PYGM), and creatine kinase (CK) during the ripening period of dry-cured ham aged up to 12 months. However, enolase and phosphoglycerate mutase (PGAM) enzymes remained visible until the end of curing in Bayonne hams (Monin et al. 1997), whereas pyruvate kinase (PK) was the only enzyme that remained constant after 11 months of curing (Soriano-Pérez et al. 2003).

Size-exclusion chromatography (SEC), performed under low-pressure, has been also used as an initial step for the analysis of dry-cured ham peptides. Peptides are fractionated according to their molecular size and the gel should have an exclusion limit smaller than the molecule of interest in order to obtain the desired separation. Thus, peptides of different molecular weight are separated within the gel matrix within the fractionation range of the gel (see Chap. 4) and then further separated through reversed-phase HPLC (see Fig. 9.1).

Figure 9.2 shows the most frequently used workflow for the identification of the complete sequence of naturally generated peptides in dry-cured ham muscle.

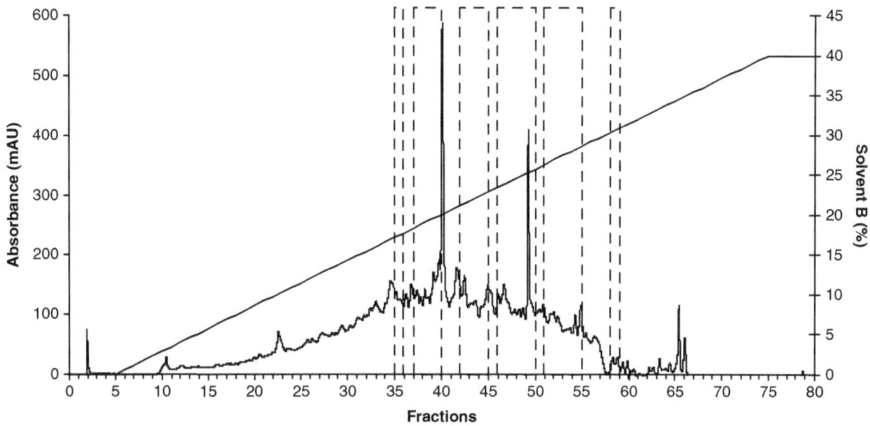

Fig. 9.1 Reversed-phase chromatographic separation of the selected pooled fractions obtained from size-exclusion chromatography. *Dotted* fractions indicate the fractions where creatine kinase fragments have been detected. *Bold line* indicates the solvent gradient of the chromatographic separation (Reprinted with permission from Mora et al., 2009a (*J. Agric. Food Chem. 57*, 8982–8988). Copyright (2009) American Chemical Society)

9.5 Mass Spectrometry Techniques for the Identification of Dry-Cured Ham Naturally Generated Peptides

9.5.1 Mass Fingerprinting (MFP)

The most frequently used strategy to study protein changes and characterize the proteomic profile of a complex mixture starts with SDS-PAGE separation. In one dimension (1D) electrophoresis proteins are isolated only according to molecular weight whereas in two dimensions (2-D) electrophoresis proteins are isolated according to both the isoelectric point and the molecular weight. Once the proteins are separated, they can be stained by using either Coomassie blue or silver staining compatible with MS, and usually digested in-gel with specific proteases such as trypsin that specifically cleave the protein on the C-terminal side of the basic amino acids arginine and lysine.

The peptides obtained after the digestion of the protein are analyzed by mass spectrometry achieving a list of peak masses. The set of measured peptide masses is the peptide mass fingerprint. This experimental mass profile is matched against the theoretical masses obtained from the *in silico* digestion at the same enzyme cleavage sites of all protein sequences in the database. The proteins in the database are then ranked according to the number of peptide masses matching their sequence within a given mass error tolerance. In order to successfully identify a peptide or protein, it is required that the masses of a certain number of fragments match with the theoretical peptide masses contained in the protein databases (Cañas et al. 2006).

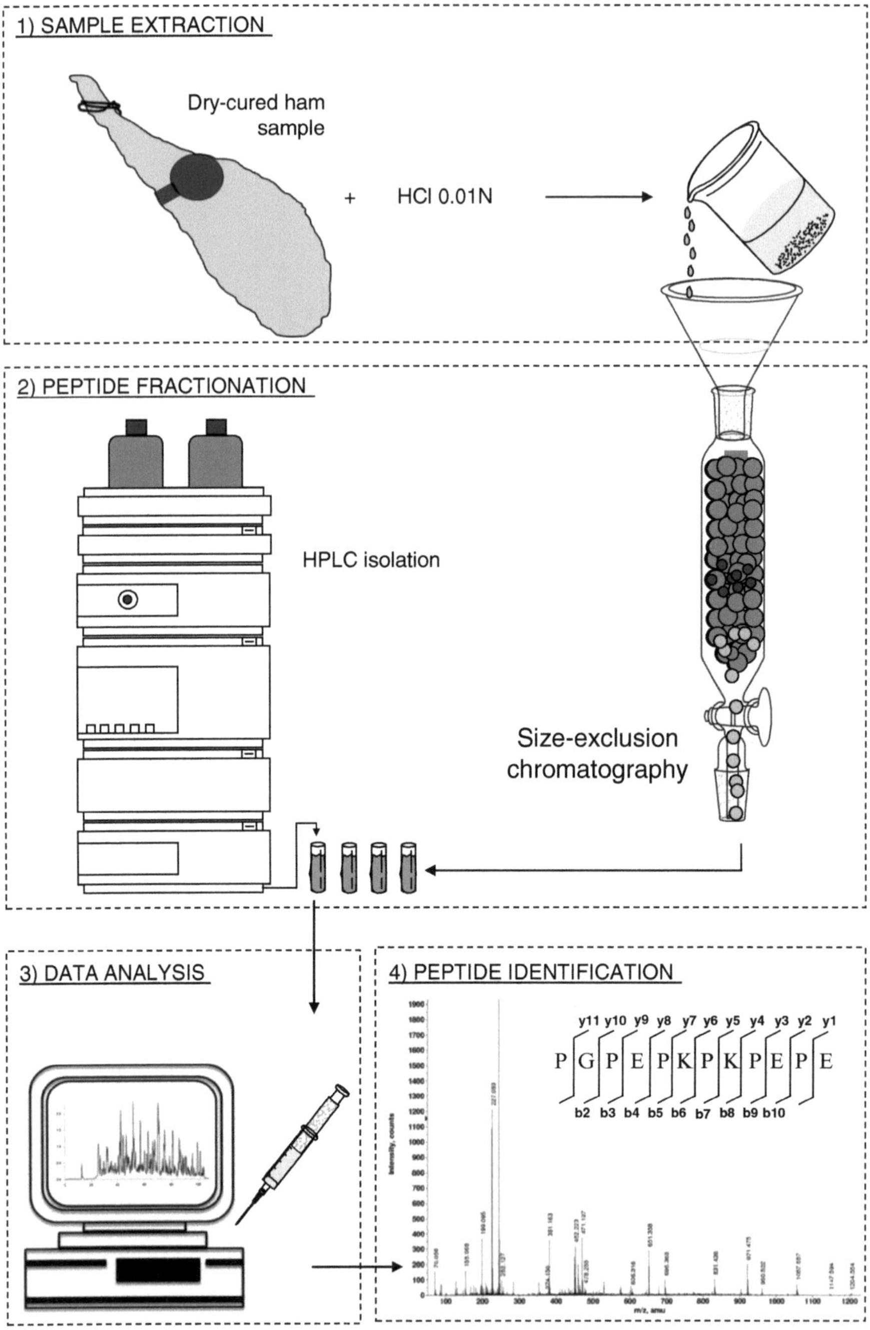

Fig. 9.2 Typical experimental workflow for the identification of the complete sequence of naturally generated peptides in dry-cured ham

This result is accepted if the data pass a statistical probability threshold of being a correct match. Proteins can be identified in this way with high-throughput compatibility and a high sensitivity even below the *f*mol range.

Two-dimensional SDS-PAGE and MALDI-TOF mass spectrometry analysis were used to study the myofibrillar and water-soluble fraction of raw ham muscles and dry-cured hams with different ripening times (Di Luccia et al. 2005). The two-dimensional maps showed the progressive disappearance of actin, tropomyosin, and myosin light chains during ripening. Some of the sarcoplasmic proteins in water extracts from pork meat markedly decreased in amount or disappeared totally during ripening. Some of the bands were digested using trypsin enzyme and subsequently analyzed by MALDI-TOF MS. Two fragments of myosin heavy chain and a new form of actin were identified in the myofibrillar fraction whereas tropomyosin α- and β-chains were detected in the water soluble protein fraction.

Sarcoplasmic proteins undergoing proteolysis during the ripening of products were also identified by MALDI-TOF mass spectrometry peptide mass fingerprinting after 2-D AUT-PAGE/SDS electrophoresis (acetic acid–urea–triton polyacrylamide gel in the first dimension and sodium dodecyl sulphate polyacrylamide gel in the second dimension). This separation technique has resulted in the identification of 13 sarcoplasmic proteins and shows higher resolution in their separation in comparison to standard 2-D IPG/SGS-PAGE as well as an easier interpretation of the 2-D maps (Picariello et al. 2006).

Despite knowledge regarding the disappearance of proteins and the generation of novel peptides, the sequences corresponding to the hydrolysis products were not analyzed in any of these studies. The major difficulty in the study of naturally generated peptides resides is the small size of these fragments that cannot be trypsin-digested because they have been hydrolyzed previously by endogenous muscle enzymes. This fact gives double difficulty: the small size of the generated peptides that are sometimes in the limit of some mass spectrometry techniques, and the impossibility of controlling the hydrolysis, having as a consequence a complex mixture of peptides from different proteins with unspecific cleavage sites. In this sense, the use of advanced proteomic techniques such as tandem mass spectrometry is essential to elucidate the sequence of these small peptides.

9.5.2 Tandem Mass Spectrometry (MS/MS)

Tandem mass spectrometry is the method of choice for protein identification in complex mixtures as well as for elucidation of the sequences of the naturally generated peptides through proteolysis.

Tandem mass spectrometry can also be performed on peptides derived by trypsin digestion and involves their fragmentation using collisionally induced dissociation (CID) to produce the characteristic spectrum that is specific for each peptide ion. The identification of the proteins is facilitated by using the generated spectra to interrogate the protein sequence databases. To make separations possible in mass spectrometry instruments, peptides have to be converted into ions and transferred into the gas phase in an ionization source. The ionization with MALDI and ESI are the most commonly employed in the analysis of peptides derived from meat and meat products. MALDI has the advantage over ESI that it is more tolerant of low concentrations of sample contaminants such as buffers and salts. One of the advantages of ESI is that ions, depending on their molecular mass and structure, may acquire multiple charges. So, the identification of peptides with higher molecular masses may be achieved using a smaller *m/z* ratio interval; also multiple charges fragment with less activation energy in the mass spectrometer.

The ions obtained in the ion source are isolated in the analyzer. The most commonly used analyzers are quadrupoles, time-of-flight instruments, and ion traps. From the point of view of resolution and reproducibility, the TOF analyzers are not as satisfactory as quadrupoles. However, certain advantages compensate for these limitations; TOF instruments are robust, fast in data acquisition, and show very good mass accuracy as well as a virtually unlimited mass range. Specialized personnel are not required, and access to the ion source is very easy. However, if the peptides of interest present molecular masses smaller than 800 Da, it is important to consider that the ions of the matrix used in MALDI ionization, commonly DBA (dihydroxy benzoic acid) or HCCA (α-Cyano-4-hydroxycinnamic acid) to ionize meat peptides, also appear as contaminants between 200 and 800 Da. They can inhibit and make the identification of peptides at that range more difficult if not impossible.

These analyzers can be combined arranging three quadrupoles, two time-of-flight, or a quadrupole and time-of-flight in tandem. There are other possible combinations of analyzers but TOF/TOT and Q/TOF have been the most commonly used in the analysis of complex mixtures of naturally generated peptides.

The first studies that focused on the elucidation of the complete sequence of meat and meat product peptides were performed using automatic Edman degradation. This is a good technique for the analysis of small peptides but it becomes tedious when the objective is to sequence longer peptides. The main disadvantage of Edman degradation is that it only can be used with pure peptides because mixtures of peptides or contaminants present in the sample could mislead the sequencing. In this sense, one peptide from glyceraldehyde-3-phosphate dehydrogenase protein, one peptide from troponin T protein, and three peptides from creatine kinase protein were isolated and identified during post-mortem aging in bovine *longissimus dorsi* muscle. Peptides generated during the aging of beef meat were isolated by HPLC and subsequently analyzed by MALDI-TOF to see the purity of the peaks. Amino acid sequencing was done by automatic Edman degradation (Stoeva et al. 2000).

Regarding meat products, a water-soluble extract of dry-cured ham was fractionated by gel filtration chromatography and fractions with the highest concentration in peptides were separated by reverse-phase and cation-exchange high-performance liquid chromatography. Small peptides of two, three, and four amino acids were sequenced from the N-terminal by automated Edman degradation (Sentandreu and Toldrá 2007b). Five peptides with structures between 151 and 381 amino acids and generated from sarcoplasmic (myoglobin and creatine kinase) and myofibrillar (troponin-I, troponin-T, and myosin light chain-2) proteins were isolated from semi-dry fermented sausages using SDS-PAGE and RP-HPLC as qualitative methodologies, and subsequently sequenced by automatic Edman degradation (Hughes et al. 2001). A study in dry-cured Parma hams resulted in the identification of a peptide with a very high degree of homology with the N-terminal part of different mammalian pyruvate kinases. The peptide was purified by means of RP-HPLC and identified by its molecular mass and amino acid sequence analysis using automatic Edman degradation (Sforza et al. 2003).

The use of modern mass spectrometry techniques with the analyzers in tandem allows a faster and more reliable identification of the peptide sequences because the obtained MS/MS spectra are matched with the theoretical sequences contained in the databases. Specific peptide sequences of the myofibrillar proteins actin (Sentandreu et al. 2007a), titin (Mora et al. 2009b), myosin light chains (Mora et al. 2011a), and troponin T (Mora et al. 2010), as well as some sarcoplasmic proteins such as creatine kinase (Mora et al. 2009a) and a group of glycolytic enzymes (Mora et al. 2011b; Sforza et al. 2003) have already been identified in dry-cured ham using ESI-Q/TOF and MALDI-TOF/TOF mass spectrometers. In this sense, Table 9.1 shows the number of peptides that have been sequenced in dry-cured ham also including their respective protein of origin.

9.6 Data Analysis

As has been previously mentioned, the identification of peptides is mostly carried out using database search approaches. This type of data analysis can also be used for the direct analysis of peptides naturally generated by endogenous enzymes (Fricker et al. 2006; Geho et al. 2006; Hardt et al. 2005a, b; Villanueva et al. 2006) although the appropriate choice of the search parameters and sequence database are crucial for successful application of this methodology. In fact, database searching of nontryptic peptides is less effective due to the lack of charge localization at the N and C termini of the naturally generated peptides. Some of the most popular search engines used in protein identification are Mascot, Sequest, and X! TANDEM, but there are more software packages designed to identify the most likely peptide sequence to match a MS/MS spectrum. Mascot from matrixscience (http://www.matrixscience.com/) is the most widely used software but the choice of its search parameters requires some user expertise, and it has to be previously evaluated with a control sample that in the case of meat peptides used to be bovine serum albumin (BSA). The use of no specific enzyme as well as the error-tolerant option to search the unmatched spectra considering possible modifications are some of the parameters to take into account in the Mascot MS/MS Ions Search when naturally generated peptides are studied.

There are many available databases, but SwissProt and NCBInr are the most commonly used in the identification of meat-derived peptides. Swiss-Prot has approximately 500,000 entries and is considered the best annotated database although it is important to consider that it is nonredundant so it is ideal for PMF

Table 9.1 Naturally generated protein fragments identified in dry-cured ham

Protein name	Accession number NCBInr	Number peptides	Sequence coverage (%)	Instrument (MS, MS/MS)	Search engine	Database	References
Small peptides (2, 3, and 4 amino acids)	–	–	–	Edman degradation	–	–	Sentandreu and Toldrá 2007b
Actin	NP_001161267	4	8	MALDI-ToF Qtrap	Mascot	UniProt	Sentandreu et al. 2007a
Myosin light chain 1 (MLC 1)	ABK55642	9	8	MALDI-ToF	Mascot	NCBInr	Mora et al. 2009b and 2011a
		137	68	nESI-LC-Q/ToF	Mascot	NCBInr	
Titin	AAD00528	5	10	MALDI-ToF nESI-LC-Q/ToF	Mascot	NCBInr	Mora et al. 2009b
Creatine kinase (CK)	NP_001123421	58	44	MALDI-ToF nESI-LC-Q/ToF	Mascot	NCBInr	Mora et al. 2009b
Slow troponin T (TnT)	Q75ZZ6	2	9	MALDI-ToF/ToF	Mascot	NCBInr	Mora et al. 2010
Fast troponin T (TnT)	BAD15381	25	18	MALDI-ToF/ToF	Mascot	NCBInr	Mora et al. 2010
Myosin light chain 2 (MLC 2)	NP_001006592	88	47	nESI-LC-Q/ToF	Mascot	NCBInr	Mora et al. 2011a
Glycogen phosphorylase (PYGM)	ABF81977	2	4	nESI-LC-Q/ToF	Mascot Paragon	NCBInr UniProt	Mora et al. 2011b
Glyceraldehyde 3-phosphate dehydrogenase (GAPDH)	ABI29187	8	8	nESI-LC-Q/ToF	Mascot Paragon	NCBInr UniProt	Mora et al. 2011b
Phosphoglycerate kinase 1 (PGK)	NP_001093402	4	7	nESI-LC-Q/ToF	Mascot Paragon	NCBInr UniProt	Mora et al. 2011b
Phosphoglycerate mutase 2 (PGAM)	CAD89670	7	47	nESI-LC-Q/ToF	Mascot Paragon	NCBInr UniProt	Mora et al. 2011b
Enolase (ENO)	NP_001037992	18	14	nESI-LC-Q/ToF	Mascot Paragon	NCBInr UniProt	Mora et al. 2011b
Piruvate kinase 3 isoform 2 (PK)	XP_001929120	2	2	nESI-LC-Q/ToF	Mascot Paragon	NCBInr UniProt	Mora et al. 2011b
		1	–	Edman degradation	–	–	
Lactate dehydrogenase (LDH)	P0039	4	5	nESI-LC-Q/ToF	Mascot Paragon	NCBInr UniProt	Mora et al. 2011b

Peptide (651.35$_{2+}$)

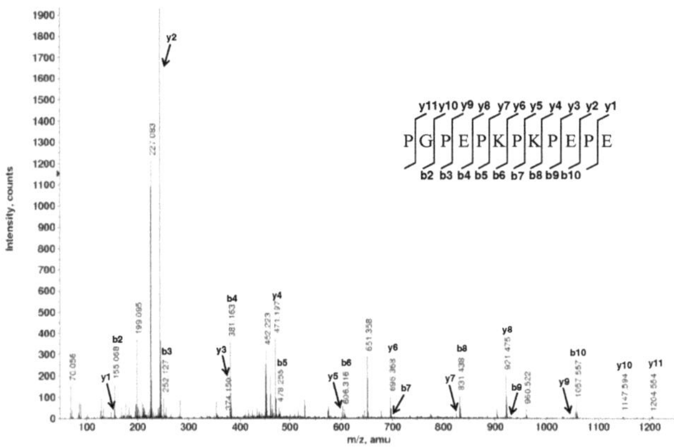

	Peptide 14 (651.35$_{2+}$)			
	b	**Seq.**	y	
1	98.06	**P**		12
2	**155.08**	**G**	**1204.62**	11
3	**252.13**	**P**	**1147.60**	10
4	**381.18**	**E**	**1050.55**	9
5	**478.23**	**P**	**921.50**	8
6	**606.32**	**K**	**824.45**	7
7	**703.38**	**P**	**696.36**	6
8	**831.47**	**K**	**599.30**	5
9	**928.52**	**P**	**471.21**	4
10	**1057.57**	**E**	374.16	3
11	1154.62	**P**	**245.11**	2
12		**E**	**148.06**	1

Fig. 9.3 MS/MS spectrum of ion 651.35^{2+} of the main peak obtained after the second reversed-phase chromatography. Peptide sequence matching the product ion spectrum is shown in capital letters, together with observed *b* and *y* ions. The spectrum is presented with its corresponding table showing the ions matched by MASCOT in *bold letters* (Reprinted with permission from Mora et al., 2009b (*J. Agric. Food Chem. 57*, 3228–3234). Copyright (2009) American Chemical Society)

searches, where the loss of one or two peptides is not very important. For MS/MS searches, the NCBInr database is a better choice. This database is large (10,000,000 entries approx.), comprehensive, and nonidentical. In the data analysis of meat-derived peptides, results obtained from both databases used to be very similar although a comparison between SwissProt and NCBInr databases revealed that the NCBInr database provided the highest number of identifications (Mora et al. 2011b). Figure 9.3 shows a MS/MS spectrum obtained with a quadrupole/time-of-flight mass spectrometer after RP-HPLC separation of a size-exclusion chromatography fraction of dry-cured ham.

9.7 Future Trends

The proteomic analysis of naturally generated peptides during the dry-cured processing of ham improves knowledge of the post-mortem proteolysis occurring during dry-cured ham processing and the assessment of the peptidases responsible for this action. The peptides generated during the proteolysis in dry-cured ham could be biomarkers of the process and, what is more interesting, they could be acting as bioactive peptides exerting either antimicrobial, antihypertensive, or any other bioactivity in the final product. In recent times, several studies have described the generation of bioactive peptides, often with cardiovascular promoting properties (usually ACE-I-inhibition or antioxidant effects) following hydrolysis of sarcoplasmic and/or myofibrillar proteins or whole meat muscle with proteolytic enzymes (Di Bernardini et al. 2011, 2012). In this sense, Sentandreu and Toldrá (2007a, b) suggested that the proteolytic action of porcine muscle dipeptidyl peptidases during the ripening period of dry-cured ham could contribute to the generation of ACE-I inhibitory peptides. The angiotensin-converting enzyme inhibitory peptides generated during the curing of meat products have been studied extensively. For example, dipeptidyl peptidases (DPP) could contribute to the generation of antihypertensive peptides among which Arg–Pro showed the strongest angiotensin-converting enzyme inhibitory activity (Jang and Lee 2005; Sentandreu and Toldrá 2007b). The use of these peptides in the development of novel meat products and their potential as ingredients of healthier food requires a better knowledge of the mechanisms that control the generation of these peptides during proteolysis and their possible biological activity, and advanced techniques of proteomics such as the mass spectrometry in tandem play a crucial role in reaching this understanding.

References

Bodwell JE, Meyer WL (1981) Purification and characterization of carboxypeptidase-A from rat skeletal-muscle. Biochemistry 10(20):2767–2777

Buscailhon S, Berdagué JL, Gandemer G, Touraille C, Monin G (1994) Effects of initial pH on compositional changes and sensory traits of French dry-cured hams. J Muscle Foods 5:257–270

Cañas B, López-Ferrer D, Ramos-Fernández A, Camafeita E, Calvo E (2006) Mass spectrometry technologies for proteomics. Brief Funct Genomic Proteomic 4:295–320

Córdoba JJ, Antequera T, Ventanas J, López-Bote C, García C, Asensio MA (1994) Hydrolysis and loss of extractability of proteins during ripening of Iberian ham. Meat Sci 37:217–227

Di Bernardini R, Rai DK, Bolton D, Kerry J, O'Neill E, Mullen AM, Harnedy P, Hayes M (2011) Isolation, purification and characterization of antioxidant peptidic fractions from a bovine liver sarcoplasmic protein thermolysin hydrolyzate. Peptides 32:388–400

Di Bernardini R, Mullen AM, Bolton D, Kerry J, O'Neill E, Hayes M (2012) Assessment of the angiotensin-I-converting enzyme (ACE-I) inhibitory and antioxidant activities of hydrolysates of bovine brisket sarcoplasmic proteins produced by papain and characterisation of associated bioactive peptidic fractions. Meat Sci 90:226–235

Di Luccia A, Picariello G, Cacace G, Scaloni A, Faccia M, Liuzzi V, Alviti G, Musso SS (2005) Proteomic analysis of water soluble and myofibrillar protein changes occurring in dry-cured hams. Meat Sci 69:479–491

Fricker LD, Lim J, Pan H, Che FY (2006) Peptidomics: identification and quantification of endogenous peptides in neuroendocrine tissues. Mass Spectrom Rev 25:327–344

Geho DH, Liotta LA, Petricoin EF, Zhao W, Araujo RP (2006) The amplified peptidome: the new treasure chest of candidate biomarkers. Curr Opin Chem Biol 10:50–55

Gianelli P, Flores M, Moya VJ, Aristoy MC, Toldrá F (2000) Effect of carnosine, anserine and other endogenous skeletal peptides on the activity of porcine muscle alanyl and arginyl aminopeptidases. J Food Biochem 24:69–78

Hardt M, Thomas LR, Dixon SE, Newport G, Agabian N, Prakobphol A, Hall SC, Witkowska HE, Fisher SJ (2005a) Toward defining the human parotid gland salivary proteome and peptidome:

identification and characterization using 2-D SDS-PAGE, ultrafiltration, HPLC, and mass spectrometry. Biochemistry 44:2885–2899

Hardt M, Witkowska HE, Webb S, Thomas LR, Dixon SE, Hall SC, Fisher SJ (2005b) Assessing the effects of diurnal variation on the composition of human parotid saliva: quantitative analysis of native peptides using iTRAQ reagents. Anal Chem 77:4947–4954

Hughes MC, Geary S, Dransfield E, McSweeney PLH, O'Neill EE (2001) Characterization of peptides released from rabbit skeletal muscle troponin-T by μ-calpain under conditions of low temperature and high ionic strength. Meat Sci 59:61–69

Jang A, Lee M (2005) Purification and identification of angiotensin converting enzyme inhibitory peptides from beef hydrolysates. Meat Sci 69:653–661

Kemp CM, Sensky PL, Bardsley RG, Buttery PJ, Parr T (2010) Tenderness: an enzymatic view. Meat Sci 84:248–256

Lametsch R, Kristensen L, Larsen MR, Therkildsen M, Oksbjerg N, Ertbjerg P (2006) Changes in the muscle proteome after compensatory growth in pigs. J Anim Sci 84:918–924

Lawrie RA, Ledward DA (2006) Lawrie's meat science, 7th edn. Woodhead Publishing, Cambridge, UK

McIntosh B, Gardiner P, McComas A (1996) Skeletal muscle: form and function. In: Berry MJ (ed) Skeletal muscle: form and function, 2nd edn. Human Kinetics, Champaign

Monin G, Marinova P, Talmant A, Martin JF, Cornet M, Lanore D, Grasso F (1997) Chemical and structural changes in dry-cured hams (Bayonne hams) during processing and effects of the dehairing technique. Meat Sci 47:29–47

Mora L, Sentandreu MA, Fraser PD, Toldrá F, Bramley PM (2009a) Oligopeptides arising from the degradation of creatine kinase in spanish dry-cured ham. J Agric Food Chem 57:8982–8988

Mora L, Sentandreu MA, Koistinen KM, Fraser PD, Toldrá F, Bramley PM (2009b) Naturally generated small peptides derived from myofibrillar proteins in serrano dry-cured ham. J Agric Food Chem 57:3228–3234

Mora L, Sentandreu MA, Toldrá F (2010) Identification of small troponin T peptides generated in dry-cured ham. Food Chem 123:691–697

Mora L, Sentandreu MA, Toldrá F (2011a) Intense degradation of myosin light chain isoforms in Spanish dry-cured ham. J Agric Food Chem 59:3884–3892

Mora L, Valero ML, Del Pino MMS, Sentandreu MA, Toldrá F (2011b) Small peptides released from muscle glycolytic enzymes during 6 dry-cured ham processing. J Proteom 74:442–450

Ouali A, Herrera-Méndez CH, Coulis G, Becila S, Boudjellal A, Aubry L, Sentandreu MA (2006) Revisiting the conversion of muscle into meat and the underlying mechanisms. Meat Sci 74:44–58

Picariello G, De Martino A, Mamone G, Ferranti P, Addeo F, Faccia M, SpagnaMusso S, Di Luccia A (2006) Proteomic study of muscle sarcoplasmic proteins using AUT-PAGE/SDS-PAGE as two-dimensional gel electrophoresis. J Chromatogr B Anal Technol Biomed Life Sci 833:101–108

Rico E, Toldrá F, Flores J (1991) Assay of cathepsin D activity in fresh pork muscle and dry-cured ham. Meat Sci 29:287–293

Rodríguez-Núñez E, Aristoy MC, Toldrá F (1995) Peptide generation in the processing of dry-cured ham. Food Chem 53:187–190

Sentandreu MA, Toldrá F (2001) Dipeptidyl peptidase activities along the processing of Serrano dry-cured ham. Eur Food Res Technol 213:83–87

Sentandreu MA, Toldrá F (2007a) Oligopeptides hydrolysed by muscle dipeptidyl peptidases can generate angiotensin-I converting enzyme inhibitory dipeptides. Eur Food Res Technol 224:785–790

Sentandreu MA, Toldrá F (2007b) Evaluation of ACE inhibitory activity of dipeptides generated by the action of porcine muscle dipeptidyl peptidases. Food Chem 102:511–515

Sentandreu MA, Armenteros M, Calvete JJ, Ouali A, Aristoy MC, Toldrá F (2007a) Proteomic identification of actin-derived oligopeptides in dry-cured ham. J Agric Food Chem 55:3613–3619

Sentandreu MA, Aubry L, Toldrá F, Ouali A (2007b) Blocking agents for ELISA quantification of compounds coming from bovine muscle crude extracts. Eur Food Res Technol 224:623–628
Sforza S, Boni M, Ruozi R, Virgili R, Marchelli R (2003) Identification and significance of the N-terminal part of swine pyruvate kinase in aged Parma hams. Meat Sci 63:57–61
Soriano-Pérez A, García-Ruiz A, Mariscal-Contreras C, Cabezudo-Ibáñez M (2003) Separation and identification of sarcoplasmic proteins from hams from three white pig crosses containing Duroc. Eur Food Res Technol 216:193–198
Stoeva S, Byrne CE, Mullen AM, Troy DJ, Voelter W (2000) Isolation and identification of proteolytic fragments from TCA soluble extracts of bovine M-longissimus dorsi. Food Chem 69:365–370
Toldrá F (1998) Proteolysis and lipolysis in flavour development of dry-cured meat products. Meat Sci 49:S101–S110
Toldrá F (2002) Dry-cured meat products. Wiley-Blackwell, Ames
Toldrá F (2004) Dry-curing. In: Jensen WK (ed) Encyclopedia of meat sciences. Elsevier, Oxford, pp 360–366
Toldrá F (2006) The role of muscle enzymes in dry-cured meat products with different drying conditions. Trends Food Sci Technol 17:164–168
Toldrá F, Etherington DJ (1988) Examination of cathepsins B, D, H and L activities in dry-cured hams. Meat Sci 23:1–7
Toldrá F, Flores M (1998) The role of muscle proteases and lipases in flavor development during the processing of dry-cured ham. Crit Rev Food Sci Nutr 38:331–352
Toldrá F, Rico E, Flores J (1993a) Cathepsin-B, cathepsin-D, cathepsin-H and cathepsin-l activities in the processing of dry-cured ham. J Sci Food Agric 62:157–161
Toldrá F, Cerveró MC, Part C (1993b) Porcine aminopeptidase activity as affected by curing agents. J Food Sci 58:724
Toldrá F, Flores M, Sanz Y (1997) Dry-cared ham flavour: enzymatic generation and process influence. Food Chem 59:523–530
Toldrá F, Aristoy MC, Flores M (2000) Contribution of muscle aminopeptidases to flavor development in dry-cured ham. Food Res Int 33:181–185
Villanueva J, Martorella AJ, Lawlor K, Philip J, Fleisher M, Robbins RJ, Tempst P (2006) Serum peptidome patterns that distinguish metastatic thyroid carcinoma from cancer-free controls are unbiased by gender and age. Mol Cell Proteom 5:1840–1852
Yamada T, Kunioka Y, Wakayama J, Aimi M, Noguchi YS, Akiyama N, Kayamori T (2004) Molecular organizations of myofibrils of skeletal muscle studied by atomic force microscopy molecular and cellular aspects of muscle contraction. In: Sugi H (ed) Springer, New York, pp 285–294

Chapter 10
Evaluation of Fish Quality and Safety by Proteomics Techniques

Carmen Piñeiro and Iciar Martinez

10.1 Introduction

Food quality is a complex concept. In marketing and economics literature, there are two main approaches to define food quality (Grunert 2005): the holistic approach, which includes within the concept of food quality "all the desirable characteristics a product is perceived to have," and the excellence approach, which views food quality as referring only to characteristics that pertain to a higher, more restrictive, or "superior" specification of the product. The holistic approach leaves wide scope for interpretation: quality can mean conforming to standards (including standards pertaining to the environment, local specialities, organic production, ethics, and even taste and smell) and it can refer to subjectively perceived quality attributes. Quality is also a factor that involves the entire production process, from raw materials, processing, and packaging up to consumption of the product.

The terms "food safety," "food security," and "food quality" should not be confused. Food safety refers to all those hazards, whether chronic or acute, that may make food injurious to the health of the consumer. Food safety is not negotiable: all food items must be safe. Food security is defined by Pinstrup-Andersen (2009) as "a situation that exists when all people, at all times, have physical,

C. Piñeiro
Institute for Marine Research (IIM), Spanish Council for Scientific Research (CSIC), C/Eduardo Cabello 6, Vigo E-36208, Spain

I. Martinez (✉)
Deptarment of Zoology and Cellular Biology, University of the Basque Country, Leioa E-48940, Spain

IKERBASQUE, Basque Foundation for Science, Bilbao E-48011, Spain
e-mail: iciar.martinez@ehu.es

F. Toldrá and L.M.L. Nollet (eds.), *Proteomics in Foods: Principles and Applications*, Food Microbiology and Food Safety 2, DOI 10.1007/978-1-4614-5626-1_10,

social, and economic access to sufficient, safe, and nutritious food that meets their dietary needs and food preferences for an active and healthy life." Food quality, as mentioned above, refers to those attributes that influence a product's value to the consumer.

According to the plain English definitions used in ISO 9000, 9001, and 9004: "The quality of something can be determined by comparing a set of inherent characteristics with a set of requirements. If those inherent characteristics meet all requirements, high or excellent quality is achieved. If those characteristics do not meet all requirements, a low or poor level of quality is achieved." According to this definition, the concept of quality is subjective and relative to how well the product fulfills the customer's expectations. Its limits are usually set by what the customer will be willing to pay and in the case of fish products, it entails aspects related to the species, size, sex, and condition, geographic origin, chemical composition (its content in vitamins, minerals, bioactive compounds, protein, and fat as well as its digestibility and desirable fatty acid composition), freshness, production method (wild, intensively or organically farmed), type of processing if applicable (frozen vs. not frozen, canned, smoked, salted, etc.), and the resulting sensory aspects (taste, texture, color, smell).

In general, however, the term "seafood quality" also includes the product's safety (Yaktine et al. 2008) and it is important to emphasize that safety and quality management of seafood may be complex due to the very large number of edible species; the fact that even farmed species are not truly domesticated (they have been commercially farmed for only a few generations); the greater variability in their environment, accessibility, and biodiversity; the difficulty in controlling their environmental conditions (temperature, pH, salinity); and the potential presence of toxic compounds in the environment.

Proteins play a crucial role in almost every biological process. In addition to providing structural support, they are responsible for an ample variety of physiological functions, including catalysis, defense, transport, and sensing, in all living systems. Whereas the genome of an organism is more or less constant and specific, its proteome (i.e., the entire complement of proteins expressed by the genome) is highly dynamic. The type and amount of proteins expressed vary not only between different types of cells in the same organism but also in a given cell type as a response to a wide diversity of stimuli and environmental factors.

The formidable capability in post-genomic analyses built up over the past few years, including the use of DNA microarrays and proteomics, underpins the ongoing research strategy. One single gene can produce a variable number of proteins whose function may be modulated by further post-translational modifications. Given that one of the major objectives of proteomics is to quantify protein levels and their dynamic changes (Hocquette et al. 2005), this approach can be used, when applied to food matrices, to improve our knowledge of the interrelationship between the changes induced during the production and processing of seafood on the protein map and the quality and safety of the product, thus opening the possibility of using proteomic techniques as robust tools to monitor the state of seafood products in a given step within the production chain.

Only complete knowledge of the exact chemical composition of a food item allows the correct evaluation of its nutritional, toxicological, and technological properties. Novel analytical technologies, including proteomics, will generate comprehensive databases of the protein composition of food items. These databases should include particularly interesting minor components that may have been overlooked by conventional methods of protein analysis, as well as changes in the protein composition originating as a consequence of applying different production and processing methods. Furthermore, proteome analysis adds the intriguing perspective of achieving a systematic overview of the metabolic routes within defined cells or tissues of food components (Pischetsrieder and Baeuerlein 2009), providing valuable information about the cellular processes that have occurred in the organism prior to slaughtering, including stress responses, inflammation, defense, apoptosis, and immunomodulation. This adds the possibility of identifying biomarkers to indicate the exposure of the organism to stresses and contaminants, in addition to helping map the chemical composition and properties of the seafood. As a consequence, proteomic analyses of seafood have the potential to tremendously increase our knowledge of the composition, safety, and effect of the production and processing methods of the raw materials and, therefore, on the impact of these variables on seafood product quality.

Further nonenzymatic post-translational protein modifications (nePTM) occur frequently during food processing and storage, rendering the food proteome even more complex. nePTMs are mostly caused by oxidation and by nonenzymatic reactions of sugars with amino acid side chains (Maillard reaction or glycation). Additionally, other nePTMs have been reported, such as condensation, elimination, or hydrolysis of side chains or breakdown of the peptide backbone. nePTMs also have significant consequences regarding the technological, nutritional, and toxicological properties of processed food and proteomic techniques have been applied to the systematic study of the formation of nePTMs in processed food items, many of which are largely unknown (Pischetsrieder and Baeuerlein 2009). The challenges of nePTM analysis are their low abundance compared to the unmodified or PTM-modified side chains and their high heterogeneity. Knowledge of the chemical structure and binding sites of nePTMs, however, is crucial to evaluate their effects on food safety and quality. The analysis of the proteome, optimally combined with the metabolome, promises to yield a systematic overview of the changes in food that are caused by changes in production parameters. More important, the recently proposed FOODOMICS approach (Herrero et al. 2012) intends to use proteome research to actually understand the molecular processes that link production parameters to food quality. Currently, proteomics is built upon the foundations of genomics, in such way that a lack of genomic information on a particular species can substantially limit success in protein identification (Graham et al. 2005) and, although sequence data from marine fish and shellfish species have significantly increased in the last decade, they are still scarce (Piñeiro et al. 2003; Forné et al. 2010). Japanese pufferfish (*Takifugu rubripes*), zebra fish (*Danio rerio*), and the estuarine fish *Fundulus heteroclitus* are the first three bony fish whose complete genome sequences became available. According to the NCBI protein database there were over 2.6

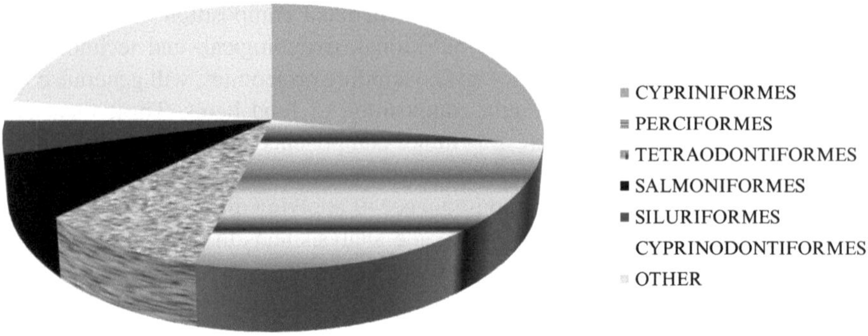

Fig. 10.1 Euteleostei orders with the highest number of protein sequences (Source NCBI protein data base)

million proteins sequenced for the subphylum Vertebrata at the beginning of 2012 of which about 14% belong to bony fish. Figure 10.1 shows the Teleostei orders with a larger number of known DNA sequences and in Table 10.1 the species of the same taxonomic class with a larger number of protein sequences available in the NCBI database are listed.

Almost 40% of the fish proteomic papers deal with issues related to aquaculture. The studies carried out thus far have clearly demonstrated the potential of proteomics to identify physiologically relevant molecules and mechanisms, biomarkers for seafood management and fish welfare, and to evaluate the impact of environmental pollution. In addition, some studies combining proteomics with genomic, metabolomic, and functional approaches provided a wider vision of the physiological functions of interest and pointed out the direction for future research (Forné et al. 2010). Interestingly, these authors stated that the reasons why proteomic techniques have not been applied to resolve problems in fishery product technology remain the same already mentioned 10 years ago by Piñeiro et al. (2003), namely that seafood proteins easily become insoluble and/or aggregate, displaying high molecular weights in addition to exhibiting a large number of isoforms and weak ionization levels.

10.2 Mapping Fish Quality by Proteomics Techniques

Research on seafood quality by means of proteomics is a complicated task: the term "seafood" refers to a large number of different species that may present tissue-, developmental stage-, and temperature-acclimation-dependent polymorphisms (Martinez et al. 1991; Watabe et al. 1992) and on the other hand, processing will further alter the protein components, thus increasing the number of spots and protein sequences with potential diagnostic value for quality determination.

The introduction of consumer perception and marketing studies enlarges the list of quality definitions previously mentioned. Santesmases (2004) distinguished between objective and subjective quality and Verdú (2003) referred to subjective or

Table 10.1 *Euteleosteii* species with the highest number of protein sequences. In *brackets* the number of sequenced proteins at the beginning of 2012 according to NCBI protein database

71,000-1,500 sequences	1,500-500 sequences	500-400 sequences	400-350 sequences	350-315 sequences
Danio rerio (70,887)	*Kryptolebias marmoratus*(1,389)	*Rutilus rutilus*(515)	*Oncorhynchus kisutch*(394)	*Pomatoschistus minutus*(346)
Tetraodon nigroviridis(28.512)	*Anoplopoma fimbria*(1,280)	*Eleutheronema tetradactylum*(497)	*Oreochromis mossambicus*(389)	*Larimichthys crocea*(345)
Oreochromis niloticus(24,076)	*Paralichthys olivaceus*(1,048)	*Oncorhynchus clarkii*(495)	*Etheostoma rufilineatum*(383)	*Gadus macrocephalus*(345)
Salmo salar(17,341)	*Larimichthys polyactis*(1,031)	*Alosa fallax*(478)	*Conger myriaster*(381)	*Mugil cephalus*(336)
Oncorhynchus mykiss(6,735)	*Epinephelus coioides*(992)	*Typhlichthys subterraneus*(459)	*Anguilla japonica*(378)	*Hypophthalmichthys nobilis*(336)
Ictalurus punctatus(4,386)	*Scleropages formosus*(883)	*Scophthalmus maximus*(449)	*Barbatula barbatula*(373)	*Epinephelus bruneus*(334)
Takifugu rubripes(2,702)	*Ctenopharyngodon idella*(856)	*Acanthemblemaria spinosa*(448)	*Rhinichthys osculus*(370)	*Denariusa bandata*(331)
Oryzias latipes(2,638)	*Oncorhynchus tshawytscha*(754)	*Acanthemblemaria aspera*(436)	*Atherina boyeri*(366)	*Anguilla anguilla*(331)
Gadus morhua(2,245)	*Ictalurus furcatus*(744)	*Gadus chalcogrammus*(430)	*Siniperca chuatsi*(360)	*Megalobrama amblycephala*(329)
Dicentrarchus labrax(,2201)	*Salvelinus alpinus*(681)	*Etheostoma simoterum*(426)	*Solea solea*(358)	*Ctenogobiops feroculus*(323)
Cyprinus carpio(1,921)	*Salmo trutta*(628)	*Oncorhynchus keta*(422)	*Perca flavescens*(356)	*Galaxias sp.* 'southern'(322)
Esox lucius(1,743)	*Sparus aurata*(603)	*Pseudorasbora parva*(422)	*Oncorhynchus kisutch*(394)	*Squalius pyrenaicus*(320)
Carassius auratus(1,539)	*Fundulus heteroclitus*(570)	*Oncorhynchus masou*(414)	*Oreochromis mossambicus*(389)	*Brevoortia tyrannus*(320)
Gasterosteus aculeatus(1,503)	*Poecilia reticulata*(529)	*Galaxias gollumoides*(410)	*Etheostoma rufilineatum*(383)	*Salmoniformes sp.*(316)
Osmerus mordax(,1493)	*Cynoglossus semilaevis*(519)	*Misgurnus anguillicaudatus*(400)	*Opsariichthys bidens*(354)	*Etheostoma caeruleum*(315)

Other taxa about 168,000

perceived quality. In the present work, we address the application of proteomics to the examination of the quality aspects inherent to the organism and already present before its harvesting and those aspects acquired or modified after its harvesting.

Quality aspects inherent to the organism would include its species, production method (wild, farmed), and food safety aspects related to exposure to environmental contaminants. Quality aspects acquired or modified after harvesting would include post-mortem changes and sensory attributes, the effect of processing and storage (cold storage, freezing, smoking), and food safety aspects related to post-harvest contaminations, mainly those of microbial origin.

10.2.1 Pre-mortem Quality and Safety Aspects of Seafood

10.2.1.1 Species and Size

Species and size are two main parameters determining the quality and price of seafood. There is an entire chapter of this book on the issue of species identification, accordingly we only mention here that in the 2DE analysis of muscle extracts, the patterns of myosin light chains (Ochiai et al. 1990; Martinez et al. 1990) and of sarcoplasmic proteins (Piñeiro et al. 1998, 2001) are always species-specific. However, given that myosin light chains and sarcoplasmic proteins are relatively abundant in the white muscle, it would be highly desirable to develop a method that, bypassing the 2DE step, would allow a direct sequencing and identification of the sample.

10.2.1.2 Production Method, Crowding, and Feeds

A second perceived quality factor is the production method, that is, whether the fish is farmed or wild, a subject which is also comprehensively dealt in another chapter of this book. Monti and co-workers (2005) showed differences between the peptide profiles of wild and farmed sea bass, especially those of some glycolytic enzymes and the parvalbumin fraction, supporting the belief that farming practices affect muscle composition (Monti et al. 2005; Eriksson and Fenyö 2005). These results were further confirmed in cod by 2DE analysis: farmed cod muscle seemed to display a different protein expression and/or a different post-mortem degradation pattern than wild cod, which was attributed to stress during cultivation, differences in post-mortem muscle conditions (e.g., pH), and/or to qualitative and quantitative differences in the expression or regulation of proteases with a role in post-mortem muscle tenderization (Martinez et al. 2007).

The muscle proteome of wild and farmed gilthead sea bream was mapped by Addis et al. (2010) who found, as expected, that the protein expression pattern in muscle was more stable than in liver (which is a more dynamic tissue) and, more interestingly, that the protein expression profiles of muscle tissue of wild and maricultured gilthead sea bream of commercial size were comparable, indicating that offshore farming in floating cages favors proper muscle tissue development and the production of high-quality fish.

Animal welfare is a highly relevant topic involving breeding and slaughtering. Farmed rainbow trout subjected to intense pre-slaughter activity displayed differences on the levels of proteins involved in energy-producing pathways and of structural proteins. Desmin in particular showed consistently lower levels in the muscle of stressed fish (Morzel et al. 2006). Regardless of whether the reduced relative amount of desmin is due to proteolysis, denaturation, or a combination of both processes, it is bound to have a negative effect on the integrity and texture of the trout muscle inasmuch as it is one of the main constituents of the cytoskeleton in muscle cells.

Veiseth-Kent et al. (2010) showed that pre-slaughter crowding in Atlantic salmon induced classic signs of both primary and secondary stress responses. It was accompanied by alterations in the amounts of 27 proteins in muscle and of 17 proteins in blood plasma, all of them involved in secondary and tertiary stress responses, including altered energy metabolism, osmotic regulation, and immune function. The main changes in the muscle of crowded salmon seemed to be increased proteolysis and/or dissociation of structural proteins (actin, myosin heavy and light chains, tropomyosin) and increased levels of enzymes involved in anaerobic energy production (creatine kinase, enolase, phosphoglycerate kinase). Apolipoprotein A-I decreased in blood plasma and the angiotensinogen complement component C3 increased. These results indicate that short-term exposure to crowding may induce changes in the immune system of salmon and explain the mechanisms causing an accelerated muscle pH decline and rigor mortis contraction.

Ochiai (2010) addressed a serious quality problem, known as "burnt meat," in the highly appreciated flesh of bluefin tuna by 2DE. He showed a high degree of protein aggregation and decomposition and mainly the absence of creatine kinase in the portion of burn muscle, in both farmed and wild specimens. "Burnt" meat lacks the characteristic bright red meat color, it has a more watery, softer texture, and it is often seen in tuna and mackerel under stress conditions when fish are caught during the spawning period in summer (Yamashita 2010). Yamashita and co-workers (2010) attributed this oxidative stress to selenium deficiency and hypoxia and indicated the relevance of the recently discovered selenoneine (with a high antioxidant ability to bind to heme proteins and protecting them from iron auto-oxidation; see Yamashita and Yamashita 2010) to prevent this quality flaw.

The amount and composition of the diet are main determinants of seafood quality. The traditional source of protein and fat in the manufacture of fish feeds has been pelagic fish stocks. Nowadays, the fact that most of these are overexploited together with ethical issues regarding the use of high-value lipid and protein sources for animal feeds rather than human food, has forced feed producers to look for alternative and more acceptable sources of nutrients most of which are of vegetable origin. Given the relevance of the liver to the general fish metabolism, most of the studies on the effect of alternative feed ingredients on fish growth have targeted this organ instead of the edible muscle tissue.

Martin et al. (2003) published a very interesting proteomics work on the effect of using soy as a protein source on the 2DE protein profile of rainbow trout liver. Fish fed diets with higher levels of soy meal displayed higher protein consumption and protein synthesis rates, increased ammonia excretion, increased activities of hepatic glutamate dehydrogenase and aspartate amino transferase, and lower efficiency of retention of synthesized protein (Martin et al. 2003). Of the approximately 800 liver protein spots whose expression pattern was examined, 33 were found to be differentially expressed according to the feed. Two structural proteins, keratin II and β-tubulin, were downregulated in fish fed the high-soy diet, suggesting an increased requirement for energy metabolism and therefore lower energy available to synthesize structural proteins in these fish. Several heat shock proteins, including at least two chaperones (HSP70 and HSP78) were also downregulated in fish fed high-soy diets. Additional differences seemed to indicate an immune response and increased emphasis on catabolism relative to anabolism in the fish fed the high-soy diet. Interestingly, a hepatic selenium-binding protein identified as a potential biomarker in this work has been previously found to increase in the presence of aryl hydrocarbons (Ishida et al. 2002). As already mentioned above, Se-containing proteins seem to be of relevance to fight oxidative stress. The authors attributed the cause for the observed altered metabolism to the copurification of antinutritional factors from soy, such as phytoestrogens, antigenic agents together with the soy meal. On the other hand, there were no apparent differences between the liver proteomes of salmon fed GM versus non-GM soy-based diets (Nini et al. 2010).

10.2.1.3 Seafood Safety: Biomarkers of Exposure to Environmental Contaminants

Classical analytical techniques used to detect many of the relevant environmental contaminants, such as polycyclic aromatic hydrocarbons (PAH), polychlorinated biphenyls (PCBs), dioxins, heavy metals, and their species (methyl mercury, arsenicum, cadmium, etc.) are time consuming, expensive, and require specialized equipment and trained personnel. Food safety and environmental monitoring are currently targeting the identification of biomarkers in order to map the exposure to contaminants and the effect that such exposure has at the molecular, tissue, and organism levels. Identification of biomarkers is obviously of high value not only to food safety but also to environmental monitoring, as indicated by Miracle and Ankley (2005) in their review about the applications of -omics techniques in the field of ecotoxicogenomics.

The effect on the proteome of juvenile Atlantic cod plasma to the exposure to crude North Sea oil spiked with alkyl phenols and polycyclic aromatic hydrocarbons was investigated by Bohne-Kjersem et al. (2009) who identified 137 proteins that were differentially expressed depending on the levels of crude oil used. The most noticeable changes, many of which occurred even at low levels, indicated alterations of fibrinolysis and the complement cascade, the immune system, fertility-linked proteins, bone resorption, fatty acid metabolism, increased oxidative stress, impaired cell mobility, and increased levels of proteins associated with apoptosis. The following proteins were identified as potential biomarkers: alpha enolase, plasminogen, alpha-1-antitrypsin, alpha-2-macroglobulin, alpha-2- anti-plasmin, prothrombin, pentraxin, tropomyosin, serotransferrin, hemopexin, Fetuin B, apolipoprotein B, and NTPase. Not surprisingly, many of the responses seemed to be linked to each other, which made the authors suggest that the use of an array of these biomarker candidates would give a better indication of adverse effects induced in the fish by oil and/or produced water compared to single biomarkers alone (Bohne-Kjersem et al. 2009).

The same research group (Meier et al. 2010) examined the response in embryos, larvae, and juvenile fish to the complex chemical mixture found in real produce water containing dispersed oil, metals, alkylphenols (APs), polycyclic aromatic hydrocarbons (PAHs), and other chemicals. In general, APs bioconcentrate in fish tissue in a dose and developmental stage-dependent manner during PW exposure and cod exposed to 1% produce water (but not to either 0.1% or 0.01%) had significantly higher levels of the biomarkers vitellogenin and CYP1A in plasma and liver, respectively (Meier et al. 2010). However, continuous exposure from egg to fry in cod to different levels of North Sea-produced water, induced large changes in the proteome also at the lower levels of 0.01% and 0.1% produce water (Bohne-Kjersem et al. 2010): the expression and modifications of myosin heavy chain were affected and those of alpha-actin, alpha-actinin, keratin K8b (S2), and cytokeratin 4 (Krt4), Hsc71 (a heat shock protein belonging to the Hsp70 family) were downregulated. In muscle, exposure changed the levels of myosin, alpha-actin, and alpha-actinin suggesting an impact of produced water on the fast skeletal muscle development, essential for somatic growth, potentially impairing general growth and development of cod fry. Furthermore, the downregulation of keratin suggests an effect on tissue integrity. In summary, myosin heavy chain, fast skeletal muscle alpha-actin, Hsc71, alpha-actinin, ATP synthase, and keratin were identified as potential biomarker candidates of the effects of produced water and 17 β-oestradiol on cod fry. These responses reflect a potential impairment of the general growth and development of cod fry with the consequent value as indicators for lowered muscle quality and exposure to safety risks. These results are of obvious interest and should be followed to document the effect of environmental contaminants on market-size fish muscle protein makeup and properties.

Surface-enhanced laser desorption/ionization (SELDI) proteomics together with genomics (heterologous cDNA arrays) were used to investigate the integrated response of rainbow trout gills to sublethal concentrations of zinc for up to 6 days' exposure (Hogstrand et al. 2002): seven proteins were unique to zinc exposure whereas four others were suppressed. A spot that remained unidentified, but suspected to be metallothionein (a low molecular weight, metal-binding peptide inducible upon exposure to metals) was upregulated. Other changes seemed to lead to promoting glycolysis and stimulating the cellular production of energy. The proteomic studies also showed an activation of an inflammatory response during zinc exposure, thus confirming the postulated immunomodulatory role for zinc, which apparently protects rainbow trout against cadmium and mercury-induced immunotoxicity (Sanchez-Dardon et al. 1999).

The effect of exposure to increasing levels of cadmium on the proteome of bastard halibut's brain was examined by Zhu et al. (2006). Among the 24 proteins identified on a 2-D-PAGE gel, 9 demonstrated a synchronous response to acute cadmium exposure, suggesting that they might represent a biomarker profile. The changes affected both cytoplasmic and mitochondrial proteins; thus creatine kinase, transcriptional regulator, and endoglucanase were upregulated and actin 1, a putative xylose repressor, transferrin, and a dehydrogenase were downregulated. The authors propose an important role for transferrine in this system as a biomarker and a key molecule in cadmium detoxification.

In summary, exposure to environmental contaminants and toxic metals induces changes in the proteome of the fish tissues examined consistent with altered metabolism, increased glycolysis and oxidative stress, possible degradation of structural proteins (myosin, actin, desmin, keratins), and upregulation of proteins involved in detoxification mechanisms (transferring, HSPs, metallothioneins). All these changes have a negative effect on the quality of the fish (mostly shown by the degradation of structural proteins) and have a clear value as seafood quality and safety indicators.

10.2.2 Quality Aspects Acquired or Modified After Harvesting

The level of denaturation of proteins in seafood has been reported to be closely related to the sensory and technological properties of the respective products (Piñeiro et al. 2003). Post-mortem changes and spoilage are the result of mainly three basic mechanisms: enzymatic autolysis, oxidation, and microbial growth. Low-temperature storage and the addition of chemicals are the methods most commonly used by the industry today to control the levels of water activity and enzymatic and oxidative reactions in order to delay product deterioration and microbial growth (Ghaly et al. 2010).

The electrophoretic and mass spectroscopic techniques used to investigate the mechanisms of post-translational protein modifications have proven useful to follow up protein modifications during food production and storage and to elucidate the relationship between the nature and structure of proteins (i.e., protein size, amino acid composition, and sequence) and their functionality (Piñeiro et al. 2003). It has become clear that protein functionality exerts a significant effect on the physical and sensory properties of seafood, highlighting the suitability of investigating relevant proteome changes in seafood products for which two-dimensional polyacrylamide gel electrophoresis (2DE-PAGE) is still the most popular technique (Piñeiro et al. 2003).

10.2.2.1 Post-mortem Changes and Sensory Attributes

Washing, bleeding, and eviscerating (in the case of fish), temperature, and degree of handling are the most relevant factors influencing the quality of seafood

(Terlouw et al. 2008; Borderías and Sánchez-Alonso 2011). The initial biochemical changes that take place in the edible tissues are due to the enzymatic breakdown of major molecules and the level of post-mortem autolysis in the muscle (with the corresponding degradation of extra- and intracellular structural proteins and variation in the composition of peptides, amino acids, and small water molecules) is a major determinant of the texture, flavor, and odor of the final products (FAO 2005). During the last few years, proteomics have been used to further understand these processes, especially in meat, where the identification of naturally generated small peptides derived from myofibrillar proteins such as myosin light chain 1, titin, and actin, have expanded the knowledge of the different proteases that influence food properties (Bendixen et al. 2011).

As far as we are aware, proteomics methodologies were initially applied to assess how variations in processing conditions during the manufacture of surimi made from pre-rigor and post-rigor cod affected the composition of the final product (Martinez et al. 1992). The results indicated an increase in protein degradation with decreasing freshness and upon the addition of Ca^{2+} salts to post-rigor, but not to pre-rigor cod, valuable information to have in order to optimize the processing parameters according to the quality of the raw material and the desired texture of the final products.

Verrez-Bagnis et al. (2001) identified a 16 kDa protein whose disappearance from the proteome of sea bass muscle with increasing post-mortem storage time indicated its potential as a biomarker for freshness in that species. In cod, Kjærsgård and Jessen (2003) identified significant changes in 11 protein spots of the partial muscle proteome (p*I* 3.5–8.0, MW 13–35 kDa): for 8 of the 9 spots whose intensity increased during the first 8 days of ice storage the increase was already significant within the first 2 h post mortem, and 2 other spots whose intensity decreased only displayed significant changes after 8 days, indicating that different biochemical processes are involved in the post-mortem changes of cod muscle.

Several gel electrophoresis and mass spectrometric studies of post-mortem changes in sea bass muscle stored at 18°C and 1°C for 5 days showed a temperature-dependent increase in proteolysis that mainly affects the myosin heavy chain and the glycolytic enzyme glyceraldehyde-3-phosphate dehydrogenase (Terova et al. 2011). Interestingly the same study shows a rapid and significant decrease in the abundance of nucleoside diphosphate kinase B and phosphoglycerate mutase 2 that appears to be temperature-independent (Terova et al. 2011). A similar preliminary study of post-mortem deterioration in ice-stored sea bream by Schiavone et al. (2008) showed that α-actin and tropomyosin were among the most stable proteins during 6 days of storage, whereas myosin light 3 (MLC3) and major histocompatibility complex Class II beta 1 proteins increased and Sec 13-like and parvalbumin significantly decreased. The relevance of these results must await the understanding of the causes for these changes. For example, an increase in the intensity of the MLC3 spot would agree with an increase in the liberation of myosin from the rigor actomyosin complex and subsequent degradation of the myosin heavy chain, which will increase the level of free MLCs. Rigor actomyosin and the intact myosin molecule are difficult to solubilize and hardly enter the 2-D gels, but MLCs are easily solubilized and visualized. These results confirm the higher susceptibility of myosin heavy chain and the stability of the light chains to proteolytic degradation shown by Martinez (1992) in frozen stored extracts of isolated actomyosin.

One of the characteristics of the proteome is that it is dynamic and one must understand the causes for the changes before these methods may be fully useful to the aquaculture and fisheries industries. The above-mentioned study by Terova et al. (2011) showed a decrease in the intensity of the parvalbumin spot upon post-mortem storage. Taking into account that parvalbumin is one of the major fish allergens, this result may confirm a previous work by Dory et al. (1998) who showed that both the number of IgE-reactive bands and the intensity of the reaction in cod extracts was greater if the fish had been stored for several days, than if the extracts had been obtained from post-rigor fish immediately after rigor mortis resolution, suggesting not a loss of allergens but their aggregation and increased allergenicity. This stresses the need to understand the reasons for the variation in intensity of the protein spots: although the formation of aggregates of parvalbumins will also induce the loss of the characteristic low molecular mass spot of these proteins, the implications for the quality and safety of the fish muscle would be opposite if the diminishing of this spot were due to the disappearance of the allergen or to its aggregation and consequent increase in potential allergenicity.

Post-mortem pH is a highly relevant variable that would affect the quality and composition of fish as well as its proteome and yet it has been largely ignored in proteome studies of seafood. It is known that post-mortem pH varies depending, for example, on the condition of the fish and the degree of stress prior to death (Roth et al. 2012). Interestingly, the only work examining the effect of the pH on post-mortem proteolytic activities showed the great impact of this variable (Wang et al. 2011). Although that work did not use proteomics, it clearly demonstrated that different proteases are active depending on the post-mortem pH, a fact illustrated by the different pattern of degradation of the myosin heavy chain. This means that the post-mortem proteome for each species will vary according to the final post-rigor pH value and that the post-mortem pH value will have more relevance on the proteome than the storage temperature, because although the changes are usually temperature-dependent in that increased storage temperature accelerates the changes, variations in the pH value will cause the appearance of different spots (i.e., originate a different proteome), due to the preferential action of different proteases.

10.2.2.2 Processing and Storage

There is a scarcity of proteomic studies addressing the modifications suffered by seafood proteins during their processing and storage, and most of them refer to chilled and frozen species of commercial value such as pufferfish, Atlantic cod, Atlantic salmon, rainbow trout, and skipjack tuna muscle. Thus, Lu and coworkers (2010) identified 21 proteins in the skeletal muscle of Japanese pufferfish using 2DE and MALDI-TOF/TOF MS with different and well-known cellular functions, and Gebriel et al. (2010) used 1D-PAGE, nanoliquid chromatography peptide fractionation, and LTQ-MS in order to achieve an Atlantic cod muscle proteome catalogue.

Specific protein modifications during processing and storage as well as the effect of lactic fermentation with *Lactobacillus* starters in Atlantic salmon muscle were studied using 2DE (Morzel et al. 2000). Their results showed that the main quantitative protein changes were due to endogenous enzymes and affected the acidic range of proteins, whereas alkaline proteins and tropomyosin were more susceptible to microbial proteases.

The production of low molecular weight peptides (<5 kDa) in post-mortem rainbow trout muscle stored in ice and their stability during cooking were studied by Bauchart et al. (2007). Post-mortem proteolysis in muscle is an important factor affecting fish texture, and low molecular weight peptides affect the taste of the products. The main peptides in trout muscle were anserine and glutathione whose concentration was almost unaffected by the 7 days of ice storage and vacuum cooking for 5 min at 70°C. MS analysis revealed the highly reproducible appearance of a limited number of not yet characterized small peptides following the treatment.

Kjærsgård and coworkers (2006a) described changes in the proteome of farmed Atlantic cod muscle stored under different cold and frozen storage temperatures using 2DE and ESI-MS/MS. Their results showed the main changes in myosin light chain, triose-phosphate isomerase, glyceraldehyde-3-phosphate dehydrogenase, aldolase A, two α-actin fragments, and nuclease diphosphate kinase B, that seemed to keep a closer relationship to the length of freezing storage than to the storage temperature.

The same research group (Kjærsgård et al. 2006b) documented how freezing storage induced an increase in the level of protein oxidation in rainbow trout muscle. These authors identified several carbonylated proteins by LC-MS/MS, including nucleoside diphosphate kinase, adenylate kinase, pyruvate kinase, actin, creatine kinase, tropomyosin, myosin light chains 1 and 2, and myosin heavy chain, and showed a reduced solubility of nucleoside diphosphate kinase in fish stored at −20°C for 2 years compared to fish stored at −80°C. The authors observed that some proteins seemed to be more susceptible to oxidation than others, and attributed these differences to their abundance, cellular localization, amino acid sequence, or biochemical function, confirming earlier findings by the same research group (Kjærsgård and Jessen 2004).

Kinoshita and coworkers (2007) identified several spots by 2DE and MALDI-TOF/TOF that corresponded to oxidized peptides in brine-frozen bonito (*Katsuwonus pelamis*) muscle defrosted and stored at 15°C for 4 days. Sequencing of five of the oxidized spots led to the identification of four as: enolase 3, aldolase, aldolase A, and L-lactate dehydrogenase A chain.

Finally, because an important source of free radicals taking part in protein oxidation is the Fenton reaction, dependent on ferrous ions present in the tissue, Pazos et al. (2011) investigated the susceptibility of sarcoplasmic and myofibrillar cod muscle proteins to in vitro catalyzed oxidation, in an attempt to identify candidates that might play a major role in the deterioration of fish quality. After 2DE separation, selected proteins whose carbonyl groups had been labeled by fluorescein-5-thiosemicarbazide (FTSC) were identified by MALDI-TOF/TOF mass spectrometry. The most vulnerable proteins to ferrous-catalyzed oxidation seemed to be lactate dehydrogenase, triosephosphate isomerase, creatine kinase, enolase, glyceraldehyde 3-phosphate dehydrogenase, nucleoside diphosphate kinase B (NDK), and phosphoglycerate mutase. Different isoforms of the last three proteins showed different susceptibilities to metal-catalyzed oxidation, indicating that post-translational modifications may change the resistance of proteins to oxidative damage. In addition, the Fe (II)/ascorbate treatment significantly increased carbonylation of mainly actin and myosin, some of whose degradation products exhibited increased carbonylation levels.

10.2.2.3 Post-harvest Food Safety: Microbial and Other Contaminations

Microbial growth is a major cause of fish spoilage resulting in the production of biogenic amines such as putrescine, histamine, and cadaverine, organic acids, sulphides, alcohols, aldehydes, and ketones with unpleasant and unacceptable off-flavors (Ghaly et al. 2010). Fresh fish stored at ambient temperature are usually spoiled by Gram-negative, fermentative bacteria, whereas chilled fish spoil under the action of psychrotolerant Gram-negative bacteria, such as *Pseudomonas* spp. and *Shewanella* spp. (Gram and Huss 2000).

Proteomics allow addressing the molecular processes underlying the physiological behavior of pathogens in food matrices, a major challenge in food microbiology. For that, it is necessary to identify both the micro-organisms present in the food and the food components that are relevant in determining the microbial stability of such foods. The latter range from small molecules (including taste compounds and food preservatives) to the macro-ingredients such as proteins, carbohydrate polymers, and fats (Havelaar et al. 2010).

Since the first bacterial genome of *Haemophilus influenza* was described by Fleischmann et al. in 1995, top-down or bottom-up proteomics approaches have allowed the fast and sensitive characterization of individual micro-organisms in mixtures (Welker 2011). Nowadays, genome sequences are available for many food-borne micro-organisms (Abee et al. 2004) as evidenced by the number of strains that are currently sequenced as reviewed by O'Flaherty and Klaenhammer (2011).

There are many studies from the early years of proteomics applying these tools to bacterial classification and to the study of common pathogenic seafood bacteria such as *Listeria, Campilobacter,* or *Staphylococcus* (Piñeiro et al. 2010; Forné et al. 2010) both under refrigeration and after high-pressure preservation procedures (Cacace et al. 2010; Bièche et al. 2011).

However, proteomics have only recently been applied to the characterization of how processing may affect the virulence of microbial pathogens: Guilbaud et al. (2008) showed that liquid smoke affected the proteomic pattern of *Listeria monocytogenes,* decreased its growth and survival, and inhibited its hemolytic potential without affecting the *hly* gene expression. These results help us understand why in spite of the apparently high prevalence of *L. monocytogenes*, particularly in lightly preserved smoked seafood, these products are rarely linked to listeriosis and support the attenuated infectious potential of *L. monocytogenes* strains isolated from the smoked salmon industry indicated by Norton et al. (2001) and Gudmundsdóttir et al. (2006).

Hazen et al. (2009) used whole-cell MALDI-TOF MS analysis to identify species and strains of *Vibrio parahaemolyticus* isolated from different geographical locations and at different times. *V. parahaemolyticus* is the leading causative agent of bacterial seafood-borne gastroenteritis in the United States, and this technique permitted the fast identification of the pathogen and its clear differentiation from the closely related *V. alginolyticus*, *V. harveyi*, and *V. campbellii.*

Böhme et al. (2010a, b) used a special extraction protocol from intact microbial cells together with MALDI-TOF-MS techniques to identify the main 26 species of seafood spoilage and pathogenic Gram-negative bacteria, including *Aeromonas hydrophila, Acinetobacter baumanii, Pseudomonas* spp., and *Enterobacter* spp. They also constructed a reference library containing the spectral fingerprints of 32 Gram-positive reference strains, including *Bacillus* spp., *Listeria* spp., *Clostridium* spp., *Staphylococcus* spp., and *Carnobacterium* spp. (Böhme et al. 2011a) and identified a variety of seafood pathogens such as *Stenotrophomonas maltophilia*, *Proteus vulgaris*, *Pseudomonas syringae*, *Pseudomonas fluorescens*, *Pseudomonas fragi*, *Bacillus cereus*, *Bacillus subtilis,* and *Serratia marcescens*, in iced and vacuum-packed mild heat-treated seafoods (Böhme et al. 2011b). The same group has recently published the first report on the identification of *Streptococcus parauberis* in seafood in general and in vacuum-packed food products in particular using the same methodology (Fernández-No et al. 2012).

Mass spectrometry (MS) is an instrumental technique that permits the identification of all types of compounds in food toxicology including seafood (Malik et al. 2010). Fernandez-No and coworkers (2011) used MS to isolate and identify the main histamine-producing bacteria from farmed turbot and blackspot sea bream and histamine production was measured by HPLC. The study revealed *Pseudomonas fragi* and *Pseudomonas syringae* to be the major histamine-forming bacteria present in farmed turbot and blackspot sea bream, respectively. Self and coworkers (2011) developed a method for the extraction of agmatine, cadaverine, histamine, phenyethylamine, putrescine, tryptamine, tyramine, and urocanic acid from canned tuna and frozen tuna loin matrices by matrix solid-phase dispersion, followed by separation and quantification of these compounds by ultrahigh-performance hydrophilic interaction chromatography (UHPLC-HILIC) with orbitrap mass spectrometric detection. This streamlined approach eliminates the need for derivatization, which has been the traditional option for liquid or gas chromatographic analysis for many of these compounds.

10.3 Challenges

The relevance of high-throughput proteomic approaches is to increase insight and understanding of how the physiology of the organism, breeding, harvesting, storage, and processing methods affect the quality and safety of ready-to eat seafood. Such understanding can then be used to optimize the entire production chain from the water to the table: from selection and breeding to processing and cooking.

There are, however, still some pitfalls associated with the techniques themselves and on how to deal best with the tremendous amount of data they generate (Pedreschi et al. 2010). Some of the limitations to the application of proteomics technologies derive from the instability of proteomes, the large number of proteins, and possible post-translational modifications, the limited detection of low abundance and highly acidic or basic proteins by 2-D gel electrophoresis, limited reproducibility, and the fact that not all proteins in a sample can be identified (Van Vliet 2011). Thus there is an urgent demand to develop sophisticated, robust, and fast analytical methodologies, capable of identifying in parallel hundreds of seafood proteins, with a variety of post-translational modifications at different expression levels (Zhang et al. 2008).

According to recent publications, the laser capture microdissection approach has been successfully applied in fish immune response studies (Prunet et al. 2012), but the use of this technique often implies the recovery of only small amounts of biological material for further proteomics analysis. Small sample size makes it difficult to detect differentially expressed proteins. Perhaps the introduction of lab-on-a-chip devices for sample processing where protein separation and trypsin digestion can take place, and directly coupling these chips to mass spectrometers, could be a good solution for the near future but it is currently a challenge (Forné et al. 2010).

Optimally analytical proteomic techniques should be combined with other "omic" approaches, particularly transcriptomic and metabolomic information together with the indispensable help of bioinformatic tools (Cifuentes et al. 2011) to map and improve seafood production, quality, and safety.

On the other hand, the high biodiversity and shortage of sequences for the large number of organisms used as seafood makes it difficult to undertake -omics strategies in the short term. There is a need to begin the compilation of useful data obtained after fully controlling all the significant variables involved, starting with the selection of relevant species on which to perform the studies under strictly controlled conditions (environmental conditions, nutrition, age, sex, stressors, contaminants, etc.). However, the results from descriptive punctual analyses without knowledge of the organism's prior history would be of only limited value and likely difficult to reproduce.

Given that most current fish farming practices seek to improve flesh quality together with the notion that white muscle is the main product for the fishing industry, a compilation of data about white muscle structure, function, and ontogeny followed by an account of the changes induced by the environment, feed, contaminants, stressors, swimming behavior, and performance related to the use of white muscle during growth from larva to adult, would be essential to fully understand the results of proteomic and other -omic studies (Videler 2011) and to devise strategies to optimize seafood breeding, harvesting, and processing.

References

Abee T, Van Schaik W, Siezen RJ (2004) Impact of genomics on microbial food safety. Trends Biotechnol 22:653–660

Addis MF, Cappuccinelli R, Tedde V, Pagnozzi D, Porcu MC, Bonaglini E, Roggio T, Uzzau S (2010) Proteomic analysis of muscle tissue from gilthead sea bream (*Sparus aurata*, L.) farmed in offshore floating cages. Aquaculture 309(1):245–252

Bauchart C, Chambon C, Mirand PP, Savary-Auzeloux I, Rémond D, Morzel M (2007) Peptides in rainbow trout (*Oncorhynchus mykiss*) muscle subjected to ice storage and cooking. Food Chem 100:1566–1572

Bendixen E, Danielsen M, Hollung K, Gianazza E, Miller I (2011) Farm animal proteomics – a review. J Proteomics 74:282–293

Bièche C, de Lamballerie M, Chevret D, Federighi M, Tresse O (2011) Dynamic proteome changes in *Campylobacter jejuni* 81-176 after high pressure shock and subsequent recovery. Ann NY Acad Sci 1189:133–138

Böhme K, Fernández-No IC, Barros-Velázquez J, Gallardo JM, Cañas B, Calo-Mata P (2010a) Comparative analysis of protein extraction methods for the identification of seafood-borne pathogenic and spoilage bacteria by MALDI-TOF mass spectrometry. Anal Method 2:1941–1947

Böhme K, Fernández-No IC, Barros-Velázquez J, Gallardo JM, Calo-Mata P, Cañas B (2010b) Species differentiation of seafood spoilage and pathogenic Gram-negative bacteria by MALDI-TOF mass fingerprinting. J Proteome Res 9:3169–3183

Böhme K, Fernández-No IC, Barros-Velázquez J, Gallardo JM, Cañas B, Calo-Mata P (2011a) Rapid species identification of seafood spoilage and pathogenic Gram-positive bacteria by MALDI-TOF mass fingerprinting. Electrophoresis 32:2951–2965

Böhme K, Fernández-No IC, Gallardo JM, Cañas B, Calo-Mata P (2011b) Safety assessment of fresh and processed seafood products by MALDI-TOF mass fingerprinting. Food Bioprocess Technol 4:907–918

Bohne-Kjersem A, Skadsheim A, Goksøyr A, Grøsvik BE (2009) Candidate biomarker discovery in plasma of juvenile cod (*Gadus morhua*) exposed to crude North Sea oil, alkyl phenols and polycyclic aromatic hydrocarbons (PAHs). Mar Environ Res 68:268–277

Bohne-Kjersem A, Bache N, Meier S, Nyhammer G, Roepstorff P, Sæle Ø, Goksøyr A, Grøsvik BE (2010) Biomarker candidate discovery in Atlantic cod *Gadus morhua* continuously exposed to North Sea produced water from egg to fry. Aquat Toxicol 96:280–289

Borderías AJ, Sánchez-Alonso I (2011) First processing steps and the quality of wild and farmed fish. J Food Sci 76:1–5

Cacace G, Mazzeo MF, Sorrentino A, Spada V, Malorni A, Siciliano RA (2010) Proteomics for the elucidation of cold adaptation mechanisms in Listeria monocytogenes. J Proteomics 73:2021–2030

Cifuentes A, Dugo P, Fanali S (2011) Advances in food analysis. J Chromatogr A 1218:7385

Dory D, Chopin C, Aimone-Gasti I, Gueant JL, Sainte-Laudy J, Moneret-Vautrin DA, Fleurence J (1998) Recognition of an extensive range of IgE-reactive proteins in cod extract. Allergy 53:42–50

Eriksson J, Fenyö D (2005) Protein identification in complex mixtures. J Proteome Res 4:387–393

FAO (2005) Fisheries and aquaculture topics. Quality of fish and fish products. Topics fact sheets. Text by Lahsen Ababouch. In: FAO fisheries and aquaculture department [online]. Rome. Updated 27 May 2005

Fernández-No IC, Böhme K, Calo-Mata P, Barros-Velázquez J (2011) Characterisation of histamine-producing bacteria from farmed blackspot seabream *Pagellus bogaraveo* and turbot *Psetta maxima*. Int J Food Microbiol 151:182–189

Fernández-No IC, Böhme K, Calo-Mata P, Cañas B, Gallardo JM, Barros-Velázquez J (2012) Isolation and characterization of *Streptococcus parauberis* from vacuum-packaging refrigerated seafood products. Food Microbiol 30:91–97

Forné I, Abián J, Cerdà J (2010) Fish proteome analysis: model organisms and non-sequenced species. Proteomics 10:858–872

Gebriel M, Uleberg K, Larssen E, Hjelle Bjørnstad A, Sivertsvik M, Møller SG (2010) Cod (*Gadus morhua)* muscle proteome cataloging using 1D-PAGE protein separation, nano-liquid chromatography peptide fractionation and linear trap quadrupole LTQ mass spectrometry. J Agric Food Chem 58:12307–12312

Ghaly AE, Dave D, Budge S, Brooks MS (2010) Fish spoilage mechanisms and preservation techniques: review. Am J Appl Sci 7:846–864

Graham DRM, Elliott ST, Van Eyk JE (2005) Broad-based proteomic strategies: a practical guide to proteomics and functional screening. J Physiol 563:1–9

Gram L, Huss HH (2000) Fresh and processed fish and shellfish. In: Lund BM, Baird-Parker AC, Gould GW (eds) The microbiological safety and quality of foods. Chapman and Hall, London, pp 472–506

Grunert KG (2005) Food quality and safety: consumer perception and demand. Eur Rev Agric Econ 32:369–391

Gudmundsdóttir S, Roche SM, Kristinsson K, Kristjánsson M (2006) Virulence of *Listeria monocytogenes* isolates from humans and smoked salmon, peeled shrimp and their processing environments. J Food Prot 69:2157–2160

Guilbaud M, Chafsey I, Pilet M, Leroi F, Prévost H, Hébraud M, Dousset X (2008) Response of *Listeria monocytogenes* to liquid smoke. J Appl Microbiol 104:1744–1753

Havelaar AH, Brul S, de Jong A, de Jonge R, Zwietering MH, Ter Kuile BH (2010) Future challenges to microbial food safety. Int J Food Microbiol 139:S79–S94

Hazen TH, Martinez RJ, Chen Y, Lafon PC, Garrett NM, Parsons MB, Sobecky PA (2009) Rapid identification of *Vibrio parahaemolyticus* by whole-cell matrix-assisted laser desorption ionization-time of flight mass spectrometry. Appl Environ Microbiol 75:6745–6756

Herrero M, Simõ C, García-Cañas V, Ibáñez E, Cifuentes A (2012) Foodomics: MS-based strategies in modern food science and nutrition. Mass Spectrom Rev 31:49–69

Hocquette J, Richardson RI, Prache S, Medale F, Duffy G, Scollan ND (2005) The future trends for research on quality and safety of animal products. Ital J Anim Sci 4:49–72

Hogstrand C, Balesaria S, Glover CN (2002) Application of genomics and proteomics for study of the integrated response to zinc exposure in a non-model fish species, the rainbow trout. Comp Biochem Physiol 133B:523–535

Ishida T, Ishii Y, Yamada H, Oguri K (2002) The induction of hepatic selenium-binding protein by aryl hydrocarbon (Ah)-receptor ligands in rats. J Health Sci 48:62–68

Kinoshita Y, Sato T, Naitou H, Ohashi N, Kumazawa S (2007) Proteomic studies on protein oxidation in bonito (*Katsuwonus pelamis*) muscle. Food Sci Technol Res 13:133–138

Kjærsgård IVH, Jessen F (2003) Proteome analysis elucidating post-mortem changes in cod (*Gadus morhua*) muscle proteins. J Agric Food Chem 51:3985–3991

Kjærsgård IVH, Jessen F (2004) Oxidation of protein in rainbow trout muscle. In: Proceedings of the 34th WEFTA conference. 184. http://www.wefta.org

Kjærsgård IVH, Nørrelykke MR, Jessen F (2006a) Changes in cod muscle proteins during frozen storage revealed by proteome analysis and multivariate data analysis. Proteomics 6: 1606–1618

Kjærsgård IVH, Nørrelykke MR, Baron CP, Jessen F (2006b) Identification of carbonylated protein in frozen rainbow trout (*Oncorhynchus mykiss*) fillets and development of protein oxidation during frozen storage. J Agric Food Chem 54:9437–9446

Lu J, Zheng J, Liu H, Li J, Chen H, Chen K (2010) Protein profiling analysis of skeletal muscle of a pufferfish, *Takifugu rubripes*. Mol Biol Rep 37:2141–2147

Malik AK, Blasco C, Picó Y (2010) Liquid chromatography-mass spectrometry in food safety. J Chromatogr A 1217:4018–4040

Martin SAM, Vilhelmsson O, Médale F, Watt P, Kaushik S, Houlihan DB (2003) Proteomic sensitivity to dietary manipulations in rainbow trout. BBA 1651:17–29

Martinez I (1992) Fish myosin degradation upon storage. In: Huss HH, Jakobsen M, Liston J (eds) Quality assurance in the fish industry. Elsevier Science, Amsterdam, pp 389–397

Martinez I, Ofstad R, Olsen RL (1990) Myosin isoforms in red and white muscles of some teleost fishes. J Muscle Res Cell Motil 11:489–495

Martinez I, Christiansen JS, Ofstad R, Olsen RL (1991) Comparison of myosin isoenzymes present in skeletal and cardiac muscles of the Arctic charr *Salvelinus alpinus* (L.). Sequential expression of different myosin heavy chains during development of the fast white skeletal muscle. Eur J Biochem 195:743–753

Martinez I, Solberg C, Lauritzen K, Ofstad R (1992) Two-dimensional electrophoretic analyses of cod (*Gadus morhua*, L.) whole muscle proteins, water-soluble fraction and surimi. Effect of the addition of CaCl2 and MgCl2 during the washing procedure. Appl Theor Electrophor 2:201–206

Martinez I, Slizyte R, Dauksas E (2007) High resolution two-dimensional electrophoresis as a tool to differentiate wild from farmed cod (*Gadus morhua*) and to assess the protein composition of klipfish. Food Chem 101:1337–1343

Meier S, Morton HC, Nyhammer G, Grøsvik BE, Makhotin V, Geffen A, Boitsov S, Kvestad KA, Bohne-Kjersem A, Goksøyr A, Folkvord A, Klungsøyr J, Svardal A (2010) Development of Atlantic cod *(Gadus morhua)* exposed to produced water during early life stages: effects on embryos, larvae and juvenile fish. Mar Environ Res 70:383–394

Miracle AL, Ankley GT (2005) Ecotoxicogenomics: linkages between exposure and effects in assessing risks of aquatic contaminants to fish. Reprod Toxicol 19:321–326

Monti G, De Napoli L, Mainolfi P, Barone R, Guida M, Marino G, Amoresano A (2005) Monitoring food quality by microfluidic electrophoresis, gas chromatography and mass spectrometry techniques: effects of aquaculture on the sea bass (*Dicentrarchus labrax)*. Anal Chem 77:2587–2594

Morzel M, Verrez-Bagnis V, Arendt EK, Fleurence J (2000) Use of two-dimensional electrophoresis to evaluate proteolysis in salmon (*Salmo salar*) muscle as affected by a lactic fermentation. J Agric Food Chem 48:239–244

Morzel M, Chambon C, Lefèvre F, Paboeuf G, Laville E (2006) Modifications of trout (*Oncorhynchus mykiss*) muscle proteins by preslaughter activity. J Agric Food Chem 54:2997–3001

Nini H, Sissener NH, Martin SAM, Cash P, Hevrøy EM, Sanden M, Hemre GI (2010) Proteomic profiling of liver from Atlantic salmon (*Salmo salar*) fed genetically modified soy compared to the near-isogenic non-GM line. Mar Biotechnol 12:273–281

Norton DM, Scarlett JM, Horton K, Sue D, Thimothe J, Boor KJ, Wiedmann M (2001) Characterization and pathogenic potential of *Listeria monocytogenes* isolates from the smoked fish industry. Appl Environ Microbiol 67:646–653

O'Flaherty S, Klaenhammer TR (2011) The impact of omic technologies on the study of food microbes. Ann Rev Food Sci Technol 2:353–371

Ochiai Y (2010) Changes in quality and denaturation of sarcoplasmic protein components in the burnt meat of bluefin tuna (*Thunnus thynnus orientalis*). Nippon Suisan Gakkaishi Jpn Ed 76:695–704

Ochiai Y, Kobayashi T, Watabe S, Hashimoto K (1990) Mapping of fish myosin light chains by two-dimensional gel electrophoresis. Comp Biochem Physiol 95B:341–345

Pazos M, Da Rocha AP, Roepstorff P, Rogowska-Wrzesinska A (2011) Fish proteins as targets of ferrous-catalyzed oxidation: identification of protein carbonyls by fluorescent labelling on two-dimensional gels and MALDI-TOF/TOF mass spectrometry. J Agric Food Chem 59:7962–7977

Pedreschi R, Maarten H, Lilley KS, Bart N (2010) Proteomics for the food industry: opportunities and challenges. CRC Crit Rev Food Sci Nutr 50:680–692

Piñeiro C, Velázquez JB, Sotelo CG, Pérez-Martín RI, Gallardo JM (1998) Two-dimensional electrophoretic study of the water-soluble protein fraction in white muscle of Gadoid fish species. J Agric Food Chem 46:3991–3997

Piñeiro C, Vázquez J, Marina AI, Barros-Velázquez J, Gallardo JM (2001) Characterization and partial sequencing of species-specific sarcoplasmic polypeptides from commercial hake species by mass spectrometry following 2-DE analysis. Electrophoresis 22:1545–1552

Piñeiro C, Barros-Velázquez J, Vázquez J, Figueras A, Gallardo JM (2003) Proteomics as a tool for the investigation of seafood and other marine products. J Proteome Res 2:127–135

Piñeiro C, Cañas B, Carrera M (2010) The role of proteomics in the study of the influence of climate change on seafood products. Food Res Int 43:1791–1802

Pinstrup-Andersen P (2009) Food security. Definition and measurement. Food Secur 1:5–7
Pischetsrieder M, Baeuerlein R (2009) Proteome research in food science. Chem Soc Rev 38:2600–2608
Prunet P, Øverli Ø, Douxfils J, Bernardini G, Kestemont P, Baron C (2012) Fish welfare and genomics. Fish Physiol Biochem 38:43–60
Roth B, Grimsbø E, Slinde E, Foss A, Stien LH, Nortvedt R (2012) Crowding, pumping and stunning of Atlantic salmon, the subsequent effect on pH and rigor mortis. Aquaculture 326–329:178–180
Sanchez-Dardon J, Voccia I, Hontela A, Chilmonczyk S, Dunier M, Boermans H, Blakley B, Fournier M (1999) Immunomodulation by heavy metals tested individually or in mixtures in rainbow trout (*Oncorhynchus mykiss*) exposed in vivo. Environ Toxicol Chem 18:1492–1497
Santesmases M (2004) Marketing. Conceptos y Estrategias. Edit. Pirámide, S.A. Madrid. 1120 pp
Schiavone R, Zilli L, Storelli C, Vilella S (2008) Identification by proteome analysis of muscle proteins in sea bream (*Sparus aurata*). Eur Food Res Technol 227:1403–1410
Self RL, Wu W, Marks HS (2011) Simultaneous quantification of eight biogenic amine compounds in tuna by matrix solid-phase dispersion followed by HPLC-orbitrap mass spectrometry. J Agric Food Chem 59:5906–5913
Terlouw EMC, Arnould C, Auperin B, Berri C, Le Bihan-Duval E, Deiss V et al (2008) Pre-slaughter conditions, animal stress and welfare: current status and possible future research. Animal 2:1501–1517
Terova G, Addis MF, Preziosa E, Pisanu S, Pagnozzi D, Biosa G, Gomati R, Bernadini G, Roggio T, Saroglia M (2011) Effects of post mortem storage temperature on sea bass (*Dicentrarchus labrax*) muscle protein degradation: analysis by 2-D DIGE and MS. Proteomics 11:2901–2910
Van Vliet E (2011) Current standing and future prospects for the technologies proposed to transform toxicity testing in the 21st century. ALTEX 28:17–44
Veiseth-Kent E, Grove H, Færgestad EM, Fjæra SO (2010) Changes in muscle and blood plasma proteomes of Atlantic salmon (*Salmo salar*) induced by crowding. Aquaculture 309:272–279
Verdú AJ (2003) Una escala multi-ítem para la medición de la calidad percibida en alimentos y bebidas. Rev Eur Dirección y Econ Empresa 12:59–76
Verrez-Bagnis V, Ladrat C, Morzel M, Noël J, Fleurence J (2001) Protein changes in post mortem sea bass (*Dicentrarchus labrax*) muscle monitored by one- and two-dimensional gel electrophoresis. Electrophoresis 22:1539–1544
Videler J (2011) An opinion paper: emphasis on white muscle development and growth to improve farmed fish flesh quality. Fish Physiol Biochem 37:337–343
Wang PA, Vang B, Pedersen AM, Martinez I, Olsen RL (2011) Post-mortem degradation of myosin heavy chain in intact fish muscle: effects of pH and enzyme inhibitors. Food Chem 124:1090–1095
Watabe S, Hwang GC, Nakaya M, Guo XF, Okamoto Y (1992) Fast skeletal myosin isoforms in thermally acclimated carp. J Biochem 111:113–122
Welker M (2011) Proteomics for routine identification of microorganisms. Proteomics 11:3143–3153
Yaktine AL, Nesheim MC, James CA (2008) Nutrient and contaminant tradeoffs: exchanging meat, poultry, or seafood for dietary protein. Nutr Rev 66:113–122
Yamashita M (2010) Stress responses of fish during catching process. In: Konno K, Ochia Y, Fukuda Y (eds) Quality control of tuna meat by optimization of fishing and handling. Koseisha-Koseikaku, Tokyo, pp 81–94
Yamashita Y, Yamashita M (2010) Identification of a novel selenium-containing compound, selenoneine, as the predominant chemical form of organic selenium in the blood of bluefin tuna. J Biol Chem 285:18134–18138
Yamashita Y, Yabu T, Yamashita M (2010) Discovery of the strong antioxidant selenoneine in tuna and selenium redox metabolism. World J Biol Chem 1:144–150
Zhang XW, Yap Y, Wei D, Chen G, Chen F (2008) Novel omics technologies in nutrition research. Biotechnol Adv 26:169–176
Zhu JY, Huang HQ, Bao XD, Lin QM, Cai ZW (2006) Acute toxicity profile of cadmium revealed by proteomics in brain tissue of *Paralichthys olivaceus*: potential role of transferrin in cadmium toxicity. Aquat Toxicol 78:127–135

Chapter 11
Farmed and Wild Fish

Maria Filippa Addis

11.1 Aquaculture and Proteomics

Aquaculture, or the farming of aquatic animals in enclosures or confinements, is an ancient practice used by man to ensure a constant supply of fresh seafood. The origin of aquaculture is credited to Chinese populations, and flourished well before 1,000 BC. The earliest document describing aquaculture, *The Classic of Fish Culture* by Shan Li, dates back to 475 BC, and reports the structure of ponds, the method of propagation of the common carp, and the growth of fry. Other sources indicate the bas-relief found in the tomb of Akihetep in Egypt, dating back to 2,500 BC, as one of the most ancient pieces of evidence of this practice. Here, a man is depicted while gathering tilapia, catfish, and other fish from a square-edged enclosure; however, whether this may or may be not be a pond is still debated. Along the centuries, men have kept fish in ponds and enclosures both for religious purposes or as reservoirs of fresh seafood, but farming techniques did not evolve substantially until the latter half of the nineteenth century and the beginning of the twentieth, when marine fish farming started to gain interest. Then, in the late 1960s, with the advent of modern technologies and materials such as plastic and fiberglass, aquaculture underwent a massive expansion, giving rise to the current widespread and extensive market. New and more efficient techniques were developed, enabling large-scale farming of freshwater and saltwater fish, from inland tanks to off-shore floating cages, and many new species were brought into culture (C.E. Nash 2011). The extensive diffusion of aquaculture enterprises was also spurred by the oil crisis, that made fish capture less economical, by the declining fish population due to uncontrolled fishing, and by the considerable increase in the global population demand for aquatic food products. In this respect, aquaculture can significantly contribute to preservation of marine biodiversity, and

M.F. Addis (✉)
Porto Conte Ricerche, Srl, S.P. 55 Porto Conte-Capo Caccia, Km 8.400,
Loc. Tramariglio 07041, Alghero (SS), Italy
e-mail: addis@portocontericerche.it

F. Toldrá and L.M.L. Nollet (eds.), *Proteomics in Foods: Principles and Applications*, Food Microbiology and Food Safety 2, DOI 10.1007/978-1-4614-5626-1_11,

promote a more rational exploitation of marine resources. Currently, aquaculture contributes to about half of the fish consumed worldwide, reaching a production of over 50 million tons in 2006 (FAO, the state of world fisheries and aquaculture 2008) (www.FAO.org), with a growth higher than any other animal food-producing sector. Recently, driven also by the oncoming market saturation for some of the most popular species, fish farmers are increasingly interested in quality improvement and certification, and in attracting consumer interest in higher-level products. The evolution of aquaculture techniques and the considerable increase in our knowledge of fish physiology, health, and nutrition can contribute to the production of fish with higher quality, safer to eat, and nutritionally more similar to wild fish, and contribute to lowering the production costs for the farmer without compromising quality.

Sensorial, textural, and nutritional properties of farmed fish depend on many factors, and the influence of farming conditions on the quality of the final product is the subject of an active and lively field of research. During farming, fish are exposed to a plethora of stimuli and environmental constraints that differ from those of their free-ranging counterparts: diet, feeding regimens, physical activity, crowding, and water quality (salinity, temperature, currents, number and variety of microbial pathogens) can be quite different from those experienced in the wild. Moreover, their extent and impact can differ significantly depending on the fish species and the farming techniques, which range from inshore recirculated tanks, to offshore cages, to rearing in coastal lakes and lagoons in semi-wild conditions. Under a nutritional perspective, investigations of quantity and composition of fatty acids have played a major role in studies assessing the impact of aquaculture techniques on fish species of economical interest. In fact, these are raised on artificial diets and have frequent access to unlimited food sources with concomitant reduced physical activity, leading to an increased amount of fat in farmed as compared to wild fish, with negative consequences on sensorial qualities and shelf-life. However, the nutritional quality of farmed fish products is generally considered to be equally elevated due to the presence of polyunsaturated fatty acids, including omega 3, which are beneficial to the heart and the circulatory system of humans. As far as proteins are concerned, farmed fish are traditionally considered as having the same quantity and quality of protein as wild fish in terms of gross composition. Nevertheless, it is now well known that farming can modify the protein expression profiles of several tissues (Addis et al. 2010a; Gornati et al. 2004a, b; Martin et al. 2001, 2003; Forné et al. 2010; Hall et al. 2003; Martinez et al. 1995; Martinez and Pettersen 1992). Being able to track and detect variations in protein expression against similar genetic backgrounds, proteomics holds considerable promise in unveiling the alterations and differences occurring between wild and farmed fish, and in understanding the underlying physiological and pathological mechanisms. The knowledge gathered by proteomics has the potential to identify and help understand these factors, and it is hoped, to provide the means to control them, enabling production of higher quality fish (Forné et al. 2010; Melamed et al. 2002; Piñeiro et al. 2003).

This chapter provides an overview of the proteomic studies that have directly assessed the impact of farming on fish tissues, focusing on muscle, liver, blood serum, kidney, and testis proteomes. Table 11.1 provides an overview of the methods applied for assessing changes in protein expression, and Table 11.2 summarizes the most relevant findings that emerged from these studies.

Table 11.1 Overview of proteomic methods used for assessing expression changes induced by farming on fish tissues

Species	Factor	Lysis conditions	Buffer conditions	Separation	Identification	References
Muscle						
S. aurata	Farming	Scalpel fragmentation and homogenization by bead beating	8 M urea, 2% CHAPS and 0.5% ampholines	1DE: SDS-PAGE 2DE: 24 cm, IEF on pH 3–10 NL strips, SDS-PAGE on 10–18% gradient gels; Coomassie and MS-compatible silver	MALDI-MS and LC-MS/MS (ESI-Q-TOF)	Addis et al. 2010a
S. aurata	Farming	Mechanical homogenization of tissue	50 mM KCl, followed by extraction in 300 mM KCl, 0.1 mM DTT, 150 mM phosphate buffer, 5 mM mgATP, dialysis and resuspension in 500 mM KCl	1DE: SDS-PAGE 2DE: IEF on pH 3–10 rods, SDS-PAGE on 15% gels; Coomassie	Peptide mapping and Western immunoblotting	Carpené et al. 1998
G. morhua	Farming	Tissue scraping and homogenization in ice	ReadyPrep Sequential Extraction Kit (Bio-Rad)	2DE: IEF on 11 cm, pH 3–10 strips, SDS-PAGE on 12.5% gels; Silver	None	Martinez et al. 2007
D. labrax	Farming	Fragmentation and mincing with a Stomacher	Water	1DE: SDS-PAGE Microfluidic separation on an Agilent 2100 Bioanalyzer	LC-MS/MS (LCQ ion trap)	Monti et al. 2005
G. morhua	Farming	Sonication	30 mM HEPES, 5 mM K_2HPO_4, 0.25 mM sucrose, 0.5 mM EDTA, 0.5% ASB14 and protease inhibitor cocktail	2DE: IEF on 13 cm, pH 5–8 strips, SDS-PAGE on 12% gels; Coomassie	LC-MS/MS (ESI-Q-TOF)	Olsson et al. 2007
Liver						
O. mykiss	Starvation	Dounce homogenization	9 M urea, 2% CHAPS, 25 mM Tris HCl pH 7.5, 3 mM EDTA, 50 mM KCl, 50 mM DTT, 2% ampholines and protease inhibitors	2E: IEF on 7 cm 4–7 strips; SDS-PAGE on 10–15% gels; Coomassie	MALDI-MS	Martin et al. 2001

(continued)

Table 11.1 (continued)

Species	Factor	Lysis conditions	Buffer conditions	Separation	Identification	References
O. mykiss	Dietary substitution	Dounce homogenization	9 M urea, 2% CHAPS, 25 mM Tris HCl pH 7.5, 3 mM EDTA, 50 mM KCl, 50 mM DTT, 2% ampholines, protease inhibitors	2DE: IEF on 7 cm 4–7 strips; SDS-PAGE on 10–15% gels; Coomassie	MALDI-MS	Martin et al. 2003
S. aurata	Handling and crowding	Ultra-Turrax homogenization and sonication	7 M urea, 2 M thiourea, 4% CHAPS buffer, 0.3% DTT, 1% ampholines, protease inhibitors	2DE: IEF on 11 cm 3–10 strips; SDS-PAGE on 12% gels; Coomassie	LC-MS/MS (ESI-Ion trap)	Alves et al. 2010
S. aurata	Temperature	Pulverization in liquid nitrogen	7 M urea, 2 M thiourea, 2% CHAPS, 80 mM DTT, 50 mM Tris HCl, pH 7.4	2DE: IEF on 24 cm, pH 3–11 NL strips, SDS-PAGE on 12.5% gradient gels; Coomassie	MALDI-MS/MS and LC-MS/MS (CapLC-QTOF)	Ibarz et al. 2012
Serum						
P. fluviatilis	Domestication	Dilution in buffer	7 M urea, 2 M thiourea, 4% CHAPS, 1% DTT, 2% ampholines	2DE: IEF on 24 cm, pH 4–7 strips; SDS-PAGE on 12.5% gels; RuBPs	LC-MS/MS (CapLC-QTOF2)	Douxfils et al. 2011
Testis						
S. senegalensis	Domestication	Mechanical distruption and sonication	Sonication in TNE buffer, addition of 1% NP-40, 1 mM PMS, and 2 mM DTT, protein precipitation and resuspension in 7 M urea, 2 M thiourea, 4% CHAPS, 30 mM Tris HCl, pH 8.5	2-D DIGE: IEF on 18 or 24 cm, pH 3–10 strips; SDS-PAGE on 12% gels	MALDI-MS	Forné et al. 2009
				CyDye labeling	IT-MS/MS (LCQ MS)	
Kidney						
S. aurata	Bacterial infections	Scalpel fragmentation and homogenization by bead beating	8 M urea, 4% CHAPS, 0.5% ampholines	2DE: IEF on 24 cm, pH 3–10 NL strips, SDS-PAGE on 10–18% gradient gels; Coomassie	MALDI-MS and LC-MS/MS (ESI-Q-TOF)	Addis et al. 2010b

Table 11.2 Summary of the most relevant findings emerged from proteome studies on farmed fish tissues

Species	Factor	Tissue	Main observations	References
G. morhua	Farming	Muscle	Differences in protein expression and protein degradation. Presence of additional spots between 100 and 50 kDa in wild, and between 45 and 35 in farmed cod Differential expression of energy metabolism enzymes with mitochondrial alterations in farmed individuals	Martinez et al. 2007; Olsson et al. 2007
D. labrax	Farming	Muscle	Overexpression of carbohydrate metabolism enzymes and underexpression of proteins involved in muscular development as a consequence of farming	Monti et al. 2005
S. aurata	Farming	Muscle	Only minor differences due to environmental factors in farmed vs wild individuals of comparable size. Significant overexpression of metabolic versus structural proteins in smaller versus larger farmed individuals	Addis et al. 2010a
S. aurata	Handling and crowding	Liver	Alterations in expression levels of proteins involved in aminoacid, carbohydrate and lipid metabolism, folding, signaling and cytoskeleton upon exposure to stress	Alves et al. 2010
O. mykiss	Dietary substitution	Liver	Increase in general catabolism and in protein turnover, downregulation of Apolipoprotein AI, downregulation of structural proteins, and alteration of nitrogen metabolism following dietary substitution with vegetable sources	Martin et al. 2003
P. fluviatilis	Domestication	Serum	Differences in abundance of the C3 complement component, transferrin and Apolipoprotein E in F4 fish versus F1 fish. Increase in production of immune response proteins in response to chronic stressors following domestication	Douxfils et al. 2011
S. aurata	Bacterial colonization	Kidney	Alterations in expression of mitochondrial enzymes and oxidative stress responses upon bacterial colonization	Addis et al. 2010b
S. senegalensis	Domestication	Testis	Alterations in protection against oxidative stress, protease inhibition mechanisms and iron and glucose metabolism in captive F1 fish versus wild fish	Forné et al. 2009

11.2 Muscle

Skeletal muscle is the edible part of fish, and is therefore the most interesting tissue from a commercial perspective. In addition, it is one of the most easily accessible for reproducible sampling and is available for analysis along the whole production chain, even in processed products. Therefore, the identification of traits enabling us to differentiate fish produced by means of aquaculture techniques in this tissue matrix would enable its monitoring "from the farm to the fork."

Muscle composition contributes strongly to quality, inasmuch as number and integrity of muscular fibers have a considerable influence on texture, elasticity, and water-holding capacity (Johnston 1999). The structural unit of muscle is the sarcomere, a supramolecular contractile assembly that accounts for about half of the total muscle proteins. This machinery is mainly composed of myosin, actin, tropomyosin, troponins, and myosin light chains, the giant proteins titin and nebulin, and several other proteins having structural or regulatory functions. Another relevant portion of the total muscle protein composition is represented by sarcoplasmic proteins, which account mainly for glycolysis and energy metabolism, providing the "fuel" for contraction. The characterization of muscle tissue proteomes and their comparative evaluation in farmed and wild fish can provide useful information on the influence exerted by farming practices, environmental variables, diet, and other factors, and assist the improvement of aquaculture productions (Reddish et al. 2008). In fact, confinement, size, diet (Kiessling et al. 1991), crowding (Gornati et al. 2004a, b), amount and intensity of physical activity (Johnston and Moon 1980; Totland et al. 1987), and water quality and temperature (Johnston et al. 1998) can all have significant effects on protein expression levels, also depending on the farming technique used, on the quality of feeds, and on dimensions and exposition to marine currents within rearing spaces. Several economically relevant farmed species have been subjected to proteomic studies of muscle tissues aimed at assessing the variations induced by farming, including Atlantic cod (*Gadus morhua*) (Martinez et al. 2007; Olsson et al. 2007), Mediterranean sea bass (*Dicentrarchus labrax*) (Monti et al. 2005), and Gilthead sea bream (*Sparus aurata*) (Addis et al. 2010a). Figure 11.1 illustrates a representative 2D-PAGE map of farmed sea bream muscle tissue.

11.2.1 Atlantic Cod: Molecular and Structural Alterations Induced by Farming

Atlantic cod (*G. morhua*) is not yet considered to be a fully domesticated species. In fact, numerous significant differences are observed between farmed and wild-captured Atlantic cod, including size and shape of the head, hepatosomatic index, skin pigmentation and flesh discoloration, also accompanied by mineral-induced melanin deposition in blood vessels located between muscle

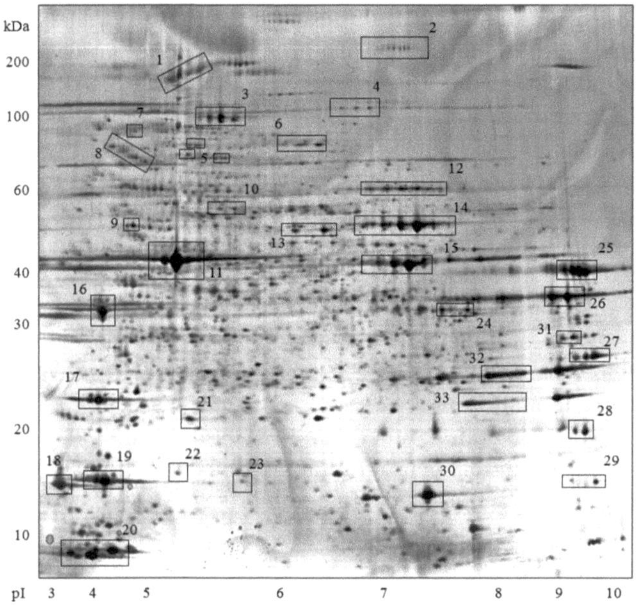

N°	Protein
1	Myosin heavy chain (MHC)
2	Myomesin
3	Myosin binding protein H
4	Muscle glycogen phosphorylase
5	MHC fragments
6	Transferrin
7	GRP78
8	Wap65
9	ATP synthase, beta subunit
10	Desmin
11	Actin, alpha skeletal muscle
12	Phosphoglucomutase 1
13	Pyruvate kinase
14	Enolase
15	Muscle-type creatine kinase M1
16	Tropomyosin 1, alpha chain
17	Myosin light chain 1
18	Myosin light chain 2
19	Myosin light chain 3
20	Parvalbumin isoforms
21	PEBP-like
22	Troponin T, white muscle, adult
23	Troponin T, white muscle, embryo
24	GAPDH
25	Fructose diphosphate aldolase
26	GAPDH
27	Phosphoglycerate mutase 2
28	Fast muscle troponin I
29	Fast muscle troponin
30	Nucleoside diphosphate kinase
31	Voltage dependent anion channel 2
32	Triosephosphate isomerase
33	Adenylate kinase

Fig. 11.1 2-D PAGE map of farmed *Sparus aurata* muscle. *Numbers* indicate the main protein groups subjected to mass spectrometry identification. Protein identities are reported on the *right*. The following conditions were applied: IEF on 24 cm, pH 3–10 IPG strips; SDS-PAGE on 8–16% polyacrylamide gradient gel; silver staining. Protein identification was performed by LC-MS/MS on a nanoLC-nanoESI-Q-TOF instrument

segments (Cooper and Midling 2007). Muscle structure and storage characteristics of farmed cod differ from those observed in wild individuals (Olsson et al. 2007). In addition, post-mortem degradation has been shown to be faster in farmed than in wild cod muscle (Ofstad et al. 1996). As the farmed and the wild cod are not yet genetically distinct, the most likely sources for these differences are environmental.

11.2.1.1 Atlantic Cod: Molecular Alterations

Martinez et al. (2007) examined the suitability of muscle proteins to differentiate farmed from wild cod by means of two-dimensional electrophoresis. Neither farming conditions nor size of fish were detailed in this work. Protein extracts were obtained from wild and farmed cod by homogenization and sequential extraction with the ReadyPrep Sequential Extraction Kit from Bio-Rad Laboratories, generating a Tris fraction and a CHAPS-urea fraction. Proteins from pools of five individuals were separated using 11-cm, pH 3–10 Immobiline strips for the first dimension, and 12.5% polyacrylamide gels for the second dimension, followed by silver staining. According to differential gel analysis, the main difference among wild and farmed fish profiles was the presence of additional spots between 100 and 50 kDa in wild fish, and between 45 and 35 kDa in farmed fish. Protein identification was not carried out; however, the authors suggested that some of these spots resulted from a differential expression induced by farming conditions, and others resulted from differences in the proteolytic degradation of high molecular weight proteins.

In general, this work indicated that a different protein expression and/or a different post-mortem degradation can be seen in farmed cod compared to wild cod muscle. As far as differential expression is concerned, this is in line with the numerous characteristics of farmed cod that deviate from those considered to be normal in the wild fish, including higher condition factor, larger liver, smaller head, and backbone malformations. Faster post-mortem deterioration in farmed fish is also a plausible option; other authors (Olsson et al. 2003) had already reported a faster post-mortem deterioration in farmed fish compared to wild fish. In farmed halibut, post-mortem degradation proceeds faster, and the flesh has lower functional properties than its wild counterpart. In the case of cod, this differential degradation is supported by the findings of Ofstad et al. (1996), who reported that farmed cod flesh is less firm and has a higher water content, lower functional properties, and a different muscle fiber structure. The factors responsible for these differences were suggested to be related to the post-mortem change of parameters such as pH, which is altered by food intake before death, or to the different abundance or expression of endogenous proteases involved in post-mortem muscle tenderization. Characterization of these proteins and their dynamics was indicated as an opportunity for improving farming conditions in order to obtain higher-quality fish with optimal muscle composition.

11.2.1.2 Atlantic Cod: Structural Alterations

In order to investigate the physiological mechanisms originating the phenotypic differences observed between farmed and wild cod, Olsson et al. (2007) performed a study using proteome analysis and electron microscopy, revealing that several metabolic disorders are present in muscle tissues of farmed Atlantic cod. Twenty farmed (2,663+/– 462 g) and nineteen wild cod (4750 +/– 870 g) were examined in this study. Protein lysates were obtained by sonication, and protein extracts were separated on 13-cm, pH 5–8 gradient strips. The second dimension was performed on 12% polyacrylamide gels, and spots were visualized using a commercial Coomassie blue stain. Upon differential analysis, 11 differentially expressed protein spots were detected between farmed and wild cod muscle, and were subjected to identification by LC-MS/MS with a Q-TOF instrument. Unfortunately, due to the lack of sequence information at the time of the analysis, only five of these produced viable data, and were identified as adenylate kinase, triosephosphate isomerase, aldolase, creatine kinase, and isocitrate dehydrogenase. All these enzymes are involved in energy metabolism, and mitochondria are the main cellular organelle where these processes take place. In keeping with these observations, ultrastructural analysis of farmed cod muscle revealed a number of intrafibrillar mitochondria showing clear cristae malformations, with ring structures formed by the inner membrane. Furthermore, significant intermyofibrillar fat deposition was observed in muscle from farmed cod, whereas no interfibrillar muscle fat was present in wild fish. Disturbances in mitochondrial–oxidative capacity in skeletal muscle can play a role in insufficient oxidation of fatty acids, and therefore in their abnormal deposition in this tissue (He et al. 2001; Vogel 2001; Petersen et al. 2003). The enzymatic alterations identified in the proteome analysis revealed the existence of alterations in energy metabolism, and supported these observations.

11.2.2 European Sea Bass: An Integrated Microfluidic Approach for Assessing Muscle Protein Profiles in Farmed Individuals

The European sea bass (*D. labrax*) is currently one of the main products of European aquaculture, and is considered to be a fish of high economical value. Monti et al. (2005) evaluated the impact of farming on its muscle proteins by using a proteomic approach based on SDS-PAGE separation followed by LC/MS-MS for protein identification, combined with relative quantification of proteins by denaturing capillary electrophoresis in farmed and wild sea bass. The authors implemented this integrated approach based on microfluidic electrophoresis and proteomic procedures in order to enable rapid detection of protein expression changes induced by aquaculture on *D. labrax* muscle.

Information on size of fish, feeding regimens, or aquaculture conditions were not reported in this work. Protein extracts were obtained by fragmentation of muscle tissues, resuspension in MilliQ water, and mincing in a Stomacher 400 circulator. Water-soluble proteins were then subjected to either SDS-PAGE or microfluidic separation and detection on an Agilent 2100 Bioanalyzer instrument. Protein 50 and Protein 200 Plus chips were used to enable separation over a wide mass range. Such analysis revealed statistically significant and reproducible differences in the protein profiles of farmed and wild muscle tissues. SDS-PAGE analysis confirmed these differences, enabling the detection of at least nine proteins with a statistically significant variation between wild and farmed *D. labrax*. Upon LC-MS/MS analysis using an LCQ ion trap instrument, identification of the differentially expressed proteins was accomplished. Among these, many enzymes involved in carbohydrate metabolism were found overexpressed in farmed sea bass, such as glyceraldehyde-3-phosphate dehydrogenase (+92%) and aldolase (+75%). In addition, the impact on creatine kinase (−45%), nucleoside diphosphate kinase B (−70%), and parvalbumin (−22%) suggested an influence of aquaculture conditions on the muscular development of farmed fish.

11.2.3 Gilthead Sea Bream: An In-depth Comparative Muscle Protein Profiling of Differently Sized Wild and Farmed Individuals

The gilthead sea bream (*S. aurata*) is the leading product of Mediterranean aquaculture (Gómez-Requeni et al. 2004), accounting for 133,026 t in 2008 (FAO – Fisheries and Aquaculture Information and Statistics Service). Although farmed for centuries, this saltwater fish has seen a considerable increase in production during the past 20 years. To obtain information on muscle protein expression profiles in this species along the production cycle, and to compare them with those seen in wild individuals, a 2DE-MS study was performed in fish from four different offshore floating cage plants in Sardinia, Italy. Interesting insights on proteome variability and on protein expression profiles associated with fish size and environmental conditions were provided. Moreover, a detailed reference 2-D map of farmed *S. aurata* muscle proteins was generated (Addis et al. 2010a).

The study was carried out on a large number of wild and farmed Gilthead sea bream, taking into account several parameters such as season, water temperature and salinity, and size of fish. Protein extracts were generated from 540 samples collected for 2 years, by fragmentation of frozen tissue, resuspension in 8-M urea–2% CHAPS buffer, and homogenization by repeated cycles of mechanical bead beating. All extracts were subjected to SDS-PAGE, and total profiles were compared by computerized densitometric analysis. Interestingly, as opposed to farmed Atlantic cod and sea bass, there were no obvious differences in muscle protein profiles that could be directly related to the farmed or wild status. Protein abundance differences were mostly either individual or related to factors other than the wild or farmed condition. Although it might appear surprising, this was not so unexpected. In a dedicated study, Carpenè et al. (1998) did not find any significant differences in the protein composition of muscle from farmed and wild sea bream, although differences in the fatty acid profiles were seen. As also observed by Martinez et al. (2007), this may be related to several factors such as the farming conditions (offshore sea cages with low rearing densities vs. crowded inland tanks), as well as the selection that occurred during the years; in fact, sea bream have been reared in captivity for a longer time, and this species might be either more adapted or best suited to be kept in captive conditions for biological or lifestyle reasons, this as opposed to Atlantic cod, for which several evident morphological alterations are still seen in captivity fish.

However, although not directly linked to the farmed or wild status, several differences were indeed present among several sample groups. To assess the identity of these proteins and to shed light on the drive for their differential expression, farmed and wild fish samples were subjected to an in-depth 2DE study followed by MALDI-MS and nanoLC-ESI Q TOF-MS. As a further classification, fish were grouped into size classes for comparison. This further analysis revealed two important sources of protein expression variability: fish size and geographic location of the farming plant. Shorter farmed (ca. 16 cm) and longer wild (ca. 26 cm) sea bream displayed significant conserved differences in their protein profiles: in smaller fish there was a higher expression of proteins involved in metabolism (sarcoplasmic proteins), whereas in larger fish there was a higher expression of proteins involved in structure and contraction (sarcomeric proteins). Interestingly, individual variations were also seen in proteins involved in seafood allergy, such as parvalbumins. Many of these observations were in line with those seen by Monti et al. (2005) in farmed sea bass. However, when differential comparison was carried out on wild and farmed fish having comparable sizes, these differences were no longer apparent. Instead, other proteins were seen to undergo statistically significant variations among groups, but in this case these were mainly associated with water temperature, such as warm temperature acclimation-related 65 kDa protein (Wap 65), to presence of infection or stressful conditions, such as transferrin, or to phosphorylation states, such as myosin binding protein H isoforms. Therefore, it must be kept in mind that in studies aimed at assessing differential expression profiles among farmed and wild fish muscle tissues, size and farming conditions are extremely significant factors to be considered in order to obtain meaningful results.

11.3 Liver

Liver functions are essential for life, as this organ performs a remarkable number of essential metabolic functions. Over 10,000 biochemical reactions are estimated to occur in this organ at any given time point, including carbohydrate, fat, and protein metabolism, storage of vitamins and minerals, and regulatory functions that control blood sugar and hormone levels. The liver is the primary organ for the synthesis of many different proteins, such as transferrin, fibrinogen, apolipoproteins, and globulins, of phospholipids and cholesterol, and of bile acid production and excretion (Gazzana and Borlak 2009). This organ is exposed to a wide range of xenobiotics and toxins, and is considerably influenced by factors such as diet, environment, and stress. All these things considered, as opposed to muscle, where less targeted studies have been performed in most cases, the consequences of aquaculture on the liver proteome have been studied by evaluating the effect of specific influencing factors. The species investigated for comparative wild/farmed liver protein expression include the rainbow trout (*Oncorhyncus mykiss*), where changes in protein profiles were studied under different feeding regimens (Martin et al. 2001, 2003), and the gilthead sea bream (*S. aurata*), where manipulation stress was evaluated (Alves et al. 2010). Figure 11.2 illustrates a representative 2-D-PAGE map of farmed sea bream liver tissue.

11.3.1 Rainbow Trout: Influence of Dietary Substitution with Vegetable Sources

Martin and coworkers (2003) applied proteomics to investigate the proteins undergoing differential expression in the liver of rainbow trout subjected to different feeding regimens, to gain insights to the biochemical processes occurring as a result of an altered

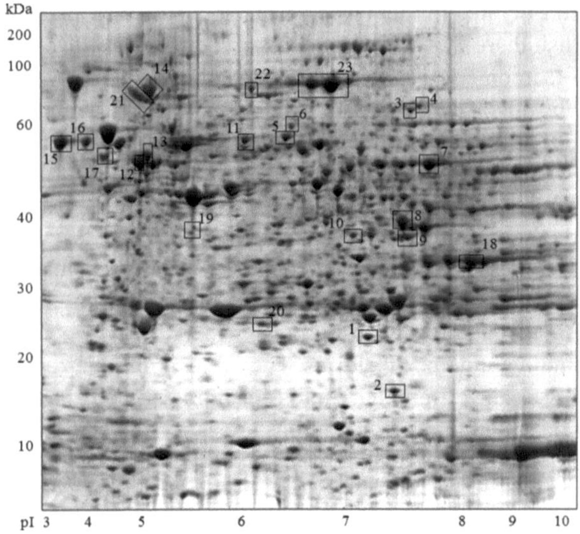

N°	Protein
1	Peroxiredoxin 1
2	Nucleoside diphosphate kinase A
3	Bifunctional purine biosynthesis protein
4	Transketolase
5	UDP-glucose dehydrogenase
6	Epoxide hydrolase 2, cytoplasmic
7	Enolase 3
8	Fructose-biphosphate aldolase
9	Cytosolic malate dehydrogenase
10	Transaldolase
11	1-acyl-sn-glycerol-3-phosphate acyltransferase
12	ATP synthase subunit beta, mitochondrial
13	Tubulin beta-1 chain
14	GRP 78
15	Calreticulin
16	Calreticulin
17	Calreticulin
18	Fructose-bisphosphate aldolase B
19	Methionine adenosyltransferase I
20	Betaine homocysteine methyltransferase
21	Wap 65
22	Iron-responsive element-binding protein 1
23	Transferrin

Fig. 11.2 2-D PAGE map of farmed *Sparus aurata* liver. *Numbers* indicate the main protein groups subjected to mass spectrometry identification. Protein identities are reported on the *right*. The following conditions were applied: IEF on 24 cm, pH 3–10 IPG strips; SDS-PAGE on 8–16% polyacrylamide gradient gel; colloidal Coomassie staining. Protein identification was performed by LC-MS/MS on a nanoLC-nanoESI-Q-TOF instrument

metabolism due to dietary substitution with vegetable sources. The two diets were formulated with ingredients commonly used in aquaculture feeds. Both contained fish meal and plant ingredients, but one (S) was higher in soybean meal, believed to induce adverse effects on different metabolic pathways, whereas the other (C) was higher in fish meal. The feeding trial was carried out for 12 weeks. Among other parameters, such as nitrogen metabolism, protein utilization, and selected enzyme activities, a 2DE analysis was performed on liver. For protein extraction, frozen tissue was homogenized in a 9-M urea–2% CHAPS buffer using a Teflon Dounce homogenizer. Soluble proteins were focused in a 7-cm pH 4–7 strip, followed by SDS-PAGE onto 10–15% gradient gels. Proteins were stained with colloidal Coomassie blue G250.

Following differential analysis of liver proteins from three fish fed diet S and three fish fed diet C, 33 protein spots were shown to be significantly altered in abundance between the diets, and 17 were successfully identified by MALDI TOF-MS. A variety of proteins including heat shock proteins (HSPs), enzymes, fatty acid binding protein, structural proteins, and proteins involved in the immune response were found to be differentially regulated. Structural proteins, such as keratin and tubulin, were downregulated in diet S, as well as HSPs. However, these displayed a differential expression of two HSP 108 isoforms. Many enzymes involved in anabolism were also downregulated, indicating an emphasis on catabolism relative to anabolism following dietary substitution with soybean. On the other hand, an increase in aldolase B and protein phosphatase A2 was detected in diet S compared to diet C, likely indicating an increase in metabolism and general turnover of proteins. Finally, an isoform of Apolipoprotein AI-1 (Apo A I) was significantly downregulated in fish fed diet S. Apo AI is involved in the removal of free cholesterol to the liver for excretion.

In conclusion, by using proteomics, this study showed for the first time that rainbow trout fed with diets containing vegetable sources undergo alterations in nitrogen metabolism and display alterations in liver protein expression profiles. This study opens the way to using this technique for designing feeds with well-balanced cost/quality ratios, and to gain insights on compatibility of feed composition with fish metabolism.

11.3.2 Gilthead Sea Bream: Influence of Handling and Crowding as Chronic Stressors

Apart from the dietary regimen, farming conditions pose many other constraints that can have a significant impact on protein expression profiles when compared to life conditions experienced in the wild. Among the most important factors, crowding and repeated handling are considered to act as significant chronic stressors. Alves and coworkers (2010) applied comparative proteomics of liver to characterize the metabolic molecular indicators of chronic stress in gilthead sea bream. Liver proteome profiles were compared in fish stressed by repetitive handling (HND), in fish crowded at high stocking densities (HSD), and in control fish (CTRL). The experiment was performed using a pooling strategy in order to minimize the effect of individual variations; to level this out, four technical replicates were analyzed for each condition (Westermeier et al. 2008). Protein extracts were obtained in a buffer containing 7-M urea–2-M thiourea–4% CHAPS buffer, and homogenization was accomplished by Ultra-Turrax treatment followed by sonication. 2DE was carried out on 11-cm pH 3–10 immobiline strips, followed by SDS-PAGE on 12% polyacrylamide gels. Staining was performed with Coomassie G250. Protein identification was carried out by LC-ESI-Ion trap MS/MS.

As a result, a total of 164 differential spots was detected, including 79 upregulated and 85 downregulated by stress; 37 were unique to handled fish (HND) and 40 were unique to crowded fish (HSD). Of these, 12 were provided a protein identification, revealing the involvement of several cellular processes related to adaptation to stressful conditions: amino acid, carbohydrate and lipid metabolism, folding, signaling, and cytoskeleton.

Concerning proteins involved in energy metabolism, the expression levels of triose phosphate isomerase, glyceraldehyde-3-phosphate dehydrogenase, alpha enolase, and pyruvate dehydrogenase were all affected by chronic stress. The authors explained this observation with the mobilization of energetic reserves involving the activation of liver glycogenolysis and gluconeogenesis to guarantee glucose supply to brain and muscles (Moyle and Cech 1996). Expression levels of fatty acid binding protein were also increased in stressed fish; this was explained as a mechanism for supporting higher rates of lipid consumption by liver cells in order to sustain the increased energetic requirements. Another observation was the decrease in calmodulin abundance upon chronic stress. The authors suggested that this might be an adaptation to chronically increased intracellular calcium levels, induced by repetitive epinephrine stimulation, aimed at preventing the triggering of apoptotic mechanisms.

Interestingly, a halving of heat-shock cognate protein 70 (hsc70) was detected in both groups subjected to chronic stress compared to the control group. HSPs are well known to be involved in the cellular response to stress (Ali et al. 2003; Basu et al. 2001; Boone and Vijayan 2002). Although belonging to the HSP class, hsc70 is constitutively expressed in cells without stress, and the observation of its downregulation is in line with previous studies (Gornati et al. 2004b; Poltronieri et al. 2007). Another downregulated protein seen in both groups was glutamine synthase. This observation was also in line with previous findings, both in fish (Hopkins et al. 1995) and in mice (Vats et al. 1999). Finally, the increase in hemoglobin expression seen in stressed fish is also in line with previous data in various fish species (Acerete et al. 2004; Biron and Benfey 1994; Montero et al. 1999; Roche and Bogé 1996), as this protein is a known indicator of response to stress due to the increase in efficiency of oxygen transportation (Bone et al. 1996; Randall et al. 1997).

In summary, liver proteomics coupled with conventional stress monitors, such as plasma cortisol levels, enabled the identification of several proteins associated with stress in fish, that may hold promise for the development of more practical tests, such as ELISAs, useful for evaluating and monitoring the existence of chronic stress factors in aquaculture plants.

11.4 Blood Serum

Blood serum (or plasma) is a biological fluid of primary importance, typically considered to be a "river" of proteins and peptides that flows through cells and tissues of the whole organism (Zhou et al. 2005). As such, it conveys and collects proteins and peptides of the organs it bathes, reflecting variations occurring in physiological and pathological conditions in distant organs and body districts. In humans and mammals, blood serum has become the "gold standard" biological matrix for discovery and diagnosis, also for its ease of sampling and the ability to acquire numerous samples during a study. In fish, the difficulties in obtaining blood samples from live animals and the stress associated with sample collection have often outweighed the practical aspects of working with serum when compared to other tissues. However, this biological fluid is frequently collected and investigated when fish are to be sacrificed, and many proteomic studies on immune response have been performed using this biological matrix. Concerning proteomic studies aimed at investigating differences between farmed and wild fish, studies are less numerous, and are also mostly focused on the immune system. However, a good example of an application of blood serum proteomics to this topic is the study by Douxfils and coworkers (2011) on the influence of the domestication process on the immune status of Eurasian perch (*Perca fluviatilis*). Figure 11.3 illustrates a representative 2-D-PAGE map of farmed sea bream serum.

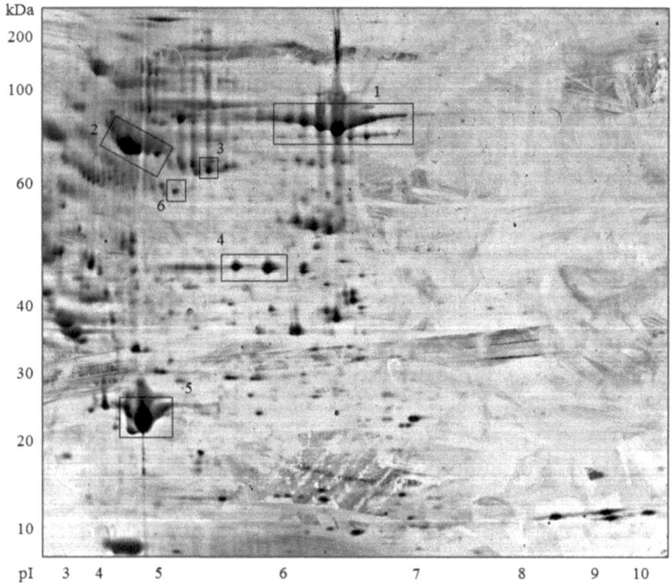

N°	Protein
1	Transferrin
2	Wap 65
3	Putative alpha-1-antitrypsin
4	Fibrinogen
5	ApoA1
6	Angiotensinogen

Fig. 11.3 2-D PAGE map of farmed *Sparus aurata* serum. *Numbers* indicate the main protein groups subjected to mass spectrometry identification. Protein identities are reported on the *right*. The following conditions were applied: IEF on 24 cm, pH 3–10 IPG strips; SDS-PAGE on 8–16% polyacrylamide gradient gel; colloidal Coomassie staining. Protein identification was performed by LC-MS/MS on a nanoLC-nanoESI-Q-TOF instrument

11.4.1 *Eurasian Perch (P. fluviatilis): Modulation of the Immune Status as a Consequence of the Genetic Selection Incurred During the Domestication Process*

The transition from the native to the captive habitat (domestication) imposes an adaptation to situations that are different from those experienced in the wild, and animals have to adapt to the new conditions (Price 1999). Therefore, a genetic selection occurs in captivity even without any intentional pressure by humans. In fish, a reduced genetic diversity has been reported in captive stocks when compared to wild populations, attributable to founder effects, inbreeding, artificial selection, and natural selection mechanisms due to survival of the fittest (Alarcon et al. 2004; Blanchet et al. 2008; Jackson et al. 2003; Koljonen et al. 2002; Pampoulie et al. 2006; Porta et al. 2006; Withler et al. 2007). In fact, natural selection in captivity acts through exclusion of the individuals unable to survive the many stressors posed by the captive conditions (Mignon-Grasteau et al. 2005). These physiological, morphological, and behavioral changes induced by domestication lead animals to be better adapted to captivity and to the presence of humans (Driscoll et al. 2009). In this context, Douxfils and coworkers (2011) combined the measurement of conventional stress indicators with microsatellite analysis and proteomic investigation of serum profiles in a wild population of Eurasian perch during their domestication process, with the aim of evaluating its influence on the stress physiology and associated immunomodulation. Specifically, the effect of chronic confinement stress on immune capacity was evaluated on two distinct generations (F1 and F4) of 1-year-old Eurasian perch juveniles reared at different stocking densities in recirculated tanks. The authors conducted the study without operating any particular morphological, physiological, or behavioral trait selection on the population under study. A combination of parameters was evaluated, including usual stress indicators, parameters related to innate and adaptive immunity, microsatellite analysis, and differential proteomic analysis of blood serum.

Proteomic experiments were performed by using pooled sera, on four replicate pools of serum per condition (unconfined F1 fish, confined F1 fish, unconfined F4 fish, confined F4 fish). Pooled sera were diluted six times and reduced in 7-M urea–2-M thiourea–4% CHAPS buffer, and proteins were subjected to isoelectric focusing after cup-loading onto rehydrated pH 4–7, 24-cm IPG strips. The second dimension was carried out in 12.5% polyacrilamide gels, and RuBPs staining (ruthenium II tris bathophenanthroline disulfonate) was performed (Rabilloud et al. 2000). Differentially expressed spots were subjected to protein identification by LC-ESI-MS/MS on a CapLC Q-TOF2.

As a result, statistically significant differences were observed in the abundance of C3 complement component, transferrin, and apolipoprotein E (ApoE) in the sera of confined F4 fish when compared to confined F1 fish. Specifically, confinement caused a decrease in transferrin abundance at day 7 in F1 fish, but not in F4 fish. Transferrin is known for its role as a positive acute phase protein, and in fish it has been demonstrated to be an activator of the macrophage antimicrobial response (Stafford and Belosevic 2003; Stafford et al. 2004). Concurrently, F4 fish showed an elevated abundance of apoE when compared to F1 fish. The authors suggested that this might provide fish with the ability to face the increased energetic demand associated with stress, as reported in the common carp *Cyprinus carpio* (Concha et al. 2003). Concerning complement C3, several differentially expressed spots produced this protein identification, although none of them matched the expected entire molecular weight, suggesting the presence of numerous fragments. Among these, a general decrease in expression was observed in F4 fish subjected to confinement stress when compared to F1 fish in the same conditions. In general, therefore, the expression level of several immune response proteins in response to chronic stressors was higher in F4 than in F1 fish, highlighting the positive effects of domestication on the immune status and therefore on the ability to fight disease.

In conclusion, by applying serum proteomics, the study by Douxfils and coworkers (2011) enabled the demonstration that, using this animal model and this farming technique, even the maintenance of fish for few generations in captivity exerts a selective pressure on the immune capacity of fish and on their ability to fight chronic stressors, although at the price of a genetic drift and the reduction of genetic diversity, allelic richness, and heterozygosity.

11.5 Kidney

In fish, the kidney is a composite organ with three different functions: excretory, hematopoietic, and endocrine. As such, it becomes an interesting organ for monitoring health and wellness of farmed fish. However, only sparse proteomic studies have been performed to assess the impact of aquaculture on the fish kidney. Figure 11.4 illustrates a representative 2-D-PAGE map of farmed sea bream kidney tissue.

11.5.1 Alterations of the Protein Expression Profile upon Bacterial Colonization of the Kidney in Gilthead Sea Bream

In a study aimed to monitor the health status of gilthead sea bream farmed in off-shore floating cages (Addis et al. 2010b), the most frequent bacteriological finding was internal organ colonization by *Moraxella* spp, a Gram-negative opportunistic bacterium. The kidney was the most affected organ in this survey, and a proteomic study was carried out in order to monitor alterations in its protein expression profile and to identify possible biomarkers linked to bacterial colonization. Protein extracts

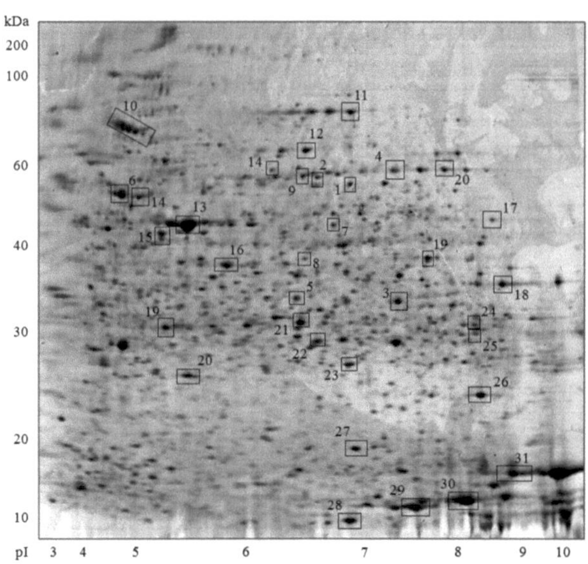

N°	Protein
1	Alanine aminotransferase
2	Aldehyde dehydrogenasa
3	Dihydrodipicolinate synthase
4	Methylmalonate semialdehyde dehydrogenase
5	Dihydrodipicolinate synthase
6	ATP synthase, beta subunit
7	S-adenosyl-homocysteine hydrolase
8	Phosphoenolpyruvate carboxykinase
9	Antiquitin
10	Wap 65
11	Transferrin
12	Transferrin
13	Beta actin
14	ATP synthase, beta subunit
15	Cytoplasmic actin
16	Beta actin
17	Acyl-CoA dehydrogenase
18	Fructose-biphosphate aldolase B
19	Electron transfer flavoprotein, alpha
20	PEBP superfamily
21	Carbonic anhydrase
22	Peroxiredoxin
23	Peroxiredoxin family protein
24	Enoyl-CoA hydratase
25	Glutathione S-transferase
26	Nucleoside diphosphate kinase
27	Cu/Zn superoxide dismutase
28	Alpha-2-globin
29	Beta globin
30	Alpha-2-globin
31	Beta globin

Fig. 11.4 2-D PAGE map of farmed *Sparus aurata* kidney. *Numbers* indicate the main protein groups subjected to mass spectrometry identification. Protein identities are reported on the *right*. The following conditions were applied: IEF on 24 cm, pH 3–10 IPG strips; SDS-PAGE on 8–16% polyacrylamide gradient gel; colloidal Coomassie staining. Protein identification was performed by LC-MS/MS on a nanoLC-nanoESI-Q-TOF instrument (Adapted from Addis et al. 2010b)

were generated from 7 positive and 28 negative kidney samples by fragmentation with a scalpel and homogenization by bead beating in 8-M urea-2% CHAPS buffer. Isoelectric focusing was carried out onto 24-cm, pH 3–10 NL IPG strips, and SDS-PAGE was performed on 8–16% polyacrylamide gradient gels, which were stained by colloidal Coomassie. Computerized image analysis identified several differential protein spots, which were identified with a combination of MALDI-MS and NanoHPLC-nanoESI-Q-TOF-MS. In particular, ten spots were found to be consistently overexpressed in proteomic maps of Moraxella-positive kidneys. Interestingly, seven of these spots corresponded to six mitochondrial enzymes, including alanine aminotransferase, aldehyde dehydrogenase, dihydrodipicolinate synthase, methylmalonate semialdehyde dehydrogenase, ATP synthase beta subunit, and Acyl-CoA-dehydrogenase. Among other proteins, peroxiredoxins, S-adenosyl-homocysteine hydrolase, wap 65, transferrin, glutathione S-transferase, carbonic anhydrase, and Cu/Zn superoxide dismutase were also identified. Mitochondrial proteins are key to many metabolic activities, and alteration of their expression is often associated with disease (Palmfeldt et al. 2009), whereas many of the other upregulated proteins are related to oxidative stress responses, infection, inflammation, or programmed cell death processes. In summary, even infection by opportunistic bacteria can lead to mitochondrial enzyme alterations in kidney tissues, producing a stressful condition potentially influencing farmed fish metabolism, and opening the way to infection by other bacterial pathogens. Although the impact on zootechnical performance remains to be demonstrated, the identification of a panel of proteins with altered expression levels might be useful as a tool for detecting and containing infections by Gram-negative pathogens before these spread to the whole farming plant.

11.6 Reproductive Organs

Another topic concerning adaptation to captivity of farmed fish, which has been addressed using proteomics, is fertility and reproduction. However, only few studies evaluated these aspects in a comparative manner between farmed fish and their wild counterparts. In this respect, Fornè et al. (2009) applied the 2-D-DIGE technology to investigate on the molecular basis for the loss of fertility consequent to captivity in Senegalese sole (*Solea senegalensis*).

11.6.1 Senegalese Sole (S. senegalensis): An Investigation of Fertility in Wild Caught Versus F1 Captive Fish by Means of Testis Proteomics

The flatfish Senegalese sole has an elevated commercial value, making its optimal reproduction in captivity of considerable interest. However, the process is not yet completely controlled; although spontaneous reproduction in fish adapted to captivity has been observed for several years (Anguis and Cañavate 2005), males from the F1 generation have a poor reproductive performance due to a reduced sperm production, sometimes resulting in the complete absence of egg fertilization (Agulleiro et al. 2006; Porta et al. 2006; Cabrita et al. 2006). Forné and coworkers applied comparative protein profiling by 2-D-DIGE to investigate the changes occurring in the testis proteome of wild-caught and F1 Senegalese sole males reared in captivity, with the aim of identifying proteins potentially involved in the low sperm production and reduced fertilization capacity of F1 males (Miura et al. 1999).

Testis samples from F1 males were collected after a treatment aimed to induce teleost spermatogenesis by means of implants containing synthetic analogues of gonadotropin-releasing hormones (F1GnRHa), a combination of GnRHa and 11-ketoandrostenedione (F1GnRHa+OA), or saline solution as a control (F1C). Testis samples were also obtained from wild-caught males at the spermiation stage (F0Mat). All testis samples were mechanically disrupted in saline buffer and then subjected to sonication. Extracted proteins were precipitated and resuspended in a pH 8.5, 7-M urea–2-M thiourea–4% CHAPS buffer. After cyanine labeling and sample mixing, samples were focused into linear pH 3–10 IPG strips, followed by SDS-PAGE into 12% polyacrylamide gels. Images were generated by scanning with a confocal laser scanner and analyzed with dedicated software. Differentially expressed spots were subjected to protein identification by analysis with a MALDI-TOF mass spectrometer, selecting the peptides to be analyzed by a following round of nanoESI IT MS/MS with an LCQ MS.

As a result, a total of 58 spots showed statistically significant differences in the analyzed groups, and 45 of them were successfully identified as the products of 31 genes. In this set of proteins, many were present in several differential spots, indicating the presence of potential differentially espressed isoforms or carrying various post-translational modifications. These included alpha-2-macroglobulin-1, hemopexin, keratin, adenosine triphosphate synthase b subunit, peroxiredoxin 6, and keratin 18, having different patterns of regulation depending on the developmental stage. It is worth noticing that such observations are facilitated and their validity is reinforced by the 2-D-DIGE technology that, combined with software analysis, provides more robust information on the presence and relative abundance of isoforms differing by isoelectric point. In total, the differential proteome analysis of testis tissue from captive F1 and wild-caught Senegalese sole males showed a significant difference in expression of 24 proteins involved in cytoskeletal organization, catabolic processes, and redox or antioxidant activities. Application of the differential proteomics approach enabled the authors of this work to discover that F1 males experience alterations in protection against oxidative stress, protease inhibition mechanisms, and iron and glucose metabolism when compared to wild fish, and these might be responsible for their impaired sperm production and reduced fertilization rates. Moreover, several new proteins not previously described in the testis of teleost fish were reported. The results of this work open the way to other investigations aimed at understanding the physiology of Senegalese sole, as well as identifying the mechanisms that impair production of viable sperm in these and other farmed teleost fish.

11.7 Concluding Remarks

The studies carried out thus far have demonstrated the potential of proteomics to identify pathways and functions influenced by farming in many fish tissues. The impact of factors such as feeding regimens, crowding, handling, temperature, growth rates, farming techniques, and many others has been characterized in most fish tissues, widening our knowledge of fish biology and providing useful tools for increasing efficiency and quality of aquaculture productions.

Nevertheless, the informative power of many differential proteomic investigations carried out on fish still suffers from the limited availability of annotated sequence data. It is hoped that in the future this problem will become less important, especially with the progress made on model organisms as well as on genome sequencing and annotation technologies. Furthermore, the integration with gene expression techniques, lipidomics, and metabolomics will enable us to fully exploit the potential of proteomics in supporting the future production of healthier and higher quality fish.

Acknowledgments The author wishes to thank Daniela Pagnozzi for the useful discussions on proteomic methods and techniques and for critical reading of the manuscript, and Stefania Ghisaura, Vittorio Tedde, and Grazia Biosa for their help with the 2-D-PAGE maps and MS identifications. The support of Prof. Sergio Uzzau and Dr. Tonina Roggio is also gratefully acknowledged.

References

Acerete L, Balasch JC, Espinosa E, Josa A, Tort L (2004) Physiological responses in Eurasian perch (*Perca fluviatilis*, L.) subjected to stress by transport and handling. Aquaculture 237:167–178

Addis MF, Cappuccinelli R, Tedde V, Pagnozzi D, Porcu MC, Bonaglini E, Roggio T, Uzzau S (2010a) Proteomic analysis of muscle tissue from gilthead sea bream (*Sparus aurata*, L.) farmed in offshore floating cages. Aquaculture 309:245–252

Addis MF, Cappuccinelli R, Tedde V, Pagnozzi D, Viale I, Meloni M, Salati F, Roggio T, Uzzau S (2010b) Influence of *Moraxella* sp. colonization on the kidney proteome of farmed gilthead sea breams (*Sparus aurata*, L.). Proteome Sci 8:50

Agulleiro MJ, Anguis V, Cañavate JP, Martínez-Rodríguez G, Mylonas CC, Cerdà J (2006) Induction of spawning of captive-reared Senegal sole (*Solea senegalensis*) using different administration methods for gonadotropin-releasing hormone agonist. Aquaculture 257:511–524

Alarcon JA, Magoulas A, Georgakopoulos T, Zouros E, Alvarez MC (2004) Genetic comparison of wild and cultivated European populations of the gilthead sea bream (*Sparus aurata*). Aquaculture 230:65–80

Ali KS, Dorgai L, Ábrahám M, Hermesz E (2003) Tissue- and stressor-specific differential expression of two hsc70 genes in carp. Biochem Biophys Res Commun 307:503–509

Alves RN, Cordeiro O, Silva T, Richard N, De Vareilles M, Marino G, Di Marco P, Rodrigues PM, Conceição LEC (2010) Metabolic molecular indicators of chronic stress in gilthead seabream (*Sparus aurata*) using comparative proteomics. Aquaculture 299:57–66

Anguis V, Cañavate JP (2005) Spawning of captive Senegal sole (*Solea senegalensis*) under a naturally fluctuating temperature regime. Aquaculture 243:133–145

Basu N, Nakano T, Grau EG, Iwama GK (2001) The effects of cortisol on heat shock protein 70 levels in two fish species. Gen Comp Endocrinol 124:97–105

Biron M, Benfey TJ (1994) Cortisol, glucose and hematocrit changes during acute stress, cohort sampling, and the diel cycle in diploid and triploid brook trout (*Salvelinus fontinalis* Mitchell). Fish Phys Biochem 13:153–160

Blanchet S, Páez DJ, Bernatchez L, Dodson JJ (2008) An integrated comparison of captive-bred and wild Atlantic salmon (*Salmo salar*): implications for supportive breeding programs. Biol Conserv 141:1989–1999

Bone Q, Marshall NB, Blaxter JHS (1996) Biology of fishes, 2nd edn. Chapman & Hall, London

Boone AN, Vijayan MM (2002) Constitutive heat shock protein 70 (HSC 70) expression in rainbow trout hepatocytes: effect of heat shock and heavymetal exposure. Comp Biochem Physiol C 132:223–233

Cabrita E, Soares F, Dinis MT (2006) Characterization of Senegal sole, *Solea senegalensis*, male broodstock in terms of sperm production and quality. Aquaculture 261:967–975

Carpené E, Martin B, Dalla LL (1998) Biochemical differences in lateral muscle of wild and farmed gilthead sea bream (*Sparus aurata* L.). Fish Physiol Biochem 19:229–238

Concha MI, Molina S, Oyarzun C, Villanueave J, Amthauer R (2003) Local expression of apolipoprotein A-1 gene and a possible role for HDL in primary defence in the carp skin. Fish Shellfish Immunol 14:259–273

Cooper M, Midling K (2007) Blood vessel melanosis: a physiological detoxication mechanism in Atlantic cod (*Gadus morhua*). Aquacult Int 15:43–54

Douxfils J, Mathieu C, Mandiki SNM, Milla S, Henrotte E, Wang N, Vandecan M, Dieu M, Dauchot N, Pigneur LM, Li X, Rougeot C, Mélard C, Silvestre F, Van Doninck K, Raes M, Kestemont P (2011) Physiological and proteomic evidences that domestication process differentially modulates the immune status of juvenile Eurasian perch (Perca fluviatilis) under chronic confinement stress. Fish Shellfish Immunol 31:1113–1121

Driscoll CA, MacDonald DW, O'Brien SJ (2009) From wild animals to domestic pets, an evolutionary view of domestication. From the Academy. In: The light of evolution III: two centuries of Darwin Sackler Colloquium, vol 106. PNAS 9971–78

Forné I, Agulleiro MJ, Asensio E, Abián J, Cerdà J (2009) 2-D DIGE analysis of Senegalese sole (*Solea senegalensis*) testis proteome in wild-caught and hormone-treated F1 fish. Proteomics 9:2171–2181

Forné I, Abián J, Cerdà J (2010) Fish proteome analysis: model organisms and non-sequenced species. Proteomics 10:858–872

Gazzana G, Borlak J (2009) An update on the mouse liver proteome. Proteome Sci 7:35

Gómez-Requeni P, Mingarro M, Calduch-Giner JA, Médale F, Martin SAM, Houlihan DF, Kaushik S, Pérez-Sánchez J (2004) Protein growth performance, amino acid utilisation and somatotropic axis responsiveness to fish meal replacement by plant protein sources in gilthead sea bream (*Sparus aurata*). Aquaculture 232:493–510

Gornati R, Terova G, Vigetti D, Prati M, Saroglia M, Bernardini G (2004a) Effects of population density on sea bass (*Dicentrarchus labrax*) gene expression. Aquaculture 230:229–239

Gornati R, Papis E, Rimoldi S, Terova G, Saroglia M, Bernardini G (2004b) Rearing density influences the expression of stress-related genes in sea bass (*Dicentrarchus labrax*, L.). Gene 341:111–118

Hall TE, Cole NJ, Johnston IA (2003) Temperature and the expression of seven muscle-specific protein genes during embryogenesis in the Atlantic cod *Gadus morhua* L. J Exp Biol 206:3187–3200

He J, Watkins S, Kelley DE (2001) Skeletal muscle lipid content and oxidative enzyme activity in relation to muscle fibre in type 2 diabetes and obesity. Diabetes 50:817–823

Hopkins TE, Wood CM, Walsh PJ (1995) Interactions of cortisol and nitrogen metabolism in the ureogenic gulf toadfish *Opsanus beta*. J Exp Biol 198:2229–2235

Ibarz A, Martín-Pérez M, Blasco J, Bellido D, de Oliveira E, Fernández-Borràs J (2012) Gilthead sea bream liver proteome altered at low temperatures by oxidative stress. Proteomics 10:963–975

Jackson TR, Martin-Robichaud DJ, Reith ME (2003) Application of DNA markers to the management of Atlantic halibut (*Hippoglossus hippoglossus*) broodstock. Aquaculture 220:245–259

Johnston IA (1999) Muscle development and growth: potential implications for flesh quality in fish. Aquaculture 177:99–115

Johnston IA, Moon TW (1980) Endurance exercise training in the fast and slow muscles of a teleost fish (*Pollachius Virens*). J Comp Physiol 135:147–156

Johnston IA, Cole NJ, Abercromby M, Vieira VLA (1998) Embryonic temperature modulates muscle growth characteristics in larval and juvenile herring *Clupea harengus*. J Exp Biol 201:623–646

Kiessling A, Storebakken T, Åsgård T, Kiessling KH (1991) Changes in the structure and function of the epaxial muscle of rainbow trout (*Oncorhynchus mykiss*) in relation to ration and age: I. Growth dynamics. Aquaculture 93:335–356

Koljonen ML, Tähtinen J, Säisä M, Koskiniemi J (2002) Maintenance of genetic diversity of Atlantic salmon (*Salmo salar*) by captive breeding programmes and the geographic distribution of microsatellite variation. Aquaculture 212:69–92

Martin SAM, Cash P, Blaney S, Houlihan DF (2001) Proteome analysis of rainbow trout (*Oncorhynchus mykiss*) liver proteins during short term starvation. Fish Physiol Biochem 24:259–270

Martin SAM, Vilhelmsson O, Médale F, Watt P, Kaushik S, Houlihan DF (2003) Proteomic sensitivity to dietary manipulations in rainbow trout. Biochim Biophys Acta 1651:17–29

Martinez I, Pettersen GW (1992) Temperature-induced precocious transitions of myosin heavy chain isoforms in the white muscle of the Arctic charr (*Salvelinus alpinus*). Bas Appl Myol 2:89–95

Martinez I, Dreyer B, Agersborg A, Leroux A, Boeuf G (1995) Effects of T3 and rearing temperature on growth and skeletal myosin heavy chain isoform transition during early development in the salmonid *Salvelinus alpinus* (L.). Comp Biochem Physiol Part B Biochem Mol Biol 122:717–725

Martinez I, Šližytė R, Daukšas E (2007) High resolution two-dimensional electrophoresis as a tool to differentiate wild from farmed cod (*Gadus morhua*) and to assess the protein composition of klipfish. Food Chem 102:504–510

Melamed P, Zhiyuan G, Hew CL, Fletcher G (2002) The potential impact of modern biotechnology on fish aquaculture. Aquaculture 204:255–269

Mignon-Grasteau S, Boissy A, Bouix J, Faure JM, Fisher AD, Hinch GN, Jensen P, Le Neindre P, Mormède P, Prunet P, Vandeputte M, Beaumont C (2005) Genetics of adaptation and domestication in livestock. Livestock Prod Sci 93:3–14

Miura T, Miura C, Ohta T, Nader MR, Todo T, Yamauchi K (1999) Estradiol-17b stimulates the renewal of spermatogonial stem cells in males. Biochem Biophys Res Comm 264:230–234

Montero D, Izquierdo MS, Tort L, Robaina L, Vergara JM (1999) High stocking density produces crowding stress altering some physiological and biochemical parameters in gilthead seabream, *Sparus aurata*, juveniles. Fish Physiol Biochem 20:53–60

Monti G, De Napoli L, Mainolfi P, Barone R, Guida M, Marino G, Amoresano A (2005) Monitoring food quality by microfluidic electrophoresis, gas chromatography, and mass spectrometry techniques: effects of aquaculture on the sea bass (*Dicentrarchus labrax*). Anal Chem 77:2587–2594

Moyle PB, Cech JJ (1996) Fishes – an introduction to ichtyology, 3rd edn. Prentice Hall, New Jersey

Nash CE (2011) The history of aquaculture. Wiley-Blackwell, Ames

Ofstad R, Egelandsdal B, Kidman S, Myklebust R, Olsen R, Hermanssona AM (1996) Liquid loss as effected by post mortem ultrastructural changes in fish muscle: cod (*Gadus morhua* L) and salmon (*Salmo salar*). J Sci Food Agric 71:301–312

Olsson GB, Friis TJ, Jensen E, Cooper M (2007) Metabolic disorders in muscle of farmed Atlantic cod (*Gadus morhua*). Aquacult Res 38:1223–1227

Olsson GB, Olsen RL, Ofstad R (2003) Post mortem structural characteristics and water-holding capacity in Atlantic halibut muscle. LWT- Food Sci Technol 36:125–133

Palmfeldt J, Vang S, Stenbroen V, Pedersen CB, Christensen JH, Bross P, Gregersen N (2009) Mitochondrial proteomics on human fibroblasts for identification of metabolic imbalance and cellular stress. Proteome Sci 28:7–20

Pampoulie C, Jörundsdóttir TD, Steinarsson A, Pétursdóttir G, Stefánsson MO, Daníelsdóttir AK (2006) Genetic comparison of experimental farmed strains and wild Icelandic populations of Atlantic cod (*Gadus morhua* L.). Aquaculture 261:556–564

Petersen KF, Befroy D, Dufour S, Dziura J, Ariyan C, Rothman DL, DiPietro L, Cline GW, Shulman GI (2003) Mitochondrial dysfunction in the elderly: possible role in insulin resistance. Science 300:1140–1142

Piñeiro C, Barros-Velázquez J, Vázquez J, Figueras A, Gallardo JM (2003) Proteomics as a tool for the investigation of seafood and other marine products. J Proteome Res 2:127–135

Poltronieri C, Maccatrozzo L, Simontacchi C, Bertotto D, Funkenstein B, Patruno M, Radaelli G (2007) Quantitative RT-PCR analysis and immunohistochemical localization of HSP70 in sea bass *Dicentrarchus labrax* exposed to transport stress. Eur J Histochem 51:125–136

Porta J, Porta JM, Martínez-Rodríguez G, Carmen AM (2006) Genetic structure and genetic relatedness of a hatchery stock of Senegal sole (*Solea senegalensis*) inferred by microsatellites. Aquaculture 251:46–55

Price EO (1999) Behavioral development in animals undergoing domestication. Appl Anim Behav Sci 65:245–271

Rabilloud T, Strub JM, Luche S, Girardet JL, van Dorsselaer A, Lunardi J (2000) Ruthenium II tris (bathophenanthroline disulfonate), a powerful fluorescent stain for detection of proteins in gel with minimal interference in subsequent mass spectrometry analysis. Proteome 1:1–14

Randall D, Burggren W, French K, Fernald R (1997) Eckert – animal physiology – mechanisms and adaptations. W.H. Freeman, New York

Reddish JM, St-Pierre N, Nichols A, Green-Church K, Wick M (2008) Proteomic analysis of proteins associated with body mass and length in yellow perch, *Perca flavescens*. Proteomics 8:2333–2343

Roche H, Bogé G (1996) Fish blood parameters as a potential tool for identification of stress caused by environmental factors and chemical intoxication. Mar Environ Res 41:27–43

Stafford JL, Belosevic M (2003) Transferrin and the innate immune response of fish: identification of a novel mechanism of macrophage activation. Dev Comp Immunol 27:539–554

Stafford JL, Wilson EC, Belosevic M (2004) Recombinant transferrin induces nitric oxide response in goldfish and murine macrophages. Fish Shellfish Immunol 17:171–185

Totland GK, Kryvi H, Jødestol KA, Christiansen EN, Tangeras A, Slinde E (1987) Growth and composition of the swimming muscle of adult Atlantic salmon *Salmo salar* L. during long-term sustained swimming. Aquaculture 66:299–313

Vats P, Mukherjee AK, Kumria MML, Singh SN, Patil SKB, Rangnathan S, Sridharan K (1999) Changes in the activity levels of glutamine synthetase, glutaminase and glycogen synthetase in rats subjected to hypoxic stress. Int J Biometeorol 42:205–209

Vogel H (2001) Mitochondrial myopathies and the role of the pathologist in the molecular era. J Neuropathol Exp Neurol 60:217–227

Westermeier R, Naven T, Höpker H (2008) Proteomics in practice: a guide to successful experimental design, 2nd edn. Wiley-VCH, Federal Republic of Germany

Withler RE, Rundle T, Beacham TD (2007) Genetic identification of wild and domesticated strains of chinook salmon (*Oncorhynchus tshawytscha*) in southern British Columbia, Canada. Aquaculture 272:S161–S171

Zhou M, Conrads TP, Veenstra TD (2005) Proteomics approaches to biomarker detection. Brief Funct Genomic Proteomic 4:69–75

Chapter 12
Fish Authentication

Mónica Carrera, Benito Cañas, and José M. Gallardo

12.1 Introduction

The intake of fish species by consumers is increasing due to the strong evidence of their positive benefits in human health. These benefits are mainly due to their high content of polyunsaturated ω-3 fatty acids proven to help in the prevention and treatment of cardiovascular, neurological, and inflammatory diseases (Hooper et al. 2006). In addition, fish are also excellent sources of micronutrients, including various vitamins (A, D, E, B1, B6, and B12) and minerals (Fe, I, P, Na, Ca, and K), and an affordable source of high-quality animal protein. A portion of 150 g of fish provides about 50–60% of the daily protein requirements for an adult (Piggot 1976). In 2007, the average annual per capita fish supply in developing countries was 9.0–15.1 kg (FAO 2010).

Attributable to this high demand, the fishery market is showing a dramatic growth in sales, producing an overexploitation of resources and the search for new or alternative fish species, which may be hazardous in an increasingly globalized market environment. Figure 12.1a,b show, respectively, the dominant species in marine fishery catches and the main seafood groups dedicated to human consumption (FAO, SOFIA 2010). Nowadays, fish can be produced in one country, processed in a second, and consumed in a third. The process of globalization has created substantial opportunities, but hand in hand with inherent risks. A very common fraudulent

M. Carrera (✉)
Institute of Molecular Systems Biology (IMSB), ETH Zürich,
Wolfgang-Pauli-Str, Zürich 16 8093, Switzerland
e-mail: carrera@imsb.biol.ethz.ch

B. Cañas
Complutense University of Madrid, Madrid, Spain

J.M. Gallardo
Marine Research Institute, CSIC,
Vigo, Pontevedra, Spain

F. Toldrá and L.M.L. Nollet (eds.), *Proteomics in Foods: Principles and Applications*, Food Microbiology and Food Safety 2, DOI 10.1007/978-1-4614-5626-1_12,

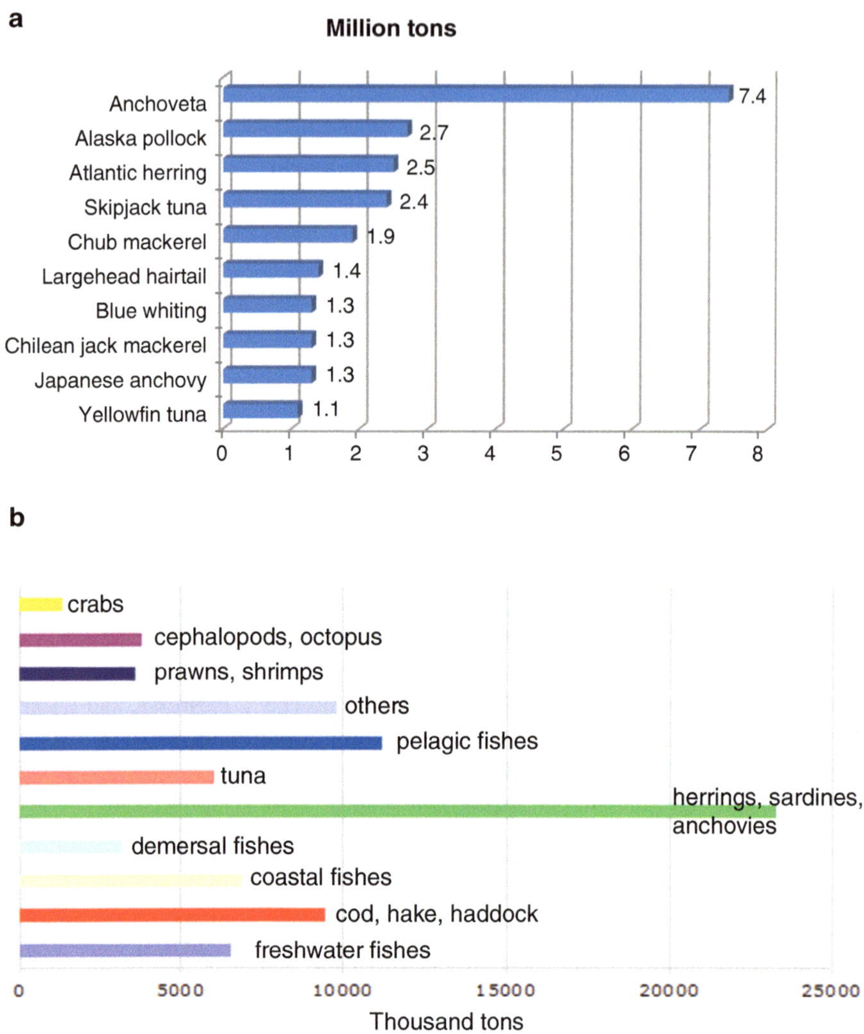

Fig. 12.1 Marine capture fisheries production: top ten species in 2008 (**a**), and the main seafood groups dedicated for human consumption (**b**) (FAO 2010)

practice is species substitution, which can be done unintentionally, but more frequently, looking for tax evasion, to launder illegally caught fish, or to sell low-priced fish species in a fraudulent substitution of more valuable higher-priced species. In addition, potential human health risks may appear as the fraudulent species can be harmful and may aggravate symptoms in sensitive human allergic patients. In fact, fish are one of the foods with more prevalence of food allergies (Sicherer and Sampson 2010).

Nowadays, consumers demand clear and reliable information about the species they eat, therefore the establishment of new standardized labeling and normative and inspection control regulations is growing continuously. The use of acceptable market names is essential for the identification of seafood because of the exceedingly great number and variety of species. Furthermore, in some cases and depending on the country, different genera share the same generic commercial name. This is the case, for example, of the generic denomination of "hake," used for commercialized species belonging to the *Merluccius* genera in Argentina, Spain, Chile, Australia, the United States, and the United Kingdom; nevertheless, other disparate species belonging to the genera *Dissostichus*, *Rexea*, or *Urophycis* are sold in Argentina, the United States, and the United Kingdom, respectively, under the same denomination of "hake." The need of common and acceptable market names for the seafood sold in interstate commerce and the need to assist manufacturers in labeling seafood products led to the publication in 1993 of *The Seafood List* by the U.S. Food and Drug Administration (U.S. FDA) to provide a source of names of several seafood species and reduce confusion within producers and consumers (Guidance for Industry 2012).

In the European Union, the need for labeling regulations has led to the Council Regulation (EC) No 104/2000 on the common organization of the markets in fishery and aquaculture products. This regulation advises that fish should be correctly labeled indicating: (1) the commercial designation of the species, (2) the production method (caught or farmed), and (3) the catch area. For this purpose, the member states draw up and publish a list of the commercial designations accepted in their territory indicating the scientific name for each species and the name in the language or languages of the member state. The indication of the catch area mentioned above is normalized in the Council Regulation (EC) No 2065/2001 in the annex following the FAO Yearbook (FAO yearbook 2000). Regulation (EC) No 104/2000 also indicates the importance of labeling seafood products with their scientific name to ensure traceability. These requirements have been implemented in each of the European states, such as Spain, where several regulations have been promulgated to assure the correct labeling and identification of seafood products (Royal Decree 1380/2002; Royal Decree 121/2004; Royal Decree 1702/2004).

To comply with all these regulations, accurate, sensitive, and fast detection methods that permit the direct authentication of fish in any food product are highly recommended. Conventional identification of unprocessed fish is done by examination of their anatomical and morphological features. However, even for marine expert biologists this is a difficult task in the case of very closely related fish species that coexist in the same catch area. This is the case, for example, of Cape hakes, *Merluccius capensis* and *Merluccius paradoxus*, two different species belonging to the Merlucciidae family with similar morphological features that overlap their geographical distribution on the South African coast. Due to this overlapping distribution, the species are caught and managed jointly and no distinction is made in stock management. In addition, and due to the development of the fishing industry, seafood products can be processed (beheaded, eviscerated, skinned, filleted, smoked, cooked, or canned), often making the identification of their external anatomical or morphological features impossible.

For all these reasons, the use of molecular tools is therefore a suitable strategy to circumvent such problems. Although over the last two decades several DNA and protein molecular markers have been developed, recent successes of proteomics methodologies make them a promising strategy for fish authentication purposes. In light of this, a comprehensive overview of the state of the art and the future of proteomics approaches for fish species authentication is given in this chapter.

12.2 Traditional Molecular Strategies Used to Assess Fish Authenticity

12.2.1 Classical Protein-Based Methods Used for Fish Authenticity

Methodologies based on the detection of biomarker proteins representative of a particular species, using mainly electrophoretic or immunological assays, have been extensively exploited for the authentication of fish species.

In this respect, sodium dodecyl sulfate-polyacrylamide gel electrophoresis (SDS-PAGE) of myofibrillar and sarcoplasmic proteins has been used for the identification of commercial fish species in both raw and processed products (Scobbie and Mackie 1988; Piñeiro et al. 1999a; Martinez et al. 2001). Using specific enzymatic staining procedures, Keenan and Saklee in 1985 published a collection of the most common enzyme patterns by SDS-PAGE for 164 different fish species. SDS-PAGE has also been used to identify those species or products subject to some thermal processing, except sterilization (Piñeiro et al. 1999a; Martinez et al. 2001).

Isoelectric focusing (IEF) is the most commonly used protein-based technique for species identification (Piñeiro et al. 2000; Tepedino et al. 2001). In fact, IEF was adopted by the Association of Official Analytical Chemists as the official validated method for species identification purposes (AOAC 1990) and the U.S. FDA offers an Internet library of IEF patterns of sarcoplasmic proteins from different fish (Regulatory Fish Encyclopedia, FDA). Urea IEF gels have been applied for discrimination of unrelated fish species in products that have been heat treated (Mackie et al. 2000; Rehbein et al. 2000). Parvalbumins are the sarcoplasmic proteins that show higher interspecific variability by IEF (Rehbein et al. 2000). Due to their thermal stability, these proteins can also be used by native IEF for the discrimination of cooked products, in addition to raw species (Carrera et al. 2010). In addition, the experiences accumulated in our lab using IEF on narrow strips of pH 4–6.5 have demonstrated the suitability of parvalbumin isoforms as species-defining markers for all commercial fish species belonging to the Merlucciidae family (Piñeiro et al. 1998; Carrera et al. 2006; Carrera 2008).

Several reports concerning the application, with authentication purposes, of the two-dimensional gel electrophoresis (2DE) to the characterization of fish proteins have been published (Piñeiro et al. 1998; Martinez and Jakobsen Friis 2004; Carrera et al. 2006). Interestingly, the application of 2DE has allowed the detection of potential species-specific proteins for the discrimination of closely related fish species, such as hake (Piñeiro et al. 1998, 2001; Carrera et al. 2006), puffer fish (Chen et al. 2004), commercial flat fish (Piñeiro et al. 1999b), and perch species (Berrini et al. 2006). 2DE database servers for different organisms are available on the Internet (World-2DPAGE List, http://world-2dpage.expasy.org/list/; Appel et al. 1996); although up to date, for Teleostei species, only one 2DE image for rainbow trout is included in the FishProm database from Aberdeen University (http://www.abdn.ac.uk/fishprom/).

The detection of potential species-specific proteins by 2DE can be further investigated by other complementary techniques such as mass spectrometry (MS). Thus, following a classical bottom-up proteomics approach, consisting of 2DE, tryptic in-gel digestion, and MS/MS analysis, representative spots or clusters of spots for the Merlucciidae family, corresponding to nucleoside-diphosphate kinase proteins (NDK) (Carrera et al. 2007), to aldolase proteins (Carrera et al. 2009), and to parvalbumin (PRVB) isoforms, were identified, characterized, and de novo sequenced (Carrera et al. 2006, 2010). Immunological methods, based on the specificity of the antigen–antibody (Ag–Ab) reaction, are particularly attractive because they combine specificity, sensitivity, and simplicity. The first efforts to produce Abs to identify fish species were described by Mairs and Sindermann in 1962, who prepared polyclonal Abs (pAbs) to discriminate clupeid species. Furthermore, pAbs against certain sarcoplasmic proteins have been developed for the discrimination of species such as sardines, salmon, trout, halibut, haddock, grouper, and Nile perch (Domínguez et al. 1997; Céspedes et al. 1999a; Asensio et al. 2003a). The production of pAbs is simple and economical. However, it presents a number of drawbacks: (1) low specificity, resulting in cross-reactivity problems, (2) limited amount of Ab, and (3) requiring continuous immunizations of new animals, leading to the existence of nonreproducible batches. These problems can be solved with the generation of monoclonal Abs (mAbs). These Abs are selected by their ability to discriminate species with high specificity and absence of cross-reactivity against other species. Also, the production of mAbs after obtaining the hybridoma is unlimited. Several mAbs have been generated against species of red snapper (Huang et al. 1995), grouper (Asensio et al. 2003b), and catfish (McNulty and Klesius 2005; Gajewski, et al. 2009). The main immunological techniques that have been employed using pAbs or mAbs are the immunodiffusion (Domínguez et al. 1997), ELISA (Huang et al. 1995; Céspedes et al. 1999a; Asensio et al. 2003a, b; McNulty and Klesius 2005) and Western blot (Zhang and Rasco 1996; Domínguez et al. 1997). Immunoassays are advantageous in that, once developed, they are easy to use, having high sensitivity and throughput, thus allowing the process of a high number of samples in a short time. However, this technology is expensive and time consuming, and is not completely exempt from potential cross-reactivity. Apart from this, immunoassay can display important limitations in the analysis of processed food because processing can greatly affect the tertiary structure of the protein and thus negatively affect recognition of the target protein by the antibody.

General limitations of these classical protein-based methods, such as being very laborious and time-consuming methodologies, may be solved with the introduction of alternative methods based on DNA amplification and hybridization.

12.2.2 Common DNA-Based Methods Used for Fish Authenticity

Analytical methods based on nucleic acid detection, mainly DNA, have achieved an outstanding position in the authentication of fishery products in recent decades (Sotelo and Pérez-Martín 2007; Rasmussen and Morrissey 2009). Compared with the above-mentioned methodologies, DNA techniques have considerably higher discriminating power as they are based on identification at the sequence level of specific DNA fragments that are unique for a particular species. Together with their sensitivity, this is why DNA-based procedures have become popular methods for unambiguous identification of fish species even for closely related species.

The analysis of specific genetic sequences can be applied to both fresh and processed products, thanks to the stable nature of the DNA molecule, its ubiquitous character, and high content in diriment information, which is not affected by variations of expression. Nuclear genes such as the 5S ribosomal DNA, 5S ribosomal RNA, internal transcribed spacer 1 (ITS1), and certain microsatellite loci have been considered for the study of phylogenetic relationships among fish (Céspedes et al. 1999b; Castillo et al. 2003; Asensio et al. 2004; Pérez and García-Vázquez 2004). Among the DNA targets, mitochondrial DNA (mtDNA) is generally preferred because of its maternal inheritance, a relatively fast evolutionary rate, and the lack of intermolecular genetic recombination. The most commonly used mtDNA markers include the cytochrome b gene (Rehbein et al. 1997; Sotelo et al. 2001; Calo-Mata et al. 2003; Chapela et al. 2007), the mtDNA control region (Quinteiro et al. 2001), and the 12S rRNA region (Comesaña et al. 2003; Zhang et al. 2006). Today, most DNA-based methods for species identification in foods consist of the highly specific amplification of one or more DNA fragments by means of polymerase chain reaction (PCR). This technique presents high potential due to its simplicity, sensibility, and specificity. In this sense, several methods have been developed in order to perform polymorphism searches such as the restriction fragment length polymorphism (PCR-RFLP) used for the identification of different species of salmon, gadoids, flatfish, and hake (Russell et al. 2000; Quinteiro et al. 2001; Sotelo et al. 2001; Pérez et al. 2004; Aranishi et al. 2005). Other techniques include the amplified fragment length polymorphism (PCR-AFLP) (Maldini et al. 2006); the single-stranded conformational polymorphism (PCR-SSCP) for the identification of tuna, salmon, flatfish, Nile perch, and hake (Colombo et al. 2005; Chapela et al. 2007); the random amplified polymorphic DNA (RAPD) for the discrimination of

Tilapia species, Nile perch, and grouper (Partis and Wells 1996; Asensio et al. 2002); and multiplex PCR for grouper fillets (Trotta et al. 2005). In addition, some examples of sequencing techniques, such as forensically informative nucleotide sequencing (PCR-FINS) have been used to identify anchovies, sardines, and hake species (Jérôme et al. 2003; Santaclara et al. 2006). The latest and more fashionable methods for fish species identification are the approaches based on real-time PCR (Sánchez et al. 2009), microarrays (Kochzius et al. 2008), and lab-on-a-chip systems (Chen et al. 2011) for their potential to identify and quantify seafood species on a large scale.

Despite all these advantages, DNA-based procedures are not exempt from some important limitations that are especially important in the case of processed foods. During the processing of fish products, disruption of the cellular integrity can occur, causing the release of hydrolytic enzymes. Together with this, heat treatment and an acid environment can negatively affect DNA integrity, reducing the length of fragments to be amplified and consequently increasing the chances of having nonspecific identifications. Another important limitation is the complexity of foods, which can yield important matrix effects that negatively affect the accuracy and robustness of results. This complicates the development of standardized protocols for DNA extraction, thus being necessary to optimize them for each particular situation to ensure that enough DNA is obtained for the analysis and inhibitors of the reduced or eliminated PCR.

Although over the last two decades, several DNA and protein molecular markers have been developed, proteomics methodologies are emerging as a promising strategy for fish authentication.

12.3 Proteomics Technologies for Fish Authentication

As a discipline, proteomics is defined as the large-scale analysis of proteins in a particular biological system at a particular time (Pandey and Mann 2000). Recent successes illustrate the role of mass spectrometry, mainly matrix-assisted laser desorption/ionization–time-of-flight (MALDI-TOF) and electrospray-ion trap (ESI-IT) mass spectrometry, as an indispensable tool for proteomics studies (Aebersold and Mann 2003). But the history of proteomics began in the 1970s with the development of 2DE, which provided the first method for displaying hundreds or thousands of proteins on a single gel (Klose 1975; O'Farrell 1975). Nowadays, bioinformatics treatment of the data has increased the scale of proteomics tools, representing a powerful strategy for a high-throughput protein and peptide identification and quantification.

Proteomics methodologies have been used for the identification of some seafood species such as mussels (López et al. 2002) and shrimp (Ortea et al. 2009), but their application on Teleostei species authentication is still scarce. An idea of the current limited impact of proteomics in any type of research involving the Teleostei group can be obtained if one considers that protein databases include 239,454 amino acid sequences from Teleostei, whereas there are more than 1,118,730 amino acid sequences reported for the Tetrapoda group (UniProtKB, February, 2012) (Fig. 12.2).

Figure 12.3 shows the proteomics pipeline and tools that are currently being used at our laboratory for fish authentication. Two consecutive phases, the discovery phase (Carrera et al. 2006, 2010) and the target-driven phase (Carrera et al. 2011) are described in detail in the following sections using as an example the identification of all the commercial fish species belonging to the Merlucciidae family.

12.3.1 Discovery Phase

In this phase, to identify new potential peptide biomarkers for the identification of fish species (i.e., Merlucciidae species), organisms whose genomes remain unsequenced, we use a classical bottom-up proteomics approach (Fig. 12.3).

Thus, potential specific proteins, according to 2DE analysis, are selected, subjected to tryptic digestion, and the recovered peptides then ionized and analyzed by means of MS. As mentioned above, different spots or clusters of spots corresponding to parvalbumin (PRVB) isoforms (11.20–11.55 kDa and p*I* 3.75–4.57 units) (Piñeiro et al. 2001; Carrera et al. 2006, 2010), nucleoside-diphosphate kinase proteins (NDK) (16.80–18.60 kDa and p*I*5.04–5.47 units) (Piñeiro et al. 2001; Carrera et al. 2007), and aldolase proteins (42–43 kDa and p*I* 6.5 units) (Carrera et al. 2009), showed noticeable qualitative interspecific differences by 2DE, and were further investigated by MS.

The capacity of the peptide mass fingerprinting (PMF) methodology by MALDI-TOF MS was ascertained by analyzing 10 closely related commercial species of the Merlucciidae family (Carrera et al. 2006). MALDI-TOF mass fingerprints of the sarcoplasmic protein PRVB defined a set of molecular fish authentication markers, relying on the presence or absence of species-specific peptide masses, providing: (1) the selective differentiation between the genus *Merluccius* and *Macruronus*; (2) the classification of the hake species in two groups according to their geographic procedence, American or Euro-African hake; and (3) the unequivocal identification of several hake species, *M. bilinearis*, *M. australis polylepsis*, *M. australis australis*, *M. productus*, *M. paradoxus,* and *M. polli*, whereas the rest of the hake species can be grouped in two clusters, comprising *M. hubbsi* and *M. gayi* in one and *M. merluccius* and *M. capensis* in the other. Due to PRVB interspecific variability and high concentration in the muscle from fish, we can forecast that this protein can be used as a good biomarker for fish species identification. The selection of PRVB as a target protein has additional importance inasmuch as it is a protein that presents a high thermostability (Kawai et al. 1992; Elsayed and Bennich 1975; Carrera et al. 2010). For that, the monitoring of peptide masses ensures an overall applicability of the method for fish authentication in both fresh and processed seafood products. A similar approach was further successfully applied for the identification of 25 different fish species (Mazzeo et al. 2008). The authors, using the same strategy previously reported for our group, characterized specific PRVB peptide masses capable of

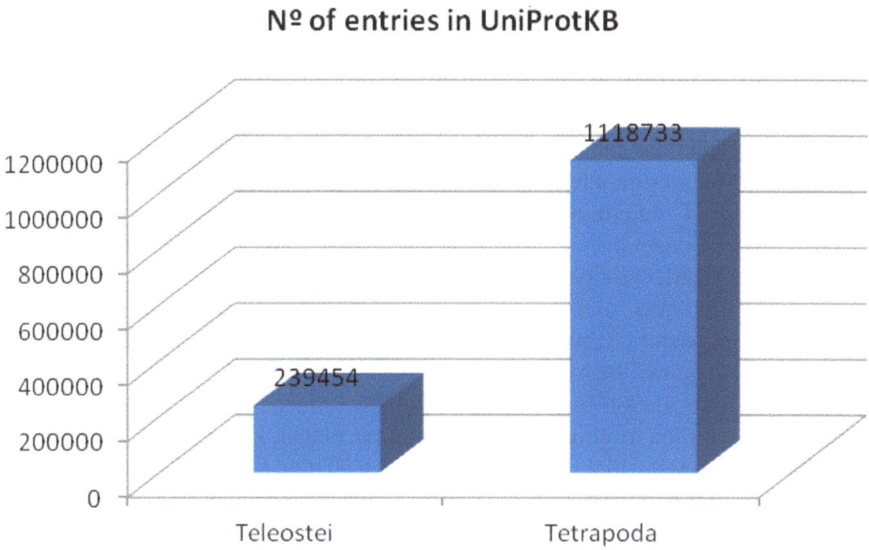

Fig. 12.2 Protein sequences annotated in UniProtKB database for Teleostei and tetrapods

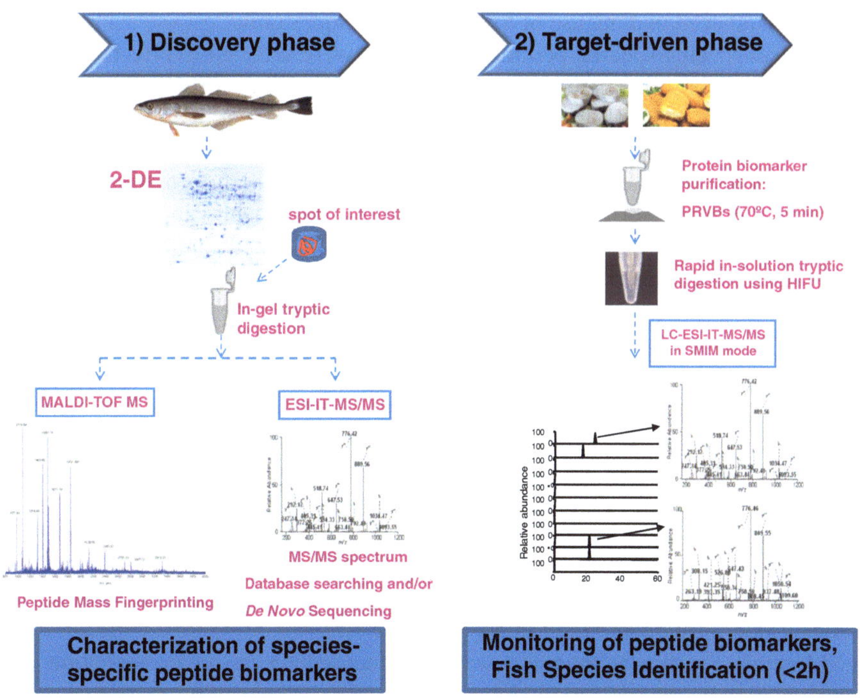

Fig. 12.3 Proteomics pipeline considered for the identification, characterization and detection of species-specific peptide biomarkers with fish authentication purposes

differentiation by MALDI-TOF MS a large number of fish species belonging to three different orders (Perciformes, Gadiformes, Pleuronectiformes).

In a subsequent study (Carrera et al. 2010), we proposed a novel strategy for the extensive characterization of all the PRVBs isoforms in all the commercial species from the Merlucciidae family (the previous 10 species and *M. senegalensis*). This strategy is based on the integration of a classical bottom-up proteomics approach with accurate M_r determination by Fourier-transform ion-cyclotron resonance (FTICR)-MS of intact proteins and selected tandem mass spectrometry (MS/MS) ion monitoring (SMIM) of peptide mass gaps. For each PRVB, mass spectra obtained by LC-ESI-IT-MS/MS from two digests (trypsin, Glu-C) were followed by database searching using Sequest (Eng et al. 1994) and *de novo* sequenced manually with the help of two programs, PEAKS (Ma et al. 2003) and DeNovoX (Thermo Electron Co.) (Scigelova et al. 2007). The deduced peptide sequences were arranged and the theoretical M_r for the resulting sequences was calculated. Experimental M_r for each PRVB was measured with high mass accuracy by FTICR-MS (0.05–4.47 ppm). The masses of several missing peptide gaps were estimated by comparing the theoretical and experimental M_r, and the MS/MS spectra corresponding to these ions were obtained by LC-ESI-IT-MS/MS in the SMIM scanning mode. Finally, all peptide sequences were combined to generate the final protein sequences. This approach allowed the complete *de novo* MS-sequencing of 25 new PRVB isoforms. This study constitutes the report accounting for the higher number of new proteins completely sequenced making use of MS-based techniques only.

Several species-specific peptide biomarkers were selected to effectively identify all the species from the Merlucciidae family. PRVBs peptide sequences with high interspecific variability, obtained after the extensive *de novo* sequencing of PRVBs previously published (Carrera et al. 2010), were used for this purpose. Eleven tryptic peptides were selected on the basis of the information that their combined presence or absence could be used to confidently identify all of the species under study. A flow diagram for the unambiguous systematic discrimination was also achieved (Carrera et al. 2011). According to this scheme, the presence/absence of several peptide biomarkers achieves: (1) identification if any member from the Merlucciidae family is present in the sample, (2) discrimination between the genera *Merluccius*, (3) classification of hake species into two groups according to their geographic distribution: American hake or Euro-African hake, and (4) finally the combination of the presence/absence of eight other peptide biomarkers allows the unambiguous identification of any specific species from the Merlucciidae family.

The selection of PRVB as the protein biomarker is particularly interesting because it is considered the major fish allergen (Elsayed and Bennich 1975), and therefore an analysis targeting this protein would have a double application, both for species identification and food safety purposes.

Moreover, particular attention was also focused on the characterization and identification by the MS-based biomarker discovery of several other peptide markers to discriminate all the commercial species belonging to the Merlucciidae family. In fact, using the same classical bottom-up proteomics approach, the characterization by *de novo* peptide sequencing of the different nucleoside diphosphate kinase (NDK) (Carrera et al. 2007) and aldolase proteins (Carrera et al. 2009), allowed the characterization of additional species-specific peptides that also can be used for fish authentication purposes.

These peptide biomarkers could be used to develop easy-to-use kits based on antibodies or monitored by MS using an innovative strategy that is described in the next phase of our pipeline.

12.3.2 Target-Driven Phase

In the second phase of the proteomics pipeline used in our laboratory for fish authentication purposes (Fig. 12.3), we developed a new targeted MS-based strategy for the fast monitoring of the species-specific peptide biomarkers found in the discovery phase (Carrera et al. 2011).

The performance of this target-driven method was established for the unequivocal identification of all commercial fish species belonging to the Merlucciidae family. It was based on (1) the purification of PRVBs by heat treatment (time 45 min), (2) their accelerated tryptic digestion using high-intensity focused ultrasound (HIFU; time 2 min), and (3) the monitoring of 11 PRVB peptide biomarkers by selected tandem mass spectrometry ion monitoring in a linear ion trap (LIT) mass spectrometer (time 60 min). Each step was individually adjusted to minimize analysis time. Thus, PRVBs, considered as the best protein biomarker for fish authentication of Merluciidae species, were purified from the sarcoplasmic extracts, taking advantage of their thermostability (Kawai et al. 1992). After treatment with heat (70ºC, 5 min), the majority of identified peptides corresponded to PRVBs (77.87%). These results demonstrated that the treatment with heat is a simple, fast, and effective procedure to purify and enrich the samples in only PRVBs. Purified PRVBs were digested with trypsin using a fast procedure by HIFU. Accelerated HIFU-tryptic digestions produced results comparable to those obtained by the conventional overnight incubation methods. Thus, the combination of a fast and easy protein purification procedure (time 45 min) with the use of HIFU for protein digestion (time 2 min) considerably simplified and reduced the time needed for sample preparation, reflected in the overall time needed for monitoring. Then, a particular combination of only 11 peptides (Table 12.1), resulting from the HIFU-assisted tryptic digestion of the thermostable proteins PRVBs, were subjected to SMIM analysis in an LIT mass spectrometer focusing the MS/MS events on the corresponding precursor ions. Once MS/MS spectra were recorded, virtual chromatograms for all the different fragments could be obtained. Tracing the highly sensitive transitions (precursor $m/z \rightarrow$ fragment m/z) for each peptide biomarker is possible for the unequivocal identification of all Merlucciidae species (Fig. 12.4). Also, the use of the SMIM mode for scanning gives the possibility of obtaining full MS/MS information necessary for the validation of the peptide biomarker sequence.

Table 12.1 Peptide biomarkers and specific transitions for the identification of all the species from the Merlucciidae family (Reprinted with permission from Carrera M, Cañas B, López-Ferrer D, Piñeiro C, Vázquez J, Gallardo JM. Fast monitoring of species-specific peptide biomarkers using high-intensity-focused-ultrasound-assisted tryptic digestion and selected MS/MS ion monitoring. Anal. Chem. 2011, 83, 5688-5695. Copyright 2011 American Chemical Society)

Biomarker code	Parvalbumin (PRVB) Peptide Sequence	SMIM Transition m/z precursor ion (z) → m/z fragment ion	Retention time (min)	S1	S2	S3	S4	S5	S6	S7	S8	S9	S10	S11	S12	Cross-reaction with proteins from other organisms by BLAST
F-MER604	LFLQVFSAGAR	604.84 (2+)→948.52 $(y''_9{}^+)$	23.00	■	■	■	■	■	■	■	■	■	■	■	■	-
G-MER517	VFGIIDQDK	517.78 (2+)→ 788.41 $(y''_7{}^+)$	20.00	■	■	■	■	■	■	■	■	■	■	■	□	-
G-MAC524	VFAIIDQDK	524.79 (2+)→ 802.43 $(y''_7{}^+)$	20.00	□	□	□	□	□	□	□	□	□	□	□	■	*Xenopus laevis*, *Xenopus tropicalis*
S-MER967	AGDSDGDGAIGVDEFAVLVK	967.97 (2+)→ 1189.68 $(y''_{11}{}^+)$	21.50	□	□	□	□	□	■	■	■	■	■	■	□	-
S-MER794	(N-Ac)A FSGILADADIAAALK	794.93 (2+)→ 1187.59 (b_2)	25.00	□	□	□	□	□	□	□	□	□	□	■	□	-
S-MER612	IGVDEFTAMLK	612.32 (2+)→ 954.46 $(y''_8{}^+)$	22.50	□	□	□	□	□	□	□	□	□	■	□	□	-
S-MER736	AEGTFTHGEFFTK	736.35 (2+)→ 966.46 $(y''_8{}^+)$	23.00	□	□	□	□	□	□	□	■	■	□	□	□	-
S-MER721	AEGTFTHGVFFTK	721.36 (2+)→ 936.49 $(y''_8{}^+)$	23.50	□	□	□	□	□	□	□	□	■	□	□	□	-
S-MER590	IGVDEFAAMVK	590.31 (2+)→ 910.43 $(y''_8{}^+)$	22.50	■	□	□	■	■	■	■	■	■	□	■	□	*Trachurus japonicus and others*[a]
S-MER973	AGDSDGDGAIGVDEWAALVK	973.46 (2+)→ 1087.57 $(y''_{10}{}^+)$	23.00	□	□	□	□	■	□	□	□	□	□	□	□	-
S-MER987	AGDSDGDGAIGVDEWAVLVK	987.48 (2+)→ 1115.61 $(y''_{10}{}^+)$	22.50	□	□	□	■	□	□	□	□	□	□	□	□	*Gadus morhua*

m/z (mass/charge).
(■) Denotes the presence of a peptide biomarker and (□) the absence.
S1 (*Merluccius merluccius*), S2 (*M.capensis*), S3 (*M. senegalensis*), S4 (*M. polli*), S5 (*M. paradoxus*), S6 (*M. hubbsi*), S7 (*M. gayi*), S8 (*M. australis polylepis*), S9 (*M. australis australis*), S10 (*M. productus*), S11(*M. bilinearis*), S12(*Macruronusspp.*).
([a]) *Sparus aurata, Fundulus grandis, Fundulus similis, Fundulus heteroclitus, Hypophthalmichthys nobilis, Paralichthys olivaceus, Theragra chalcogramma.*

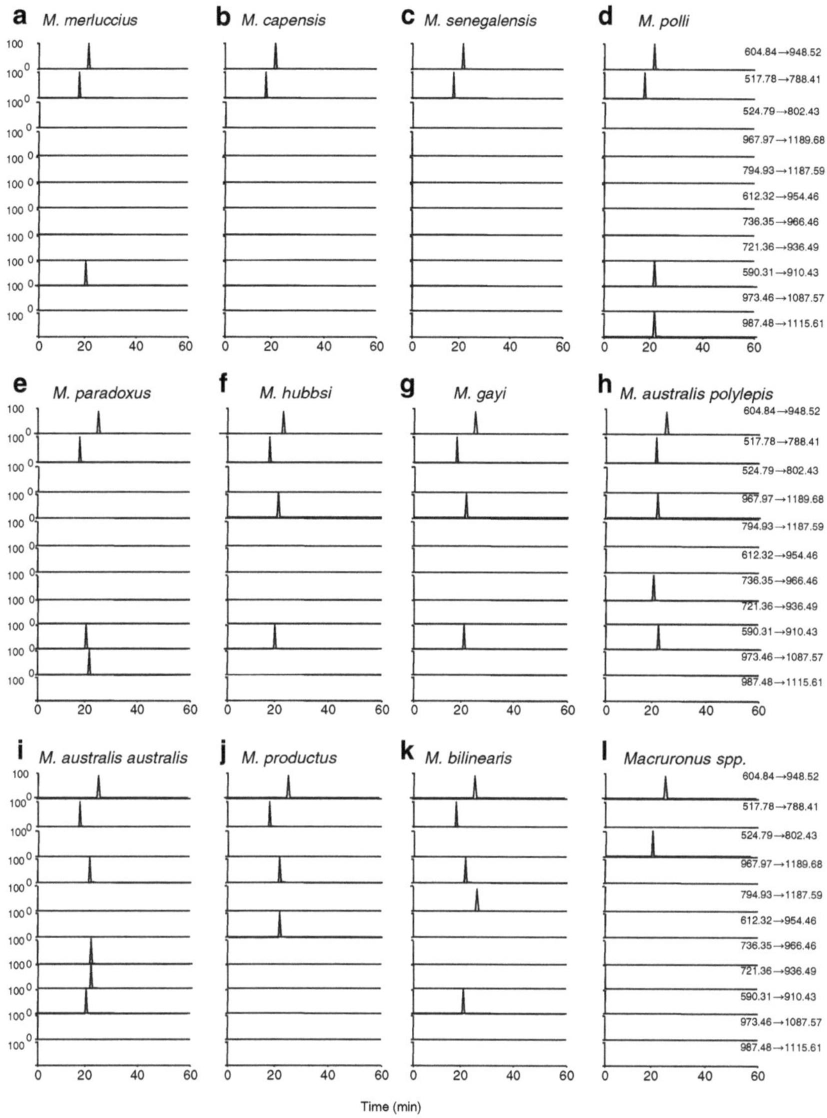

Fig. 12.4 Reference SMIM traces for each Merlucciidae species, plotting the corresponding canonical transition for each PRVB tryptic peptide biomarker (Reprinted with permission from Carrera M, Cañas B, López-Ferrer D, Piñeiro C, Vázquez J, Gallardo JM. Fast monitoring of species-specific peptide biomarkers using high-intensity-focused-ultrasound-assisted tryptic digestion and selected MS/MS ion monitoring. Anal. Chem. 2011, 83, 5688-5695. Copyright 2011 American Chemical Society)

An additional validation step using commercial fish products successfully demonstrated the applicability of this new targeted strategy for the fast detection of mislabeling practices in both fresh and processed fish products.

With this new strategy, we demonstrated that all relevant fish species belonging to the Merlucciidae family present in any seafood product can be unequivocally identified in less than 2 h.

This workflow constitutes the fastest method for peptide biomarker monitoring and its application to food quality control provides authorities with a rapid and effective method for food authentication and traceability to guarantee quality and safety to consumers.

12.4 Concluding Remarks and Future Outlook

As we discussed in this chapter the potentiality of the proteomics pipeline developed in our laboratory for fish species identification is noticeable. The two consecutive steps (discovery phase and target-driven phase) allow the identification and characterization of species-specific peptides that can be monitored by MS allowing the unequivocal and fast identification of fish species in any seafood product. Currently this pipeline is also being successfully applied in our laboratory for the identification of other species belonging to the order Decapoda (Ortea et al. 2009) and to develop a rapid and reliable method for bacterial identification in foodstuffs (Böhme et al. 2011). Nevertheless, we consider that its application field is not restricted to food authentication purposes and that may offer new opportunities to the food science sector such as the detection of allergens, the characterization of bioactive peptides, the study of the effects caused by processing and storing on food proteins, and so on. In addition, we consider that the procedures and results obtained using this proteomics pipeline may be stored in a centralized and web-accessible open source that supports their dissemination and their potential applicability to further food science projects.

References

Aebersold R, Mann M (2003) Mass spectrometry-based proteomics. Nature 422:198–207

AOAC (1990) Official method of analysis. In: Association of official analytical chemists, Helrich, K. (Ed.). 15th Edn., vol. 2, Arlington, Virginia, USA

Appel RD, Bairoch A, Sanchez JC, Vargas JR, Golaz O, Pasquali C, Hochstrasser DF (1996) Federated 2-DE database: a simple means of publishing 2-DE data. Electrophoresis 17:540–546

Aranishi F, Okimoto T, Izumi S (2005) Identification of gadoid species (Pisces, Gadidae) by PCR-RFLP analysis. J Appl Genet 46:69–73

Asensio L, González I, Fernández A, Rodríguez MA, Lobo E, Hernández PE, García T, Martín R (2002) Application of random amplified polymorphic DNA (RAPD) analysis for identification of grouper (*Epinephelus guaza*), wreck fish (*Polyprion americanus*), and Nile perch (*Lates niloticus*) fillets. J Food Prot 65:432–435

Asensio L, González I, Rodríguez MA, Mayoral B, López-Calleja I, Hernández PE, García T, Martín R (2003a) Identification of grouper (*Epinephelus guaza*), wreck fish (*Polyprion americanus*), and Nile perch (*Lates niloticus*) fillets by polyclonal antibody-based enzyme linked immunosorbent assay. J Agric Food Chem 51:1169–1172

Asensio L, González I, Rodríguez MA, Mayoral B, López-Calleja I, Hernández PE, García T, Martín R (2003b) Development of a specific monoclonal antibody for grouper (*Epinephelus guaza*) identification by an indirect enzyme-linked immunosorbent assay. J Food Prot 66:886–889

Asensio L, González I, Rodríguez MA, Hernández PE, García T, Martín R (2004) PCR-ELISA for the semiquantitative detection of Nile perch (*Lates niloticus*) in sterilized fish muscle mixtures. J Agric Food Chem 52:4419–4422

Berrini A, Tepedino V, Borromeo V, Secchi C (2006) Identification of freshwater fish commercially labelled "perch" by isoelectric focusing and two-dimensional electrophoresis. Food Chem 96:163–168

Böhme K, Fernández-No IC, Barros-Velázquez J, Gallardo JM, Cañas B, Calo-Mata P (2011) Rapid species identification of seafood spoilage and pathogenic Gram-positive bacteria by MALDI-TOF mass fingerprinting. Electrophoresis 32:2951–2965

Calo-Mata P, Sotelo CG, Pérez-Martín RI, Rehbein H, Hold GL, Russell VJ, Pryde S, Quinteiro J, Rey-Méndez M, Rosa C, Santos AT (2003) Identification of gadoid fish species using DNA-based techniques. Eur Food Res Technol 217:259–264

Carrera M (2008) Proteómica bottom-up y top-down de organismos poco secuenciados. Secuenciación *de novo* de péptidos biomarcadores para la identificación de especies de la familia Merlucciidae. Tesis Doctoral, Vigo University

Carrera M, Cañas B, Piñeiro C, Vázquez J, Gallardo JM (2006) Identification of commercial hake and grenadier species by proteomic analysis of the parvalbumin fraction. Proteomics 6:5278–5287

Carrera M, Cañas B, Piñeiro C, Vázquez J, Gallardo JM (2007) *De novo* mass spectrometry sequencing and characterization of species-specific peptides from nucleoside diphosphate kinase B for the classification of commercial fish species belonging to the family Merlucciidae. J Proteome Res 6:3070–3080

Carrera M, Barros L, Cañas B, Gallardo JM (2009) Discrimination of South African commercial fish species (*Merluccius capensis* and *Merluccius paradoxus*) by LC-MS/MS analysis of the protein aldolase. J Aquat Food Prod Tech 18:67–78

Carrera M, Cañas B, Vázquez J, Gallardo JM (2010) Extensive de novo sequencing of new parvalbumin isoforms using a novel combination of bottom-up proteomics, accurate molecular mass measurement by FTICR-MS and selected MS/MS ion monitoring. J Proteome Res 9:4393–4406

Carrera M, Cañas B, López-Ferrer D, Piñeiro C, Vázquez J, Gallardo JM (2011) Fast monitoring of species-specific peptide biomarkers using high-intensity-focused-ultrasound-assisted tryptic digestion and selected MS/MS ion monitoring. Anal Chem 83:5688–5695

Castillo AG, Martínez JL, García-Vázquez E (2003) Identification of Atlantic hake species by a simple PCR-based methodology employing microsatellite loci. J Food Prot 66:2130–2134

Céspedes A, García T, Carrera E, González I, Fernández A, Asensio L, Hernández PE, Martín R (1999a) Indirect enzyme-linked immunosorbent assay for the identification of sole (*Solea solea*), European plaice (*Pleuronectes platessa*), flounder (*Platichthys flesus*), and Greenland halibut (*Reinhardtius hippoglossoides*). J Food Prot 62:1178–1182

Céspedes A, García T, Carrera E, González I, Fernández A, Hernández PE, Martín R (1999b) Identification of sole (*Solea solea*) and Greenland halibut (*Reinhardtius hippoglossoides*) by PCR amplification of the 5S rDNA gene. J Agric Food Chem 47:1046–1050

Chapela MJ, Sánchez A, Suárez MI, Pérez-Martín RI, Sotelo CG (2007) A rapid methodology for screening hake species (*Merluccius* spp.) by single-stranded conformation polymorphism analysis. J Agric Food Chem 55:6903–6909

Chen TY, Shiau CY, Wei CI, Hwang DF (2004) Preliminary study on puffer fish proteome species identification of puffer fish by two-dimensional electrophoresis. J Agric Food Chem 52:2236–2241

Chen S, Zhang J, Chen W, Xu D, Zhou Y (2011) Species identification of grouper and snapper in Taiwan Strait using polymerase chain reaction-restriction fragment length polymorphism analysis and lab-on-a-chip system. Se Pu 29:677–680

Colombo F, Mangiagalli G, Renon P (2005) Identification of tuna species by computer-assisted and cluster analysis of PCR-SSCP electrophoretic patterns. Food Control 16:51–53

Comesaña AS, Abella P, Sanjuan A (2003) Molecular identification of five commercial flatfish species by PCR-RFLP analysis of a 12S rRNA gene fragment. J Sci Food Agric 83:752–759

Council Regulation (EC) No 104/2000 of 17 December 1999 on the common organization of the markets in fishery and aquaculture products. Off J Eur Com L 17:22
Council Regulation (EC) No 2065/2001 of 22 October 2001 laying down detailed rules for the application of Council Regulation (EC) No 104/2000 as regards informing consumers about fishery and aquaculture products. Off J Eur Com, L:278, 6
Domínguez E, Pérez MD, Puyol P, Calvo M (1997) Use of immunological techniques for detecting species substitution in raw and smoked fish. Z Lebensm Unters Forsch A 204:279–281
Elsayed S, Bennich H (1975) The primary structure of allergen M from cod. Scand J Immunol 4:203–208
Eng JK, McCormack AL, Yates JR III (1994) An approach to correlate tandem mass spectral data of peptides with amino acid sequences in a protein database. J Am Soc Mass Spectrom 5:976–989
FAO, SOFIA (2010) The state of world fisheries and aquaculture 2010. ISSN 1020–5489
FAO yearbook (2000) Fishery statistics. Catches. Vol 86/1
Gajewski KG, Chen YT, Hsieh YH (2009) Production and characterization of monoclonal antibodies specific to pangasius catfish, basa and tra. J Food Sci 74:C241–C247
Guidance for Industry (2012) The seafood list – FDA's guide to acceptable market names for seafood sold in interstate commerce
Hooper L, Thompson RL, Harrison RA, Summerbell CD, Ness AR, Moore HJ, Worthington HV, Durrington PN, Higgins JP, Capps NE, Riemersma RA, Ebrahim SB, Davey SG (2006) Risk and benefits of omega 3 fats for mortality, cardiovascular disease, and cancer: systematic review. Br Med J 332:752–760
Huang TS, Marshall MR, Kuo KJ, Otwell WS, Wei CI (1995) Development of monoclonal antibodies for red snapper (*Lutjanus campechanus*) identification using enzyme linked immunosorbent assay. J Agric Food Chem 43:2301–2307
Jérôme M, Lemaire C, Verrez-Bagnis V, Etienne M (2003) Direct sequencing method for species identification of canned sardine and sardine-type products. J Agric Food Chem 51:7326–7332
Kawai Y, Uematsu S, Shinano H (1992) Effect of heat-treatment on some physicochemical properties and emulsifying activity of carp sarcoplasmic protein. Nippon Suisan Gakkai 58:1327–1331
Keenan CP, Shaklee JB (1985) Electrophoretic identification of raw and cooked fish fillets and other marine products. Food Tech Australia 37:117–128
Klose J (1975) Protein mapping by combined isoelectric focusing and electrophoresis of mouse tissues. A novel approach to testing for induced point mutations in mammals. Humangenetik 26:231–243
Kochzius M, Nölte M, Weber H, Silkenbeumer N, Hjörleifsdottir S, Hreggvidsson GO, Marteinsson V, Kappel K, Planes S, Tinti F, Magoulas A, Garcia-Vázquez E, Turan C, Hervet C, Campo Falgueras D, Antoniou A, Landi M, Blohm D (2008) DNA microarrays for identifying fishes. Mar Biotechnol 10:207–217
López JL, Marina A, Álvarez G, Vázquez J (2002) Application of proteomics for fast identification of species-specific peptides from marine species. Proteomics 2:1658–1665
Ma B, Zhang K, Hendrie C, Liang C, Li M, Doherty-Kirby A, Lajoie G (2003) PEAKS: powerful software for peptide *de novo* sequencing by tandem mass spectrometry. Rapid Commun Mass Spectrom 17:2337–2342
Mackie I, Craig A, Etienne M, Jérôme M, Fleurence J, Jessen F, Smelt A, Kruijt A, Malmheden Yman I, Ferm M, Martinez I, Pérez-Martín R, Piñeiro C, Rehbein H, Kündiger R (2000) Species identification of smoked and gravad fish products by sodium dodecylsulphate polyacrylamide gel electrophoresis, urea isoelectric focusing and native isoelectric focusing: a collaborative study. Food Chem 71:1–7
Mairs DF, Sindermann CJ (1962) A serological comparison of five species of Atlantic clupeoid fishes. Biol Bull 123:330–343
Maldini M, Marzano FN, Fortes GG, Papa R, Gandolfi G (2006) Fish and seafood traceability based on AFLP markers: elaboration of a species database. Aquaculture 261:487–494
Martinez I, Jakobsen Friis TJ (2004) Application of proteome analysis to seafood authentication. Proteomics 4:347–354

Martinez I, Jakobsen Friis T, Seppola M (2001) Requirements for the application of protein sodium dodecyl sulfate-polyacrylamide gel electrophoresis and randomly amplified polymorphic DNA analyses to product speciation. Electrophoresis 22:1526–1533

Mazzeo MF, de Giulio B, Guerriero G, Ciarcia G, Malorni A, Russo GL, Siciliano RA (2008) Fish authentication by MALDI-TOF mass spectrometry. J Agric Food Chem 56:11071–11076

McNulty ST, Klesius PH (2005) Development of an indirect enzyme-linked immunoabsorbent assay using a monoclonal antibody to identify *Ictalurus* sp. fillets. Aquacult Res 36:1279–1284

O'Farrell PH (1975) High resolution two-dimensional electrophoresis of proteins. J Biol Chem 250:4007–4021

Ortea I, Cañas B, Gallardo JM (2009) Mass spectrometry characterization of species-specific peptides from arginine kinase for the identification of commercially relevant shrimp species. J Proteome Res 8:5356–5362

Pandey A, Mann M (2000) Proteomics to study genes and genomes. Nature 405:837–846

Partis L, Wells RJ (1996) Identification of fish species using random amplified polymorphic DNA (RAPD). Mol Cell Probes 10:435–441

Pérez J, García-Vázquez E (2004) Genetic identification of nine hake species for detection of commercial fraud. J Food Prot 67:2792–2796

Pérez M, Álarez C, Balado M, Cabado AG, Vieites JM, Presa P (2004) Identification of South Atlantic hakes (*Merluccius australis* and *Merluccius hubbsi*) in processed foods by PCR-RFLPs of cytochrome b gene. J Aquat Food Product Technol 13:59–67

Piggot GM (1976) New approaches to marketing fish. In: Altschul AM (ed) New protein foods. Academic, New York, pp 1–37

Piñeiro C, Barros-Velázquez J, Sotelo CG, Pérez-Martín RI, Gallardo JM (1998) Two dimensional electrophoretic study of the water-soluble protein fraction in white muscle of gadoid fish species. J Agric Food Chem 46:3991–3997

Piñeiro C, Barros-Velázquez J, Pérez-Martín RI, Martínez I, Jacobsen T, Rehbein H, Kündiger R, Mendes R, Etienne M, Jerome M, Craig A, Mackie IM, Jessen F (1999a) Development of a sodium dodecyl sulfate-polyacrylamide gel electrophoresis reference method for the analysis and identification of fish species in raw and heat-processed samples: a collaborative study. Electrophoresis 20:1425–1432

Piñeiro C, Barros-Velázquez J, Sotelo CG, Gallardo JM (1999b) The use of two dimensional electrophoresis in the characterization of the water-soluble protein fraction of commercial flat fish species. Z Lebensm Unters Forsch A 208:342–348

Piñeiro C, Barros-Velázquez J, Pérez-Martín RI, Gallardo JM (2000) Specific enzyme detection following isoelectric focusing as a complimentary tool for the differentiation of related Gadoid fish species. Food Chem 70:241–245

Piñeiro C, Vázquez J, Marina AI, Barros-Velázquez J, Gallardo JM (2001) Characterization and partial sequencing of species-specific sarcoplasmic polypeptides from commercial hake species by mass spectrometry following two-dimensional electrophoresis. Electrophoresis 22:1545–1552

Quinteiro J, Vidal R, Izquierdo M, Sotelo CG, Chapela MJ, Pérez-Martín RI, Rehbein H, Hold GL, Russel VJ, Pryde SE, Rosa C, Santos AT, Rey-Méndez M (2001) Identification of hake species (*Merluccius* genus) using sequencing and PCR-RFLP analysis of mitochondrial DNA control region sequences. J Agric Food Chem 49:5108–5114

Rasmussen RS, Morrissey MT (2009) Application of DNA-based methods to identify fish and seafood substitution on the commercial market. Compr Rev Food Sci Food Safety 8:118–154

Regulatory Fish Encyclopedia, FDA (2011) (http://www.fda.gov/Food/FoodSafety/Product-Specific Information/Seafood/RegulatoryFishEncyclopediaRFE/ucm219129.htm)

Rehbein H, Kress G, Schmidt T (1997) Application of PCR-SSCP to species identification of fishery products. J Sci Food Agric 74:35–41

Rehbein H, Kündiger R, Pineiro C, Perez-Martin RI (2000) Fish muscle parvalbumins as marker proteins for native and urea isoelectric focusing. Electrophoresis 21:1458–1463

Royal Decree 1380/2002 of Dec 20 (BOE No. 3, 3 Jan 2003) from Ministry of Agriculture, Fisheries and Food, Spain

Royal Decree 1702/2004 of Jul 16 (BOE No. 172, 17 Jul 2004) from Ministry of Agriculture, Fisheries and Food, Spain

Royal Decree 121/2004 of Jan 23 (BOE No. 31, 5 Feb 2004) from Ministry of Agriculture, Fisheries and Food, Spain

Russell VJ, Hold GL, Pryde SE, Rehbein H, Quinteiro J, Rey-Méndez M, Sotelo CG, Pérez-Martín RI, Santos AT, Rosa C (2000) Use of restriction fragment length polymorphism to distinguish between salmon species. J Agric Food Chem 48:2184–2188

Sánchez A, Quinteiro J, Rey-Méndez M, Pérez-Martín RI, Sotelo CG (2009) Identification of European hake species (*Merluccius merluccius*) using real-time PCR. J Agric Food Chem 57:3397–3403

Santaclara FJ, Cabado AG, Vieites JM (2006) Development of a method for genetic identification of four species of anchovies: *E. encrasicolus*, *E. anchoita*, *E. ringens* and *E. japonicus*. Eur Food Res Technol 223:609–614

Scigelova M, Maroto F, Dufresne C, Vázquez J (2007) High-throughput *de novo* sequencing. http://www.thermo.com/

Scobbie AE, Mackie IM (1988) The use of sodium dodecylsulphate polyacrylamide gel electrophoresis is fish species identification -a procedure suitable for cooked and raw fish. J Sci Food Agricul 44:343–351

Sicherer SH, Sampson HA (2010) Food allergy. J Allergy Clin Immunol 125(2 Suppl 2):S116–S125

Sotelo CG, Pérez-Martín RI (2007) Fish and seafood authentication. In: *Authentication of food and wine* (SE Ebeler, GR Takeoka, P Winterhalter, eds.), pp. 126–137. ACS Symposium Series 952, Washington (USA)

Sotelo CG, Calo-Mata P, Chapela MJ, Pérez-Martín RI, Rehbein H, Hold GL, Russell VJ, Pryde S, Quinteiro J, Izquierdo M, Rey-Méndez M, Rosa C, Santos AT (2001) Identification of flatfish (Pleuronectiforme) species using DNA-based techniques. J Agric Food Chem 49:4562–4569

Tepedino V, Berrini A, Borromeo V, Gaggioli D, Cantoni C, Manzoni P, Secchi C (2001) Identification of commercial fish species belonging to the orders pleuronectiformes and gadiformes: library of isoelectric focusing patterns. J AOAC Int 84:1600–1607

Trotta M, Schönhuth S, Pepe T, Cortesi ML, Puyet A, Bautista JM (2005) Multiplex PCR method for use in real-time PCR for identification of fish fillets from grouper (*Epinephelus* and *Mycteroperca* species) and common substitute species. J Agric Food Chem 53:2039–2045

Zhang Q, Rasco BA (1996) Quantitative Western blotting method for determining the species of origin in surimi. IFT annual meeting: book of abstracts. p. 146. ISSN 1082–1236

Zhang J, Huang H, Cai Z, Huang L (2006) Species identification in salted products of red snappers by semi-nested PCR-RFLP based on the mitochondrial 12S rRNA gene sequence. Food Control 17:557–563

Chapter 13
Proteomics in Milk and Milk Processing

Qiang Zhang and Catherine J. Carpenter

13.1 Introduction

Milk is a key source of food for mammals to feed their young offspring. In addition, cow's milk, a major agroeconomic product, continues to be consumed by humans in various forms throughout their lifetimes. Milk proteins provide essential nutrition for health, growth, and development. Although many are thoroughly digested in the gut to provide vital amino acids, other proteins are either partially or minimally broken down and exert functionality related to the structures of their digestive products (Knip et al. 2010; Castell et al. 1997; Sanderson and Walker 2000). In the example of human milk proteins, some facilitate the digestion and uptake of other nutrients in milk (Hamosh 2001), and others defend against pathogenic bacteria and viruses, help bolster the acquired immune system (Hamosh 2001; Goldman 1993; Schack-Nielsen and Michaelsen 2007), influence cognitive development (Schack-Nielsen and Michaelsen 2007; Daniels and Adair 2005), affect metabolic development (Abbott et al. 1996; Frid et al. 2005; Pal and Ellis 2011; Zivkovic et al. 2011), regulate food intake and satiety (Luhovyy et al. 2007), and aid the development (Karhumaa et al. 2001) and maturation of the gastrointestinal tract (GIT) (Sanderson and Walker 2000; Kanwar and Kanwar 2009; Heird et al. 1984; Harmsen et al. 2000). Despite the increased understanding of the various health and developmental benefits of milk proteins, there is little information available regarding the underlying molecular origins of their effects. Even less information can be found on the development of milk function over the course of lactation where milk proteins may aid differently in newborns and young

Q. Zhang, Ph.D. (✉)
Mead Johnson Nutrition Pediatric Nutrition Institute,
2400 West Lloyd Expressway, Evansville, IN 47721, USA
e-mail: qiang.zhang@mjn.com

C.J. Carpenter, Ph.D.
Department of Chemistry and Biochemistry, University of California-Santa Barbara,
Santa Barbara, CA 93106, USA

F. Toldrá and L.M.L. Nollet (eds.), *Proteomics in Foods: Principles and Applications*, Food Microbiology and Food Safety 2, DOI 10.1007/978-1-4614-5626-1_13,

infants. Numerous efforts have been carried out to identify the "secret ingredients" that may contribute to the above-mentioned benefits conferred by milk proteins (Séverin and Wenshui 2005; Walker 2010; Lönnerdal 2004). One approach is to apply proteomic tools to explore milk proteomes and, it is hoped, from there researchers can gain more insight into biological processes, signaling pathways, nutritional values, and novel functional food ingredients that are novel or whose importance in milk is not yet understood. In the example of human milk, numerous proteomic and nonproteomic studies (Casado et al. 2009; D'Alessandro et al. 2010) have been carried out in the past decade, investigating the proteins in milk whey and milk-fat-globule membrane (MFGM) (Hettinga et al. 2011; Lu et al. 2011; Liao et al. 2011a, b; Mangé et al. 2008; Palmer et al. 2006; Cavaletto et al. 2004; Fortunato et al. 2003). The number of proteins characterized in human milk has increased steadily over the years. Despite this advancement in human milk proteomics, the total number of proteins identified in human milk remains somewhat disappointing, considering that over 1,000 proteins have been identified in blood serum, another human fluid, using a similar experimental setup (Hanash et al. 2008). The deficit in identified human-milk proteins can be mainly ascribed to the lack of approaches that can adequately unmask low-abundance proteins in the presence of the high-abundance ones. Compared to blood serum, milk is even more notoriously dominated by major proteins. Owing to the technical difficulty, there is still no comprehensive view of the milk proteome. As a result, the biology and functions of milk proteins remain incompletely understood, thus inhibiting the further exploration of the dietary benefits that milk can provide.

Our laboratory has since shown that protein fractionation with larger-than-normal sample loads provides a means to explore low-abundance proteins in milk. Using sample loads in the range of a few milligrams and mixed-bed ion-exchange (IEX) protein fractionation, we were able to identify 297 proteins in cow's milk (Le et al. 2011). We recently applied this larger-than-normal load approach to protein fractionation using SDS-PAGE. Upon combining the IEX and SDS-PAGE data, we were able to identify over 600 proteins in cow's milk. A similar setup was applied to human milk and nearly 1,000 proteins were identified (Gao et al. 2012).

Defining the milk proteome and, equally important if not more so, how it changes during the course of lactation are key steps towards an improved understanding of milk biology and function. Obtaining this information has the potential to provide novel insights in the areas of dairy science and food processing. For instance, although the complement system found in blood is well known to play an important role in host defense, its corresponding role in milk is still uncertain (Ogundele 2001). The defense system has the potential to damage host tissue; however, its regulation and potential impact on breastfeeding women is largely undefined. Milk proteins are also observed to promote gastrointestinal development (Sanderson and Walker 2000; Lönnerdal 2004), however, the protein candidates that account for the maturation of the GIT in young infants remain an open question.

The establishment of human and cow's milk proteomes can be used to reveal common functions conferred by cow's milk proteins and their human counterparts.

In addition, the comparison can provide guidance for future experiments to better our understanding of the mechanisms that regulate milk expression and factors that influence milk production and quality. As the comprehensive milk proteome and its changes in protein abundances during lactation become more apparent, it is evident that milk proteins are more than a premium source of amino acids, but an expansive repertoire of functional food ingredients. Our better understanding has the potential to motivate improvement in industrial processing to preserve milk protein functions better. Moreover, it may guide manufacturers in developing improved functional foods based on dairy products. In this chapter, we highlight recent findings in milk proteomes for both humans and cows that can be used to bolster our understanding of novel milk functions. With the improved understanding of milk proteins, we further describe the use of proteomics in characterizing heat-induced protein modifications during industrial processes, which may reduce the nutritional value and function of milk proteins. Finally, we discuss the use of proteomic analysis that can guide the optimization of industrial processing conditions and selection of milk materials.

13.2 Milk Proteome

13.2.1 Protein Constituents in Milk Secretome

Through centrifugation, milk can be separated into MFGM, soluble whey, and casein micelles. MFGM is typically obtained by centrifuging whole milk at low speed at approximately 2,000×g for 20 min at 4°C. The remaining component of milk is termed skim milk, which contains whey proteins and caseins in a micellar form. Skim milk can be further fractionated into soluble milk whey and a casein pellet using ultracentrifugation typically at 100,000×g for 1 h at 4°C (Dewettinck et al. 2008; Keenan 2001). The whey consists of proteins, carbohydrates, and other nutrients that are soluble in water. The casein pellet is made predominantly of casein proteins, minerals, insoluble salts, leukocytes, and cell debris which may slough off from mammary epithelium during lactation or milk production. The relative abundance of soluble whey proteins and micellar caseins changes during lactation. In cow's milk, the whey protein/casein ratios invert from approximately 80:20 in colostrum to 20:80 in mature milk. In human milk, the whey/casein ratios decrease from 90:10 in early lactation to 50:50 in late lactation (Kunz and Lonnerdal 1992).

13.2.2 Functional Milk Proteins

In addition to their innate nutritional value, milk proteins have been implicated as a primary source of functional foods. In general, functional foods are those that are capable of providing specific health benefits above or beyond their nutritional value. Human MFGM (hMFGM) and bovine MFGM (bMFGM) have been the subject of numerous studies in the past 50 years and have received increasing attention lately. MFGM proteins are present at low concentrations in milk, approximately 1–4% of total protein content in milk, with whey proteins and soluble caseins constituting the rest (Cavaletto et al. 2008). Although MFGM proteins would probably have less classical nutritional value as a source of essential amino acids compared to the major protein constituents of milk whey and caseins, they have been found to exert various beneficial functions in the defense mechanisms of anti-inflammation, antitumor, immunoregulation, and in lowering blood cholesterol (Dewettinck et al. 2008; Riccio 2004; Spitsberg 2005; Peterson et al. 2001). Cow's whey proteins are a premium source of essential amino acids compared to many other edible proteins (Smithers 2008). They were found to have antitumor effects against colon, breast, and prostate cancers (Gill and Cross 2000). Indeed, back in the seventeenth century, cow's whey was known to have beneficial effects in host defense, including sepsis, wound healing, and control of "stomach disease" (Smithers 2008). Whey proteins, from both humans and cows, were also found to promote cell proliferation and growth, a fact that is often ascribed to the growth factors and cytokines present in milk whey (Belford et al. 1995). Peptides derived from cow's casein, for example, phosphopeptides and β-casomorphins, have already found interesting applications both as dietary supplements and pharmaceutical preparations (Schlimme and Meisel 1995).

13.2.3 Milk Secretion

Milk production and secretion is a complex process that warrants more basic research to further our understanding of these processes. Such understanding may provide important insights into lactogenesis (milk production), which may, in turn, be used to fine-tune milk volume, fat content, and protein quality in both breastfeeding women and lactating farm animals. The synthesis and secretion of milk by mammary epithelial cells is a highly ordered process that involves two distinct pathways. The aqueous fraction of milk, which contains water-soluble whey proteins and casein micelles, is secreted through an exocytic pathway involving the endoplasmic reticulum (ER), Golgi complex, and secretory vesicles (De Matteis and Luini 2008). The fat fraction of milk is comprised primarily of droplets of triacylglycerols that are surrounded by plasma membranes. The fat droplets are thought to be first released from ER membranes into cytosol as triacylglycerol-rich microlipid droplets, which are partially coated by fat-soluble proteins. Microdroplets fuse with each other and give rise to larger cytoplasmic lipid droplets, which are unidirectionally transported to the apical plasma membrane, enveloped by plasma membrane and released via a budding process (Heid and Keenan 2005). The membrane forms a protective coating surrounding the lipid globules, which allows the dispersion of fat and fat-soluble contents in the milk plasma. The plasma-membrane-enveloped lipid droplets are termed MFGM and the proteins secreted along with the milk fat are termed MFGM proteins. Proteomic tools were used to understand the mechanism of milk secretion. In a mouse study, proteomic analysis suggested that MFGM and its proteins originated from the ER in mammary gland epithelial cells (Wu et al. 2000). An in vitro study of murine mammary epithelial cells suggested that actin cytoskeleton may be involved during milk secretion (Desrivieres et al. 2003).

13.3 Current Status of Milk Proteome Exploration

13.3.1 MFGM Proteome

The MFGM proteome and its protein function are not completely understood. Relatively little information is available regarding the change of MFGM protein levels and the corresponding regulatory mechanism during lactation. The deficiency of such information can be in part ascribed to the lack of cell lines that can secrete milk (Heid and Keenan 2005). Traditional biochemical approaches often address one protein at a time. A proteomic approach, in contrast, yields the identities of many proteins in a single experiment. MFGM represents a diverse source of proteins and we highlight recent analysis of the MFGM proteome which may shed more light onto some of the signaling and secretory pathways used by the mammary gland. In addition, we describe some novel health benefits conferred by milk proteins.

The first extensive proteomic investigation of MFGM was carried out by Reinhardt and Lippolis using SDS-PAGE fractionation of proteins (Reinhardt and Lippolis 2006). A total of 116 proteins were identified in bMFGM. A quantitative approach using isobaric tags for relative and absolute quantitation (iTRAQ) followed by strong cation exchange (SCX) fractionation of tryptic peptides was further applied to measure the developmental changes of bMFGM proteome between day-one colostrum and day-seven milk (Reinhardt and Lippolis 2008). Among the 138 bMFGM proteins identified, 26 are upregulated in colostrum and 19 upregulated in milk with a change greater than or equal to a factor of two. Various bMFGM proteins associated with lipid transport, synthesis, and secretion, such as acyl-CoA synthetase, lanosterol synthase, lysophosphatidic acid acyltransferase, and fatty-acid binding protein were found upregulated in day-7 milk compared to day-1 colostrum. A number of the colostrum-upregulated proteins are associated with biological functions in lipid transport, synthesis, and secretion. In contrast, mucin 1 and 15 were found to be upregulated in day-1 colostrum compared to day-7 milk. Proteins involved in vesicle transport and protein trafficking were also found to exert various regulatory changes between colostrum and milk, which may have implications in milk production and secretion. Using a two-dimensional gel electrophoresis (2DE) workflow, a total of 95 protein spots were characterized for bMFGM. A number of proteins were found to have the same molecular weight but with different isoelectric points (pIs), which implied post-translational modifications for MFGM proteins (Vanderghem et al. 2008). A similar proteomic setup using 2DE and isoelectric focusing (IEF) was employed by Fong et al. (2007).

Using a 2-D-LC approach, namely SCX fractionation of tryptic peptides derived from in vitro protein tryptic digestion followed by reversed-phase LC/MS separation,

Affolter and his coworkers characterized 133 proteins from buttermilk protein (BMP) concentration, a bMFGM-derived industrial product (Affolter et al. 2009). Label-free relative quantification between BMP and whey protein concentrate showed representative proteins uniquely enriched in each of the two sample types. Fatty-acid binding protein, butyrophilin, xanthine dehydrogenase/oxidase, lactadherin, and adipophilin are well-characterized bMFGM proteins and were found at higher abundance in BMP. Similar to the workflow used by Reinhardt and Lippolis, a recent study used SDS-PAGE to fractionate bMFGM and further expanded the proteome to a total of 232 proteins (Hettinga et al. 2011). For human milk, Charlwood et al. used 2DE to separate hMFGM proteins and characterized the major species present (Charlwood et al. 2002). Fortunato et al. used a similar setup and analyzed over 100 protein spots (Fortunato et al. 2003). A recent study used a complementary peptide ligand library (CPLL) to enrich low-abundance proteins and identified 191 proteins in hMFGM (Liao et al. 2011a). A number of identified proteins were found to be related to lipid metabolism and homeostasis. The developmental change of the hMFGM proteome was explored with a number of proteins found to be upregulated in either early or late lactation. In addition to human and cow, the proteomic analysis of MFGM has also been applied to goat, sheep, and other farm animals (Pisanu et al. 2011; Cebo et al. 2012, 2010; Cunsolo et al. 2011a, b). The secretion mechanism of MFGM suggests that it may be rich in cellular proteins during milking. The 200 or so MFGM proteins identified thus far likely constitute a small portion of the entire MFGM proteome. Further exploratory work is needed to gain a more complete picture of MFGM proteomes in various species.

13.3.2 Whey Proteomes

Historically, cow's whey was essentially considered a nuisance by cheese and casein manufacturers. Over time, scientific and technological advancements have diverted whey proteins and other ingredients from the waste stream to a valuable dairy raw material, and help set the foundation for the modern use of whey in the food and related industries. Details of the key developmental steps in the whey industry can be found in the review by Smithers (2008). Bovine whey proteins have an exceptional nutritional value in terms of being a source of essential amino acids. In addition, whey proteins have been widely recognized as an important source of functional ingredients that are involved in a variety of biological processes. A key step in understanding the physiological benefits of whey proteins as a functional food is to determine their molecular identities. Using 2DE, Smolenski identified over 30 whey proteins in cow's milk and a number of them are related to host defense function (Smolenski et al. 2007). Enrichment of low-abundance milk proteins using CPLL led to 149 proteins being identified from cow's whey (D'Amato et al. 2009). Using SCX protein fractionation, 244 proteins were identified in whey-derived protein concentrate (Affolter et al. 2009). Further advancement in proteomics technology has allowed more proteins to be identified in cow's whey in recent years. In our group, we used mixed-bed IEX to fractionate cow's whey proteins and identified 293 proteins (Le et al. 2011). The comparison of colostrum and mature milk indicates a qualitative consistency of proteome over the course of lactation. Recent progress in protein fractionation has further allowed us to expand the cow's milk proteome to over 600 proteins (to be published).

While the cow's whey proteome has been substantially expanded, a parallel effort has been carried out for human whey. With the depletion of high-abundance proteins using immunoprecipitation, Palmer et al. identified 151 proteins in human whey (Palmer et al. 2006). A similar approach using CPLL to enrich low-abundance proteins allowed the identification of 115 human whey proteins (Liao et al. 2011b). A number of proteins were found to have higher abundance in early lactation, for example, α-1-antitrypsin, carbonic anhydrase, galectin-3-binding protein, lactadherin, lipoprotein lipase, and tenascin. In contrast, CD14, fatty-acid binding protein, lysozyme C, and several others were found to have higher abundance during late lactation. In our group, we found that low-abundance proteins can be measured with a larger-than-normal sample load. With this setup, we expanded the proteome of human whey to 976 proteins (Gao et al. 2012). The more in-depth understanding of the human whey proteome allows us to delineate several biological processes and signaling pathways that are novel or whose importance in milk was not previously understood.

13.3.2.1 Immunoglobulins

Host defense is probably the most well-understood functional role conferred by whey proteins. Milk whey contains a variety of antimicrobial proteins that function to both safeguard the lactating mammary gland and provide protection to suckling infants while their immune systems are still immature. Consistent with the defensive roles that milk provides through immunological processes, gene ontology (GO) analysis revealed that milk proteins related to immune system processes are overrepresented when compared to the entire human genome (Fig. 13.1a) (Gao et al. 2012). Among various protective milk proteins, immunoglobulins (Igs) in whey are probably the most well-studied ones. They combat pathogenic infections and enhance gut health during milk intake (Field 2005). For example, immunoglobulin A (IgA) takes a secretory form in milk which is termed secretory IgA (sIgA). sIgA can fight against pathogenic infections in the intestinal lumen in addition to providing its direct nutritional value as a source of amino acids. In contrast to sIgA, another immunoglobulin identified in human whey, immunoglobulin G (IgG) binds to antigens that may be found in the intestinal lumen and transfers them across the epithelial layer via binding to the epithelial neonatal Fc receptor (FcRn), which recognizes the Fc region of IgG, a process termed transcytosis (Rojas and Apodaca 2002). Antigens so transported are then released into the basal propria underlying epithelial layer, which subsequently induce B-cell proliferation, immune activation, or tolerance (Rojas and Apodaca 2002). Igs were successfully identified in several proteomic analyses (Hettinga et al. 2011; Le et al. 2011; Gao et al. 2012; Smolenski et al. 2007). Indeed, proteomics revealed novel expression patterns of Igs. A proteomic approach can be used to measure the changes of expression levels for a given protein under

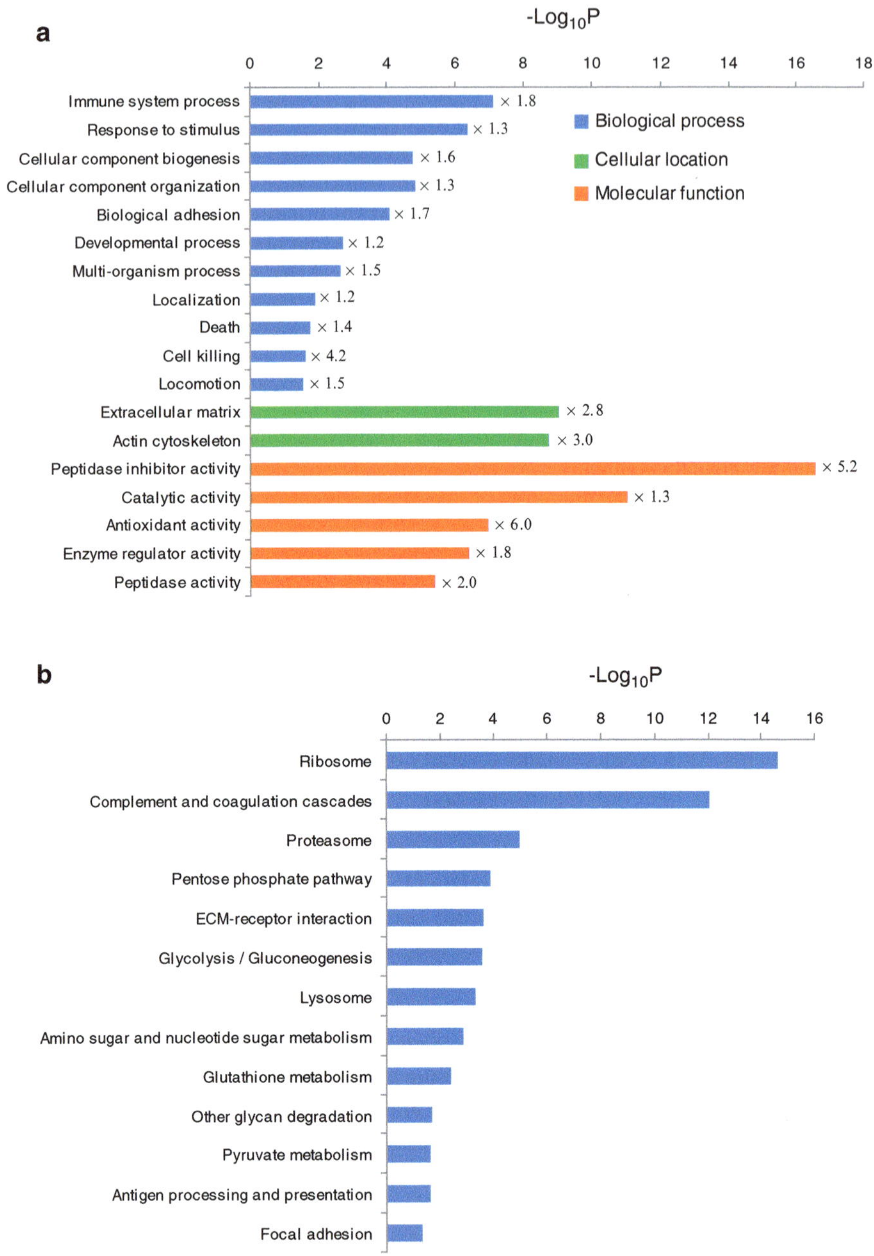

Fig. 13.1 Enrichment analysis. (**a**) Gene ontology (GO) distributions for the human milk proteome in biological processes (*blue*), cellular location (*green*) and molecular function (*orange*). For each category, the factor of enrichment compared to the human average is shown adjacent to data bar. (**b**) KEGG pathways for the human milk proteome. Proteins involved in pentose, glucose, pyruvate, amino/nucleotide sugar and glycan metabolism are all significantly enriched (Data from Gao et al. 2012)

various conditions, for example, between early and mature lactation. Our group found that sIgA and immunoglobulin M (IgM) are more abundant in early lactation, whereas IgG is more abundant in late lactation (Fig. 13.2). The difference in expression between sIgA and IgG suggests a transformation in milk's function from supporting direct pathogen-killing as conferred by sIgA in early lactation towards an antigen intake facilitated by IgG that aims to develop an infant's own immunity as lactation progresses. Interestingly, levels of IgG in newborns are comparable to those of their mothers because of transplacental transport of maternal IgG. Infants delay production of their own IgG until about six months after birth. The combination of late production of endogenous IgG and catabolism of maternal IgG leads to a transient deficiency of IgG in infants during the period from birth to one year of age (Murphy 2012). The increasing supply of IgG from transitional milk to mature milk may provide a means to complement the synchronous decrease in IgG during early infancy. The increased understanding of developing immune functions may be applied to the manufacture of functional food with optimal defense function at various stages of human development.

13.3.2.2 Complement System

In addition to measuring expressional changes of a given protein, proteomic analysis can also be utilized to compare relative abundances among proteins. Higher protein concentration is associated in general with a higher number of MS/MS events (spectral counts) (Zhang et al. 2011). The spectral counts of each protein were used to assess relative protein abundances in milk. Using this approach, we found that sIgA presents in milk at a higher abundance than IgG and IgM (Fig. 13.2). The approach was also used to explore the relative abundance of other proteins involved in the complement system. The complement system is an important component of the immune system that protects the body against pathogens. There are two major pathways that can activate the complement system. In what is known as the "classical" pathway, the complement system works in conjunction with antibodies that recognize the surface of pathogens. In the "alternative" pathway, the microbial surfaces directly activate the complement system and the presence of antibodies is not required (Murphy 2012). We found that the spectral counts span a wide range for the complement system proteins identified, from approximately 6,000 counts for C3 to 2 counts for C1S, suggesting a likely broad dynamic range of concentration for these species. KEGG pathway analysis (Huang et al. 2009a, b) reveals that the complement cascade is among the most significantly enriched pathways in human milk (Fig. 13.1b). Among various complement components, C1r, C1s, and C2 are specific to the classical pathway whereas CFB, CFI, and CFH are characteristic of the alternative pathway. Our proteomic analysis showed higher abundance for CFB, CFI, and CFH than for C1r, C1s, and C2 (Fig. 13.3). The higher levels of CFB, CFI, and CFH in relation to C1 and C2, suggest a predominant role for the alternative pathway compared to the C1- and C2-mediated classical pathway.

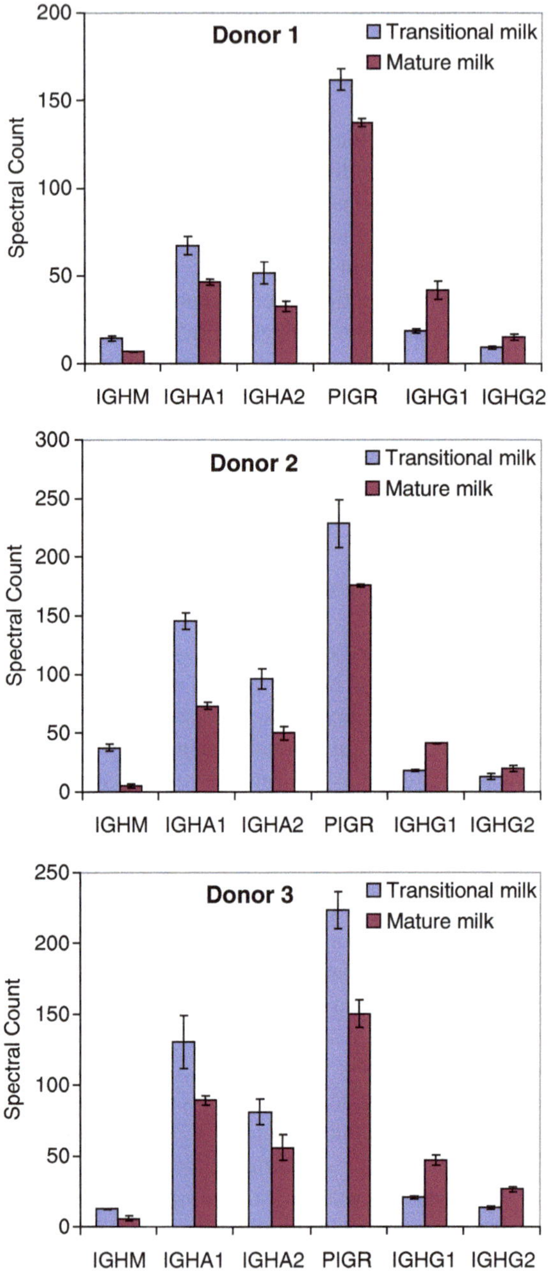

Fig. 13.2 Relative-abundance distribution of IgM (IGHM), sIgA (IGHA1, IGHA2 and PIGR) and IgG (IGHG1 and IGHG2) in transitional milk and mature milk from three individual donors (Data from Gao et al. 2012)

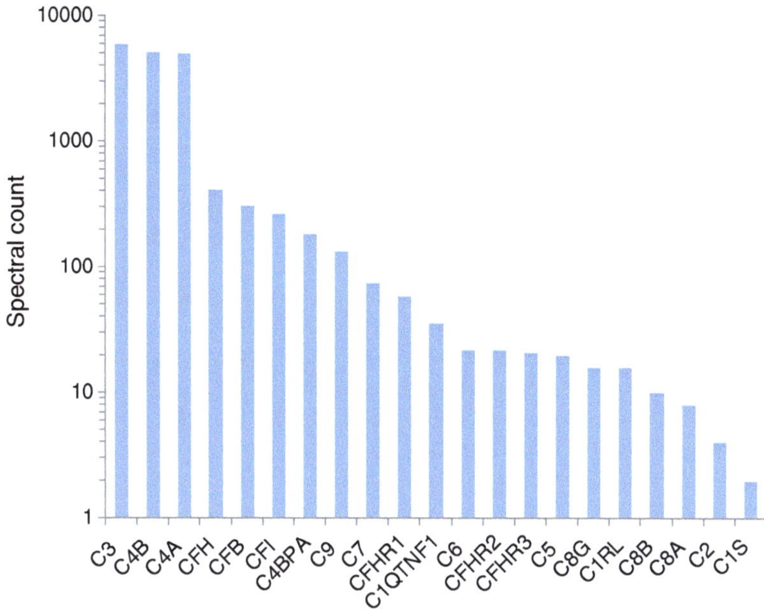

Fig. 13.3 Spectral-count distribution of complement components and complement component-related proteins in milk (Data from Gao et al. 2012)

13.3.2.3 Transepithelial Transport of Milk Proteins

Another key feature in developing functional-food concepts is to assess the transport of functional ingredients during intestinal uptake. *Intracellular transcytosis* is an important process in both milk protein secretion by mammals and uptake by their offspring. As mentioned above, both the mammary secretion of sIgA and the intestinal uptake of IgG occur via transcytosis. Another mechanism for nutrient release and uptake is via *intercellular transepithelial transport*. Albumin and several other proteins in milk were thought to be derived from blood serum during lactation. The observation of these proteins in milk, however, indicates that, in fact, they undergo intercellular transepithelial transport in the mammary gland during milk production. Intestinal transepithelial transport of small peptides was previously found in cell cultures, demonstrating the possible uptake of the nutrient species (Satake et al. 2002; Shimizu et al. 1997). Immune factors (e.g., lymphocytes) were also found to transverse the neonatal intestine (Hanson et al. 2003). An important class of milk proteins belongs to those often found in the extracellular matrix (ECM). ECM proteins act as a structural support for cells and connective tissue (Hynes 2009). In addition, cellular signaling occurs through ECM proteins that play a vital role in the differentiation, proliferation, and migration of cells through adhesion receptors, for example, integrins (Hynes 2009; Morrison and Cutler 2010). Our group found that proteins with a cellular location designation of ECM are significantly overrepresented in human milk compared to the entire human genome (Fig. 13.1a) (Gao et al. 2012).

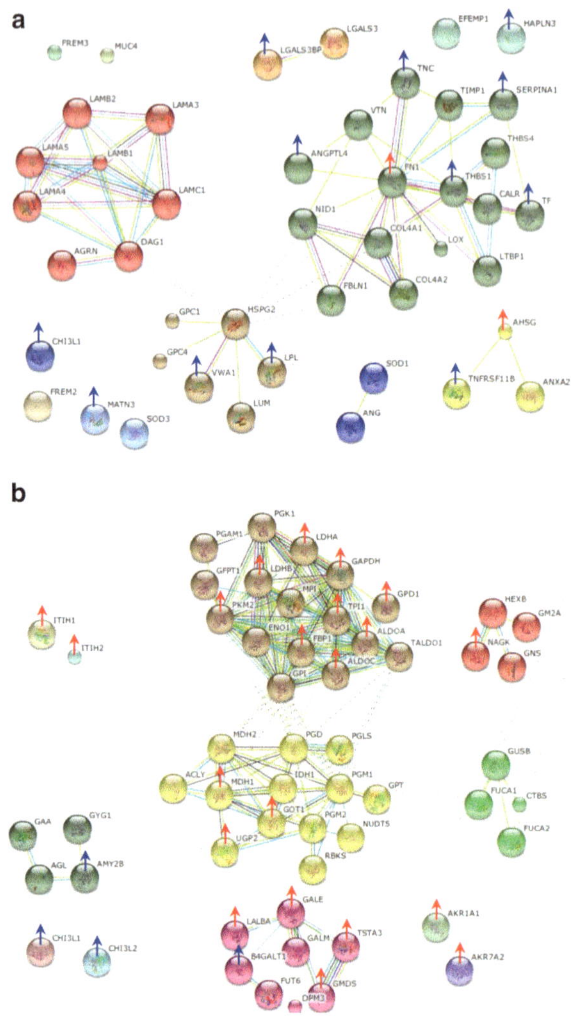

Fig. 13.4 Analysis of interaction networks of (**a**) extracellular matrix proteins and (**b**) carbohydrate metabolism proteins using STRING. Densely connected networks were clustered and color-coded. Proteins upregulated in transitional milk and mature milk are labeled with *blue* and *red arrows*, respectively (Data from Gao et al. 2012)

We further noticed that many milk ECM proteins have extraordinarily high abundance in early lactation (Fig. 13.4a) (Gao et al. 2012). Matrix metalloproteinases (MMPs) cleave ECM to allow side branching of mammary alveoli during pregnancy, which probably leads to the enrichment of maternal ECM proteins in the mammary gland (Wiseman and Werb 2002). Such side branching was thought to halt when lactation begins. The observation of high levels of ECM proteins in milk during early lactation implies that ECM proteins can cross the mammary-gland epithelium, probably by transepithelial transport in a manner similar to the transport of serum albumin into milk. The rapid decrease of milk ECM proteins after early lactation supports the idea that ECM degradation in the mammary gland halts when lactation begins. Based on this indication that ECM proteins cross the epithelium in the mammary glands of the mother, we conjectured that, in the infant, ECM proteins may transport across intestinal epithelial cells to reach the basal propria underneath where they can perform their functions (Gao et al. 2012).

13.3.2.4 Cytokines and Chemokines

In addition to contributing to developmental functions through ECM-receptor interactions, the ECM also acts as a reservoir of cytokines that are vital in developmental processes (Hynes 2009). Milk is known to contain a diverse array of cytokines and chemokines (Field 2005). Typical cytokines include interleukins (ILs) and growth factors, such as transforming growth factor β (TGF-β) (Bottcher et al. 2003, 2000a, b; Bryan et al. 2006; Hawkes et al. 2001). Although cytokines are used in protecting the host, they are also transferred by the immune system to newborns and infants to help them defend against infection. In the example of milk cytokines (Field 2005), they contribute to important immunoregulatory functions, including regulation of sIgA secretion and maturation of the infant's intestinal immune system (Sanderson and Walker 2000; Bottcher et al. 2000a; Donnet-Hughes et al. 2000). With a proteomic approach, a number of cytokines, cytokine receptors, and chemokines have been newly identified in human milk (Gao et al. 2012). In the examples of tumor necrosis factor ligand superfamily members 11B, –13 (TNFSF11B, –13) and nicotinamide phosphoribosyltransferase (NAMPT), GO analysis shows that they play an important role in immune cell development. In cow's whey, a proteomic approach allowed for the identification of the low-abundance protein TGF-β2 and its receptor (Le et al. 2011).

13.3.2.5 Regulation of Milk Nutrients

In recent decades, it has been suggested that milk proteins play additional functional roles in the regulation and development of the immune system, maturation of the gastrointestinal tract, and in glucose homeostasis (Schack-Nielsen and Michaelsen 2007; Abbott et al. 1996; Frid et al. 2005; Pal and Ellis 2010; German et al. 2002). Perhaps an obvious question to ask about milk as a functional food source is, "Can the constituent proteins of milk regulate milk's own nutritional content?" In an indication that this is possible, changes in the abundance of MFGM proteins involved in lipid metabolism may play a role in regulating fat content in milk (Reinhardt and Lippolis 2008). Our proteomic observations also suggest a possible feedback control mechanism conferred by whey proteins in regulating milk carbohydrate contents (Gao et al. 2012). It is known that milk is rich in carbohydrate content that increases from transitional milk to mature milk (Jenness 1979). Pathway analysis of milk proteins showed a diverse array of them are involved in carbohydrate metabolism (Fig. 13.1b) with many being upregulated in mature milk (Fig. 13.4b). The upregulation in mature milk of metabolic proteins involved in various carbohydrate pathways reflects the transforming metabolic activities in breastfeeding women who provide increased milk carbohydrate content to their babies during later lactation.

13.4 Digestion and Bioavailability of Milk Proteins

Milk immunoglobulins and immunoregulatory proteins are active in the digestive tract of newborns. In the example of bovine milk immunoglobulins, IgG, IgA, and IgM, over 20% of ingested immunoglobulins were found to survive gastric digestion in humans and retain their antimicrobial activities (Roos et al. 1995). Several cow's milk proteins were also found to be digestion-resistant in the human digestive tract (Castell et al. 1997; Harris et al. 2010; Clare et al. 2003; Lönnerdal 2003; Mahe et al. 1996; Sorva et al. 1994). In our group, we utilized a proteomics approach to measure the digestion of milk proteins using an in vitro model mimicking the human stomach. The kinetics of protein degradation was determined for more than 200 proteins. Various quantities of proteins were identified after excessive exposure to digestion conditions, which suggests the common theme of incomplete hydrolysis of milk proteins during gastric digestion (to be published).

There is growing evidence of the functional activities exerted by intact or poorly broken down proteins, however, milk also contains a large number of bioactive peptides with various biological activities (Clare and Swaisgood 2000; Ricci-Cabello et al. 2012; Nagpal et al. 2011). The peptides can either exist naturally in milk or be derived from protein precursors via hydrolysis. Probably through receptor binding, transcytosis, or transepithelial transport, these peptides may act as potential physiological modulators during the gastrointestinal digestion of milk. For example, apelin peptides have been thought to contribute to the gastrointestinal development of newborns and were recently identified in cow's colostrum and milk (Mesmin et al. 2011). Several comprehensive reviews covering many types of milk-derived bioactives have been published (Pihlanto and Korhonen 2003; Korhonen and Pihlanto 2007, 2003). Although details of the activation mechanism remain unknown, proteolytic digestion under gastric conditions has also been proposed as a means to produce bioactive components such as partially digested but still active proteins and peptides (Gobbetti et al. 2002). Proteomic analysis of cow's milk and human milk revealed that milk proteins are overpopulated by proteases (Le et al. 2011; Gao et al. 2012). In vitro models suggest that some cytokines and chemokines remain inactive and can be proteolytically activated during gastric digestion (Calhoun et al. 1999). Recently, milk TGF-β protein was found to be retained during in vitro digestion and is thus able to improve gut integrity (Hering et al. 2011).

13.5 Milk Processing

13.5.1 Heat-Induced Protein Modifications

Milk whey is a valuable protein source in that it confers a high content of essential amino acids in a ratio similar to the required composition for human nutrition. Moreover, some proteins are resistant to gastric digestion and contribute to a wide variety of physiological activities in the intestinal lumen, blood stream, and organ tissues. Cow's whey proteins have received particular attention because whey is a major human food and agroeconomic product. During industrial processing, however, liquid whey is usually heated to achieve sterile liquid products or to obtain milk powder, for example, whey protein concentrate (WPC), using high-temperature spray-drying. The heat treatment can impact the nutritional value of whey proteins in terms of their amino acid bioavailability and protein and peptide function (Mauron 1990).

The major whey proteins α-lactalbumin and β-lactoglobulin have been used as models to explore the effects of thermal processing on the reaction of milk proteins with lactose (Jones et al. 1998; Meltretter et al. 2007; Morgan et al. 1997). Several groups have established that the nonenzymatic Maillard reaction is the primary modification that occurs to milk proteins during thermal processing (Luz Sanz et al. 2007; Monaci and van Hengel 2007; Fenaille et al. 2004; Meltretter and Pischetsrieder 2008). In the reaction, the carbonyl group of milk lactose can react with lysine residues in proteins, leading to the formation of Amadori products (Acharya and Manning 1980). The Amadori products can undergo further reactions, leading to a wide variety of chemical structures, namely, advanced glycation end-products (AGEs). Additional chemical changes induced by heat treatment include lactulosyl-lysine degradation into advanced glycol-oxidation products, such as N-carboxymethyl-lysine, oxalic acid monolysinylamide, pentosidine, and galactosyl-β-pyranone (Meltretter et al. 2007; Arena et al. 2010). More recently, protein crosslinking of α- and β-caseins and deamidation of α_{s1}-casein were also observed during storage of milk samples (Holland et al. 2011). In Amadori products, blocked lysine residues are inaccessible by enzymatic digestion in the gastrointestinal tract and may, in that way, influence protein digestibility (Seiquer et al. 2006; Carbonaro 2006). Similar to the heat-induced lactosylation of lysine residues, *in vitro* enzymatic protein glycosylation is a common motif observed in raw milk proteins. For example, of 976 whey proteins we identified in human milk, over 300 are classified as glycoproteins based on GO analysis. Reduced digestibility of glycoproteins compared to nonglycosylated proteins may make them a source of a wide variety of bioactive peptides that can exert beneficial biological functions during food protein intake.

13.5.2 Proteomic Characterization of Protein Modifications

During industrial processes, the degrees of thermal treatment increase as one goes from pasteurization (75–85°C for 15–30 s) to ultra-high temperature (UHT) sterilization (138–145°C for 2–5 s) to spray-drying (van Boekel 1998). The extent of Maillard reaction occurrence is an important metric in assessing the quality of milk proteins under various heat processes. The extent of protein lactosylation could depend largely on the thermal procedures used during industrial milk processing. The capability of proteomics in determining protein modifications has long stimulated interest in using it to assess the effects of heat treatment on protein lactosylation and milk quality. Various proteomic approaches measuring protein lactosylation have been used to evaluate the thermal history of bovine milk proteins and to differentiate their commercial quality (Losito et al. 2007; Siciliano et al. 2000; Czerwenka et al. 2006; Meltretter et al. 2009; Carbonaro 2004).

One approach is to enrich lactosylated protein motifs and subject them to proteomic investigation. Affinity chromatography containing *m*-aminophenylboronic-acid-bound agarose was used to selectively trap lactosylated peptides and characterize their lactosylation sites (Arena et al. 2010). After breaking down milk proteins into peptides by endoproteinase LysC, the immobilized boronic acid moiety on the agarose beads was used to recognize *cis*-diol groups on lactosylated peptides, which were subsequently characterized by mass spectrometry. A number of lactosylation sites were identified among high-abundance milk proteins with markedly increased lactosylation. The extent increased from pasteurized milk to UHT-treated milk to milk powder, a trend consistent with a greater extent of heat treatment that each of the sample types received. Indeed, it has been demonstrated that the lactosylated protein forms account for as much as 3%, 30%, and 70% of the β-lactoglobulin content in pasteurized, UHT-treated, and spray-dried milk samples, respectively (Monaci and van Hengel 2007; Leonil et al. 1997). Hong et al. used trypsin in their bottom-up proteomic approach to identify proteins in milk-based infant formula (Hong et al. 2007). Trypsin has been used to probe unmodified lysine residues because it selectively cleaves unmodified N-termini at lysine residues, but not at lactosylated sites. The analysis led to a total of 154 peptides assigned to 31 proteins being characterized in infant formula. The observations suggest that a number of protein domains remain unmodified after extensive heat treatment.

Protein modifications can also take place during storage when elevated temperatures are encountered. Proteomics tools were applied to investigate protein cross-linking, deamidation, and lactosylation in UHT-treated milk samples during storage (Holland et al. 2011; Le et al. 2012). Milk samples can be subjected to various heat conditions during storage, transportation, or changes in geographic locations. Stabilities of milk samples at various storage temperatures and durations were examined. At elevated temperature, for example 40°C, extended storage time leads to a higher degree of protein cross-linking of α- and β-caseins, deamidation of α_{s1}-casein, and lactosylation of major whey proteins, which is presumably due to the accelerated thermodynamics at modestly elevated thermal conditions compared to mild conditions.

2DE has been demonstrated to be a powerful tool in visualizing changes that occur to milk proteins under various processes and conditions (Chevalier et al. 2009). Proteins that undergo physicochemical modifications can exhibit different behavior in their electrophoresis separations. Decreased resolution in protein IEF was observed for milk samples with extended storage time (Holland et al. 2011). Despite its successful application in distinguishing native and modified milk proteins, the 2DE approach presents problems in handling and automation. Czerwenka et al. used reversed-phase LC to assess modifications of milk proteins (Czerwenka et al. 2006). A decrease in retention time was observed with an increasing number of lactose units attached to a protein. Attachment of the highly hydrophilic lactose to a lysine residue increases protein hydrophilicity and hence reduces the retention time of proteins in reversed-phase chromatography.

Optimization of industrial processing conditions and careful selection of raw materials can effectively reduce Maillard reactions in processed milk (Cattaneo et al. 2009). To further preserve the nutrient value of whey proteins, the development of new industrial processes would be required, from raw materials to finished products, that use less heat treatment while achieving the sterile requirements of making safe food products. Affolter et al. used proteomic fingerprints to distinguish dairy products based on different sources of ingredients (Affolter et al. 2009). Absolute quantification of targeted peptides was achieved by spiking samples with stable-isotope-labeled peptide internal standards. The authors suggested that the absolute quantification of peptides derived from in vitro tryptic digestion may be used to guide industrial processes throughout raw material selection, processing, formulation, quality control, and storage. Furthermore, they suggest it could be used in assessing the digestibility, bioactivity, and bioavailability of functional food ingredients that are derived from milk proteins.

13.6 Future Prospects

The recent advancements in milk proteomics can provide a platform for future searches for milk proteins that play vital physiological roles in health. Compelling evidence suggests that proteomics can provide guidance in manufacturing premium-quality protein ingredients based on dairy milk. Significant contributions to this area are likely to come from in-depth proteomic studies of milk from human and various animal species. For example, an improved understanding of human milk will provide a reference point in developing functional food derived from farm animals. Despite the fact that the mammary gland remains extraordinarily active throughout lactation, a thorough proteomic analysis quantifying milk proteins at multiple lactational stages is not yet available. Proteomic measurements of milk subproteomes that may cross the intestinal epithelium via transcytosis or transepithelial transport should help clarify the bioavailability of milk proteins. The application of quantitative approaches, such as stable isotope labeling by amino acids in cell culture (SILAC), remains very limited for milk proteomics. Analysis of stable-isotope-labeled intestinal epithelial cell lines would provide a means to differentiate proteins either being taken up from foods or synthesized internally by intestinal epithelial cells and may provide insights to underlying molecular mechanisms in cellular responses when stimulated by various food sources.

References

Abbott RD, Curb JD, Rodriguez BL, Sharp DS, Burchfiel CM, Yano K (1996) Effect of dietary calcium and milk consumption on risk of thromboembolic stroke in older middle-aged men. The Honolulu Heart Program. Stroke 27(5):813–818

Acharya AS, Manning JM (1980) Amadori rearrangement of glyceraldehyde-hemoglobin Schiff base adducts. A new procedure for the determination of ketoamine adducts in proteins. J Biol Chem 255(15):7218–7224

Affolter M, Grass L, Vanrobaeys F, Casado B, Kussmann M (2009) Qualitative and quantitative profiling of the bovine milk fat globule membrane proteome. J Proteomics 73(6):1079–1088

Arena S, Renzone G, Novi G, Paffetti A, Bernardini G, Santucci A, Scaloni A (2010) Modern proteomic methodologies for the characterization of lactosylation protein targets in milk. Proteomics 10(19):3414–3434

Belford DA, Rogers ML, Regester GO, Francis GL, Smithers GW, Liepe IJ, Priebe IK, Ballard FJ (1995) Milk-derived growth factors as serum supplements for the growth of fibroblast and epithelial cells. In Vitro Cell Dev Biol Anim 31(10):752–760

Bottcher MF, Jenmalm MC, Garofalo RP, Bjorksten B (2000a) Cytokines in breast milk from allergic and nonallergic mothers. Pediatr Res 47(1):157–162

Bottcher MF, Jenmalm MC, Bjorksten B, Garofalo RP (2000b) Chemoattractant factors in breast milk from allergic and nonallergic mothers. Pediatr Res 47(5):592–597

Bottcher MF, Jenmalm MC, Bjorksten B (2003) Cytokine, chemokine and secretory IgA levels in human milk in relation to atopic disease and IgA production in infants. Pediatr Allergy Immunol 14(1):35–41

Bryan DL, Forsyth KD, Gibson RA, Hawkes JS (2006) Interleukin-2 in human milk: a potential modulator of lymphocyte development in the breastfed infant. Cytokine 33(5):289–293

Calhoun DA, Lunoe M, Du Y, Staba SL, Christensen RD (1999) Concentrations of granulocyte colony-stimulating factor in human milk after in vitro simulations of digestion. Pediatr Res 46(6):767–771

Carbonaro M (2004) Proteomics: present and future in food quality evaluation. Trends Food Sci Tech 15:209–216

Carbonaro M (2006) Application of two-dimensional electrophoresis for monitoring gastrointestinal digestion of milk. Amino Acids 31(4):485–488

Casado B, Affolter M, Kussmann M (2009) OMICS-rooted studies of milk proteins, oligosaccharides and lipids. J Proteomics 73(2):196–208

Castell JV, Friedrich G, Kuhn CS, Poppe GE (1997) Intestinal absorption of undegraded proteins in men: presence of bromelain in plasma after oral intake. Am J Physiol Gastr L 273 ((1 36–1)):G139–G146

Cattaneo S, Masotti F, Pellegrino L (2009) Liquid infant formulas: technological tools for limiting heat damage. J Agric Food Chem 57(22):10689–10694

Cavaletto M, Giuffrida MG, Conti A (2004) The proteomic approach to analysis of human milk fat globule membrane. Clin Chim Acta 347(1–2):41–48

Cavaletto M, Giuffrida MG, Conti A (2008) Milk fat globule membrane components–a proteomic approach. Adv Exp Med Biol 606:129–141

Cebo C, Caillat H, Bouvier F, Martin P (2010) Major proteins of the goat milk fat globule membrane. J Dairy Sci 93(3):868–876

Cebo C, Rebours E, Henry C, Makhzami S, Cosette P, Martin P (2012) Identification of major milk fat globule membrane proteins from pony mare milk highlights the molecular diversity of lactadherin across species. J Dairy Sci 95(3):1085–1098

Charlwood J, Hanrahan S, Tyldesley R, Langridge J, Dwek M, Camilleri P (2002) Use of proteomic methodology for the characterization of human milk fat globular membrane proteins. Anal Biochem 301(2):314–324

Chevalier F, Hirtz C, Sommerer N, Kelly AL (2009) Use of reducing/nonreducing two-dimensional electrophoresis for the study of disulfide-mediated interactions between proteins in raw and heated bovine milk. J Agric Food Chem 57(13):5948–5955
Clare DA, Swaisgood HE (2000) Bioactive milk peptides: a prospectus. J Dairy Sci 83(6):1187–1195
Clare DA, Catignani GL, Swaisgood HE (2003) Biodefense properties of milk: the role of antimicrobial proteins and peptides. Curr Pharm Design 9(16):1239–1255
Cunsolo V, Muccilli V, Saletti R, Foti S (2011a) Review: applications of mass spectrometry techniques in the investigation of milk proteome. Eur J Mass Spectrom (Chichester, Eng) 17(4):305–320
Cunsolo V, Muccilli V, Fasoli E, Saletti R, Righetti PG, Foti S (2011b) Poppea's bath liquor: the secret proteome of she-donkey's milk. J Proteomics 74(10):2083–2099
Czerwenka C, Maier I, Pittner F, Lindner W (2006) Investigation of the lactosylation of whey proteins by liquid chromatography-mass spectrometry. J Agric Food Chem 54(23):8874–8882
D'Alessandro A, Scaloni A, Zolla L (2010) Human milk proteins: an interactomics and updated functional overview. J Proteome Res 9(7):3339–3373
D'Amato A, Bachi A, Fasoli E, Boschetti E, Peltre G, Sénéchal H, Righetti PG (2009) In-depth exploration of cow's whey proteome via combinatorial peptide ligand libraries. J Proteome Res 8(8):3925–3936
Daniels MC, Adair LS (2005) Breast-feeding influences cognitive development in Filipino children. J Nutr 135(11):2589–2595
De Matteis MA, Luini A (2008) Exiting the golgi complex. Nat Rev Mol Cell Biol 9(4):273–284
Desrivieres S, Prinz T, Castro-Palomino Laria N, Meyer M, Boehm G, Bauer U, Schafer J, Neumann T, Shemanko C, Groner B (2003) Comparative proteomic analysis of proliferating and functionally differentiated mammary epithelial cells. Mol Cell Proteomics 2(10):1039–1054
Dewettinck K, Rombaut R, Thienpont N, Le TT, Messens K, Van Camp J (2008) Nutritional and technological aspects of milk fat globule membrane material. Int Dairy J 18(5):436–457
Donnet-Hughes A, Duc N, Serrant P, Vidal K, Schiffrin EJ (2000) Bioactive molecules in milk and their role in health and disease: the role of transforming growth factor-beta. Immunol Cell Biol 78(1):74–79
Fenaille F, Morgan F, Parisod V, Tabet JC, Guy PA (2004) Solid-state glycation of beta-lactoglobulin by lactose and galactose: localization of the modified amino acids using mass spectrometric techniques. J Mass Spectrom 39(1):16–28
Field CJ (2005) The immunological components of human milk and their effect on immune development in infants. J Nutr 135(1):1–4
Fong BY, Norris CS, MacGibbon AKH (2007) Protein and lipid composition of bovine milk-fat-globule membrane. Int Dairy J 17(4):275–288
Fortunato D, Giuffrida MG, Cavaletto M, Garoffo LP, Dellavalle G, Napolitano L, Giunta C, Fabris C, Bertino E, Coscia A, Conti A (2003) Structural proteome of human colostral fat globule membrane proteins. Proteomics 3(6):897–905
Frid AH, Nilsson M, Holst JJ, Bjorck IM (2005) Effect of whey on blood glucose and insulin responses to composite breakfast and lunch meals in type 2 diabetic subjects. Am J Clin Nutr 82(1):69–75
Gao X, McMahon RJ, Woo JG, Davidson BS, Morrow AL, Zhang Q (2012) Temporal changes in milk proteomes reveal developing milk functions. J Proteome Res 11:3897–3907
German JB, Dillard CJ, Ward RE (2002) Bioactive components in milk. Curr Opin Clin Nutr 5(6):653–658
Gill HS, Cross ML (2000) Anticancer properties of bovine milk. Br J Nutr 84(Suppl 1):S161–S166
Gobbetti M, Stepaniak L, De Angelis M, Corsetti A, Di Cagno R (2002) Latent bioactive peptides in milk proteins: proteolytic activation and significance in dairy processing. Crit Rev Food Sci 42(3):223–239
Goldman AS (1993) The immune system of human milk: antimicrobial, antiinflammatory and immunomodulating properties. Pediatr Infect Dis J 12(8):664–671
Hamosh M (2001) Bioactive factors in human milk. Pediatr Clin North Am 48(1):69–86
Hanash SM, Pitteri SJ, Faca VM (2008) Mining the plasma proteome for cancer biomarkers. Nature 452(7187):571–579

Hanson LA, Korotkova M, Lundin S, Haversen L, Silfverdal SA, Mattsby-Baltzer I, Strandvik B, Telemo E (2003) The transfer of immunity from mother to child. Ann N Y Acad Sci 987:199–206

Harmsen HJM, Wildeboer-Veloo ACM, Raangs GC, Wagendorp AA, Klijn N, Bindels JG, Welling GW (2000) Analysis of intestinal flora development in breast-fed and formula-fed infants by using molecular identification and detection methods. J Pediatr Gastroenterol Nutr 30(1):61–67

Harris P, Johannessen KM, Smolenski G, Callaghan M, Broadhurst MK, Kim K, Wheeler TT (2010) Characterisation of the anti-microbial activity of bovine milk ribonuclease4 and ribonuclease5 (angiogenin). Int Dairy J 20(6):400–407

Hawkes JS, Bryan DL, Neumann MA, Makrides M, Gibson RA (2001) Transforming growth factor beta in human milk does not change in response to modest intakes of docosahexaenoic acid. Lipids 36(10):1179–1181

Heid HW, Keenan TW (2005) Intracellular origin and secretion of milk fat globules. Eur J Cell Biol 84(2–3):245–258

Heird WC, Schwarz SM, Hansen IH (1984) Colostrum-induced enteric mucosal growth in beagle puppies. Pediatr Res 18(6):512–515

Hering NA, Andres S, Fromm A, van Tol EA, Amasheh M, Mankertz J, Fromm M, Schulzke JD (2011) Transforming growth factor-beta, a whey protein component, strengthens the intestinal barrier by upregulating claudin-4 in HT-29/B6 cells. J Nutr 141(5):783–789

Hettinga K, van Valenberg H, de Vries S, Boeren S, van Hooijdonk T, van Arendonk J, Vervoort J (2011) The host defense proteome of human and bovine milk. PLoS One 6(4):e19433

Holland JW, Gupta R, Deeth HC, Alewood PF (2011) Proteomic analysis of temperature-dependent changes in stored UHT milk. J Agric Food Chem 59(5):1837–1846

Hong SS, Park JH, Kwon SW (2007) Determination of proteins in infant formula by high-performance liquid chromatography-electrospray tandem mass spectrometry. J Chromatogr B Analyt Technol Biomed Life Sci 845(1):69–73

Huang DW, Sherman BT, Lempicki RA (2009a) Systematic and integrative analysis of large gene lists using DAVID bioinformatics resources. Nat Protoc 4(1):44–57

Huang DW, Sherman BT, Lempicki RA (2009b) Bioinformatics enrichment tools: paths toward the comprehensive functional analysis of large gene lists. Nucleic Acids Res 37(1):1–13

Hynes RO (2009) The extracellular matrix: not just pretty fibrils. Science 326(5957):1216–1219

Jenness R (1979) The composition of human milk. Semin Perinatol 3(3):225–239

Jones AD, Tier CM, Wilkins JP (1998) Analysis of the Maillard reaction products of beta-lactoglobulin and lactose in skimmed milk powder by capillary electrophoresis and electrospray mass spectrometry. J Chromatogr A 822(1):147–154

Kanwar JR, Kanwar RK (2009) Gut health immunomodulatory and anti-inflammatory functions of gut enzyme digested high protein micro-nutrient dietary supplement-Enprocal. BMC Immunol 10(7):1–19

Karhumaa P, Leinonen J, Parkkila S, Kaunisto K, Tapanainen J, Rajaniemi H (2001) The identification of secreted carbonic anhydrase VI as a constitutive glycoprotein of human and rat milk. Proc Natl Acad Sci USA 98(20):11604–11608

Keenan TW (2001) Milk lipid globules and their surrounding membrane: a brief history and perspectives for future research. J Mammary Gland Biol Neoplasia 6(3):365–371

Knip M, Virtanen SM, Seppa K, Ilonen J, Savilahti E, Vaarala O, Reunanen A, Teramo K, Hamalainen AM, Paronen J, Dosch HM, Hakulinen T, Akerblom HK (2010) Dietary intervention in infancy and later signs of beta-cell autoimmunity. N Engl J Med 363(20):1900–1908

Korhonen H, Pihlanto A (2003) Food-derived bioactive peptides–opportunities for designing future foods. Curr Pharm Design 9(16):1297–1308

Korhonen H, Pihlanto A (2007) Technological options for the production of health-promoting proteins and peptides derived from milk and colostrum. Curr Pharm Design 13(8):829–843

Kunz C, Lonnerdal B (1992) Re-evaluation of the whey protein/casein ratio of human milk. Acta Paediatr 81(2):107–112

Le A, Barton LD, Sanders JT, Zhang Q (2011) Exploration of bovine milk proteome in colostral and mature whey using an ion-exchange approach. J Proteome Res 10(2):692–704

Le TT, Deeth HC, Bhandari B, Alewood PF, Holland JW (2012) A proteomic approach to detect lactosylation and other chemical changes in stored milk protein concentrate. Food Chem 132:655–662

Leonil J, Molle D, Fauquant J, Maubois JL, Pearce RJ, Bouhallab S (1997) Characterization by ionization mass spectrometry of lactosyl beta-lactoglobulin conjugates formed during heat treatment of milk and whey and identification of one lactose-binding site. J Dairy Sci 80(10):2270–2281

Liao Y, Alvarado R, Phinney B, Lonnerdal B (2011a) Proteomic characterization of human milk fat globule membrane proteins during a 12 month lactation period. J Proteome Res 10(8):3530–3541

Liao Y, Alvarado R, Phinney B, Lonnerdal B (2011b) Proteomic characterization of human milk whey proteins during a twelve-month lactation period. J Proteome Res 10(4):1746–1754

Lönnerdal B (2003) Nutritional and physiologic significance of human milk proteins. Am J Clin Nutr 77(6):1535s–1536s

Lönnerdal B (2004) Human milk proteins: key components for the biological activity of human milk. Adv Exp Med Biol 554:11–25

Losito I, Carbonara T, Monaci L, Palmisano F (2007) Evaluation of the thermal history of bovine milk from the lactosylation of whey proteins: an investigation by liquid chromatography-electrospray ionization mass spectrometry. Anal Bioanal Chem 389(7–8):2065–2074

Lu J, Boeren S, de Vries SC, van Valenberg HJ, Vervoort J, Hettinga K (2011) Filter-aided sample preparation with dimethyl labeling to identify and quantify milk fat globule membrane proteins. J Proteomics 75(1):34–43

Luhovyy BL, Akhavan T, Anderson GH (2007) Whey proteins in the regulation of food intake and satiety. J Am Coll Nutr 26(6):704S–712S

Luz Sanz M, Corzo-Martinez M, Rastall RA, Olano A, Moreno FJ (2007) Characterization and in vitro digestibility of bovine beta-lactoglobulin glycated with galactooligosaccharides. J Agric Food Chem 55(19):7916–7925

Mahe S, Roos N, Benamouzig R, Davin L, Luengo C, Gagnon L, Gausserges N, Rautureau J, Tome D (1996) Gastrojejunal kinetics and the digestion of [15N]beta-lactoglobulin and casein in humans: the influence of the nature and quantity of the protein. Am J Clin Nutr 63(4):546–552

Mangé A, Bellet V, Tuaillon E, Van de Perre P, Solassol J (2008) Comprehensive proteomic analysis of the human milk proteome: contribution of protein fractionation. J Chromatogr B 876(2):252–256

Mauron J (1990) Influence of processing on protein quality. J Nutr Sci Vitaminol (Tokyo) 36(Suppl 1):S57–S69

Meltretter J, Pischetsrieder M (2008) Application of mass spectrometry for the detection of glycation and oxidation products in milk proteins. Ann N Y Acad Sci 1126:134–140

Meltretter J, Seeber S, Humeny A, Becker CM, Pischetsrieder M (2007) Site-specific formation of Maillard, oxidation, and condensation products from whey proteins during reaction with lactose. J Agric Food Chem 55(15):6096–6103

Meltretter J, Birlouez-Aragon I, Becker CM, Pischetsrieder M (2009) Assessment of heat treatment of dairy products by MALDI-TOF-MS. Mol Nutr Food Res 53(12):1487–1495

Mesmin C, Fenaille F, Becher F, Tabet JC, Ezan E (2011) Identification and characterization of apelin peptides in bovine colostrum and milk by liquid chromatography-mass spectrometry. J Proteome Res 10(11):5222–5231

Monaci L, van Hengel AJ (2007) Effect of heat treatment on the detection of intact bovine beta-lactoglobulins by LC mass spectrometry. J Agric Food Chem 55(8):2985–2992

Morgan F, Leonil J, Molle D, Bouhallab S (1997) Nonenzymatic lactosylation of bovine beta-lactoglobulin under mild heat treatment leads to structural heterogeneity of the glycoforms. Biochem Biophys Res Commun 236(2):413–417

Morrison B, Cutler ML (2010) The contribution of adhesion signaling to lactogenesis. J Cell Commun Signal 4(3):131–139

Murphy KP (2012) Janeway's immunobiology, 8th edn. Garland Science, London, pp 48–71, 527–528

Nagpal R, Behare P, Rana R, Kumar A, Kumar M, Arora S, Morotta F, Jain S, Yadav H (2011) Bioactive peptides derived from milk proteins and their health beneficial potentials: an update. Food Funct 2(1):18–27
Ogundele MO (2001) Role and significance of the complement system in mucosal immunity: particular reference to the human breast milk complement. Immunol Cell Biol 79(1):1–10
Pal S, Ellis V (2010) The chronic effects of whey proteins on blood pressure, vascular function, and inflammatory markers in overweight individuals. Obesity (Silver Spring) 18(7):1354–1359
Pal S, Ellis V (2011) Acute effects of whey protein isolate on blood pressure, vascular function and inflammatory markers in overweight postmenopausal women. Br J Nutr 105(10):1512–1519
Palmer DJ, Kelly VC, Smit AM, Kuy S, Knight CG, Cooper GJ (2006) Human colostrum: identification of minor proteins in the aqueous phase by proteomics. Proteomics 6(7):2208–2216
Peterson JA, Scallan CD, Ceriani RL, Hamosh M (2001) Structural and functional aspects of three major glycoproteins of the human milk fat globule membrane. Adv Exp Med Biol 501:179–187
Pihlanto A, Korhonen H (2003) Bioactive peptides and proteins. Adv Food Nutr Res 47:175–276
Pisanu S, Ghisaura S, Pagnozzi D, Biosa G, Tanca A, Roggio T, Uzzau S, Addis MF (2011) The sheep milk fat globule membrane proteome. J Proteomics 74(3):350–358
Reinhardt TA, Lippolis JD (2006) Bovine milk fat globule membrane proteome. J Dairy Res 73(4):406–416
Reinhardt TA, Lippolis JD (2008) Developmental changes in the milk fat globule membrane proteome during the transition from colostrum to milk. J Dairy Sci 91(6):2307–2318
Ricci-Cabello I, Herrera MO, Artacho R (2012) Possible role of milk-derived bioactive peptides in the treatment and prevention of metabolic syndrome. Nutr Rev 70(4):241–255
Riccio P (2004) The proteins of the milk fat globule membrane in the balance. Trends Food Sci Tech 15(9):458–461
Rojas R, Apodaca G (2002) Immunoglobulin transport across polarized epithelial cells. Nat Rev Mol Cell Biol 3(12):944–955
Roos N, Mahe S, Benamouzig R, Sick H, Rautureau J, Tome D (1995) 15N-labeled immunoglobulins from bovine colostrum are partially resistant to digestion in human intestine. J Nutr 125(5):1238–1244
Sanderson IR, Walker WA (2000) Development of the gastrointestinal tract. B. C. Decker, Hamilton, Ontario L8N 3K7, pp 83–102, 147–164, 227–260
Satake M, Enjoh M, Nakamura Y, Takano T, Kawamura Y, Araî S, Shimizu M (2002) Transepithelial transport of the bioactive tripeptide, Val-Pro-Pro, in human intestinal Caco-2 cell monolayers. Biosci Biotechnol Biochem 66(2):378–384
Schack-Nielsen L, Michaelsen KF (2007) Advances in our understanding of the biology of human milk and its effects on the offspring. J Nutr 137(2):503S–510S
Schlimme E, Meisel H (1995) Bioactive peptides derived from milk proteins. Structural, physiological and analytical aspects. Nahrung 39(1):1–20
Seiquer I, Diaz-Alguacil J, Delgado-Andrade C, Lopez-Frias M, Munoz Hoyos A, Galdo G, Navarro MP (2006) Diets rich in Maillard reaction products affect protein digestibility in adolescent males aged 11–14 y. Am J Clin Nutr 83(5):1082–1088
Séverin S, Wenshui X (2005) Milk biologically active components as nutraceuticals: review. Crit Rev Food Sci 45(7–8):645–656
Shimizu M, Tsunogai M, Arai S (1997) Transepithelial transport of oligopeptides in the human intestinal cell, Caco-2. Peptides 18(5):681–687
Siciliano R, Rega B, Amoresano A, Pucci P (2000) Modern mass spectrometric methodologies in monitoring milk quality. Anal Chem 72(2):408–415
Smithers GW (2008) Whey and whey proteins-from 'gutter-to-gold'. Int Dairy J 18(7):695–704
Smolenski G, Haines S, Kwan FYS, Bond J, Farr V, Davis SR, Stelwagen K, Wheeler TT (2007) Characterisation of host defence proteins in milk using a proteomic approach. J Proteome Res 6(1):207–215
Sorva R, Makinen-Kiljunen S, Juntunen-Backman K (1994) β-Lactoglobulin secretion in human milk varies widely after cow's milk ingestion in mothers of infants with cow's milk allergy. J Allergy Clin Immunol 93(4):787–792

Spitsberg VL (2005) Invited review: bovine milk fat globule membrane as a potential nutraceutical. J Dairy Sci 88(7):2289–2294

van Boekel MAJS (1998) Effect of heating on Maillard reactions in milk. Food Chem 62(4):403–414

Vanderghem C, Blecker C, Danthine S, Deroanne C, Haubruge E, Guillonneau F, De Pauw E, Francis F (2008) Proteome analysis of the bovine milk fat globule: enhancement of membrane purification. Int Dairy J 18(9):885–893

Walker WA (2010) Mead Johnson Symposium: functional proteins in human milk: role in infant health and development. J Pediatr 156(2 Suppl):S1–S2

Wiseman BS, Werb Z (2002) Stromal effects on mammary gland development and breast cancer. Science 296(5570):1046–1049

Wu CC, Howell KE, Neville MC, Yates Iii JR, McManaman JL (2000) Proteomics reveal a link between the endoplasmic reticulum and lipid secretory mechanisms in mammary epithelial cells. Electrophoresis 21(16):3470–3482

Zhang Q, Faca V, Hanash S (2011) Mining the plasma proteome for disease applications across seven logs of protein abundance. J Proteome Res 10(1):46–50

Zivkovic AM, German JB, Lebrilla CB, Mills DA (2011) Human milk glycobiome and its impact on the infant gastrointestinal microbiota. Proc Natl Acad Sci USA 108(Suppl 1):4653–4658

Chapter 14
Cheese Processing

Aldo Di Luccia, Michele Faccia, Caterina Incoronato, Filomena Inglese, Carmela Lamacchia, Sara Lamparelli, Mariaconsiglia Occidente, and Donato Matassino

14.1 Some Aspects of Cheese Milk

Cheese is a "biologically active" food, and its quality depends heavily on raw milk quality (chemical–microbiological composition and enzymatic activity) and processing (Fig. 14.1). In fact, milk is the source both of substrates (carbohydrates, fat, and protein) and enzymes which are at the basis of the complex phenomenon responsible for the organoleptic features of each cheese.

14.1.1 Composition

Starting from the consciousness that the major part of milk is used for cheese production, the milk properties associated with cheese making (lactoprotein variants, colloidal calcium phosphate content, degree of titratable acidity) play a prime role. Aptitude to coagulation is a primary requisite; for the evaluation of this aptitude the

A. Di Luccia
Department of SAFE, University of Foggia, via Napoli, 25, Foggia, Italy

ConSDABI, Piano Cappelle, Benevento, Italy

M. Faccia
Department of DIBCA, University of Bari, via Amendola, Bari, Italy

C. Incoronato • F. Inglese • M. Occidente • D. Matassino (✉)
ConSDABI, Piano Cappelle, Benevento, Italy
e-mail: matassinod@consdabi.org

C. Lamacchia • S. Lamparelli
Department of SAFE, University of Foggia,
via Napoli, 25, Foggia, Italy

F. Toldrá and L.M.L. Nollet (eds.), *Proteomics in Foods: Principles and Applications*, Food Microbiology and Food Safety 2, DOI 10.1007/978-1-4614-5626-1_14,

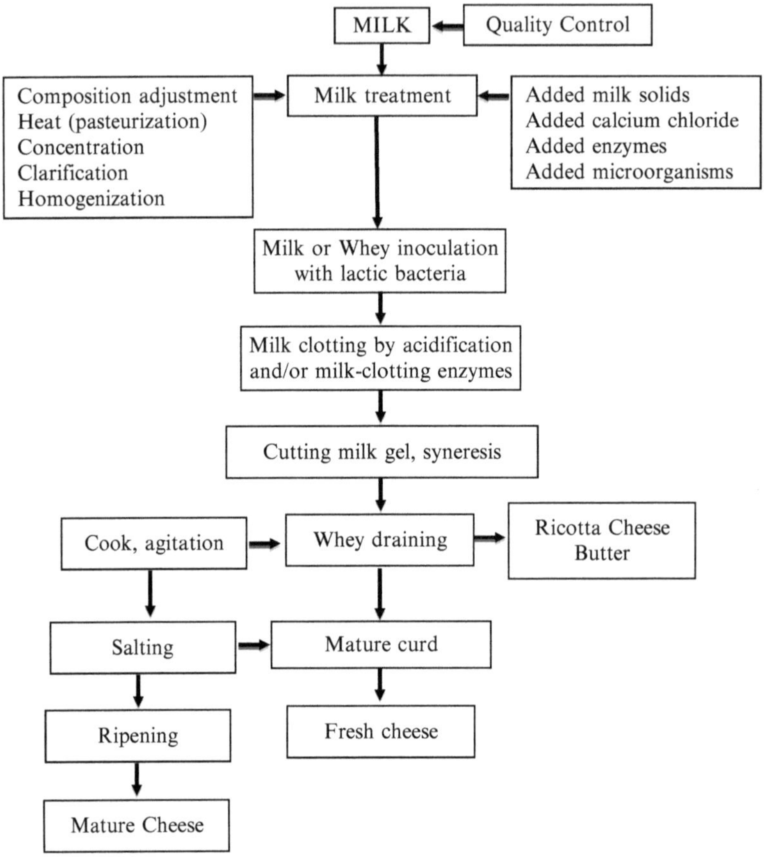

Fig. 14.1 Schematic representation of cheese manufactures and general phases of hard, soft and fresh cheeses

lactodynamographic technique provides useful information through the estimation of some parameters[1] (Matassino et al. 2007a).

Some of the aspects mentioned above can be addressed in the context of selection programs aimed at improving cheese-making ability. One of them is the improving "casein fraction" through the selection of carriers of genetic variants associated with milk particularly suited to cheese-making. As is known, milk of subjects with BB genotype at k-casein and beta-lactoglobulin *loci* shows a better cheese-making quality and provides a higher cheese yield (Aleandri et al. 1990; Mariani et al. 1995).

[1] Lactodynamographic parameters: (*r*)—renneting time in minutes; (*K*20)—rate of curd formation measured by time taken from lactodynamographic output to reach the amplitude of 20 mm; (*a*30)—curd firmness measured 30 min after rennet addition. In addition to previous lactodynamographic parameters it is increasingly useful to employ $a_{1/2r}$ which provides information about curd firmness at one half of renneting time.

The progress in molecular biology methods gives impetus to the promotion of research and expansion of knowledge of milk processing through a new approach to the study.

The application of a proteomic strategy to cheese milk has a general purpose summarized in the following points: (1) identification of the heterogeneity degree (qualitative/quantitative) of the milk protein fraction as a result of lactoprotein *loci* polymorphism and/or post-translational modifications, (2) study of the relationship between milk protein polymorphism and dairy quality, and (3) identification of protein fragments from "caseoma" as an expression of enzyme activities.

Among these objectives, the detection of lactoprotein genetic polymorphism (casein fractions: α_{s1}-, α_{s2}-, β-, and κ-CN; serum proteins: α-lactalbumin and β-lactoglobulin) assumes a role of primary importance. For example, the application of the analysis system based on ELISA provides tangible benefits to the operators of the dairy industry. A practical development is the monitoring of cheese milk and in particular of k-casein content. This application has to be considered as an innovative and useful tool for farmers because it could allow—quickly and at low cost—the diversification of the destination of the milk, addressing the better milk towards more profitable productive lines.

The "dairy quality," due to intrinsic (species, breed, individual) and extrinsic (feed, health, lactation period, and technologies) factors, is represented not only by lactoprotein fraction, but also by fat/protein ratio, state of colloidal phase (i.e., mineralization of the casein micelles), and ionic calcium availability as well as microbiological parameters, enzymatic activity, and its chemism.

The fat/protein ratio influences yield and cheese quality and different optimum values are required depending on the type of cheese. Its standardization to the required level can be achieved both by adding/removing fat and by increasing protein content: in the latter the use of microfiltration at a 0.1-μm cutoff is widely diffused, because it allows concentration on casein (Beckman et al. 2010). Addition of milk powder or powdered milk proteins is also performed, but it is not allowed for PDO cheeses: the illegal use can be detected by chromatographic techniques (Pellegrino et al. 2010).

The mineralization of the casein micelles and the availability of ionic calcium have an important role in milk coagulation and rheological properties of the curd (Lucey and Fox 1993). Intact and well-mineralized micelles in cheese milk, together with adequate presence of Ca^{++} or addition of $CaCl_2$ (e.g., 0.01%), increases the cross-link between casein micelles increasing the strength of the curd (Zhong et al. 2007).

14.1.2 Microbiological Quality

As is known, cheese milk microbiological quality requires a low total bacterial count (TBC) and influences the aptitude to coagulation; nevertheless, an adequate presence of favorable micro-organisms has to be guaranteed to favor cheese making. When milk TBC is too high, high-temperature short-time pasteurization (HTST, i.e., 72–75°C for 15 s) is performed. Such a process could give rise to two consequences: recovery of some whey proteins in the cheese matrix with consequent yield increase (an extreme recovery can impair the curd firmness), or the flattening of the microflora (its reconstitution is required through the use of natural[2] or commercial selected starter, to assure correct acidification and proteolysis during ripening process).

14.1.3 Proteolytic Activity

The enzymatic activity of the milk, due to both apical membrane or cytoplasm of the secretory cells and raw milk, plays a key role in cheese production (Kelly et al. 2006). The indigenous proteinases, originated from animal blood, pass into epithelial secretory cells of the mammary gland and/or are released from lysosomes of somatic cells. Plasmin—the main endogenous enzyme—is an alkaline serine protease and originates from the blood, whereas cathepsin D is a lysosomal aspartyl protease released from somatic cells (O'Driscoll et al. 1999; Larsen et al. 2006). Both enzymes affect the cheese-making yield, as they reduce the concentration of the intact casein present in milk and play a key role in cheese ripening.

Milk somatic cell count (SCC) is known to increase with mastitis, an inflammatory reaction of the udder, commonly resulting from bacterial infection. Owing to this disease an increase of polymorphonucleated leukocytes afflux from blood into the milk occurs; the latter have caseinolytic and lipolytic activity (Grieve and Kitchen 1985). The most important proteinases associated with mastitis are plasmin and cathepsins, whose activity increases in milk with high SCC (Kelly and McSweeney 2003).

14.1.4 Some Biochemical Aspects

Milk transformation in cheese is a complex and concomitant action of endogenous and exogenous enzymes. These latter derive from rennet. Rennet is an exogenous enzyme containing as a major component, chymosin, an aspartyl protease with 88–94% milk clotting activity (MCA), and pepsin, a serine protease with 6–12% MCA. Rennet coagulates milk through a specific action of chymosin on breakdownbv site 105 Phe-106 Met of κ-casein delivering the C-terminus glycomacropeptide (GMP) in the whey and the N-terminus, para-κ-casein, remains linked to the micelles. As a result of renneting, a micelle destabilization occurs and calcium-sensitive caseins (α_s and β) form a calcium paracaseinate gel including whey and fat globules.

Other kinds of aspartyl or cisteyl proteinase coagulants can substitute the rennet such as: (1) biotechnological calf chymosin from several microorganisms, (2) bovine pepsin, or (3) proteinases from *R. meihei*, *R. pusillus,* and *C. parasitica* fungi. Moreover, proteases from a number of plants including fig (*Ficus carica*), cardoon (*Cynara cardunculus* L.), paw paw (*Carica papaya*), pineapple (*Ananas sativa*), and

[2] Natural starters are a very complex association of micro-organisms used for producing traditional cheese in artisanal dairies. These micro-organisms strongly contribute to the peculiar organoleptic features of these products, whereas selected starters give raise to "nonspecific" cheeses.

castor oil seeds (*Ricinus communis*) also have been used to coagulate milk (Garg and Johri 1994; Faccia et al. 2012). However, these coagulants have nonspecific proteolytic activity on caseins (Sousa and Malcata 2002; Silva et al. 2003).

As mentioned above, primary proteolysis occurs in milk and cheese due to milk endogenous enzymes, mainly plasmin and cathepsin D (Kelly et al. 2006), and then by rennet enzymes (Grappin et al. 1985; Fox and McSweeney 1997).

Furthermore, in the case of cold storage milk, enzymes also delivered by psychrotrophic bacteria concur to primary proteolysis.

Plasmin cleaves proteins on the carboxyl side of L-Lys and L-Arg residues (Grufferty and Fox 1988), with a preference for L-Lys. Plasmin preferentially hydrolyzes β-CN and with a slower action α_{s1}-, α_{s2}-caseins, but has little or no activity on the whey proteins α-lactalbumin and β-lactoglobulin (Cassens et al. 1999). Three C-terminus fragments are generated from hydrolysis of bovine β-CN: γ_1 f(29–209), γ_2 f(106–209), and γ_3 f(108–209) (Green and Foster 1974; Eigel 1977); the concentration of these components increases both in nonlactating cows and milk containing a superior threshold limit somatic cell count. γ-casein fraction can be considered a "quality molecular marker" inasmuch as a significant positive correlation between γ-casein fragments and cheese ripening was observed ("Parmigiano Reggiano" and "Grana Padano") (Addeo et al. 1994; Gaiaschi et al. 2000).

The proteomic approach allows the individuation in river buffalo milk of a further fragment named γ_4 f(69–209), due to a substitution of Gln at position 68 of the β-CN primary structure with Lys (Trieu-Cuot and Addeo 1981; Somma et al. 2008; Di Luccia et al. 2009).

In sheep and goat, the three fragments derived from β-CN hydrolysis are f(29–207), f(106–207), and f(108–207).

Fragments of α_{s1}-CN, termed λ-CN, were generated by plasmin and identified by Aimutis and Eigel (1982). These authors also established that the estimated molecular weight of λ-CN ranged from 5,000 to 6,000 Da.

O'Flaherty (1997) and Kelly et al. (2006) found peptides in the λ-casein fraction of good quality bulk milk that were generated by plasmin cleavage at Phe^{23}–Phe^{24}, Phe^{24}–Val^{25} Lys^{34}–Glu^{35} Val^{37}–Asn^{38} sites.

Plasmin, B, D, H cathepsins, elastase, as well as amino- and carboxypeptidases have been shown to originate 20 different peptides from α_{s1}- and β-caseins (Larsen et al. 2010).

D cathepsin, chymosin, and psychrotrophic microbial enzymes have similar specificity in hydrolyzing proteins at cleavage sites that involve neutral and/or polar amino acids. Chymosin is known also to cleave α_{s1}-CN at several sites; initially it gives rise to α_{s1}-I (f25 to 199) and later to α_{s1}-II (f25 to 169) and further products (Fox 1989).

The primary cleavage site of chymosin on α_{s2}-casein is at Phe^{88}–Tyr^{89}; nevertheless, further possible cleavage sites are Tyr^{95}–Leu^{96}, Gln^{97}–Tyr^{98}, Tyr^{98}–Leu^{99}, Phe^{163}-–Leu^{164}, Phe^{174}–Ala^{175}, and Tyr^{179}–Leu^{180} (McSweeney et al. 1994).

Cathepsin D, similarly to chymosin, produces the glycomacropeptide, κ-CN f(106–169), by the enzymatic cleavage of the Phe^{105}–Met^{106} bond (McSweeney et al. 1995); furthermore, two other cleavage sites of cathepsin D on κ-casein have been identified: Leu^{32}–Ser^{33} and Leu^{79}–Ser^{80}. These two enzymes also hydrolyze α_{s1}- at

Phe^{24}–Phe^{25} and α_{s2}-caseins at Leu^{99}–Tyr^{100}, Leu^{123}–Asn^{124}, Leu^{180}–Lys^{181}, and Thr^{182}-Val^{183} (Larsen et al. 1996).

Secondary proteolysis essentially occurs during ripening cheese when fragments from primary proteolysis are mainly hydrolyzed by endo- and exopeptidases of microflora present in the cheese. Biochemical mechanisms and the formation of small peptides and amino acids are well reviewed by Sousa et al. (2001), Upadhyay et al. (2004) and McSweeney (2004). These authors established that indigenous milk enzymes (mainly plasmin and cathepsins), and residual rennet (chymosin and plasmin) produce water-insoluble and soluble peptides. Over a longer period of time, indigenous enzymes also produce peptides responsible for the taste of cheeses (Visser 1993). Hence, the action of miscellaneous cysteine, aspartyl, and serine proteases on casein fraction determines many peptides that are generated by the breakdown of bonds among neutral, aromatic, and basic amino acids. Therefore, the diversity and number of peptides depend on casein heterogeneity (genetic variants, phosphorylation, and glycosylation degree), effectiveness of enzyme activity due to technological parameters (pH, temperature, and humidity) and on typical and/or selected microflora that grow in the milk.

14.2 Proteomic Approach to Cheese

Proteomics represent an important milestone of the post-genomic era, especially in light of the view that current methods addressed towards gene expression analysis (e.g., RNA-Seq, RNA sequencing), although able to identify directly any changes that take place "downstream of the transcription," fail to provide information about post-translational modification that, instead, could have a significant role in defining the quality characteristics of a food. Indeed, continuous scientific advances undeniably highlight the importance of proteomics in the complex phenomenon of cheese production.

During the conversion of milk in cheese, biochemical, chemical, and physical events occur. Proteolysis is one of the most important biochemical events that occurs through manufacture and ripening of cheese. Proteolysis contributes substantially to taste and odor, through generation of large polypeptides, and a range of oligo and small peptides, and including free amino acids and their degradation products (Mallatou et al. 2004; Pappa et al. 2006; Upadhyay et al. 2004), such as Strecker's aldehyde (Dunn and Lindsay 1985; Avsar et al. 2004; Curtin and McSweeney 2004). Polypeptide and peptide fragments derived from proteolysis of casein fractions have been utilized mainly to establish the ripening time of cheese and cheese authenticity. For these purposes electrophoretic, chromatographic, and immunochemical detection techniques have been employed (Addeo et al. 1990; Gaiaschi et al. 2000, 2001; Senocq et al. 2002; Veloso et al. 2004; Matassino et al. 2007b, 2008; Matassino 2011).

Proteomic techniques are potentially appropriate: (1) to identify products of primary proteolysis and large peptides from early stages of secondary proteolysis, characterizing the extent and nature of cheese proteolysis (Alli et al. 1998; Gouldsworthy et al. 1996; Pappa et al. 2008; Piraino et al. 2007); and (2) to provide evidence of fraud and/or sophistication in order to authenticate cheeses.

14.2.1 *Primary Proteolysis*

Primary proteolysis can be defined as the first action of endogenous and rennet enzymes in caseins that give rise to large insoluble peptides detectable by SDS-PAGE in a range of estimated molecular weights ranging from 200 to 6 kDa. 2-DGE alone was extensively used to characterize cheeses (Trieu-Cuot and Gripon 1982; Marshall and Williams 1988; Molina et al. 2002) and it is plausible that it will be dominant for some period of time. Nevertheless, this technique has to be coupled with analytical strategies at high resolution in order to obtain a better separation of slightly different protein isoforms. It also has to be opportunely integrated to deepen the knowledge of post-translational modifications. At present, global proteomic studies to characterize cheeses are few and involved: teleme cheese (Pappa et al. 2008), cheddar (Hinz et al. 2012), and Swiss-type cheese (Jardin et al. 2012).

A typical Italian "pasta filata" cheese studied by proteomic approach was "Caciocavallo" of Italian autochthonous grey cattle (ex Podolian) (Matassino et al. 2008).

From these studies it emerged that 2-DGE maps compared at different times of ripening (1, 3, 6, and 12 months) showed a marked dynamism in casein fractions, expressed by the different number of spots according to the time of ripening. All spots were identified by chemical-assisted fragmentation-post source decay (CAF-PSD) analyses. By this procedure, six different fragments from β-CN, due to plasmin action, were identified. These fragments, common to all 2-DGE maps were grouped into four classes (termed A, B, C, and D; Fig. 14.2). In particular, the spots belonging to the A group (including those of γ_1-CN) arose from primary proteolysis of β-CN and their volume decreased during ripening in concomitance with an increase of spot volume belonging to the B and C groups (including γ_2- and γ_3-CN, respectively); this increase was ascribable to the primary proteolysis on β-CN and to the secondary proteolysis on proteins of the A group. The volume variation of the proteins included in the three groups could, hence, represent a useful indicator of Caciocavallo ripening time, as evidenced by other authors for Grana Padano cheese (Gaiaschi et al. 2001) and Comté cheese (Dupont et al. 2003). Moreover, it is worth noting the presence and evolution in time of peptides, characterized by a low molecular weight and pI assigned to the D group (Fig. 14.2); the latter were detectable at the first ripening and disappeared through ripening time.

A typical Italian Pecorino characterized by the proteomic approach, was "Pecorino of Laticauda." This study showed that proteolysis degree was influenced by the initial average weight of cheese moulds, independent of ripening time; indeed, cheese moulds with a starting weight of about 3 kg showed a more marked casein hydrolysis, in particular of β-CN, in comparison with moulds with a starting weight of about 5 kg (Matassino et al. 2008); this behavior was also demonstrated by Faccia et al. (2003) for "Canestrato Pugliese" ovine cheese.

Recently, Matassino (2010), analyzing pecorino cheeses from Bagnolese and Laticauda ovine breeds has individuated a molecular marker, termed "ConSDABI," identified by MALDI-TOF and CAF-PSD, as an α_{s1}-CN fragment, that is differently

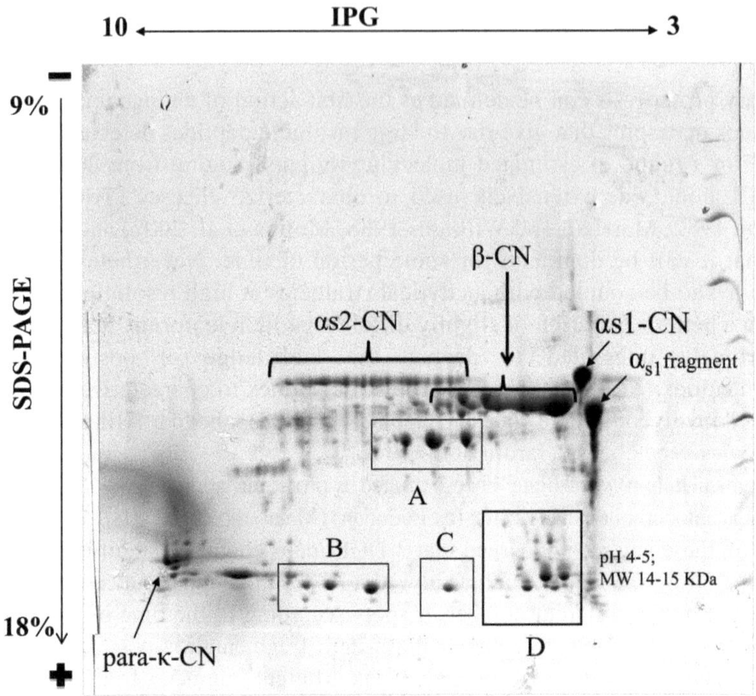

Fig. 14.2 Two dimensional map of insoluble fraction from "Caciocavallo" cheese at 1 month of ripening. Subset indicated: (**a**) γ1-CN; (**b**) γ2-CN; (**c**) γ3-CN and (**d**) a mixture of fragments from secondary proteolysis of β- and α_{s1}-CN

present in the two ovine breeds independently of ripening time. In particular, pecorino of Bagnolese has shown a significantly higher amount of the above-cited fragment at any ripening time ($p<0.001$ at 3 and 6 months; $p<0.01$ at 9 months) in comparison with Pecorino of Laticauda (Fig. 14.3). Therefore, the ConSDABI marker can be considered (1) a potential "breed" marker, or (2) a potential "process" marker.

Furthermore, another molecular marker, called ALMI, has been identified as able to discriminate fresh and conserved (frozen or from abroad) curds, both employed for mozzarella cheese production (Faccia et al. 2011).

14.2.2 Secondary Proteolysis

A further proteolysis due to endogenous enzymes, rennet, and microbial enzymes gives rise to water-soluble small peptides. These peptides, identified and structurally characterized by high-performance liquid chromatography coupled mass spectrometry (LC/MS), allow us to define the different enzymatic activities of cheese as well as their specificity (Addeo et al. 1992, 1994; Ferranti et al. 1997; Singh et al. 1994; Gagnaire et al. 2001). Gagnaire et al. (2001), analyzing proteolytic

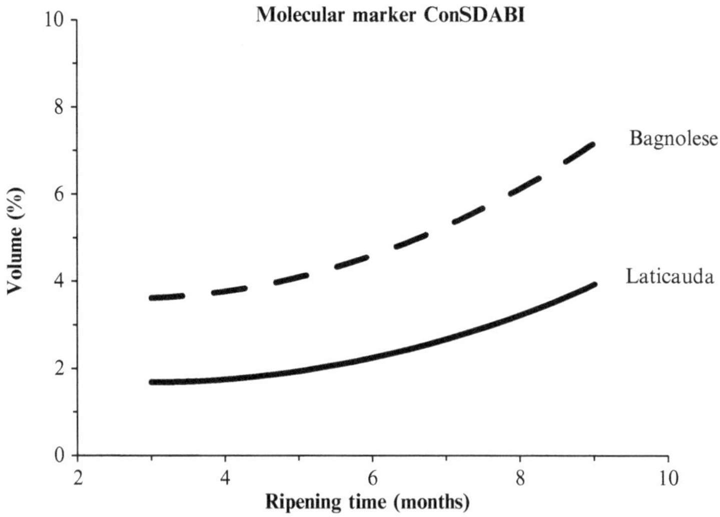

Fig. 14.3 Trend of molecular marker ConSDABI through ripening phase in relation to the variation percentage of volume intensity. *Dashed line* indicates pecorino of Bagnolese, *continuous line* indicates pecorino of Laticauda

changes occurring during Emmental cheese ripening by LC/tandem MS, identified 89 peptides: 51 arose from α_{s1}-casein, 28 from β-casein, 9 from α_{s2}-casein (9), and 1 from κ-casein. Using the same approach, Gagnaire et al. (2011) identified 123 peptides in Ragusano cheese: 72 arising from β-casein, 34 from α_{s1}-casein, and 17 from α_{s2}-casein. All the authors showed that the number of peptides, including the mineral carrier as well as antihypertensive and immunomodulating peptides and phosphopeptides, dramatically decreased throughout ripening. Other authors (Pappa et al. 2008), employing LC/MS with a tandem mass spectrometry QTOF to identify peptides extracted from water-soluble fraction (especially peptides from β- and α_{s1}-CN), were able to distinguish the milk source (cow, goat, and sheep) used for teleme cheese manufacture supporting the potential of proteomic techniques to describe the origin and development of this processed food product.

14.3 Conclusions

The overall proteomic analysis of cheeses, concerning primary and secondary proteolysis, shows that the endogenous and rennet enzymes act mainly during primary proteolysis on β- and α_{s1}-CN originating oligo and small peptides, which are the substrate of endo- and eso-peptidase enzymes during secondary proteolysis. To investigate primary proteolysis the separation strategy based on two-dimensional gel electrophoresis, has been proved to be the most suitable; high-performance liquid chromatography can be applied to secondary proteolysis studies. The study of primary proteolysis is a powerful tool especially to individuate frauds and/or sophistication useful for cheese authenticity. The analysis of secondary proteolysis is a powerful tool to investigate the evolution of proteolysis during ripening process.

The differences observed in all cheeses analyzed demonstrated that the proteolysis and therefore the peculiarity and quality of cheeses depend on milk characteristics and on various parameters of a flowchart of cheese production. Whatever change occurs in these parameters is reflected in qualitative or quantitative modification of casein fragments from primary and secondary proteolysis. Considering the wide assortment of cheese produced in the world, many different peptides, that are generated through the ripening process for each type of cheese, will exist. We could consider each cheese as a Rosetta Stone on which are inscribed the peptide sequences obtained from proteomic studies performed by LC/MS-MS. These peptidic hieroglyphs, once deciphered, will allow a better understanding of the relationship among biochemical, chemical, and physical phenomena and technological phases of cheese production.

References

Addeo F, Moio L, Chianese L, Stingo C, Resmini P, Berner I, Krause I, Di Luccia A, Bocca A (1990) Use of plasmin to increase the sensitivity of the detection of bovine milk in ovine cheese by gel isoelectric focusing of γ2-caseins. Milchwissenschaft 45(11):708–711

Addeo F, Mauriello R, Moio L, Laezza P, Chianese L, Di Luccia A (1992) Ovine casein variant identification using electrophoretic, immunochemical and chromatographic techniques. Milchwissenschaft 47:283–287

Addeo F, Chianese L, Sacchi R, Musso SS, Ferranti P, Malorni A (1994) Characterization of the oligopeptides of Parmigiano-Reggiano cheese. J Dairy Res 61: 365–374

Aimutis WR, Eigel WN (1982) Identification of λ-casein as plasmin-derived fragments of bovine α_{s1}-casein. J Dairy Sci 65:175–181

Aleandri R, Buttazzoni LG, Schneider JC, Caroli A, Davoli R (1990) The effects of milk protein polymorphisms on milk components and cheese-producing ability. J Dairy Sci 73:241–255

Alli I, Okoniewska M, Gibbs BF, Konishi Y (1998) Identification of peptides in cheddar cheese by electrospray ionization mass spectrometry. Int Dairy J 8:643–649

Avsar YK, Karagul-Yuceer Y, Drake MA, Singh TK, Yoon Y, Cadwallader KR (2004) Characterization of nutty flavor in cheddar cheese. J Dairy Sci 87:1999–2010

Beckman SL, Zulewska J, Newbold M, Barbano DM (2010) Production efficiency of micellar casein concentrate using polymeric. spiral-wound microfiltration membranes. J Dairy Sci 93:4506–4517

Cassens PWJR, Visser S, Gruppen H, Voragen AGJ (1999) β-Lactoglobulin hydrolysis. 1. Peptide composition and functional properties of hydrolysates obtained by the action of plasmin, trypsin, and *Staphylococcus aureus* V8 protease. J Agric Food Chem 47:2973–2979

Curtin AC, McSweeney PLH (2004) Catabolism of amino acids in cheese during ripening. In: Fox PF, Mcsweeney PLH, Cogan TM, Guinee TP (eds) Cheese chemistry, physics and microbiology vol 1., Elsevier, London, pp 435–454

Di Luccia A, Picariello G, Trani A, Alviti G, Loizzo P, Faccia M, Addeo F (2009) Occurrence of β-casein fragments in cold stored and curdled river buffalo (*Bubalus Bubalis*, L.) milk. J Dairy Sci 92:1319–1329

Dunn HC, Lindsay RC (1985) Evaluation of the role of microbial strecker-derived aroma compounds in unclean-type flavors of cheddar cheese. J Dairy Sci 68:2859–2874

Dupont D, Rolet-Repecaud O, Senocq D (2003) A new approach to monitoring proteolysis phenomena using antibodies specifically directed against the enzyme site on its substrate. Anal Biochem 317:240–246

Eigel WN (1977) Formation of γ1-A2, γ2-A2 and γ3-A2 caseins by in vitro proteolysis of β-casein A2 with bovine plasmin. Int J Biochem 8:187–192

Faccia M, Gambacorta G, Liuzzi VA, Alviti G, Di Luccia A (2003) Influence of cheese weight and type of rennet on composition and proteolysis of "Canestrato Pugliese" cheese II. Chromatographic characterization of soluble nitrogen. Ital J Food Sci 15(1):75–84

Faccia M, Loizzo P, Trani A, Gagliardi R, Lamparelli S, Scrocchi P, Di Luccia A (2011) Un approccio analitico alla rivelazione dell'uso di cagliate conservate nella produzione della mozzarella vaccina. Scienza e Tecnica Lattiero-Casearia 62(4):233–240

Faccia M, Picariello G, Trani A, Loizzo P, Gambacorta G, Lamacchia C, Di Luccia A (2012) Proteolysis of Cacioricotta cheese made from goat milk coagulated with caprifig (*Ficus carica sylvestris*) or calf rennet. Eur Food Res Technol 234:527–533

Ferranti P, Itolli E, Barone F, Malorni A, Garro G, Laezza P, Chianese L, Migliaccio F, Stingo V, Addeo F (1997) Combined high resolution chromatographic techniques (FPLC and HPLC) and mass spectrometry-based identification of peptides and proteins in Grana Padano cheese. Lait 77:683–697

Fox PF (1989) Proteolysis during cheese manufacture and ripening. J Dairy Sci 72:1379–1400

Fox PF, McSweeney PLH (1997) Rennets: their role in milk coagulation and cheese ripening. In: Law BA (ed) Microbiology and biochemistry of cheese and fermented milk. Chapman and Hall, Blackie Academic &Professional, London, pp 1–249

Gagnaire V, Mollé D, Herrouin M, Léonil J (2001) Peptides identified during emmental cheese ripening: origin and proteolytic systems involved. J Agric Food Chem 49(9):4402–4413

Gagnaire V, Carpino S, Pediliggieri C, Jardin J, Lortal S, Licitra G (2011) Uncommonly thorough hydrolysis of peptides during ripening of Ragusano cheese revealed by tandem mass spectrometry. J Agric Food Chem 59(23):12443–12452

Gaiaschi A, Beretta B, Poiesi C, Conti A, Giuffrida MG, Galli CL, Restani P (2000) Proteolysis of α_s-casein as a marker of Grana Padano cheese ripening. J Dairy Sci 83:2733–2739

Gaiaschi A, Beretta B, Poiesi C, Conti A, Giuffrida MG, Galli CL, Restani P (2001) Proteolysis of β-casein as a marker of Grana Padano cheese ripening. J Dairy Sci 84:60–65

Garg SK, Johri BN (1994) Rennet: current trends and future research. Food Rev Int 10:313–355

Gouldsworthy AM, Leaver J, Banks JM (1996) Application of a mass spectrometry sequencing technique for identifying peptides present on cheddar cheese. Int Dairy J 6:781–790

Grappin R, Rank TC, Olson NE (1985) Primary proteolysis of cheese proteins during ripening. J Dairy Sci 68:531–540

Green ML, Foster PMD (1974) Comparison of the rates of proteolysis during ripening of cheddar cheeses made with calf rennet and swine pepsin as coagulants. J Dairy Res 41:269–282

Grieve PA, Kitchen BJ (1985) Proteolysis in milk: the significance of proteinases originating from milk leucocytes and a comparison of the action of leucocyte, bacterial and natural milk proteinases on casein. J Dairy Res 52:101–112

Grufferty MB, Fox PF (1988) Milk alkaline proteinase. J Dairy Res 55:609–630

Hinz K, O'Connor PM, O'Brien B, Huppertz T, Ross RP, Kelly AL (2012) Proteomic study of proteolysis during ripening of cheddar cheese made from milk over a lactation cycle. J Dairy Res 27:1–9

Jardin J, Mollè D, Piot M, Lortal S, Gagnaire V (2012) Quantitative proteomic analysis of bacterial enzymes released in cheese during ripening. Int J Food Microbiol 155:19–28

Kelly AL, McSweeney PLH (2003) Indigenous proteolytic enzymes in milk. In: McSweeney PLH, Fox PF (eds) Advanced dairy chemistry I, proteins, 2nd edn. Kluwer Academic-Plenum, New York, pp 495–544

Kelly AL, O'Flaherty F, Fox PF (2006) Review. Indigenous proteolytic enzymes in milk: a brief overview of the present state of knowledge. Int Dairy J 16:563–572

Larsen LB, Benfeldt C, Rasmussen LK, Petersen TE (1996) Bovine milk procathepsin D and cathepsin D: coagulation and milk protein degradation. J Dairy Res 63:119–130

Larsen LB, McSweeney PLH, Hayes MG, Andersen JB, Ingvartsen KL, Kelly AL (2006) Variation in activity and heterogeneity of bovine milk proteases with stage of lactation and somatic cell count. Int Dairy J 16:1–8

Larsen LB, Hinz K, Jørgensen ALW, Møller HS, Wellnitz O, Bruckmaier RM, Kelly AL (2010) Proteomic and peptidomic study of proteolysis in quarter milk after infusion with lipoteichoic acid from *Staphylococcus aureus*. J Dairy Sci 93:5613–5626

Lucey JA, Fox PF (1993) Importance of calcium and phosphate in cheese manufacture: a review. J Dairy Sci 76:1714–1724

Mallatou H, Pappa EC, Boumba VA (2004) Proteolysis in Teleme cheese made from ewes', goats' or a mixture of ewes and goats' milk. Int Dairy J 14:977–987

Mariani P, Summer A, Anghinetti A, Senese C, Di Gregorio P, Rando A, Serventi P (1995) Effetti dell'allele αs1-Cn G sulla ripartizione percentuale delle caseine α s1, α s2, βe k in vacche di razza Bruna. Ind Latte 31(4):3–13

Marshall T, Williams KM (1988) High resolution two-dimensional electrophoresis and sodium dodecyl sulphate-polyacrylamide gel electrophoresis of cheese proteins after rapid solubilisation. Electrophoresis 9(3):143–147

Matassino D (2010) Presentazione dei risultati Unità operativa 'Tracciabilità e Rintracciabilità'. Convegno "Presentazione dei Risultati del Progetto Selmol", organizzato da: Istituto Sperimentale 'Lazzaro Spallanzani', Associazione Italiana Allevatori (AIA), Consorzio di Ricerca e Sperimentazione degli Allevatori (CRSA), Accademia dei Georgofili, Ministero per le Politiche Agricole, Alimentari e Forestali (Mipaaf), Firenze, 15–16 settembre. www.selmol.eu

Matassino D (2011) Rintracciabilità di razza e autenticazione di prodotto mediante approccio proteomico. In: Marchitelli C, Ranieri MS, Negrini R (ed) I Georgofili Quaderni 2010 III, supplemento a 'I Georgofili', Atti dell'Accademia dei Georgofili, Anno 2010, Serie VIII- vol 7 (186. dall'inizio). Polistampa, Firenze, pp 305–309

Matassino D, Barone CMA, Di Luccia A, Incoronato C, Inglese F, Marletta D, Occidente M, Roncada P (2007a) Genomica e proteomica funzionali. In: I Georgofili – Quaderni 2006 –I. Società Editrice Fiorentina, Firenze, pp 201–354

Matassino D, Gigante G, Inglese F, Serluca G, Di Luccia A (2007b) Proteomic approach to the study of proteolysis in Podolian caciocavallo cheese. Ital J Anim Sci 6 (suppl. 1):453

Matassino D, Gigante G, Inglese F, Romagnuolo F, Di Luccia A (2008) Safeguard of dairy products for a multifunctional rural development of Apennine bioterritories. Proteomics of 'Pecorino di Laticauda'. Preliminary results. In: 3rd symposium of the mediterranean livestock industry network "strategies of the dairy sector in the mediterranean basin". Bari, 16–17 settembre 2008

McSweeney PLH (2004) Biochemistry of cheese ripening. Int J Dairy Technol 57:127–144

McSweeney PLH, Olson NF, Fox PF, Healy A (1994) Proteolysis of bovine αs2-casein by chymosin. Z Lebensm Unters Forsch 119:429–432

McSweeney PLH, Fox PF, Olson NF (1995) Proteolysis of bovine caseins by cathepsin D—preliminary observations and comparison with chymosin. Int Dairy J 5:321–336

Molina E, Ramos M, Amigo L (2002) Characterisation of the casein fraction of Ibérico cheese by electrophoretic techniques. J Sci Food Agric 82:1240–1245

O'Driscoll BM, Rattray FP, McSweeney PLH, Kelly AL (1999) Protease activities in raw milk determined using a synthetic heptapeptide substrate. J Food Sci 64:606–611

O'Flaherty F (1997) Characterisation of some minor caseins and protease peptones of bovine milk. M.Sc. thesis. National University of Ireland, Cork, Ireland

Pappa EC, Kandarakis IG, Zerfiridis GK, Anifantakis EM, Sotirakoglou K (2006) Influence of starter cultures on the proteolysis of Teleme cheese made from different types of milk. Lait 86:273–290

Pappa EC, Robertson JA, Rigby NM, Mellon F, Kandarakis I, Mills ENC (2008) Application of proteomic techniques to protein and peptide profiling of Teleme cheese made from different types of milk. Int Dairy J 18:605–614

Pellegrino L, Cattaneo S, Masotti F, Psathas G (2010) Detection of milk powder and caseinates in Halloumi cheese. J Dairy Sci 93:3453–3460

Piraino P, Upadhyay VK, Rossano A, Riccio P, Parente E, Kell AL (2007) Use of mass spectrometry to characterize proteolysis in cheese. Food Chem 101:964–972

Senocq D, Dupont D, Rolet-Repecaud O, Levieux D (2002) ELISA for monitoring the cleavage of β-casein at site Lys28-Lys29 by plasmin during Comt. cheese ripening. J Dairy Res 69:491–500
Silva SV, Allmere T, Malcata FX, Andrén A (2003) Comparative studies on the gelling properties of cardosins extracted from Cynara cardunculus and chymosin on cow's skim milk. Int Dairy J 13:559–564
Singh TK, Fox PF, Hojrup P, Healy A (1994) A scheme for the fractionation of cheese nitrogen and identification of principal peptides. Int Dairy J 4:111–122
Somma A, Ferranti P, Addeo F, Mauriello R, Chianese L (2008) Peptidomic approach based on combined capillary isoelectric focusing and mass spectrometry for the characterization of the plasmin primary products from bovine and water buffalo β-casein. J Chromatogr A 1192:294–300
Sousa MJ, Malcata FX (2002) Advances in the role of a plant coagulant (*Cynara cardunculus*) in vitro and during ripening of cheeses from several milk species. Lait 82:151–170
Sousa MJ, Ard Y, McSweeney PLH (2001) Advances in the study of proteolysis during cheese ripening. Int Dairy J 11:327–345
Trieu-Cuot P, Addeo F (1981) Occurrence of gamma casein in buffalo milk. J Dairy Res 48:311–317
Trieu-Cuot P, Gripon J-C (1982) A study of proteolysis during Camembert cheese ripening using isoelectric focusing and two-dimensional electrophoresis. J Dairy Res 49:501–510
Upadhyay VK, McSweeney PLH, Magboul AAA, Fox PF (2004) Proteolysis in cheese during ripening. In: Fox PF, McSweeney PLH, Cogan TM, Guinee TP (eds) Cheese chemistry, physics and microbiology, vol 1. Elsevier, London, pp 391–433
Veloso ACA, Teixeira N, Peres AM, Mendonc A, Ferreira IMPLVO (2004) Evaluation of cheese authenticity and proteolysis by HPLC and urea–polyacrylamide gel electrophoresis. Analytical nutritional and clinical methods. Food Chem 87:289–295
Visser S (1993) Symposium: proteolytic enzymes and cheese ripening. Proteolytic enzymes and their relation to cheese ripening and flavor: an overview. J Dairy Sci 76:329–350
Zhong Q, Daubert CR, Velev OD (2007) Physicochemical variables affecting the rheology and microstructure of rennet casein gels. J Agric Food Chem 55(7):2688–2697

Chapter 15
Lactic Acid Bacteria in Fermented Foods

Françoise Rul, Monique Zagorec, and Marie-Christine Champomier-Vergès

15.1 Introduction

Lactic acid bacteria (LAB) are the main agents ensuring the safety and organoleptic properties of fermented foods. LAB encompass more than 10 genera which include a large number of species (above 100). However, the amount of LAB species used as starters for food fermentation is limited. In the dairy industry, *Lactobacillus delbrueckii ssp. bulgaricus* and *Streptococcus thermophilus* are the only two species involved in yogurt production, whereas *Lactococci* (*L. lactis*, *L. cremoris*, *L. diacetylactis*) and other *Lactobacilli* (*L. helveticus*, *L. acidophilus*, *L. casei*) are also involved in cheese making. In fermented meat products, *Lactobacillus sakei*, *Lactobacillus curvatus*, *Lactobacillus plantarum*, *Pediococcus pentosaceus*, and *Pediococcus acidilactici* are the main starters. *Oenococcus oeni* is the emblematic wine starter. In addition, the natural microflora of the components used for food fermentation or present in the final fermented products may also encompass a great diversity of LAB species. However, most of the scientific studies have focused on starters. Concerning proteomics, its use to study the adaptation or functions of starters in fermented products is mainly restricted to laboratory conditions or to a few model matrices. This limitation results from a major technical problem: food matrices, whatever their animal origin (milk, meat) or vegetal origin are rich in proteins and peptides that may interfere with the study of LAB involvement. Indeed, many proteomic analyses dealing with food refer to the food

F. Rul • M. Zagorec • M.-C. Champomier-Vergès (✉)
INRA, UMR1319 Micalis, Jouy-en-Josas, F-78350, France

AgroParisTech, UMR Micalis, Jouy-en-Josas, F-78350, France
e-mail: marie-christine.champomier-verges@jouy.inra.fr

M. Zagorec
INRA, UMR 1014 Secalim, 44307 Nantes, France

LUNAM Université, Oniris, Nantes, 44307, France

F. Toldrá and L.M.L. Nollet (eds.), *Proteomics in Foods: Principles and Applications*, Food Microbiology and Food Safety 2, DOI 10.1007/978-1-4614-5626-1_15,

matrix rather than to the microbes present in it (D'Alessandro and Zolla 2012; Sentandreu and Sentandreu 2011; Molina et al. 2002). In many fermented products, the biomass represented by bacteria is negligible when compared to the proteins and peptides from the food components, and proteomic analyses of these products therefore do not require a previous removal of bacteria. As examples of such studies to investigate meat quality, see Vallejo-Cordoba et al. (2010) and Nakamura et al. (2010). In the present chapter we present studies dealing with a proteomic approach that allows exploring the ability of lactic acid bacteria to grow in fermented food models, their important functions for the quality of the products, and their ability to resist the harsh conditions encountered during fermentation processes. A few proteomics examples related to food spoilage and food safety, or to bacteria fingerprinting are also presented.

15.2 Generic Analyses in Laboratory Model Conditions: Adaptability to Conditions Mimicking the Food Processes

15.2.1 Introduction

In nature, the ability to respond quickly to stress is essential for survival. Indeed, under adverse conditions, bacteria set up cellular responses and defense mechanisms that significantly improve their chances of successful adaptation to harsh and/or sudden environmental changes. LAB, as do other bacteria, have to face various and often simultaneous stresses during starter handling and storage and also during food processes, in particular physicochemical stresses such as freezing, cold, heat, extreme pH, osmotic pressure, oxidative agents, or high hydrostatic pressure (HHP). A better comprehension of stress resistance is necessary to understand the adaptative response to these unfavorable conditions and thus to predict the potential functions that would be essential for the survival/maintenance of starters under such industrial conditions. This would allow and rationalize the preparation of adapted strains and their improvement for specific industrial applications. Table 15.1 proposes a synthetic summary of proteomic studies reported in the literature, and dealing with LAB starters of dairy products (*L. lactis*, *L. delbrueckii ssp. bulgaricus*, *L. helveticus*, *L. casei*, *L. acidophilus*, *S. thermophilus*), sourdough (*L. sanfranciscensis*), sausage (*L. sakei*), or a starter that can be found in all of these products (*L. plantarum*). In these articles, authors investigated adaptation of the bacteria to conditions encountered during fermentation processes. Table 15.1 summarizes only proteins whose expression was upregulated, representing therefore functions and regulatory mechanisms necessary for bacteria to overcome these stressful conditions.

We analyzed these data to search for: (1) proteins commonly upregulated under several stressful conditions and/or upregulated by several species thus representing a global stress signature, (2) species-specific responses, and (3) stress-specific responses.

15.2.2 Global Stress Signatures

From all data summarized in Table 15.1, general stress proteins are among the most important families of proteins characterizing the bacterial response to various stress conditions: these proteins are induced whatever the stress or the food matrix, and in all species. As an example, several of the Clp proteins both respond to different stresses in one bacterial species, but also respond to the same stress in different bacterial species: ClpL is upregulated after both oxidative-, heat-, cold- and lactose-starvation stresses in *S. thermophilus*. ClpP responds to lactose starvation (which provokes growth arrest) in *S. thermophilus*, and also to the entry in the stationary phase in *L. plantarum*. It also responds to a cold shock in both *S. thermophilus* and *L. acidophilus*. It is known that a pre-adaptation to one stress condition helps bacterial cells to become resistant to other stresses. This reflects the multiple stresses that can be encountered during food processes. In addition, this may explain the multiple protections ensured by stress-response proteins. For example, a pre-adaptation of *L. delbrueckii* ssp. *bulgaricus* to acid confers a better cryotolerance.

Another relevant feature in Table 15.1 is that both general stress proteins and proteins described as specific for a stress, such as oxidative stress-response proteins, actually respond to a large panel of stressful conditions. For instance, Dpr is indeed upregulated after an oxidative stress in *S. thermophilus*, but also after acid stress in this bacterium and after conditions leading to growth arrest in *S. thermophilus*, *L. plantarum*, and *L. helveticus*.

Proteins involved in carbohydrate transport and metabolism generally respond to conditions reflecting growth rate such as entry to late exponential or stationary phases, lactose starvation, or decrease of growth temperature. Some of those, however, are also upregulated after various stresses. In addition, upregulation of proteins involved in nitrogen metabolism is mostly observed under conditions of growth arrest (entry to late exponential or stationary phases, lactose starvation) in different LAB species. A similar, although less pronounced, tendency is observed for proteins involved in nucleotide transport and metabolism. This certainly shows that a central metabolism is required for bacteria switching from optimal conditions for fast growth, to harsher conditions. This implies that bacteria should not be considered as dormant when they enter a famine state after feast conditions.

Interestingly, upregulation of proteins involved in translation or transcription arises after various stress conditions, rather than under growth-arrest conditions. In addition, these upregulated proteins are also typical of the response of *L. sakei* and also, to a lesser extent of *L. sanfranciscensis*, after HHP treatment.

15.2.3 Species-Specific Responses

Among the data available in the literature, only a few species comparisons can be performed. Growth under acidic conditions or acid stress is among the conditions that have been best described in lactic acid bacteria. Indeed, LAB producing

Table 15.1 Synthetic overview of proteins up-regulated by different starters grown under various conditions. The LAB species studied are used for various food types: sourdough, dairy products (yogurt, cheese), meat (sausage) or for all of these. The type of stress applied is listed, as well as the experimental conditions. Proteins are listed according to their functional category

Protein name	Stress[a]	Food type	Species	Experimental conditions[b]	References
General stress proteins					
Asp	Cold-shock	Meat	*L. sakei*	4 °C (CDM)	(Marceau et al. 2004)
ClpB	Acid	Dairy	*L. lactis*	2 h, pH 5.0 (SA)	(Budin-Verneuil et al. 2005)
ClpC	Acid		*L. lactis*	2 h, pH 5.0 (SA)	(Budin-Verneuil et al. 2005)
ClpE	Acid		*L. lactis*	pH 5.5 (GSA)	(Frees et al. 2003)
				2 h, pH 5.0 (SA)	(Budin-Verneuil et al. 2005)
ClpL	Acid		*L. bulgaricus*	40 min, pH 4.9 (MRS)	(Fernandez et al. 2008)
	Oxidative		*S. thermophilus*	180 min, 1 mM H_2O_2 (MRS)	(Arena et al. 2006)
	Heat-shock		*S. thermophilus*	20 and 180 min, 50 °C (MRS)	(Arena et al. 2006)
	Cold-shock		*S. thermophilus*	180 min, 20 °C (MRS)	(Arena et al. 2006)
	Lactose starvation		*S. thermophilus*	180 min, no lac (MRS)	(Arena et al. 2006)
	HHP	Sourdough	*L. sanfranciscensis*	60 min 80 MPa at 30 MPa/min (MRS)	(Vogel et al. 2005)
				80 MPa at 30 MPa/min (MRS)	(Hörmann et al. 2006)
ClpP	Acid	Dairy	*L. lactis*	pH 5.5 (GSA)	(Frees et al. 2003)
	Oxidative		*S. thermophilus*	180 min, 1 mM H_2O_2 (MRS)	(Arena et al. 2006)
	Cold-shock		*S. thermophilus*	20 and 180 min, 20 °C (MRS)	(Arena et al. 2006)
			L. acidophilus	8 h, 26 °C (MRS)	(Wang et al. 2005)
	Lactose starvation		*S. thermophilus*	20 and 180 min, no lac (MRS)	(Arena et al. 2006)
	Stationary phase	All	*L. plantarum*	Late-stationary/mid-log phase (MRS)	(Cohen et al. 2006)
	HHP	Sourdough	*L. sanfranciscensis*	80 MPa at 30 MPa/min (MRS)	(Hörmann et al. 2006)
CspA and B	Cold-shock	Dairy	*S. thermophilus*	20 and 180 min, 20 °C (MRS)	(Arena et al. 2006)
	HHP	Sourdough	*L. sanfranciscensis*	25 to 200 MPa/1 h (H medium)	(Drews et al. 2002)
	Co-culture[c]	Sourdough	*L. sanfranciscensis*	18 h (WFH)	(Di Cagno et al. 2007)
CspE	Acid	Dairy	*L. lactis*	2 h, pH 5.0 (M17)	(Budin-Verneuil et al. 2005)
DnaK	Acid	Dairy	*L. bulgaricus*	30 min, pH 4.75 (CDM)	(Lim et al. 2000)
				40 min, pH 4.9 (MRS)	(Fernandez et al. 2008)

			L. lactis	30 min, pH 5.5 (M17)	(Hartke et al. 1996)
				pH 5.5 (GSA)	(Frees et al. 2003)
				2 h, pH 5.0 (SA)	(Budin-Verneuil et al. 2005)
	Oxidative		*S. thermophilus*	180 min, 1 mM H_2O_2 (MRS)	(Arena et al. 2006)
	Heat-shock		*L. bulgaricus*	15 min, 65 °C (CDM)	(Gouesbet et al. 2002)
			L. lactis	43 °C (GSA)	(Kilstrup et al. 1997)
			S. thermophilus	20 and 180 min, 50 °C (MRS)	(Arena et al. 2006)
	Decreasing T° gradient		*L. helveticus*	55–20 °C (whey)	(Di Cagno et al. 2006)
	Osmotic		*L. lactis*	2.5 % NaCl (GSA)	(Kilstrup et al. 1997)
				NaCl 5 M (CDM)	(Zhang et al. 2010)
	Stationary phase		*L. lactis*	(RSM and SA)	(Larsen et al. 2006)
	HHP	Meat	*L. sakei*	10 min, 400 MPa (MRS)	(Jofré et al. 2007)
		Sourdough	*L. sanfranciscensis*	80 MPa at 30 MPa/min (MRS)	(Hörmann et al. 2006)
Gls24-like	Acid	Dairy	*S. thermophilus*	20 and 180 min, pH 5.0 (MRS)	(Arena et al. 2006)
	Cold-shock		*S. thermophilus*	20 and 180 min, 20 °C (MRS)	(Arena et al. 2006)
		Meat	*L. sakei*	4 °C (CDM)	(Marceau et al. 2004)
	Oxidative	Dairy	*S. thermophilus*	20 and 180 min, 1 mM H_2O_2 (MRS)	(Arena et al. 2006)
	Lactose starvation		*S. thermophilus*	20 and 180 min, no lac (MRS)	(Arena et al. 2006)
	HHP	Meat	*L. sakei*	400 MPa 10 min (MRS)	(Jofré et al. 2007)
Glycoprotein endopeptidase		Dairy	*L. acidophilus*	8 h, 26 °C (MRS)	(Wang et al. 2005)
GroEL	Acid		*L. bulgaricus*	30 min, pH 4.75 (CDM)	(Lim et al. 2000)
				24 h, uncontrolled pH (MRS)	(Silva et al. 2005)
				40 min, pH 4.9 (MRS)	(Fernandez et al. 2008)
			L. lactis	30 min, pH 5.5 (M17)	(Hartke et al. 1996)
				pH 5.5 (GSA)	(Frees et al. 2003)
				2 h, pH 5.0 (SA)	(Budin-Verneuil et al. 2005)

(continued)

Table 15.1 (continued)

Protein name	Stress[a]	Food type	Species	Experimental conditions[b]	References
			S. thermophilus	20 and 180 min, pH 5.0 (MRS)	(Arena et al. 2006)
	Heat-shock		*L. bulgaricus*	15 min, 65 °C (CDM)	(Gouesbet et al. 2002)
			L. lactis	43 °C (GSA)	(Kilstrup et al. 1997)
			S. thermophilus	Emmental processing (milk)	(Guimont et al. 2002)
	Cold-shock		*S. thermophilus*	180 min, 20 °C (MRS)	(Arena et al. 2006)
	Decreasing T° gradient		*L. helveticus*	55–20 °C (whey)	(Di Cagno et al. 2006)
	Osmotic		*L. lactis*	2.5 % NaCl (GSA)	(Kilstrup et al. 1997)
	Oxidative		*S. thermophilus*	20 and 180 min, 1 mM H_2O_2 (MRS)	(Arena et al. 2006)
	Lactose starvation		*S. thermophilus*	20 and 180 min, no lac (MRS)	(Arena et al. 2006)
	Stationary phase		*L. lactis*	(RSM and SA)	(Larsen et al. 2006)
	Cheese processing		*S. thermophilus*	Emmental processing (milk)	(Guimont et al. 2002)
	HHP	Sourdough	*L. sanfranciscensis*	60 min 80 MPa at 30 MPa/min (MRS)	(Vogel et al. 2005)
GroES	Acid	Dairy	*L. bulgaricus*	30 min, pH 4.75 (CDM)	(Lim et al. 2000)
				24 h, uncontrolled pH (MRS)	(Silva et al. 2005)
				40 min, pH 4.9 (MRS)	(Fernandez et al. 2008)
			L. lactis	pH 5.5 (GSA)	(Frees et al. 2003)
			S. thermophilus	180 min, pH 5.0 (MRS)	(Arena et al. 2006)
	Heat-shock		*L. lactis*	43 °C, (GSA)	(Kilstrup et al. 1997)
			S. thermophilus	20 and 180 min, 50 °C (MRS)	(Arena et al. 2006)
	Osmotic		*L. lactis*	2.5 % NaCl (GSA)	(Kilstrup et al. 1997)
	Co-culture[c]	Sourdough	*L. sanfranciscensis*	18 h (WFH)	(Di Cagno et al. 2007)
				80 MPa at 30 MPa/min (MRS)	(Hörmann et al. 2006)
GrpE	Acid	Dairy	*L. bulgaricus*	40 min, pH4.9 (MRS)	(Fernandez et al. 2008)
				30 min, pH 5.25/pH6 (WP+YE)	(Streit et al. 2008)
		Sourdough	*L. sanfranciscensis*	60 min pH5.0 (MRS)	(De Angelis et al. 2001)
	Heat-shock	Dairy	*S. thermophilus*	180 min, 50 °C (MRS)	(Arena et al. 2006)
	Lactose starvation		*L. casei*	24 h, no lac (semi-defined medium)	(Hussain et al. 2009)
	Stationary phase	All	*L. plantarum*	Late-stationary/mid-log phase (MRS)	(Cohen et al. 2006)

HSP	Stationary phase	All	*L. plantarum*	Late-stationary/mid-log phase (MRS)	(Cohen et al. 2006)
	Acid	Dairy	*L. bulgaricus*	24 h, uncontrolled pH (MRS)	(Silva et al. 2005)
			L. lactis	pH 5.5 (GSA)	(Frees et al. 2003)
			S. thermophilus	2 h, pH 4.9 (M17)	(Gonzalez-Marquez et al. 1997)
	Decreasing T° gradient		*L. helveticus*	55–20 °C (whey)	(Di Cagno et al. 2006)
	Heat-shock		*S. thermophilus*	Emmental processing (milk)	(Guimont et al. 2002)
	HHP	Sourdough	*L. sanfranciscensis*	80 MPa at 30 MPa/min (MRS)	(Hörmann et al. 2006)
Ssp21	Acid		*L. lactis*	pH 5.5 (GSA)	(Frees et al. 2003)
Usp	Cold-shock	Meat	*L. sakei*	4 °C (CDM)	(Marceau et al. 2004)
UspA-like	Acid	Dairy	*L. lactis*	2 h, pH 5.0 (M17)	(Budin-Verneuil et al. 2005)
Oxidative stress response proteins					
AhpC	Acid	Dairy	*L. lactis*	2 h, pH 5.0 (SA)	(Budin-Verneuil et al. 2005)
Catalase	Stationary phase	All	*L. plantarum*	Late-stationary/mid-log phase (MRS)	(Cohen et al. 2006)
Dpr	Acid	Dairy	*S. thermophilus*	20 and 180 min, pH 5.0 (MRS)	(Arena et al. 2006)
	Lactose starvation		*S. thermophilus*	20 min, no lac (MRS)	(Arena et al. 2006)
	Stationary phase	All	*L. plantarum*	Late-stationary/mid-log phase (MRS)	(Cohen et al. 2006)
	Oxidative	Dairy	*S. thermophilus*	20 and 180 min, 1 mM H_2O_2 (MRS)	(Arena et al. 2006)
	Decreasing T° gradient		*L. helveticus*	55–20 °C (whey)	(Di Cagno et al. 2006)
GshR	Oxidative		*S. thermophilus*	20 and 180 min, 1 mM H_2O_2 (MRS)	(Arena et al. 2006)
	Stationary phase	All	*L. plantarum*	Late-stationary/mid-log phase (MRS)	(Cohen et al. 2006)
MsrA	Cold-shock	Meat	*L. sakei*	4 °C (CDM)	(Marceau et al. 2004)
Nox	Oxidative	Dairy	*S. thermophilus*	20 and 180 min, 1 mM H_2O_2 (MRS)	(Arena et al. 2006)
Ohr/Osm	Cold-shock	Meat	*L. sakei*	4 °C (CDM)	(Marceau et al. 2004)
Oxidoreductase	Acid	Dairy	*L. bulgaricus*	40 min, pH 4.9 (MRS)	(Fernandez et al. 2008)

(continued)

Table 15.1 (continued)

Protein name	Stress[a]	Food type	Species	Experimental conditions[b]	References
	Stationary phase	All	*L. plantarum*	Late-stationary/mid-log phase (MRS)	(Cohen et al. 2006)
SodA	Oxidative	Dairy	*S. thermophilus*	180 min, 1 mM H_2O_2 (MRS)	(Arena et al. 2006)
	Acid		*L. lactis*	2 h, pH 5.0 (M17 and SA)	(Budin-Verneuil et al. 2005)
SufB and C	Oxidative		*S. thermophilus*	20 and 180 min, 1 mM H_2O_2 (MRS)	(Arena et al. 2006)
ThrX	Oxidative		*S. thermophilus*	20 and 180 min, 1 mM H_2O_2 (MRS)	(Arena et al. 2006)
	Decreasing T° gradient		*L. helveticus*	55–20 °C (whey)	(Di Cagno et al. 2006)
	HHP	Sourdough	*L. sanfranciscensis*	80 MPa at 30 MPa/min (MRS)	(Hörmann et al. 2006)
Tpx	Acid	Dairy	*L. lactis*	2 h, pH 5.0 (SA)	(Budin-Verneuil et al. 2005)
Carbohydrate transport and metabolism proteins					
AdhE	Acid	Dairy	*L. lactis*	2 h, pH 5.0 (M17)	(Budin-Verneuil et al. 2005)
Als	Acid		*L. lactis*	2 h, pH 5.0 (SA)	(Budin-Verneuil et al. 2005)
BglA	Acid		*L. lactis*	2 h, pH 5.0 (M17 and SA)	(Budin-Verneuil et al. 2005)
CcpA	Stationary phase	Dairy	*L. lactis*	Stationary (RSM and SA)	(Larsen et al. 2006)
Eno	Decreasing T° gradient		*L. helveticus*	55–20 °C (whey)	(Di Cagno et al. 2006)
	Co-culture[c]	Sourdough	*L. sanfranciscensis*	18 h (WFH)	(Di Cagno et al. 2007)
FbaA	Acid	Dairy	*L. bulgaricus*	40 min, pH 4.9 (MRS)	(Fernandez et al. 2008)
GapA1	Acid		*L. bulgaricus*	24 h, uncontrolled pH (MRS)	(Silva et al. 2005)
			L. lactis	2 h, pH 5.0 (SA)	(Budin-Verneuil et al. 2005)
	Heat-shock		*L. lactis*	43 °C (GSA)	(Kilstrup et al. 1997)
	Decreasing T° gradient		*L. helveticus*	55 °C to 20 °C (whey)	(Di Cagno et al. 2006)
	Lactose starvation		*S. thermophilus*	20 and 180 min, no lac (MRS)	(Arena et al. 2006)
	HHP	Meat	*L. sakei*	10 min, 400 MPa (MRS)	(Jofré et al. 2007)
GalE	Lactose starvation	Dairy	*S. thermophilus*	180 min, no lac (MRS)	(Arena et al. 2006)
	Stationary	All	*L. plantarum*	Late-stationary/mid-log phase (MRS)	(Cohen et al. 2006)
	Late exponential	Dairy	*S. thermophilus*	Late/early exponential (milk)	(Hervé-Jimenez et al. 2008)
GalK	Stationary phase	All	*L. plantarum*	Late-stationary/mid-log phase (MRS)	(Cohen et al. 2006)
	Late exponential	Dairy	*S. thermophilus*	Late/early exponential (milk)	(Hervé-Jimenez et al. 2008)
GalU	Acid		*L. lactis*	2 h, pH 5.0 (SA)	(Budin-Verneuil et al. 2005)

	Lactose starvation		*S. thermophilus*	180 min, no lac (MRS)	(Arena et al. 2006)
GlmS	Late exponential		*S. thermophilus*	Late/early exponential (milk)	(Hervé-Jimenez et al. 2008)
	Acid		*L. lactis*	2 h, pH 5.0 (M17 and SA)	(Budin-Verneuil et al. 2005)
Gnp	Stationary phase	All	*L. plantarum*	Late-stationary/mid-log phase (MRS)	(Cohen et al. 2006)
GlpD	Cold shock	Meat	*L. sakei*	4°C (CDM)	(Marceau et al. 2004)
	Stationary phase	Dairy	*L. lactis*	Stationary (SM)	(Larsen et al. 2006)
HasC	Acid	Dairy	*L. lactis*	2 h, pH 5.0 (M17)	(Budin-Verneuil et al. 2005)
HHP	Sourdough		*L. sanfranciscensis*	80 MPa at 30 MPa/min (MRS)	(Hörmann et al. 2006)
	Co-culture[c]		*L. sanfranciscensis*	18 h (WFH)	(Di Cagno et al. 2007)
Hpr	Co-culture[c]		*L. sanfranciscensis*	18 h (WFH)	(Di Cagno et al. 2007)
Ldh	Acid	Dairy	*L. bulgaricus*	24 h, uncontrolled pH (MRS)	(Silva et al. 2005)
			L. lactis	2 h, pH 5.0 (SA)	(Budin-Verneuil et al. 2005)
			S. thermophilus	180 min, pH 5.0 (MRS)	(Arena et al. 2006)
	Lactose starvation		*L. casei*	24 h, no lac (semi-defined medium)	(Hussain et al. 2009)
	Co-culture[c]	Sourdough	*L. sanfranciscensis*	18 h (WFH)	(Di Cagno et al. 2007)
MtlD	Stationary phase	All	*L. plantarum*	Late-stationary/mid-log phase (MRS)	(Cohen et al. 2006)
MtlR	Acid	Dairy	*L. lactis*	2 h, pH 5.0 (SA)	(Budin-Verneuil et al. 2005)
PdhC	Acid	Dairy	*L. lactis*	2 h, pH 5.0 (M17)	(Budin-Verneuil et al. 2005)
PfkA	Cold-shock		*S. thermophilus*	180 min, 20 °C (MRS)	(Arena et al. 2006)
	Lactose starvation		*L. casei*	24 h, no lac (semi-defined medium)	(Hussain et al. 2009)
Pgk	Acid		*L. bulgaricus*	24 h, uncontrolled pH (MRS)	(Silva et al. 2005)
			L. lactis	2 h, pH 5.0 (SA)	(Budin-Verneuil et al. 2005)
	Heat-shock		*L. lactis*	43 °C (GSA)	(Kilstrup et al. 1997)
	Lactose starvation		*L. casei*	24 h, no lac (semi-defined medium)	(Hussain et al. 2009)
	Co-culture[c]	Sourdough	*L. sanfranciscensis*	18 h (WFH)	(Di Cagno et al. 2007)
Pgm	Co-culture[c]		*L. sanfranciscensis*	18 h (WFH)	(Di Cagno et al. 2007)
Pmi	Lactose starvation	Dairy	*L. casei*	24 h, no lac (semi-defined medium)	(Hussain et al. 2009)
	Stationary phase	All	*L. plantarum*	Late-stationary/mid-log phase (MRS)	(Cohen et al. 2006)
	Acid		*L. lactis*	2 h, pH 5.0 (M17)	(Budin-Verneuil et al., 2005)

(continued)

Table 15.1 (continued)

Protein name	Stress[a]	Food type	Species	Experimental conditions[b]	References
Pox	Stationary phase		*L. plantarum*	Late-stationary/mid-log phase (MRS)	(Cohen et al. 2006)
Pta	Acid	Dairy	*L. lactis*	2 h, pH 5.0 (M17)	(Budin-Verneuil et al. 2005)
Pts9AB	Stationary phase		*L. plantarum*	Late-stationary/mid-log phase (MRS)	(Cohen et al. 2006)
PtsI	Heat-shock	Dairy	*L. bulgaricus*	15 min, 65 °C (CDM)	(Gouesbet et al. 2002)
	Late exponential		*S. thermophilus*	Late/early exponential (milk)	(Hervé-Jimenez et al. 2008)
PycA	Acid		*L. lactis*	2 h, pH 5.0 (M17 and SA)	(Budin-Verneuil et al. 2005)
Pyk	Acid		*L. bulgaricus*	24 h, uncontrolled pH (MRS)	(Silva et al. 2005)
			L. lactis	2 h, pH 5.0 (SA)	(Budin-Verneuil et al. 2005)
	Cold-shock		*L. acidophilus*	8 h, 26 °C (MRS)	(Wang et al. 2005)
	Heat-shock		*L. lactis*	43 °C (GSA)	(Kilstrup et al. 1997)
	Decreasing T° gradient		*L. helveticus*	55–20 °C (whey)	(Di Cagno et al. 2006)
RbsK	HHP	Sourdough	*L. sanfranciscensis*	25–200 MPa/h (H medium)	(Drews et al. 2002)
				60 min 80 MPa at 30 MPa/min (MRS)	(Vogel et al. 2005)05
				80 MPa at 30 MPa/min (MRS)	(Hörmann et al. 2006)
RmlB	Acid	Dairy	*L. lactis*	2 h, pH 5.0 (M17)	(Budin-Verneuil et al. 2005)
RmlC	Lactose starvation		*L. casei*	24 h, no lac (semi-defined medium)	(Hussain et al. 2009)
Rpe	Lactose starvation		*L. casei*	24 h, no lac (semi-defined medium)	(Hussain et al. 2009)
Sugar hydrolase	HHP	Sourdough	*L. sanfranciscensis*	80 MPa at 30 MPa/min (MRS)	(Hörmann et al. 2006)
ScrB	Late exponential	Dairy	*S. thermophilus*	Late/early exponential (milk)	(Hervé-Jimenez et al. 2008)
Tpi	Acid		*L. bulgaricus*	24 h, uncontrolled pH (MRS)	(Silva et al. 2005)
Translation – transcription					
ArgS	Acid		*L. lactis*	2 h, pH 5.0 (M17 and SA)	(Budin-Verneuil et al. 2005)
DNA topo III	Decreasing T° gradient		*L. helveticus*	55–20 °C (whey)	(Di Cagno et al. 2006)
EF-Tu,	Acid		*S. thermophilus*	20 and 180 min, pH 5.0 (MRS)	(Arena et al. 2006)
			L. lactis	2 h, pH 5.0 (SA)	(Budin-Verneuil et al. 2005)
	Cold-shock		*S. thermophilus*	180 min, 20 °C (MRS)	(Arena et al. 2006)
	HHP	Sourdough	*L. sanfranciscensis*	25–200 MPa/1 h (H medium)	(Drews et al. 2002)
				60 min 80 MPa at 30 MPa/min (MRS)	(Vogel et al. 2005)

EF-Ts	Acid	Dairy	*S. thermophilus*	180 min, pH 5.0 (MRS)	(Arena et al. 2006)
	Heat-shock		*S. thermophilus*	20 and 180 min, 50 °C (MRS)	(Arena et al. 2006)
EF-G	Acid		*S. thermophilus*	180 min, pH 5.0 (MRS)	(Arena et al. 2006)
	Heat-shock		*S. thermophilus*	20 and 180 min, 50 °C (MRS)	(Arena et al. 2006)
EngA	Co-culture[c]	Sourdough	*L. sanfranciscensis*	18 h (WFH)	(Di Cagno et al. 2007)
Fmt	Stationary phase	Dairy	*L. lactis*	Stationary (RSM and SA)	(Larsen et al. 2006)
GatA	Acid		*L. lactis*	2 h, pH 5.0 (M17 and SA)	(Budin-Verneuil et al. 2005)
GltX	Acid		*L. lactis*	2 h, pH 5.0 (M17)	(Budin-Verneuil et al. 2005)
IleS	Co-culture[d]/late exponential		*S. thermophilus*	Late/early exponential, (milk)	(Hervé-Jimenez et al. 2009)
L5, S10P, S10, L7/L12, L31	Co-culture[c]	Sourdough	*L. sanfranciscensis*	18 h (WFH)	(Di Cagno et al. 2007)
L7/L12	Decreasing T° gradient	Dairy	*L. helveticus*	55–20 °C (whey)	(Di Cagno et al. 2006)
LysS	Acid		*L. lactis*	2 h, pH 5.0 (M17)	(Budin-Verneuil et al. 2005)
NusG	HHP	Meat	*L. sakei*	10 min, 400 MPa (MRS)	(Jofré et al. 2007)
PheT	Acid	Dairy	*L. lactis*	2 h, pH 5.0 (SA)	(Budin-Verneuil et al. 2005)
	Osmotic		*L. lactis*	NaCl 5 M (CDM)	(Zhang et al. 2010)
ProS	Acid		*L. lactis*	2 h, pH 5.0 (M17)	(Budin-Verneuil et al. 2005)
ValS	Co-culture[d]/late exponential		*S. thermophilus*	Late/early exponential, (milk)	(Hervé-Jimenez et al. 2009)
RplJ	HHP	Meat	*L. sakei*	10 min, 400 MPa (MRS)	(Jofré et al. 2007)
RpoD	Acid	Dairy	*L. lactis*	2 h, pH 5.0 (M17)	(Budin-Verneuil et al. 2005)
RpsA	Acid		*L. lactis*	2 h, pH 5.0 (M17)	(Budin-Verneuil et al. 2005)
RpsB	Stationary phase		*L. lactis*	Stationary (RSM and SA)	Stationary (RSM and SA)
	Acid		*L. lactis*	2 h, pH 5.0 (M17)	(Budin-Verneuil et al. 2005)
RpsF	HHP	Meat	*L. sakei*	10 min, 400 MPa (MRS)	(Jofré et al. 2007)
Rrf	HHP		*L. sakei*	10 min, 400 MPa (MRS)	(Jofré et al. 2007)
TypA	Acid	Dairy	*L. lactis*	2 h, pH 5.0 (SA)	(Budin-Verneuil et al. 2005)
Nucleotide transport and metabolism					
Adk	HHP	Meat	*L. sakei*	10 min, 400 MPa (MRS)	(Jofré et al. 2007)
GuaA	Acid	Dairy	*L. lactis*	2 h, pH 5.0 (M17)	(Budin-Verneuil et al. 2005)
GuaB	Stationary phase	Dairy	*L. lactis*	Stationary (RSM and SA)	(Larsen et al. 2006)

(continued)

Table 15.1 (continued)

Protein name	Stress[a]	Food type	Species	Experimental conditions[b]	References
PurA	Stationary phase		*L. lactis*	Stationary (RSM and SA)	(Larsen et al. 2006)
PurE	Acid		*L. bulgaricus*	30 min, pH 5.25/pH 6 (WP+YE)	(Streit et al. 2008)
PurH	Osmotic		*L. lactis*	5 M NaCl (CDM)	(Zhang et al. 2010)
PurR	Co-culture[d]/late exponential		*S. thermophilus*	Late/early exponential, (milk)	(Hervé-Jimenez et al. 2009)
PyrE	Stationary phase	All	*L. plantarum*	Late-stationary/mid-log phase (MRS)	(Cohen et al. 2006)
PyrC	Stationary phase		*L. plantarum*	Late-stationary/mid-log phase (MRS)	(Cohen et al. 2006)
PyrG	Acid	Dairy	*L. lactis*	2 h, pH 5.0 (M17)	(Budin-Verneuil et al. 2005)
Nitrogen metabolism					
Asd2	Stationary phase		*L. plantarum*	Late-stationary/mid-log phase (MRS)	(Cohen et al. 2006)
AspC	Acid	Dairy	*L. lactis*	2 h, pH 5.0 (SA)	(Budin-Verneuil et al. 2005)
CysM1/M2	Late exponential		*S. thermophilus*	Late/early exponential (milk)	(Hervé-Jimenez et al. 2008)
CysD	Late exponential		*S. thermophilus*	Late/early exponential (milk)	(Hervé-Jimenez et al. 2008)
CysK	Lactose starvation		*L. casei*	24 h, no lac (semi-defined medium)	(Hussain et al. 2009)
DapB	Stationary phase	All	*L. plantarum*	Late-stationary/mid-log phase (MRS)	(Cohen et al. 2006)
GlnQ	Lactose starvation	Dairy	*S. thermophilus*	180 min, no lac (MRS)	(Arena et al. 2006)
	Heat-shock		*S. thermophilus*	Emmental processing (milk)	(Guimont et al. 2002)
GltB and D	Acid		*L. lactis*	2 h, pH 5.0 (SA)	(Budin-Verneuil et al. 2005)
IlvB	Co-culture[d]/late exponential		*S. thermophilus*	Late/early exponential, (milk)	(Hervé-Jimenez et al. 2009)
	Late exponential		*S. thermophilus*	Late/early exponential (milk)	(Hervé-Jimenez et al. 2008)
IlvC	Co-culture[d]/late exponential		*S. thermophilus*	Late/early exponential, (milk)	(Hervé-Jimenez et al. 2009)
LivF	Lactose starvation		*S. thermophilus*	180 min, no lac (MRS)	(Arena et al. 2006)
LeuB	Co-culture[d]/late exponential		*S. thermophilus*	Late/early exponential, (milk)	(Hervé-Jimenez et al. 2009)
MetA	Late exponential		*S. thermophilus*	Late/early exponential (milk)	(Hervé-Jimenez et al. 2008)
MetE	Late exponential		*S. thermophilus*	Late/early exponential (milk)	(Hervé-Jimenez et al. 2008)
MetK	Acid		*L. lactis*	2 h, pH 5.0 (SA)	(Budin-Verneuil et al. 2005)
PepC	Acid		*L. lactis*	2 h, pH 5.0 (SA)	(Budin-Verneuil et al. 2005)
PepD1	Acid		*L. bulgaricus*	40 min, pH 4.9 (MRS)	(Fernandez et al. 2008)
PepM	Acid		*L. lactis*	2 h, pH 5.0 (SA)	(Budin-Verneuil et al. 2005)

PepO	Acid		*L. lactis*	2 h, pH 5.0 (SA)	(Budin-Verneuil et al. 2005)
Pro dehydrogenase	Co-culture[c]	Sourdough	*L. sanfranciscensis*	18 h (WFH)	(Di Cagno et al. 2007)
SerA	Late exponential	Dairy	*S. thermophilus*	Late/early exponential (milk)	(Hervé-Jimenez et al. 2008)
ThrA1	Stationary phase	All	*L. plantarum*	Late-stationary/mid-log phase (MRS)	(Cohen et al. 2006)
Urease	Acid	Dairy	*S. thermophilus*	180 min, pH 5.0 (MRS)	(Arena et al. 2006)
YugD (protease)	Acid		*L. lactis*	2 h, pH 5.0 (SA)	(Budin-Verneuil et al. 2005)
Fatty acid metabolism					
Acc	Acid		*L. bulgaricus*	40 min, pH 4.9 (MRS)	(Fernandez et al. 2008)
Cfa	Stationary phase	All	*L. plantarum*	Late-stationary/mid-log phase (MRS)	(Cohen et al. 2006)
	Acid	Dairy	*L. lactis*	2 h, pH 5.0 (M17 and SA)	(Budin-Verneuil et al. 2005)
EstA	Heat-shock		*S. thermophilus*	Emmental processing (milk)	(Guimont et al. 2002)
FabI	Acid		*L. bulgaricus*	40 min, pH 4.9 (MRS)	(Fernandez et al. 2008)
	Stationary phase	All	*L. plantarum*	Late-stationary/mid-log phase (MRS)	(Cohen et al. 2006)
Cell division/DNA metabolism					
DivIVA	Acid	Dairy	*S. thermophilus*	180 min, pH 5.0 (MRS)	(Arena et al. 2006)
FtsA	Acid		*L. lactis*	2 h, pH 5.0 (SA)	(Budin-Verneuil et al. 2005)
Stress-induced DNA binding protein	Stationary phase	All	*L. plantarum*	Late-stationary/mid-log phase (MRS)	(Cohen et al. 2006)
Single-stranded	HHP	Meat	*L. sakei*	10 min, 400 MPa (MRS)	(Jofré et al. 2007)
DNA binding protein	Acid	Dairy	*L. bulgaricus*	30 min, pH 5.25/pH 6 (WP+YE)	(Streit et al. 2008)
	Decreasing T° gradient		*L. helveticus*	55–20°C (whey)	(Di Cagno et al. 2006)
Other functions					
Adenosyl methyl transferase	Co-culture[c]	Sourdough	*L. sanfranciscensis*	18 h (WFH)	(Di Cagno et al. 2007)
Aldo/ketoreductase	Co-culture[c]	Sourdough	*L. sanfranciscensis*	18 h (WFH)	(Di Cagno et al. 2007)
ArcA	Acid	Dairy	*L. lactis*	2 h, pH 5.0 (M17)	(Budin-Verneuil et al. 2005)
ArcB	Acid	Dairy	*L. lactis*	2 h, pH 5.0 (M17 and SA)	(Budin-Verneuil et al. 2005)

(continued)

Table 15.1 (continued)

Protein name	Stress[a]	Food type	Species	Experimental conditions[b]	References
ATPase H+	Acid		*S. thermophilus*	20 and 180 min, pH 5.0 (MRS)	(Arena et al. 2006)
ATPA	Acid		*L. lactis*	2 h, pH 5.0 (M17 and SA)	(Budin-Verneuil et al. 2005)
AtpF	Acid		*L. lactis*	2 h, pH 5.0 (SA)	(Budin-Verneuil et al. 2005)
B6 vitamin synthesis	Acid		*L. bulgaricus*	30 min, pH 5.25/pH 6 (WP+YE)	(Streit et al. 2008)
BusAA	Acid		*L. lactis*	2 h, pH 5.0 (SA)	(Budin-Verneuil et al. 2005)
EraL	Acid		*L. lactis*	2 h, pH 5.0 (M17)	(Budin-Verneuil et al. 2005)
FolP	Co-culture[d]/late exponential		*S. thermophilus*	Late/early exponential, (milk)	(Hervé-Jimenez et al. 2009)
GidA	Acid		*L. lactis*	2 h, pH 5.0 (M17)	(Budin-Verneuil et al. 2005)
HipO1	Late exponential		*S. thermophilus*	Late/early exponential (milk)	(Hervé-Jimenez et al. 2008)
LuxS	Acid		*L. bulgaricus*	30 min, pH 5.25/pH 6 (WP+YE)	(Streit et al. 2008)
			L. lactis	15–25 min pH 4.5 (GSA)	(Frees et al. 2003)
				2 h, pH 5.0 (SA)	(Budin-Verneuil et al. 2005)
NadE	Stationary phase		*L. lactis*	Stationary (RSM and SA)	(Larsen et al. 2006)
PanE	Late exponential		*S. thermophilus*	Late/early exponential (milk)	(Hervé-Jimenez et al. 2008)
PpiA	Acid		*L. bulgaricus*	30 min, pH 5.25/pH 6 (WP+YE)	(Streit et al. 2008)
PstF	Acid		*L. lactis*	2 h, pH 5.0 (SA)	(Budin-Verneuil et al. 2005)
RecA	Acid		*L. lactis*	2 h, pH 5.0	(Budin-Verneuil et al. 2005)
Response regulator 11	Stationary phase	All	*L. plantarum*	Late-stationary/mid-log phase (MRS)	(Cohen et al. 2006)
Response regulator 01	Co-culture[d]/late exponential	Dairy	*S. thermophilus*	Late/early exponential, (milk)	(Hervé-Jimenez et al. 2009)
TuaH	Decreasing T° gradient		*L. helveticus*	55–20 °C (whey)	(Di Cagno et al. 2006)

[a]Decreasing T° gradient: incubation conditions from 55 °C to 20 °C, mimicking the manufacture or natural whey starter. *HHP* High hydrostatic pressure

[b]*CDM* chemically defined medium, *H medium* Homohiochii medium (Kleynmans et al. 1989), *MRS* MRS medium (De Man et al. 1960), *RSM* Reconstituted Skim Milk, *SA* chemically defined medium, GSA 1 % glucose SA medium (Jensen and Hammer 1993), *WFH* Wheat Flour Hydrolysate (wheat flour+yeast-extract+glucose+maltose), *WP* whey powder, *YE* Yeast extract, *No lac* no lactose in the medium

[c]Co-culture with *L. plantarum* DC400, *L. brevis* CR13 or *L. rossiae* A7

[d]Co-culture with *S. thermophilus* and *L. bulgaricus*

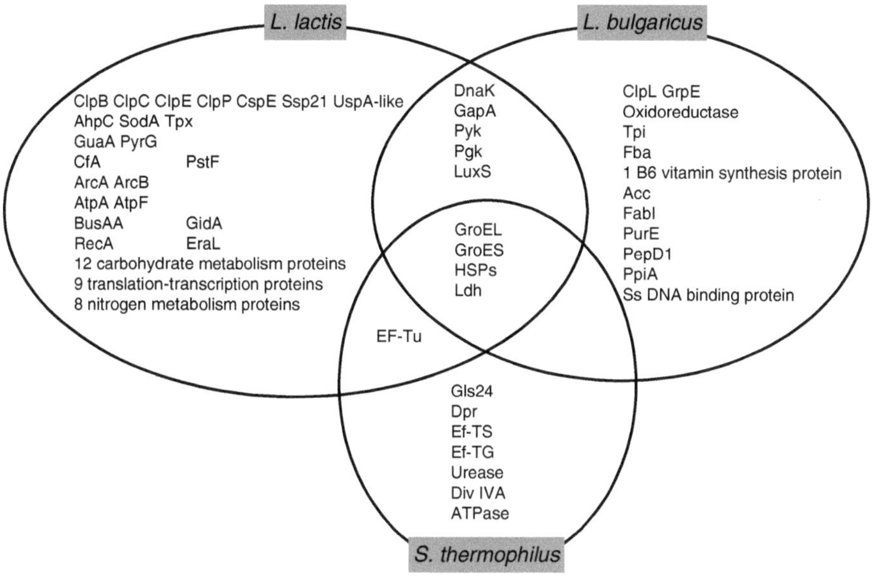

Fig. 15.1 Venn diagram representing the response of *S. thermophilus, L. lactis,* and *L. bulgaricus* to acid conditions. The up-regulated proteins common to several species or species-specific are presented

lactic acid and low pH is one characteristic of fermented products. For that reason, many proteomic studies report the effect of low pH in several starter species. In Fig. 15.1, the response of three dairy LAB (*L. lactis, L. delbrueckii* ssp. *bulgaricus*, and *S. thermophilus*) to one stress, acidic conditions are compared in order to detect if a species-specific response exists. *L. lactis* is the most responsive species, with 61 proteins upregulated followed by *L. delbruecki* ssp. *bulgaricus* with 21 over-expressed proteins and *S. thermophilus* with 12 over-expressed proteins. Strikingly, a substantial species-specific answer is observed. Indeed, only four upregulated proteins are shared by the three species: three general stress proteins (GroES, GroEL, HSP) and lactate dehydrogenase. *S. thermophilus* and *L. delbruecki* ssp. *bulgaricus* have no other common upregulated protein, *L. delbruecki* ssp. *bulgaricus* and *L. lactis* have one, and *L. lactis* and *L. delbruecki* ssp. *bulgaricus* share five. In *S. thermophilus*, the functions induced after an acid shock may contribute to the intracellular pH homeostasis with the upregulation of H^+-ATPase responsible for proton extrusion, and urease able to produce NH_3. These results clearly indicate that each species has a specific protein pattern. Each species over-expresses similar functional categories, but the targeted proteins are specific for each bacterium.

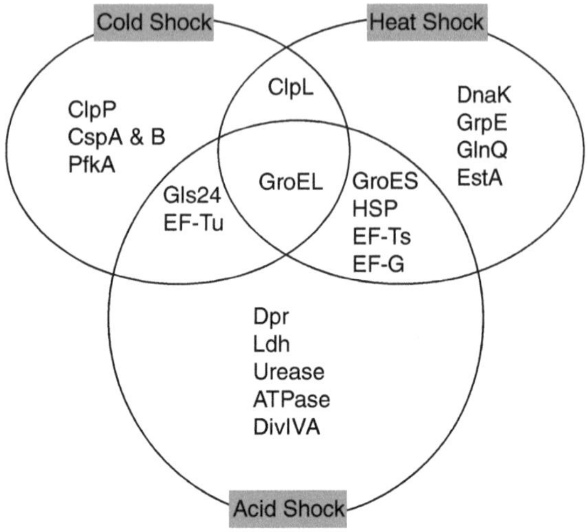

Fig. 15.2 Venn diagram representing the response of *S. thermophilus* to three different stresses. The up-regulated proteins common to several stresses or stress specific are presented

In Table 15.1 we can also observe that, in the oxidative stress-response protein category, *L. lactis* over-expresses AhpC, SodA, and Tpx, whereas *L. delbrueckii ssp. bulgaricus* and *S. thermophilus* over-express only an oxidoreductase and Dpr, respectively.

Also, *L. lactis* showed upregulation of more glycolysis proteins when compared to other species: under acidic stress conditions, *L. lactis* over-expresses twelve proteins whereas *L. delbrueckii* ssp. *bulgaricus* and *S. thermophilus* over-express, respectively, six and one proteins of the carbohydrate metabolism category. In conclusion, for a given stress, a species-specific response indeed exists.

15.2.4 Stress-Specific Responses

Regarding data presented in Table 15.1, we also considered the different stresses in order to evaluate a possible stress-specific response in LAB. As an example, we considered three stresses: heat, cold, or acid shock and compared the response of one LAB species, *S. thermophilus* to each of these three stresses. As shown in Fig. 15.2, only one general stress protein (GroEL) is upregulated following each of the three stresses. Some proteins are common to the response to two different stresses. Indeed, four proteins (HSP, EF-Ts, EF-G, GroES) are upregulated under either heat or acid conditions; two other proteins (Gls24, EF-Tu) are upregulated either by cold or acid shock. Only one, ClpL, is common to heat and cold shock response. It thus appears that, in addition to a common central stress response, *S. thermophilus* elicits different protein sets in order to adapt to each environmental modification.

Regarding oxidative stress response, *S. thermophilus* over-expresses only stress-response proteins, five being general stress proteins, and the seven remaining specifically targeting the oxidative metabolism. Compared to other stresses applied, in which several functional protein categories are generally upregulated, this is the sole example of such a response focused on only a given functional category.

Finally, we noticed that during growth at low temperature two proteins characteristic of oxidative stress response (MsrA and OhrA) were upregulated in *L. sakei,* whereas in *S. thermophilus* the response was mainly characterized by induction of general stress proteins, in particular the cold shock proteins CspA and B.

Concerning non-dairy LAB, HHP treatment has also been studied. Indeed HHP can be used as a preservative treatment, alternative to filtration or heat sterilization that is not applicable in some solid and raw food matrices (dry sausage, sourdough). After an HHP treatment, *L. sakei* and *L. sanfranciscensis* show quite different responses. Both over-express general stress proteins but different ones. The only common protein upregulated after HHP treatment in both bacteria is DnaK.

15.3 Analyses in Food Matrices

As mentioned above, proteomics studies of LAB behavior in food products has been mainly limited by the high protein content of food materials. A very few in situ studies have been reported, mainly concerning cheese. Yvon et al. (2011) have compared the ability in cheese making of two *L. lactis* strains with different properties. One strain was more proteolytic, as evaluated by growth in cheese matrix, amino acid production, and proteolytic enzymatic activities. For the two strains, regarding protein expression pattern, functional categories affected by growth in cheese were mainly related to acid stress response and amino acid starvation, indicating that strains have to cope with these stressful environmental conditions when grown in cheese. However, differences were observed between the two strains. A stronger growth limitation for one strain was linked with a lower proteolytic activity. This study underlines the importance of proteomics for evaluating dynamics of LAB population in cheese and consequent technological properties enlightening that *L. lactis* cheese adaptation depends on proteolytic activities.

Regarding the cheese process, several LAB can be involved. The Emmental cheese ecosystem, for instance, is composed of complex microflora including different LAB species (*L. helveticus, L. bulgaricus, S. thermophilus*) and *Propionibacterium freundenreichii.* During cheese ripening, bacterial proteins can be released due to bacterial cell lysis. These proteins contribute to cheese organoleptic and textural properties. Gagnaire et al. (2004) have performed a proteomic study in order to evaluate the respective part of each flora in the ripening process. Proteins released in the soluble fraction, after exclusion of milk and cheese proteins by chromatography, have been analyzed by two-dimensional gel electrophoresis. They could identify 21 proteins from *S. thermophilus*, 17 from *L. delbruecki* ssp. *bulgaricus,* and 8 from *P. freundenrichii.* These proteins were allocated to different functional categories such as stress proteins, DNA repair, or oxido-reduction, indicating a stress cellular status of the LAB during cheese ripening. Several peptidases were also identified, some being attributed to *S. thermophilus* and others to *L. delbruecki* ssp.

bulgaricus, enlightening the respective roles of these two species in the ripening process.

Proteomic analysis (MALDI-MS) has also been successfully used to monitor the extent of bacterial digestion of milk proteins during yogurt production (Fedele et al. 1999). This study showed that: (1) changes in milk protein profiles were due to the action of the two yogurt LAB *S. thermophilus* and *L. delbruecki* ssp. *bulgaricus*, (2) casein hydrolysis varied among the strains tested, and (3) proteolysis was relevant when milk was fermented with mixtures of the two yogurt bacterial species, probably because of synergistic phenomena. This tool could be employed to determine the effectiveness of different yogurt LAB ratios for yogurt production, in terms of production of peptides that can bear probiotic properties.

Recently, a new methodology has been applied to the proteomic analysis of cheese ripening (Jardin et al. 2012). The ITRAQ (isobaric tagging for relative and absolute quantification) method allows identification and quantification of proteins in a single LC-MS/MS run. It was used to follow Emmental-type cheese during ripening, by analyzing proteins present in the aqueous phase. By this method, the authors could overcome differences in the dynamics range between milk and bacterial proteins, and could identify proteins from both origins. Proteins from milk did not show significant increase in the aqueous fraction. Bacterial proteins identified were issued from *L. helveticus* or *S. thermophilus*. A major increase in their concentration was observed between day 20 and day 69 of ripening. These bacterial proteins were mainly assigned to the functional category of stress proteins. This study thus confirmed the previous finding of a stress status for LAB during cheese ripening. It also opens fields for a simultaneous analysis of cheese and bacterial proteins and thus a better global understanding of cheese making.

No such deep in situ comprehensive approach has been reported for meat, even though some studies have attempted to approach meat matrix conditions. Fadda et al. (2010) have studied the response of *L. sakei* grown in a chemically defined medium after addition of either myofibrillar or sarcoplasmic proteins. Most of the proteins showing modified expression were observed in the presence of myofibrillar extract (16 proteins) compared to sarcoplasmic protein addition (6 proteins). Most proteins were less expressed in the presence of the meat extracts. General stress proteins were downregulated whereas proteins of energy metabolism, pyrimidine metabolism, and translation were upregulated indicating that *L. sakei* is really adapted to the presence of the meat substrate. In other words, this means that meat is indeed the favorite habitat of *L. sakei* and that laboratory media constitute a stressing environment. This situation is not observed with cheese bacteria that show a stress status during the cheese process.

15.4 Food Safety: Biogenic Amine Production

Proteomics have also been used as a tool either to detect or study biogenic amine production, or even for the fingerprinting of biogenic amine-producing strains or species. Although many fermented foods may be confronted with the risk of biogenic amine production, wine and dry fermented sausages are the most critical. In addition, fish products, fermented or not, are also susceptible to host bacteria able to produce biogenic amines such as histamine, tyramine, or putrescine. A proteomic analysis, combining two-dimension gel electrophoresis to MALDI-TOF and MS-MS *de novo* sequencing, was used to determine proteins synthesized by two *Lactobacillus* strains (*Lactobacillus* sp. 30a ATCC33222 and a *Lactobacillus* sp. strain isolated from an amine-contaminated wine) under various laboratory growth conditions (Pessione et al. 2005). The results helped to identify proteins involved in amino acid transport and conversion to biogenic amines, but could not confirm the hypothesis that amino acid accumulation induces biogenic amine synthesis, nor that this synthesis is a bacterial response to medium acidification.

15.5 Identification, Fingerprinting, or Characterization of Bacterial Diversity by Proteomics

The MALDI-TOF technique can be used to fingerprint bacteria (Giebel et al. 2010) including biogenic amine producers. As an example, the MALDI-TOF spectra obtained on bacterial extracts from *Pseudomonas* strains isolated from fish products gave different fingerprints when obtained from biogenic amine producing or non-producing strains (Fernández-No et al. 2011). A similar approach was used aimed at identifying bacterial species present in seafood products, including the LAB species *Carnobacterium divergens*, *Carnobacterium gallinarum*, and *Carnobacterium maltaromaticum* and also two *Staphylococcus* species (*S. epidermitis* and *S. xylosus*) which are naturally present or used as starter for the fermentation of sausage (Böhme et al. 2011). These examples show that the pattern of MALDI-TOF spectra obtained from food bacteria can be used as a tool to differentiate strains or genus, or even to identify new isolates.

Proteomics have also been used as a tool for investigating LAB natural diversity. The meat-borne bacterium *L. sakei* has been described as a species showing high genomic diversity and 10 genotypic groups could be described (Chaillou et al. 2009). Ten strains representing these clusters have been studied by a proteomic approach based on two-dimensional gel electrophoresis. This revealed a difference of up to 20% in the number of expressed proteins detected in gels, with specific strain protein patterns. Moreover, it was observed that the ten strains could split into two clusters, based on the pattern displayed by the four isoforms of the glyceraldehyde-3P-dehydrogenase (GapA) on gels. These two isoform families were associated with the two *L. sakei* subspecies previously reported in the literature as *L. sakei* subsp. *sakei* and *L. sakei* subsp. *carnosus*.

Regarding the expression of this diversity, the metabolism of *L. sakei* strains has also been explored by proteomics in order to evaluate the potential of strains for technological applications. McLeod et al. (2010) have compared ten *L. sakei* food isolates, when grown on either glucose or ribose as the carbon source, these two sugars being present in meat. A common regulation by ribose was observed for all strains but the study also pointed out differences between strains. For all strains, a total of ten proteins was upregulated after growth on ribose and six proteins were downregulated on glucose. A commercial starter strain and a putative biopreservative strain exhibited a different regulation in the utilization of these two carbon sources. One strain, isolated from fermented fish showed a higher level of expression of stress proteins when compared to the other strains.

Guillot et al. (2003) compared the 2-D gel electrophoresis profiles of two *L. lactis* strains. Apart from the whole set of glycolytic enzymes that were present in both strains, at conserved positions, this study revealed an important protein polymorphism. For strain NCDO763, a dairy strain belonging to *cremoris* subspecies, 26 proteins were not present in the *L. lactis* subsp. *lactis* IL1403 strain map. Among these, there were enzymes involved in lactose assimilation and amino acid metabolism (peptidases, amino acid synthases) that are of particular relevance for growth on milk and thus for cheese making.

These reports indicate that the proteomic approach can also be a useful tool for exploring bacterial diversity, in association with other methods such as genomics, to go deeper into LAB strain performance analysis for a better understanding of strain-specific technological properties.

15.6 Conclusion

Proteomic analyses of LAB from fermented foods have shown that bacteria use different strategies to face the various conditions encountered during the food production processes. When confronted with the same stress, different species will modify the protein they synthesize in different but similar ways: the functional categories that are affected by a specific stress are often similar, but the proteins that are upregulated after this stress may vary from species to species. The degree of similarity or difference between strains varies with the kind of stress or growth condition tested.

However, most of the proteomic results presented to date in this field issue from two-dimensional gel electrophoresis followed by mass spectrometry protein identification. One should not forget that the method mostly refers to cytoplasmic proteins, in a limited pI and MW window, and that only the emerging part of the iceberg is accessible to analysis. This narrow point of view is, however, informative about bacterial functions that have to be taken into consideration to understand and improve the production of fermented foods.

In addition, recent LAB whole genome sequencing studies have revealed large intraspecies diversity in both genome size and content. This means that stress response observed with one strain may be different in another strain. Whole mechanisms of bacterial adaptation to fermented food conditions are yet to be totally elucidated. Proteomics have allowed, nevertheless, the indication of some marker proteins that may be useful for the selection of well-adapted efficient starters for industrial applications.

References

Arena S, D'Ambrosio C, Renzone G, Rullo R, Ledda L, Vitale F, Maglione G, Varcamonti M, Ferrara L, Scaloni A (2006) A study of *Streptococcus thermophilus* proteome by integrated analytical procedures and differential expression investigations. Proteomics 6(1):181–192

Böhme K, Fernández-No IC, Barros-Velázquez J, Gallardo JM, Cañas B, Calo-Mata P (2011) Rapid species identification of seafood spoilage and pathogenic Gram-positive bacteria by MALDI-TOF mass fingerprinting. Electrophoresis 32:2951–2965

Budin-Verneuil A, Pichereau V, Auffray Y, Ehrlich D, Maguin E (2005) Proteome characterization of acid tolerance response in *Lactococcus lactis* MG1363. Proteomics 5(18): 4794–4807

Chaillou S, Daty M, Baraige F, Dudez A-M, Anglade P, Jones R, Alpert C-A, Champomier-Vergès M-C, Zagorec M (2009) Intraspecies genomic diversity and natural population structure of the meat-borne lactic acid bacterium *Lactobacillus sakei*. Appl Environ Microbiol 75(4):970–980

Cohen DPA, Renes J, Bouwman FG, Zoetendal EG, Mariman E, de Vos WM, Vaughan EE (2006) Proteomic analysis of log to stationary growth phase *Lactobacillus plantarum* cells and a 2-DE database. Proteomics 6(24):6485–6493

D'Alessandro A, Zolla L (2012) We are what we eat: food safety and proteomics. J Proteome Res 11(1):26–36

De Angelis M, Bini L, Pallini V, Cocconcelli PS, Gobbetti M (2001) The acid-stress response in *Lactobacillus sanfranciscensis* CB1. Microbiology 147(7):1863–1873

De Man JC, Rogosa M, Sharpe ME (1960) A medium for the cultivation of lactobacilli. J Appl Bacteriol 23(1):130–135

Di Cagno R, De Angelis M, Limitone A, Fox PF, Gobbetti M (2006) Response of *Lactobacillus helveticus* PR4 to heat stress during propagation in cheese whey with a gradient of decreasing temperatures. Appl Environ Microbiol 72(7):4503–4514

Di Cagno R, De Angelis M, Limitone A, Minervini F, Simonetti MC, Buchin S, Gobbetti M (2007) Cell-cell communication in sourdough lactic acid bacteria: a proteomic study in *Lactobacillus sanfranciscensis* CB1. Proteomics 7(14):2430–2446

Drews O, Weiss W, Reil G, Parlar H, Wait R, Görg A (2002) High pressure effects step-wise altered protein expression in *Lactobacillus sanfranciscensis*. Proteomics 2(6):765–774

Fadda S, Anglade P, Baraige F, Zagorec M, Talon R, Vignolo G, Champomier-Vergès M-C (2010) Adaptive response of *Lactobacillus sakei* 23K during growth in the presence of meat extracts: a proteomic approach. Int J Food Microbiol 142(1–2):36–43

Fedele L, Seraglia R, Battistotti B, Pinelli C, Traldi P (1999) Matrix-assisted laser desorption/ionization mass spectrometry for monitoring bacterial protein digestion in yogurt production. J Mass Spectrom 34(12):1338–1345

Fernandez A, Ogawa J, Penaud S, Boudebbouze S, Ehrlich D, van de Guchte M, Maguin E (2008) Rerouting of pyruvate metabolism during acid adaptation in *Lactobacillus bulgaricus*. Proteomics 8(15):3154–3163

Fernández-No IC, Böhme K, Calo-Mata P, Barros-Velázquez J (2011) Characterisation of histamine-producing bacteria from farmed blackspot seabream (*Pagellus bogaraveo*) and turbot (*Psetta maxima*). Int J Food Microbiol 151(2):182–189

Frees D, Vogensen FK, Ingmer H (2003) Identification of proteins induced at low pH in *Lactococcus lactis*. Int J Food Microbiol 87(3):293–300

Gagnaire V, Piot M, Camier B, Vissers JPC, Jan G, Léonil J (2004) Survey of bacterial proteins released in cheese: a proteomic approach. Int J Food Microbiol 94(2):185–201

Giebel R, Worden C, Rust SM, Kleinheinz GT, Robbins M, Sandrin TR (2010) Microbial fingerprinting using matrix-assisted laser desorption ionization time-of-flight mass spectrometry (MALDI-TOF MS) applications and challenges. Adv Appl Microbiol 71:149–184

Gonzalez-Marquez H, Perrin C, Bracquart P, Guimont C, Linden G (1997) A 16 kDa protein family overexpressed by *Streptococcus thermophilus* PB18 in acid environments. Microbiology 143:1587–1594

Gouesbet G, Jan G, Boyaval P (2002) Two-dimensional electrophoresis study of *Lactobacillus delbrueckii* subsp. *bulgaricus* thermotolerance. Appl Environ Microbiol 68(3):1055–1063

Guillot A, Gitton C, Anglade P, Mistou M-Y (2003) Proteomic analysis of *Lactococcus lactis*, a lactic acid bacterium. Proteomics 3(3):337–354

Guimont C, Chopard MA, Gaillard JL, Chamba JF (2002) Protein composition of a strain of *Streptococcus thermophilus* grown in milk at 42°C and in emmental making thermal conditions. Sci Aliment 22(1–2):59–66

Hartke A, Bouché S, Giard J-C, Benachour A, Boutibonnes P, Auffray Y (1996) The lactic acid stress response of *Lactococcus lactis* subsp. *lactis*. Curr Microbiol 33(3):194–199

Hervé-Jimenez L, Guillouard I, Guedon E, Gautier C, Boudebbouze S, Hols P, Monnet V, Rul F, Maguin E (2008) Physiology of *Streptococcus thermophilus* during late stage of milk fermentation with special regard to sulfur amino-acid metabolism. Proteomics 8(20):4273–4286

Hervé-Jimenez L, Guillouard I, Guedon E, Boudebbouze S, Hols P, Monnet V, Maguin E, Rul F (2009) Post-genomic analysis of *Streptococcus thermophilus* cocultivated in milk with *Lactobacillus delbrueckii* subsp. *bulgaricus*: involment of nitrogen, purine, and iron metabolism. Appl Environ Microbiol 75(7):2062–2073

Hörmann S, Scheyhing C, Behr J, Pavlovic M, Ehrmann M, Vogel RF (2006) Comparative proteome approach to characterize the high-pressure stress response of *Lactobacillus sanfranciscensis* DSM 20451(T). Proteomics 6(6):1878–1885

Hussain MA, Knight MI, Britz ML (2009) Proteomic analysis of lactose-starved *Lactobacillus casei* during stationary growth phase. J Appl Microbiol 106(3):764–773

Jardin J, Mollé D, Piot M, Lortal S, Gagnaire V (2012) Quantitative proteomic analysis of bacterial enzymes released in cheese during ripening. Int J Food Microbiol 155(1–2):19–28

Jensen PR, Hammer K (1993) Minimal Requirements for Exponential Growth of *Lactococcus lactis*. Appl Environ Microbiol 59(12):4363-4366

Jofré A, Champomier-Vergès M-C, Anglade P, Baraige F, Martín B, Garriga M, Zagorec M, Aymerich T (2007) Protein synthesis in lactic acid and pathogenic bacteria during recovery from a high pressure treatment. Res Microbiol 158(6):512–520

Kilstrup M, Jacobsen S, Hammer K, Vogensen F (1997) Induction of heat shock proteins DnaK, GroEL, and GroES by salt stress in *Lactococcus lactis*. Appl Environ Microbiol 63(5):1826–1837

Kleynmans WP, Heinzl U, Hammers H (1989) *Lactobacillus suebicus* sp.nov., an obligately heterofermentative *lactobacillus* species isolate from fruit mashes. Syst Appl Microbiol 11: 267–271

Larsen N, Boye M, Siegumfeldt H, Jakobsen M (2006) Differential expression of proteins and genes in the lag phase of *Lactococcus lactis* subsp. *lactis* grown in synthetic medium and reconstituted skim milk. Appl Environ Microbiol 72:1173–1179

Lim EM, Ehrlich SD, Maguin E (2000) Identification of stress-inducible proteins in *Lactobacillus delbrueckii* subsp. *bulgaricus*. Electrophoresis 21(12):2557–2561

Marceau A, Zagorec M, Chaillou S, Méra T, Champomier-Vergès M-C (2004) Evidence for involvement of at least six proteins in adaptation of *Lactobacillus sakei* to cold temperatures and addition of NaCl. Appl Environ Microbiol 70(12):7260–7268

McLeod A, Zagorec M, Champomier-Vergès M-C, Naterstad K, Axelsson L (2010) Primary metabolism in *Lactobacillus sakei* food isolates by proteomic analysis. BMC Microbiol 10:120

Molina E, Ramos M, Amigo L (2002) Characterisation of the casein fraction of Iberico cheese by electrophoretic techniques. J Sci Food Agric 82(10):1240–1245

Nakamura R, Nakamura R, Nakano M, Arisawa K, Ezaki R, Horiuchi H, Teshima R (2010) Allergenicity study of EGFP-transgenic chicken meat by serological and 2D-DIGE analysis. Food Chem Toxicol 48(5):1302–1310

Pessione E, Mazzoli R, Giuffrida MG, Lamberti C, Garcia-Moruno E, Barello C, Conti A, Giunta C (2005) A proteomic approach to studying biogenic amine producing lactic acid bacteria. Proteomics 5(3):687–698

Sentandreu MA, Sentandreu E (2011) Peptide biomarkers as a way to determine meat authenticity. Meat Sci 89(3):280–285

Silva J, Carvalho AS, Ferreira R, Vitorino R, Amado F, Domingues P, Teixeira P, Gibbs PA (2005) Effect of the pH of growth on the survival of *Lactobacillus delbrueckii* subsp. *bulgaricus* to stress conditions during spray-drying. J Appl Microbiol 98(3):775–782

Streit F, Delettre J, Corrieu G, Béal C (2008) Acid adaptation of *Lactobacillus delbrueckii* subsp. *bulgaricus* induces physiological responses at membrane and cytosolic levels that improves cryotolerance. J Appl Microbiol 105(4):1071–1080

Vallejo-Cordoba B, Rodríguez-Ramírez R, González-Córdova AF (2010) Capillary electrophoresis for bovine and ostrich meat characterisation. Food Chem 120(1):304–307

Vogel RF, Pavlovic M, Hörmann S, Ehrmann MA (2005) High pressure-sensitive gene expression in *Lactobacillus sanfranciscensis*. Braz J Med Biol Res 38(8):1247–1252

Wang Y, Delettre J, Guillot A, Corrieu G, Béal C (2005) Influence of cooling temperature and duration on cold adaptation of *Lactobacillus acidophilus* RD758. Cryobiology 50(3):294–307

Yvon M, Gitton C, Chambellon E, Bergot G, Monnet V (2011) The initial efficiency of the proteolytic system of *Lactococcus lactis* strains determines their responses to a cheese environment. Int Dairy J 21(5):335–345

Zhang Y, Zhang Y, Zhu Y, Mao S, Li Y (2010) Proteomic analyses to reveal the protective role of glutathione in resistance of *Lactococcus lactis* to osmotic stress. Appl Environ Microbiol 76(10):3177–3186

Chapter 16
Wine Quality

Marzia Giribaldi

16.1 Introduction

One of the most ancient legends about the origins of wine tells of a Persian princess who tried to poison herself so as to put an end to her love pangs. So she drank some juice from the bottom of a jar containing rotten grapes, and, instead of being killed by them, she forgot all her pain, and soon thereafter she regained her lover's heart. The King himself tried the prodigious drink, and began to spread its fame throughout his kingdom. Many other legends regarding the prodigious powers of wine can be found in literature, and even in the main world religions, of which the episode of Noah's drunkenness in the Old Testament is an example.

The domestication of the grapevine can be traced back to the third millennium BC: since then, the cultivation of *Vitis* ssp. has expanded to all the continents, with the exception of Antarctica. According to the International Organization of Vine and Wine (O.I.V. Report 2008), in 2007 the surface area given over to grape cultivation accounted for almost 7.8 million hectares, Europe being the main producer, transformer, and consumer of grapes for winemaking. Among the products derived from grapes, wine is the most significant. Wine production touched 266 million hL in 2007 (O.I.V. Report 2008), showing a sharp decrease (more than 20%) since the beginning of the 1980s. This decrease has been accompanied by increasing attention on the part of consumers towards wine quality. Fraud concerning adulterated wines (Frank in 2007 reported that up to 5% of the wine sold in secondary markets could be counterfeit) leads to important economic losses in the wine trade, mostly due to the producers rightly or wrongly losing credibility, as well as causing severe safety alarms.

M. Giribaldi (✉)
National Research Council – Institute of Sciences of Food Production,
via Leonardo da Vinci 44-10095, Grugliasco, TO, Italy
e-mail: marzia.giribaldi@ispa.cnr.it

F. Toldrá and L.M.L. Nollet (eds.), *Proteomics in Foods: Principles and Applications*, Food Microbiology and Food Safety 2, DOI 10.1007/978-1-4614-5626-1_16,

The importance of the wine trade and the growing attention paid by modern consumers towards food quality has raised interest in recent years among the scientific community in defining quality markers for wine. The quality of a food is not unambiguous: it is made up of safety issues, organoleptic features, technological aspects, and nutritional requirements, among other characteristics. All these features have been investigated in wines, and almost all of them have been studied with proteomic means, mostly in the last two decades.

The scientific reviews on wine protein analysis published to date can be grouped into two main categories: those devoted to methodological issues (Moreno-Arribas et al. 2002; Flamini and De Rosso 2006; Curioni et al. 2008; Le Bourse et al. 2010) and those focused on the application of proteomics to specific wine issues (Ferreira et al. 2002; Giribaldi and Giuffrida 2010; Pedreschi et al. 2010; D'Alessandro and Zolla 2012). Moreno-Arribas and colleagues mainly described the methods used for the extraction, quantification, and separation of wine proteins (Moreno-Arribas et al. 2002). Interestingly, mass spectrometry (MS) in their report was briefly described as a tool for identifying proteins, and only a few years later the review published by Flamini and De Rosso (2006) focused on its use for the separation, quantification, and identification of grape and wine proteins, thus paving the way for the leading role acquired by MS in the analysis of foods and beverages in recent years. In 2002, Ferreira and collaborators revised the state of the art on wine proteins by providing an overview of their origin and function, and focused on the turbidity issues caused by them, including possible solutions to remove them. More recently, Curioni et al. (2008) and Le Bourse et al. (2010) provided updated reviews of the latest trends in the analytical techniques used for the study of grape juice and wine proteins. In 2010, Giribaldi and Giuffrida updated the proteomic studies published since 2005, covering aspects of grape physiology and grape berry ripening, as well as protein function in wines (Giribaldi and Giuffrida 2010). In the same year, wine proteomics was included in a review devoted to the application of proteomics to various important food industry sectors (Pedreschi et al. 2010). The most recent survey on the issue is the review by D'Alessandro and Zolla (2012) on proteomics applications in the field of wine safety and traceability.

The present chapter briefly and exhaustively describes the papers that recount proteomic means to study wine proteins in recent decades. However, the authors strongly recommend consulting these reviews in order to achieve a more complete understanding of wine protein science.

16.2 Methodological Aspects in Wine Proteomics

In recent decades, the application of new techniques to the characterization of wine proteins has brought the number of wine proteins identified by proteomics to over 100 (D'Amato et al. 2011). Traditional techniques, such as electrophoresis and chromatography, originally allowed for the detection of the most abundant proteins (accounting for only about 30% of the total protein species in wine, according to D'Amato and collaborators).

One major step towards achieving these goals was made in 2007, when two independent studies, one supported by a French–Italian consortium (Jaillon et al. 2007) and the other by an Italian–American initiative (Velasco et al. 2007), published the genome sequence of one non-cultivated highly homozygous and of one cultivated highly heterozygous Pinot Noir clone, respectively. The annotation of the grapevine genome is still in progress, and although it may sometimes complicate the interpretation of research findings, the availability of genome sequences has significantly boosted grape and wine proteomics.

16.2.1 Quantification of Protein Content in Wines

Typically, protein concentrations have been reported in a range from 15 to 230 mg/L (Monteiro et al. 2001; Ferreira et al. 2002; Waters et al. 2005). Proteins are thus considered as minor components of both white and red wines, with little nutritional relevance. Several techniques are currently available for total protein quantification in foods and beverages, but wine is typically rich in compounds that may interfere with normal quantification techniques, such as phenolics, ethanol, and organic acids (Marchal et al. 1997; Moreno-Arribas et al. 2002; Le Bourse et al. 2010).

The standard protein quantification method in the food industry is based on the evaluation of total proteins by conversion of nitrogen measured by Kjeldahl assay (multiplying nitrogen content by 6.25) (AOAC method 960.52), but this method is known to cause frequent overestimations of wine protein content (Vidigal et al. 2012). The reliable quantification of wine proteins may have an impact on the stability of the wine itself, as they are major causes of white wine clouding (Waters et al. 2005). The most common methods used for protein quantification in wines to date are based on spectrophotometric detections, such as the Bradford (Bradford 1976), Lowry (Lowry et al. 1951), Biuret (Gornall et al. 1949), or Smith (Smith et al. 1985) tests, which have often been used due to their ease and speed (Vidigal et al. 2012). Nevertheless, the presence of interfering compounds and the absence of standard wine proteins may lead to erroneous quantification with these methods (Moreno-Arribas et al. 2002; Le Bourse et al. 2010, 2011; Vidigal et al. 2012).

In recent years, Vincenzi et al. (2005) and Smith et al. (2011) have provided comparisons between the more widely used protein quantification methods and developed improved procedures for protein recovery and quantification in white and red wines. Vincenzi et al. (2005) concluded that potassium dodecyl sulphate (KDS) precipitation coupled with Smith's assay gave the most accurate results, consistent with those obtained by densitometric quantification of SDS-PAGE protein bands. Smith et al. (2011) concluded that, in red wines, protein precipitation with cold trichloroacetic acid/acetone and quantification based on Bradford's assay absorbance using a yeast mannoprotein invertase standard gave results similar to those obtained by micro-Kjeldahl analysis.

Some authors have suggested using HPLC-based techniques to achieve reliable quantification of wine proteins (Peng et al. 1997; Pocock and Waters 2006).

Marangon et al. (2009) developed a quantification method for protein fractions separated by hydrophobic interaction chromatography based on comparison with the HPLC peak area of two standard proteins (cytochrome c and bovine serum albumin, BSA), but this method may suffer from limitations due to the use of non-wine standard proteins (Le Bourse et al. 2010). To circumvent this major drawback, Le Bourse et al. (2011) published a method for purifying grape juice class IV chitinase and thaumatin-like (TL) proteins by liquid chromatography. The pure protein fractions were then used to build reliable calibration curves for ultra-HPLC and for ELISA quantification of these proteins in different grape juices and wine samples, thus providing a valuable tool for future oenological studies.

The densitometric quantification of wine protein bands from SDS-PAGE gels (after staining with Coomassie Brilliant Blue) was frequently used in the past (Marchal et al. 2000; Hsu and Heatherbell 1987a, b). Recently, Sauvage et al. (2010) and Dufrechou et al. (2010, 2012) reported a quantification method based on absorption/desorption of proteins with excess bentonite, a clay with protein absorption capacity commonly used in wine fining, followed by SDS-PAGE, image analysis, and quantification of wine protein bands using a standard BSA band. This method, although it is said to provide an estimate of total wine proteins, fails to take into account low and high molecular weight proteins, which are not visualized in the SDS-PAGE gel. Moreover, it relies on the staining intensity of a non-wine protein, BSA, similarly to most of the previously cited techniques, loaded in a single concentration on the gel, and it thus fails to build a standard curve, with a regression and a correlation coefficient. In addition, similarly to all the methods based on Coomassie Brilliant Blue protein staining, including the Bradford method, the composition of some proteins may not give a linear response (Fountoulakis et al. 1992).

The most recent advance in the field of protein quantification in white wines is the modified Lowry assay coupled with solid-phase extraction recently developed by Vidigal et al. (2012). The method is based on the retention of proteins in the solid support, nitrilotriacetic acid Superflow beads (Qiagen), charged by Cu^{2+}, and on the absorbance at 500 nm after addition of Folin–Ciocalteu's reagent. Results from Vidigal's work represent a significant advance with respect to current methods in terms of sample volume required, poor pre-processing before analysis, and automation (Vidigal et al. 2012). The limitations due to the use of non-wine standard protein should be evaluated carefully, and further investigations into its reliability for wine glycoprotein quantification are needed.

16.2.2 Methods Used in Wine Protein Characterization

As brilliantly summarized in several reviews (Moreno-Arribas et al. 2002; Flamini and De Rosso 2006; Curioni et al. 2008; Le Bourse et al. 2010), the main techniques used for grape and wine proteome analysis can be grouped into chromatography, electrophoresis, and MS-based methods.

Electrophoresis and 2-D-electrophoresis have been widely used in the past and are still in use to characterize the protein fraction of wines, often in association with chromatographic techniques. Early works mainly used electrophoretic techniques, in the form of both native gel electrophoresis (Bayly and Berg 1967; Pueyo et al. 1993; Moreno-Arribas et al. 1999) and denaturing gel electrophoresis (Yokotsuka et al. 1991; Waters et al. 1992, 1993; Pueyo et al. 1993; Dorrestein et al. 1995; Santoro 1995; Marchal et al. 1996). Wine proteins have often been separated according to their isoelectric point as a preparatory or analytical step (Murphey et al. 1989; Yokotsuka et al. 1991; Pueyo et al. 1993; Dawes et al. 1994; Marchal et al. 1996; Luguera et al. 1998). The coupling of isoelectrofocusing and denaturing electrophoresis led to detailed screening of wine proteins ever since its early application to wine proteomics (Hsu and Heatherbell 1987a, b; Hsu et al. 1987). Protein immunoblotting has been widely used as a tool for investigating the origin and the structural similarity of wine proteins from different varieties (Hsu and Heatherbell 1987a, b; Hsu et al. 1987; Marchal et al. 1998; Monteiro et al. 1999; Ferreira et al. 2000; Monteiro et al. 2001, 2003a, b; Dambrouck et al. 2003; Manteau et al. 2003; Monteiro et al. 2007). In more recent years, the coupling of electrophoresis with protein identification by mass spectrometry has allowed researchers to unravel the complexity of the wine proteome in different conditions, and thus improved wine science (Okuda et al. 2006; Cilindre et al. 2008; Wigand et al. 2009; Sauvage et al. 2010; Vincenzi et al. 2011). Moreover, the introduction of enrichment technologies, such as the use of combinatorial peptide ligand libraries (CPLL), has increased the detection of low-abundance constitutive or contaminating proteins in wines (Cereda et al. 2010; D'Amato et al. 2010, 2011).

Capillary electrophoresis has been applied in the past to wine protein analysis (Moine Ledoux et al. 1992; Luguera et al. 1997, 1998; Dizy and Bisson 1999), and one recent example of its potential is represented by varietal differentiation recently obtained by high-performance capillary electrophoresis of wine proteins and shikimic acid quantification by Chabreyrie et al. (2008).

Chromatographic approaches used in wine proteomics for protein purification, separation, and characterization include FPLC (Waters et al. 1992, 1993; Dawes et al. 1994; Dorrestein et al. 1995; Waters et al. 1995; Luguera et al. 1998; Monteiro et al. 1999, 2001, 2003a, b, 2007; Esteruelas et al. 2009), HPLC (Tyson et al. 1981; Santoro 1995; Yokotsuka and Singleton 1997; Girbau et al. 2004), size exclusion chromatography (Pellerin et al. 1993; Gonçalves et al. 2002), affinity chromatography (Pellerin et al. 1993; Waters et al. 1993; Marchal et al. 1996; Gonçalves et al. 2002; Vanrell et al. 2007), and more recently hydrophobic interaction chromatography (Falconer et al. 2010; Marangon et al. 2009, 2011a, b), hydrophilic interaction chromatography, titanium dioxide enrichment, and hydrazide chemistry enrichment (Palmisano et al. 2010).

Mass spectrometry and N-terminal sequencing have greatly increased our understanding, allowing the identification of several wine proteins. In recent years, MS has become a useful tool for wine traceability, allowing for varietal fingerprinting and contaminant detection in both red and white wines (Szilágyi et al. 1996; Weiss et al. 1998; Kwon 2004; Catharino et al. 2006; Carpentieri et al. 2007; Chambery et al. 2009; Monaci et al. 2010, 2011; Simonato et al. 2011; Tolin et al. 2012; Nunes-Miranda et al. 2012).

16.3 The Origin of Wine Proteins

One of the main aspects investigated by wine proteomics has been the elucidation of the origin of wine proteins. Early investigations declared wine proteins to be derived exclusively from the grape berry (Luguera et al. 1998; Ruiz-Larrea et al. 1998; Ferreira et al. 2000). One reason could be the fact that, according to almost all the reports, grape-derived proteins, and especially the class named pathogenesis-related (PR) proteins (Linthorst 1991), represent the vast majority of the protein components found in all studied wines (Waters et al. 1996; Monteiro et al. 2001; Okuda et al. 2006; Wigand et al. 2009; Dufrechou et al. 2010; Sauvage et al. 2010; Vincenzi et al. 2011; Dufrechou et al. 2012). According to other hypotheses, the poor visualization of yeast proteins from wines may be due to the revelation method used, as their sugar moiety makes them poorly detectable by Coomassie and silver-based stains (Waters et al. 1993; Dambrouck et al. 2003; Wigand et al. 2009).

Currently, most authors agree on the mixed origin of wine proteins, as yeast-deriving proteins (mostly cell wall components) have been demonstrated to occur in wines by several methodological approaches, such as chromatography (Marchal et al. 1996; Yokotsuka and Singleton 1997; Monteiro et al. 2001; Gonçalves et al. 2002; Palmisano et al. 2010), electrophoresis (Waters et al. 1993; Dupin et al. 2000; Kwon 2004; Cilindre et al. 2008; Wigand et al. 2009; D'Amato et al. 2011; Marangon et al. 2011a, b), immunostaining (Monteiro et al. 2001; Dambrouck et al. 2003), and mass spectrometry (Simonato et al. 2011; Tolin et al. 2012). The functions of parietal yeast mannoproteins include adsorption of ochratoxin A, combination with phenolic compounds, increased growth of malolactic bacteria, inhibition of tartrate salt crystallization, interaction with the yeast-derived superficial film (*flor*) formed in the manufacture of sherry-type products, reinforcement of aromatic components, and wine enrichment during aging on fine lees (Caridi 2006; Blasco et al. 2011). Their major roles, being haze-protective factors in white wines and foam aids in sparkling wines, are detailed in the following chapter.

Other fermentative agents in wines include several types of bacteria. The presence of bacterial proteins in wines used to be excluded by immunostaining (Dambrouck et al. 2003) until recently. Simonato et al. (2011) and Tolin et al. (2012), by means of LC-MS/MS profiling, were the first, to the author's knowledge, to detect one 60 kDa chaperonin from *Oenococcus oeni,* a lactic acid bacterium involved in malolactic fermentation.

Fungal pathogens are responsible for considerable economic losses for winemakers, and the costs of field prevention against their occurrence are a major expense for agriculture. The presence of proteins deriving from fungal infection of grape clusters on the vine has been demonstrated by Western blot and electrophoresis in recent years (Kwon 2004; Cilindre et al. 2007, 2008). One recent investigation of a commercial Valpolicella red wine revealed the presence of proteins from several fungal pathogens (D'Amato et al. 2011).

16.4 The Role of Proteins in Wines

Although they are minor constituents in wines, proteins are widely recognized to have a significant impact on wine quality. They are thought to contribute to wine taste and body (Jones et al. 2008), and to the foaming properties of sparkling wines (Vanrell et al. 2007; Blasco et al. 2011; Coelho et al. 2011). They are known to be detrimental for wine quality on some occasions, causing turbidity in white wines (Waters et al. 1992, 2005). Moreover, some of the wine proteins have been found to be allergenic for some susceptible individuals (Pastorello et al. 2003; Vassilopoulou et al. 2007; Giribaldi and Giuffrida 2010; Gonzalez-Quintela et al. 2011).

Profiling by nano-HPLC/tandem MS of a Sauvignon Blanc wine led to the identification of 20 major proteins, including several yeast proteins (Kwon 2004). The identified grape-derived proteins were vacuolar invertase, TL proteins, class IV endochitinase, and β-glucanase (Kwon 2004). The 2DE pattern of one Chardonnay wine confirmed the occurrence of several vacuolar invertase protein spots, as well as the presence of PR proteins such as osmotins and thaumatins, and detected for the first time one low molecular weight lipid transfer protein (Okuda et al. 2006). The presence of these proteins has since been confirmed in red and white wines by other published reports (Sauvage et al. 2010; Dufrechou et al. 2012; Lambri et al. 2012), with minor differences, such as the detection of grape ripening-related proteins and of PR4 proteins (chitin-binding proteins) in some cases (Cilindre et al. 2008; Esteruelas et al. 2009; Marangon et al. 2009; Wigand et al. 2009).

The application of more sensitive techniques such as direct MS analysis, or the use of enrichment strategies such as CPLL (Righetti et al. 2011) and glycopeptide enrichment (Palmisano et al. 2010), has helped to identify several other low-abundance proteins present in red (Simonato et al. 2011; Tolin et al. 2012) and white wines (D'Amato et al. 2011). Although there is detailed knowledge of the type of proteins that wine may contain, there is still little understanding of the role they may play in wine, especially with regard to low-abundance proteins.

16.4.1 Heat-Unstable Proteins and Haze Formation

Wine proteins may cause a common white wine defect called "casse protéique." During bottle storage, occasional extremes of temperature may lead to protein aggregation and flocculation, which causes turbidity (Waters et al. 2005). A haze or deposit in bottled wine can reduce or invalidate its commercial value, and winemakers usually perform fining treatments, typically with bentonite, to avoid this turbidity. The occurrence of this defect led to early studies on wine proteins (Koch and Sajak 1959; Moretti and Berg 1965; Bayly and Berg 1967; Hsu and Heatherbell 1987b; Waters et al. 1992).

Although total protein quantity may have an impact on the probability of haze development in white wines (Mesquita et al. 2001), not all wine protein fractions seem to share the tendency to flocculate, as some are more heat-labile than others (Moretti and Berg 1965; Bayly and Berg 1967; Hsu and Heatherbell 1987b; Hsu et al. 1987; Moine Ledoux et al. 1992; Waters et al. 1992). One recent application of CPLL to the soluble fraction and sediment of one white wine revealed a very limited overlap between the two types of proteins (D'Amato et al. 2011), thus contributing to the hypothesis of differential haze-forming tendencies for different wine proteins.

The heat-unstable protein fraction is mainly made up of grape PR proteins (Waters et al. 1996; Esteruelas et al. 2009; Sauvage et al. 2010; Marangon et al. 2011a, b; Vincenzi et al. 2011; Dufrechou et al. 2012). These proteins are able to persist through the winemaking process (Vincenzi et al. 2011), mainly due to their resistance to proteolysis and to their stability at acid pH (Linthorst 1991). The major contributors to natural wine haze to date have been identified as β-glucanases, class IV chitinases, and TL proteins (Waters et al. 1996; Esteruelas et al. 2009; Falconer et al. 2010; Marangon et al. 2011b; Sauvage et al. 2010; Dufrechou et al. 2012). These proteins were characterized recently for their haze-forming tendency and absorbance by bentonite (Sauvage et al. 2010). A progressive sensitivity to heat-induced precipitation, and a concomitant increased susceptibility to bentonite absorption, was found for β-glucanases, class IV chitinases, and a fraction of TL proteins, with invertases and the other fraction of TLs being less affected. These results confirmed previous findings on the thermal stability of purified chitinase, invertase, and TL protein, which reported that chitinase is the major player in heat-induced wine haze formation, probably due to its low melting temperature (Falconer et al. 2010).

The aggregation kinetics in white wines at different temperatures were determined by dynamic light-scattering experiments (Dufrechou et al. 2010). At low temperature (40 °C), aggregation took place during the heating phase, whereas at higher temperatures (60 °C and 70 °C) protein aggregation mainly developed during the cooling phase. Results confirmed the differential heat sensitivity of diverse TL protein fractions and the haze-forming tendency at low temperatures of β-glucanases and class IV chitinases (Sauvage et al. 2010), but a temperature lower than that reported in model wine solutions (Falconer et al. 2010) was found to be necessary for invertase aggregation and precipitation. Recently, Marangon and co-workers (2011b) found that the natural haze of white wine consisted mainly of class IV chitinase, with the contribution of β-glucanase and, for the first time, of one yeast cell wall glucantransferase. Using a model wine solution, they found that haze in the presence of purified chitinase and TL protein was formed only when sulphate was present. Similar results were found in a protein-free wine added with isolated chitinase, TL proteins, and sulphate. As reported for invertase by Dufrechou et al. (2010), the wine proteins tested by Marangon and co-workers (2011b) had lower unfolding temperatures in real wine than in model wine solutions.

Because different wines with different haze potential usually contain very similar protein fractions (Ferreira et al. 2000; Monteiro et al. 2001; Wigand et al. 2009), one or more unknown non-proteinaceous wine component(s) (termed X factors) are thought to be needed to cause visible haze formation (Mesquita et al. 2001; Waters et al. 2005; Batista et al. 2009). Candidate factors that may play a modulating role in wine haze formation include the sulfate anion (Pocock et al. 2007; Marangon et al. 2011a, b), pH value (Batista et al. 2009; Dufrechou et al. 2012), ionic strength (Dufrechou et al. 2010, 2012; Marangon et al. 2011a), phenolic compounds (Waters et al. 1995; Marangon et al. 2010; Esteruelas et al. 2011), and organic acids (Batista et al. 2010). To date, the identity of the X factor remains unclear.

Marangon et al. (2011a) studied the impact of ionic strength and sulfate upon thermal aggregation of purified grape chitinases and TL proteins in a model wine solution. They reported that, although TL proteins are not very susceptible to ionic strength changes, chitinase isoforms behave differently, one being precipitated above 21 mM, the other above 100 mM. Sulphate, even at low concentration, increased the instability of both chitinase isoforms, and it had no effect on TLs. Very recently, Dufrechou et al. (2012) published a report on the effects of ionic strength, pH, and temperature on wine protein instability, using both model and real wines. By screening aggregation kinetics, they proposed a model for heat-induced haze formation which includes a balance between pH-induced unfolding, leading to conformational changes responsible for colloidal aggregation of wine proteins at low pH, and heat-induced unfolding, leading to denaturation and aggregation at higher temperatures.

The validity of the experiments on wine protein instability involving the analysis of wine model solutions containing organic acids has been recently questioned: Batista et al. (2010) found a dramatic reduction in the haze potential of wine proteins when measured in the presence of organic acids normally encountered in wines. This reduction was also observed in real wines when added with organic acids. They suggested this phenomenon may be linked to the removal of considerable amounts of phenolics, which are apparently involved in protein haze formation (Waters et al. 1995; Marangon et al. 2010; Esteruelas et al. 2011).

Glycosylated proteins are known to play an important role in wine turbidity, as they may interact with tannins, polyphenols, and other proteins (Siebert et al. 1996). Moine-Ledoux et al. (1992) showed that wines aged on yeast lees were less prone to haze formation and were stabilized by the addition of less bentonite than wines aged without lees. Subsequently, they were able to demonstrate that this protection from haze was due to a 32-kDa fragment of glycosylated yeast invertase (Moine-Ledoux and Dubourdieu 1999). Dupin et al. (2000) proposed a competitive mechanism between yeast mannoprotein and wine proteins for unknown wine components, otherwise required for the formation of large insoluble aggregates of denatured protein (the X factor?). Other glycoproteins showing haze-protective activity include whole yeast invertase (Moine-Ledoux and Dubourdieu 1999), Arabinogalactan proteins (Waters et al. 1994b; Pellerin et al. 1993), and high molecular weight yeast mannoproteins (Waters et al. 1993; Waters et al. 1994a). To date, several studies have demonstrated that modified yeast strains overproducing mannoproteins significantly contributed to improved white wine stability (Brown et al. 2007; Gonzalez-Ramos et al. 2008). The recent screening of wine glycoproteins published by Palmisano and co-workers (2010) may increase the possibilities for the characterization of yeast and grape glycoproteins, which may have a technical application in the reduction of white wine haziness. To this intent, one predictive assay for wine haze tendency based on the separate recovery and quantification of wine proteins and glycoproteins was recently developed (Fusi et al. 2010). The authors showed that protein content and glycoprotein concentrations are different in wine, whereas their electrophoretic patterns are almost superimposable. They also demonstrated a straightforward connection between their assay and prediction of haze as measured by traditional assays.

16.4.2 Foam Formation and Stability

Wine proteins have been shown to play an important role in the sparkling wine industry because they are known to promote foam formation and stability. A positive correlation between protein concentration and foam formation in sparkling wines has been reported since the earliest studies (Brissonet and Maujean 1993; Malvy et al. 1994; Andres-Lacueva et al. 1996; Marchal et al. 1996; Luguera et al. 1997, 1998). The occurrence of protein degradation in sparkling wines has been shown to reduce their foamability. Dambrouck et al. (2005) found that a significant decrease in both the total protein and the grape invertase contents of Champagne-base wines was correlated with the loss of wine foaming properties.

Several studies have investigated the detrimental effect on foam stability of infection by fungal pathogens, such as *Botrytis cinerea.* The reduction in foamability registered in the presence of botrytized grapes was due to fungal proteases able to significantly hydrolyze wine proteins (Girbau et al. 2004; Marchal et al. 2006; Cilindre et al. 2007, 2008).

Vanrell and colleagues demonstrated that the use of bentonite fining treatments on sparkling wines caused a significant reduction in foam formation and stability. This effect was due to the registered depletion of all the protein fractions by bentonite, except for the high molecular mass fraction, which probably contains glycoproteins and polysaccharides (Vanrell et al. 2007).

As previously described, during alcoholic fermentation and aging on lees, glycosylated proteins (mannoproteins) are released by the yeasts. These proteins were recently shown to have potential foam-active properties in wine and also in beers (Blasco et al. 2011). Very recently, one experiment on molecular reconstituted model sparkling wines demonstrated that foam height and foam stability increased exponentially with the concentration of high molecular weight mannoproteins (Coelho et al. 2011).

Due to the contribution to foam formation and stability of some wine proteins, especially the high molecular weight glycosylated proteins, the investigation of fining methods other than bentonite, with a more selective removal capacity, is still one of the major needs of the oenological industry.

16.5 Wine Proteins as a Tool for Traceability

16.5.1 Varietal/Geographical Differentiation

The transformation of the wine market into a global market has pushed producers and legislators towards the approval of protection policies for several high-quality wines. To this end, labeling policies have been created throughout the world. The two main types of labeling policy are derived from the American and the French model. In the first model, wines are labeled according to the most abundant grape variety used (minimum 75%). In the French policy, the system of Protected Designations of Origin, geographical criteria are as important as varietal ones. Regardless of the system used, wines bearing protected labels are considered of higher quality and are generally more expensive than non-labeled wines. It is thus not surprising to see efforts being made by scientists to develop new techniques to prove wine authenticity. Most of these techniques rely on DNA typing (Siret et al. 2000; García-Beneytez et al. 2002), biochemical characterization of both volatile and non-volatile compounds (Rebolo et al. 2000; Moret et al. 1994), and analysis of stable isotopes (Day et al. 1995; Di Paola-Naranjo et al. 2011).

In more recent years, the study of compounds that can be used in grape traceability has expanded to nitrogenous compounds, including proteins. The first steps towards varietal differentiation of wines based on protein profiling were taken by Pueyo et al. (1993), who found differences in the native electrophoresis patterns of musts obtained from different grape varieties, and by Moreno-Arribas et al. (1999), who analyzed 41 musts made from a mixture of grapes from large vineyards and were able to group them according to the grape variety using the same approach.

In 2002, Rodriguez-Delgado and collaborators used capillary gel electrophoresis of wine proteins in order to differentiate between different wines from the Canary Islands, and found that, although similar, the relative amounts of specific protein fractions allowed differentiation among them, due to the different grape varieties used, the soil in which the vines grew, and the climatic conditions. One recent application of high-performance capillary electrophoresis to varietal differentiation of still white wines, based on protein profiling coupled with shikimic acid quantification, has been proposed (Chabreyrie et al. 2008). Comparison of the SDS-PAGE patterns of commercial red, rosé, and white wines from different varieties revealed great similarities among the analyzed wines, although some differences could be found (Wigand et al. 2009). The protein band identified as lipid transfer protein, for example, was not detected in most of the commercial red wines, although it was fully detected in the Dornfelder red wine, less in the rosé wine, and not in white wines, probably due to the shorter contact times between wine and skins (Wigand et al. 2009).

MS analysis of wine proteins has been proposed as a tool for wine authentication since 1996, when Szilágyi and colleagues published their results on the application of MALDI for distinguishing wines and musts. MALDI and SELDI (surface-enhanced laser desorption/ionization) were then used for the fingerprinting of proteins in different wines (Weiss et al. 1998), and ESI-MS was used on directly infused musts and wines, and proved to be able to reveal the addition of unfermented must or sugar (Catharino et al. 2006). MALDI-TOF-MS profiling of peptides obtained by tryptic digestion has been recently proposed as a tool for differentiating high-quality white wines from the Campania region (Chambery et al. 2009).

To promote the use of MS profiling of wine proteins as a tool for differentiation, more methodological and technical evidence is needed. To this end, one recent study has been published by Nunes-Miranda et al. (2012), taking into account the type of matrix, the number of bottles of white wine, the number of technical replicates, and the number of spots, as well as the classification algorithm used. In their report, the best conditions for the reliable profiling of unprocessed wine proteins were found to be the use of α-Cyano-4-hydroxycinnamic acid matrix, mixed 0.75:1 with analyzed wine, with three spots from five different bottles of each wine as minimum requirements, with the Bayes-Net algorithm performing the best in these conditions (Nunes-Miranda et al. 2012). Although expensive both in terms of costs and time, more studies in this direction are urgently needed to move from the field of research to real applications of MS for proteins and peptides in the field of varietal/geographical differentiation of wines.

16.5.2 Detection of Contaminating Proteins

One major issue for wine traceability in the last decade has been the detection of residual proteins deriving from fining treatments. Some winemakers usually add protein-based fining agents (milk casein, egg ovalbumin, fish gelatin, gluten) in order to reduce or eliminate potential sediments of grape and yeast proteins during long-term bottle storage (D'Alessandro and Zolla 2012). These fining proteins may cause severe problems for wine commercialization, as most of them are potentially allergenic, and are now subjected to mandatory labeling. The techniques used to date for detection of contaminating proteins of animal and/or plant origin in wines mainly relied on the antibody/antigen reaction, such as ELISA (Rolland et al. 2008; Weber et al. 2009; Lacorn et al. 2011) and Western blotting (Weber et al. 2009). The detection limit of these methods is often considered too high (100 μg/L for Weber et al. 2009, 2010). One recent clinical work by Vassilopoulou et al. (2011) reported that, although no allergen was detected by traditional methods in the fined wines, positive skin prick test reactions and basophil activation to the treated wines were observed in the majority of patients with allergy to milk, egg, or fish, correlating with the concentration of the fining agents used.

Mass spectrometry has been applied in recent years to the detection of these fining proteins to wines. Capillary LC combined with ESI-Q-TOF-MS was used by Monaci et al. (2010) for the detection of caseins in white wines, with a declared limit of detection (50 mg/L) which is still much higher than approved ELISA methods. Very recently, one commercial ELISA kit has been validated for detection of caseins in white wine with a declared detection limit as low as 1 ppm (Restani et al. 2011).

Mass spectrometry has also been used for the detection of gluten-derived proteins in red wines by LC-MS/MS analysis (Simonato et al. 2011), and the method proved to be significantly more sensitive (LOD: 1 mg/L) than the usual ELISA methods (LOD: 50 mg/L). The same research approach has recently been used for egg protein detection in red wines, and again in this case proved to be more effective than immunochemical methods, achieving an LOD of 5 mg/L of egg white (Tolin et al. 2012).

The application by Cereda et al. in 2010 of the CPLL to white wines allowed the detection of amounts of added caseins as low as 1 μg/L. The same research group performed screening of commercial Italian red wines using the same approach, and was able to detect the use of milk proteins for red wine fining instead of the expected occurrence of egg ovalbumin (D'Amato et al. 2010). Major criticisms of the cited method for the detection of allergenic fining proteins in wines are the poor quantitative results, mainly due to the limited dynamic range of electrophoresis and staining techniques. Nevertheless, their significant improvement in the detection limit of contaminating proteins in wines have boosted the chances of preventing frauds that can seriously damage consumer health.

16.6 Concluding Remarks

Wine proteomics has recently achieved new relevance, and the number of surveys devoted to oenological aspects influenced by the wine proteins has exponentially increased since the accomplishment of grape genome sequencing. Nevertheless, more efforts towards absolute protein quantification and standardization of the methods are currently needed in the field of wine proteomics, particularly for its use in quality assessment and in traceability.

References

Andres-Lacueva C, Lopez-Tamames E, Lamuela-Raventos RM, Buxaderas S, de la Torre-Boronat MC (1996) Characteristics of sparkling base wines affecting foam behavior. J Agric Food Chem 44:989–995

AOAC International (2010) Official methods of analysis of AOAC International, Method 960.52

Batista L, Monteiro S, Loureiro VB, Teixeira AT, Ferreira RB (2009) The complexity of protein haze formation in wines. Food Chem 112:169–177

Batista L, Monteiro S, Loureiro VB, Teixeira AT, Ferreira RB (2010) Protein haze formation in wines revisited. The stabilising effect of organic acids. Food Chem 122:1067–1075

Bayly FC, Berg HW (1967) Grape and wine proteins of white wine varietals. Am J Enol Vitic 17:18–32

Blasco L, Viñas M, Villa TG (2011) Proteins influencing foam formation in wine and beer: the role of yeast. Int Microbiol 14:61–71

Bradford MM (1976) Rapid and sensitive method for the quantitation of microgram quantities of protein utilizing the principle of protein-dye binding. Anal Biochem 72:248–254

Brissonet F, Maujean A (1993) Characterization of foaming proteins in a champagne base wine. Am J Enol Vitic 44:297–301

Brown SL, Stockdale VJ, Pettolino F, Pocock KF, de Barros LM, Williams PJ, Bacic A, Fincher GB, Hoj PB, Waters EJ (2007) Reducing haziness in white wine by overexpression of *Saccharomyces cerevisiae* genes YOL155c and YDR055w. Appl Microbiol Biotechnol 73:1363–1376

Caridi A (2006) Enological functions of parietal yeast mannoproteins. Antonie Van Leeuwenhoek 89:417–422

Carpentieri A, Marino G, Amoresano A (2007) Rapid fingerprinting of red wines by MALDI mass spectrometry. Anal Bioanal Chem 389:969–982

Catharino RR, Cunha IB, Fogaça AO, Facco EM, Godoy HT, Daudt CE, Eberlin MN, Sawaya AC (2006) Characterization of must and wine of six varieties of grapes by direct infusion electrospray ionization mass spectrometry. J Mass Spectrom 41:185–190

Cereda A, Kravchuk AV, D'Amato A, Bachi A, Righetti PG (2010) Proteomics of wine additives: mining for the invisible via combinatorial peptide ligand libraries. J Proteomics 73:1732–1739

Chabreyrie D, Chauvet S, Guyon F, Salagoïty MH, Antinelli JF, Medina B (2008) Characterization and quantification of grape variety by means of shikimic acid concentration and protein fingerprint in still white wines. J Agric Food Chem 56:6785–6790

Chambery A, del Monaco G, Di Maro A, Parente A (2009) Peptide fingerprint of high quality Campania white wines by MALDI-TOF mass spectrometry. Food Chem 113:1283–1289

Cilindre C, Castro AJ, Clément C, Jeandet P, Marchal R (2007) Influence of *Botrytis cinerea* infection on Champagne wine proteins (characterized by two-dimensional electrophoresis/immunodetection) and wine foaming properties. Food Chem 103:139–149

Cilindre C, Jégou S, Hovasse A, Schaeffer C, Castro AJ, Clément C, Van Dorsselaer A, Jeandet P, Marchal R (2008) Proteomic approach to identify champagne wine proteins as modified by *Botrytis cinerea* infection. J Proteome Res 7:1199–1208

Coelho E, Rocha SM, Coimbra MA (2011) Foamability and foam stability of molecular reconstituted model sparkling wines. J Agric Food Chem 59:8770–8778

Curioni A, Vincenzi S, Flamini R (2008) Proteins and peptides in grape and wine. In: Flamini R (ed) Hyphenated techniques in grape and wine chemistry. Wiley, Chichester, pp 249–287

D'Alessandro A, Zolla L (2012) We are what we eat: food safety and proteomics. J Proteome Res 11:26–36

D'Amato A, Kravchuk AV, Bachi A, Righetti PG (2010) Noah's nectar: the proteome content of a glass of red wine. J Proteomics 73:2370–2377

D'Amato A, Fasoli E, Kravchuk AV, Righetti PG (2011) *Mehercules, adhuc Bacchus*! The debate on wine proteomics continues. J Proteome Res 10:3789–3801

Dambrouck T, Marchal R, Marchal-Delahaut L, Parmentier M, Maujean A, Jeandet P (2003) Immunodetection of proteins from grapes and yeast in a white wine. J Agric Food Chem 51:2727–2732

Dambrouck T, Marchal R, Cilindre C, Parmentier M, Jeandet P (2005) Determination of the grape invertase content (using PTA-ELISA) following various fining treatments versus changes in the total protein content of wine. relationships with wine foamability. J Agric Food Chem 53:8782–8789

Dawes H, Boyes S, Keene J, Heatherbell D (1994) Protein instability of wines: influence of protein isoelectric point. Am J Enol Vitic 4:319–326

Day MP, Zhang B, Martin GJ (1995) Determination of the geographical origin of wine using joint analysis of elemental and isotopic composition. II—differentiation of the principal production zones in France for the 1990 vintage. J Agric Food Chem 67:113–123

Di Paola-Naranjo RD, Baroni MV, Podio NS, Rubinstein HR, Fabani MP, Badini RG, Inga M, Ostera HA, Cagnoni M, Gallegos E, Gautier E, Peral-Garcia P, Hoogewerff J, Wunderlin DA (2011) Fingerprints for main varieties of Argentinean wines: *terroir* differentiation by inorganic, organic, and stable isotopic analyses coupled to chemometrics. J Agric Food Chem 59:7854–7865

Dizy M, Bisson LF (1999) White wine protein analysis by capillary zone electrophoresis. Am J Enol Vitic 50:120–127

Dorrestein E, Ferreira RB, Laureano O, Teixeira AR (1995) Electrophoretic and FPLC analysis of soluble proteins in four Portuguese wines. Am J Enol Vitic 46:235–242

Dufrechou M, Sauvage FX, Bach B, Vernhet A (2010) Protein aggregation in white wines: influence of the temperature on aggregation kinetics and mechanisms. J Agric Food Chem 58:10209–10218

Dufrechou M, Poncet-Legrand C, Sauvage FX, Vernhet A (2012) Stability of white wine proteins: combined effect of pH, ionic strength, and temperature on their aggregation. J Agric Food Chem 60:1308–1319

Dupin I, McKinnon BM, Strockdale VJ, Ryan C, Boulay M, Markides AJ, Jones GP, Williams PJ, Waters EJ (2000) *Saccharomyces cerevisiae* mannoproteins that protect wine from protein haze: their release during fermentation and lees contact and a proposal for their mechanism of action. J Agric Food Chem 48:3098–3105

Esteruelas M, Poinsaut P, Sieczkowski N, Manteau S, Fort MF, Canals JM, Zamora F (2009) Characterization of natural haze protein in Sauvignon white wine. Food Chem 113:28–35

Esteruelas M, Kontoudakis N, Gil M, Fort MF, Canals JM, Zamora F (2011) Phenolic compounds present in natural haze protein of Sauvignon white wine. Food Res Int 44:77–83

Falconer RJ, Marangon M, Van Sluyter SC, Neilson KA, Chan C, Waters EJ (2010) Thermal stability of thaumatin-like protein, chitinase, and invertase isolated from Sauvignon blanc and Semillon juice and their role in haze formation in wine. J Agric Food Chem 58:975–980

Ferreira RB, Monteiro S, Piçarra-Pereira MA, Tanganho MC, Loureiro VB, Teixeira AR (2000) Characterization of the proteins from grapes and wines by immunological methods. Am J Enol Vitic 51:22–28

Ferreira RB, Piçarra-Pereira MA, Monteiro S, Loureiro VB, Teixeira AR (2002) The wine proteins. Trends Food Sci Tech 12:230–239

Flamini R, De Rosso M (2006) Mass spectrometry in the analysis of grape and wine proteins. Expert Rev Proteomics 3:321–331

Fountoulakis M, Juranville J-F, Manneberg M (1992) Comparison of the Coomassie brilliant blue, bicinchoninic acid and Lowry quantitation assays, using non-glycosylated and glycosylated proteins. J Biochem Biophys Methods 24:265–274

Frank M (2007) Counterfeit bottles multiply as global demand for collectible wine surges. Wine Spectator. 31 Jan–28 Feb:14

Fusi M, Mainente F, Rizzi C, Zoccatelli G, Simonato B (2010) Wine hazing: a predictive assay based on protein and glycoprotein independent recovery and quantification. Food Control 21:830–834

García-Beneytez E, Moreno-Arribas MV, Borrego J, Polo MC, Ibáñez J (2002) Application of a DNA analysis method for the cultivar identification of grape musts and experimental and commercial wines of *Vitis vinifera* L. using microsatellite markers. J Agric Food Chem 50:6090–6096

Girbau T, Stummer BE, Pocock KF, Baldock GA, Scott ES, Waters EJ (2004) The effect of *Uncinula necator* (powdery mildew) and *Botrytis cinerea* infection of grapes on the levels of haze-forming pathogenesis-related proteins in grape juice and wine. Aust J Grape Wine Res 10:125–133

Giribaldi M, Giuffrida MG (2010) Heard it through the grapevine: proteomic perspective on grape and wine. J Proteomics 73:1647–1655

Gonçalves F, Heyraud A, de Pinho MN, Rinaudo M (2002) Characterization of white wine mannoproteins. J Agric Food Chem 50:6097–6101

Gonzalez-Quintela A, Gomez-Rial J, Valcarcel C, Campos J, Sanz ML, Linneberg A, Gude F, Vidal C (2011) Immunoglobulin-E reactivity to wine glycoproteins in heavy drinkers. Alcohol 45:113–122

Gonzalez-Ramos D, Cebollero E, Gonzalez R (2008) A recombinant *Saccharomyces cerevisiae* strain overproducing mannoproteins stabilizes wine against protein haze. Appl Environ Microbiol 74:5533–5540

Gornall AG, Bardawill CJ, David MM (1949) Determination of serum proteins by means of the biuret reaction. J Biol Chem 177:751–766

Hsu JC, Heatherbell DA (1987a) Isolation and characterization of soluble proteins in grapes, grape juice, and wine. Am J Enol Vitic 38:6–10
Hsu JC, Heatherbell DA (1987b) Heat-unstable proteins in wine. I. Characterization and removal by bentonite fining and heat treatment. Am J Enol Vitic 38:11–16
Hsu JC, Heatherbell DA, Flores JH, Watson BT (1987) Heat-unstable proteins in wine. II. Characterization and removal by ultrafiltration. Am J Enol Vitic 38:17–22
Jaillon O, Aury JM, Noel B, Policriti A, Clepet C, Casagrande A, Choisne N, Aubourg S, Vitulo N, Jubin C, Vezzi A, Legeai F, Hugueney P, Dasilva C, Horner D, Mica E, Jublot D, Poulain J, Bruyère C, Billault A, Segurens B, Gouyvenoux M, Ugarte E, Cattonaro F, Anthouard V, Vico V, Del Fabbro C, Alaux M, Di Gaspero G, Dumas V, Felice N, Paillard S, Juman I, Moroldo M, Scalabrin S, Canaguier A, Le Clainche I, Malacrida G, Durand E, Pesole G, Laucou V, Chatelet P, Merdinoglu D, Delledonne M, Pezzotti M, Lecharny A, Scarpelli C, Artiguenave F, Pè ME, Valle G, Morgante M, Caboche M, Adam-Blondon AF, Weissenbach J, Quétier F, Wincker P, French-Italian Public Consortium for Grapevine Genome Characterization (2007) The grapevine genome sequence suggests ancestral hexaploidization in major angiosperm phyla. Nature 449:463–467
Jones PR, Gawel R, Francis IL, Waters EJ (2008) The influence of interactions between major white wine components on the aroma, flavour and texture of model white wine. Food Qual Pref 19:596–607
Koch J, Sajak E (1959) A review and some studies on grape protein. Am J Enol Vitic 10:114–123
Kwon SW (2004) Profiling of soluble proteins in wine by nano-high-performance liquid chromatography/tandem mass spectrometry. J Agric Food Chem 52:7258–7263
Lacorn M, Gößwein C, Immer U (2011) Determination of residual egg white proteins in red wines during and after fining. Am J Enol Vitic 62:382–385
Lambri M, Dordoni R, Giribaldi M, Violetta MR, Giuffrida MG (2012) Heat-unstable protein removal by different bentonite labels in white wines. LWT Food Sci Technol 46:460–467
Le Bourse D, Jégou S, Conreux A, Villaume S, Jeandet P (2010) Review of preparative and analytical procedures for the study of proteins in grape juice and wine. Anal Chim Acta 667:33–42
Le Bourse D, Conreux A, Villaume S, Lameiras P, Nuzillard JM, Jeandet P (2011) Quantification of chitinase and thaumatin-like proteins in grape juices and wines. Anal Bioanal Chem 401:1541–1549
Linthorst HJM (1991) Pathogenesis-related proteins of plants. Crit Rev Plant Sci 10:123–150
Lowry OH, Rosebrough NJ, Farr AL, Randall RJ (1951) Protein measurement with Folin phenol reagent. J Biol Chem 193:265–275
Luguera C, Moreno-Arribas V, Pueyo E, Polo MC (1997) Capillary electrophoretic analysis of wine proteins. Modifications during the manufacture of sparkling wines. J Agric Food Chem 45:3766–3770
Luguera C, Moreno-Arribas V, Pueyo E, Bartolomé B, Polo MC (1998) Fractionation and partial characterization of protein fractions present at different stages of the production of sparkling wines. Food Chem 63:465–471
Malvy J, Robillard B, Duteurtre B (1994) Influence des proteínes sur le comportement de la mousse des vins de Champagne. Sci Aliments 14:87–98
Manteau S, Lambert B, Jeandet P, Legendre L (2003) Changes in chitinase and thaumatin-like pathogenesis-related proteins of grape berries during the Champagne winemaking process. Am J Enol Vitic 54:267–272
Marangon M, Van Sluyter SC, Haynes PA, Waters EJ (2009) Grape and wine proteins: their fractionation by hydrophobic interaction chromatography and identification by chromatographic and proteomic analysis. J Agric Food Chem 57:4415–4425
Marangon M, Vincenzi S, Lucchetta M, Curioni A (2010) Heating and reduction affect the reaction with tannins of wine protein fractions differing in hydrophobicity. Anal Chim Acta 660:110–118
Marangon M, Sauvage F-X, Waters EJ, Vernhet A (2011a) Effects of ionic strength and sulfate upon thermal aggregation of grape chitinases and thaumatin-like proteins in a model system. J Agric Food Chem 59:2652–6262
Marangon M, Van Sluyter SC, Neilson KA, Chan C, Haynes PA, Waters EJ, Falconer RJ (2011b) Roles of grape thaumatin-like protein and chitinase in white wine haze formation. J Agric Food Chem 59:733–740

Marchal R, Bouquelet S, Maujean A (1996) Purification and partial biochemical characterization of glycoproteins in champenois Chardonnay wine. J Agric Food Chem 44:1716–1722

Marchal R, Seguin V, Maujean A (1997) Quantification of interferences in the direct measurement of proteins in wines from the champagne region using the Bradford method. Am J Enol Vitic 48:303–309

Marchal R, Berthier L, Legendre L, Marchal-Delahaut L, Jeandet P, Maujean A (1998) Effects of *Botrytis cinerea* infection on the must protein electrophoretic characteristics. J Agric Food Chem 46:4945–4949

Marchal R, Chaboche D, Marchal-Delahaut L, Gerland C, Gandon JP, Jeandet P (2000) Detection and quantification of lysozyme in champagne wines. J Agric Food Chem 48:3225–3321

Marchal R, Warchol M, Cilindre C, Jeandet P (2006) Evidence for protein degradation by *Botrytis cinerea* and relationships with alteration of synthetic wine foaming properties. J Agric Food Chem 54:5157–5165

Mesquita PR, Piçarra-Pereira MA, Monteiro S, Loureiro VB, Teixeira AR, Ferreira RB (2001) Effect of wine composition on protein stability. Am J Enol Vitic 52:324–330

Moine Ledoux V, Dulau L, Dubourdieu D (1992) Interprétation de l'amélioration de la stabilité protéique des vins au cours de l'élevage sur lies. J Int Sci Vigne Vin 26:239–251

Moine-Ledoux V, Dubourdieu D (1999) An invertase fragment responsible for improving the protein stability of dry white wines. J Sci Food Agric 79:537–543

Monaci L, Losito I, Palmisano F, Visconti A (2010) Identification of allergenic milk proteins markers in fined white wines by capillary liquid chromatography-electrospray ionization-tandem mass spectrometry. J Chromatogr A 1217:4300–4305

Monaci L, Losito I, Palmisano F, Godula M, Visconti A (2011) Towards the quantification of residual milk allergens in caseinate-fined white wines using HPLC coupled with single-stage Orbitrap mass spectrometry. Food Addit Contam 28:1304–1314

Monteiro S, Piçarra-Pereira MA, Tanganho MC, Rente JP, Loureiro VB, Teixeira AR, Ferreira RB (1999) Preparation of polyclonal antibodies specific for wine proteins. J Sci Food Agric 79:772–778

Monteiro S, Piçarra-Pereira MA, Mesquita PR, Loureiro VB, Teixeira A, Ferreira RB (2001) The wide diversity of structurally similar wine proteins. J Agric Food Chem 49:3999–4010

Monteiro S, Barakat M, Piçarra-Pereira MA, Teixeira AR, Ferreira RB (2003a) Osmotin and thaumatin from grape: a putative general defense mechanism against pathogenic fungi. Phytopathology 93:1505–1512

Monteiro S, Piçarra-Pereira MA, Teixeira AR, Loureiro VB, Ferreira RB (2003b) Environmental conditions during vegetative growth determine the major proteins that accumulate in mature grapes. J Agric Food Chem 51:4046–4053

Monteiro S, Piçarra-Pereira MA, Loureiro VB, Teixeira AR, Ferreira RB (2007) The diversity of pathogenesis-related proteins decreases during grape maturation. Phytochemistry 68:416–425

Moreno-Arribas MV, Cabello F, Polo MC, Martín-Alvarez PJ, Pueyo E (1999) Assessment of the native electrophoretic analysis of total grape must proteins for the characterization of *Vitis vinifera* L. cultivars. J Agric Food Chem 47:114–120

Moreno-Arribas MV, Pueyo E, Polo MC (2002) Analytical methods for the characterization of proteins and peptides in wines. Anal Chim Acta 458:63–75

Moret I, Scarponi G, Cescon P (1994) Chemometric characterization and classification of five venetian white wines. J Agric Food Chem 42:1143–1153

Moretti RH, Berg HW (1965) Variability among wines to protein clouding. Am J Enol Vitic 16:69–78

Murphey JM, Spayd SE, Powers JR (1989) Effect of grape maturation on soluble protein characteristics of Gewürztraminer and White Riesling juice and wine. Am J Enol Vitic 40:199–207

Nunes-Miranda JD, Santos HM, Reboiro-Jato M, Fdez-Riverola F, Igrejas G, Lodeiro C, Capelo JL (2012) Direct matrix assisted laser desorption ionization mass spectrometry-based analysis of wine as a powerful tool for classification purposes. Talanta. doi:10.1016/j.talanta.2012.01.017

Okuda T, Fukui M, Takayanagi T, Yokotsuka K (2006) Characterization of major stable proteins in Chardonnay wine. Food Sci Technol Res 12:131–136

Organisation Internationale de la Vigne et du Vin (2008) World Vitivinicultural statistics 2007, supplement to the OIV Report

Palmisano G, Antonacci D, Larsen MR (2010) Glycoproteomic profile in wine: a 'sweet' molecular renaissance. J Proteome Res 9:6148–6159

Pastorello EA, Farioli L, Pravettoni V, Ortolani C, Fortunato D, Giuffrida MG, Perono Garoffo L, Calamari AM, Brenna O, Conti A (2003) Identification of grape and wine allergens as an endochitinase 4, a lipid-transfer protein, and a thaumatin. J Allergy Clin Immunol 111:350–359

Pedreschi R, Hertog M, Lilley KS, Nicolaï B (2010) Proteomics for the food industry: opportunities and challenges. Crit Rev Food Sci Nutr 50:680–692

Pellerin P, Waters E, Brillouet J-M (1993) Characterization of two arabinogalactan-proteins from red wine. Carbohydr Polym 22:187–192

Peng Z, Pocock KF, Waters EJ, Francis IL, Williams PJ (1997) Taste properties of grape (*Vitis vinifera*) pathogenesis-related proteins isolated from wine. J Agric Food Chem 45:4639–4643

Pocock KF, Waters EJ (2006) Protein haze in bottled white wines: how well do stability tests and bentonite fining trials predict haze formation during storage and transport? Aust J Grape Wine Res 12:212–220

Pocock KF, Alexander GM, Hayasaka Y, Jones PR, Waters EJ (2007) Sulfate: a candidate for the missing essential factor that is required for the formation of protein haze in white wine. J Agric Food Chem 55:1799–1807

Pueyo E, Dizy M, Polo C (1993) Variety differentiation of must and wines by means of protein fraction. Am J Enol Vitic 44:255–260

Rebolo S, Peña RM, Latorre MJ, García S, Botana AM, Herrero C (2000) Characterisation of Galician (NW Spain) Ribeira Sacra wines using pattern recognition analysis. Anal Chim Acta 417:211–220

Restani P, Uberti F, Tarantino C, Ballabio C, Gombac F, Bastiani E, Bolognini L, Pavanello F, Danzi R (2011) Validation by a collaborative interlaboratory study of an ELISA method for the detection of caseinate used as a fining agent in wine. Food Anal Methods 2011:1–7

Righetti PG, Fasoli E, Boschetti E (2011) Combinatorial peptide ligand libraries: the conquest of the 'hidden proteome' advances at great strides. Electrophoresis 32:960–966

Rodriguez-Delgado MA, Malovana S, Montelongo FJ, Cifuentes A (2002) Fast analysis of proteins in wines by capillary gel electrophoresis. Eur Food Res Technol 214:536–540

Rolland JM, Apostolou E, de Leon MP, Stockley CS, O'Hehir RE (2008) Specific and sensitive enzyme-linked immunosorbent assays for analysis of residual allergenic food proteins in commercial bottled wine fined with egg white, milk, and non grape-derived tannins. J Agric Food Chem 56:349–354

Ruiz-Larrea F, Lopez R, Santamaria M, Ruiz MC, Zarazaga M, Gutierrez AR, Torres C (1998) Soluble proteins and free amino nitrogen content in must and wine of cv. Viura in La Rioja. Vitis 37:139–142

Santoro M (1995) Fractionation and characterization of must and wine proteins. Am J Enol Vitic 46:250–254

Sauvage FX, Bach B, Moutounet M, Vernhet A (2010) Proteins in white wines: thermo-sensitivity and differential adsorbtion by bentonite. Food Chem 118:26–34

Siebert KJ, Carrasco A, Lynn PY (1996) Formation of protein-polyphenol haze in beverages. J Agric Food Chem 44:1997–2005

Simonato B, Mainente F, Tolin S, Pasini G (2011) Immunochemical and mass spectrometry detection of residual proteins in gluten fined red wine. J Agric Food Chem 59:3101–3110

Siret R, Boursiquot JM, Merle MH, Cabanis JC, This P (2000) Toward the authentication of varietal wines by the analysis of grape (*Vitis vinifera* L.) residual DNA in must and wine using microsatellite markers. J Agric Food Chem 48:5035–5040

Smith PK, Krohn RI, Hermanson GT, Mallia AK, Gartner FH, Provenzano MD, Fujimoto EK, Goeke NM, Olson BJ, Klenk DC (1985) Measurement of protein using bicinchoninic acid. Anal Biochem 150:76–85

Smith MR, Penner MH, Bennett SE, Bakalinsky AT (2011) Quantitative colorimetric assay for total protein applied to the red wine Pinot noir. J Agric Food Chem 59:6871–6876

Szilágyi Z, Vas G, Mády G, Vékey K (1996) Investigation of macromolecules in wines by matrix-assisted laser desorption/ionization time-of-flight mass spectrometry. Rapid Commun Mass Spectrom 10:1141–1143

Tolin S, Pasini G, Curioni A, Arrigoni G, Masi A, Mainente F, Simonato B (2012) Mass spectrometry detection of egg proteins in red wines treated with egg white. Food Control 23:87–94

Tyson PJ, Luis ES, Day WR, Walker B, Lee TH (1981) Estimation of soluble protein in must and wine by high-performance liquid chromatography. Am J Enol Vitic 32:241–243

Vanrell G, Canals R, Esteruelas M, Fort F, Canals JM, Zamora F (2007) Influence of the use of bentonite as a ridding agent on foam quality and protein fraction of sparkling wines (Cava). Food Chem 104:148–155

Vassilopoulou E, Zuidmeer L, Akkerdaas J, Tassios I, Rigby NR, Mills EN, van Ree R, Saxoni-Papageorgiou P, Papadopoulos NG (2007) Severe immediate allergic reactions to grapes: part of a lipid transfer protein-associated clinical syndrome. Int Arch Allergy Immunol 143:92–102

Vassilopoulou E, Karathanos A, Siragakis G, Giavi S, Sinaniotis A, Douladiris N, Fernandez-Rivas M, Clausen M, Papadopoulos NG (2011) Risk of allergic reactions to wine, in milk, egg and fish-allergic patients. Clin Transl Allergy 1:10

Velasco R, Zharkikh A, Troggio M, Cartwright DA, Cestaro A, Pruss D, Pindo M, Fitzgerald LM, Vezzulli S, Reid J, Malacarne G, Iliev D, Coppola G, Wardell B, Micheletti D, Macalma T, Facci M, Mitchell JT, Perazzolli M, Eldredge G, Gatto P, Oyzerski R, Moretto M, Gutin N, Stefanini M, Chen Y, Segala C, Davenport C, Demattè L, Mraz A, Battilana J, Stormo K, Costa F, Tao Q, Si-Ammour A, Harkins T, Lackey A, Perbost C, Taillon B, Stella A, Solovyev V, Fawcett JA, Sterck L, Vandepoele K, Grando SM, Toppo S, Moser C, Lanchbury J, Bogden R, Skolnick M, Sgaramella V, Bhatnagar SK, Fontana P, Gutin A, Van de Peer Y, Salamini F, Viola R (2007) A high quality draft consensus sequence of the genome of a heterozygous grapevine variety. PLoS One 2:e1326

Vidigal SSMP, Tóth IV, Rangel AOSS (2012) Determination of total protein content in white wines by solid phase spectrometry in a SI-LOV system. Talanta. doi:10.1016/j.talanta.2011.12.028 (in press)

Vincenzi S, Mosconi S, Zoccatelli G, Pellegrina CD, Veneri G, Chignola R, Peruffo A, Curioni A, Rizzi C (2005) Development of a new procedure for protein recovery and quantification in wine. Am J Enol Vitic 56:182–187

Vincenzi S, Marangon M, Tolin S, Curioni A (2011) Protein evolution during the early stages of white winemaking and its relations with wine stability. Aust J Grape Wine Res 17:20–27

Waters EJ, Wallace W, Williams PJ (1992) Identification of heat-unstable wine proteins and their resistance to peptidases. J Agric Food Chem 40:1514–1519

Waters EJ, Wallace W, Tate ME, Williams PJ (1993) Isolation and partial characterization of a natural haze protective factor from wine. J Agric Food Chem 41:724–730

Waters EJ, Pellerin P, Brillouet JM (1994a) A *Saccharomyces* mannoprotein that protects wine from protein haze. Carbohydr Polym 23:185–191

Waters EJ, Pellerin P, Brillouet JM (1994b) A wine arabinogalactan-protein that reduces heat-induced wine protein haze. Biosci Biotechnol Biochem 58:43–48

Waters EJ, Peng Z, Pocock KF, Williams PJ (1995) Proteins in white wine, I: procyanidin occurrence in soluble proteins and insoluble protein hazes and its relationship to protein instability. Aust J Grape Wine Res 1:86–93

Waters EJ, Shirley NJ, Williams PJ (1996) Nuisance proteins of wine are grape pathogenesis-related proteins. J Agric Food Chem 44:3–5

Waters EJ, Alexander G, Muhlack R, Pocock KF, Colby C, O'Neill BK, Høj PB, Jones P (2005) Preventing protein haze in bottled white wine. Aust J Grape Wine Res 11:215–225

Weber P, Steinhart H, Paschke A (2009) Determination of the bovine food allergen casein in white wines by quantitative indirect ELISA, SDS – PAGE, Western Blot and immunostaining. J Agric Food Chem 57:8399–8405

Weber P, Steinhart H, Paschke A (2010) Characterization, antigenicity and detection of fish gelatine and isinglass used as processing aids in wines. Food Addit Contam Part A Chem Anal Control Expo Risk Assess 27:273–282

Weiss KC, Yip T-T, Hutchens TW, Bisson LF (1998) Rapid and sensitive fingerprinting of wine proteins by matrix-assisted laser desorption/ionization time-of-flight (MALDI-TOF) mass spectrometry. Am J Enol Vitic 49:231–239

Wigand P, Tenzer S, Schild H, Decker H (2009) Analysis of protein composition of red wine in comparison with rosé and white wines by electrophoresis and high-pressure liquid chromatography-mass spectrometry (HPLC-MS). J Agric Food Chem 57:4328–4333

Yokotsuka K, Ebihara T, Sato T (1991) Comparison of soluble proteins in juice and wine from Koshu grapes. J Ferment Bioeng 71:248–253

Yokotsuka K, Singleton VL (1997) Glycoproteins: characterization in a hybrid grape variety (Muscat Bailey A) juice, fermenting must, and resultant red wine. Am J Enol Vitic 48:100–114

Chapter 17
Eggs

Gustavo Martos, Elena Molina, and Iván López-Expósito

17.1 Introduction

Unfertilized hen's eggs are commonly employed in the food industry because of their technological properties, such as foaming, gelling, and emulsification, and their high nutritional value. In addition, eggs are an abundant dietary source of vitamins (A, D, E, and K), saturated and unsaturated fatty acids, carotenoids, minerals, and proteins (Burley and Vadehra 1989). Egg functionality in food products is correlated with its chemical composition, especially its protein content. Individual components of hen's egg are also associated with numerous biological functions, which include antimicrobial activity, protease inhibitory action, vitamin-binding properties, and antigenic or immunogenic characteristics (Li-Chan et al. 1995).

The avian egg (Fig. 17.1) is formed in the mature hen (*Gallus gallus*) by a reproductive system composed of an ovary and oviduct. The yolk is formed in the follicular sac by the deposition of continuous layers of yolk material. When the yolk matures, the follicular sac ruptures and releases the yolk, which quickly reaches the magnum section of the oviduct, where the dense portion of the albumen (thick egg white) is added. Subsequently, two shell membranes form to contain the yolk and dense white loosely until the rest of the albumen is added in the oviduct shell gland. Here the shell is formed and also two cordlike structures, the chalazae, maintaining the yolk centered in the egg. In the last portion of the oviduct, the vagina, a thin protein coating called "bloom" is applied to the shell to keep harmful bacteria or dust from entering the eggshell pores.

G. Martos • E. Molina • I. López-Expósito (✉)
Instituto de Investigación en Ciencias de la Alimentación, CIAL (CSIC-UAM),
C/Nicolás Cabrera 9, Madrid 28049, Spain
e-mail: ivan.lopez@csic.es

F. Toldrá and L.M.L. Nollet (eds.), *Proteomics in Foods: Principles and Applications*, Food Microbiology and Food Safety 2, DOI 10.1007/978-1-4614-5626-1_17,

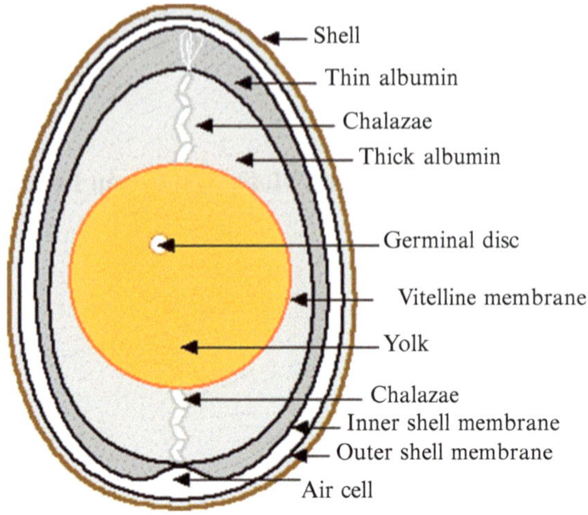

Fig. 17.1 Schematic representation of the different parts of an avian egg

Egg protein content is approximately 13%, 17% in the yolk, and 11% in the white (Burley and Vadehra 1989). Egg proteins may serve as a food source for the developing embryo, function as shock-absorber, antimicrobial barrier, or exert other biological activities. Knowledge concerning the detailed protein composition of hen's egg can increase the potential applications in the food industry and can also enhance our understanding of various biological processes. Furthermore, eggs are a potential source of bioactive peptides (Miguel and Aleixandre 2006) and may become medically important as carriers of therapeutic proteins (Kawabe et al. 2006).

To date, several approaches have been made to separate and identify the proteins located in hen's egg. Some of those include chromatographic techniques (Awade and Efstathiou 1999; Miguel et al. 2005), electrophoresis and immunochemical methods (Kitabatake et al. 1988; Desert et al. 2001; Holen and Elsayed 1990), and isoelectric focusing (Beeley 1971). Recently, mass spectrometry has represented a quantum leap in egg protein identification, leading to the finding of over 100 proteins both in the egg yolk (Mann and Mann 2008; Farinazzo et al. 2009) and white (Guerin-Dubiard et al. 2006; Raikos et al. 2006; D'Ambrosio et al. 2008; Mann and Mann 2011).

17.2 Egg White Proteins

Avian egg white is a high-viscosity solution of about 11% protein and glycoprotein. It functions as a shock-absorber, keeps the yolk in place, constitutes an antimicrobial barrier, and provides water, protein, and other nutrients to the developing embryo. In addition to these biological roles, it is an inexpensive source of high-quality protein for food industries, contains proteins of pharmaceutical interest, and also proteins that have found widespread use in biomedical research and protein chemistry (Stevens 1991; Mine and Kovacs-Nolan 2004; Anton et al. 2006; Mine 2007).

Before the 1990s, only 13 proteins were often referenced in egg white, some of which were not even fully characterized. Classical biochemical techniques, such as chromatographic and electrophoretic separations, together with Edman sequence analysis, usually run into the difficulty of the unfavorable egg white composition, with only three proteins—ovalbumin, ovotransferrin and ovomucoid—accounting for 75% of total protein content. Moreover, proteins in egg white have very different molecular masses, ranging from 7 kDa for ovosecretoglobin to 8,000 kDa for soluble ovomucin, and extreme pI values, from 3.9 for ovoglycoprotein to 10.7 for lysozyme (Guerin-Dubiard et al. 2006). Additionally, numerous proteins present abundant post-translational modifications, such as high degrees of glycosylation and phosphorylated residues. However, the irruption of mass spectrometry (MS) techniques has overcome many of these difficulties and has greatly improved our knowledge of the egg white proteome over the last years. The major recent proteomic advances are discussed in the following sections.

17.2.1 DE – Mass Spectrometry

2-D-electrophoresis (2DE) allows the separation of egg white proteins according to their isoelectric point (pI) and molecular weight (Mw). Egg white is usually manually separated from the yolk and gently homogenized. Then it is sonicated in a detergent-containing buffer and proteins are precipitated with acetone and resuspended in a rehydration buffer to apply on immobilized pH gradient (IPG) strips of variable pH ranges (Gorg et al. 2000). Focused IPG strips are reduced and alkylated prior to being run on the SDS-PAGE slabs for the molecular weight separation. Proteins are visualized by colorimetric or fluorescent staining (Fig. 17.2) and spots are excised from the gel and subjected to a tryptic digestion following a standard protocol (e.g., Shevchenko et al. 2006).

For MALDI-TOF MS analysis, tryptic digests can be directly mixed with a saturated matrix solution of α-cyano-4-hydroxycinnamic acid (CHCA) and loaded onto a sample plate. Protein identification is carried out by peptide mass fingerprinting (PMP), considering in the sequence database search the following modifications: fixed carbamidomethylation of cysteines (covalent addition of a carbamidomethyl group (57.07 Da) due to the reduction and alkylation steps during the equilibration of IPG strips prior to the second dimension) and variable methionine oxidation. Using this methodology, Raikos et al. (2006) identified five proteins, namely, ovalbumin, conalbumin (ovotransferrin), clusterin, activin receptor II-A, and the hypothetical protein FLJ 10305, out of 13 spots detected in the egg white 2-D gel. According to the authors, some of the limitations for further protein identification

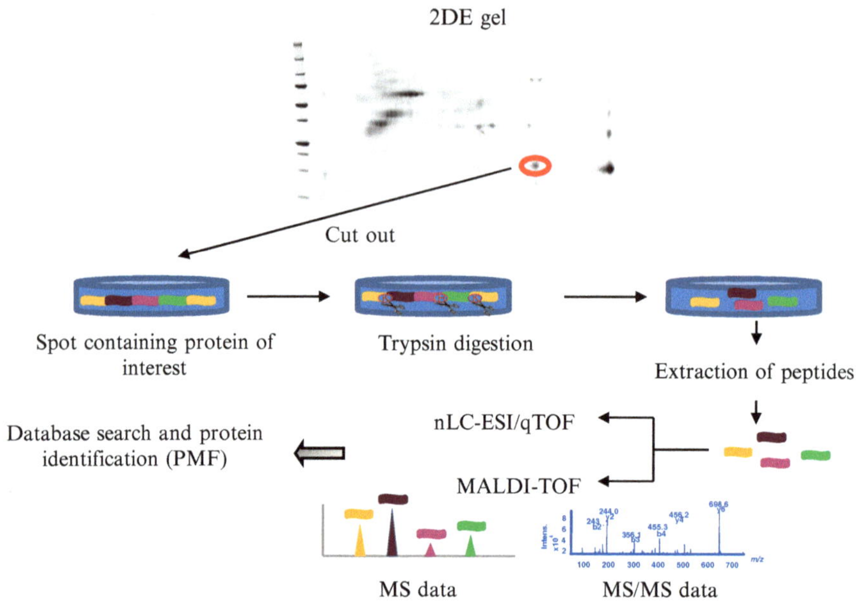

Fig. 17.2 Standard protocol for protein analysis by 2-DE and mass spectrometry

were that the precipitation method could have resulted in the loss of several proteins from the 2-D map, that proteins such as lysozyme present a pI which is not within the range of the IPG strips used in the study, and that glycosylation may interfere with the software identification due to the different molecular masses of the sugar-bearing peptides.

Guerin-Dubiard et al. (2006) carried out multiple 2DE separations of egg white proteins employing various pH ranges (3–10, 4–7, and 5.5–6.7). A total of 69 spots was detected and analyzed by MALDI-TOF MS. Eleven spots were successfully identified by PMF alone after standard trypsinolysis. These corresponded to well-known proteins such as ovalbumin, ovotransferrin, and ovoinhibitor. For most protein spots, the authors performed internal peptide sequencing by nanoliquid chromatography (nLC)-MS/MS. In this methodology, tryptic hydrolysates are concentrated and desalted on a C18 pre-column and the nanoscale LC eluent from the analytical column is sent to the nanoelectrospray ionization source of a quadrupole/time-of-flight (qTOF) mass spectrometer. The high sensitivity, resolution, and accuracy of the qTOF instrument produced very informative MS/MS spectra. These were analyzed using the Mascot search engine that led to the confirmation of PMF MALDI-TOF analysis and the unambiguous identification of 55 protein spots. Despite the high number of identifications, only 16 egg white proteins were found, among which two, namely vitelline membrane protein VMO-1 and Tenp (a potential antimicrobial protein), had not been detected in egg white before. Again, some limitations were the extreme molecular weight or pI of some egg white proteins, such as ovomucin (Mw > 8,000 kDa), ovomacroglobulin (Mw = 780 kDa) or avidin (pI = 10). Moreover, some unidentified spots could correspond to proteins that are not indexed yet in the international protein databanks.

Despite its limitations, the 2DE-MS approach presents some advantages with respect to other methodologies that do not make use of a previous bidimensional electrophoretic separation. 2DE reinforces identification by confirming Mw and pI matches of analyzed spots. Furthermore, numerous isoforms can be observed in the gel allowing for tentative identification of polymeric, glycosylated, and phosphorylated isoforms. Thus, Guerin-Dubiard et al. (2006) were able to recognize the diphosphorylated, monophosphorylated, and nonphosphorylated ovalbumin isoforms, as well as some polymeric forms of higher molecular weight. Five isoforms found for ovotransferrin were attributed to different glycoforms, as has been shown for human transferrin. Additionally, an apparent glycosylated and a dimeric form of lysozyme were identified in two protein spots.

17.2.2 1D-PAGE – Mass Spectrometry

The results from Raikos et al. (2006) and Guerin-Dubiard et al. (2006) indicated that the known protein inventory of egg white proteins was far from being complete. A major breakthrough came with the use of LC-FT-ICR (Fourier-transform ion cyclotron resonance) MS, performing MS/MS and MS^3 experiments. Mann (2007) identified 78 chicken egg white proteins by analyzing 18 sections of a 1D-PAGE gel containing the separated egg white proteins. Recently, the number of identified proteins rose up to 202 (Mann and Mann 2011) using a similar methodology that employed an orbitrap mass analyzer. ICR and orbitrap analyzers are coupled after a linear ion trap, which first accumulates and stores ions, in hybrid high-performance mass spectrometers. These instruments have unsurpassed resolving power, sequencing speed, mass accuracy, and sensitivity. The masses are not resolved in space or time as with other techniques but only by sensing the frequency that each ion produces as it rotates or oscillates in an electromagnetic field.

The workflow in these studies starts with the egg white separation from the yolk, homogenization, and lyophilization. Proteins are then subjected to SDS-PAGE and the gel is cut into several sections for in-gel reduction, carbamidomethylation, and digestion with trypsin (Wilm et al. 1996). Eluted peptides are cleaned using C18 tips (Rappsilber et al. 2003) and analyzed by the nLC-FT (ICR or orbitrap) mass spectrometer.

This approach has hitherto been the most successful, leading to the identification of numerous proteins for the first time in the egg white and further characterization of previously reported ones. For instance, a protein was detected presenting 62% sequence identity with mammalian acyloxyacyl hydrolases, responsible for the cleavage of acyl chains from bacterial lipopolysaccharides (LPS). Another newly detected egg white component was lymphocyte antigen 86 (MD-1), also involved in triggering cellular responses to LPS. Several other proteins identified for the first time in egg white had been reported to occur in chicken plasma and cerebrospinal fluid, such as serum albumin, various immunoglobulins, or apolipoproteins. This suggests that those are proteins generally present in body fluids, including the oviduct fluid.

The abundance of identified proteins can be estimated by calculating the exponentially modified protein abundance index (emPAI). This parameter gives a rough estimation of the protein concentration. It is calculated as $10^{PAI} - 1$, where PAI is the number of observed unique parent ions divided by the number of theoretically observable ions (Ishihama et al. 2005). For instance, an emPAI value of 99 allowed the statement that ovosecretoglobulin, the first nonmammalian secretoglobulin identified (Mann 2007), had a similar abundance to that of cystatin (emPAI = 99) or avidin (emPAI = 138).

It is interesting to note that some typical major yolk residents, such as apovitellinin-I, vitellogenin-1 to −3, and apolipoprotein B, were also found among the egg white components. This was ascribed to the fact that residual proteins not taken up by the egg cell may be liberated from the ovary together with the egg and comigrate into the oviduct, mixing with egg white proteins secreted in the magnum section. In addition, many low-abundance proteins detected are proteins normally found in intracellular compartments, suggesting that they may have reached the oviduct fluid as by-products of the secretion of major egg white proteins or could also derive from damaged leaky cells of the epithelium lining the oviduct. These data indicate a decrease in the finding of new proteins among secreted egg white proteins, becoming more difficult to identify additional functional and original egg white components and suggesting that a deep, albeit not complete, proteome characterization has now been reached.

Despite the tremendous identification power of the methodology consisting of 1D-PAGE and nano-LC separation coupled with current FT-ICR and orbitrap MS instruments, it is not exempt from some limitations. For example, the same egg proteins were detected in different distant sections of the gel, which could be attributed to limited solubility in the electrophoresis sample buffer, aggregation upon unfolding, or partial degradation by egg white proteases before the electrophoretic separation. Other discrepancies between relative mobility in the gel and predicted mass may be due to protein complex formation. Poor solubility of some egg white proteins in the electrophoresis buffer may prevent them from being subsequently detected as well. Furthermore, undetected proteins may also be encoded by the estimated 5–10% of the chicken genome that has not been sequenced yet (Hillier et al. 2004). In addition, many egg white proteins are glycosylated and phosphorylated and peptides bearing such modifications need specialized methods to be detected.

There are some known egg white enzymatic activities that are difficult to associate with specific database entries, either because they correspond to the nonsequenced genome region or because they may be attributed to identified but poorly characterized proteins. Enzymatic activities such as acid and alkaline phosphatase, β-N-acetylhexoxaminidase, α-manosidase, catalase, or ribonuclease (Seko et al. 1997) have not been found a corresponding protein yet.

17.2.3 *Combinatorial Peptide Ligand Libraries: Egg White Proteins*

The critical composition of the egg white proteome, which makes the identification of minor hidden species an arduous task, has led some investigators to employ the combinatorial peptide ligand libraries (CPLL), also known as equalizer beads technology. This novel affinity approach was introduced in 2005 by Thulasiraman et al. and was claimed to equalize the proteins in a biological sample by selective binding to highly diverse but specific binding sites introduced by hexapeptides. The equalizer beads feature a combinatorial library from hexameric peptide ligands on their surface so that there virtually exists a bead that specifically binds to every protein in the sample. Thus, the equalization would result from major proteins saturating all of the available ligands and all minor proteins would be captured by their respective specific ligands. Unfortunately, a few years later Keidel et al. (2010) proved the equalizing mechanism wrong, demonstrating that beads interact with protein mixtures according to a general hydrophobic binding mechanism, where diversity in surface ligands plays a negligible role. Nonetheless, for reduction of complexity, dynamic range, and high-abundant, albeit hydrophilic, proteins from complex mixtures, the application of beads with certain hydrophobic surface properties may be useful.

D'Ambrosio et al. (2008) employed two solid-phase combinatorial peptide libraries (known as ProteoMiner and its carboxylated version) for the analysis of egg white proteins. The initial sample was obtained by manually separating the white, homogenization, and addition of a protease inhibitor cocktail. After equilibration in PBS, the sample was mixed overnight with the ProteoMiner library and the supernatant (proteins not bound to the beads) was placed in contact with the carboxylated library and incubated overnight. Captured proteins by both libraries were eluted with three sequential solutions. Eluates were run on a SDS-PAGE gel and gel lanes were cut into 15 slices after the electrophoretic separation for trypsinization and protein identification by μLC-ESI-IT-MS/MS experiments (Fig. 17.3). The identified unique gene products amounted to a total of 148, a value that doubled the egg white proteins found a year before by Mann (2007) but inferior to the number of proteins subsequently identified by the same author using the orbitrap mass spectrometer (Mann and Mann 2011).

Protein qualitative comparison between the starting egg white sample and the bead-treated sample was also performed by 2DE. It is clear that the peptide library technology improves subsequent MS identification by reducing the abundance dynamic range of the proteins in the sample, leading to a higher number of detectable species. However, one major egg white protein, ovotransferrin, escaped capture by the libraries together and was found only in the starting material. This could be attributed to the low hydrophobicity and large size of the protein, which might preclude sufficient hydrophobic interaction with the bead-bound hexapeptides.

It has been suggested that, given the higher sensitivity of the new generation mass spectrometers, the investigation of the egg white proteome would no longer

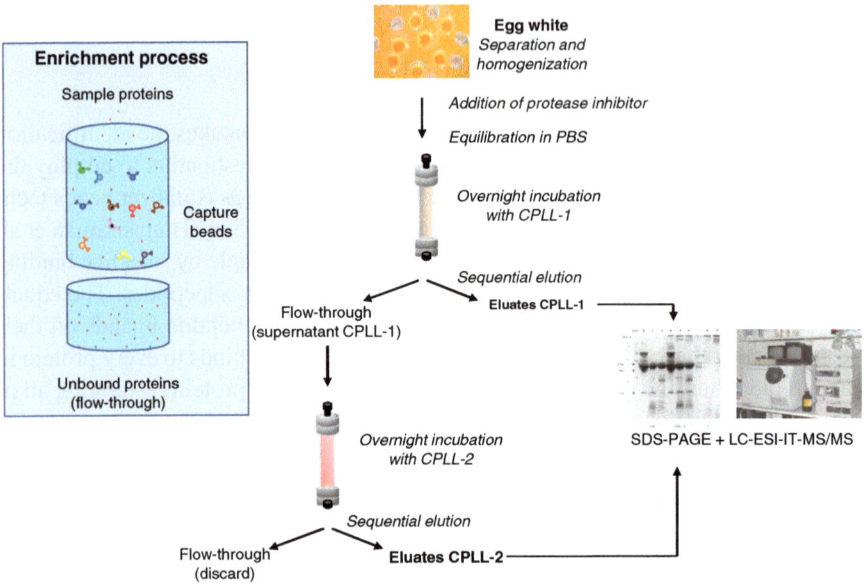

Fig. 17.3 Scheme of the use of combinatorial ligand libraries (CPLL) for the study of the chicken egg white proteome. CPLL-1 and CPLL-2 correspond to Proteominer peptide library and its carboxylated version, respectively

need to rely on enrichment technologies that modify the protein relative content in an unpredictable way and render impossible any quantification. However, considering the obvious improvement achieved using peptide libraries, it could be argued that a potential combined use of this technology, together with novel orbitrap spectrometers (as in Mann and Mann 2011) could lead to a much greater number of detected proteins, including many yet undiscovered ones. In fact, only few of the new egg white proteins identified using the peptide libraries (D'Ambrosio et al. 2008) were also detected in the orbitrap MS study.

17.3 Egg Yolk Proteins

The chicken egg yolk nourishes the developing embryo with lipids, proteins, vitamins, and minerals. It also accumulates maternal antibodies providing the first immune defense of the embryo. Apart from its biological role, the egg yolk is widely used in food and cosmetics industries as a binding agent, emulsifier, and natural antioxidant. It has been used as a source of bioactive peptides (Anton et al. 2006; Mine and Kovacs-Nolan 2006) and certain therapeutic applications have also been suggested (Hartmann and Wilhelmson 2001; Kovacs-Nolan et al. 2005).

The egg yolk consists of approximately 48% water, 33% lipids, and 17% protein (Burley and Vadehra 1989). Many yolk constituents are synthesized in the liver, transported with the blood to the ovary, and incorporated into the egg cell by receptor-mediated endocytosis. Hen egg yolk protein content is about 68% low-density lipoproteins (LDL), 16% high-density lipoproteins (HDL), 10% livetins, and 4% phosvitins. Livetins (α or serum albumin, β and λ or Ig λ) are the main constituents of the plasma water-soluble fraction whereas LDL, HDL, and phosvitins constitute the granular or globular fraction. All yolk LDL apoproteins, except apovitellenin I, derive from the blood very low density lipoprotein (VLDL), apolipoprotein B (Apo B), which is enzymatically cleaved during its transfer into the yolk. HDL apoproteins (the lipovitellins) together with the phosvitins and the yolk plasma glycoproteins YGP40 and YGP42, derive from the cleavage of the vitellogenins, which are constituents of blood HDL particles. In addition to these major proteins, egg yolk has been reported to contain many enzymes, often characterized by their activity only (Burley and Vadehra 1989; Seko et al. 1997).

17.3.1 1D-PAGE: Mass Spectrometry

Mass spectrometry techniques have also represented a breakthrough for protein identification in egg yolk. Using the 1D-PAGE/LC-MS approach, Mann and Mann (2008) were able to identify 119 proteins from hen egg yolk, 86 of which had not been detected in yolk previously. For the analysis of egg yolk proteins, sample preparation entails a previous separation step of the granular and plasma fractions, as described by Ahn et al. (2006). Proteins of every fraction were then run in a SDS-PAGE gel, lanes cut into several slices, trypsinized, and peptides were eluted and cleaned using C18 tips before analysis by nLC-FT-ICR-MS.

Most identified proteins were found in both fractions, but usually with different abundances, as deduced from their emPAI values. In addition to the highest abundant proteins, yolk contained proteases and protease inhibitors, antioxidative enzymes such as superoxide dismutase and glutathione peroxidase, vitamin- and cobalamin-binding proteins, and many other serum proteins. Furthermore, many egg white proteins, such as cystatin, ovalbumin-related protein Y, or lysozyme were also found in the yolk, apart from the already reported major egg white proteins ovalbumin, ovotransferrin, and ovomucoid (Williams 1962; Saito and Martin 1966; Sugimoto et al. 1984). In contrast to the egg white, the yolk contained very few components of basement membrane or by-products of secretion processes, which authors ascribed to the receptor-mediated endocytosis mechanism required for intracellular accumulation of yolk components. Therefore, most yolk proteins are probably recruited from blood, although some may be produced by granulosa cells surrounding the oocyte.

Various enzymes or binding proteins previously reported to occur in the yolk (Burley and Vadehra 1989; Seko et al. 1997) have not found a corresponding protein, which could be missing in the 5–10% chicken genome portion that has not been sequenced yet (Hillier et al. 2004). Alternatively, these enzymatic activities may be attributable to one of the identified but insufficiently characterized proteins.

17.3.2 *Combinatorial Peptide Ligand Libraries: Egg Yolk Proteins*

As in the egg white, the combinatorial peptide ligand library (CPLL), ProteoMiner, containing hexapeptides terminating with a primary amine (CPLL-1), or modified with a terminal carboxyl (CPLL-2) or tertiary amine group (CPLL-3), has been employed for the deep analysis of the yolk plasma proteins. Using this technology, Farinazzo et al. (2009) doubled the number of proteins (up to 255) found in egg yolk plasma, as compared to the previous study of Mann and Mann (2008). However, inasmuch as this methodology is based on solid-phase adsorption, it is not amenable to particulate matter treatment and, therefore, could not be applied to the yolk granular fraction. Nonetheless, the authors achieved unprecedented identification results by subsequently treating yolk plasma proteins with the three versions of the CPLL.

Interestingly, it was suggested that the main force driving the bead capture capacity was the charge of the hexapeptide terminus. Thus, CPLL-1 captured the vast majority of proteins, as they are mostly charged negatively under the buffer conditions (at physiological pH two-thirds of all proteins in any proteome bear a net negative charge because they have pI values below pH 7). The carboxylated variant, CPLL-2, bound approximately an additional 20% proteome, nonredundant with CPLL-1 and mainly corresponding to cationic proteins (pI > 7). The third library, CPLL-3, captured essentially the same proteins adsorbed by CPLL-1, reinforcing the argument that binding is primarily modulated by the charge of the terminus. These data should be considered in conjunction with the results of Keidel et al. (2010), which portrayed a general nonspecific hydrophobic interaction between the CPLL-1 hexapeptides and the proteins, where diversity of hexapeptide amino acid composition plays a negligible role. Taken together, hydrophobic and electrostatic interactions with the terminal group would determine the binding, the latter being mostly responsible for the specific capture of anionic or cationic proteins. Still, some drawbacks of the CPLL technology comprise highly hydrophilic proteins, which are likely to be excluded from the capture event, as well as other proteins designated as aberrant, because they behave in an unpredictable way, resulting in either an abnormally low or high adsorption onto the beads.

Despite the limitations of misbehaving proteins, the CPLL treatment was very successful, given the fact that the egg yolk protein list was increased by about 170% (from 115 to 316 total). The biological significance and role of these new species is yet to be investigated.

17.4 Eggshell Proteins

Eggshells are crucial structures for avian species. They shelter the embryo from the microbial and physical environment, regulate the exchange of water and gases during extra-uterine development, and provide calcium to the growing embryo once the yolk stores are depleted. Eggshells are formed during movement of the yolk/ovum complex along the oviduct (isthmus section and uterus) by producing a multilayered mineral–organic composite (Arias et al. 2003). It mostly consists of three layers: the innermost eggshell double membrane, the calcified layer, and the cuticle (Mann et al. 2008).

17.4.1 Eggshell Membrane Proteins

Intact eggshell membranes are essential for shell calcification in laying hens (Baumgartner et al. 1978; Chowdhury 1990; Arias et al. 1997), providing at the same time a barrier to prevent inward mineralization. The low solubility of eggshell membranes hampers their analysis, although diverse studies have shed light on their composition. They have been shown to contain collagens, proteoglycans (Arias and Fernandez 2001), a few proteins that also occur in the egg white and calcified eggshell layer, such as lysozyme and ovotransferrin (Nys et al. 2001, 2004), β-N-acetylglucosaminidase (Ahlborn et al. 2006), and a cystein-rich protein named CREMP (Kodali et al. 2011) that could contribute to the extensive cross-linking of the membranes' proteinaceous fibers.

17.4.2 Eggshell Matrix Proteins

The mineralized shell consists primarily of calcite, the most stable polymorph of calcium carbonate. Its formation is largely influenced by an organic matrix composed of proteins, glycoproteins, and proteoglycans, designated "eggshell matrix proteins," which are progressively incorporated from the uterine fluid during calcification. Eggshell matrix proteins can be classified in three groups: egg white proteins also present in the eggshell, ubiquitous proteins that are found in many tissues, and eggshell-specific matrix proteins. Among the egg white proteins are ovalbumin, lysozyme, ovotransferrin, and clusterin (Hincke 1995; Hincke et al. 2000; Gautron et al. 2001b; Mann et al. 2003). Osteopontin, a glycophosphoprotein found in the bone and other hard tissues, is an example of a ubiquitous protein that occurs in the eggshell matrix (Chien et al. 2009). Due to the capacity of mammalian renal osteopontin to inhibit calcium oxalate crystal formation (Shiraga et al. 1992), it has been suggested that this protein may modulate the speed of calcium carbonate precipitation during eggshell formation (Nys et al. 2004). Eggshell-specific matrix proteins are unique to the shell calcification and have been proposed to regulate the mineralization process and defend against microbial colonization. However, the specific role of many of these proteins, primarily ovocleidins and ovocalyxins, needs to be further elucidated (Rose and Hincke 2009).

The analysis of eggshell matrix proteins usually involves a decalcification step with EDTA or acetic acid. This treatment yields soluble and insoluble constituents, the former being much easier to study. A complementary source for these proteins has been the uterine fluid that contains the precursors of matrix proteins in their functional and native forms prior to incorporation into eggshell. Several approaches were initially employed that led to the identification of a variety of eggshell matrix proteins. Those included amino acid microsequencing of electroblotted bands of proteins separated by SDS-PAGE, Western blotting with specific antibodies, and N-terminal sequencing of shell matrix proteins purified by liquid chromatography (Hincke et al. 1995; Gautron et al. 1997, 2001a, b). Other eggshell matrix proteins were characterized by a combination of molecular cloning, immunochemistry, and bioinformatics (Hincke et al. 1999; Gautron et al. 2007).

The irruption of high-throughput mass spectrometry methods led to a dramatic increase in the number of identified eggshell proteins. Mann et al. (2006) comprehensively analyzed the acid-soluble fraction of the eggshell calcified layer by LC-FT-ICR tandem mass spectrometry. Over 500 proteins were identified, including the most abundant proteins that were already known (mentioned above). It was proposed that many of them would be remnants of previous stages of egg formation or intracellular proteins released by breakdown of the cells lining the oviduct. Therefore, many proteins present in the uterine fluid during the calcification process would become assimilated into the eggshell in a nonspecific manner. However, eggshell-specific proteins are indeed highly likely to be relevant to eggshell function.

Insoluble eggshell matrix proteins following EDTA treatment were studied by Miksik et al. (2010), who employed an eggshell fractionation protocol consisting of limited cleavage with cynogen bromide and gradual precipitation with sodium chloride. Subsequent reduction, alkylation, and trypsin digestion steps were performed on samples prior to analysis by HPLC–MS/MS utilizing an ESI-IT mass spectrometer. A total of 28 proteins was identified, including a few proteins from the cuticle. Collagens I and III and an abundant protein similar to Kunitz-like protease inhibitor were found for the first time in the eggshell.

17.4.3 Eggshell Cuticle Proteins

One and a half hours before oviposition (egg expulsion) mineralization stops and an organic thin cuticle is deposited onto the eggshell surface. The cuticle covers the calcified layer and fills the entry to its pores, creating a barrier that prevents dehydration of the interior, bacterial penetration of pores, and limits microbial colonization of the egg's surface (Board and Halls 1973; Ruiz and Lunam 2000). Eggs with absent or incomplete cuticle are more prone to bacterial contamination (Sparks and Board 1984). The cuticle mainly consists of proteins (>85%) and possesses two layers: the hydroxyapatite-mineralized inner layer and the outer nonmineralized layer.

Rose-Martel et al. (2012) carried out a comprehensive proteomic study of the outermost cuticle layer of the avian egg. Protein extraction was performed by manually massaging the egg surface with a solution containing 1% SDS, 2 mM DTT. The solubilized proteins were concentrated and separated by SDS-PAGE. The gel was sliced into 13 sections for reduction, alkylation, and trypsin digestion of the protein bands. Extracted peptides were analyzed by nLC-ESI-MS/MS in a linear ion trap LTQ mass spectrometer. A total of 47 proteins was identified, many of them with known antimicrobial activity. In addition to ovocalyxin-32 (OCX-32), lysozyme, and ovotransferrin, the authors detected cystatin, a bactericidal protein against Gram-positive and Gram-negative bacteria and ovoinhibitor, which possesses inhibitory activity against fungal and bacterial proteases. In addition, the protein similar to Kunitz-like protease inhibitor was detected as a major constituent (along with OCX-32, ovalbumin, and lysozyme). A potential role was suggested for this protein in inhibiting bacterial and fungal proteases, as do other members of the family of serine protease inhibitors. Finally, proteins involved in the lipid metabolism and transport were also identified, such as apolipoprotein A-IV and vitellogenin I and II.

17.5 Egg-Related Proteomic Applications

17.5.1 Detection of Egg Allergens

After cow's milk, hen's egg allergy is the second most common food allergy in infants and young children (Sicherer and Sampson 2006). Currently, the most efficient approach for egg allergy is total avoidance of the offending compound. However, the omnipresence of egg-derived components in cooked or manufactured food products renders the approach difficult (Allen et al. 2007), and inadvertent exposure may lead to life-threatening anaphylactic responses. Stringent labeling regulations need to be enforced to provide full information about egg components in food products to ensure safety consumption for allergic individuals. The MS detection of egg allergenic proteins in food seems a promising approach inasmuch as it allows accurate detection at very low concentration levels.

Lee and Kim (2010) compared immunochemical, DNA-based, and MS-based methods to detect egg allergens in foods such as mayonnaise and biscuit. Egg white allergens ovalbumin and ovotransferrin, and egg yolk allergen α-livetin were successfully identified by LC-qTOF-MS. However, they were not able to identify the immunodominant egg white allergen ovomucoid, possibly because their sample preparation method failed to denature ovomucoid, avoiding its further tryptic digestion. However, ovomucoid detection should be feasible, as it has already been identified by LC-MS/MS using an in-solution digestion protocol that included the reduction and carbamidomethylation of the sample, dissolution in urea containing buffer, and digestion with lysyl endopeptidase prior to trypsin digestion (Mann 2007).

Heick et al. (2011a, b) developed an analytical LC-QqQ-MS/MS (triple quadrupole mass analyzer) method capable of simultaneously identifying seven allergenic foods (milk, egg, soy, peanut, hazelnut, walnut, and almond) in bread, before and after baking. The authors implemented the choice of marker peptides from all foods and the method showed superior sensitivity to ELISA for the detection of egg in the raw and baked product, with only one of the four ELISA test kits assayed being able to recognize egg in the baked matrix. Therefore, the mass spectrometric-based method is also a useful tool for egg allergen screening in processed foods.

17.5.2 Egg White Thinning During Storage

Prolonged storage of eggs at ambient temperature produces a loss in viscosity of egg white known as egg white thinning. Thinned egg white is more prone to microbial infections and is regarded as a sign of loss of quality that leads to staleness. In order to investigate the biochemical changes responsible for the natural thinning of thick egg whites, Omana et al. (2010) carried out a proteomic study that investigated the changes in the abundance of egg white proteins during storage for 10, 20, 30, and 40 days. Using 2DE, they detected several spots corresponding to proteins that significantly changed their abundance during storage. The identification was performed by excising spots of interest, in-gel trypsin digestion, and extraction of peptides for LC-qTOF MS analysis. Identified proteins were ovalbumin, clusterin, ovoinhibitor, ovotransferrin, and prostaglandin D2 synthase.

The increase in pH during egg storage is known to produce spatial unfolding of egg white proteins (Koehler 1974). Omana and coworkers suggested that the increase in the abundance of clusterin, a protein known to interact with partly folded proteins (Guerin-Dubiard et al. 2006), could stabilize partially unfolded proteins arising from the storage, thus causing egg white to lose its viscous nature. Furthermore, it was noticed that ovalbumin and clusterin underwent proteolysis during storage, because several spots corresponding to these proteins presented migration rates indicative of lower molecular weight than the native protein. The authors suggested that degradation of these proteins, due to proteolytic enzymes or as a result of the pH increase, could also contribute to the egg white thinning phenomenon.

References

Ahlborn GJ, Clare DA, Sheldon BW, Kelly RW (2006) Identification of eggshell membrane proteins and purification of ovotransferrin and beta-NAGase from hen egg white. Protein J 25:71–81

Ahn DU, Lee SH, Singam H, Lee EJ, Kim JC (2006) Sequential separation of main components from chicken egg yolk. Food Sci Biotechnol 15:189–195

Allen CW, Campbell DE, Kemp AS (2007) Egg allergy: are all childhood food allergies the same? J Paediatr Child Health 43:214–218

Anton M, Nau F, Nys Y (2006) Bioactive egg components and their potential uses. Worlds Poult Sci J 62:429–438

Arias JL, Fernandez MS (2001) Role of extracellular matrix molecules in shell formation and structure. Worlds Poult Sci J 57:349–357

Arias J, Cataldo M, Fernandez M, Wu J, Kessi E (1997) Effect of beta-aminoproprionitrile on the eggshell mineralisation. Br Poult Sci 38:351–356

Arias J, Wiff J, Fuenzalida V, Fernandez M (2003) Molecular regulation of avian eggshell biomineralization. In: Kobayashi I, Ozawa H (eds) Biomineralization: formation, diversity, evolution and application. Tokai University Press, Japan, pp 13–17

Awade AC, Efstathiou T (1999) Comparison of three liquid chromatographic methods for egg-white protein analysis. J Chromatogr B Biomed Sci Appl 723:69–74

Baumgartner S, Brown DJ, Salevsky E Jr, Leach RM Jr (1978) Copper deficiency in the laying hen. J Nutr 108:804–811

Beeley JG (1971) Ovomucoid heterogeneity examined by flat bed isoelectric focusing. Biochim Biophys Acta 230:595–598

Board R, Halls N (1973) The cuticle: a barrier to liquid and particle penetration of the shell of the hen's egg. Br Poult Sci 14:69–97

Burley RW, Vadehra DV (1989) The avian egg: chemistry and biology. Wiley, New York

Chien YC, Hincke MT, McKee MD (2009) Avian eggshell structure and osteopontin. Cells Tissues Organs 189:38–43

Chowdhury S (1990) Shell membrane system in relation to lathy-rogen toxicity and copper deficiency. Worlds Poult Sci J 46:156–169

D'Ambrosio C, Arena S, Scaloni A, Guerrier L, Boschetti E, Mendieta ME et al (2008) Exploring the chicken egg white proteome with combinatorial peptide ligand libraries. J Proteome Res 7:3461–3474

Desert C, Guerin-Dubiard C, Nau F, Jan G, Val F, Mallard J (2001) Comparison of different electrophoretic separations of hen egg white proteins. J Agric Food Chem 49:4553–4561

Farinazzo A, Restuccia U, Bachi A, Guerrier L, Fortis F, Boschetti E et al (2009) Chicken egg yolk cytoplasmic proteome, mined via combinatorial peptide ligand libraries. J Chromatogr A 1216:1241–1252

Gautron J, Hincke MT, Nys Y (1997) Precursor matrix proteins in the uterine fluid change with stages of eggshell formation in hens. Connect Tissue Res 36:195–210

Gautron J, Hincke MT, Mann K, Panheleux M, Bain M, McKee MD et al (2001a) Ovocalyxin-32, a novel chicken eggshell matrix protein. Isolation, amino acid sequencing, cloning, and immunocytochemical localization. J Biol Chem 276:39243–39252

Gautron J, Hincke MT, Panheleux M, Garcia-Ruiz JM, Boldicke T, Nys Y (2001b) Ovotransferrin is a matrix protein of the hen eggshell membranes and basal calcified layer. Connect Tissue Res 42:255–267

Gautron J, Murayama E, Vignal A, Morisson M, McKee MD, Rehault S et al (2007) Cloning of ovocalyxin-36, a novel chicken eggshell protein related to lipopolysaccharide-binding proteins, bactericidal permeability-increasing proteins, and plunc family proteins. J Biol Chem 282:5273–5286

Gorg A, Obermaier C, Boguth G, Harder A, Scheibe B, Wildgruber R et al (2000) The current state of two-dimensional electrophoresis with immobilized pH gradients. Electrophoresis 21:1037–1053

Guerin-Dubiard C, Pasco M, Molle D, Desert C, Croguennec T, Nau F (2006) Proteomic analysis of hen egg white. J Agric Food Chem 54:3901–3910

Hartmann C, Wilhelmson M (2001) The hen's egg yolk: a source of biologically active substances. Worlds Poult Sci J 57:13–28

Heick J, Fischer M, Kerbach S, Tamm U, Popping B (2011a) Application of a liquid chromatography tandem mass spectrometry method for the simultaneous detection of seven allergenic foods in flour and bread and comparison of the method with commercially available ELISA test kits. J AOAC Int 94:1060–1068

Heick J, Fischer M, Popping B (2011b) First screening method for the simultaneous detection of seven allergens by liquid chromatography mass spectrometry. J Chromatogr A 1218:938–943

Hillier LW, Miller W, Birney E, Warren W, Hardison RC, Ponting CP et al (2004) Sequence and comparative analysis of the chicken genome provide unique perspectives on vertebrate evolution. Nature 432:695–716

Hincke MT (1995) Ovalbumin is a component of the chicken eggshell matrix. Connect Tissue Res 31:227–233

Hincke MT, Tsang CP, Courtney M, Hill V, Narbaitz R (1995) Purification and immunochemistry of a soluble matrix protein of the chicken eggshell (ovocleidin 17). Calcif Tissue Int 56:578–583

Hincke MT, Gautron J, Tsang CP, McKee MD, Nys Y (1999) Molecular cloning and ultrastructural localization of the core protein of an eggshell matrix proteoglycan, ovocleidin-116. J Biol Chem 274:32915–32923

Hincke MT, Gautron J, Panheleux M, Garcia-Ruiz J, McKee MD, Nys Y (2000) Identification and localization of lysozyme as a component of eggshell membranes and eggshell matrix. Matrix Biol 19:443–453

Holen E, Elsayed S (1990) Characterization of four major allergens of hen egg-white by IEF/SDS-PAGE combined with electrophoretic transfer and IgE-immunoautoradiography. Int Arch Allergy Appl Immunol 91:136–141

Ishihama Y, Oda Y, Tabata T, Sato T, Nagasu T, Rappsilber J et al (2005) Exponentially modified protein abundance index (emPAI) for estimation of absolute protein amount in proteomics by the number of sequenced peptides per protein. Mol Cell Proteomics 4:1265–1272

Kawabe Y, Naka T, Ando-Noumi N, Matsumoto H, Ono K, Nishijima K et al (2006) Transport of human immunoglobulin G and Fc-fusion proteins to chicken egg yolk. J Biosci Bioeng 102:518–523

Keidel EM, Ribitsch D, Lottspeich F (2010) Equalizer technology–equal rights for disparate beads. Proteomics 10:2089–2098

Kitabatake N, Ishida A, Doi E (1988) Physicochemical and functional properties of hen ovalbumin dephosphorylated by acid phosphatase(food & nutrition). Agric Biol Chem 52:967–973

Kodali VK, Gannon SA, Paramasivam S, Raje S, Polenova T, Thorpe C (2011) A novel disulfide-rich protein motif from avian eggshell membranes. PLoS One 6:e18187

Koehler HH (1974) Physicochemical appraisal of changes in egg white during storage. J Agric Food Chem 22:288–293

Kovacs-Nolan J, Phillips M, Mine Y (2005) Advances in the value of eggs and egg components for human health. J Agric Food Chem 53(22):8421–8431

Lee JY, Kim CJ (2010) Determination of allergenic egg proteins in food by protein-, mass spectrometry-, and DNA-based methods. J AOAC Int 93:462–477

Li-Chan ECY, Powrie WD, Nakai S (1995) The chemistry of eggs and egg products. In: Stadelman WJ, Cotterill OJ (eds) Egg science and technology, 4th edn. The Haworth Press, New York, pp 105–175

Mann K (2007) The chicken egg white proteome. Proteomics 7:3558–3568

Mann K, Mann M (2008) The chicken egg yolk plasma and granule proteomes. Proteomics 8:178–191

Mann K, Mann M (2011) In-depth analysis of the chicken egg white proteome using an LTQ Orbitrap Velos. Proteome Sci 9:7

Mann K, Gautron J, Nys Y, McKee MD, Bajari T, Schneider WJ et al (2003) Disulfide-linked heterodimeric clusterin is a component of the chicken eggshell matrix and egg white. Matrix Biol 22:397–407

Mann K, Macek B, Olsen JV (2006) Proteomic analysis of the acid-soluble organic matrix of the chicken calcified eggshell layer. Proteomics 6:3801–3810

Mann K, Olsen JV, Maäoek B, Gnad F, Mann M (2008) Identification of new chicken egg proteins by mass spectrometry-based proteomic analysis. Worlds Poult Sci J 64:209–218

Miguel M, Aleixandre A (2006) Antihypertensive peptides derived from egg proteins. J Nutr 136:1457–1460

Miguel M, Manso MA, López-Fandiño R, Ramos M (2005) Comparative study of egg white proteins from different species by chromatographic and electrophoretic methods. Eur Food Res Technol 221:542–546

Miksik I, Sedlakova P, Lacinova K, Pataridis S, Eckhardt A (2010) Determination of insoluble avian eggshell matrix proteins. Anal Bioanal Chem 397:205–214

Mine Y (2007) Egg proteins and peptides in human health–chemistry, bioactivity and production. Curr Pharm Des 13:875–884
Mine Y, Kovacs-Nolan J (2004) Biologically active hen egg components in human health and disease. J Poult Sci 41:1–29
Mine Y, Kovacs-Nolan J (2006) New insights in biologically active proteins and peptides derived from hen egg. Worlds Poult Sci J 62:87–95
Nys Y, Gautron J, Mckee M, Garcia-Ruiz J, Hinke M (2001) Biochemical and functional characterization of eggshell matrix proteins in hens. Worlds Poult Sci J 57:401–413
Nys Y, Gautron J, Garcia-Ruiz J, Hinke M (2004) Avian eggshell mineralisation: biochemical and functional characterization of matrix proteins. Comptes Rendues Palevol 3:549–562
Omana DA, Liang Y, Kav NN, Wu J (2010) Proteomic analysis of egg white proteins during storage. Proteomics 11:144–153
Raikos V, Hansen R, Campbell L, Euston SR (2006) Separation and identification of hen egg protein isoforms using SDS-PAGE and 2-D gel electrophoresis with MALDI-TOF mass spectrometry. Food Chem 99:702–710
Rappsilber J, Ishihama Y, Mann M (2003) Stop and go extraction tips for matrix-assisted laser desorption/ionization, nanoelectrospray, and LC/MS sample pretreatment in proteomics. Anal Chem 75:663–670
Rose ML, Hincke MT (2009) Protein constituents of the eggshell: eggshell-specific matrix proteins. Cell Mol Life Sci 66:2707–2719
Rose-Martel M, Du J, Hincke MT (2012) Proteomic analysis provides new insight into the chicken eggshell cuticle. J Prot 75:2697–2706
Ruiz J, Lunam C (2000) Ultra structural analysis of the eggshell: contribution of the individual calcified layers and the cuticle to hatchability and egg viability in broiler breeders. Br Poult Sci 41:584–592
Saito Z, Martin WG (1966) Ovalbumin and other water-soluble proteins in avian yolk during embryogenesis. Can J Biochem 44:293–301
Seko A, Juneja LR, Yamamoto T (1997) Hen eggs – their basic and applied science. CRC Press, Boca Raton
Shevchenko A, Tomas H, Havlis J, Olsen JV, Mann M (2006) In-gel digestion for mass spectrometric characterization of proteins and proteomes. Nat Protoc 1:2856–2860
Shiraga H, Min W, VanDusen WJ, Clayman MD, Miner D, Terrell CH et al (1992) Inhibition of calcium oxalate crystal growth in vitro by uropontin: another member of the aspartic acid-rich protein superfamily. Proc Natl Acad Sci U S A 89:426–430
Sicherer SH, Sampson HA (2006) Food allergy. J Allergy Clin Immunol 117:S470–S475
Sparks N, Board R (1984) Cuticle, shell porosity and water intake through hen's eggshells. Br Poult Sci 25:267–276
Stevens L (1991) Egg white proteins. Comp Biochem Physiol B 100:1–9
Sugimoto Y, Hanada S, Koga K, Sakaguchi B (1984) Egg-yolk trypsin inhibitor identical to albumen ovomucoid. Biochim Biophys Acta 788:117–123
Thulasiraman V, Lin S, Gheorghiu L, Lathrop J, Lomas L, Hammond D et al (2005) Reduction of the concentration difference of proteins in biological liquids using a library of combinatorial ligands. Electrophoresis 26:3561–3571
Williams J (1962) A comparison of conalbumin and transferrin in the domestic fowl. Biochem J 83:355–364
Wilm M, Shevchenko A, Houthaeve T, Breit S, Schweigerer L, Fotsis T et al (1996) Femtomole sequencing of proteins from polyacrylamide gels by nano-electrospray mass spectrometry. Nature 379:466–469

Chapter 18
Fruits and Vegetables

Peter M. Bramley

18.1 Introduction

Studies on the protein content of fruits and vegetables have increased significantly in recent years, partly due to the advances in proteomic techniques, as described in Sect. 18.1 of this chapter. In particular, forensic investigations of fruit and vegetable products, with respect to allergens, genetic modification, and pathogen infection have been reported for a range of species, as has the use of proteomics to investigate the molecular physiology of fruit and vegetable development and ripening and post-harvest changes, with a view to improving quality and yield. Although the protein and amino acid content of fruits is low (typically ca 1% of fresh weight), proteins play key roles in fruit development, maturation, and ripening as well as protecting against damage in storage. In some fruits such as apple, avocado, and tomato, the protein content increases during ripening, reflecting the importance of certain proteins in fruit development. In those vegetables containing seeds (e.g., beans, peas, pulses), some 30% of the fresh weight can be from proteins, due to the storage proteins in seeds (Jenks and Bebeli 2011). In addition, increases in certain amino acids such as glutamic acid are responsible for the characteristic flavors of ripe fruits and vegetables. The proteomes of GM and non-GM crops are discussed in Chap. 31 of this book, and allergens are described in Chap. 30.

This review focuses upon commercially important fruits and vegetables, including tomato, citrus, pepper, peach, potato, strawberry, and pear, with respect to the molecular physiology of fruit development and ripening, post-harvest physiology and abiotic stress responses, improvements to taste and flavor, and the proteomes of different

P.M. Bramley (✉)
School of Biological Sciences, Royal Holloway, University of London,
Egham, Surrey, TW20 0EX, UK
e-mail: p.bramley@rhul.ac.uk

F. Toldrá and L.M.L. Nollet (eds.), *Proteomics in Foods: Principles and Applications*, Food Microbiology and Food Safety 2, DOI 10.1007/978-1-4614-5626-1_18,

varieties and cultivars, as well as those grown under organic and conventional horticulture. Improvements in the extraction and purification of proteins from recalcitrant fruits and vegetables have been reviewed by Song and Braun (2008).

18.2 The Molecular Physiology of Fruit Development and Ripening

Fruit development and ripening are processes unique to plants and are important in the human diet, especially from fleshy species, as it is the ripe fruit that we consume. Fruit development can generally be considered to occur in four phases: fruit set, a period of rapid cell division, a cell expansion phase, and finally ripening/maturation. Following fruit set, the cell division phase gradually shifts into the cell expansion phase. The rate and duration of cell division varies among fruits and also among tissues within a fruit. In tomato, for example, the cell division phase lasts approximately 7–10 days, whereas cell expansion lasts 6–7 weeks. Ripening occurs synchronously with seed and embryo maturation. In fleshy fruits it is designed to make the fruit appealing to animals that eat the fruit as a means for seed dispersal. Ripening involves the softening, increased juiciness and sweetness, and color changes of the fruit. Fleshy fruits are designated as either climacteric or nonclimacteric (Fig. 18.1). The former are characterized by a burst of ethylene production and increased respiration rate at the start of ripening, whereas nonclimacteric fruit ripen without the need for this phytohormone (Adams-Phillips et al. 2004). This separation is not linked to taxonomic groups, as species belonging to the same family; for example, tomato and pepper are climacteric and nonclimacteric, respectively. In addition, some fruit, such as the tomato, apple, banana, plum, chili pepper, pear, and peach can ripen off the parent plant, whereas others, such as blackberry, strawberry, cherry, lemon, and orange, cannot and must remain on the parent to enable full ripening.

Given these complexities, it is unsurprising that considerable efforts have gone into understanding the molecular physiology and regulation of these processes, especially at the genetic and epigenetic levels (Giovannoni 2004; Seymour et al. 2008). Much of this work has been aimed at crop improvement, with respect to taste, texture, nutrient content, flavor, yield, and appearance of ripe fruits in the diet. Metabolite profiling of fruit has shown the changes that occur to the metabolome during the ripening process, many of which are responsible for the characteristic color changes and smell of fresh ripe produce (e.g. Mounet et al. 2009; Zanor et al. 2009; Vogel et al. 2010). These datasets for tomato have been collated in the Tomato Functional Genomics Database (TFGD, Fei et al. 2011). More recently, proteomic studies have been reported to further our knowledge at the molecular level (reviewed by Renaut et al. 2006; Pedreschi et al. 2010; Palma et al. 2011). Taken together these experimental approaches are yielding important information about the developmental, biochemical, physiological, and structural differences that affect the development and ripening of fruit. Such knowledge should lead to agronomic improvements to crops. In this section, proteomic studies on tomato, citrus, pepper, potato, stone

Fig. 18.1 Climacteric and non-climacteric fruit of nutritional interest (Taken from Palma et al. 2011, with permission)

fruit, strawberry, and peach are discussed. A typical protocol used to investigate the proteomics of fruit ripening is shown in Fig. 18.2 (Palma et al. 2011).

18.2.1 Tomato

Tomato (*Solanum lycopersicum*) fruit ripening has been the most extensively studied of all fleshy fruits and often serves as a model for the genetic, physiology, and metabolic changes during ripening. It has been subjected to detailed genomics

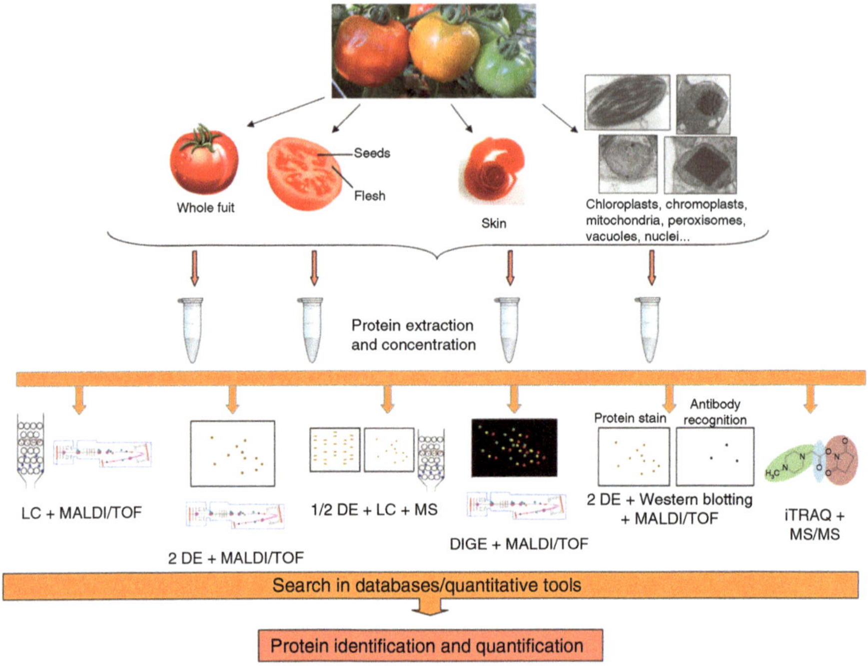

Fig. 18.2 Strategy to investigate the proteomics of fruit ripening. Fruits at different ripening stages can be studied as whole fruits, but also as separate components (flesh, seeds, skin), or at a subcellular level. Once proteins have been extracted, partially purified and characterized, they are subjected to proteomic tools. The interrogation of the results with databases leads to the identification of proteins and the use of reliable approaches can lead to the quantification of proteins, which are useful for comparative proteomic analyses (Taken from Palma et al. 2011, with permission)

(Adams-Phillips 2004; Giovannoni 2004; Seymour et al. 2008; Barone et al. 2009), transcriptomic (e.g., Baxter et al. 2005), metabolomic (Tikunov et al. 2005; Carrari and Fernie 2006; Zanor et al. 2009; DiLeo et al. 2011) and proteomic studies (e.g., Cánovas et al. 2004), and increasingly combinations of all omic technologies in a systems biology approach (Fraser et al. 2009; Mounet et al. 2009; Karlova et al. 2011; Osorio et al. 2011). There are abundant species, cultivars, GM, and introgression lines available for study, many of which are available from the Tomato Genetics Resource Center (http://tgrc.ucdavis.edu/), with the Sol Genomics Network (http://solgenomics.net/) being a repository for much of this information.

An early study by Schuch and co-workers (1989), using MALDI-TOF MS/MS studied the proteins in fruit at three stages: unripe, medium, and fully ripe. They identified two proteins, of 34 and 44 kDa that increased during ripening. These were found to be pectin esterase and heterotrimeric GTP binding protein fragment. These proteins were suggested as ripening-specific protein markers.

Using more advanced techniques, Kok and co-workers (2008) identified 655 protein spots by 2DE, with 53 changing during fruit ripening. Of these, 26 increased, and the remainder decreased. As expected, gene expression changes preceded those of protein synthesis. A detailed study of the cherry tomato proteome throughout ripening (six stages) revealed 1,791 spots, with 8% being variable throughout fruit development and ripening (Faurobert et al. 2007). Clustered correlation analysis identified groups of proteins with similar profiles. Proteins associated with amino acid metabolism and protein synthesis were found to increase during cell division and then decrease. During cell expansion, those related to photosynthesis and cell wall formation increased transiently, whereas during fruit development and ripening there was an increase in proteins linked to carbon compounds, carbohydrate metabolism, and oxidative processes. In addition, stress proteins and those linked to fruit senescence were increased. Therefore, this study was able to prove the identity of proteins that would be expected to change during fruit development and ripening, based upon the phenotype of the fruit. A more detailed list is shown in Table 18.1. The presence of stress proteins was also found in ripe fruit of two tomato ecotypes (Rocco et al. 2006). Proteome changes due to abiotic stress are discussed in Sect. 18.3, below.

During ripening, increases in free amino acids are found in tomato fruit, especially in cultivated varieties, with glutamate being the principal free amino acid (Boggio et al. 2000). Free glutamic acid is a major factor in the so-called "umami" (savoriness) taste of foods. Higher levels of glutamate have been detected in ripe tomato fruit, with concomitant increases in the aminating activity of glutamate dehydrogenase (α-ketogutarate to glutamate) and of 4-amino butyrate amino transferase (Sorrequieta et al. 2010).

Proteomic studies on chromoplasts from ripe fruit have been successful in identifying proteins that are specific to the chromoplast. Barsan et al. (2010) identified 998 proteins, 209 of which had not been previously listed in plastidial databases. Of these, all the lipoxygenase family was identified, together with those for chlorophyll degradation, lipid metabolism, and trafficking. Perhaps surprisingly, the enzymes of the Calvin cycle and oxidative pentose phosphate pathway were also found.

18.2.2 Pepper

There have been limited studies on protein changes during pepper ripening. Lee and co-workers (2006) investigated protein changes in the placental tissue of pungent and nonpungent fruit during ripening, using 2DE and LC MS/MS. They found 37 of 2,600 protein spots were exclusively in pungent fruit or increased significantly in these fruit during ripening. Of these, 22 were identified and grouped into eight classes, including enzymes for capsaicin biosynthesis, fatty acid metabolism, the Calvin cycle, and defense mechanisms.

One hundred and fifty-one proteins of ripe chromoplasts of bell pepper have been identified by MS/MS. These include carbohydrate and amino acid metabolism enzymes. Two proteins directly associated with carotenoid synthesis in the ripening fruit were found: capsanthin–capsorubin synthase and fibrillin (Siddique et al. 2006).

Table 18.1 Examples of changes to protein levels during fruit ripening

Fruit and protein category	Protein	Change
Tomato (Faurobert et al. 2007)		
C compounds and CHO metabolism	Acid β–fructo furanosidase	+
	α-Galactosidase	+
Photosynthesis/respiration	PSII O_2 evolving complex	–
	3-Phosphoglycerate dehydrogenase	–
	$NADP^+$-Specific isocitrate dehydrogenase	+
Protein synthesis/storage	eIF-5A 4	+, then–
Protein fate/degradation	Peptide Met sulfoxide reductase E4	+
	Cysteine protease	+
Secondary metabolism	Similar to lipocalin	+
Stress response	HSP 20, HSc 70, sHSP	+
Detoxification	Ascorbate peroxidase	+
Electron transport	Ripening protein E8	+
Transport	Mitochondrial ATPase β subunit	–
	Vacuolar H^+ ATPase	–
Citrus (Yun et al. 2010)		
Carbon compounds	Aconitase	+
	Alcohol dehydrogenase	+
	6-Phosphogluconate dehydrogenase	+
Amino acid metabolism	Chorismate synthase 1	–
	Cysteine synthase	–
Protein translation	β-Tubilin	+
	Translation initiation factor 1A	–
Stress	HSP 70	+
Strawberry (Bianco et al. 2009)		
Stress	HSP 60, 70	+
C metabolism and respiration	Triose phosphate isomerase	–
	Alcohol dehydrogenase	+
Secondary metabolism	Chalcone synthase and isomerase	+
	Flavonone 3-hydroxylase	+
	Naringenin 7-glucoside synthase	+
Protein synthesis	eIF5A	+

+ increase, – decrease, *CHO* carbohydrate, *HSP* heat stress protein, *HSc* heat shock cognate protein

18.2.3 Potato

A comprehensive study on the potato tuber proteome over the course of tuberization, tuber development, and sprouting revealed five protein expression patterns, with 150 proteins showing highly significant differences in abundance over the period. Another 50 proteins showed smaller changes. From the 150, 59 were identified. Tuber development was found to be accompanied by the accumulation of patatin isoforms as well as defense and disease proteins. Proteins associated with energy production, carbohydrate metabolism, and protein processing decreased in maturation (Lehesranta et al. 2006).

18.2.4 Stone Fruit

In order to assess the correct state of harvest for stone fruits such as plums, peaches, and nectarines, protein that increase as the fruit mature were assessed by Abdi and colleagues (2002). The consequences of an incorrect judgment of maturity are economically serious. Harvesting too early results in a significant loss of yield, as stone fruit grow rapidly during the last few days of maturation and often do not develop acceptable flavors. Harvesting too late, however, results in the fruit having a short shelf-life. Four proteins, named Z1, Z2, Y, and X were found to appear just before optimum harvest and were shown to be a more accurate reflection of the correct harvesting time than more traditional measures such as skin color and softening. The authors suggest that measurement of these proteins could be used in a test kit by horticulturists.

18.2.5 Strawberry

A proteome reference map of strawberry at various stages of ripening has been created using a wide range of high-throughput proteomic techniques, including 2DE, MudPIT, and DIGE (Bianco et al. 2009). It shows a wide range of biological processes that change during ripening, including energy and carbohydrate metabolism, signal transduction, stress responses, and transcription, concomitant with constitutive and differentially expressed proteins. The authors have related these changes and the functions of the identified proteins to quality traits such as flavor, antioxidant levels, and macronutrients.

18.2.6 Citrus

Using label-free LC MS/MS, Katz and co-workers (2010, 2011) were able to identify changes to key metabolic enzymes during citrus development. The level of invertase remained constant, but an invertase inhibitor was upregulation during fruit maturation. In addition, sucrose phosphate synthase and sucrose 6-phosphate phosphatase were increased, causing higher levels of sucrose in the juice sac in late fruit development. Protein profiles of the Moro and Cadenera cultivars at ripening revealed 64 differentially expressed proteins on 2DE (Muccilli et al. 2009). Most of these were related to sugar metabolism and increased in the Moro cultivar, whereas stress-related proteins were more abundant in Cadenera. Enzymes of the anthocyanin biosynthetic pathway were also more prominent in Moro, which has a dark red fruit.

18.3 Proteome Changes Due to Abiotic Stress

In order to survive under environmental stresses, plants rely upon the plasticity of the proteome to maintain the functionality of the cell. These stresses can be biotic or abiotic, with the latter including drought, salinity, high and low temperatures, freezing, and chemical toxins such as heavy metals. In extreme cases, crop yields can be severely affected, with reductions of up to 75%.

Families of proteins associated with the plant's response to stress have been identified. They may be newly synthesized, accumulate in the cells, or even decrease in response to abiotic stresses. They are involved with a variety of processes, including signaling, host defense mechanisms, carbohydrate metabolism, amino acid metabolism, and translation. A number of these proteins are families of heat shock (heat stress) proteins (HSPs; Wang et al. 2004; Timperio et al. 2008; Takáč et al. 2011). An understanding of these events is of practical and basic importance, as it may lead to improvements in plant defenses (Afroz et al. 2011) and reduce yield losses (Salekdeh and Komatsu 2007). In addition, the genes of stress-related proteins that have specific beneficial functions could be transformed into other crops.

Examples of the types of proteins that have been identified as stress-related are shown in Table 18.2. It seems that plants adapt in similar ways to different stresses, but are expressed in different parts of the cell (Fig. 18.3). One frequent effect is the production of HSPs that are induced by several environmental stresses. The term HSP is somewhat misleading, as they are also produced in response to nonthermal stress. It is likely that under these circumstances, their roles are different from their molecular chaperone role in thermal responses (Timperio et al. 2008). As proteomic technologies advance, the number of proteins found to change on abiotic stress increases. It remains to be seen whether all of these changes are linked by a common regulator.

18.4 Post-Harvest Physiology: Changes to the Proteome During Storage

Post-harvest physiology refers to effects of handling and storage of horticultural produce following harvest. Between harvest and consumption of horticultural crops, especially soft fruits, losses in quality and quantity occur. In developed countries, these can be 5–25%, but in undeveloped countries, they are typically between 20% and 50% (Kader 2002). Biological factors that influence such losses include respiration, ethylene production, compositional changes, growth and development, transpiration, physiological breakdown, and biotic stresses caused by plant pathogens. Therefore, storage conditions must minimize these deleterious changes to extend the lifespan of the produce without affecting nutritional quality and taste. Such storage conditions vary between species. The use of proteomics in this field is relatively new, but is being used increasingly to study the influence of protein profiles on the

Table 18.2 Examples of stress-related proteins that accumulate in higher plants subjected to abiotic stress

Abiotic stress	Species	Protein	Reference
Light	*Arabidopsis*	HSP70	Rossell et al. 2002; Kimura et al. 2003
	Rice	HSPs	Yang et al. 2007
	Panax ginseng	smHSPs	Nam et al. 2005
Salt	Rice	GST, enolase	Salekdeh et al. 2002; Dooki et al. 2006; Nohzadeh Malakshah et al. 2007
	Lemna minor	HSP70	Ireland et al. 2004
High temperature	Rice	High and low Mr HSPs	Miernyk 1997; Lee et al. 2007
	Tomato	HSPC1, C2, C3, C4	Polenta et al. 2007
Cold	Rice	85 proteins	Yan et al. 2006
	Pea	20 proteins	Taylor et al. 2005
	Poplar	COR proteins	Renaut et al. 2004
	Wheat	27 proteins	Kamal et al. 2010
	Grass pea	15 proteins	Chattopadhyay et al. 2011
Freezing			
	Rye	Antifreeze proteins	Marentes et al. 1993
Drought	Chestnut	smHSPs, HSP70	Lopez-Matas et al. 2004
	Wheatgrass	LEA proteins, osmotin	Gazanchian et al. 2007
	Sugar beet	HSP17	Hajheidari et al. 2005
	Medicago truncatala	Methionine synthase	Larrainzar et al. 2007
	Arabidopsis	HSP70	Lin et al. 2001
Heavy metals	Rice	192 proteins	Pandey et al. 2010
	Lemna minor	HSP70	Ireland et al. 2004
	Arabidopsis	HSP70, HSP60	Sarry et al. 2006
	Rice	HSP70, HchA	Ahsan et al. 2007

HSP heat shock protein, *smHSP* small HSP, *COR* cold regulated, *GST* glutathione S transferase, *ASR1* abscisic acid stress ripening, *LEA* late embryogenesis abundant, *HchA* chaperone protein

quality and longevity of harvested fruits and vegetables. Generally, qualitative and quantitative changes are found in proteins, both enzymic and nonenzymic, involved in respiration, protein synthesis and degradation, and defense mechanisms to respond to stress. In this respect some of the changes are similar to those described in Sect. 18.3, above, on abiotic stress.

A proteomic study on the juice sacs of *Citrus reticulate* using 2DE and MALDI/TOF/TOF revealed changes to 74 proteins when stored post-harvest at 4°C (Yun et al. 2010). Of these, 27 were cytosolic and the majority related to changes in dehydration and respiration. Under anaerobic conditions, the fruit showed changes of at least 1.5-fold in 33 proteins. Of these in the peel, 64% were stress related with 6% linked to energy production, whereas in the juice, 38% of those affected were from energy-producing pathways (Shi et al. 2008). These changes are presumably responsible for the off-odors of acetaldehyde and ethanol in deteriorating fruit.

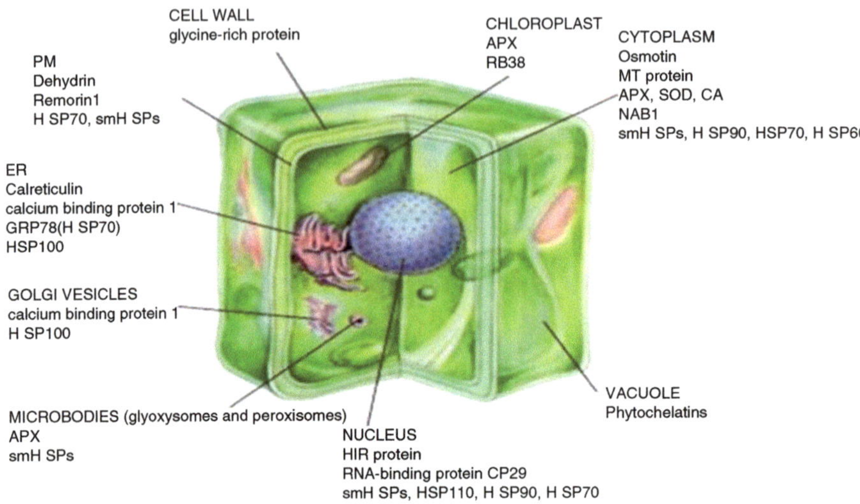

Fig. 18.3 Localization of the major stress-related proteins in the plant cell. Protein belonging to the specific class and those belonging to the common response pathway upon stresses are listed together (Taken from Timperio et al. 2008, with permission)

Kiwi fruit are often stored in an atmosphere of ozone in order to reduce microbial spoilage, respiration, and ethylene oxidation. Using 1D SDS-PAGE and mass spectrometry Minas et al. (2012) have shown that such an atmosphere causes alteration to 102 proteins, induces protein carbonylation, stimulates antioxidant and antifree radical activities, and improves post-harvest quality. Changes to O_2/CO_2 concentrations for storing pears causes impaired respiration and activation of defense mechanisms, as shown by 2DE LC ESI MS/MS (Pedreschi et al. 2008). Valencia oranges, on the other hand, are heat treated post harvest to induce defense mechanisms such as lignin biosynthesis and phytoalexin formation. These reduce chilling injury. Analysis of proteins post heat treatment found 61 changes to protein levels, with 28 of these identified by MALDI/TOF. They comprised proteins involved in defense, cell rescue, and virulence (Perotti et al. 2011). Peaches are also heat treated in this way to enhance defense mechanisms. Proteomic analysis found 35 different post-harvest expressed proteins, of which 43% were stress responses, 17 cell structure, 7 the glycolytic pathway, and 3 ripening and senescence (Zhang et al. 2011). The chilling of peach fruit has been studied using DIGE. Several proteins, including polygalacturonase, dehydrins, pectin methyl esterase, catalase, and NADP-dependent isocitrate dehydrogenase were used to distinguish healthy from chill-damaged fruit (Nilo et al. 2010).

Chilling tomatoes post harvest affects the cell wall proteins by increasing β–galactosidase and reducing polygalacturonase protein levels (Rugkong et al. 2010). Page and co-workers (2010) have investigated the chilling tolerance of two tomato lines. They found that the firm genotype at harvest was the most sensitive to cold storage. Cold storage increased smHSPs and freezing tolerance, but decreased acid invertase and polygalacturonase.

There are fewer studies of this type reported for vegetables. However, Owiti and co-workers (2011) have studied changes to the cassava proteome. Cassava has a very short storage life of some 24–48 h after harvest (Sanchez et al. 2006). Extending this life span would increase the value of the crop significantly. Owiti et al. (2011) have produced a proteome map and analyzed the quantitative increases in proteins post harvest using iTRAQ. They discovered 67 and 170 proteins increased early and late after harvest, respectively. They included linamarase (β-D-glucosidase), glutamate-rich protein, β-1, 3-glucanase, pectin methyl esterase, and proteins of signaling pathways. This variety suggests highly regulated and complex metabolic changes that, once understood, could be used in biotechnology to reduce deterioration. iTRAQ was also used to identify changes to the proteome of potato tubers during cold storage over a five-month period (Yang et al. 2011). Of the 4,463 proteins identified, only 46 showed differential changes (±1.41) during storage at 5°C. These included enzymes involved in starch breakdown and sucrose formation (α-amylase, sucrose synthase) as well as chaperonins and HSPs.

18.5 Proteome Differences Between Cultivars and Varieties

Proteomics has the potential to be used as a screen for identifying the differences between fruits or vegetables from different species, wild relatives, ecotypes, varieties, cultivars, and ecotypes. To do so requires robust databases of protein profiles from crops that enable diagnostic proteins to be identified and quantified. In turn, the differences can assist in the discovery of the genetic diversity of species and so on, with a view to breeding key traits, such as disease and pest resistance, into elite lines.

A good example of this approach is the potato. Potato has a very large number of commercial cultivars (Bradshaw and Mackay 1994). Most of these are tetraploids of *Solanum tuberosum*, but many have commercially desirable characteristics introgressed from wild species. In order to distinguish the differences at the protein level and identify mutations, 2DE of several genotypes (landraces, varieties, and GM lines) was used to create a reference map of the potato tuber (Lehesranta et al. 2005). Thirty-two non-GM potato lines, comprising 21 named tetraploid cultivars, 8 landraces, and 3 diploids were analyzed. Over 1,900 polypeptides from all genotypes were found, with significant qualitative and quantitative differences between them. The diploid genotypes could be distinguished from the tetraploids by PCA and large differences in the isoforms of patatin were found. HPLC ESI MS/MS was used to identify many proteins that contribute to diseases and defense responses, the glycolytic pathway, sugar metabolism, and protein storage.

A smaller study, on two ecotypes of tomato, has also shown differences in the proteomes. Fruit from the cultivars Ailsa Craig and San Marzano were harvested at three stages of maturation and their protein profiles compared by 2DE, followed by PMF MS/MS. At the breaker stage of fruit ripening there was a 57% overlap in their proteomes, showing a partial conservation of the genotype, but also allowing them to be distinguished from each other (Rocco et al. 2006).

A third example is that of peach cultivars. Fruit mesocarp from two cultivars, one melting flesh and one nonmelting flesh, was analyzed by 2DE and LC-ESI MS/MS during fruit ripening. Differences in 53 protein spots were found between the two cultivars at the same ripening stages, of which 35 were identified. Sucrose synthase and α–amylase were increased in the nonmelting flesh cultivar, suggesting this was the cause of the higher tissue integrity and functionality. Such a discovery could help in marker-assisted breeding programs (Prinsi et al. 2011).

18.6 The Proteomes of Crops Grown by Organic and Conventional Cultivation

The debate on the possible benefits of organically grown crops over those cultivated conventionally has spurred a number of studies on the comparative proteomics of varieties grown under both regimes. Lester and Saftner (2011) report that organic crops often have higher dry matter, ascorbic acid, phenolic, and sugar, but lower moisture, nitrate, and protein contents and yields than conventionally grown crops. There have been relatively few reports on the comparative proteomics of crops grown under conventional and organic methods. Lehesranta et al. (2007) quantified the effects of different farming systems with organic, low input, and conventional production and key components of the systems (fertility management, crop protection practices, and rotational designs) on the protein profiles of potato tubers, determined by 2DE and subsequent protein identification using HPLC-ESI-MS/MS. Principal component analysis revealed that only fertility management practices (organic vs. mineral fertilizer) had a significant effect of protein composition. Of the 1,100 tuber proteins separated by 2DE, only 160 were quantifiably different, with 17 at higher levels in tubers grown conventionally and 143 more abundant in the organic fertilizer regime. Those identified by MS are involved in protein synthesis and turnover, carbon and energy metabolism, and defense responses, suggesting that organic fertilization leads to increased stress responses in the potato tuber. These differences, however, are less than those found in an earlier study on different varieties (Lehesranta et al. 2005).

Comparisons of cabbage and carrot proteomes from crops gown under conventional and two organic systems (slurry and green manure) also showed some differences in the amounts of proteins in the head and root, respectively. However, in both cases the number of such proteins was low, with 58 of 1,300 and 68 of 1,800 being different in cabbage and carrot, respectively (Nawrocki et al. 2011). However, the studies on potato, carrot, and cabbage do show that proteome analysis can accurately identify differences between such cropping regimes.

Protein profiles of wheat grains, grown on organic and conventional plots, showed differences in 25 proteins of which 16 are candidates for diagnostic proteins to verify the authenticity of organic grains (Zorb et al. 2009).

18.7 Future Perspectives

There is no doubt that the technological advances in proteomics have been used to good effect with fruits and vegetables, allowing us to understand better the molecular changes that occur during fruit and vegetable development and ripening, as well as the more deleterious changes to the proteome from post-harvest storage and abiotic stress. It is now important to utilize a systems approach to study these events. Combining proteomics with metabolomics and genomics will do this, even for the more intractable crops. In addition, further improvements to protein extraction, separation, and identification will expand our databases of plant proteomes. In turn, these studies will allow plant breeding of crops to be improved for yield, flavor, and nutritional content.

References

Abdi N, Holford P, McGlasson B (2002) Application of two-dimensional gel electrophoresis to detect proteins associated with harvest maturity in stone fruit. Postharvest Biol Technol 26:1–13

Adams-Phillips L, Barry C, Giovannoni J (2004) Signal transduction systems regulating fruit ripening. Trends Plant Sci 9:331–338

Afroz A, Ali GM, Mir A, Komatsu S (2011) Application of proteomics to investigate stress-induced proteins for improvement in crop protection. Plant Cell Rep 30:745–763

Ahsan N, Lee DG, Lee SH, Kang KY, Lee JJ, Kim PJ, Yoon HS, Kim JS, Lee BH (2007) Excess copper induced physiological and proteomic changes in germinating rice seeds. Chemosphere 67:1182–1193

Barone A, Chiusano ML, Ercolano MR, Giuliano G, Grandillo S, Frusciante L (2009) Structural and functional genomics of tomato. Intl J Plant Genet 2008:1–12

Barsan C, Sanchez-Bel P, Rombaldi C, Egea I, Rossignol M, Kuntz M, Zouine M, Latche A, Bouzayen M, Pech J-C (2010) Characteristics of the tomato chromoplast revealed by proteomic analysis. J Exp Bot 61:2413–2431

Baxter CJ, Sabar M, Quick WP, Sweetlove LJ (2005) Comparison of changes in fruit gene expression in tomato introgression lines provides evidence of genome-wide transcriptional changes and reveals links to mapped QTLs and described traits. J Exp Bot 56:1591–1604

Bianco L, Lopez L, Scalone AG, Di Carli M, Desiderio A, Benvenuto E, Perrotta G (2009) Strawberry proteome characterization and its regulation during fruit ripening and in different genotypes. J Proteomics 72:586–607

Boggio SB, Palatnik JF, Heldt HW, Valle EM (2000) Changes in amino acid composition and nitrogen metabolizing enzymes in ripening fruits of *Lycopersicon esculentum* Mill. Plant Sci 159:125–133

Bradshaw JE, Mackay GR (1994) Breeding strategies for clonally propagated potatoes. In: Bradshaw JE, Mackay GR (eds) Potato genetics. CAB International Press, Wallingford, pp 467–497

Cánovas FM, Dumas-Gaudot E, Recorbet G, Jorrin J, Mock H-P, Rossignol M (2004) Plant proteome analysis. Proteomics 4:285–298

Carrari F, Fernie AR (2006) Metabolic regulation underlying tomato fruit development. J Exp Bot 57:1883–1897

Chattopadhyay A, Subba P, Pandey A, Bhushan D, Kumar R, Datta A, Chakraborty S, Chakraborty N (2011) Analysis of the grasspea proteome and identification of stress-responsive proteins upon exposure to high salinity, low temperature, and abscisic acid treatment. Phytochemistry 72:1293–1307

DiLeo MV, Strahan GD, den Bakker M, Hoekenga OA (2011) Weighted correlation network analysis (WGCNA) applied to the tomato fruit metabolome. PLoS ONE 6:e26683

Dooki AD, Mayer-Posner FJ, Askari H, Zaiee AA, Salekdeh GH (2006) Proteomic responses of rice young panicles to salinity. Proteomics 6:6498–6507

Faurobert M, Mihr C, Bertin N, Pawlowski T, Negroni L, Sommerer N, Causse M (2007) Major proteome variations associated with cherry tomato pericarp development and ripening. Plant Physiol 143:1327–1346

Fei Z, Joung J-G, Tang X, Zheng Y, Huang M, Lee JM, McQuinn R, Tieman DM, Alba R, Klee HJ, Giovannoni JJ (2011) Tomato functional genomics database: a comprehensive resource and analysis package for tomato functional genomics. Nucl Acids Res 39:D1156–D1163

Fraser PD, Enfissi EM, Bramley PM (2009) Genetic engineering of carotenoid formation in tomato fruit and the potential application of systems and synthetic biology approaches. Archiv Biochem Biophys 483:196–204

Gazanchian A, Hajheidari M, Khoshkholgh Sima N, Hosseini SG (2007) Proteome response of *Elymus elongatum* to severe water stress and recovery. J Exp Bot 58:291–300

Giovannoni JJ (2004) Genetic regulation of fruit development and ripening. Plant Cell 16:S170–S180

Hajheidari M, Abdollahian-Noghabi M, Askari H, Heidari M, Sadeghian SY, Ober ES, Salekdeh GH (2005) Proteome analysis of sugar beet leaves under drought stress. Proteomics 5:950–960

Ireland HE, Harding SJ, Bonwick GA, Jones M, Smith CJ, Williams JHH (2004) Evaluation of heat shock protein 70 as a biomarker of environmental stress in *Fucus serratus* and *Lemna minor*. Biomarkers 9:139–155

Jenks MA, Bebeli PJ (2011) Breeding for fruit quality. Wiley-Blackwell, Oxford

Kader AA (2002) Post harvest technology of horticultural crops, 3rd edn. University of California Agriculture and Natural Resources, Oakland

Kamal AHM, Kim KH, Shin KH, Choi JS, Baik BK, Tsujimoto H, Heo HY, Park CS, Woo SH (2010) Abiotic stress responsive proteins of wheat grain determined using proteomics technique. Aust J Crop Sci 4:196–208

Karlova R, Rosin FM, Busscher-Lange J, Parapunova V, Do PT, Fernie AR, Fraser PD, Baxter C, Angenent GC, de Maagd RA (2011) Transcriptome and metabolite profiling show that APETALA2a is a major regulator of tomato fruit ripening. Plant Cell 23:923–941

Katz E, Fon M, Eigenheer RA, Phinney BS, Fass JN, Lin D, Sadka A, Blumwald E (2010) A label-free differential quantitative mass spectrometry method for the characterization and identification of protein changes during citrus fruit development. Proteome Sci 8:68–87

Katz E, Hwan Boo K, Youn Kim H, Eigenheer RA, Phinney BS, Shulaev V, Negre-Zakharov F, Sadka A, Blumwald E (2011) Label-free shotgun proteomics and metabolite analysis reveal a significant metabolic shift during citrus fruit development. J Exp Bot 62:5367–5384

Kimura M, Yamamoto YY, Seki M, Sakurai T, Sato M, Abe T, Yoshida S, Manabe K, Shinozaki K, Matsui M (2003) Identification of *Arabidopsis* genes regulated by high light–stress using cDNA microarray. Photochem Photobiol 77:226–233

Kok EJ, Lehesranta SJ, van Dijk JP, Helsdingen JR, Dijksma WTP, Van Hoef AMA, Koistinen KM, Karenlampi SO, Kuiper HA, Keijer J (2008) Changes in gene and protein expression during tomato ripening – consequences for the safety assessment of new crop plant varieties. Food Sci Technol Int 14:503–518

Larrainzar E, Wienkoop S, WeckwerthW LR, Arrese-Igor C, González M (2007) *Medicago truncatula* root nodule proteome analysis reveals differential plant and bacteroids responses to drought stress. Plant Physiol 144:1495–1507

Lee JM, Kim S, Lee JY, Yoo EU, Cho MC, Cho MR, Kim B-D, Bahk YY (2006) A differentially expressed proteomic analysis in placental tissues in relation to pungency during the pepper fruit development. Proteomics 6:5248–5259

Lee D-G, Ahsan N, Lee S-H, Kang KY, Bahk JD, Lee I-J, Lee B-H (2007) A proteomic approach in analyzing heat-responsive proteins in rice leaves. Proteomics 7:3369–3383

Lehesranta SJ, Davies HV, Shepherd LVT, Nunan N, McNicol JW, Auriola S, Koistinen KM, Suomalainen S, Kokko HI, Karenlampi SO (2005) Comparison of tuber proteomes of potato varieties, landraces, and genetically modified lines. Plant Physiol 138:1690–1699

Lehesranta SJ, Davies HV, Shepherd LVT, Koistinen KM, Massat N, Nunan N, McNicol JW, Kärenlampi SO (2006) Proteomic analysis of the potato tuber life cycle. Proteomics 6:6042–6052

Lehesranta SJ, Koistinen KM, Massat N, Davies HV, Shepherd LVT, McNicol JW, Cakmak I, Cooper J, Luck L, Karenlampi SO, Leifert C (2007) Effects of agricultural production systems and their components on protein profiles of potato tubers. Proteomics 7:597–604

Lester GE, Saftner RA (2011) Organically versus conventionally grown produce: common production inputs, nutritional quality, and nitrogen delivery between the two systems. J Agric Food Chem 59:10401–10406

Lin BL, Wang JS, Liu HC, Chen RW, Meyer Y, Barakat A, Delseny M (2001) Genomic analysis of the Hsp70 superfamily in *Arabidopsis thaliana*. Cell Stress Chaperones 16:201–208

Lopez-Matas MA, Nunez P, Soto A, Allona I, Casado R, Collada C, Guevara M-A, Aragoncillo C, Luis GL (2004) Protein cryoprotective activity of a cytosolic small heat shock protein that accumulates constitutively in chestnut stems and is up-regulated by low and high temperatures. Plant Physiol 134:1708–1717

Marentes E, Griffith M, Mlynarz A, Brush RA (1993) Proteins accumulate in the apoplast of winter rye leaves during cold acclimation. Physiol Plant 87:499–507

Miernyk JA (1997) The 70 kDa stress-related proteins as molecular chaperones. Trends Plant Sci 2:80–87

Minas IS, Tanou G, Belghazi M, Job D, Manganaris GA, Molassiotis A, Vasilakakis M (2012) Physiological and proteomic approaches to address the active role of ozone in kiwifruit postharvest ripening. J Exp Bot (in press)

Mounet F, Moing A, Garcia V, Petit J, Maucourt M, Deborde C, Bernillon S, Le Gall G, Colquhoun I, Defernez M, Giraudel L, Rolin D, Rothan C, Lemaire-Chamley M (2009) Gene and metabolite regulatory network analysis of early developing fruit tissues highlights new candidate genes for the control of tomato fruit composition and development. Plant Physiol 109:1505–1528

Muccilli V, Licciardell C, Fontanini D, Russo DP, Cunsolo V, Saletti R, Recupero GR, Foti S (2009) Proteome analysis of *Citrus sinensis* L. (Osbeck) flesh at ripening time. J Proteomics 73:134–152

Nam MH, Kim SI, Liu JR, Yang DC, Lim YP, Kwon KH, Yoo JS, Park YM (2005) Proteomic analysis of Korean ginseng (*Panax ginseng* C.A. Meyer). J Chromatogr B 815:147–155

Nawrocki A, Thorup-Kristensen K, Jensen ON (2011) Quantitative proteomics by 2DE and MALDI MS/MS uncover the effects if organic and conventional cropping methods on vegetable products. J Proteomics 74:2810–2825

Nilo R, Saffie C, Lilley K, Baeza-Yates R, Cambiazo V, Campos-Vargas R, González M, Meisel LA, Retamales J, Silva H, Orellana A (2010) Proteomic analysis of peach fruit mesocarp softening and chilling injury using difference gel electrophoresis (DIGE). BMC Genet 11:43–63

Nohzadeh Malakshah S, Habibi Rezaei M, Heidari M, Salekdeh GH (2007) Proteomic reveals new salt responsive proteins associated with rice plasma membrane. Biosci Biotechnol Biochem 71:2144–2154

Osorio S, Alba R, Damasceno CMB, Lopez-Casado G, Lohse M, Zanor MI, Tohge T, Usadel B, Rose JKC, Fei Z, Giovannoni JJ, Fernie AR (2011) Systems biology of tomato fruit development: combined transcript, protein, and metabolite analysis of tomato transcription factor (nor, rin) and ethylene receptor (Nr) mutants reveals novel regulatory interactions. Plant Physiol 157:405–425

Owiti J, Grossmann J, Gehrig P, Dessimoz C, Laloi C, Hansen MB, Gruissem W, Vanderschuren H (2011) iTRAQ-based analysis of changes in the cassava root proteome reveals pathways associated with post-harvest physiological deterioration. Plant J 67:145–156

Page D, Gouble B, Valot B, Bouchet JP, Callot C, Kretzschmar A, Causse M, Renard CMCG, Faurobert M (2010) Protective proteins are differentially expressed in tomato genotypes differing for their tolerance to low-temperature storage. Planta 232:483–500

Palma JM, Corpas FJ, del Río LA (2011) Proteomics as an approach to the understanding of the molecular physiology of fruit development and ripening. J Proteomics 74:1230–1243

Pandey A, Rajamani U, Verma J, Subba P, Chakraborty N, Datta A, Chakraborty S, Chakraborty N (2010) Identification of extracellular matrix proteins of rice (*Oryza sativa* L.) involved in dehydration-responsive network: a proteomic approach. J Proteome Res 9:3443–3464

Pedreschi R, Hertog M, Robben J, Noben J-P, Nicolaï B (2008) Physiological implications of controlled atmosphere storage of 'Conference' pears (*Pyrus communis* L.): a proteomic approach. Postharvest Biol Technol 50:110–116

Pedreschi R, Hertog M, Lilley KS, Nicolas B (2010) Proteomics for the food industry: opportunities and challenges. Crit Rev Food Sci Nutr 50:680–692

Perotti VE, del Vecchio HA, Sansevich A, Meier G, Bello F, Cocco M, Garrán SG, Anderson C, Vázquez D, Podestá FE (2011) Proteomic, metabolomic, and biochemical analysis of heat treated Valencia oranges during storage. Postharvest Biol Technol 62:97–114

Polenta GA, Calvete JJ, González CB (2007) Isolation and characterization of the main small heat shock proteins induced in tomato pericarp by thermal treatment. FEBS J 274:6447–6455

Prinsi B, Negri AS, Fedeli C, Morgutti S, Negrini N, Cocucci M, Espen L (2011) Peach fruit ripening: a proteomic comparative analysis of the mesocarp of two cultivars with different flesh firmness at two ripening stages. Phytochemistry 72:1251–1262

Renaut J, Lutts S, Hoffmann L, Hausman JF (2004) Responses of poplar to chilling temperatures: proteomic and physiological aspects. Plant Biol 6:81–90

Renaut J, Hausman JF, Wisniewski ME (2006) Proteomics and low-temperature studies: bridging the gap between gene expression and metabolism. Physiol Plant 126:97–109

Rocco M, D'Ambrosio C, Arena S, Faurobert M, Scaloni A, Marra M (2006) Proteomic analysis of tomato fruits from two ecotypes during ripening. Proteomics 6:3781–3791

Rossell JB, Wilson IW, Pogson BJ (2002) Global changes in gene expression in response to high light in *Arabidopsis*. Plant Physiol 130:1109–1120

Rugkong A, Rose JKC, Lee SJ, Giovannoni JJ, O'Neill MA, Watkins CB (2010) Cell wall metabolism in cold-stored tomato fruit. Postharvest Biol Technol 57:106–113

Salekdeh GH, Komatsu S (2007) Crop proteomics: aim at sustainable agriculture of tomorrow. Proteomics 7:2976–2996

Salekdeh GH, Siopongco J, Wade LJ, Ghareyazie B, Bennett J (2002) Proteomic analysis of rice leaves during drought stress and recovery. Proteomics 2:1131–1145

Sanchez T, Chavez AL, Ceballos H, Rodriguez-Amaya DB, Nestel P, Ishitani M (2006) Reduction or delay of post-harvest physiological deterioration in cassava roots with higher carotenoid content. J Sci Food Agric 86:634–639

Sarry JE, Kuhn L, Ducruix C, Lafaye A, Junot C, Hugouvieux V, Jourdain A, Bastien O, Fievet JB, Vailhen D, Amekraz B, Moulin C, Ezan E, Garin J, Bourguignon J (2006) The early responses of *Arabidopsis thaliana* cells to cadmium exposure explored by protein and metabolite profiling analyses. Proteomics 6:2180–2198

Schuch W, Bird CR, Ray J, Smith CJS (1989) Control and manipulation of gene expression during tomato fruit ripening. Plant Mol Biol 13:303–311

Seymour G, Poole M, Manning K, King GJ (2008) Genetics and epigenetics of fruit development and ripening. Curr Opin Plant Biol 11:58–63

Shi JX, Chen S, Gollop N, Goren R, Goldschmidt EE, Porat R (2008) Effects of anaerobic stress on the proteome of citrus fruit. Plant Sci 175:478–486

Siddique MA, Grossmann J, Gruissem W, Baginsky S (2006) Proteome analysis of bell pepper (*Capsicum annuum* L.) chromoplasts. Plant Cell Physiol 47:1663–1673

Song J, Braun G (2008) Application of proteomic techniques to fruits and vegetables. Curr Proteomics 5:191–201

Sorrequieta A, Ferraro G, Boggio SB, Valle EM (2010) Free amino acid production during tomato fruit ripening: a focus on L-glutamate. Amino Acids 10:1523–1532

Takáč T, Pechan T, Šamaj J (2011) Differential proteomics of plant development. J Proteomics 74:577–588

Taylor NL, Heazlewood JL, Day DA, Millar AH (2005) Differential impact of environmental stresses on the pea mitochondrial proteome. Mol Cell Proteomics 4:1122–1133

Tikunov Y, Lommen A, de Vos CHR, Verhoeven HA, Bino RJ, Hall RD, Bovy AG (2005) A novel approach for nontargeted data analysis for metabolomics. Large-scale profiling of tomato fruit volatiles. Plant Physiol 139:1125–1137

Timperio AM, Egidi MG, Zolla L (2008) Proteomics applied on plant abiotic stresses: role of heat shock proteins (HSP). J Proteomics 71:391–411

Vogel JT, Tieman DM, Sims CA, Odabasi AZ, Clark DG, Klee HJ (2010) Carotenoid content impacts flavour acceptability in tomato (*Solanum lycopersicum*). J Sci Food Agric 90:2233–2240

Wang W, Vinocur B, Shoseyov O, Altman A (2004) Role of plant heat-shock proteins and molecular chaperones in the abiotic stress response. Trends Plant Sci 9:244–252

Yan SP, Zhang QY, Tang ZC, Su WA, Sun WN (2006) Comparative proteomic analysis provides new insights into chilling stress responses in rice. Mol Cell Proteomics 5:484–496

Yang P, Chen H, Liang Y, Shen S (2007) Proteomic analysis of de-etiolated rice seedlings upon exposure to light. Proteomics 7:2459–2468

Yang Y, Qiang X, Owsiany K, Zhang S, Thannhauser TW, Li L (2011) Evaluation of different multidimensional LC-MS/MS pipelines for isobaric tags for relative and absolute quantitation (iTRAQ)-based proteomic analysis of potato tubers in response to cold storage. J Prot Res 10:4647–4660

Yun Z, Li W, Pan Z, Xu J, Cheng Y, Deng X (2010) Comparative proteomics analysis of differentially accumulated proteins in juice sacs of ponkan (*Citrus reticulata*) fruit during postharvest cold storage. Postharvest Biol Technol 56:189–201

Zanor MI, Rambla J-L, Chaib J, Steppa A, Medina A, Granell A, Fernie AR, Causse M (2009) Metabolic characterization of loci affecting sensory attributes in tomato allows an assessment of the influence of the levels of primary metabolites and volatile organic contents. J Exp Bot 60:2139–2154

Zhang L, Yu Z, Jiang L, Jiang J, Luo H, Fu L (2011) Effect of post-harvest heat treatment on proteome change of peach fruit during ripening. J Proteomics 74:1139–1145

Zorb C, Betsche T, Lagenkamper G (2009) Search for diagnostic proteins to prove authenticity of organic wheat grains (*Triticum aestivum* L.). J Agric Food Chem 57:2932–2937

Chapter 19
Wheat Grain Proteomics for the Food Industry

Angéla Juhász, Paula Moolhuijzen, Matthew Bellgard, Rudi Appels, and Frank Békés

19.1 Introduction

The content of this chapter focuses on the analysis of protein in wheat grain because this provides the major source of protein utilized in the food industry. The novel aspects in this area that are of particular interest are the advances in wheat and barley genomics (reviewed in Appels et al. 2011) which now provide a basis for a whole-of-proteome approach to the analysis of the wheat grain. Studies reviewing food proteomics related to aspects of wheat and covering plant proteomics for cereal research can be found in Thiellement et al. 1999; Kersten et al. 2002; Rakwal and Agrawal 2003; Peck 2005; Agrawal and Rakwal 2006; Komatsu and Yano 2006. The potential in new methodologies is also considered in these reviews, most of which focus mainly on rice proteomics as a valuable reference for wheat and barley grain protein studies. It is now feasible to identify many wheat and barley proteins in 2-D gel electrophoresis and mass spectroscopy (MS) analysis based on homology between wheat/barley expressed genes and rice/Arabidopsis genes.

A. Juhász (✉)
Agricultural Institute, Centre for Agricultural Research of the Hungarian Academy of Sciences, Martonvásár 2462, Hungary
e-mail: juhasz.angela@agrar.mta.hu

P. Moolhuijzen • M. Bellgard • R. Appels
Centre for Comparative Genomics, Murdoch University, Western Australia 6150, Australia

F. Békés
FBFD PTY LTD,
Beecroft, NSW 2119, Australia

F. Toldrá and L.M.L. Nollet (eds.), *Proteomics in Foods: Principles and Applications*, Food Microbiology and Food Safety 2, DOI 10.1007/978-1-4614-5626-1_19,

Wheat seeds comprise three main parts: a protective seed shell, a small embryo, and a starchy nutrient reservoir. Due to the abundance of a range of proteins they have been of particular interest for the analysis in different gel-based and gel-free proteomic approaches. Wheat seed storage proteins serve as deposits of nitrogen for germination, and have a significant impact on human nutrition. Many aspects of the processing, assembly, deposition, and degradation of grain proteins are now well understood. Proteome-level approaches used in seed science are summarized in Fig. 19.1.

Proteome mining mainly deals with identification and characterization of proteins from different seed organelles or subproteomes utilizing the differences observed in their structure and solubility. Analysis of subproteomes, mainly those of cereal endosperms or germ protein fractions, serve as an important base for studies that relate to end-use quality studies. Functional proteomics aims to characterize signaling and metabolic pathways, function of protein complexes, and protein activities during developmental processes such as seed development, dormancy, or germination in order to interpret environmental and chemical affects on grain composition. Spatiotemporal analysis of expressed proteins form a subgroup of the functional proteomics studies and relies especially on the outputs from DNA- and RNA-based studies. Expression studies enhance our understanding of seed responses to different abiotic stresses, pathogen attack, or symbiotic interactions, including the effects of gene silencing or introduction of additional genes. The major challenge of structural proteomics is to assign predicted or in-silico modeled protein structure data to protein sequences obtained from genome and transcriptome sequencing projects. Structural proteomic studies of cereal seeds are used to predict the presence of epitopes important to food products, plant–pathogen interactions, to identify receptor functions by the analysis of ligand–protein interactions, or to analyze enzyme complexes. Structural proteomics builds on computational proteomics, utilizing the potential of data analysis and 3D modeling. The collection and evaluation of data from protein databases combined with gel-based, MS-based, or immunoanalytical computational analyses can deliver outputs in drug discovery, personalized medical treatments, and food safety analyses. The analyses of post-translational modifications, which modulate the activity of many seed proteins and the modified "subproteomes" that result, such as glycosylated, phosphorylated proteins or modified membrane proteins, can now be studied using mass spectrometry. Post-translational modifications leading to polymer formation through intermolecular disulfide binding is especially important in the structure–function relationship analysis of wheat gluten for food products.

19.2 The Proteome Complement of the Wheat Grain

Methodologies used in wheat seed proteomics include 2-D-GE based proteomics for relatively quantitative analysis for comparing small numbers of samples. The combination of 2-D-GE and mass spectrometry analyses, such as MALDI-TOF-MS or

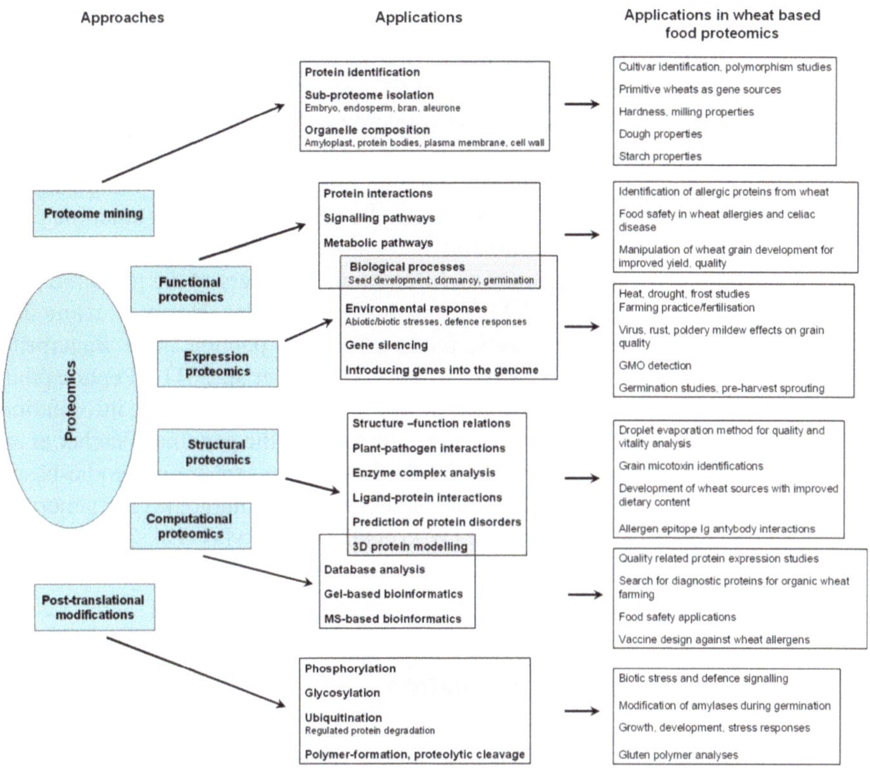

Fig. 19.1 Proteome-level studies in analysing the wheat gain are summarized with a particular focus on the translation of these studies to application in the food and related industries

ESI-MS, can resolve the seed proteome to 3,000 spots as demonstrated by Dupont et al. (2011) in their very extensive analysis. This approach can also identify many post-translational modifications by comparing protein sequence with physical properties.

Proteins with extreme hydrophobicity, mass, or pIs can be underrepresented in standard approaches because of solubility (for review, see Santoni et al. 2000; Görg et al. 2004) or other technical limitations. To identify these extreme proteins, gel-free proteomic approaches, such as multidimensional protein identification technology (MudPIT; Link et al. 1999; Washburn et al. 2001), have been explored. MudPIT separates peptides by a strong cation exchange phase in the first dimension followed by reverse-phase chromatography in the second dimension. Comparisons of MudPIT and 2D-GE indicate that these technologies are complementary (Koller et al. 2002; Katavic et al. 2006). Another gel-free MS approach is the isotope-coded affinity tags (ICAT; Gygi et al. 1999), used for quantitative comparison of proteomes.

Although peptide mass fingerprinting is a sensitive high-throughput method, it is not always easy to identify the plant proteins or genes by fingerprinting because the complete nucleotide sequence of the genome of various species, including wheat, is not yet available and the protein database does not have enough information for the identification of unknown proteins. In addition, plant proteins are often post-translationally modified which means peptide–mass fingerprinting initially needs to deal with a subpopulation of peptides that are not modified. Improved sample preparation methods for the peptide mapping are important to develop for MS analysis of wheat (Woo et al. 2002). Due to the high proline and glutamine content observed in grain storage proteins the usual enzymatic digestion protocols did not result in sufficient accuracy to differentiate members of the prolamin protein families, such as HMW or LMW glutenins and gliadins. However, using different combinations of proteinases, more distinctive peptide mass fingerprint datasets can be generated (Vensel et al. 2005, Dupont et al. 2011). Peptides that are distinctive for specific proteins are the key to acquiring validating information for the expression of a gene that has been annotated in the genome (Futcher et al. 1999; Gygi et al. 1999; Appels et al. 2012). In order to analyze proteome-based protein expression on a large scale, mass spectroscopy technologies are generally considered to be the most appropriate (Woo et al. 2002; Skylas et al. 2005; Appels et al. 2012).

19.3 Wheat Seed Protein Isolation

Seed tissues contain a large number of diverse proteins, with different chemical characteristics which have resulted in the development of special extraction and purification approaches that retain the quality of proteins for analysis on 2-D gels or by MS (Branlard and Bancel 2007). For cereal grains the major storage components are starch and different carbohydrates, which have a detrimental effect on the extractability of proteins as result of seed proteins being present as aggregates or coupled to other compounds such as starch granules or cell wall elements. Compared to other plant tissues, the extraction of seed proteins is first required to break the hard structure of mature seed using grinding or milling. This results in solubility characteristics of proteins that are dependent on particle size and the homogeneity in size distribution of the ground material. Extraction methods in seed proteomics can be divided into two main types: using universal protocols for the main protein groups, including classes of water/salt soluble proteins, moderately hydrophobic proteins, enriched hydrophobic proteins, and membrane proteins; and more specific cereal proteins targeting the water, salt, dilutes–alcohol, and light acid/base classes of soluble fractions (Görg et al. 2007).

Fresh, developing seed tissues need the use of liquid nitrogen during isolation to rapidly freeze the tissues before grinding. The presence of some contaminating compounds, such as phenolics, phosphatases, lipids, salt ions, nucleic acids, insoluble materials, and proteolytic enzymes requires the use of special chemicals. Proteolytic enzymes are intrinsically involved in many aspects of plant physiology and development, ranging from the mobilization of storage proteins during seed germination to the initiation of cell death and senescence programs (Loukas 2002; Creemers 2002; Schaller 2004). Post-translational modification of proteins, for protein assembly, for the activation of prohormones to hormones, and for maturation and control, the activities of enzymes are processes carried out by grain components present in small amounts. Different mixtures of protease inhibitors, tricloroacetic acid (TCA) precipitation, or boiling in anionic detergent solution are often used to avoid protein degradation in seed tissue extracts. Loss of function of enzymes or membrane proteins can be avoided using native nondenaturating conditions during the extraction (Fido et al. 2004). Presence of phenolic compounds generally has a detrimental effect on the separation of proteins resulting in horizontal and vertical streaks on 2D-GE. Precipitation of proteins with TCA followed by rinsing with ice cold acetone + TCA will help to eliminate phenolic compounds present in fresh tissues. Starch or complex carbohydrates can also seriously reduce protein extraction, and is a primary problem in proteomic analyses of maize or rice seeds. Advances in the methodologies for cereal proteomics provides the basic translating research output to the food industry.

19.4 The Wheat Endosperm Proteome

The structure of the wheat grain can be categorized simply into three distinct fractions, the outer layers, the germ, and the endosperm. The endosperm plays the most important role from a human consumption point of view and is the fraction studied in most detail (Dunbar 1985; Skylas et al. 2001; Amiour et al. 2002; Islam et al. 2002; Dupont et al. 2011; Tasleem-Tahir et al. 2012). The development of wheat seeds has been studied by 2D-GE where developing and mature seed proteomes were compared (Skylas et al. 2000; Debiton et al. 2011; Hurkman et al. 2009; Tasleem-Tahir et al. 2012). In these studies 250 proteins were identified in developing endosperm (Vensel et al. 2005). Another study separated whole wheat endosperm amyloblasts or amyloblast membrane proteins during development and protein were identified by LC-MS/MS (Andon et al. 2002).

The proteins located in the endosperm determine the characteristics of elasticity, extensibility, and gas-holding ability of the wheat dough. Particularly important are the proteins within the gluten complex, and the low and high molecular weight glutenin subunit proteins (LMW-GS and HMW-GS; Shewry et al. 2009; Islam et al. 2012). Within the polymeric gluten fraction, the A-group (M_r range of 80,000–120,000 Da) corresponds to the HMW-GS (Payne et al. 1979). The major group, the B-subunits, consists of the typical LMW-GS that are basic proteins with a M_r range from 42,000 to 51,000 Da. The minor C-subunits are a mixture of proteins belonging either to LMW-GS or to γ- and α-gliadins with an odd number of cysteine residues (Shewry and Tatham 1997; Payne et al. 1979).

Gliadins are the most diverse group of storage proteins in wheat and are characterized by their solubility in 70% aqueous alcohol. This group of proteins shows extensive polymorphism across wheat genotypes. Flour also contains smaller amounts of other storage proteins such as globulins and triticins, and proteins such as amylase and protease inhibitors that may protect against insects and fungi. A range of enzymes exists in very small amounts. These nongluten proteins encompass smaller percentages of the seed protein content; however, they play a crucial role in different metabolic processes during grain development, ripening, and desiccation and can have a relevant role in defense mechanisms against biotic and abiotic stresses (Shewry and Halford 2002; Shewry et al. 2002; Shewry 2009; Skylas et al. 2000). Several early studies demonstrated the utility of 2D-GE to visualize the total complement of flour proteins (Payne et al. 1985; Wrigley 1970), and the combination of 2D-GE and MS techniques improves the analytical capability (Skylas et al. 2005; Mamone et al. 2009; Ferranti et al. 2007). In addition to gliadins and glutenins some minor storage protein groups with similar characteristics to the traditional prolamin groups were also identified. The detailed analysis of farinins (avenin-like proteins), purinins (low molecular weight gliadins), and triticins has been carried out in the cultivar Butte 86 by Dupont et al. (2011).

The water-soluble albumins and salt-soluble globulins of wheat grain comprise mainly metabolic and protective proteins whereas at least some of the insoluble structural proteins may be extracted with diluted acid or alkalines in the glutenin fraction. Albumins and globulins, each accounting for approximately 10% of total flour proteins, are known to be soluble enzymes and proteins soluble in polar solution, respectively. More than 2,000 spots can be extracted from flour with salt solution. Numerous studies have been carried out for the identification and mapping of wheat endosperm proteins (Merlino et al. 2009; Skylas et al. 2000, 2005; Islam et al. 2003a, b; Gao et al. 2009). Studies on kernel responses to heat stress have enabled identification of many of the enzymes in wheat kernels (Majoul et al. 2003, 2004; Skylas et al. 2000; Dupont et al. 2006). However, the majority of soluble proteins still need to be identified and mapped.

Amphiphilic proteins are detergent-soluble membrane proteins (Amiour et al. 2002) and together with gluten proteins play an important role in wheat quality. Some of them, such as puroindolines, are lipid binding proteins, and are strongly linked to dough foaming properties and to fine crumb texture. Most of our knowledge about this protein class comes from the work of Amiour et al. (2002, 2003) who carried out proteomic analysis and genetic mapping using 81 F9 lines, progeny obtained from an interspecific cross "W7984"x"Opata", a cross intensively characterized in the International Triticeae Mapping Initiative (ITMI*map*). Sixty-four spots were inherited as Mendelian trait loci (MTLs) and their corresponding genes located on 11 chromosomes. Image analysis of 2-D gels allowed quantification of protein variation and 35 protein quantitative traits (PQLs) were identified. Many of these proteins displayed similarities in their sequences to proteins in databases and particularly proteins associated with membranes. Some amphiphilic proteins were located between the protein matrix and the starch granule of wheat endosperm, and could thus, due to their diversity and quantity, influence grain hardness and consequently dough properties

19.5 The Amyloplast Proteome

Studies of specific plant tissues and organelles, such as the chloroplast (Kleffmann et al. 2004; Peltier et al. 2000; Zybailov et al. 2009) and chloroplast membrane (Ferro et al. 2003; Rolland et al. 2003) have been studied in significantly more detail than the amyloplast, not only because it is more common in the plant kingdom but also because amyloplasts are difficult to isolate in the laboratory due to fragility of the amyloplast membrane in contrast to the density of the resident starch granules (Denyer and Pike 2008). The first proteomic investigation of wheat amyloplast by Andon et al. (2002) was carried out on whole amyloplasts and purified amyloplast membranes. From the total number of 171 proteins, 108 were identified from whole amyloplasts and 63 from purified amyloplast membranes. A total of 46% of the proteins could be identified using sequence data from the rice genome database. The largest class of proteins with a known function (24%) was found to be starch synthesis and storage enzymes. This figure includes two peptides from starch branching enzymes, one from globulin, three from glycogen starch synthase precursors, 10 from various glutenins, and 46 peptides from 11 different isoforms of gamma gliadin. It was evident from this study that most of the proteins in the amyloplast membrane and whole amyloplast were unknown.

The view of the importance of starch in relation to the overall quality of wheat has changed significantly. For many years starch was considered to play only a minor role in wheat product quality. It has been only recently recognized that the contribution of starch to flour-processing quality is important (Rahman et al. 2000). Starch has a dominant contribution in determining the quality of noodle products and it is a major contributor to the nutritive value of wheat-based products (Regina et al. 2006). Progress in improving our understanding of starch-related quality research has been reviewed by Wrigley et al. (2009) and Park et al. (2009).

The protocol for the extraction and 2D-GE of wheat starch granule-associated proteins developed by Bancel et al. (2010) consists of several steps: (1) washing and sonication to remove interfering compounds (storage proteins) from the surface of the granules, (2) scanning electron microscopy (SEM) observations to monitor purification and granule swelling, and (3) appropriate protein extraction and solubilization to obtain enough proteins for proteomic analysis. For quantitative proteomic analysis of proteins associated with wheat starch granules, a two-step protein extraction protocol was developed including TCA/acetone precipitation and phenol extraction. With this protocol, proteins were extracted from wheat starch granules and solubilized, and 2D-GE protein maps analyzed (Bancel et al. 2010). The majority of the spots associated with starch granules were identified by peptide mass fingerprinting and MS/MS and functionally classified into carbohydrate metabolism and stress defense. Affinity electrophoresis has been utilized in the identification of enzyme-type proteins (Kosar-Hashemi et al. 2006), especially the protein-related starch biosynthetic (Konik-Rose et al. 2007; Ahmed et al. 2007), and hydrolyzing enzymes and their inhibitors (Wu et al. 2011). The affinity substrates are glucan (corn starch or amylopectin) added to the native gel polymerization mixture followed by staining methods developed by Nishi et al. (2001).

19.6 Proteome of Wheat Germ

Wheat germ and bran are by-products of the milling industry, and there is a growing interest in the possibilities of making better use of these potentially valuable food resources, especially in the baking and dietary products industries. The composition of wheat germ is of nutritional significance and the commercial milling process is very efficient in separating the endosperm from other anatomical parts of the grain, including germ. Wheat germ is the embryonic tissue of the wheat grain, and detailed knowledge of its proteome is necessary to define the processes that occur during germination, especially pre-harvest sprouting. The germ fraction is predominantly made up of the mature wheat embryo, which is poised to undergo significant developmental changes once sufficient water is available to allow biochemical processes to take place.

The proteome of wheat germ has been characterized in detail by Mak et al. (2006) using gel-based 2-D proteomics tools. In total, 612 protein spots were excised from the pH 4.0–7.0 and 6.0–9.0 gel maps for PMF analysis. About two-thirds of the proteins identified were in functional categories associated with processes involved in activating growth and development of the embryo such as transcription, translation, energy, and general metabolism, transport, cell division, and signaling processes. A major class of proteins present in the germ fraction represents stress-associated proteins, which in contrast are poorly represented in the endosperm. About 25% of the protein spots identified belonged to protein groups of either unclassified function or represent proteins that were considered to be hypothetical proteins in the genomic database. Membrane proteins number about 20–40% of the total proteins in cells, however, their poor solubility and inherent hydrophobicity results in difficulties during extraction (Stevens and Arkin 2000). Increased solubilization of denatured membrane proteins was achieved by the use of surfactants such as SDS, which is incompatible with isoelectrofocusing (IEF) (Nouwens et al. 2000). Therefore, this type of protein may be underrepresented in the germ proteome maps. Approximately 20% of the proteins identified in the germ were enzymes, which are represented in all six Enzyme Commission (EC) classification groups, namely oxydoreductases, transferases, hydrolases, lyases, isomerases, and ligases, whereas only two EC classification groups, oxidoreductases and isomerases, were expressed in the mid-development endosperm.

The protein composition of germinating wheat seeds has been analyzed in the first three days of germination (Mak et al. 2009). All identified proteins involved in transcription, transporter activities, or secondary mechanisms have shown significant decreases in their abundance. Decreasing levels of expression were also identified for some proteins involved in the energy-related mechanism, protein synthesis, and degradation, as well as proteins of signal transduction, stress-related mechanisms, and metabolism. The proteins that increased in abundance during the germination period have diverse roles including energy, protein degradation, protein folding, and in the cytoskeleton.

19.7 Proteome of Wheat Bran

Bran has an important function as a protective barrier for the grain. The grain is a potential food for insects, fungi, and bacteria, and it is also exposed to many environmental stresses, thus the bran must have properties that protect it from all these factors. Proteomic analysis of wheat bran and bran tissue fractions have been reported by Jerkovicz et al. 2010 and Laubin et al. 2008, and spatial and temporal expression of peripherial layer proteins were followed by Tasleem-Tahir et al. (2011). Collection of pure bran tissue fractions is limited by the strong bonds between the various bran tissue layers and endosperm in dry grain. Thus, a method to obtain bran layers free from contaminants, such as adjacent tissue and endosperm, is required to provide a suitable sample for proteomic analysis. The outermost component of the grain is the bran, which is composed of about seven distinct tissue layers. The outermost layers of the bran are collectively called the pericarp. The pericarp consists of the epidermis, hypodermis, cross-cells, and finally the tube cells. The next three tissue layers are the testa, nucellar, and aleurone. The aleurone tissue is in contact with the endosperm and it is the only bran tissue layer that is still alive and functional at the cellular level in mature grain and is critical during germination (Antoine et al. 2003, 2004). Proteomic analysis of bran tissue fractions from wheat grain revealed the location and distribution of many common plant defense-related proteins, which appear to be specific to certain tissue layers within the bran. Around 80% of proteins identified within the outer layers provide resistance to fungal and bacterial colonization and so fulfill a general defensive role rather than targeting specific biotic stresses. Defense-and stress-related proteins were also identified in the inner fraction, however, they comprised only around 5% of the protein array. The major protein in the inner fraction was 7S globulin storage protein, which is likely to be involved in defense against oxalate-secreting fungi and also to serve in protecting itself during grain development when there is potentially a high oxalate level (Dunwell et al. 2000). Some of the major classes of defense proteins are termed pathogenesis-related proteins (PR), and include PR-1, PR-2 (1, 3-beta-glucanases), PR-3 (chitinases), PR-4 (wheatwin), and PR-5 (thaumatin-like proteins) (Selitrennikoff 2001; Desmond et al. 2006).

Studies of the aleurone layer of the seed have been stimulated by its nutritional and health benefits. The AL (aleurone) provides proteins rich in lysine and also contains several vitamins (B1, B2, B3, B6, B9, and E) and minerals (P, K, Mg, Mn, and Fe). Although biochemical studies have already been conducted to distinguish and characterize the different layers of wheat bran (Antoine et al. 2004), until recently this was not the case for protein composition of the wheat AL.

Developmental processes of peripherial layers have also been monitored by Tasleem-Tahir et al. (2011). More than 200 proteins of grain peripheral layers (inner pericarp, hyaline, testa, and aleurone layer) were identified and classified in 16 different functional categories. Study of the protein expression over time allowed identification of five main profiles and four distinct phases of development. Wheat grain peripheral layers at early stages are the site of metabolic activity, photosynthesis, and all the metabolic pathways linked to the reactive oxygen species (ROS) production and detoxification. Composite expression curves indicated that there was a shift from metabolic processes, translation, transcription, and ATP interconversion towards storage and defense processes. Protein synthesis, protein turnover, signal transduction, membrane transport, and biosynthesis of secondary metabolites were the mediating functions of this shift.

19.8 Application of Wheat Proteomics in Fundamental and Applied Wheat Research

Proteomic research activities on wheat and the possible and/or already existing application areas of proteomics in wheat research have been reviewed by Skylas et al. (2005). Dough quality, starch properties, puroindoline, chromosomal location of coding genes, effects of growth environment, insect damage to wheat grains, and genotype identification were included by Skylas et al. (2005). More recent reviews have been published summarizing the different techniques of proteomics both in plant biology and food science and are provided in Table 19.1.

The development of techniques to improve extraction and separation and to identify proteins and peptides is facilitating functional proteomics and the analysis of subproteomes from small amounts of starting material, such as seed tissues (reviewed in Jorrin et al. 2007; Miernyk and Hajduch 2011; Finnie et al. 2011). The combination of proteomics with structural and functional analysis is increasingly applied to target subsets of proteins. These “next-generation” proteomics studies differentiate between spatial subproteomes and functional subproteomes and contribute to understanding the processes controlling cereal grain development and nutritional and processing characteristics as well as the changes caused by biotic (Kav et al. 2007) and abiotic (Hashiguchi et al. 2010) stresses. The importance of abiotic stresses caused by global climate change has particular significance in the future of agriculture and food industries. Advances in crop proteomics provide the basis for interpreting changes in the grain as a result of climatic factors including photosynthetic stress, air pollutants, thermal stress including heat and cold, and osmotic stress (drought, salt, flooding), and metal stress to help to eliminate or to reduce the negative effects of these factors on the food production chain (Finnie et al. 2011).

Reviews on the use of proteomics in food technology (Carbonaro 2004; Mamone et al. 2009; Gaso-Sokac et al. 2010; Pischetsrieder and Baeuerlein 2009) focus on

Table 19.1 Techniques of proteomics and examples in seed biology and wheat based food proteomics

Approach	Applications	Examples for applications in wheat based food proteomics	References
Proteome mining	*Protein identification*	Cultivar identification, polymorphism studies	Cornish et al. 2001; Skylas et al. 2001; Wrigley 2002; Wrigley et al. 2003; Wu et al. 2011; Yahata et al. 2005; Dupont et al. 2011
		Wheat lines with chromosome deletions	Islam et al. 2002, 2003b
		Primitive wheats as gene sources	Islam et al. 2003a; Wu et al. 2011
	Sub-proteome isolation (embryo, endosperm, bran, aleurone)	Hardness, milling properties	Amiour et al. 2002, 2003; Branlard et al. 2003a; Lesage et al. 2012
		Dough properties	Liu et al. 2009, 2010; Dworschak et al. 1998; Ghirardo et al. 2005; Muccilli et al. 2005; Salt et al. 2005
		Starch properties	Hurkman et al. 2008; Tetlow et al. 2008
	Organelle composition (amyloplast, protein bodies, plasma membrane, cell wall)	Starch granule development	Andon et al. 2002; Balmer et al. 2006a; Bechtel and Wilson 2003
Functional proteomics	*Protein interactions*	Identification of allergic proteins from wheat	Akagawa et al. 2007; De Angelis et al. 2008, 2010; Larré et al. 2011; Ruoppolo et al. 2003; Sotkovsky et al. 2008; Yang et al. 2011; Kasarda et al. 2008
		Food safety applications in wheat allergies, celiac disease	De Angelis et al. 2006; Gobbetti et al. 2008; Haraszi et al. 2011; Heick et al. 2011; Matsuo et al. 2004
	Signalling pathways		
	Metabolic pathways	Starch synthesys analysis	Balmer et al. 2006a, b; Dupont 2008

(continued)

Table 19.1 (continued)

Approach	Applications	Examples for applications in wheat based food proteomics	References
Expression proteomics	*Biological processes (seed development, dormancy, germination)*	Manipulation of wheat grain development for improved yield, quality	Altenbach et al. 2011;
		Germination studies, pre-harvest sprouting	Kamal et al. 2009; Guo et al. 2011
	Environmental responses (abiotic/biotic stresses, defence responses)	Heat, drought, frost studies Farming practice/fertilistaion	Finnie et al. 2004; Horváth-Szanics et al. 2006; Hurkman et al. 2009; Laino 2007; Laino et al. 2010; Majoul et al. 2004; Miernyk and Hajduch 2011; Hashiguchi et al. 2010; Neilson et al. 2010; Sancho et al. 2008; Skylas et al. 2002, 2005
		Virus, rust, poldery mildew effects on grain quality	Altenbach et al. 2011; Flate et al. 2005
			Antoine et al. 2003; Blein et al. 2002; Desmond et al. 2006; Dunwell et al. 2000; Gane et al. 1998; Gorlach et al. 1996; Kav et al. 2007; Laino 2007; Laino et al. 2010; Mak et al. 2006; Mellersh et al. 2002; Niranjan-Raj et al. 2006; Selitrennikoff 2001; Wang et al. 2005; Zhou et al. 2005
	Gene silencing *Introducing genes into the genome*	GMO detection	Debiton et al. 2011; Gobaa et al. 2007; Guo et al. 2011; Horvath-Szanics et al. 2006; Islam et al. 2003a; Di Luccia et al. 2005;

Structural proteomics	*Structure-function relations*	Droplet evaporation method for quality and vitality analysis	Kokornaczyk et al. 2011
	Plant-pathogen interactions	Grain micotoxin identifications	Berthiller et al. 2005; Bhatnagar et al. 2008
	Enzyme complex analysis		
	Ligand protein interactions	Development of wheat sources with improved dietary content (functional foods)	Pollet et al. 2010; Laubin et al. 2008; Merlino et al. 2009
	Prediction of protein disorders		
	3D protein modelling	Allergen epitope Ig antybody interactions	
Computational proteomics	*3D protein modelling*	Allergen epitope Ig antybody interactions	
	Database analysis	Quality related proteins and their expression	Juhász et al. 2011; Altenbach et al. 2011
		Food safety applications	Heick et al. 2011
	Gel-based bioinformatics	Search for diagnostic proteins for organic wheat farming	Zörb et al. 2006, 2009
		Identification wheat allergens	Akagawa et al. 2007; Denery-Papini et al. 2007; Larré et al. 2011; Yang et al. 2011, Saha and Raghava 2006a, b
	MS-based bioinformatics	Determination of wheat quality based on MS analysis of wheat seed	Gottlieb et al. 2002; Herrero et al. 2011; Altenbach et al. 2011; Ghirardo et al. 2005; Dupont et al. 2011
Post-translational modifications	*Phosphorilation*	Biotic stress signalling, defence signalling	Rampitsch et al. 2010; Bykova et al. 2011
		Starch branching enzyme activity analyses	Tetlow et al. 2004
	Glycolysation	Modification of glutenins	Bollecker et al. 1998
	Ubiquitination	Growth, development, stress response analyses	Wilson et al. 2005; Wong et al. 2004
	Polymer formation proteolytic cleavage	Gluten polymer analyses	Beasley et al. 2002; Gao et al. 1992; Lindsay et al. 2000; Weegels et al. 1997

applications related to the characterization and standardization of raw materials – including areas important for selecting parents for crossing in breeding (glutenin allele identification) or regulating the entire wheat chain (variety identification) process development, and detection of batch-to-batch variations and quality control of the final product. Aspects of food safety, especially regarding biological and microbial safety and the use of genetically modified foods are also important targets.

In wheat the knowledge of starch biosynthesis has led to developing wheat varieties with better nutritional properties to produce lower calorie cereal-based foods. The understanding of the interrelationships among metabolic pathways helps to break the inverse relationship between harvest yield and protein content, in order to maintain a high level of production with better quality source material for the industry. Improved knowledge about the protective mechanism of the plant and the effects of biotic stresses on the protein composition has a direct relationship to food safety. The better understanding of the effects of genetic and environmental factors on the expression and deposition of different proteins directly lead to the production of wheat varieties with better end-use quality, with higher nutritive value, and with less harm for sensitive individuals.

Proteomic analysis of amphiphilic proteins and albumins/globulins has been applied to discover metabolic pathways and to define the physiological responses caused by natural or manmade mutations and/or introducing or silencing genes in the genome. Proteomic analysis carried out by Horváth-Szanics et al. (2006).

Both qualitative and quantitative aspects of the protein composition related to prolamin proteins and amphiphilic proteins are directly related to the end-use quality to the wheat-based products. Therefore their investigation can lead directly to quality improvement. The schema of Mamone et al. (2009) illustrating how proteomics can be applied in quality-related wheat research is an excellent example of this integrated strategy.

19.9 Some Examples of Proteomics in Wheat Research

19.9.1 Uncoupling Starch and Protein Accumulation in the Grain

Understanding the molecular mechanisms that regulate the balance between starch and protein synthesis could aid in efforts to uncouple the inverse relationships between starch and protein synthesis. Breeding for increased grain yield tends to increase starch more than protein, resulting in flour with a lower protein percentage, and there seems to be a limit to maximum grain protein content in high yielding varieties (Oury and Godin 2007; Rharrabti et al. 2001) although there are exceptions. When wheat is grown under cool, well-watered conditions, it is possible to obtain high grain yield together with high protein concentration by supplying nitrogen after anthesis (Altenbach et al. 2003; Banziger et al. 1994; Brown and Petrie 2006). However, additional nitrogen alters grain amino acid content and there is a negative relationship between protein concentration and the concentration of essential amino acids such as lysine, cysteine, and methionine (Rharrabti et al. 2001). There is also a trade-off between grain yield and protein content under conditions of drought and/or high temperature, which reduce accumulation of starch more than protein (Altenbach et al. 2003, Hawker and Jenner 1993). This is partly because high temperature reduces transcript levels and enzyme activity of starch synthase (Hurkman et al. 2003). High-protein flours produced under conditions of high temperature were also reported to have increased proportions of glutamine- and proline-rich proteins and reduced proportions of cysteine- and methionine-rich proteins (Dupont et al. 2006).

The number of proteins in amyloplast fractions has been arranged into 18 proposed metabolic and biosynthetic pathways (Balmer et al. 2006a; Dupont 2008). The enzymes are arranged into maps that include enzyme and compound names in a single view and suggest that amyloplasts play a central role in the process of transforming the sucrose, glutamine, and other amino acids into the array of proteins, lipids, nucleic acids, and carbohydrates needed to sustain the intense accumulation of starch and protein in the developing grain. Until recently they have been discussed mainly in terms of their role in starch biosynthesis, whereas their role in synthesis of other endosperm components has not been considered in detail.

Starch is produced by the majority of higher plant species inside plastids and represents a major storage product of many of the seeds and storage organs produced agriculturally and used for human consumption as well as many important industrial applications. The starch granule is a complex polymeric structure with a hierarchical order, allowing efficient packing of large amounts of glucose into a water-insoluble form, and is composed of two distinct types of polymer: amylose and amylopectin. The significant progress in the understanding of the biosynthesis of amylose and amylopectin as well as the granule mass formation and the role of different enzymes of starch synthesis and defining the functional properties of the product are summarized in several excellent reviews (Thompson 2000; Ball and Morell 2003).

To date, information on how coordination between amylopectin-synthesizing enzymes is achieved is sparse. Recent work in wheat endosperm amyloplasts suggests protein phosphorylation is involved in modulating the catalytic activity of some key enzymes (the SBEII class) and their ability to form physical interactions with other starch-metabolizing enzymes (Tetlow et al. 2004). Analysis of transgenic Arabidopsis and potato (*Solanum tuberosum*) plants also indicates a role for 14-3-3 proteins in the regulation of SS activity in the formation of assimilatory (transient) starch in leaves, presumably through protein–protein interactions (Sehnke et al. 2001; Zuk et al. 2005). Investigation of protein–protein interactions among enzymes of amylopectin biosynthesis in developing wheat endosperm by Tetlow et al. (2008) provided direct evidence for the existence of protein complexes involved in amylopectin biosynthes.

19.9.2 Understanding Grain Hardness and Milling Properties of the Grain: The "Puroindoline Story"

The puroindolines have been identified as a component defining hardness (a measure of the resistance to deformation under applied stress) of grain (reviewed in Turnbull and Rahman 2002). Hardness and vitreousness are separate traits linked by overlapping quality trait loci under genetic control at the 5D Ha locus (Turnbull and Rahman 2002). The environment affects the expression of both, but does not change the ranking of cultivars relative to each other for this trait. Modifying genes account for the range of hardness within hard or soft classes. It is the expression of these genes that is thought to be important in determining milling quality. Hardness has been found to relate mostly to the surface properties of the starch particles of the endosperm. Different proteins interacting on the granular membrane play an essential role in this process (Shewry et al. 2009). The puroindolines are basic lipid-binding proteins, rich in cysteine, of about 13,000 Da (Gautier et al. 1994; Morris 2002; Jones et al. 2006). They belong to the 2S albumin superfamily of proteins and are encoded by genes at two loci on the short arm of chromosome 5D: *Pina-D1* (two alleles, *a* and *b*) and *Pinb-D1* (many alleles, *a–g*, *l*, *p*, *q*) (Morris 2002; Jones et al. 2006; Wanjugi et al. 2008). Bread wheat generally contains two types of puroindolines differing slightly in size. Soft wheat possesses both puroindoline a and puroindoline b, due to the allele combination *Pina-D1a* and *Pinb-D1a*. As their starch granules are loosely attached to the protein matrix, soft wheat crushes easily, producing largely intact starch granules and fine flour. On the other hand, the starch granules of hard wheat are tightly bound to the protein matrix, requiring greater milling energy and producing coarser flour with higher levels of starch damage.

Until recently the cellular functions of puroindolines during grain development remained unknown. To gain further insight into their biological function, Lesege et al. (2012) carried out a comparative proteome analysis of two near-isogenic lines (NILs) of bread wheat *Triticum aestivum* L. cv. Falcon differing in the presence or absence of the puroindoline-a gene (Pina) and kernel hardness. Proteomes of the two NILs were compared at four developmental stages of the grain for the metabolic albumin/globulin fraction and the Triton-extracted amphiphilic fraction. Proteome variation showed that, during grain development, folding proteins and stress-related proteins were more abundant in the hard line compared with the soft one. These results, taken together with ultrastructural observations showing that the formation of the protein matrix occurred earlier in the hard line, were consistent with an earlier endosperm cell death in the hard NIL than in the soft one. Quantification of the albumin/globulin fraction and amphiphilic proteins at each developmental stage strengthened this hypothesis as a plateau was revealed from the 500°C day stage in the hard NIL whereas synthesis continued in the soft one clearly. Puroindolines differentiate between the hard and soft grain class, but explain only around 30% of the within-class variations of endosperm texture. The intensive research activity on looking for other contributors to grain hardness variation has identified significant QTLs for this trait outside the *Hardness* locus on chromosomes 1 and 5 (Groos et al. 2006), individually explaining ~15% of the variation of hardness whereas Bettge and Morris (2000) found that the variation in hardness related to growing conditions was largely accounted for by the amount and composition of pentosans.

19.9.3 Glutenin Allelic Composition: The Key Factor of the Strength and Extensional Properties of the Dough

Relating the protein composition to certain quality traits by statistical means is a frequently used methodology to relate structure/composition to functionality in cereal science. The classic tool used by breeders is the Payne score (Payne 1987; Payne et al. 1987) providing a single number to estimate dough strength from the HMW glutenin allelic composition. This approach has been extended to include LMW glutenin alleles in similar mathematical formulas (Gupta et al. 1991; Cornish et al. 2006; Békés et al. 2006; Eagles et al. 2002). By the application of sophisticated statistical approaches, the wheat simulator of Eagles et al. (2002), and the protein scoring system (PSS) of Békés et al. (2006), it is possible to describe the effects of both HMW and LMW GS alleles on dough strength and extensibility, individually and the pairwise interactions among the alleles. Applications of the models indicate that the approach of relating allelic composition to quality attributes is possible with careful data selection and applying robust mathematical tools; the genetic potential of a line, with a certain combination of alleles on the six glutenin coding loci, can be meaningfully predicted where both the contribution of the individual alleles and their pairwise interactions play equally important roles (Békés et al. 2006. Cornish et al. 2006; Branlard et al. 2003b; Baracskai et al. 2011).

The genetic makeup of gluten proteins, the information about the allelic composition of glutenin and gliadin alleles, therefore becomes essential information in wheat breeding. Large databases containing the allelic composition of wheat varieties grown around the world are available to search for certain combinations of alleles holding specific quality attributes (Metakovsky et al. 2006; Wrigley et al. 2006a, b). The MALDI-TOF-based proteome analysis of wheat proteins has been found to provide a sensitive, reproducible, and high-throughput methodology for defining the required allelic composition (Liu et al. 2010) and can serve the needs of breeding programs. The technologies can now resolve the *b, u, al*, and *br* alleles of the Bx7-type subunits coded by the *Glu-B1* locus (Gao et al. 2012). The amounts of x-type subunits in both the *al* and *br* alleles are overexpressed but although this extra amount of HMW-GS subunit in the *al* allele provides extra dough strength, the presence of the *br* allele has a negative effect on dough strength, speculated to be caused by a tyrosine–cystein mutation (Gao et al. 2012).

Despite the abundance of the LMW-GS, they have received much less attention than the HMW-GS, probably due to their complexity, heterogeneity, and comigration with gliadins in SDS-PAGE (Appelbee et al. 2009; D'Ovidio and Masci 2004). In the SDS-PAGE system, utilizing gliadins as indicators provided an indirect way to define LMW-GS alleles (Jackson et al. 1996). MALDI-TOF-MS is currently the most efficient method to analyze these proteins and requires only 4–5 min per sample. It is a high-throughput technology for analyzing wheat gluten proteins (Dworschak et al. 1998; Ghirardo et al. 2005; Muccilli et al. 2005; Liu et al. 2009). In addition, simple, rapid, and sensitive PCR approaches to assay specific genes directly have proven to be useful for identifying LMW-GS composition in common wheat (Appelbee et al. 2009; Zhang et al. 2004; Wang et al. 2009, 2010). The comprehensive comparison of the results derived from using proteomic or PCR-based methods is discussed by Liu et al. 2010.

19.9.4 Wheat Grain Proteins as Allergens

The unusual amino acid composition and primary structure of prolamins have nutrition- and health-related consequences. Consumption of gluten containing food can cause health problems for a significant number of people who consume foods derived from wheat, rye, barley, and oats. The general public in most Western countries is now aware of the potential adverse effects of cereals containing gluten with reports appearing in the lay press (Braly and Hogganm 2002; Ford 2008; Wangen 2009) promoting gluten-free diets. Many of these reports, however, fail to draw attention to the importance of appropriate diagnosis or defining the nature of the gluten “intolerance” an individual may have.

The terms wheat-related intolerance (sensitivity) and wheat allergy are often used interchangeably. However, it is generally accepted that wheat sensitivity is specifically defined as the negative reaction in individuals to wheat with clinical manifestations such as chronic urticaria, gastrointestinal problems (i.e., irritable bowel syndrome, eosinophilic colitis, ulcerative colitis), depression, eczema, and low blood iron levels, which can sometimes take days to manifest. In contrast, wheat allergy is a negative reaction where symptoms appear rapidly following exposure to macromolecules such as proteins and includes acute anaphylactic hypersensitivity, baker’s asthma (Houba et al. 1998), and the autoimmune enteropathy, celiac disease (CD) (van Heel and West 2006). Celiac disease (CD) is an inflammatory disorder of the upper small intestine triggered by the ingestion of wheat, rye, barley, and possibly oat products. Wheat sensitivity affects many more people (about 15% of human population) than wheat allergy (about 1–4% population). In the early 1970s, the gluten fraction, specifically the gliadins were reported to be responsible for celiac disease, anaphylactic hypersensitivity, and baker’s asthma. Numerous studies have led to the identification of wheat prolamin proteins containing toxic epitopes responsible for CD, from both the gluten and soluble proteins of wheat.

The storage proteins of CD toxic cereals have been classified into three groups: a high molecular weight (HMW) group; a medium molecular weight (MMW) group; and a low molecular weight (LMW) group, the latter being the major group (Tatham and Shewry 2008; Wieser and Koehler 2008). The detailed analysis of these groups of proteins led to the identification of the toxic epitopes, the mechanism of toxicity, and the effects of genetic background of CD to be understood (reviewed by Anderson and Wieser 2006; Catassi and Fasano (2008), Wieser and Koehler 2008; van Heel and West 2006). In contrast, wheat sensitivity (Tanabe et al. 1996, Sandiford et al. 1997; Matsuo et al. 2004; Battais et al. 2005a, b) is not well defined and much less is known about the basics of this complex disorder.

Regarding wheat allergies, the flour-soluble fraction consisting of albumins/globulins (A/G) has also been reported to contain IgE-binding proteins (Weiss et al. 1997; Varjonen et al. 2000; Takizawa et al. 2001). Proteins of the α-amylase/trypsin inhibitor family (Armentia 1994), (Sotkovsky et al. 2008), β-amylase (Pastorello et al. 2007), peroxidases (Pastorello et al. 2007; Sanchez-Monge et al. 1997), LTP1 (Pastorello et al. 2007; Palacin et al. 2007, 2010), and serpins (Pastorello et al. 2007; Sander et al. 2001; Akagawa et al. 2007) were reported to be allergens in food or respiratory allergies. However, it is noteworthy that the reactivity of the proteins within each family is dependent on their isoforms and that IgE obtained from patients recognizes the allergens with variable sensitivity and specificity. Pastorello et al. (2007), using two-dimensional electrophoresis followed by immunoblotting, identified new allergens in this A/G fraction, including granule-bound starch synthases, glyceraldehyde-3P-deshydrogenase, and puroindolines. Until now, members of 14 families of allergens belonging to the A/G fraction and their involvement in wheat allergies has been studied (Tatham and Shewry 2008; Battais et al. 2008; DeAngelis et al. 2010).

Demands raised by food safety issues and food-processing applications involve analytical and labeling issues in relation to the gluten content of food (Wieser 2008; Haraszi et al. 2011; Heick et al. 2011). The activities include finding formulations with functionality similar to wheat gluten (Arendt et al. 2008) and looking for less harmful wheat varieties and wheat relatives (Klockenbring et al. 2001; Nakamura et al. 2005; Spaenij-Dekking et al. 2005; van Herpen et al. 2006; Denery-Papini et al. 2007) as well as methods and technologies to reduce the level of harmful compounds in wheat-based products such as the use of different enzymes (Malandain 2005; Mitea, et al. 2008; Stenman et al. 2010), and/or sour-dough technology (Di Cagno et al. 2004; Decock and Capelle 2005; De Angelis et al. 2006; Ehren et al. 2008; Ganzle et al. 2008, Gobbetti et al. 2008, Goesaert et al. 2008).

The methodology for the identification of allergenic proteins by using the proteomic approach is referred to as "allergenomics" (Yagami et al. 2004; Akagawa et al. 2007). In allergenomics, total proteins in allergen sources are solubilized with a strong anionic detergent and urea and effectively resolved with 2-D electrophoresis. Subsequently, IgE-reactive proteins are detected by immunoblotting using allergic patients' sera. Candidate allergens can be identified by their site-specific degradation and subsequent mass spectroscopic analysis of the fragmented peptides and database search.

Liquid chromatography–tandem mass spectrometry methodologies are considered to be the method of the future assessing gluten-free or low-gluten products (Haraszi et al. 2011; Heick et al. 2011). This method deals with the levels of harmful wheat proteins and the levels of certain nonwheat ingredients used in wheat baked products. Detailed protein analysis of bread products by MALDI-TOF and 2D-GE protein separation followed by MS/MS protein identification has shown that the lupin conglutin proteins are clearly differentiated during the bread-making process into salt-soluble (alfa conglutins) and salt-insoluble classes (beta conglutins) (Islam et al. 2012). These changes in solubility, and the identification of new entities using MALDI-TOF, reflect the wheat–lupin protein interactions that await further study (Islam et al. 2012), especially in relation to physical attributes of lupin proteins such as the higher thermal stability of α-conglutins compared to β-conglutins. In their analysis of different commercially produced wheat starches Kasarda et al. (2008) showed that a proteome level analysis provided a useful "quality control" to identify their gluten contamination through the a-gliadin content.

The combination of component analyses with high-throughput bioinformatic analysis, has a great potential to identify allergen candidates and to develop vaccines against them (Mora and Telford 2010). Spurred by a growing market and the realization that strict gluten exclusion is difficult and often only partially successful, therapies for CD are rapidly evolving (Tye-Din and Anderson 2008). Clinical trials are under way to assess drugs and treatments targeted at reducing paracellular permeability in the small intestine, further digesting gluten in the stomach by way of prolyl, and glutaminyl endopeptidases, or cysteine endoproteases, and inhibiting migration of T cells to the inflamed intestine through mediation by chemokine receptor 9 (CCR9) (Osorio et al. 2012). Other drugs under consideration are inhibitors of intestinal tTG and peptide binding to HLA-DQ2. Agents that qualitatively alter the pathogenic T-cell response to gluten by enhancing regulatory T cells offer the prospect of sustained immune tolerance to gluten. A peptide-based therapeutic vaccine is being developed to test this potential in CD (Anderson et al. 2006). For the latter approach, the most successful immunologic studies now rely upon synthetic peptides or recombinant proteins. Computer-assisted search strategies for toxic components of gluten have been reasonably comprehensive, using public databases such as Genbank (http://www.ncbi.nlm.nih.gov/genbank/). In the past, functional studies have focused on only eight distinct recombinant α-gliadins (from the 58 α-gliadin genes in Genbank) and five γ-gliadins (from 48 genes), but recent studies involved ω-gliadins, LMW and HMW glutenin subunits, even hordeins, or secalins. (Camarca et al. 2009).

19.9.5 Wheat Allergen Candidate Identification for Vaccine Design

Immunoinformatics is the application of informatics techniques to the immunity system for the prediction of epitope targets and vaccine design. This is also referred to as "computer aided vaccine design (CAVD) or computational vaccinology" (Flower and Doytchinova 2002; Sollner et al. 2010). An epitope is the complementary protein region in the antibody (Ab) antigen (Ag) binding sites in Ag–Ab reaction. The epitope sequence in this region can be either conformational (discontinuous) or linear presenting (Huang and Honda 2010, Kulkarni-Kale et al. 2005). The requirements for epitope prediction for vaccine design include the facts that the epitope occurs on the surface of the protein and is more flexible than the rest of the protein.

The epitope region requires a high degree of exposure to the solvent and the amino acids making the epitope charged and hydrophilic.

The main methods for the identification of epitopes currently in use are:

1. The Parker Hydrophilic Prediction Method (Parker et al. 1986) utilized three parameters: hydrophilicity, Janin's scale accessibility, and the flexibility of Karplus and Schultz (1985). The hydrophilicity parameter was calculated using HPLC from the retention coefficient of model synthetic peptides. The surface profile was determined by summing the parameters for each residue of a seven-residue segment and assigning the sum to the fourth residue; this is one of the most useful prediction algorithms (Parker et al. 1986).
2. Surface Accessibility Prediction (Emini et al. 1985).
3. Beta-Turn Prediction (Chou and Fasman 1978).
4. Exposed surface, polarity, and antigenic propensity of polypeptide chains have been correlated with the location of continuous epitopes. (Kolaskar's semi-empirical method uses physiological properties of amino acid residues and the frequencies of occurrence of amino acids in experimentally known epitopes (Kolaskar and Tongaonkar 1990)).

B-cell epitopes are recognized by antibodies of the immune system and are used in the design of vaccines and diagnostics tests. It is therefore of interest to develop improved methods for predicting B-cell epitopes (Larsen et al. 2006; Moolhuijzen 2011).

Peptide epitopes from gliadin and high molecular weight subunits have been identified as associated with wheat-dependent–exercise-induced-anaphylaxis (WDEIA) (Tatham and Shewry 2008). Immunological databases and analysis resources are available to identify candidate epitope peptides; these have been reviewed for high-throughput analysis accessibility (Moolhuijzen 2011; Salimi et al. 2010) (Table 19.2).

To identify wheat vaccine gene candidates, the *Triticum aestivum* gene index (TaGI) (http://compbio.dfci.harvard.edu/tgi/gi/tagi/searching/xpress_search.html) dataset of expressed genes was screened for wheat proteins with candidate epitopic peptide regions that bind B cell IgE.

1. Translate the TaGI into protein sequences.
2. Screen predicted protein sequences for B-cell continuous epitopic residues greater than six amino acids in length.
3. Validate epitopic protein sequences against known protein domains.
4. Partition proteins identified with epitopes into known and unknown protein domain groups.
5. Identify those genes expressed in grain (protein families that may trigger allergic reactions).

Table 19.2 Summary of B-cell epitope prediction tools reviewed

Site	Prediction	URL	HTP
ABCPred	B-cell	http://www.imtech.res.in/raghava/abcpred/	No
Antigenic	B-cell	http://emboss.sourceforge.net/apps/release/5.0/emboss/apps/antigenic.html	Yes
BCEPred	Linear B-cell	http://www.imtech.res.in/raghava/bcepred/index.html	No
BCIPep	B-cell blast database	http://www.imtech.res.in/raghava/bcipep/data.html	Yes
Bepipred B-cell epitopes	Predict location of linear B-cell epitopes	http://www.cbs.dtu.dk/services/BepiPred/	Yes
CEP	Conformational epitope prediction server B-cell	http://bioinfo.ernet.in/cep.htm (no longer available)	No
IEDB	B-cell and Tcell helper epitope tools	http://tools.immuneepitope.org/main/html/bcell_tools.html	No

6. Cluster sequences to identify conserved sequence domains within families of proteins.
7. Group classes of epitope sequences using SDAP. These known epitopic peptides are a positive control to validate the data analysis strategy.
8. Screen wheat candidate epitope peptides against the human proteome and delete entries with more than 70% positive similarity to avoid an allergic reaction in humans to the vaccine.

In the TaGI there exist 216,452 tentative consensus sequences and 12,784 (5.9%) transcripts containing predicted linear B-cCell peptides (greater than six residues in length). To identify the known and unknown protein families the 12,784 B-cell transcripts can be translated into protein sequences and searched against the NCBI conserved domain database using RPS-Blast. All sequences with good alignment to known transcripts (<1e–10) were classified as known protein families. The number of known protein families' transcripts found equaled 8,601 and clustered into (1,363 clusters, 973 singletons) (examples shown in Fig. 19.2). The number of transcripts of unknown protein families equaled 4,183 and clustered into 965 clusters, 2,321 singletons.

19.10 Conclusion

The extensive studies on the wheat proteome reviewed in this chapter provide a good foundation for integrating knowledge about specific genes expressed in the form of proteins and the genes annotated within the wheat genome DNA sequence as this is established through an extensive international effort (http://www.wheatgenome.org).

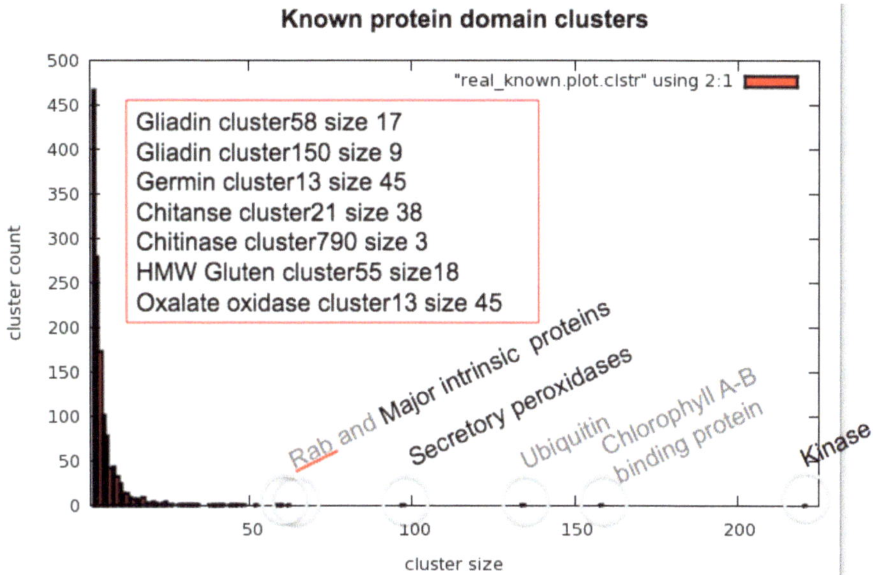

Fig. 19.2 A plot of the number of clustered transcripts with predicted B-cell epitopes versus the cluster size, the larger clusters and circled in grey and the annotation in black are known allergen proteins. The boxed insert lists the other known allergen clusters and the number of transcripts (size of the cluster)

The correct assignment of gene sequences within the genome is fundamental to integrating the DNA-encoded information with the proteome. Large-scale sampling of peptides and the process of relating these to gene sequences in the genome is a key cross-referencing of the proteome back to the hypothesized gene sequences in order to validate gene assignments. The analysis of the wheat proteome with the view to application in the analysis of foods is advancing quickly as the result of progress in the human and yeast fields. The srmatlas project (Tian et al. 2012; http://www.srmatlas.org) is focused on selected reaction monitoring (SRM) mass spectrometry assays (Anderson and Hunter 2006; Lange et al. 2008; Picotti et al. 2009) for all human proteins (http://www.srmatlas.org/). The SRM mass spectrometry assays utilize a triple quadrupole mass spectrometer to identify a peptide(s) associated with a protein of interest in an iterative process eventually to provide highly accurate identification and quantification. The SRM assays can detect multiple targets in complex samples. The SRM initiative complements the human protein atlas (http://www.proteinatlas.org/) which is focused on antibody-based proteomics. Version 8.0 of the human protein atlas, released in May 2011, included 11,260 genes with protein expression profiles based on 14,506 antibodies (Ponten et al. 2011). These advances provide a valuable vision for the analysis of the wheat proteome. The iMOP (http://www.imop.uzh.ch) process is now in place to facilitate the transfer of the experience of colleagues within the Human Proteome Organization (HUPO) to analyzing the proteomes of other organisms (Jones et al. 2012; Orchard et al. 2012).

The selected classes of wheat proteins reviewed in detail in this chapter can be detected in food products (Islam et al. 2012; Mamone et al. 2009) using standard denaturing extraction techniques and already provide valuable information to be used in the whole wheat chain from pre-breeding through to process and product development, process control, and quality assurance. The latter application, diagnostics for identity validation and predicting the presence of toxic/allergic epitopes in foodstuffs, is a key issue for human health and also for the future of the wheat industry. The expansion of detection technologies to specific peptides using SRM mass spectrometry assays would greatly facilitate the association of particular proteins with quality attributes of foods. The advanced, high-throughput, mass spectroscopic detection protocols can also facilitate the screening of breeding lines and cultivars to provide a basis for targeting particular wheat lines in a breeding program to specific food products. The detailed and high-resolution analyses of the grain proteome by Dupont et al. (2011) provide the basis for assaying specific proteins associated with quality attributes. The well-established example discussed in this chapter relates to the association of a specific complement of HMW glutenin subunit proteins with bread quality and the high-throughput detection of these "fingerprints" using MALDI-TOF. Many other classes of proteins have been or will be associated with particular end-product attributes and form the basis for an essential suite of tools for characterizing food components and end-products. In these developments centers for carrying out the high-resolution and high-throughput technologies and the associated bioinformatics play a critical role in data capture and management, metadata capture, minimum descriptions for experiments, analysis, and integration (e.g., Barga et al. 2011; Bellgard 2005; Rhee et al. 2006; Roessner et al. 2011; Taylor et al. 2008). A particular bioinformatics challenge is the integration of disparate data and information sources from the new -omics technologies through to old legacy databases and/or spreadsheets and fieldbooks that contain information such as phenotypic or epidemiological data that are not necessarily in easily exchangeable data formats (Goble and Stevens 2008).

References

Agrawal GK, Rakwal R (2006) Rice proteomics: a cornerstone for cereal food crop proteomes. Mass Spectrum Rev 25:1–5

Ahmed J, Ramaswamy H S, Ayad A, All, Alvarez P (2007) Effect of high-pressure treatment on rheological, thermal and structural changes in Basmati rice flour slurry. J Cereal Sci 46:148–156

Akagawa M, Handoyo T, Ishii T, Kumazawa S, Morita N, Suyamam K (2007) Proteomic analysis of wheat flour allergens. J Agric Food Chem 55:6863–6870

Altenbach SB, Dupont FM, Kothari KM, Chan R (2003) Temperature water and fertilizer influence the timing of key events during grain development in a US spring wheat. J Cereal Sci 37:9–20

Altenbach SB, Tanaka CK, Hurkman WJ, Whitehand LC, William H, Vensel WH, Dupont FM (2011) Differential effects of a post-anthesis fertilizer regimen on the wheat flour proteome determined by quantitative 2-DE. Proteome Sci 9(1–13):46

Amiour N, Merlino M, Leroy P, Branlard G (2002) Proteomic analysis of amphiphilic proteins of hexaploid wheat kernels. Proteomics 2:632–641
Amiour N, Merlino M, Leroy P, Branlard G (2003) Chromosome mapping and identification of amphiphilic proteins of hexaploid wheat kernels Theor. Appl Genet 108:62–72
Anderson L, Hunter CL (2006) Quantitative mass spectrometric multiple reaction monitoring assays for major plasma proteins. Mol Cell Proteomics 5:573–588
Anderson RP, Wieser H (2006) Medical applications of gluten-composition knowledge. In: Wrigley CW, Békés F, Bushuk W (eds) Gliadin and glutenin the unique balance of wheat quality. AACCI Press, St Paul, pp 387–409
Anderson RP, Hill AVS, Jewell DP (2006) Diagnosis of celiac disease using gliadin epitopes. US Patent 7,144,569 B1
Andon NL, Hollingworth S, Koller A, Greenland A, Haynes PA (2002) Proteomic characterization of wheat amyloplasts using identification of proteins by tandem mass spectrometry. Proteomics 2:1156–1168
Antoine C, Peyon S, Mabille F, Lapierre C, Bouchet B, Abecassis J, Rouau X (2003) Individual contribution of grain outer layers and their cell wall structure to the mechanical properties of wheat bran. J Agric Food Chem 51:2026–2033
Antoine C, Peyron S, Lullien-Pellerin V, Abecassis J, Rouau X (2004) Wheat bran tissue fractionation using biochemical markers. J Cereal Sci 39:387–393
Appelbee MJ, Mekuria GT, Nagasandra V, Bonneau JP, Eagles HA, Eastwood RF, Mather DE (2009) Novel allelic variants encoded at the Glu-D3 locus in bread wheat. J Cereal Sci 49:254–261
Appels R, Barsby T, Risacher T, Bekes F (2011) Linking the genome to phenotypes in wheat: advances in technologies and concepts. In: Bonjean AP, Angus WJ, van Ginkel M (eds) The world wheat book, 2nd edn. Lavoisier, Paris/New York, pp 709–748
Appels R, Barrero R, Bellgard M (2012) Advances in biotechnology and linking outputs to variation in complex traits: Plant and animal genome meeting January 2012, Functional & Integrative Genomics, 12: DOI: 10.1007/s10142-012-0270-7
Arendt EK, Renzetti S, Moore MM (2008) Novel approaches in the design of gluten-free cereal products. Food Sci Technol 22:43–46
Armentia A (1994) Allergens associated with baker's asthma. Allergy 49:906
Ball SG, Morell MK (2003) From bacterial glycogen to starch: understanding the biogenesis of the plant starch granule. Annu Rev Plant Biol 54:207–233
Balmer Y, Vensel WH, Dupont FM, Buchanan BB, Hurkman WJ (2006a) Proteome of amyloplasts isolated from developing wheat endosperm presents evidence of broad metabolic capability. J Exp Bot 57:1591–1602
Balmer Y, Venselm WH, Cai N, Manieriem W, Schurmann P, Hurkman WJ, Buchanan BB (2006b) A complete ferredoxin-thioredoxin system regulates fundamental processes in amyloplasts. Proc Natl Acad Sci USA 103:2988–2993
Bancel E, Rogniaux H, Debiton C, Chambon C, Branlard G (2010) Extraction and proteome analysis of starch granule-associated proteins in mature wheat kernel. J Proteome Res 9:3299–3310
Banziger M, Feil B, Stamp P (1994) Competition between nitrogen accumulation and grain growth for carbohydrates during grain filling of wheat. Crop Sci 34:440–446
Baracskai I, Balázs G, Liu L, Ma W, Oszvald M, Newberry M, Tömösközi S, Láng L, Bedő Z, Békés F (2011) A retrospective analysis of HMW and LMW glutenin alleles of cultivars bred in Martonvásár. Hungary Cereal Res Comm 39:225–236
Barga R, Howe B, Beck D, Bowers S, Dobyns W, Haynes W, Higdon R, Howard C, Roth C, Stewart E, Welch D, Kolker E (2011) Bioinformatics and data-intensive scientific discovery in the beginning of the 21st century. OMICS 15:199–201
Battais F, Coucoux P, Popineau Y, Kanny G, Moneret-Vautrin DA, Denery-Papini S (2005a) Food allergy to wheat: differences in immunoglobulin E-binding proteins as a function of age or symptoms. J Cereal Sci 42:109–117
Battais F, Mothes T, Moneret-Vautrin DA, Pineau F, Kanny G, Popineau Y (2005b) Identification of IgE-binding epitopes on gliadins for patients with food allergy to wheat. Allergy 60:815–821

Battais F, Richard C, Jacquenet S, Denery-Papini S, Moneret-Vautrin DA (2008) Wheat grain allergies: an update on wheat allergens. Eur Ann Allergy Clin Immunol 40:67–76
Beasley HL, Uthayakumaran S, Stoddard FL, Partidge SJ, Daqiq L, Chong P, Békés F (2002) Synergistic ans additive effects of three HMW-GS loci. II. Effects on wheat dough functionality and end-use quality. Cereal Chem 79:301–307
Bechtel DB, Wilson JD (2003) Amyloplast formation and starch granule development in hard red winter wheat. Cereal Chem 80:175–183
Békés F, Keméeny S, Morell MK (2006) An integrated approach to predicting end-product quality of wheat. Eur J Agron 25:155–162
Bellgard MI (2005) Bioinformatics from comparative genomic analysis through to integrated systems. In: Ruvinsky A, Marshall-Graves J (eds) Mammalian genomics. CABI, Cambridge, MA, pp 393–409
Berthiller F, Dall'Asta C, Schuhmacher R, Lemmens M, Adam G, Krska R (2005) Masked mycotoxins: determination of a deoxynivalenol glucoside in artificially and naturally contaminated wheat by liquid chromatography-tandem mass spectrometry. J Agric Food Chem 53:3421–3425
Bettge AD, Morris FM (2000) Relationships among grain hardness, pentosan fractions, and end-use quality of wheat. Cereal Chem 77:241–247
Bhatnagar D, Rajasekaran K, Payne GA, Brown RL, Yu J, Cleveland TE (2008) The 'omics' tools: genomics, proteomics, metabolomics and their potential for solving the aflatoxin contamination problem. World Mycotoxin J 1:3–12
Blein JP, Coutos-Thevenot P, Marion D, Ponchet M (2002) From elicitins to lipid-transfer proteins: a new insight in cell signalling involved in plant defence mechanisms. Trends Plant Sci 7:293–296
Bollecker SSJ, Kaiser KP, Köhler P, Wieser H, Schofield JD (1998) Re-examination of the glycosylation of high M_r Subunits of wheat glutenin. J Agric Food Chem 46:4814–4823
Braly J, Hogganm R (2002) Dangerous grains why gluten cereal grains may be hazardous to your health. Penguin Group, New York
Branlard G, Bancel E (2007) Protein extraction from cereal seeds. In: Thiellement H, Zivy M, Damerval C, Méchin V (eds) Plant proteomics: methods and protocols, vol 355, Methods in molecular biology. Humana Press, Totowa, pp 15–25
Branlard G, Amiour NI, Gaborit T, Herbette S, Dardevet M, Marion D (2003a) Diversity of puroindolines as revealed by two-dimensional electrophoresis. Proteomics 3:168–174
Branlard G, Dardevet M, Amiour N, Igrejas G (2003b) Allelic diversity of HMW and LMW glutenin subunits and omega-gliadins in French bread wheat (*Triticum aestivum* L.). Genet Resour Crop Evol 50:669–679
Brown BD, Petrie S (2006) Irrigated hard winter wheat response to fall spring and late season applied nitrogen. Field Crops Res 96:260–268
Bykova NV, Hoehn B, Rampitsch C, Banks TW, Stebbing JA, Fan T, Knox RE (2011) Redox-sensitive proteome and antioxidant strategies in wheat seed dormancy control. Proteomics 11:865–882
Camarca A, Anderson RP, Mamone G, Fierro O, Facchiano A, Costantini S, Zanzi D, Sidney J, Auricchio S, Sette A, Troncone R, Gianfrani C (2009) Intestinal T cell responses to gluten peptides are largely heterogeneous: implications for a peptide-based therapy in celiac disease. J Immunol 182:4158–4166
Carbonaro M (2004) Proteomics: present and future in food quality evaluation. Trends Food Sci Technol 15:209–216
Catassi C, Fasano A (2008) Celiac disease. In: Gluten-free cereal products and beverages Ed EK Arendt F and Dal Bello. Academic, San Diego, pp 1–28
Chou PY, Fasman GD (1978) Prediction of the secondary structure of proteins from their amino acid sequence. Adv Enzymol Relat Areas Mol Biol 47:45–148
Cornish GB, Siriamornpun S, Skylass D, Békés F, Wrigley CW, Wooton M (2001) HMW and LMW glutenin subunit and gliadin protein markers in genetic maps. Aust J Agric Res 52:1161–1171
Cornish GB, Békés F, Eagles HA, Payne PI (2006) Prediction of dough properties for bread wheats. In: Wrigley CW, Békés CF, Bushuk W (eds) Gliadin and glutenin. The unique balance of wheat quality. AACC Press, St. Paul, pp 243–280

Creemers J (2002) Proteolytic enzymes as therapeutic targets – keystone symposium. I Drugs 5:216–219
D'Ovidio R, Masci S (2004) The low-molecular-weight glutenin subunits of wheat gluten. J Cereal Sci 39:321–339
De Angelis M, Coda R, Silano M (2006) Fermentation by selected sourdough lactic acid bacteria to decrease the intolerance to rye and barley flours. J Cereal Sci 43:301–314
De Angelis M, Minervini F, Caputo L, Cassone A, Coda R, Calasso MP (2008) Proteomic analysis by two-dimensional gel electrophoresis and starch characterization of *Triticum turgidum* L. var. durum cultivars for pasta making. J Agric Food Chem 56:8619–8628
De Angelis M, Di Cagno R, Minervini F, Rizzello CG, Gobbetti M (2010) Two-dimensional electrophoresis and IgE-mediated food allergy. Electrophoresis 31:2126–2136
Debiton C, Merlino C, Chambon C, Bancel E, Decourteix D, Planchot V, Branlard G (2011) Analyses of albumins globulins and amphiphilic proteins by proteomic approach give new insights on waxy wheat starch metabolism. J Cereal Sci 53:160–169
Decock P, Capelle S (2005) Bread technology and sourdough technology. Trends Food Sci Technol 16:113–120
Denery-Papini S, Lauriere M, Branlard G, Morisset M, Pecquet C, Choudat D (2007) Influence of the allelic variants encoded at the Gli-B1 locus responsible for a major allergen of wheat on IgE reactivity for patients suffering from food allergy to wheat. J Agric Food Chem 55:799–805
Denyer K, Pike M (2008) Isolation of amyloplasts. In: Bonifacino JS, Dasso M, Harford JB, Lippincott-Schwartz J, Yamada KM (eds) Current protocols in cell biology. Wiley, New York, pp 3281–3285
Desmond OJ, Edgar CI, Manners JM, Maclean DJ, Schenk PM, Kazan K (2006) Methyl jasmonate induced gene expression in wheat delays symptom development by the crown rot pathogen Fusarium pseudograminearum. Physiol Mol Plant Pathol 67:171–179
Di Cagno R, De Angelis M, Auricchio S (2004) A sour-dough bread made from wheat and non-toxic flours and started with selected lactobacilli is tolerated in Celiac Sprue. Appl Environ Microbiol 70:1088–1096
Di Luccia A, Lamacchia C, Fares C, Padalino L, Mamone G, La Gatta B, Gambacorta G, Faccia M, Di Fonzo N, La Notte E (2005) A proteomic approach to study protein variation in GM durum wheat in relation to technological properties of semolina. Ann Chim (Rome) 95:405–414
Dunbar BD, Bundman DS, Dunbar BS. (1985) Identification of cultivarspecific proteins of winter wheat (*T. aestivum L.*) by high resolution twodimensional polyacrylamide gel electrophoresis and color-based silver stain. Electrophoresis 6:39–43
Dunwell JM, Khuri S, Gane PJ (2000) Microbial relatives of the seed storage proteins of higher plants: conservation of structure and diversification of function during evolution of the cupin superfamily. Microbiol Mol Biol Rev 64:153–179
Dupont FM (2008) Metabolic pathways of the wheat (*Triticum aestivum*) endosperm amyloplast revealed by proteomics. BMC Plant Biol 8:39
Dupont FM, Hurkman WJ, Vensel WH, Chan R, Lopez R, Tanaka CK, Altenbach SB (2006) Differential accumulation of sulfur-rich and sulfur-poor wheat flour proteins is affected by temperature and mineral nutrition during grain development. J Cereal Sci 44:101–112
Dupont FM, Vensel WH, Tanaka CK, Hurkman WJ, Altenbach SB (2011) Deciphering the complexities of the wheat flour proteome using quantitative two-dimensional electrophoresis, three proteases and tandem mass spectrometry. Proteome Sci 9:10
Dworschak RG, Ens W, Standing KG, Preston KR, Marchylo BA, Nightingale MJ, Stevenson SG, Hatcher DW (1998) Analysis of wheat gluten proteins by matrix-assisted laser desorption/ionization mass spectrometry. J Mass Spectrom 33:429–435
Eagles HA, Hollamby GJ, Gororo NN, Eastwood RF (2002) Estimation and utilisation of glutenin gene effects from the analysis of unbalanced data from wheat breeding programs. Aust J Agric Res 53:367–377
Ehren J, Govindarajan S, Morón B, Minshull J, Khosla C (2008) Protein engineering of improved prolyl endopeptidases for celiac sprue therapy. Protein Eng Des Sel 21:699–707

Emini EA, Hughes JV, Perlow DS, Boger J (1985) Induction of hepatitis A virus-neutralizing antibody by a virus-specific synthetic peptide. J Virol 55:836–839

Ferranti P, Mamone G, Picariello G, Addeo F (2007) Mass spectrometry analysis of gliadins in celiac disease. J Mass Spectrom 42:1531–1548

Ferro M, Salvi D, Brugiere S, Miras S, Kowalski S, Louwagie M, Garin J, Joyard J, Rolland N (2003) Proteomics of the chloroplast envelope membranes from *Arabidopsis thaliana*. Mol Cell Proteomics 2:325–345

Fido RJ, Mills CEN, Rigby NM, Shewry PR (2004) Protein extraction from plant tissues. In: Cutler P (ed) Methods in molecular biology, vol 244, 2nd edn, Protein purification protocols. Humana Press, Totowa

Finnie C, Steenholdt T, Noguera OR, Knudsen S, Larsen J, Brinch-Pedersen H, Holm PB, Olsen O, Svensson B (2004) Environmental and transgene expression effects on the barley seed proteome. Phytochemistry 65:1619–1627

Finnie C, Sultan A, Grasser KD (2011) From protein catalogues towards targeted proteomics approaches in cereal grains. Phytochemistry 72:1145–1153

Flate NES, Hollung K, Ruud L, Sogn T, Færgestad EM, Skarpeid HJ, Magnus EM, Uhlen AK (2005) Combined nitrogen and sulphur fertilisation and its effect on wheat quality and protein composition measured by SE-FPLC and proteomics. J Cereal Sci 41:357–369

Flower DR, Doytchinova IA (2002) Immunoinformatics and the prediction of immunogenicity. Appl Bioinformatics 1:167–176

Ford R (2008) The gluten syndrome is wheat causing you harm? RRS Global LT, Christchurch

Futcher B, Latter GI, Monardo P, McLaughlin CS, Garrels JI (1999) A sampling of the yeast proteome. Mol Cell Biol 19:7357–7368

Gane PJ, Dunwell JM, Warwicker J (1998) Modeling based on the structure of vicilins predicts a histidine cluster in the active site of oxalate oxidase. J Mol Evol 46:488–493

Ganzle MG, Loponen J, Gobbetti M (2008) Proteolysis in sourdough fermentation: mechanism and potential for improved bread quality. Trends Food Sci Technol 19:513–521

Gao L, Ng PKW, Bushuk W (1992) Structure of glutenin based on farinograph and electrophoretic results. Cereal Chem 69:452–455

Gao L, Wang A, Li X, Dong K, Wang K, Appels R, Ma W, Yan Y (2009) Wheat quality related differential expressions of albumins and globulins revealed by two-dimensional difference gel electrophoresis (2-D DIGE). J Proteomics 73:279–296

Gao X, Appelbee MJ, Mekuria GT, Chalmers KJ, Mather DE (2012) A second 'overexpression' allele at the Glu-B1 high-molecular-weight glutenin locus of wheat: sequence characterisation and functional effects. Theor Appl Genet 124:333–343

Gaso-Sokac D, Kovac S, Josic D (2010) Application of proteomics in food technology and food biotechnology: process development, quality control and food safety. Food Technol Biotechnol 48:284–295

Gautier M-F, Aleman ME, Guirao A, Marion D, Joudrier P (1994) *Triticum aestivum* puroindolines two basic cystine-rich seed proteins: cDNA sequence analysis and developmental gene expression. Plant Mol Biol 25:43–57

Ghirardo A, Sorensen HA, Petersen M, Jacobsen S, Sondergaard I (2005) Early prediction of wheat quality, analysis during grain development using mass spectrometry and multivariate data analysis. Rapid Commun Mass Spectrom 19:525–532

Gobaa S, Brabant C, Kleijer G, Stamp P (2007) Effect of the 1BL.1RS translocation and of the Glu-B3 variation on fifteen quality tests in a doubled haploid population of wheat. J Cereal Sci 41:181–192

Gobbetti M, De Angelis M, Di Cagno R, Rizzello CG (2008) Sourdough/lactic acid bacteria. In: EK Arendt F, Dal B (eds) Gluten-free cereal products and beverages. Academic Press, San Diego, USA pp 267–288

Goble C, Stevens R (2008) State of the nation in data integration for bioinformatics. J Biomed Informatics 41:687–693

Goesaert H, Courtin CM, Delcour JA (2008) Use of enzymes in the production of cereal-based functional foods and food ingredients. In: EK Arendt F, Dal B (eds) Gluten-free cereal products and beverages. Academic, San Diego, pp 237–266

Görg A, Weiss W, Dunn MJ (2004) Current two-dimensional electrophoresis technology for proteomics. Proteomics 4:3665–3685

Görg A, Klaus A, Lück C, Weiland F, Weiss W (2007) Two-dimensional electrophoresis with immobilized pH gradients for proteome analysis. A laboratory manual. http://www.wzw.tum.de/proteomik

Gorlach J, Volrath S, Knauf-Beiter G, Hengy G, Beckhove U, Kogel KH, Oostendorp M, Staub T, Ward E, Kessmann H (1996) Benzothiadiazole, a novel class of inducers of systemic acquired resistance, activates gene expression and disease resistance in wheat. Plant Cell 8:629–643

Gottlieb DM, Schultz J, Petersen M, Nesic L, Jacobsen S, Sondergaard I (2002) Determination of wheat quality by mass spectrometry and multivariate data analysis. Rapid Commun Mass Spectrom 16:2034–2039

Groos C, Bervas E, Charmel G (2006) Genetic analysis of grain protein content, grain hardness and dough rheology in a hard×hard bread wheat progeny. J Cereal Sci 40:93–100

Guo H, Zhang Y, Li Y, Ren J, Wang X, Niu H, Yin J (2011) Identification of changes in wheat seeds proteome in response to anti-trx s gene. PLoS ONE I 6, 7 I e222SS

Gupta RB, Békés F, Wrigley CW (1991) Prediction of physical dough properties from glutenin subunit composition in bread wheats: correlation studies. Cereal Chem 68:328–333

Gygi SP, Rochon Y, Franza BR, Aebersold R (1999) Correlation between protein and mRNA abundance in yeast. Mol Cell Biol 19:1720–1730

Haraszi R, Chassaigne H, Macquet A, Ulberth F (2011) Analytical methods for detection of gluten in food – last method developments in support to the legislations on labelling of foodstuffs. J AOAC 94:1–20

Hashiguchi A, Ahsan N, Komatsu S (2010) Proteomics application of crops in the context of climatic changes. Food Res Int 43:1803–1813

Hawker JS, Jenner CF (1993) High temperature affects the activity of enzymes in the committed pathway of starch synthesis in developing wheat endosperm. Aust J Plant Physiol 20:197–209

Heick J, Fischer M, Kerbach S, Tamm U, Popping B (2011) Application of a liquid chromatography tandem mass spectrometry method for the simultaneous detection of seven allergenic foods in flour and bread and comparison of the method with commercially available ELISA test kits. J AOAC 94:1–9

Herrero M, Simó C, García-Canas V, Ibanez E, Cifuentes A (2011) Foodomics: MS-based strategies in modern food science and nutrition. Mass Spectrom Rev. doi:10.1002/mas.20335

Horváth-Szanics E, Szabó Z, Janáky T, Pauk J, Gy H (2006) Proteomics as an emergent tool for identification of stress-induced proteins in control and genetically modified wheat lines. Chromatographia 63:S143–S147

Houba RD, Heederik G, Doekes G (1998) Wheat sensitization and work-related symptoms in the baking industry are preventable. An epidemiologic study. Am J Resp Critical Care Med 158:1499–1503

Huang J, Honda W (2006) CED: a conformational epitope database. BMC Immunol 7:7

Hurkman WJ, McCue KF, Altenbach SB, Korn A, Tanaka CK, Johnson EL, Bechtel DB, Anderson OD, DuPont FM (2003) Expression of genes for starch biosynthesis is regulated by high temperature in developing wheat endosperm. Plant Sci 164:873–881

Hurkman WJ, Vensel WH, DuPont FM, Altenbach SA, Buchanan BB (2008) Endosperm and amyloplast proteomes of wheat grain Chapter 14. In: Agrawa GK, Rakwai GK (eds) Plant proteomics technologies strategies and applications. Wiley, London, pp 23–47

Hurkman WJ, Vensel WH, Charlene CH, Tanaka K, Whitehand L, Altenbach SB (2009) Effect of high temperature on albumin and globulin accumulation in the endosperm proteome of the developing wheat grain. J Cereal Sci 49:12–23

Islam N, Woo SH, Tsujimoto H, Kawasaki H, Hirano H (2002) Proteome approaches to characterize seed storage proteins related to ditelocentric chromosomes in common wheat. Proteomics 2:1146–1155

Islam N, Tsujimoto H, Hirano H (2003a) Proteome analysis of diploid tetraploid and hexaploid wheat: towards understanding genome interaction in protein expression. Proteomics 3:549–557

Islam N, Tsujimoto H, Hirano H (2003b) Wheat proteomics: relationship between fine chromosome deletion and protein expression. Proteomics 3:307–316

Islam S, Ma W, Yan G, Békés F, Appels R (2012) Modifying processing and health attributes of wheat bread through changes in composition genetics and breeding. In: Bread making Improving quality, 2nd edn. SP Cauvain. CRC Press, Boston, pp 259–296

Jackson EA, Morel MH, Sontag-Strohm T, Branlard G, Metakovsky EV, Redaelli R (1996) Proposal for combining the classification systems of alleles of Gli-1 and Glu-3 loci in bread wheat (*Triticum aestivum* L.). J Genet Breed 50:321–336

Jerkovicz A, Kriegel AM, Bradner JR, Atwell BJ, Roberts TH, Willows RD (2010) Strategic distribution of protective proteins within bran layers of wheat protects the nutrient-rich endosperm. Plant Physiol 152:1459–1470

Jones B, Morris C, Békés F, Wrigley CW (2006) Proteins that complement the roles of gliadin and glutenin. In: Wrigley CW, Békés F, Bushuk W (eds) Gliadin and glutenin: the unique balance of wheat quality. AACC Int, St Paul, pp 413–446

Jones AME, Ruedi Aebersold R, Ahrens CH, Rolf A, Katja Baerenfaller K, Baker M, Bendixen E, Briggs S, Philip Brownridge P-B, Erich Brunner E, Daube M, Deutsch EW, Ueli Grossniklaus U, Joshua Heazlewood J, Hengartner MO, Hermjakob H, Jovanovic M, Lawless C, Lochnit G, Martens L, Ravnsborg C, Sabine P, Schrimpf SP, Shim Y-H, Subasic D, Tholey A, van Wijk K, von Mering C, Weiss M, Zheng X (2012) The HUPO initiative on model organism proteomes, iMOP. Proteomics 12:340–345

Jorrin V, Maldonado AM, Castillejo MA (2007) Plant proteome analysis: a 2006 update. Proteomics 7:2947–2962

Juhász A, Makai S, Sebestyén E, Tamás L, Balázs E (2011) Role of conserved non-coding regulatory elements in LMW glutenin gene expression. PLoS ONE 6(12):e29501

Kamal AHM, Kim KH, Shin DH, Seo HS, Shin KH, Park CS, Heo HY, Woo SH (2009) Proteomics profile of pre-harvest sprouting wheat by using MALDI-TOF mass spectrometry. Plant Omics J 2:110–119

Karplus PA, Schulz GE (1985) Prediction of chain flexibility in proteins – a tool for the selection of peptide antigens. Naturwissenschaften 72:212–213

Kasarda DD, Dupont FM, Vensel WH, Altenbach SB, Lopez R, Tanaka CK, Hurkman WJ (2008) Surface-associated proteins of wheat starch granules: suitability of wheat starch for celiac patients. J Agric Food Chem 56:10292–10302

Katavic V, Agrawal GK, Hajduch M, Harris SL, Thelen JJ (2006) Protein and lipid composition analysis of oil bodies from two Brassica napus cultivars. Proteomics 6:4586–4598

Kav NNV, Srivastava S, Yajima W, Sharma N (2007) Application of proteomics to investigate plant-microbe interactions. Curr Proteomics 4:28–43

Kersten B, Bürkle L, Kuhn EJ, Giavalisco P, Konthur Z, Lueking A, Walter G, Eickhoff H, Schneider U (2002) Large-scale plant proteomics. Plant Mol Biol 48:133–141

Kleffmann T, Russenberger D, von Zychlinski A, Christopher W, Sjolander K, Gruissem W, Baginsky S (2004) The *Arabidopsis thaliana* chloroplast proteome reveals pathway abundance and novel protein functions. Curr Biol 14:354–362

Klockenbring T, Boese IA, Bauer IR, Goerlich R (2001) Comparative investigations of wheat and spelt cultivars: IgA IgE IgG 1 and IgG4 binding characteristics. Food Agric Immunol 13:171–181

Kokornaczyk MO, Dinelli G, Marotti I, Benedettelli S, Nan D, Betti L (2011) Self-organized crystallization patterns from evaporating droplets of common wheat grain leakages as a potential tool for quality analysis. Scientific World J 11:1712–1725

Kolaskar AS, Tongaonkar PC (1990) A semi-empirical method for prediction of antigenic determinants on protein antigens. FEBS Lett 276:172–174

Koller A, Washburn MP, Lange BM, Andon NL, Deciu C, Haynes PA, Hays L, Schieltz D, Ulaszek R, Wei J et al (2002) Proteomic survey of metabolic pathways in rice. Proc Natl Acad Sci USA 99:11969–11974

Komatsu S, Yano H (2006) Update and challenges on proteomics in rice. Proteomics 6:4057–4068

Konik-Rose C, Thistleton J, Chanvrier H, Tan I, Halley P, Gidley M, Kosar-Hashemi B, Wang H, Larroque O, Ikea J, McMaugh S, Regina A, Rahman A, Morell MK, Li Z (2007) Effects of starch synthase IIa gene dosage on grain protein and starch in endosperm of wheat. Theor Appl Genet 115:1053–1065

Kosar-Hashemi B, Irwin JA, Higgins J, Rahman S, Morell MM (2006) Isolation, identification and characterization of starch-interacting proteins by 2-D affinity electrophoresis. Electrophoresis 27:1832–1839

Kulkarni-Kale U, Bhosle S, Kolaskar AS (2005) CEP: a conformational epitope prediction server. Nucleic Acids Res. 33(Web Server issue):W168–171

Laino P (2007) The effects of the abiotic and biotic stresses on the wheat proteome. Thesis, Univ. Tuscia, Viterbo, Italy

Laino P, Shelton D, Finnie C, De Leonardis AM, Mastrangelo AM, Svensson B, Lafiandra D, Masci S (2010) Comparative proteome analysis of metabolic proteins from seeds of durum wheat (cv.Svevo) subjected to heat stress. Proteomics 10:2359–2368

Lange V, Malmstrom JA, Didion J, King NL, Johansson BP, Schafer J, Rameseder J, Wong C-H, Deutsch EW, Brusniak M-Y, Lange V, Malmstrom JA, Didion J, King NL, Johansson BP, Schafer J, Rameseder J, Wong C-H, Deutsch EW, Brusniak M-Y, Buhlmann P, Bjorck L, Domon B, Aebersold R (2008) Targeted quantitative analysis of *Streptococcus pyogenes* virulence factors by multiple reaction monitoring. Mol Cell Proteomics 7:1489–1500

Larré C, Lupi R, Gombaud G, Brossard C, Branlard G, Moneret-Vautrin DA, Rogniaux H, Denery-Papini S (2011) Assessment of allergenicity of diploid and hexaploid wheat genotypes: identification of allergens in the albumin/globulin fraction. J Proteomics 74:1279–1289

Larsen JE, Lund O, Nielsen M (2006) Improved method for predicting linear B-cell epitopes. Immunome Res 2:2

Laubin B, Lullien-Pellerin V, Nadaud I, Gaillard-Martinie B, Chambon C, Branlard G (2008) Isolation of the wheat aleurone layer for 2-D electrophoresis and proteomics analysis. J Cereal Sci 48:709–714

Lesage VS, Merlino M, Chambon C, Bouchet B, Marion D, Branlard G (2012) Proteomes of hard and soft near-isogenic wheat lines reveal that kernel hardness is related to the amplification of a stress response during endosperm development. J Exp Bot 63:1001–1011

Lindsay MP, Tamás L, Appels R, Skerritt JH (2000) Direct evidence that number and location of cysteine residues affect glutenin polymer structure. J Cereal Sci 31:32–333

Link AJ, Eng J, Schieltz DM, Carmack E, Mize GJ, Morris DR, Garvik BM, Yates JR III (1999) Direct analysis of protein complexes using mass spectrometry. Nat Biotechnol 17:676–682

Liu L, Wang AL, Appels R, Ma JH, Xia XC, Lan P, He ZH, Bekes F, Yan YM, Ma WJ (2009) A MALDI-TOF based analysis of high molecular weight glutenin subunits for wheat breeding. J Cereal Sci 50:295–301

Liu L, Ikeda TM, Branlard G, Pena RJ, Rogers WJ, Lerner SE, Kolman MA, Xia X, Wang L, Ma W, Appels R, Yoshida H, Wang A, Yan Y, He Z (2010) Comparison of low molecular weight glutenin subunits identified by SDS-PAGE, 2-DE, MALDI-TOF-MS and PCR in common wheat. BMC Plant Biol 10:124

Loukas A (2002) Proteolytic enzymes as therapeutic targets – keystone symposium. Targeting ICE ACE IDrugs 5:220–221

Majoul T, Bancel E, Triboi E, BenHamida J, Branlard G (2003) Proteomic analysis of the effect of heat stress on hexaploid wheat grain: characterisation of heat responsive proteins from total endosperm. Proteomics 3:175–183

Majoul T, Bancel E, Triboi E, BenHamida J, Branlard G (2004) Proteomic analysis of the effect of heat stress on hexaploid wheat grain: characterisation of heat responsive proteins from non prolamins fraction. Proteomics 4:505–513

Mak Y, Skylas DJ, Willows R, Connolly A, Cordwell SJ, Wrigley CW, Sharp PJ, Copeland L (2006) A proteomic approach to the identification and characterisation of protein composition in wheat germ Funct Integr Genomics 6:322–337
Mak Y, Willows RD, Roberts TH, Wrigley CW, Sharp PJ, Copeland L (2009) Germination of wheat: a functional proteomics analysis of the embryo. Cereal Chem 86:281–289
Malandain H (2005) Transglutaminases: a meeting point for wheat allergy celiac disease and food safety. Eur Ann Allergy Clin Immunol 37:397–403
Mamone G, De Caro S, Di Luccia A, Addeo F, Ferranti P (2009) Proteomic-based analytical approach for the characterization of glutenin subunits in durum wheat. J Mass Spectrometry 44:1709–1723
Matsuo H, Morita E, Tatham AS, Morimoto K, Horikawa T, Osuna H (2004) Identification of the IgE-binding epitope in omega-5 gliadin a major allergen in wheat-dependent exercise-induced anaphylaxis. J Biol Chem 279:12135–12140
Mellersh DG, Foulds IV, Higgins VJ, Heath MC (2002) H_2O_2 plays different roles in determining penetration failure in three diverse plant-fungal interactions. Plant J 29:257–268
Merlino M, Leroy P, Chambon C, Branlard G (2009) Mapping and proteomic analysis of albumin and globulin proteins in hexaploid wheat kernels (*Triticum aestivum* L). Theor Appl Gen 118:1321–1337
Metakovsky EV, Branlard G, Graybosch RA, Békés F, Cavanagh CR, Wrigley CW, Bushuk W (2006) The gluten composition of wheat varieties and genotypes. Part I. Gliadin composition table. www.aaccnet.org
Miernyk JA, Hajduch B (2011) Seed proteomics. J Proteomics 74:389–400
Mitea C, Havenaar R, Drijfhout JW, Edens L, Dekking L, Koning F (2008) Efficient degradation of gluten by a prolyl endoprotease in a gastrointestinal model: implications for coeliac disease. Gut 57:25–32
Moolhuijzen P (2011) Bioinformatics approaches for functional predictions in diverse informatics environments. BSc Thesis, Murdoch University, Perth, WA, Australia
Mora M, Telford JL (2010) Genome-based approaches to vaccine development. J Mol Med 88:143–147
Morris CF (2002) Puroindolines: the molecular basis of wheat grain hardness. Plant Mol Biol 48:633–647
Muccilli V, Cunsolo V, Saletti R, Foti S, Masci S, Lafiandra D (2005) Characterization of B- and C-type low molecular weight glutenin subunits by electrospray ionization mass spectrometry and matrix-assisted laser desorption/ionization mass spectrometry. Proteomics 5:719–728
Nakamura A, Tanabe S, Watanabe J, Makino T (2005) Primary screening of relatively less allergenic wheat varieties. J Nutr Sci Vitaminol Tokyo 51:204–206
Neilson KA, Gammulla CG, Mirzaei M, Imin N, Haynes PA (2010) Proteomic analysis of temperature stress in plants. Proteomics 10:828–845
Niranjan-Raj SN, Sarosh BR, Shetty HS (2006) Induction and accumulation of polyphenol oxidase activities as implicated in development of resistance against pearl millet downy mildew disease. Funct Plant Biol 33:563–571
Nishi A, Nakamura Y, Tanaka N, Satoh H (2001) Biochemical and genetic effects of amylose-extender mutation in rice endosperm. Plant Physiol 127:459–472
Nouwens AS, Cordwell SJ, Larsen MR, Molloy MP, Gillings M, Willcox MDP, Walsh BJ (2000) Complementing genomics with proteomics: the membrane subproteome of Pseudomonas aeruginosa. Electrophoresis 21:3797–3809
Orchard S, Albar JP, Deutsch EW, Eisenacher M, Binz PA, Martinez-Bartolomé S, Vizcaino JA, Hermjakob H (2012) From proteomics data representation to public data flow: a report on the HUPO-PSI workshop. Proteomics 12:351–355
Osorio C, Wen N, Gemini R, Zemetra R, von Wettstein D, Rustgi S (2012) Targeted modification of wheat grain protein to reduce the content of celiac causing epitopes. Funct Integr Genomics. doi:10.1007/s10142-012-0287-y
Oury F-X, Godin C (2007) Yield and grain protein concentration in bread wheat: how to use the negative relationship between the two characters to identify favourable genotypes? Euphytica 157:45–57

Palacin A, Bartra J, Munoz R, Diaz-Perales A, Valero A, Salcedo G (2010) Anaphylaxis to wheat flour-derived foodstuffs and the lipid transfer protein syndrome: apotential role of wheat lipid transfer protein Tri a 14. Int Arch Allergy Immunol 152:178–183

Palacin A, Quirce S, Armentia A, Fernandez-Nieto M, Pacios LF, Asensio T (2007) Wheat lipid transfer protein is a major allergen associated with baker's asthma. J Allergy Clin Immunol 120:1132–1138

Park S-H, Wilson J, Seabourn JD (2009) Starch granule size distribution of hard red winter and hard red spring wheat: its effects on mixing and breadmaking quality. J Cereal Sci 49:98–105

Parker JM, Guo D, Hodges RS (1986) New hydrophilicity scale derived from high-performance liquid chromatography peptide retention data: correlation of predicted surface residues with antigenicity and X-ray-derived accessible sites. Biochemistry 25:5425–5432

Pastorello EA, Farioli L, Conti A, Pravettoni V, Bonomi S, Iametti S (2007) Wheat IgE-mediated food allergy in European patients: alpha-amylase inhibitors lipid transfer proteins and low-molecular-weight glutenins allergenic molecules recognized by double-blind placebo-controlled food challenge. Int Arch Allergy Immunol 144:10–22

Payne PI (1987) Genetics of wheat storage proteins and effect of allelic variation on bread-making quality. Annu Rev Plant Physiol Plant Mol Biol 38:141–153

Payne PI, Corfield KG, Blackman JA (1979) Identification of a high-molecular-weight subunit of glutenin whose presence correlates with bread-making quality in wheats of related pedigree. Theor Appl Genet 55:153–159

Payne PI, Holt LM, Jarvis MG, Jackson EA (1985) Two-dimensional fractionation of the endosperm proteins of bread wheat (*Triticum aestivum*): biochemical and genetic studies. Cereal Chem 62:319–326

Payne PI, Nightingale MA, Krattiger AF, Holt LM (1987) The relationships between HMW glutenin subunit composition and the bread-making quality of British-grown wheat-varieties. J Sci Food Agric 40:51–65

Peck SC (2005) Update on proteomics in Arabidopsis. Where do we go from here? Plant Physiol 138:591–599

Peltier JB, Friso G, Kalume DE, Roepstorff P, Nilsson F, Adamska I, van Wijk KJ (2000) Proteomics of the chloroplast: systematic identification and targeting analysis of lumenal and peripheral thylakoid proteins. Plant Cell 12:319–341

Picotti P, Bodenmiller B, Mueller LN, Domon B, Aebersold R (2009) Full dynamic range proteome analysis of S. cerevisiae by targeted proteomics. Cell 138:795–806

Pischetsrieder M, Baeuerlein R (2009) Proteome research in food science. Chem Soc Rev 8:2600–2608

Pollet A, Lagaert S, Kulminskaya A, Delcour JA, Courtin CM (2010) Mutagenesis and subsite mapping underpin the importance for substrate specificity of the aglycon subsites of glycoside hydrolase family 11 xylanases. Biochim Biophys Acta Proteins Proteomics 1804:977–985

Ponten F, Schwenk JM, Asplund A, EdqvistP HD (2011) The human protein Atlas as a proteomic resource for biomarker discovery (Review). J Intern Med 270:428

Rahman S, Li Z, Batey IL, Cochrane MP, Appels R, Morell MK (2000) Genetic alteration of starch functionality in wheat. J Cereal Sci 31:91–110

Rakwal R, Agrawal GK (2003) Rice proteomics: current status and future perspectives. Electrophoresis 24:3378–3389

Rampitsch C, Subramaniam R, Djuric-Ciganovic S, Bykova NV (2010) The phosphoproteome of Fusarium graminearum at the onset of nitrogen starvation. Proteomics 10:124–140

Regina A, Bird AR, Li Z, Rahman S, Mann G, Chanliaud E, Berbezy P, Topping D, Morell MK (2007) Bioengineering cereal carbohydrates to improve human health. Cereal Food World 52:182–187

Regina A, Bird A, Topping D, Bowden S, Freeman J, Barsby T, Kosar-Hashemi B, Li Z, Rahmand S, Morell MM (2006) High-amylose wheat generated by RNA interference improves indices of large-bowel health in rats. PNAS 103:3547–3551

Rharrabti Y, Elhani S, Martos-Nunez V, Garcia del Moral L (2001) Protein and lysine content grain yield and other technological traits in durum wheat under Mediterranean conditions. J Agric Food Chem 49:3802–3807
Rhee SY, Dickerson J, Xu D (2006) Bioinformatics and its applications in plant biology. Annu Rev Plant Biol 57:335–360
Roessner U, Nahid A, Hunter A, Bellgard M (2011) Metabolomics—the combination of analytical chemistry, biology and informatics. In: Moo-Young M (ed) Comprehensive biotechnology, vol 1, 2nd edn. Academic, Burlington, pp 447–459
Rolland N, Ferro M, Seigneurin-Berny D, Garin J, Douce R, Joyard J (2003) Proteomics of chloroplast envelope membranes. Photosynth Res 78:205–230
Ruoppolo M, Orru S, D'Amato A, Francese S, Rovero P, Marino G, Esposito C (2003) Analysis of transglutaminase protein substrates by functional proteomics. Protein Sci 12:1290–1297
Saha S, Raghava GP (2006a) Prediction of continuous B-cell epitopes in an antigen using recurrent neural network. Proteins 65:40–48
Saha S, Raghava GP (2006b) AlgPred: Prediction of allergenic proteins and mapping of IgE epitopes. Nucleic Acids Res 34:W202-W209, Web server issue
Salimi N, Fleri W, Peters B, Sette A (2010) Design and utilization of epitope-based databases and predictive tools. Immunogenetics 62:185–196
Salt LJ, Robertson JA, Jenkins JA, Mulholland F, Mills C (2005) The identification of foam-forming soluble proteins from wheat (Triticum aestivum) dough. Proteomics 5:1612–1623
Sanchez-Monge R, Garcia-Casado G, Lopez-Otin C, Armentia A, Salcedo G (1997) Wheat flour peroxidase is a prominent allergen associated with baker's asthma. Clin Exp Allergy 27:1130–1137
Sancho AI, Gillabert M, Tapp H, Shewry PR, Skeggs PK, Mills EN (2008) Effect of environmental stress during grain filling on the soluble proteome of wheat (*Triticum aestivum*) dough liquor. J Agric Food Chem 56:5386–5393
Sander I, Flaggem A, Mergetm R, Halderm TM, Meyer HE, Baur X (2001) Identification of wheat flour allergens by means of 2-dimensional immunoblotting. J Allergy Clin Immunol 107:907–991
Sandiford CP, Tatham AS, Fido R, Welch JA, Jones MG, Tee RD (1997) Identification of the major water/salt insoluble wheat proteins involved in cereal hypersensitivity. Clin Exp Allergy 27:1120–1129
Santoni V, Molloy M, Rabilloud T (2000) Membrane proteins and proteomics: un amour impossible? Electrophoresis 21:1054–1070
Schaller A (2004) A cut above the rest: the regulatory function of plant proteases. Planta 220:183–197
Sehnke PC, Chung HJ, Wu K, Ferl RJ (2001) Regulation of starch accumulation by granule-associated plant 14-3-3 proteins. Proc Natl Acad Sci USA 98:765–770
Selitrennikoff CP (2001) Antifungal proteins. Appl Environ Microbiol 67:2883–2894
Shewry PR (2009) Wheat. J Exp Bot 60:1537–1553
Shewry PR, Halford NG (2002) Cereal seed storage proteins: structures, properties and role in grain utilization. J Exper Bot 53:947–958
Shewry PR, Tatham AS (1997) Disulfide bounds in wheat gluten proteins. J Cereal Sci 25:207–227
Shewry PR, Halford NG, Belton P, Tatham AS (2002) The structure and properties of gluten: an elastic protein from wheat grain. Phil Trans Royal Soc B Bio Sci 357:133–142
Shewry PR, D'Ovidio R, Lafiandra D, Jenkins JA, Mills NF, Békés F (2009) In: Khan K, Shewry PR (eds) Wheat chemistry and technology. AACC Press, St Paul, pp 223–298
Skylas DJ, Mackintosh JA, Cordwell ST, Basseal DJ, Walsh BJ, Harry J, Blumenthal C, Copeland L, Wrigley CW, Rathmell W (2000) Proteome approach to the characterisation of protein composition in the developing and mature wheat-grain endosperm. J Cereal Sci 32:169–188
Skylas DJ, Copeland L, Rathmell W, Wrigley CW (2001) The wheat-grain proteome as a basis for more efficient cultivar identification. Proteomics 1:1542–1546
Skylas DJ, Cordwell SJ, Hains PG, Larsen MR, Basseal DJ, Walsh BJ, Blumenthal C, Rathmell WG, Copeland L, Wrigley CW (2002) Heat shock of wheat during grain filling: characterisation of proteins associated with heat-tolerance using a proteome approach. J Cereal Sci 35:175–188
Skylas DJ, Van Dyk D, Wrigley CW (2005) Proteomics of wheat grain. J Cereal Sci 41:165–179

Sollner J, Heinzel A, Summer G, Fechete R, Stipkovits L, Szathmary S, Mayer B (2010) Concept and application of a computational vaccinology workflow. Immunome Res 6(Suppl 2):S7

Sotkovsky P, Hubalek M, Hernychova L, Novak P, Havranova M, Setinova I (2008) Proteomic analysis of wheat proteins recognized by IgE antibodies of allergic patients. Proteomics 8:1677–1691

Spaenij-Dekking L, Kooy-Winkelaar Y, van Veelen P, Drijfhoutm JW, Jonker H, van Soest L (2005) Natural variation in toxicity of wheat: potential for selection of nontoxic varieties for celiac disease patients. Gastroenterology 129:797–806

Stenman SM, Lindfors K, Venäläinen JI, Männisto PT, Auriola S, Kaukovirta-Norja A, Jantunen A, Laurila K, Qiao SW, Sollid LM, Mauriala T, Mäki M, Kaukinen K (2010) Degradation of coeliac disease-inducing rye secalin by germinating cereal enzymes: diminishing toxic effects in intestinal epithelial cells. Clin Exp Immunol 161:242–249

Stevens TJ, Arkin IT (2000) Do more complex organisms have a greater proportion of membrane proteins in their genomes? Proteins Struct Funct Bioinform 39:417–420

Takizawa T, Arakawa H, Tokuyama K, Morikawa A (2001) Identification of allergen fractions of wheat flour responsible for anaphylactic reactions to wheat products in infants and young children. Int Arch Allergy Immunol 125:51–56

Tanabe S, Arai S, Yanagiharam Y, Mita H, Takahas-him K, Watanabe M (1996) A major wheat allergen has a Gln-Gln-Gln-Pro-Pro motif identified as an IgE-binding epitope. Biochem Biophys Res Commun 219:290–293

Tasleem-Tahir A, Nadaud I, Girousse C, Martre P, Marion D, Branlard G (2011) Proteomic analysis of peripheral layers during wheat grain development. Proteomics 11:371–379

Tasleem-Tahir A, Nadaud I Chambon C and Branlard G (2012) Expression profiling of starchy endosperm metabolic proteins at 21 stages of wheat grain development. J Proteome Res 11: 2754–2773

Tatham AS, Shewry PR (2008) Allergy to wheat and related cereals. Clin Exp Allergy 38:1712–1726

Taylor CF, Field D, Sansone SA, Aerts J, Apweiler R, Ashburner M, Ball CA, Binz PA, Bogue M, Booth T, Brazma A, Brinkman RR, Michael Clark A, Deutsch EW, Fiehn O, Fostel J, Ghazal P, Gibson F, Gray T, Grimes G, Hancock JM, Hardy NW, Hermjakob H, Julian RK Jr, Kane M, Kettner C, Kinsinger C, Kolker E, Kuiper M, Le Novère N, Leebens-Mack J, Lewis SE, Lord P, Mallon AM, Marthandan N, Masuya H, McNally R, Mehrle A, Morrison N, Orchard S, Quackenbush J, Reecy JM, Robertson DG, Rocca-Serra P, Rodriguez H, Rosenfelder H, Santoyo-Lopez J, Scheuermann RH, Schober D, Smith B, Snape J, Stoeckert CJ Jr, Tipton K, Sterk P, Untergasser A, Vandesompele J, Wiemann (2008) Promoting coherent minimum reporting guidelines for biological and biomedical investigations: the MIBBI project. Nat Biotechnol 26: 889–889

Tetlow IJ, Wait R, Lu Z, Akkasaeng R, Bowsher CG, Esposito S, Kosar-Hashemi B, Morell MK, Emes MJ (2004) Protein phosphorylation in amyloplasts regulates starch branching enzyme activity and protein-protein interactions. Plant Cell 16:694–708

Tetlow IJ, Kim G, Beisel KG, Cameron S, Makhmoudova S, Liu F, Bresolin NSR, Morell MM, Emes MJ (2008) Analysis of protein complexes in wheat amyloplasts reveals functional interactions among starch biosynthetic enzymes. Plant Physiol 146:1878–1891

Thiellement H, Bahrman N, Damerval C, Plomion C, Rossignol M, Santoni V, de Vienne D, Zivy M (1999) Proteomics for genetic and physiological studies in plants. Electrophoresis 20:2013–2026

Thompson DB (2000) On the non-random nature of amylopectin branching. Carbohydrate Polymers 43:223–239

Tian Q, Price ND, Hood L (2012) Systems cancer medicine: towards realization of predictive, preventive, personalized, and participatory (P4) medicine. J Internal Med. doi:10.1111/j.1365-2796.2011.02498.x, Accepted Article

Turnbull KM, Rahman S (2002) Endosperm texture in wheat. J Cereal Sci 36:327–337

Tye-Din J, Anderson A (2008) Immunopathogenesis of celiac disease. Curr Gastroenterol Rep 10:458–465

van Heel DA, West J (2006) Recent advances in coeliac disease. Gut 55:1037–1046 van Herpen TW, Goryunova SV, van der Schoot J, Mitreva M, Salentijn E, Vorst O (2006) Alpha-gliadin genes from the A B and D genomes of wheat contain different sets of celiac disease epitopes. BMC Genomics 7:1

Varjonen E, Vainio E, Kalimo K (2000) Antigliadin IgE-indicator of wheat allergy in atopic dermatitis. Allergy 55:386–391

Vensel WH, Tanaka CK, Cai N, Wong JH, Buchanan BB, Hurkman WJ (2005) Developmental changes in the metabolic protein profiles of wheat endosperm. Proteomics 5:1594–1611

Wang Y, Yang L, Xu H, Li Q, Ma Z, Chu C (2005) Differential proteomic analysis of proteins in wheat spikes induced by Fusarium graminearum. Proteomics 5:4496–5003

Wang LH, Zhao XL, He ZH, Ma W, Appels R, Pena RJ, Xia XC (2009) Characterization of low-molecular-weight glutenin subunit Glu-B3 genes and development of STS markers in common wheat (*Triticum aestivum* L.). Theor Appl Genet 118:525–539

Wang LH, Li GY, Pena RJ, Xia XC, He ZH (2010) Development of STS markers and establishment of multiplex PCR for Glu-A3 alleles in common wheat (*Triticum aestivum* L.). J Cereal Sci 51:305–312

Wangen S (2009) Healthier without wheat a new understanding of wheat allergies celiac disease and non-celiac gluten intolerance. Innate Health, Sieattle

Wanjugi HW, Martin JM, Giroux MJ (2008) Influence of puroindolines A and B individually and in combination on wheat milling and bread traits. Cereal Chem 84:540–547

Washburn MP, Wolters D, Yates JR, Washburn MP, Wolters D, Yates JR III (2001) Large-scale analysis of the yeast proteome by multidimensional protein identification technology. Nat Biotechnol 19:242–247

Weegels PL, Hamer RJ, Schofield JD (1997) Depolymerisation and re-polymerisation of wheat glutenin during dough processing. II. Changes in composition. J Cereal Sci 25:155–163

Weiss W, Huber G, Engel KH, Pethran A, Dunn MJ, Gooley AA (1997) Identification and characterization of wheat grain albumin/globulin allergens. Electrophoresis 18:826–833

Wieser H (2008) Detection of gluten. In: Arendt EK, Dal Bello F (eds) Gluten-free cereal products and beverages. Academic, San Diego, pp 47–80

Wieser H, Koehler P (2008) The biochemical basis of CD. Cereal Chem 85:1–13

Wilson ID, Barker GLA, Chungui L, Coghill JA, Beswick RW, Lenton JR, Edwards KJ (2005) Alteration of the embryo transcriptome of hexaploid wheat (*Triticum aestivum* cv. Mercia) during maturation and germination. Funct Integr Genomics 5:144–154

Wong JH, Cai N, Balmer Y, Tanaka CK, Vensel WH, Hurkman WJ, Buchanan BB (2004) Thiredoxin targets of developing wheat seeds identified by complementary proteomic approaches. Phytochemistry 65:1629–1640

Woo SH, Fukuda M, Islam N, Takaoka M, Kawasaki H, Hirano H (2002) Efficient peptide mapping and its application to identify embryo proteins in rice proteome analysis. Electrophoresis 23:647–654

Wrigley CW (1970) Protein mapping by combined gel electrofocusing and electrophoresis: application to the study of genotypic variations in wheat gliadins. Biochem Genet 4:504–516

Wrigley CW (2002) Testing for grain quality at the gene level. Cereal Foods World 47:156–157

Wrigley CW, Batey IL, Skylas DJ, Sharp PJ (2003) Grain quality assessment using proteomics and genomics. Food Australia 55:143–146

Wrigley CW, Békés F, Cavanagh CR, Martinov S, Bushuk W (2006a) The gluten composition of wheat varieties and genotypes. Part II. Composition table for the LMW subunits of glutenin. www.aaccnet.org

Wrigley CW, Békés F, Cavanagh CR, Martinov S, Bushuk S (2006b) The gluten composition of wheat varieties and genotypes. Part III. Composition table for the HMW subunits of glutenin. www.aaccnet.org

Wrigley CW, Asenstorfer R, Batey IL, Cornish GB, Day L, Mares D, Mrva K (2009) The biochemical and molecular basis of wheat quality, Chapter 21. In: Carver B (ed) Wheat: science and trade. Wiley-Blackwell, Ames, pp 495–520

Wu MJ, McKay S, Howes NK, Hegedus E, Chin J (2011) Polymorphism and pedigree analysis of b-amylase iso-enzymes in Australian wheat. J Cereal Sci 53:362–370

Yagami T, Haishima Y, Tsuchiya T, Tomotaka-Yagami A, Kano H, Matsunaga K (2004) Proteomic analysis of putative latex allergens. Int Arch Allergy Immunol 135:3–11

Yahata E, Maruyama-Funatsuki W, Nishio Z, Tabiki T, Takata K, Yamamoto Y, Tanida M, Saruyama H (2005) Wheat cultivar-specific proteins in grain revealed by 2-DE and their application to cultivar identification of flour. Proteomics 5:3942–3953

Yang L, Tian D, Luo Y, Zhang R, Ren C, Zhou X (2011) Proteomics-based identification of storage metabolic and allergenic proteins in wheat seed from 2-DE gels. African J Agric Res 6:808–816

Zhang W, Gianibelli MC, Rampling LR, Gale KR (2004) Characterisation and marker development for low molecular weight glutenin genes from Glu-A3 alleles of bread wheat (*Triticum aestivum* L.). Theor Appl Genet 108:1409–1419

Zhou W, Kolb FL, Riechers DE (2005) Identification of proteins induced or upregulated by Fusarium head blight infection in the spikes of hexaploid wheat (*Triticum aestivum*). Genome 48:770–780

Zörb C, Langenkamper G, Betsche T, Niehaus K, Barsch A (2006) Metabolite profiling of wheat grains (*Triticum aestivum* L.) from organic and conventional agriculture. J Agric Food Chem 54:8301–8306

Zörb C, Betsche T, Langenkamper G (2009) Search for diagnostic proteins to prove authenticity of organic wheat grains (*Triticum aestivum* L.). J Agric Food Chem 57:2932–2937

Zuk M, Weber R, Szopa J (2005) 14-3-3 Protein down-regulates key enzyme activities of nitrate and carbohydrate metabolism in potato plants. J Agric Food Chem 53:3454–3460

Zybailov B, Friso G, Kim J, Rudella A, Rodriguez VR, Asakura Y, Sun Q, van Wijk KJ (2009) Large scale comparative proteomics of a chloroplast Clp protease mutant reveals folding stress altered protein homeostasis and feedback regulation of metabolism. Mol Cell Proteomics 8:1789–1810

Chapter 20
Proteomics and Applications to Food Science in Rice

Pingfang Yang, Mohammad-Zaman Nouri, and Setsuko Komatsu

Abbreviations

ABA	Abscisic acid
GA	Gibberellic acid
TCA	Tricarboxylic acid
ROS	Reactive oxygen species
JA	Jasmonic acid

20.1 Introduction

Rice, wheat, maize, and soybean provide more than half the calories consumed by the world's population. Rice production in Asia has been doubled since 1961 based on the breeding of new rice cultivars with an intensive cultivation system. Furthermore, as the global climate changes, high temperature, flooding, and drought have become the most important environmental factors that can influence the yield and quality of crops. In Asia, over 25 million ha were affected by drought. In addition to being a very important agriculture crop, rice is useful as a model plant for

P. Yang
Key Laboratory of Plant Germplasm Enhancement and Speciality Agriculture, Wuhan Botanical Garden, Chinese Academy of Science, Wuhan 430074, China

M.-Z. Nouri
Rice Research Institute of Iran-Deputy of Mazandaran, Amol 46191-91951, Iran

S. Komatsu (✉)
National Institute of Crop Science,
2-1-18 Kannondai, Tsukuba 305-8518, Japan
e-mail: skomatsu@affrc.go.jp

F. Toldrá and L.M.L. Nollet (eds.), *Proteomics in Foods: Principles and Applications*, Food Microbiology and Food Safety 2, DOI 10.1007/978-1-4614-5626-1_20,

biological research because it has a smaller genome than those of other cereals (Devos and Gale 2000). The Rice Genome Sequencing Project (The International Rice Genome Sequencing Project 2005) produced a map-based, high-quality sequence that covers 95% of the 389 Mb genome of rice, including virtually all of the euchromatin and two complete centromeres. The annotation of the rice genome has progressed rapidly (The Rice Annotation Project 2007), and most of the predicted genes are supported by full-length cDNAs (Kikuchi et al. 2003). Since the completion of the sequencing of the rice genome, the challenge for the plant-research community has been to identify the function and regulation of rice genes.

The major purpose of rice cultivation is to produce high quality and yield grains that are consumed as the staple food in Asia especially in the southeast area. In addition, seeds also play a central role in the rice lifecycle. Acquiring high quality and viable seeds is a prerequisite for rice cultivation. Not surprisingly, rice seed especially grain filling and seed germination study is one of the hot areas in rice biology. Furthermore, uncovering the regulatory mechanism of rice grain filling and seed germination is also important for rice breeding and production. With the advent of proteomic technologies, large-scale protein profiling has been applied to elucidate the mechanism of these two complex processes, which showed that the regulation happened in different pathways at different levels. Abiotic stresses including drought, flood, high or low temperature, and salinity will seriously affect the grain filling process and hence decrease the yield and quality, and they can also inhibit rice seed germination. In this review, rice seed proteomics and rice proteomics against environmental stresses are summarized and future perspectives of food science in rice are also discussed.

20.2 Rice Seed Proteomics

20.2.1 Proteomics of Grain Filling Period

Carbohydrates, proteins, and oils are three major reserves that accumulate in plant seeds (Weber et al. 2005). Compared with other plant species, such as soybean and oilseeds, rice predominantly accumulates carbohydrates. In mature rice grain, starch accounts for about 85% of the total dry mass (Ruuska et al. 2002). The starch is mainly synthesized in endosperm during a period of seed development generally defined as grain filling. In rice, grain filling roughly refers the stage between 6 and 20 days after flowering (Xu et al. 2008). During this stage, seeds experience dramatic morphological, physiological, and metabolic changes (Rubel et al. 1972). Yield and nutrition, which are the two major economic characteristics of rice grains, are largely determined by the synthesis and storage of starches, proteins, and minerals during grain filling. Manipulation of different biosynthetic pathways can significantly change these characteristics (Ye et al. 2000), which indicates a close correlation of gene expression in different pathways. Expression analyses at the mRNA level have identified many genes that are preferentially expressed during rice grain filling and participate in starch and protein synthesis and storage (Zhu et al. 2003; Liu et al. 2010; Gaur et al. 2011). Gene expression at the RNA level is not always consistent with that at the protein level (Greenbaum et al. 2003; Watson et al. 2003; Schmidt et al. 2007); furthermore, information about post-translational modification could not be acquired through RNA. Undoubtedly, study at the protein level will provide more direct evidence, since the biosynthetic pathways are catalyzed by different enzymes. Proteomic analyses about seed filling have been carried out in many different plant species, such as *Medicago truncatula* (Gallardo et al. 2003, 2007; Repetto et al. 2008), soybean (Agrawal et al. 2008), oilseeds (Hajduch et al. 2005, 2006), barley (Finnie et al. 2002), wheat (Laino et al. 2010), maize (Mechin et al. 2007), and rice (Xu et al. 2008). All these studies revealed that the proteins involved in cellularization and cell division decreased, and the proteins involved in cell expansion and reserve accumulation increased during seed filling.

The most comprehensive proteomic study about rice grain filling to date was conducted by Xu et al. (2008). In their study, they identified 345 differentially displayed proteins among which 45% were involved in different metabolic pathways and 20% were involved in protein biosynthesis and destination. The starch synthesis related proteins increased at the beginning of the grain filling stage and peaked at the time when the filling was almost finished (Xu et al. 2008; Liu et al. 2010), which can help to enhance starch synthesis and accumulation. Because the accumulation of starch can result in an anoxic condition, there is a transition from central carbon metabolism to fermentation during the late stage of grain filling, which was also reported in maize (Mechin et al. 2007). The starch biosynthesis, constituents, starch granule structure, and arrangement can all affect grain quality (Myers et al. 2000; Kang et al. 2005; Yamakawa et al. 2007). Starch is composed of two major constituents: unbranched amylose (linear a-1, 4-polyglucans) and branched amylopectin (a-1, 6-branched polyglucans). A suitable ratio between amylose and amylopectin is one of the determinants for rice grain quality. Mutants that lead to low amylose content (<2%) will set seeds with opaque endosperm and hence poor quality (Dung et al. 2000; Isshiki et al. 2000). It was reported that the starch biosynthesis related enzymes exhibited maximum activities at 15 days after flowering (Liu et al. 2010). The longer the activities exist, the better the grains are.

MADS-box transcription factor OsMADS6 was found regulating the expression of ADP-glucose pyrophosphorylase genes that encode the rate-limiting step enzyme in starch synthesis pathway, whereas the expression of OsMADS6 is subjected to an epigenetic regulation (Zhang et al. 2010a). Phytohormone cytokinins can enhance, whereas abscisic acid (ABA), ethylene and methyl-jasmonic acid (JA) can negatively affect the rice grain filling (Kim et al. 2009; Zhang et al. 2010b; Zhu et al. 2011). ABA and ethylene suppress the expression of starch synthesis genes, which can result in poor filled grains of the inferior spikelets (Zhu et al. 2011). Methyl-JA could induce the biosynthesis of ABA and hence suppress the expression of starch synthesis genes (Kim et al. 2009). Drought can reduce the grain filling through the effect of methyl-JA (Kim et al. 2009). Transcription factor AP37 can reverse the effect of drought to grain filling (Oh et al. 2009). Delay the action of ABA and

Table 20.1 The differentially displayed proteins involved in different metabolic pathways during seed filling in rice, rapeseed and soybean

	Rice[a]	Rapeseed[b]	Soybean[c]
Sugar conversion	16	12	8
Starch synthesis	23	4	1
Lipid and sterols metabolism	6	13	10
Protein synthesis	13	3	17
Glycolysis	21	20	16
Alcoholic fermentation	9	N/A	2
Pentose phosphate	2	5	N/A
TCA pathway	3	5	4
Amino acid metabolism	14	21	14
Nitrogen and sulfur metabolism	6	1	4
Nucleotide metabolism	8	1	7
Secondary metabolism	11	10	7

[a]Data are from Xu et al. (2008)
[b]Data from Agrawal et al. (2008)
[c]Data from Hajduch et al. (2006)

ethylene might be an effective way to fulfill the grains and hence increase the yield. All these data give us a clue that nuclear proteomic analysis might help us to further understand the mechanism of grain filling. Unfortunately, there has been no such report in rice until now. Proteins involved in transcription regulation, RNA processing and transport, chromatin modifying, RNA interference, and RNA-directed DNA methylation have been found stored in the nucleus at the early stage of seed filling in *Medicago truncatula* (Repetto et al. 2008). These data implied that active gene expression regulation happened at different levels during seed development. In the nucleus, RNA binding proteins play pivotal roles in RNA processing; in cytoplasmic, RNA transportation, localization, translation, storage, and degradation also need RNA binding proteins (Dreyfuss et al. 2002; Fedoroff 2002). Except for those well-known RNA binding proteins, many metabolism-related proteins were also found to be cytosolic RNA binding proteins (Doroshenk et al. 2009). Why should these metabolic proteins bind to RNAs is still an open question.

In addition to starch, proteins account for 6–10% of the dry mass and are important for the nutrition, cooking, and brewing quality of rice grains (Bressani et al. 1971; Hamaker 1994). Rice storage proteins include glutelins, prolamins, globulins, and albumins to name several. Studies in *Medicago truncatula* showed that many ribosomal proteins were stored in the nucleus in preparation for the following intensive protein synthesis (Repetto et al. 2008). These data are consistent with the fact that the stocking proteins are the major reserves (about 40%) in *Medicago truncatula*. Table 20.1 shows the number of differentially displayed proteins involved in different biosynthetic pathways during the seed filling in rice, rapeseed, and soybean. It is clear that different plants synthesize and accumulate different reserves during seed filling. How the plant balances the synthesis of different reserves is very important because common carbon precursors are usually shared by different biosynthetic pathways. A RNA binding protein was found to bind to glutelin and prolamine RNAs in vivo. This protein is distributed to multicellular compartments, and might play a role in the expression of storage protein genes (Crofts et al. 2010). Proteomic studies showed that two proteins, lipoxygenase and SBPs, were over-represented in developing seeds of legume plants such as soybean (Agrawal et al. 2008) and *Medicago trunctula* (Gallardo et al. 2003) and underrepresented in non-legume plants such as rapeseed (Hajduch et al. 2006), wheat (Vensel et al. 2005), maize (Mechin et al. 2007), and rice (Xu et al. 2008). These two proteins might contribute to the accumulation of stocking proteins in legume seeds. A comprehensive comparison among the proteomic data of the developing seeds from different plant species will be helpful.

20.2.2 Proteomics During Germination

Higher plants produce their progeny mainly through seeds. Regulation of seed germination is critical for the plant lifecycle. By definition, germination refers to the physiological process that starts from the uptake of water by the dry seeds and ends at the protrusion of the radicle (Bewley 1997). Water uptake during the whole process of germination could be divided into three phases. Phase I is a rapid water absorption stage that is followed by phase II. During phase II, the water absorption is almost stopped. Phase III is another rapid water uptake stage, which can only come into true after the completion of germination. Dormant seeds cannot enter into phase III without breaking the dormancy. Rice seed is not dormant. It will germinate once suitable conditions are available. Several studies have been conducted in *Arabidopsis* and rice to analyze the genome expression profiles and their changes at both the transcriptomic and proteomic levels during seed germination. These studies showed that genes related to translation, protein degradation, and cell wall modification were preferentially expressed during germination (Holdsworth et al. 2008). Although transcription is not required for germination, translation is absolutely necessary, inhibition of translation will totally block the germination (Rajjou et al. 2004; He et al. 2011).

Before germination, the dry mature seeds are quiescent. Upon imbibition, seeds will rapidly resume metabolic activity and hence experience dramatic physiological and biochemical changes (Salon et al. 1988; Ehrenshaft and Brambl 1990; Botha et al. 1992). Recovery of the metabolism will result in the mobilization of the reserves, which can provide enough substrates and energy for the following seedling growth. In rice endosperm, starch is organized into semicrystalline arrays and stored in the form of granules (Zeeman et al. 2002, 2004a). The first step of starch degradation should be attacking the granule surface. The only enzyme to catalyze this step in plants is alpha-amylase. In germinating cereal endosperm, alpha-amylases hydrolyze a-1,4 linkages within the glucose polymers exposed on the surface of the starch granules, which will release soluble glucans for further degradation

(Smith et al. 2005). The second step is debranching that is catalyzed by isoamylase. This enzyme hydrolyzes the a-1,6 linkages of the branching polymers. The first two steps of degradation will produce linear and soluble glucans that could be further degraded through two alternative pathways. The first one can release glucose 1-phosphate and is catalyzed by glucan phosphorylase (Zeeman et al. 1998); the second one is catalyzed by beta-amylase and produces maltose that is then converted to glucose 1-phosphate and glucose by maltose phosphorylase (Lao et al. 1999). It was reported that the linear glucans were degraded through the second pathway in chloroplasts (Zeeman et al. 2004b).

The enzymes involved in two pathways were all detected in the germinating rice seeds (He et al. 2011), which implies that starch might be degraded through both ways during germination. Except for alpha-amylase which is dramatically increased during germination, other starch degradation-related enzymes were constant in this process (Yang et al. 2007; Kim et al. 2008). The degradation of starch during germination should have been programmed during the seed's development and maturation. In cereal seeds, alpha-amylase is mainly produced in the aleurone layer, and induced by the phytohormone gibberellic acid (GA) (Mitsunaga et al. 1994). Thus the degradation of starch is heavily dependent on GA. To the contrary, proteomic analysis in *Arabidopsis* GA-deficient mutants showed that GA has no effect on the mobilization of storage lipids and proteins, but it can affect the cell wall loosening (Gallardo et al. 2002). Phytohormones ABA and GA play antagonistic roles in regulating seed germination. ABA is involved in dormancy maintenance, whereas GA promotes germination (Finch-Savage and Leubner-Metzger 2006). Proteomic analysis showed that ABA can suppress the expression of those GA-responsive proteins during rice seed germination (Kim et al. 2008). The alpha-amylase is also induced by sulfuric acid in rice seeds (Mitsunaga et al. 2007), which provides a clue that treatment with sulfuric acid might help rice seed germination. After alpha-amylase production it is transferred into endosperm where it degrades the stored starch. Proteomic analysis showed that alpha-amylase can exist both full length and as fragments in cereal seeds (Bak-Jensen et al. 2007). Whether both states are functional is still an open question.

The product of starch degradation is glucose 1-phosphate. It could be further degraded through glycolysis and the tricarboxylic acid (TCA) cycle. All the enzymes involved in the central carbon metabolism exist in the dry mature seeds (Yang et al. 2007; He et al. 2011). Upon imbibition, their activities are resumed, and the metabolic pathways are recovered. Because of the compact structure of seeds, the endosperm tissue is anaerobic. At the early stage of germination, anaerobic respiration is the major source of ATPs. One round of glycolysis can only produce two ATPs, which is not enough at all for the germination, so most of the glycolytic enzymes are upregulated to increase the availability of ATPs (Yang et al. 2007). The fermentation will help to remove the pyruvate along with TCA cycle (He et al. 2011). Upon the protrusion of radicle, the aerobic respiration increases dramatically. So at the late stage of germination, the TCA cycle will proceed rapidly and act as the major source of ATPs (Kim et al. 2008).

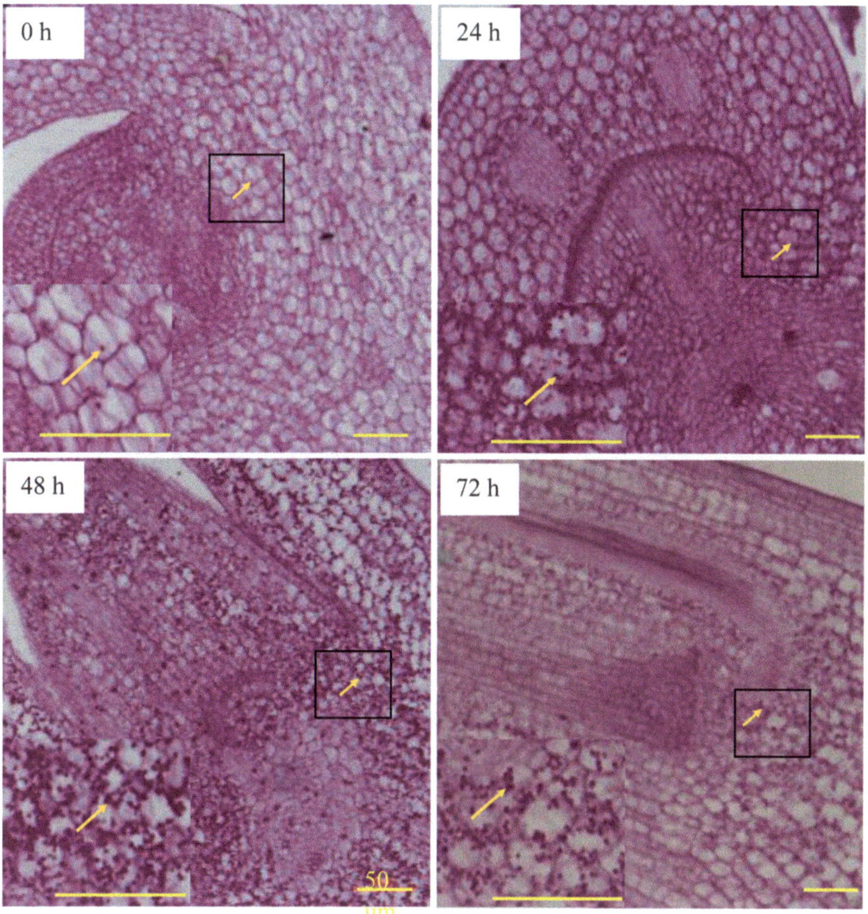

Fig. 20.1 PAS stained sections of the rice embryos showing the accumulation of starch granules in the cells around the vascular tissue during the germination. (Bars = 50 um; He et al. Proteomics, 2011)

During germination, the central metabolism not only functions as an energy resource but also provides precursors for the biosynthetic pathways. The fast starch degradation will result in over accumulation of monosaccharides and other precursors. These precursors could not be efficiently utilized by the biosynthetic pathways that largely occur post-germination. These precursors and monosaccharides might be converted into polysaccharides again in the embryos, and be prepared for the following seedling establishment. All the enzymes related to gluconeogenesis and starch biosynthesis were detected constantly in the germinating rice seeds. Gradual accumulation of starch granules was observed in the cells around the bundle tissue in the embryo during germination (Fig. 20.1; He et al. 2011). Incubation of the dissected rice embryos with the media containing sugars proved that these starch granules are newly synthesized in the embryos (Matsukura et al. 2000). In addition to starch, stored proteins are the other major reserves that experience degradation during germination. The amino acids derived from the degradation are used for the synthesis of the functional proteins. It seems that the degradation is mainly executed by the protease rather than the proteasome system (He et al. 2011).

Long-time storage of the dry mature seeds as well as fast metabolism in germination can both result in the accumulation of reactive oxygen species (ROS) that play an important role in seed germination regulation (Bailly 2004). ROS might act as signaling molecules that promote the germination. Previous reports have shown that nitric oxide can release the seeds from dormancy and promote germination in several different plant species (Zhang et al. 2003; Bethke et al. 2006; Zhao et al. 2007). In *Arabidopsis*, this radical can induce the catabolism of ABA during the first hours of imbibition, and hence release the dormancy (Liu et al. 2009). ROS can also alter protein structure and function through either direct or indirect ways. Protein carbonylation is one of the most important post-translational modifications under oxidative stress (Nystrom 2005). Carbonylation can provoke protein degradation, which might be bifunctional. On the positive side, it can promote the degradation of those storage proteins and germination inhibitory proteins, however, it can be fatal by targeting those physiologically important proteins (Job et al. 2005). ROS can also affect protein function through altering the redox status of the cysteinyl residues (Spadaro et al. 2010). To minimize the impeditive effects of ROS, rice has evolved fine mechanisms to regulate the redox homeostasis during germination. Many ROS scavenging enzymes such as SOD and APX were detected to be increased in the germinating rice seeds (Yang et al. 2007; Kim et al. 2008; He et al. 2011). These enzymes function sequentially to detoxify the superoxide anions into water, and hence protect the cell from oxidative damage. Proteins such as thioredoxin can help to regulate the redox status of cysteinyl residue (Yano and Kuroda 2006).

Although we have obtained abundant knowledge about rice grain filling and seed germination through proteomic analyses, there is still a long way to go to get a full understanding about these complicated processes. Considering that rice is an important crop, our study should be conducted under conditions very close to the real ones for agriculture production. Subcellular proteomic analysis about the nuclear proteins, RNA binding proteins, and protein post-translational modification will also be helpful for us in elucidating the mechanisms of these complex processes.

20.3 Rice Proteomics Against Environmental Stress

20.3.1 Proteomics of Yield Affected by Environmental Stress

Among cereals, rice has diverse varieties adopted to grow in a wide range of environmental conditions from foothills to deepwater in lowland areas. Rice yield is a quantitative trait affected by several factors including genetic, morphological, physiological, and environmental conditions. Analysis of yield trait is usually performed by evaluation of the main constituting factors of the grain yield. The traits of panicle number per hill, filled spikelet per panicle, and grain weight are the main components of yield in rice. Although these characters are mainly controlled genetically, their performance is varied according to the growth situation and environmental stresses. Drought, salinity, low and high temperature, flooding, and anoxia are important environmental stresses. Yield components are highly affected by the stresses, which ultimately reduce the crop yield (Boonjung and Fukai 1996; Davatgar et al. 2009; Liu and Bennett 2011). Each of the main components of yield can be affected by various factors. For instance, flag leaf, peduncle, and anther are reported to play a critical role on spikelet fertility and grain weight of rice under drought stress (Liu et al. 2005). Therefore, evaluation of the plant characteristics related to the seed setting and fertility at the reproductive stage gives better insight to understanding the regulatory mechanism of environmental factors.

Morphological evaluations of rice indicated that filled spikelet per panicle and weights of grain were significantly reduced under drought and salinity (Dooki et al. 2006; Liu et al. 2005; Mushtaq et al. 2008). Temperature stresses such as heat and cold could also significantly reduce the number of fertile spikelets during anthesis (Jagadish et al. 2007; Sato et al. 2011). Reduction in spikelet fertility and seed setting can directly affect the rice yield . To deeply evaluate the molecular mechanisms of rice stress response, several strategies were adopted. Proteomics have the potential ability to clarify the effects of environmental stresses on growth, development, and ultimately yield of rice. Specifically, evaluation of rice proteome at the reproductive stage under environmental stress and analysis of yield components will be helpful.

Fertility of the spikelet is the most sensitive yield component of rice under environmental stresses such as drought (Liu and Bennett 2011), salinity (Dooki et al. 2006), and cold (Hayashi et al. 2004). Increasing the number of unfilled spikelet per panicle due to spikelet sterility can highly reduce the yield. Figure 20.2 shows that exposing rice to salinity stress (8 ds/m) could cause leaf folding and spikelet sterility. A comparative proteome analysis of rice anther between drought-tolerant and drought-sensitive genotypes at the heading stage indicated that two groups of proteins (reversible and irreversible) were significantly regulated by drought stress and rewatering (Liu and Bennett 2011). The number of irreversible proteins regulated by rewatering in a drought-sensitive cultivar was much higher than that in a drought-tolerant cultivar. Furthermore, cysteine protease, anther wall degradation protein, were downregulated only in the drought-sensitive cultivar. According to the proteomics results, irreversible effects of drought interrupted anther development in sensitive cultivar and caused poor anther dehiscence and lower pollen density (Liu and Bennett 2011). Therefore, dehydration at the heading stage affects the process of rice anthesis and rewatering cannot fully compensate the deleterious effects.

Extreme temperatures negatively affect the reproductive stage of rice growth. Low temperature can cause pollen sterility, which is considered as one of the main reasons for yield reduction. Cold could repress the transportation of entire sucrose, which might result in the production of nonviable pollen (Oliver et al. 2005). A proteomic study of mature anther under cold stress indicated that the proteins

Fig. 20.2 Effects of salinity on spikelet sterility of rice. Rice (*Oryza sativa* L., Landrace of Tarom) seedlings were transplanted in pot and irrigated normally (*up*) or subjected to salinity (8 ds/m) from the second week after transplanting (*bottom*)

related to metabolism and stress response were differentially regulated (Hayashi et al. 2004). Deficiency in anther dehiscence and the presence of nonviable pollen are the main causes of spikelet sterility that directly affect the number of filled grains and yield.

Proper elongation of the peduncle (uppermost internode) that leads to full exsertion of panicle from the flag leaf sheath is controlled genetically. However, some genotypes such as male sterile lines with WA cytoplasm are disabled to exsert the panicle fully even in normal growth conditions (Gangashetti et al. 2004). Yang et al. (2006) have studied the peduncle proteome at the milky stage of rice seed ripening. According to their results, RuBisCO was not accumulated in the peduncle which indicates low photosynthesis activity. Proteins related to the metabolism, signaling, and stress resistance were highly accumulated, emphasizing a high physiological and stress-resistant activity in the peduncle (Yang et al. 2006). Panicle exsertion failure is one of the limiting factors for obtaining high grain yield in rice, which usually happens under drought stress. In a proteomic study of peduncle under drought stress, proteins involved in cell division and elongation, signaling, and lignin biosynthesis were differentially expressed. Expression of six ABA-responsive proteins under drought stress indicated that hormonal interactions might cause the failure of peduncle elongation (Muthurajan et al. 2011). In as much as all of the trapped spikelets in the leaf sheath will be sterile, it can significantly reduce the yield of rice.

Grain weight is one of the important economical characteristics of rice. The grain filling in rice is genetically controlled. According to a transcriptomic analysis, genes involved in carbohydrate biosynthesis and starch metabolism such as starch biosynthesis, ADP-glucose pyrophosphorylase, sucrose-P synthase, and invertase were expressed higher in the panicle of high-yielding hybrid rice variety than in its parental lines (Wei et al. 2009). Accumulation of storage proteins in the seeds not only supports seed germination, but also improves seed quality as a source of food. A proteomic survey indicated that several storage proteins including different globulins, prolamins, and glutelins start to accumulate in rice seeds at 14 days after anthesis (Koller et al. 2002). Environmental stresses reduce endosperm cell division in rice seed. Furthermore, the stresses negatively affect the efficiency of photosynthesis, which in turn alters the source–sink relationship and remobilization processes. Decrease in the capacity of reserve accumulation and shortage of assimilates are major causes of reduction in grain filling (Setter and Flannigan 2001; Sreenivasulu et al. 2007). A proteome study at different stages of rice growth indicated that rice leaf has a major role in mobilization of nutrients to the grains and higher photosynthetic activity in the leaf causes a higher rate of grain filling (Zhao et al. 2005). Therefore, any changes in source and sink capacity induced by environmental stresses can directly affect the grain filling.

20.3.2 Proteomics for Identification of Stress Tolerance Mechanisms

Minimizing yield loss is the final purpose of stress-related studies. Numerous factors affect the level of rice tolerance to environmental stimuli. These factors can be roughly classified into three major groups including physiological status of the plant, stress situation, and genetic background. It is well known that plant physiological status is affected by several factors such as fertilizer supply, developmental stage, and so on. Application of too much fertilizer may negatively affect the tolerance of rice to stresses. It was reported that spikelet sterility under cold stress was correlated with nitrogen supply; rice pollen grains would be physiologically damaged by low temperature if the plant were grown under high nitrogen conditions (Hayashi et al. 2004). Identification of the external factors that can affect the level of stress tolerance will help us to understand the mechanisms of stress tolerance and hence improve the yield of rice under stressful conditions. The developmental stage of the plant is another physiological factor that determines the level of stress tolerance. Neilson et al. (2010) have reviewed the differences of stress responses in plants at the vegetative and reproductive stages. They concluded that the early stage of male gametophyte development was more susceptible to cold, drought, salinity, heat, and flooding stresses in rice and several other important crops

(Neilson et al. 2010). Susceptibility of rice at the reproductive stage to abiotic stress was also confirmed by physiological and proteomic studies (Dooki et al. 2006; Davatgar et al. 2009; Hayashi et al. 2004; Liu and Bennett 2011). Most of the abiotic stresses could cause infertility of the spikelet, which is a typical sign of stress damage on plant growth.

The intensity, duration, and type of stress can also affect plant tolerance. Han et al. (2009) have studied the effects of various high temperature stresses from 35°C to 45°C on the proteome of rice seedlings. According to their results, increasing the temperature could induce more proteins that were involved in various protection machineries and antioxidative pathways. Similar results could also be obtained by changing the stress duration. A proteomic study of rice seedlings under various drought stress duration indicated that protein expression pattern under prolonged drought stress was different from that exposed to a short period of stress (Ali and Komatsu 2006). According to the proteome analysis of rice under various stress intensity or duration, plant cells might adjust themselves to respond to the changes of stressful conditions.

The type of abiotic stress is another factor that can affect plant response. Environmental stimuli are varied from water-related stresses and ion toxicity to temperature stresses and several other growth-limiting factors. A considerable number of proteomic studies were performed to explain the negative effects of major environmental stresses such as drought (Ali and Komatsu 2006), salinity (Abbasi and Komatsu 2004), cold (Cui et al. 2005), heat (Han et al. 2009), flooding, and anoxia (Shingaki-Wells et al. 2011) on rice at the vegetative stage. Although responsive proteins to various abiotic stresses were classified and their functions were described (Kang et al. 2010; Salekdeh and Komatsu 2007), less attention has been paid to the mechanism of stress tolerance of rice at the reproductive stage.

Major physiological and cellular responses of rice to drought, salinity, cold, heat, and flooding or anoxia were classified and the main differentially expressed protein groups were summarized based on a review of the literature (Table 20.2). Most of the negative effects of drought, salinity, cold, and heat were related to the disturbance of spikelet fertility and seed filling, which was described in Sect. 20.3.1. However, comparison of cellular responses and differentially regulated proteins under abiotic stresses indicated that damages to the photosynthetic machinery were the common responses to most of the stresses. Downregulation of photosynthetic electron transport, photosystem I and II, and RuBisCO activity under drought, salinity, cold, and heat stress have already been reported (Ali and Komatsu 2006; Zinn et al. 2010). In fact, reduction of grain weight and yield under abiotic stress could be ascribed to the decrease in photosynthetic efficiency and the alteration of the source–sink relationship and remobilization processes (Sreenivasulu et al. 2007). Reduction of photosynthetic efficiency reduces assimilation, which in turn affects the grain filling.

The genetic background is the most important factor that determines the tolerance of plant to abiotic stresses. If not all, the majority of plant responses to stresses are genetically controlled. A tolerant variety usually has specific cellular mechanisms that may not be found in susceptible plants. Comparison between tolerant and

Table 20.2 Summary of effects of environmental stresses on cellular functions and protein regulations at reproductive stage of rice

		Protein expression/cellular function	
Stress	Physiological/cellular response at reproductive stage	Increase	Decrease
Drought	Rapid stomatal closure, reduction of photosynthetic activity, electrolyte leakage, leaf folding and growth inhibition, decrease in endosperm cell division, inhibition of anthesis, spikelet desiccation, abnormal pollination and pollen germination, incomplete panicle exsersion, spikelet sterility	1-ROS scavengers 2-late embryogenesis abundant proteins 3-actin depolymerizing factor	1-RuBisCO 2-Photosynthetic electron transport 3-isoflavone reductase-like protein
Salinity	Reduction in panicle length and spikelet number per panicle, inhibition of cell elongation, wilting and dying of old leaves and necrosis of young leaves, reduction in photosynthetic activity, spikelet sterility	1-ROS scavengers 2-activation of Na^+/H^+ exchange 3-lipid peroxidation and electrolyte leakage	1-RuBisCO 2-metabolic activity of the cell 3-chloroplast activity
Cold	Enhance in lignification, disruption of sugar metabolism in the tapetum, flower abortion, pollen and ovule infertility, increasing levels of protective osmolytes, energy shortfall due to mitochondrial dysfunction, spikelet sterility	1-ROS scavengers 2-alcohol dehydrogenase 3-biosynthesis of cell wall components	1-RuBisCO 2-Protein disulfide isomerase 3-H^+-transporting ATP synthase
Flooding/anoxia	Carbohydrate and energy shortfall, slow photosynthesis, increase in cell wall extensibility and cell elongation, slow O_2 and CO_2 influx	1-alcohol dehydrogenase 2-expression of α-amylase 3-expansin family proteins	1-oxidative phosphorylation 2-Starch content 3-Total sugar
Heat	Affect on respiration, damage to the photosynthetic apparatus, increase in ion leakage and lipid peroxidation, spikelet sterility	1-ROS scavengers 2-expression of small HSPs 3-protection related proteins	1-RuBisCO 2-photosystem I and II activity 3-starch synthesis-related proteins

susceptible or wild-type and transgenic plants using proteomic approaches can be very helpful in elucidating the mechanism of stress tolerance in rice. Results of comparative proteomic analyses in rice under abiotic stresses at the reproductive stage indicated specific cellular mechanisms including accumulation of more APX and higher antioxidant capacity in a salt-tolerant genotype (Dooki et al. 2006); better cell membrane integrity, less lipid peroxidation, and increased stomatal closure; and ABA sensitivity in drought-tolerant varieties (Choudhary et al. 2009; Hu et al. 2006); higher capacity to generate ATP in the absence of oxygen through fermentative metabolism in flood-tolerant genotypes (Lasanthi-Kudahettige et al. 2007); and less ABA accumulation in cold-tolerant varieties (Ji et al. 2011). These mechanisms are parts of stress-responsive pathways that help to improve the tolerance of rice varieties to abiotic stresses.

20.4 Future Prospective of Food Science in Rice

Taken together, environmental stresses are inevitable in rice production and elucidation of the cellular mechanisms of stress response and tolerance would help to reduce the deleterious effects of stresses. Proteomics has the potential ability to resolve this issue through analyzing the stress-related cellular responses. Although the effects of major environmental stresses including drought, salinity, cold, heat, flooding, and anoxia have been extensively studied in rice, our knowledge about the impact of stresses on yield and yield-related traits is still insufficient. Furthermore, exploration of the molecular mechanisms in rice responding to several different stresses that occurred simultaneously is still a challenge. It can raise hope for the development of new tolerant varieties of rice to produce acceptable yields under stress conditions to answer these questions.

Whereas proteomics can be used in the cloning of unique genes by means of differential analysis, the technique is also useful in studying post-translational modifications that can affect the activity and binding of a protein and alter its role within the cell. By the use of the completely analyzed gene and the analysis of the functions of the encoded proteins, proteomics is expected to have a marked impact on breeding new lines of rice. In the near future, a tight correction should be attainable between proteomics data and other data from comprehensive analyses. These data will provide future in-depth characterization of the proteome of rice and its subcellular compartments, and have the potential to identify new targets for improvement in rice crops.

References

Abbasi FM, Komatsu S (2004) A proteomic approach to analyze salt-responsive proteins in rice leaf sheath. Proteomics 4:2072–2081

Agrawal GK, Hajduch M, Graham K, Thelen JJ (2008) In-depth investigation of the soybean seed-filling proteome and comparison with a parallel study of rapeseed. Plant Physiol 148:504–518

Ali GM, Komatsu S (2006) Proteomic analysis of rice leaf sheath during drought stress. J Proteome Res 5:396–403
Bailly C (2004) Active oxygen species and antioxidants in seed biology. Seed Sci Res 14:93–107
Bak-Jensen KS, Laugesen S, Ostergaard O, Finnie C, Roepstorff P, Svensson B (2007) Spatio-temporal profiling and degradation of alpha-amylase isozymes during barley seed germination. FEBS J 274:2552–2565
Bethke PC, Libourel IG, Reinöhl V, Jones RL (2006) Sodium nitroprusside, cyanide, nitrite, and nitrate break *Arabidopsis* seed dormancy in a nitric oxide-dependent manner. Planta 223:805–812
Bewley JD (1997) Seed germination and dormancy. Plant Cell 9:1055–1066
Boonjung H, Fukai S (1996) Effects of soil water deficit at different growth stages on rice growth and yield under upland conditions. 2. Phenology, biomass production and yield. Field Crop Res 48:47–55
Botha FC, Potgieter GP, Botha AM (1992) Respiratory metabolism and gene expression during seed germination. J Plant Growth Regul 11:211–224
Bressani R, Elias LG, Juliano BO (1971) Evaluation of the protein quality and milled rices differing in protein content. J Agric Food Chem 19:1028–1034
Choudhary MK, Basu D, Datta A, Chakraborty N, Chakraborty S (2009) Dehydration-responsive nuclear proteome of rice (*Oryza sativa* L.) illustrates protein network, novel regulators of cellular adaptation, and evolutionary perspective. Mol Cell Proteomics 8:1579–1598
Crofts AJ, Crofts N, Whitelegge JP, Okita TW (2010) Isolation and identification of cytoskeleton-associated prolamine mRNA binding proteins from developing rice seeds. Planta 231:1261–1276
Cui S, Huang F, Wang J, Ma X, Cheng Y, Liu J (2005) A proteomic analysis of cold stress responses in rice seedlings. Proteomics 5:3162–3172
Davatgar N, Neishabouri MR, Sepaskhah AR, Soltani A (2009) Physiological and morphological responses of rice (*Oryza sativa* L.) to varying water stress management strategies. Int J Plant Prod 3:19–32
Devos MK, Gale DM (2000) Genome relationships: the grass model in current research. Plant Cell 12:637–643
Dooki AD, Mayer-Posner FJ, Askari H, Zaiee AA, Salekdeh GH (2006) Proteomic responses of rice young panicles to salinity. Proteomics 6:6498–6507
Doroshenk KA, Crofts AJ, Morris RT, Wyrick JJ, Okita TW (2009) Proteomic analysis of cytoskeleton-associated RNA binding proteins in developing rice seed. J Proteome Res 8:4641–4653
Dreyfuss G, Kim VN, Kataoka N (2002) Messenger-RNA-binding proteins and the messages they carry. Nat Rev Mol Cell Biol 3:195–205
Dung LV, Mikami I, Amano E, Sano Y (2000) Study on the response of dull endosperm 2–2, du2-2, to two Wx alleles in rice. Breed Sci 50:215–219
Ehrenshaft M, Brambl R (1990) Respiration and mitochondrial biogenesis in germinating embryos of maize. Plant Physiol 93:295–304
Fedoroff NV (2002) RNA-binding proteins in plants: the tip of an iceberg. Curr Opin Plant Biol 5:452–459
Finch-Savage WE, Leubner-Metzger G (2006) Seed dormancy and the control of germination. New Phytol 171:501–523
Finnie C, Melchior S, Roepstorff P, Svensson B (2002) Proteome analysis of grain filling and seed maturation in barley. Plant Physiol 129:1308–1319
Gallardo K, Job C, Groot SP, Puype M, Demol H, Vandekerckhove J, Job D (2002) Proteomics of *Arabidopsis* seed germination. A comparative study of wild-type and gibberellin-deficient seeds. Plant Physiol 129:823–837
Gallardo K, Le Signor C, Vandekerckhove J, Thompson RD, Burstin J (2003) Proteomics of *Medicago truncatula* seed development establishes the time frame of diverse metabolic processes related to reserve accumulation. Plant Physiol 133:664–682
Gallardo K, Firnhaber C, Zuber H, Héricher D, Belghazi M, Henry C, Küster H, Thompson R (2007) A combined proteome and transcriptome analysis of developing *Medicago truncatula*

seeds: evidence for metabolic specialization of maternal and filial tissues. Mol Cell Proteomics 6:2165–2179

Gangashetti MG, Jena KK, Shenoy VV, Freeman WH (2004) Inheritance of elongated uppermost internode (*EUI*) and identification of a RAPD marker linked to *eui* gene in rice. Curr Sci 87:469–475

Gaur VS, Singh US, Kumar A (2011) Transcriptional profiling and in silico analysis of Dof transcription factor gene family for understanding their regulation during seed development of rice *Oryza sativa* L. Mol Biol Rep 38:2827–2848

Greenbaum D, Colangelo C, Williams K, Gerstein M (2003) Comparing protein abundance and mRNA expression levels on a genomic scale. Genome Biol 4:117

Hajduch M, Ganapathy A, Stein JW, Thelen JJ (2005) A systematic proteomic study of seed filling in soybean: establishment of highresolution two-dimensional reference maps, expression profiles, and an interactive proteome database. Plant Physiol 137:1397–1419

Hajduch M, Casteel JE, Hurrelmeyer KE, Song Z, Agrawal GK, Thelen JJ (2006) Proteomic analysis of seed filling in *Brassica napus*. Developmental characterization of metabolic isozymes using high-resolution two dimensional gel electrophoresis. Plant Physiol 141:32–46

Hamaker BR (1994) The influence of rice protein on rice quality. In: Marshall WE, Wadsworth JI (eds) Rice science and technology. Dekker, New York, pp 177–194

Han F, Chen H, Li XJ, Yang MF, Liu GS, Shen SH (2009) A comparative proteomic analysis of rice seedlings under various high-temperature stresses. Biochim Biophys Acta 1794:1625–1634

Hayashi T, Yamaguchi T, Nakayama K, Komatsu S, Koike S (2004) High nitrogen conditions enhance the cooling damage to pollen in rice plants: proteome analysis of mature anthers. In: Fischer T et al (eds) New directions for a diverse planet. Proceedings for the 4th international crop science congress, Brisbane

He D, Han C, Yao J, Shen S, Yang P (2011) Constructing the metabolic and regulatory pathways in germinating rice seeds through proteomic approach. Proteomics. doi:10.1002/pmic.201000598

Holdsworth MJ, Finch-Savage WE, Grappin P, Job D (2008) Post-genomics dissection of seed dormancy and germination. Trends Plant Sci 13:7–13

Hu H, Dai M, Yao J, Xiao B, Li X, Zhang Q, Xiong L (2006) Overexpressing a NAM, ATAF, and CUC (NAC) transcription factor enhances drought resistance and salt tolerance in rice. Proc Natl Acad Sci USA 103:12987–12992

Isshiki M, Nakajima M, Satoh H, Shimamoto K (2000) Dull rice mutants with tissue-specific effects on the splicing of the waxy pre-mRNA. Plant J 23:451–460

Jagadish SV, Craufurd PQ, Wheeler TR (2007) High temperature stress and spikelet fertility in rice (*Oryza sativa* L.). J Exp Bot 58:1627–1635

Ji X, Dong B, Shiran B, Talbot MJ, Edlington JE, Hughes T, White RG, Gubler F, Dolferus R (2011) Control of abscisic acid catabolism and abscisic acid homeostasis is important for reproductive stage stress tolerance in cereals. Plant Physiol 156:647–662

Job C, Rajjou L, Lovigny Y, Belghazi M, Job D (2005) Patterns of protein oxidation in *Arabidopsis* seeds and during germination. Plant Physiol 138:790–802

Kang HG, Park S, Matsuoka M, An G (2005) White-core endosperm floury endosperm-4 in rice is generated by knockout mutations in the C-type pyruvate orthophosphate dikinase gene (OsPPDKB). Plant J 42:901–911

Kang S, Chen S, Da S (2010) Proteomics characteristics of rice leaves in response to environmental factors. Front Biol 5:246–254

Kikuchi S, Satoh K, Kawagashira N et al (2003) Collection mapping, and annotation of over 28000 cDNA clones from japonica rice. Science 301:376–379

Kim ST, Kang SY, Wang Y, Kim SG, Hwang DH, Kang KY (2008) Analysis of embryonic proteome modulation by GA and ABA from germinating rice seeds. Proteomics 8:3577–3587

Kim EH, Kim YS, Park SH, Koo YJ, Choi YD, Chung YY, Lee IJ, Kim JK (2009) Methyl jasmonate reduces grain yield by mediating stress signals to alter spikelet development in rice. Plant Physiol 149:1751–1760

Koller A, Washburn MP, Lange BM, Andon NL, Deciu C, Haynes PA, Hays L, Schieltz D, Ulaszek R, Wei J, Wolters D, Yates JR 3rd (2002) Proteomic survey of metabolic pathways in rice. Proc Natl Acad Sci USA 99:11969–11974

Laino P, Shelton D, Finnie C, De Leonardis AM, Mastrangelo AM, Svensson B, Lafiandra D, Masci S (2010) Comparative proteome analysis of metabolic proteins from seeds of durum wheat (cv. Svevo) subjected to heat stress. Proteomics 10:2359–2368

Lao NT, Schoneveld O, Mould RM, Hibberd JM, Gray JC, Kavanagh TA (1999) An *Arabidopsis* gene encoding a chloroplast-targeted â-amylase. Plant J 20:519–527

Lasanthi-Kudahettige R, Magneschi L, Loreti E, Gonzali S, Licausi F, Novi G, Beretta O, Vitulli F, Alpi A, Perata P (2007) Transcript profiling of the anoxic rice coleoptile. Plant Physiol 144:218–231

Liu JX, Bennett J (2011) Reversible and irreversible drought-induced changes in the anther proteome of rice (*Oryza sativa* L.) genotypes IR64 and moroberekan. Mol Plant 4:59–69

Liu JX, Raveendran M, Mushtaq R, Bennett J (2005) Proteomic analysis of drought-responsiveness in rice: OsADF5. In: Tuberosa R, Phillips RL, Gale M (eds) In the wake of the double helix: from the green revolution to the gene revolution. Avenue Media, Bologna, pp 251–262

Liu Y, Shi L, Ye N, Liu R et al (2009) Nitric oxide-induced rapid decrease of abscisic acid concentration is required in breaking seed dormancy in *Arabidopsis*. New Phytol 183:1030–1042

Liu X, Guo T, Wan X, Wang H, Zhu M, Li A, Su N, Shen Y, Mao B, Zhai H, Mao L, Wan J (2010) Transcriptome analysis of grain-filling caryopses reveals involvement of multiple regulatory pathways in chalky grain formation in rice. BMC Genomics 11:730

Matsukura C, Saitoh T, Hirose T, Ohsugi R, Perata P, Yamaguchi J (2000) Sugar uptake and transport in rice embryo. Expression of companion cell-specific sucrose transporter (*OsSUT1*) induced by sugar and light. Plant Physiol 124:85–93

Mechin V, Thevenot C, Le Guilloux M, Prioul J-L, Damerval C (2007) Developmental analysis of maize endosperm proteome suggests a pivotal role for pyruvate orthophosphate dikinase. Plant Physiol 143:1203–1219

Mitsunaga S, Tashiro T, Yamaguchi J (1994) Identification and characterization of gibberellin-insensitive mutants selected from among dwarf mutants of rice. Theor Appl Genet 87:705–712

Mitsunaga S, Kobayashi M, Fukui S, Fukuoka K, Kawakami O, Yamaguchi J, Ohshima M, Mitsui T (2007) Alpha-amylase production is induced by sulfuric acid in rice aleurone cells. Plant Physiol Biochem 45:922–925

Mushtaq R, Katiyar S, Bennett J (2008) Proteomic analysis of drought stress-responsive proteins in rice endosperm affecting grain quality. J Crop Sci Biotechnol 11:227–232

Muthurajan R, Shobbar ZS, Jagadish SV, Bruskiewich R, Ismail A, Leung H, Bennett J (2011) Physiological and proteomic responses of rice peduncles to drought stress. Mol Biotechnol 48:173–182

Myers AM, Morell MK, James MG, Ball SG (2000) Recent progress toward understanding biosynthesis of the amylopectin crystal. Plant Physiol 122:989–997

Neilson KA, Gammulla CG, Mirzaei M, Imin N, Haynes PA (2010) Proteomic analysis of temperature stress in plants. Proteomics 10:828–845

Nystrom T (2005) Role of oxidative carbonylation in protein quality control and senescence. EMBO J 24:1311–1317

Oh SJ, Kim YS, Kwon CW, Park HK, Jeong JS, Kim JK (2009) Overexpresson of the transcription factor AP37 in rice improves grain yield under drought conditions. Plant Physiol 150:1368–1379

Oliver SN, Van Dongen JT, Alfred SC, Mamun EA, Zhao X, Saini HS, Fernandes SF, Blanchard CL, Sutton BG, Geigenberger P, Dennis ES, Dolferus R (2005) Cold-induced repression of the rice anther-specific cell wall invertase gene *OSINV4* is correlated with sucrose accumulation and pollen sterility. Plant Cell Environ 28:1534–1551

Rajjou L, Gallardo K, Debeaujon I, Vandekerckhove J, Job C, Job D (2004) The effect of a-amanitin on the *Arabidopsis* seed proteome highlights the distinct roles of stored and neosynthesized mRNAs during germination. Plant Physiol 134:1598–1613

Repetto O, Rogniaux H, Firnhaber C, Zuber H, Küster H, Larré C, Thompson R, Gallardo K (2008) Exploring the nuclear proteome of *Medicago truncatula* at the switch towards seed filling. Plant J 56:398–410

Rubel A, Rinne RW, Canvin DT (1972) Protein, oil, and fatty-acid in developing soybean seeds. Crop Sci 12:739–741

Ruuska SA, Girke T, Benning C, Ohlrogge JB (2002) Contrapuntal networks of gene expression during *Arabidopsis* seed filling. Plant Cell 14:1191–1206
Salekdeh GH, Komatsu S (2007) Crop proteomics: aim at sustainable agriculture of tomorrow. Proteomics 7:2976–2996
Salon C, Raymond P, Pradet A (1988) Quantification of carbon fluxes through the tricarboxylic acid cycle in early germinating lettuce embryos. J Biol Chem 263:12278–12287
Sato Y, Masuta Y, Saito K, Murayama S, Ozawa K (2011) Enhanced chilling tolerance at the booting stage in rice by transgenic over expression of the ascorbate peroxidase gene, OsAPXa. Plant Cell Rep 30:399–406
Schmidt MW, Houseman A, Ivanov AR, Wolf DA (2007) Comparative proteomic and transcriptomic profiling of the fission yeast *Schizosaccharomyces pombe*. Mol Syst Biol 3:79
Setter TL, Flannigan BA (2001) Water deficit inhibits cell division and expression of transcripts involved in cell proliferation and endoreduplication in maize endosperm. J Exp Bot 52:1401–1408
Shingaki-Wells RN, Huang S, Taylor NL, Carroll AJ, Zhou W, Millar AH (2011) Differential molecular responses of rice and wheat coleoptiles to anoxia reveal novel metabolic adaptations in amino acid metabolism for tissue tolerance. Plant Physiol 156:1706–1724
Smith AM, Zeeman SC, Smith SM (2005) Starch degradation. Annu Rev Plant Biol 56:73–98
Spadaro D, Yu B, Spoel SH, Chu C et al (2010) The redox switch: dynamic regulation of protein function by cysteine modifications. Physiol Plant 138:360–371
Sreenivasulu N, Sopory SK, Kavi Kishor PB (2007) Deciphering the regulatory mechanisms of abiotic stress tolerance in plants by genomic approaches. Gene 388:1–13
The International Rice Genome Sequencing Project (2005) The map-based sequence of the rice genome. Nature 436:793–800
The Rice Annotation Project (2007) Crated genome annotation of *Oryza sativa* ssp. Japonica and comparative genome analysis with *Arabidopsis thaliana*. Genome Res 17:175–183
Vensel WH, Tanaka CK, Cai N, Wong JH, Buchanan BB, Hurkman WJ (2005) Developmental changes in the metabolic protein profiles of wheat endosperm. Proteomics 5:1594–1611
Watson BS, Asirvatham VS, Wang L, Sumner LW (2003) Mapping the proteome of barrel medic (*Medicago truncatula*). Plant Physiol 131:1104–1123
Weber H, Borisjuk L, Wobus U (2005) Molecular physiology of legume seed development. Annu Rev Plant Biol 56:253–279
Wei G, Tao Y, Liu G, Chen C, Luo R, Xia H, Gan Q, Zeng H, Lu Z, Han Y, Li X, Song G, Zhai H, Peng Y, Li D, Xu H, Wei X, Cao M, Deng H, Xin Y, Fu X, Yuan L, Yu J, Zhu Z, Zhu L (2009) A transcriptomic analysis of superhybrid rice LYP9 and its parents. Proc Natl Acad Sci USA 106:7695–7701
Xu SB, Li T, Deng ZY, Chong K, Xue Y, Wang T (2008) Dynamic proteomic analysis reveals a switch between central carbon metabolism and alcoholic fermentation in rice filling grains. Plant Physiol 148:908–925
Yamakawa H, Hirose T, Kuroda M, Yamaguchi T (2007) Comprehensive expression profiling of rice grain filling-related genes under high temperature using DNA microarray. Plant Physiol 144:258–277
Yang P, Liang Y, Shen S, Kuang T (2006) Proteome analysis of rice uppermost internodes at the milky stage. Proteomics 6:3330–3338
Yang P, Li X, Wang X, Chen H, Chen F, Shen S (2007) Proteomic analysis of rice (*Oryza sativa*) seeds during germination. Proteomics 7:3358–3368
Yano H, Kuroda M (2006) Disulfide proteome yields a detailed understanding of redox regulations: a model study of thioredoxin-linked reactions in seed germination. Proteomics 6:294–300
Ye X, Al-Babili S, Kloti A, Zhang J, Lucca P, Beyer P, Potrykus I (2000) Engineering the provitamin A (beta-carotene) biosynthetic pathway into (carotenoid-free) rice endosperm. Science 287:303–305
Zeeman SC, Northrop F, Smith AM, Rees T (1998) A starch-accumulating mutant of *Arabidopsis* thaliana deficient in a chloroplastic starch hydrolyzing enzyme. Plant J 15:357–365

Zeeman SC, Pilling E, Tiessen A, Kato L, Donald AM, Smith AM (2002) Starch synthesis in *Arabidopsis*. Granule synthesis, composition, and structure. Plant Physiol 129:516–529

Zeeman SC, Smith SM, Smith AM (2004a) The breakdown of starch in leaves. New Phytol 163:247–261

Zeeman SC, Thorneycroft D, Schupp N, Chapple A, Weck M et al (2004b) Plastidial a-glucan phosphorylase is not required for starch degradation in *Arabidopsis* leaves but has a role in the tolerance of abiotic stress. Plant Physiol 135:849–858

Zhang H, Shen W, Xu L (2003) Effect of nitric oxide on the germination of wheat seeds and its reactive oxygen species metabolisms under osmotic stress. Acta Bot Sin 45:901–905

Zhang H, Chen T, Wang Z, Yang J, Zhang J (2010a) Involvement of cytokinins in the grain filling of rice under alternate wetting and drying irrigation. J Exp Bot 61:3719–3733

Zhang J, Nallamilli BR, Mujahid H, Peng Z (2010b) OsMADS6 plays an essential role in endosperm nutrient accumulation and is subject to epigenetic regulation in rice (*Oryza sativa*). Plant J 64:604–617

Zhao C, Wang J, Cao M, Zhao K, Shao J, Lei T, Yin J, Hill GG, Xu N, Liu S (2005) Proteomic changes in rice leaves during development of field-grown rice plants. Proteomics 5:961–972

Zhao M, Tian Q, Zhang W (2007) Nitric oxide synthase-dependent nitric oxide production is associated with salt tolerance in *Arabidopsis*. Plant Physiol 144:206–217

Zhu T, Budworth P, Chen W, Provart N, Chang HS, Guimil S, Su W, Estes B, Zou G, Wang X (2003) Transcriptional control of nutrient partitioning during rice grain filling. Plant Biotechnol J 1:59–70

Zhu G, Ye N, Yang J, Peng X, Zhang J (2011) Regulation of expression of starch synthesis genes by ethylene and ABA in relation to the development of rice inferior and superior spikelets. J Exp Bot. doi:10.1093/jxb/err088

Zinn KE, Tunc-Ozdemir M, Harper JF (2010) Temperature stress and plant sexual reproduction: uncovering the weakest links. J Exp Bot 61:1959–1968

Chapter 21
Beer Proteomics

Gianluca Picariello, Chiara Nitride, Gianfranco Mamone, Takashi Iimure, Francesco Addeo, and Pasquale Ferranti

21.1 Introduction

Beer is the most consumed alcoholic beverage in the world. It is also believed to be the oldest fermented beverage as the production of certain primeval forms of beer can be traced back to Mesopotamian writings that are nearly 5,000 years old. The process of making beer is known as brewing.

Barley (*Hordeum vulgare* L.), and to a much less extent other cereals, are the basic raw material for brewing. They are also the primary source of enzymes that, through the saccharification of starch, produce the sugars that are fermented by *Saccharomyces* sp. yeasts. *Hordeum vulgare* belongs to the *Triticeae* subfamily of the *Poaceae*, which includes important small-grain cereal species such as wheat and rye. Barley is an extremely adaptable cereal grain that can be cultivated in a wide climatic zone ranging from subarctic to subtropical latitudes. It was one of the

G. Picariello (✉) • F. Addeo • P. Ferranti
Istituto di Scienze dell'Alimentazione – Consiglio Nazionale delle Ricerche (CNR), Via Roma 64, 83100, Avellino, Italy

Dipartimento di Scienza degli Alimenti, Università di Napoli "Federico II", Parco Gussone, Portici, Napoli 80055, Italy
e-mail: picariello@isa.cnr.it

C. Nitride
Dipartimento di Scienza degli Alimenti, Università di Napoli "Federico II", Parco Gussone, Portici, Napoli 80055, Italy

G. Mamone
Istituto di Scienze dell'Alimentazione – Consiglio Nazionale delle Ricerche (CNR), Via Roma 64 83100, Avellino, Italy

T. Iimure
Bioresources Research and Development Department, Sapporo Breweries Ltd, 37-1, Nittakizaki, Ota, Gunma 370-0393, Japan

F. Toldrá and L.M.L. Nollet (eds.), *Proteomics in Foods: Principles and Applications*, Food Microbiology and Food Safety 2, DOI 10.1007/978-1-4614-5626-1_21,

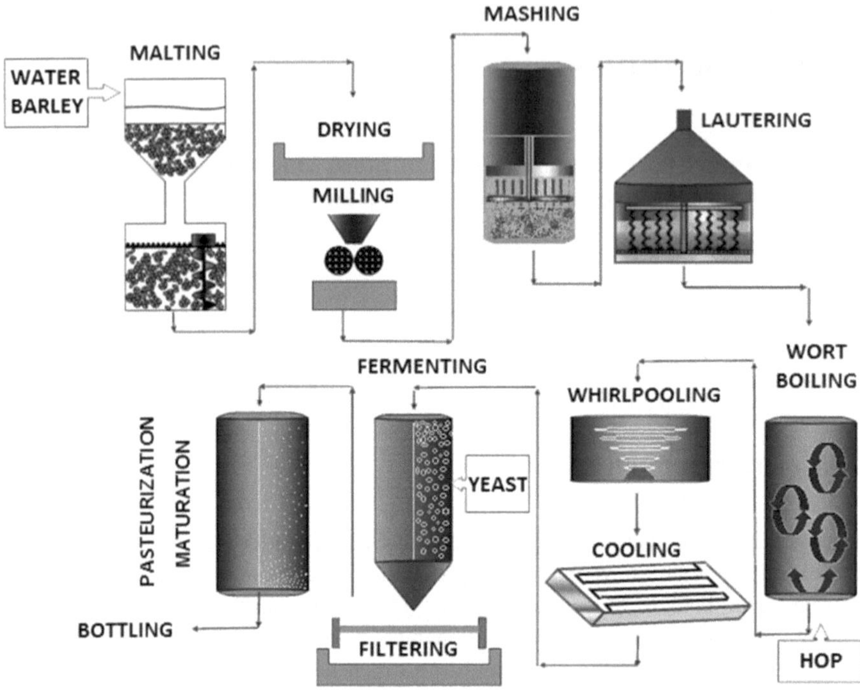

Fig. 21.1 Schematic and simplified representation of the brewing process

first agricultural species to be domesticated and is currently ranked the fifth most grown crop in the world (Baik and Ullrich 2008). Aside from the increased interest in reintroducing it into the human diet due to its nutritional potential, the largest use of barley is for malting purposes by the brewing industry. Two varieties of barley are used for malting: two-row and six-row depending on the arrangement of the kernels around the spike shaft. Two-row barley is the generally preferred variety, as it yields malt with higher extracts, lighter color, lower protein levels, and is claimed to have a more refined flavor than six-row barley. However, six-row barely has a slightly higher diastatic power (roughly a measure of the activity of the starch-converting enzymes) than two-row. To balance the effect of the high protein content of six-row barley, brewers usually cut malted grains with unmalted grains such as corn or rice.

21.2 The Brewing Process

A conventional brewing process is schematized in Fig. 21.1. The flowchart highlights the production of a so-called lager beer with bottom fermenting yeast. There are many variants of the brewing process depending on the specific type of beer.

In contrast, ale and stout beers are produced with top-fermenting yeasts, and the process differs somewhat from that outlined for the lager. Regardless, malted barley is almost always the main source of both starch and enzymes.

21.2.1 Malting

Malting is a controlled process of germination that is realized to induce physical and biochemical modifications to cereal grains. These changes are then stabilized by grain drying and kilning. Malting consists of three main steps: (1) steeping: grains are soaked in water for 2 or 3 days to allow them to increase their moisture level from 12 to at least 40–45%; (2) germination: in this phase, the embryo increases its diameter by approximately 1/3 and germination begins to maintain embryo growth. Germination is interrupted by drying after 5–6 days, when radicles develop and are approximately 3/4 of the length of the seed; and (3) kilning: modifications are stabilized by a mild to heavy thermal treatment (65–220°C). The development of melanoidins causes the seeds to become dark as a consequence of the Maillard reactions. Several studies have confirmed that kilning has only a marginal effect on hydrolytic enzymes and that these enzymes retain the great part of their activity. The hydrolytic activity of the grain enzymes is explicated in the later stages of the brewing process (e.g., mashing).

21.2.2 Grinding Malt

The milling of malt and of other eventual cereal adjuncts represents one of the most critical steps in the brewing process and influences certain parameter processes such as pressure at lautering. The purpose of milling is to split the husk and reduce the starchy endosperm to a consistency fine enough to allow easy access for water solubilization (Bamforth 2003). Milling can be a dry or a wet process (Eblinger 2009). In this latter case, the hydration of the husk reduces the risk of its damage during milling.

21.2.3 Mashing

Mashing is the water-solubilization of malt components by a combination of enzymatic, physical, and chemical processes. Grist is efficiently mixed with water at 55–60°C to gelatinize the starch, thereby facilitating its enzymatic degradation. Water is added into the mash tun at a proportion of approximately 2.5–3.5- L/Kg malt. The temperature increment during mashing is characterized by specific thermal rests during which a series of biochemical events occurs. (1) β-glucanase rest (37–45°C). β-glucanase and starch-debranching glycosidases are mobilized to hydrolyze β-glucans and starch, respectively. The activity of these enzymes is critical in assisting the activity of amylases to produce fermentable sugars and to lower the viscosity of the system (Aehle 2007). (2) Acid rest wherein phytase (optimum 30–53°C) catalyzes the demolition of insoluble phytin, a complex organic phosphate containing both calcium and magnesium, to phytic acid with a consequent reduction of the wort pH (up to 5.0–5.5) and precipitation of calcium phosphate (Aehle 2007). In modern processes, this step is generally omitted because the wort can be alternatively acidified by adding a low amount of malt that has been preacidified by lactic acid or by exploiting the action of lactic acid bacteria (Lambic beers). (3) Protein rest (45–55°C) where proteases extensively degrade proteins to peptides and amino acids. By facilitating the mobilization of endosperm polysaccharides, protein rest affects the level of malt modification. In addition, the proteolysis allows beer clarification and supplies a nitrogen substrate fundamental for the growth involved in fermenting. A "proteolytic compromise" has to be achieved in order to guarantee the persistence of proteins that stabilize foam and to remove haze-forming polypeptides. (4) Saccharification rest where α-amylase partially hydrolyzes amylose to dextrin that is an unfermentable sugar that remains intact in beer, influencing its heaviness. Amylose also produces maltose because of the action of β-amylase that is released from the aleuronic layer during the protein rest. Maltose is a fermentable disaccharide that directly affects the alcohol degree of beer. These two enzymes have different optimal temperatures: the optimal temperature for β-amylase is approximately 58–63°C, and that for α-amylase is approximately 10°C higher. Considering that the activity of the different proteases is highly dependent on temperature, the adjustment of this parameter is the primary control mechanism of the mashing process for conferring the desired qualitative traits to beer.

21.2.4 Lautering

By lautering, the liquid wort is separated from the residual spent grain (e.g., husk). Lautering starts with the mashout step that consists in raising the temperature of the mash (around 77°C) to arrest the enzymatic conversion of starches and make the mash and wort fluid. As the mash is added to the lauter tun, the spent grain begins functioning as a sand filter that captures mash debris. Once a bed of the grain husks is established, the wort is drawn from the bottom of the tun and is recirculated to the top of the bed until the first wort reaches the desired clarity. To drain the wort completely from the mash and to maximize the extraction of sugars, the grain bed is uniformly distributed (sparging) and sprinkled with hot water (Briggs et al. 2004).

21.2.5 Wort Boiling

Wort boiling can be divided into two steps: hot holding and evaporation. Different chemical reactions take place during hot holding such as (1) the Maillard reactions occur at an appreciable rate at approximately 80°C and are responsible for the neo-generation of flavor and color substance; (2) enzyme inactivation; and (3) the wort sterilization that eliminates any organisms that might compromise yeast fermentation. Proteins and protein–tannin complexes tend to coagulate and precipitate, thus clarifying wort. An optimal pH for protein coagulation is below pH 5.2 and can be achieved by wort acidification.

Hops (*Humulus lupulus*) are added to the wort during boiling (150–600 g/hL) in two steps, at the beginning and at the end of the process. Hopping consists in the release of soluble α- and β-acids. α-acids (such as humulone and compounds structurally correlated) are isomerized (mainly to *cis*- and *trans*-isohumulone) and have a positive effect on yeast fermentation; β-acids, such as myrcene, are not isomerized and contribute to a beer's aromatics and bitterness. The wort (90°C) is cooled in a heat exchanger. The final temperature depends on the desired beer type: 10 or 20°C for top and bottom fermenting beer, respectively. The sedimentation of suspended particles is carried out by gravitational pull and can be promoted by fining agents. After a period of approximately 3 days, the beer is filtered or centrifuged. The removal of wort solids is usually accomplished by "whirlpooling".

21.2.6 Fermentation

Yeasts turn wort into beer by fermenting sugars produced by the demolition of starch. Ethanol is produced by the yeast metabolism of carbohydrates under anaerobic or low-oxygen conditions. Top-fermented beers are most commonly produced with *Saccharomyces cerevisiae*. Top-cropping yeasts, also termed "ale yeast" such as *S. cerevisiae*, are so-called because they form a foam at the top of the wort bulk during fermentation. Bottom-cropping yeasts such as *S. pastorianus*, formerly known as *S. carlsbergensis*, are typically used to produce lager-type beers, although they can also produce ale-type beers. These yeasts ferment well at low temperatures.

21.2.7 Pasteurization and Bottling

Pasteurization is the last step in brewing and can occur before or after bottling. In flash pasteurization, the beer flows through a heat exchanger that typically raises the temperature to 72°C. Beer is bottled in previously sterilized bottles. In pasteurization after bottling, a tunnel pasteurizer is used. The higher the temperature, the more rapidly microorganisms are destroyed; however, the beer flavor changes significantly as a function of temperature. The most significant of these changes are due to oxidation that is occasionally prevented by incorporating antioxidants, such as vitamin C.

21.3 From Barley Grain to Beer: The Protein Path

21.3.1 Barley Proteins

If minor amounts of yeast-derived enzymes are excepted, the majority of beer proteins are derived from barley or other cereals used for brewing. Barley grain proteins are comprised of a complex mixture of polypeptides that are commonly classified by the extractability characteristics according to the procedure introduced by Osborne (1895). Albumins and globulins are heterogeneous protein groups soluble in water and in saline aqueous solution, respectively. They are structurally correlated nonstorage components of cell walls or metabolic enzymes and are primarily located in the aleuronic layer and embryo. The most abundant of these proteins can accumulate in relatively large amounts and are thus enrolled as secondary storage reserves. It is important to note that pathogenesis-related protein factors constitute a significant amount of albumin and globulins.

The 60–70% ethanol-soluble fraction of barley grain contains hordeins, a class of major storage proteins that reside in the starchy endosperm and account for at least half of the total protein content. Hordeins are a complex polymorphic mixture of proteins coded by multigene families. They can be divided into four groups according to electrophoretic mobility: B-, C-, D- and γ-hordeins. B- (30–45 kDa) and C- (45–75 kDa) hordeins are the dominating subtypes, accounting for 70–80% and 10–12% respectively, whereas D- (105 kDa) and γ-hordeins (35–40 kDa) are minor components from a quantitative standpoint. The B-hordeins are usually subdivided further into B1, B2, and B3 subfamilies (Skerritt and Janes 1992). Osborne coined the term *prolamins* for the alcohol-soluble proteins of barley and the homologue counterparts found in other cereals to emphasize their high content of proline and amine nitrogen. Barley glutelins were initially thought to constitute a distinct protein family, extractable only in denaturing and sulfhydryl-reducing buffers. They were later characterized as gene products structurally related to prolamins, classifiable primarily as B- and D-hordeins (Skerritt and Janes 1992). Thus, those extracted in aqueous alcohol are properly the monomeric C- and B-hordeins. The D- and remaining B-hordeins (sulfur-rich) are in part associated by disulfide bridges and, in the presence of water, self-organize in gel-like protein structures that resemble those of wheat gluten. It has been hypothesized that within these glutenlike aggregates, the D-hordeins (high MW subunits) form a backbone that binds to B-hordeins (low MW subunits) via intermolecular disulphides (Moonen et al. 1987). The relative ratio of the hordein subtypes is determined by a number of factors.

The influence of genetic and climatic factors and growth conditions on the expression of characteristic hordein patterns, although not yet fully elucidated, assumes particular relevancy.

During seed germination, hordeins are heavily degraded to supply substrates for the protein synthesis in the growing embryo. Similarly, during the malting and mashing processes, hordeins are hydrolyzed up to small peptide fragments or even amino acids. Hordein mobilization is necessary for hydrolases to access starch grain to release fermentable carbohydrates. By determining the accessibility to starch B-, C-, and D-hordeins affect the diastatic power of malt, thereby justifying why the total hordein grain content is negatively associated with malting quality (Smith and Simpson 1983). In contrast, taking into account the level of hydrolases, total protein content is directly correlated to diastatic power. The amount of soluble proteins versus the total protein amount, referred to as the "Kolbach index," is used by maltsters as a reliable predictor for potential starch conversion. In general, the total hordein content alone is to be considered an inadequate predictor of malting quality, primarily because the relationship is strictly cultivar-specific. Considering that the assessment of quantifiable indices of malting quality is of huge interest for brewers for the purpose of malt selection and breeding programs, several research efforts have attempted to link specific hordein patterns to the malting quality. For instance, a putative direct causal relationship has been suggested between the amount of D-hordein and the malt extract (Howard et al. 1996) that, however, has not been definitively accepted based on investigations that exploited pairs of near-isogenic barley lines both with and without D-hordeins (Brennan et al. 1998; Molina-Cano et al. 2001). Conflicting observations have also been recorded when the B- to C-hordeins ratio has been correlated with the activity of amylases and to malting quality (Qi et al. 2006). The wide genotypic and environment-induced variability of the hordein expression will require further studies to establish comprehensive relationships that consistently link malt quality factors and specific hordein levels.

21.3.2 Protein Modification During Malting and Mashing

The conversion of barley into beer represents one of the most ancient and also the most intricate examples of applied enzymology. Barley components undergo complex and extensive transformations during the malting, mashing, and fermenting processes.

Barley proteins in particular are modified in different ways and to a variable extent during each of these processes. The degree of modification depends on several factors including the function of proteins, their location in the caryopsis, their structure, and other physical and chemical characteristics, including water solubility.

More than 40 endopeptidases have been identified in malt, and they have been grouped into four general classes: cysteine, metallo, aspartic, and serine proteinases

(Zhang and Jones 1995). Cysteine endoproteases are the most active category of proteases in germinating barley grain, accounting for up to 90% of the total activity in the degradation of prolamins (Enari 1986).

The cysteine endoprotease activity in barley is due primarily to two endoproteinases (EP), EP-A (37 kDa) and EP-B (30 kDa), that exhibit a different substrate specificity. During germination, EP-A and EP-B are secreted into the endosperm from the scutellar epithelium and aleurone, in response to gibberellic acid (Koehler and Ho 1990). Additional active proteases in malt that belong to the four classes have been identified by direct isolation and characterization or inferred by zymographic investigations relying on opportunely designed inhibition studies (Jones 2005). Proteolysis is controlled by a very delicate equilibrium with a high impact on malting quality and, hence, on the final quality of the beer. This equilibrium is modulated by the action of a number of protease inhibitors (Jones 2005) and also by hydrolytic events that can degrade useful metabolic enzymes such as barley β-amylase (Schmitt and Marinac 2008).

Reducing or oxidizing agents may affect the activity of proteases. Structural or conformational features of the protein substrates may be altered, rendering them more or less extractable or available for hydrolytic action. Reportedly, the activity of malt proteases is significantly enhanced by reducing agents and lowered by oxidizing agents (Jones and Budde 2003). Overall, this reduction has to be ascribed to a reductive depolymerization of hordeins, especially those that are sulfur-rich, which can acquire a certain degree of solubility once released by their aggregated forms. The comparison of hordein extractability in reducing and nonreducing conditions from unmalted barley, barley, and spent grain, has demonstrated that disulfide bonds undergo cleavage and B- and D-hordeins are extensively proteolyzed during malting. However, the complete hordein repertory does not undergo digestion during malting. In contrast, mashing probably induces the formation of novel disulfide bonds and promotes hordein aggregation, ultimately resulting in their removal from the low-alcohol solution in the wort and the beer. The mashing temperature is critical in determining the proteolytic equilibrium because it affects the activity of proteases and solubility of the protein components. Relatively small amounts of intact proteins survive the remainder of the harsh brewing process that generally includes wort boiling, fermentation, and filtering.

It is well established that the most abundant proteins in beer are the Z-barley protein and nonspecific lipid transfer protein (ns-LTP) (Curioni et al. 1995). Z-barley protein, a ~43 kDa hydrophobic protein, is found in two isoforms: Z4 (80%) and Z7 (20%). Similarly two ns-LTPs occur in beer: ns-LTP1 of 9.7 kDa and ns-LTP2 of 7.0 kDa. Either these two proteins undergoes heavy nonenzymatic glycation (Maillard reactions) and acylation, which are likely also to occur upon malt kilning, or they undergo at least a partial unfolding during wort boiling. Interestingly, ns-LTP is almost foam-inactive in the native form, whereas it becomes foam-active after thermal denaturation/glycation. Although the survival of small amounts of protease/amylase inhibitors and other metabolic proteins has been well established, the persistence of intact hordeins in beer is still under discussion. The possible presence of hordeins or hordein-derived low-sized peptides has several technological and immunological implications. Their identification in beer represents an analytical challenge that is stimulating the development and the application of up-to-date high-resolution strategies, including several proteomic and peptidomic approaches.

21.3.3 The Characterization of Beer Proteins: Classical Analytic Methodologies and the Advent of Proteomics

Proteins play a primary role in determining the properties of malt and beer. They affect flavor, color, foam formation and stability, mouth feel, and colloidal stability with the latter having a direct effect on shelf-life. There appears to be some disagreement within the literature data regarding the average protein content of beer, most likely because of the great varieties of commercialized beer worldwide. The protein amounts reported cover a rather broad range. Whereas protein-derived nitrogenous, as also including amino acids and very short peptides, is estimated to fall in the range 2–5 g/L, the actual polypeptide fraction of beer is around 0.5 g/L. The polypeptide fraction of beer is extremely heterogeneous as it encompasses hydrolytic peptides constituted by a few amino acid residues as well as intact proteins of at least ~100 kDa. For several decades, many classical biochemical analytic methodologies have addressed the characterization of proteins in beer. As early as in 1978, Sørensen and Ottesen used gel filtration chromatography to divide beer proteins in two main categories: those with estimated MW of 44 kDa and those with an estimated MW of 10 kDa. It is worth noting that they started their large-scale purification with 100 l of beer. The chromatographic fractions further analyzed by isoelectrofocusing, exhibited many isoforms migrating at distinct isoelectric points (pI). Later on, it was demonstrated that such a spreading was a consequence of the Maillard-type glycation that masks a variable number of lysyl residues, affecting the protein charge (Hejgaard and Kaersgaar 1983). The same two protein fractions were observed by SDS-PAGE analysis. To identify proteins in beer, opportune antibodies were raised against the barley albumin and globulin fractions. Immunochemical experiments made it possible to identify barley Z (Hejgaard and Kaersgaar 1983) and ns-LTP (Sørensen et al. 1993). ELISA immunoassays were also developed to quantify protein Z4 and ns-LTP to evaluate the impact of these proteins on beer foam quality (Evans et al. 1999). Further in-depth successive single-protein characterizations have demonstrated that during the boiling stages, barley ns-LTP is heavily glycated, resistant to hydrolysis, and, in part, denatured (Jégou et al. 2000, 2001).

Malt-derived hordeins were initially recognized as primary haze-forming factors using electrophoretic separation and immunochemical detection (Asano et al. 1982). In a classic article, Ellis et al. (1990) detected a high amount of potentially harmful hordeins and gliadinlike epitopes in both malt and beer, thereby precluding beer consumption by individuals with celiac disease. Later on, immunochemical assays that were performed throughout the brewing process using antibodies raised against the specific hordein subfamilies detected large fragments of B- and C-hordeins in beer that, however, had not been further characterized (Sheehan and Skerritt 1997).

Although a definitive estimation of hordein in beer samples still remains out of reach, the above determinations undoubtedly suffered from several limitations, primarily due to a lack of specificity and an inability to handle large dynamic ranges of protein abundance. Thus, there remains an incompatibility in the data regarding the occurrence of intact or large-sized hordein fragments in beer, and from several standpoints, it has remained undefined for many years. In fact, the identification of low-abundance proteins and more reliable proof of the survival of hordeins in beer were delayed until the development of adequate technological platforms.

The introduction of advanced techniques, such as two-dimensional electrophoresis (2DE) expanded the study of the beer proteins to include the tracing of the pathway of their "natural selection" in the brewing process. Barley grain proteomes were among the first protein sets profiled by 2DE, when the term "proteome" did not yet exist. The investigations by the group of Angelika Görg primed a more systematic study of the biochemical events underlying the brewing process, and they provided remarkable evidence of the successful application of 2DE separation techniques to proteins in seeds (Weiss et al. 1992; Görg et al. 1992a, b). Pioneering silver-stained 2DE analyses of beer proteins showed a pattern dominated by polypeptides in the range of 34–43 kDa (Marshall and Williams 1987) that were mainly ascribed to Z4-barley, as successively confirmed (Pressi et al. 1993).

The development and progressive improvement of the matrix-assisted laser desorption ionization (MALDI) and electrospray (ESI) mass spectrometry (MS) ushered in the proteomic era by enabling the high-throughput analysis of complex proteomes. Over the last decade, the 2DE separation and the identification of protein spots by peptide mass mapping (also referred to as "peptide mass fingerprinting") supported by bioinformatic tools have come to be considered "classic proteome approaches." These approaches have followed the development of the protein through all of its stages of production. Proteomic analyses have been performed on barley grain, (Østergaard et al. 2004), malt (Bak-Jensen et al. 2004), wort (Iimure et al. 2012a), and more recently on beer (Fasoli et al. 2010; Iimure et al. 2010; Perrocheau et al. 2005; Picariello et al. 2011; Colgrave et al. 2012). The barley grain proteome has undergone several comprehensive untargeted and descriptive analyses. Such analyses have definitively identified a large number of the major metabolic proteins as well as their actual or putative role in the process of seed germination. However, a great part of the more recent proteomic research has evolved by primarily addressing questions of functional proteomics. They therefore target the specific topics of development, physiopathology, and quality of crop plants through the detection of differentially expressed protein subsets. Thus, proteomic analysis of embryo, aleurone, and starchy endosperm has provided insight into the biochemical events that occur at various stages of seed development, from grain filling to germination. The dynamic evolution of the barley proteome referred to as "the spatiotemporal barley proteomics" has been reviewed in detail, with focus on the structural characterization of the thioredoxin system and specific amylases (Bønsager et al. 2007;

Finnie and Svensson 2009). The proteome of the mature barley grain is clearly dominated by the so-called "pathogenesis-related (PR) proteins." PR proteins have been evolutionarily selected to endure proteolysis and to remain soluble at a low pH. Due to their structural properties, these proteins are barely affected by the extreme conditions of vacuolar, apoplastic, or intercellular spaces where they are localized and are thus able to survive the harsh brewing treatments (Gorjanović 2009). Therefore, not surprisingly, the brewing processes select the PR proteins, that are ultimately among the most abundant proteins in beer.

Proteomic analysis has also been utilized as a powerful and reproducible system for the characterization of transgenic genotypes designed to increase the tolerance of barley to varying stress conditions, compared to wild-type barley. The heterologous expression of specific genes can produce unexpected modifications of the proteome. These can be effectively explored by the 2DE-MS approach (Finnie et al. 2004).

The comparison among 2DE maps of water-soluble barley malt and beer directly visualized the modifications that occur through brewing involving the protein patterns (Perrocheau et al. 2005). This study along with successive investigations demonstrated wort to have a very simplified protein pattern compared to water-soluble malt. This is due to many of the non-PR proteins not surviving reduction, heat treatments, and fermentation and because they are proteolyzed or removed from the wort bulk because of aggregation (Jin et al. 2011a). Related to this aspect, recent proteome analysis of the wort boiling process has furnished evidence of a partial aggregation and precipitation of Z4-serpin that was found to be covalently associated with small proteolytic fragments of barwin (Iimure et al. 2012a). As expected, wort boiling gradually causes protein unfolding, although it does not substantially change the protein pattern (Jin et al. 2009a). However, the conformational transitions concerning both secondary and tertiary protein structures have an effect on the final quality of beer.

Recent comparative studies agree on the massive removal of hordeins from wort, contingent on their undertaking of either proteolysis or aggregation. This and other aspects related to the protein changes that occur during malting and brewing await a response that is to be provided through the use of proteomic-based platforms. Thus, the research in this field is still fervent, fueled by the increased potential of the functional proteomic studies to improve control of the brewing process significantly.

21.3.4 Beer Proteomics

Considering the critical influence of polypeptides in determining foam firmness, chill haze formation, colloidal stability, mouth feel and palatability, the comprehensive characterization of the beer proteome has important technological implications. As with the analysis of complex proteomes, the potential for classical methodologies of protein analysis in the high-throughput analysis of beer have been largely surpassed by the upgraded proteomic approaches, especially those based on MS identification and quantification.

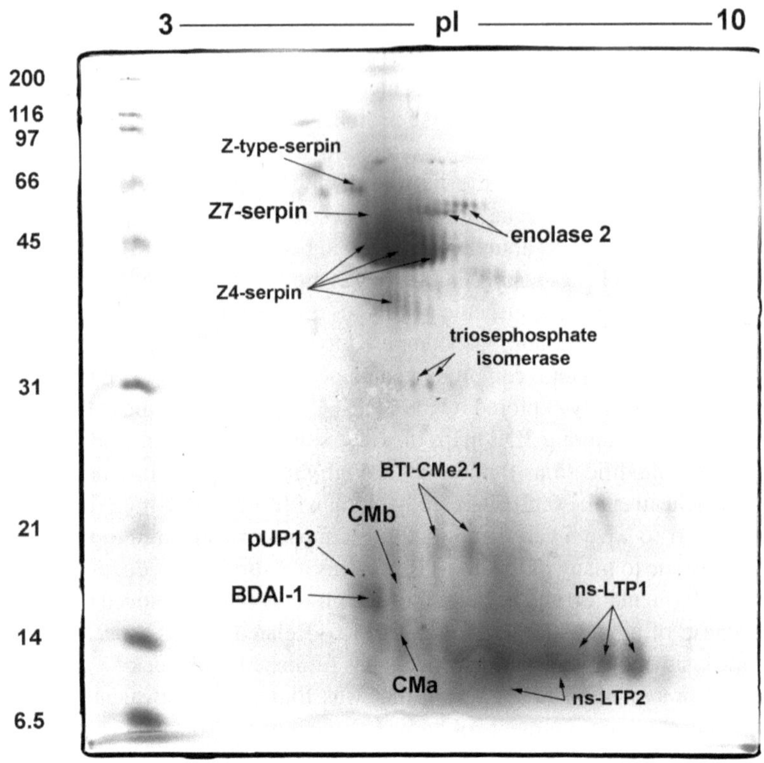

Fig. 21.2 Two-dimensional electrophoresis (2DE) map of a representative barley-malt Italian lager beer. Protein spots have been assigned by image comparison with analogue 2DE maps from the literature and by peptide mass mapping

Figure 21.2 shows a typical immobilized 3–10 pH gradient 2-D IPG/SDS-PAGE separation of proteins from a lager beer precipitated by 12% trichloroacetic acid. Protein spots were assigned by image comparison with analogous maps from the literature and by MALDI-TOF MS peptide mapping. In agreement with previous studies (Perrocheau et al. 2005; Iimure et al. 2010), the protein pattern is largely dominated by the barley chymotrypsin inhibitors Z4-serpin and its related Z7 isoform that produced a broad set of spots centered around 43 kDa and pI 5.6. Ns-LTP1 along with minor amounts of ns-LTP2 prevailed at the lower molecular weights. Common poor focusing of alkaline proteins causes ns-LTP1 to exhibit rather diffuse spots (pI = 8.2 for ns-LTP1 and 7.0 for ns-LTP2). As expected (Hejgaard and Kaersgaar 1983), the protein spots of Z4-serpin and ns-LTPs were spread over a broad pI range due to multiple nonenzymatic glycation. A part trace amounts of glycolytic enzymes (i.e., enolase 2 and triosephosphate isomerase) released by *Saccharomyces cerevisiae* at a late stage of the brewing process, the remaining proteins in the map have ascertained or putative functions as protease/amylase inhibitors.

The structural traits related to their function, including primary and secondary structure and folding, provide them with particular stability, thereby justifying their persistence in beer. From a technical viewpoint, the separation and identification of beer proteins is complicated by the concomitance of several factors, such as the heavy nonenzymatic glycation and, primarily, the wide dynamic range of concentrations. The "deep" proteome of beer is characterized by a number of gene products that occur in a very low abundance or migrate at extreme pI or MW values and are barely represented in the classical 2DE maps. These components have a considerable impact on the technological properties of beer that are detailed in a subsequent section. Therefore, with the aim of enlarging the protein annotation and more accurate detection, approaches have been specifically designed to analyze the beer proteome. To increase resolution in the separation of minor protein spots, two separate 2DE maps were obtained for each beer sample, with a pI in the ranges 4–7 and 6–9, respectively. This strategy was particularly successful for detecting and characterizing the alkaline proteins and allowed for the comparison of the protein cartography of 11 beer samples brewed on a pilot scale from different barley cultivars and with different levels of malt modification. This analysis singled out a substantial number of protein spots. For example, 85 out of 199 protein spots, ascribable to 12 gene products, were identified by peptide mass mapping (Iimure et al. 2010). Notably, a "core" of typical beer proteins is substantially shared by all of the beer samples examined, regardless of the barley cultivar brewed and the technological process. The "core" of the barley-derived proteins of beer is schematized in Table 21.1.

The beer proteins enriched through silica-gel adsorption were analyzed by SDS-PAGE followed by MALDI-TOF mass fingerprinting and nanoflow ESI-MS/MS (Picariello et al. 2011). A wheat-derived 16.4 protein, the aveninlike protein A1, that shares homology with γ3-hordein was thusly identified. This protein had already been detected by a gel-free proteomic analysis (Weber et al. 2009) and was successively confirmed as a relatively highly represented beer polypeptide (Colgrave et al. 2012).

The concern regarding the elevated range of concentrations has been challenged through the use of the combinatorial peptide ligand library (CPLL) technology (ProteoMiner). ProteoMiner technology relies on treatment of protein extracts with a library of resin bead-bound hexapeptides that covers a wide range of physicochemical properties. Based on their sequence, proteins exhibit an affinity towards specific hexapeptides and compete in binding to them. In theory, high-abundance proteins quickly saturate their ligands and the excess washes out. In contrast, when sufficiently high starting volumes of the protein source are treated, low-abundance proteins are concentrated on the relevant binding sites. Proteins are then eluted and analyzed by MS-based techniques. This technology has recently been demonstrated to be especially powerful in nonlinearly compressing very complex proteomes (Fonslow et al. 2011) as it produces a remarkable enrichment of the low- and very-low-abundance polypeptides in spite of a partial depletion of the more largely represented polypeptides. After removal of interfering polyphenols with polyvinylpolypyrrolidone, proteins from a 330-mL bottle of beer were enriched with different homemade CPLL. Captured proteins were separated by SDS-PAGE, digested by trypsin in gel and

Table 21.1 Proteins most commonly identified in beer by proteomic analysis

Protein	Accession no.	MW	Theoretical pI
Z4 serpin	gi|1310677	43.3	5.72
Z7 serpin	gi|75282567	42.8	5.45
ns-LTP 1	gi|128376	9.7	8.19
ns-LTP 2	gi|128377	7.0	6.98
D-hordein	gi|1167498	75.0	8.00
B3-hordein	gi|123459	30.2	7.74
B1-hordein	gi|18929	31.5	6.54
γ1-hordein	gi|123464	32.7	8.12
γ3-hordein	gi|288709	33.2	6.70
Avenin-like protein A1[a]	gi|122244839	16.3	8.24
α-amylase/trypsin inhibitor CMd	gi|585291	16.1	5.24
α-amylase inhibitor BDAI-1	gi|123970	16.4	5.36
α-amylase inhibitor BMAI-1	gi|2506771	15.8	6.36
Trypsin inhibitor CMe	gi|1405736	16.1	7.50
Trypsin inhibitor BTI-CMe3	gi|2707924	16.3	7.51
α-amylase/trypsin inhibitor CMb	gi|585290	16.5	5.77
α-amylase/trypsin inhibitor CMa	gi|585289	15.5	5.87
Trypsin/amylase inhibitor pUP13	gi|225102	14.7	5.35
Hordoindoline A	gi|7671688	13.2	8.60
Hordoindoline B1	gi|75172332	13.2	8.56
Hordoindoline B2	gi|374258553	13.2	8.73
Barwin	gi|114832	13.7	7.76

[a]By homology from *Triticum aestivum*

analyzed by tandem MS. Thus, the protein inventory of beer was enlarged to 20 different barley protein families. Due to the use of minor amounts of corn in brewing, two maize proteins were additionally identified in the examined beer sample. Interestingly, as many as 40 gene products derived from the *Saccharomyces* species were detected and identified. The majority of these *Saccharomyces*-derived proteins, although intuitively expected, had not been previously detected in beer. The exploration of the "deep" beer proteome evidenced the substantial absence of hordeins, except for traces of γ3-hordein, and supports the findings of other reports (Perrocheau et al. 2005; Picariello et al. 2011). Conversely, recent gel-free shotgun HPLC-ESI-MS/MS investigations suggested that hordeins in beer do occur at a certain level, implying that beer is potential harmful to those with celiac disease (Weber et al. 2009; Colgrave et al. 2012). The entire protein extract is digested with trypsin before nanoflow LC-electrospray (ESI) and MS analysis, as a general procedure of the gel-free shotgun proteomic approach. Despite a further increase of the analytical complexity of the protein system, the multiplication of protein-derived fragments enhances the probability of detection and confident identification of at least one peptide of the parent protein. In addition, MS techniques are particularly sensitive for the analysis of peptides and MS/MS also enables the sequencing of protein-specific motifs. The shotgun analysis makes it possible to extend the proteome coverage to the less-abundant components, overcoming the majority of the limitations imposed by the 2DE electrophoresis analysis, such as those related to the protein concentration range and to the extreme values of MW and pI.

Thus far, only two articles report on the gel-free shotgun proteomic analysis of beer (Weber et al. 2009; Colgrave et al. 2012). The research by Colgrave et al. (2012) compared 60 commercial beer samples and included beer brewed with genetically modified barley grains carrying mutations that prevented the accumulation of B- and C-hordeins. In addition, the hordein-derived fragments were relatively quantified monitoring a set of "proteotypic" peptides through a triple quadrupole multiple reaction monitoring (MRM) MS approach.. "Proteotypic" peptides represent "easy to monitor" analytical surrogates, whose MRM transitions are established by opportune explorative analyses -of the parent proteins. In both of the shotgun proteomic analyses the unfractionated proteinaceous material, including that at low MW, was digested with trypsin and analyzed by HPLC/ESI-Q-TOF-MS. Therefore, this procedure does not allow inference as to whether hordein-related tryptic peptides were derived from intact/large-sized fragments of hordeins or from short water-soluble peptides generated by the hydrolytic events of mating and brewing. However, in a specific trial focused on the integrity of beer proteins, a D-hordein (75 kDa) was identified in beer exclusively in a 7–17 kDa SDS-PAGE band, demonstrating that the hordein traces detected could indeed be small or large-sized water-soluble fragments (Hao et al. 2006). Regardless, the immunological potential of beer remains a concern that merits an in-depth discussion in a later section of this chapter.

It is envisaged that the suite of proteins found in beer will be completed in the very near future. This is contingent upon the completion of the *Hordeum vulgare* genome sequencing that is in its final stages. The increased application of methods for the quantitative assessment of the beer proteome will enable a strict control of its processing parameters, improving and standardizing quality and providing the desired features to this ancient yet modern beverage.

21.3.5 Beer Peptidomics

The proteolytic events that occur during malting and mashing release a massive amount of peptide fragments into the bulk solution. For instance, well-modified malt contains less than half the amount of hordeins of the original barley (Celus et al. 2006). Many of the peptides generated from insoluble proteins can indeed be water-soluble and are found in the final product of beer. Several size-exclusion chromatography analyses have shown that no less than 60–70% of the total amount of beer polypeptides consist of hydrolytic fragments arising from barley proteins (Sørensen and Ottesen 1978; Osman et al. 2003; Picariello et al. 2011). These observations supported the concept that beer most likely contained peptides deriving from hordein. In addition, it has previously been believed that the proline-rich peptides were involved in determining some quality traits of beer, such as foam

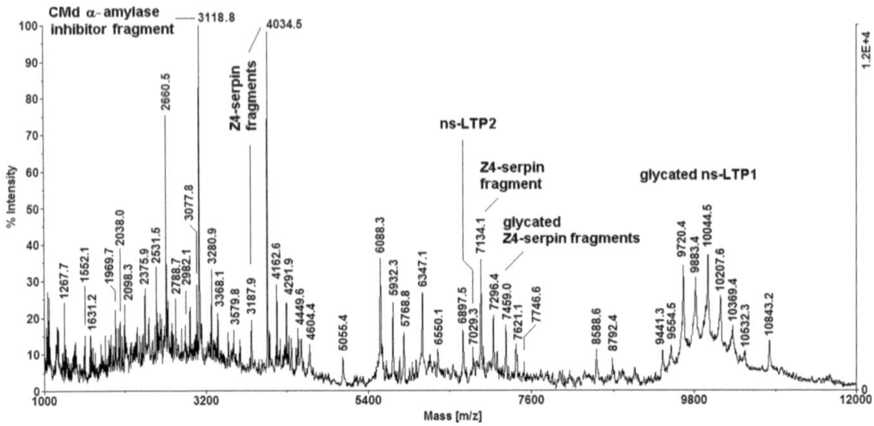

Fig. 21.3 MALDI-TOF MS spectrum of the low molecular weight polypeptides extracted from an Italian lager beer. Peptides were separated from the high molecular weight proteins by size exclusion chromatography and purified with a C18 reversed phase cartridge prior to analysis. The spectrum clearly, but not exhaustively, evidences the high complexity of the peptide fraction of beer (The Figure has been adapted from Mamone et al. 2011. Expert Rev. Proteomics 8(1):95–115 with the permission of Expert Reviews Ltd.)

stabilization and, especially, haze formation (Kauffman et al. 1994; Kaverva et al. 2005). Despite their importance, up to a few years ago, none of these peptides had been identified and no direct evidence had been collected to support the hypothesis that hordeins in beer are not completely degraded into amino acids. This paucity must be ascribed to the extreme analytical complexity of the low-MW fraction of beer that severely challenges the classical methodologies of peptide characterization. The MALDI-TOF MS analysis, shown in Fig. 21.3, demonstrates the highly heterogeneous nature of beer peptides. In fact, the peptide signals cover a virtually continuous range of molecular mass from 1 to 12 kDa. The peptides' nontryptic nature, the absolute lack of cleavage specificity, and the extensive nonenzymatic glycation further complicated the identification process. The newer HPLC-MS and MS/MS techniques are beginning to address such complexity with two recent independent reports targeting the peptide fraction of beer. Several medium-sized peptides (2.5–7.2 kDa) were tentatively assigned to fragments of Z4-serpin, ns-LTP1, and CM α-amylase/trypsin inhibitors using single-stage MALDI-TOF-MS analysis (Picariello et al. 2011; Mamone et al. 2011). The same authors used nanoflow HPLC-ESI Q-TOF MS/MS to sequence more than 20 small peptide sequences (MW up to 1,661 Da). Remarkably, this method definitively identified several hordein-derived peptides in beer for the first time. The simultaneous presence of unhydrolyzed ns-LTPs, known to be highly resistant to proteases, along with their hydrolytic fragments suggest that metabolic proteins are at least partly unfolded during brewing, and thereby become susceptible to proteolytic enzymes. The study by Colgrave et al. (2012) confirmed the presence of hordein-derived peptides in beer and enlarged the inventory of the detected species.

Although the parent proteins are much less represented in barley than their B- and C- cognates, the fragments of D- and γ-hordeins appeared as some of the most represented peptides. In particular, the shortage of peptides arising from C-hordeins is most likely dependent on the water insolubility of their repeated octapeptide motifs, PQQPFPQQ (Colgrave et al. 2012). Fragments of the aveninlike proteins A were also detected. Regardless, due to the complexity emphasized above, the beer peptides are still far from being comprehensively catalogued.

21.4 Technological Aspects of Beer Proteomics and Peptidomics

Beer proteins contribute to several of beer's qualities such as foam stability and haze formation. These two qualities are of particular importance for both brewers and consumers. Foam stability is influenced by proteins (Dale and Young 1987, Siebert and Knudson 1989), hop iso-α-acids (Simpson and Hughes 1994), nonstarch polysaccharides (Evans et al. 1999), fatty acids (Kobayashi et al. 2002), and metal ions (Evans and Sheehan 2002). Haze formation in beer is due to the interaction between haze-active proteins and polyphenols (Asano et al. 1982; Siebert et al. 1996). The relationship of protein factors to foam stability and haze formation have been extensively analyzed. Recent advancements in proteomic techniques, such as high resolution 2DE coupled with MS, have facilitated a more detailed investigation of the role of specific proteins in affecting foam stability and haze formation. In addition, novel proteins related to these quality traits have been identified.

Nevertheless, it has become clear that the structural basis of both foam-stabilization and haze formation are not affected by only a single protein but by the combination of several interacting polypeptides as well as the formation of heterocomplexes, such as the protein–polyphenol type. Therefore, the soon forthcoming proteomics-related "interactomics" studies will affect the exploration of these mechanisms.

Preliminary proteomic investigations have demonstrated that the protein profile of beer depends on the barley cultivar, and on the malting and brewing conditions (Okada et al. 2008; Iimure et al. 2010; van Nierop et al. 2004). Several targeted proteomic approaches have recently been developed for relating the pattern of beer proteins to its character traits and process parameters. For instance, an investigation based on anion exchange chromatography coupled with MALDI-TOF MS has demonstrated that the glycation degree is reproducibly process-dependent. Thus, the monitoring of glycated ns-LTP1 has been suggested as a means to establish the progress of protein conversion during the brewing process (Bobalova and Chmelik 2007; Laštovi ková et al. 2010). The characterization of glycated ns-LTP1 for the purpose of validating molecular indicators of the brewing process has been advanced by the application of a shotgun approach based on the 2-D chromatography fractionation of proteins and MALDI-MS/MS identification of tryptic peptides (Petry-Podgórska et al. 2010). By focusing on the technological aspects of beer, the study revealed that proteomics differential analyses are required for optimizing raw materials and process parameters for higher-quality beer. To this purpose, the correlation of character traits with the phenotypic expression of signature proteins will complement the genetic-based assessments, such as the analyses carried out by Robinson et al. (2007) and Iimure et al. (2011). These assessments in turn attempt to plan innovative barley breeding programs.

21.4.1 Foam-Stabilizing Proteins

Foam is one of beer's most important characteristics as it influences consumers' purchasing decisions with respect to the beer freshness, proper carbonation, and mouthfeel. Therefore, brewers attempt to optimize this quality by controlling the content of surface-active proteins. Protein Z (mainly protein Z4) and ns-LTP1 have undergone extensive investigation for their relationship with beer foam stability. Beer protein Z and ns-LTP1 were previously characterized by several methods such as enzyme-linked immunosorbent assays (ELISA), SDS-PAGE, and correlation analysis with foam stability (Evans et al. 1999; Leiper et al. 2003a).

The need of a more detailed characterization of process-modified Z and ns-LTP1 proteins has led to the recent application of proper proteomic strategies. Iimure et al. (2012a) analyzed the protein Z spots of sweet wort, boiled wort, and trub on 2DE gels by MALDI-TOF MS and LC-ESI MS/MS. They revealed that during wort boiling, protein Z was precipitated by binding comparatively small-sized specific fragments derived from barwin, a sweet wort protein. In addition, confirmation that beer protein Z4 had several modifications, such as glycation and partial digestion was also inferred by the characterization of the peptide fraction. To establish the role of Z-protein in foam-stability, 2DE maps of beers brewed with protein Z4 and/or protein Z7 deficient barley mutants were recently compared. Regular beers showed protein Z spots of approximately 60, 43, and 39 kDa (Iimure et al. 2012b). In contrast, the 60 and 39 kDa spots were not observed in the beer samples obtained from a protein Z4 deficient mutant. The foam stability between the beer samples was comparable, suggesting that the contribution of Z proteins to beer foam stability is not greater than that of other beer proteins. Perrocheau et al. (2006) used MS to analyze purified glycated ns-LTP1 from both malt and beer. They recorded the reduction of disulfide bonds and potential sulfhydryl exchange assessed by labeling the free thiol groups of water-soluble, heat-stable proteins from barley, malt, and beer and analyzed them using 2DE. Free cysteines were detected in the ns-LTP1 spots for malt and beer, but not barley. The degree of ns-LTP1 modification after heating (96 or 105°C) was analyzed by assessing the change in ns-LTP1 spot locations on 2DE (van Nierop et al. 2004). Taking into consideration that ns-LTP1 is foam-active in its unfolded conformation, these studies outlined the mechanism through which wort boiling temperature may affect foam stability. Hao et al. (2006) enlarged the annotation of foam-stabilizing proteins to include 24 gene products. The proteins in beer foam were then specifically analyzed by SDS-PAGE followed by MS and database search.

Several investigators have contributed to the distinction between foam-positive and foam-negative proteins using proteomic approaches. Okada et al. (2008) analyzed three fractions of beer samples by 2DE, including beer whole proteins, salt-precipitated proteins, and proteins concentrated from beer foam. By comparing the 2DE gel images, they identified barley dimeric alpha-amylase inhibitor-I (BDAI-1) as a foam-positive protein. Iimure et al. (2008) detected an additional foam-related protein by analyzing beer samples with different malt contents, including an all barley-malt beer, a beer brewed with malt and adjuncts (67% malt in the total raw material), and a low-malt beer (<24% of malt in the total raw material). Two beer samples with different foam stabilities were examined. A 2DE spot at approximately 12 kDa was lower in the beers with higher foam stability and higher in the beers with poorer foam stability. This spot was identified by MS and database search as the yeast thioredoxin, a yeast intracellular protein (Gan 1991; Pedrajas et al. 1999). This identification suggested that it could be a foam-negative protein. Additional glycolytic enzymes such as enolase and triose-phosphate isomerase, which are also intracellular proteins, were identified in beer by several investigators using proteomic techniques (Fasoli et al. 2010; Iimure et al. 2010; Perrocheau et al. 2005; Picariello et al. 2011). It has been suggested that the physiological conditions of yeast cells affect not only foam (Dreyer et al. 1983) but also flavor stability (Guido et al. 2004). The yeast-derived proteins might serve as markers to determine optimal yeast physiological conditions for stabilizing both foam and flavor.

Hordein-derived polypeptides have been suggested to be active components in beer foam fractions (Evans et al. 2003; Sheehan and Skerritt 1997). However, low molecular mass peptides derived from the aveninlike protein A as well as B-, C-, D-, and γ-hordeins were only recently ascertained by LC-MS/MS analysis (Picariello et al. 2011; Colgrave et al. 2012). The relative contents of peptides arising from different hordein subtypes depend on the beer sample. In any case, the relationship between beer hordein-derived polypeptides and foam stability remains contentious (Leiper et al. 2003a; Evans et al. 2003). Proteome approaches, such as LC-MS/MS including quantitative MRM approaches, should enable us to correlate hordein subfamilies and their derived peptides with foam stability.

21.4.2 Haze Formation

Haze formation is one of the most serious problems affecting beer quality. Consumers generally judge a beer that is hazy as stale (Yasui et al. 1998). Beer proteins and polyphenols contribute to haze formation (Asano et al. 1982; Siebert et al. 1996). Proline-rich hordeins have long been believed to influence haze formation directly (Asano et al. 1982). Silica gel absorbs proline-rich proteins, such as hordein; thus, brewers add silica gel during beer filtration to prevent colloidal beer haze (Leiper et al. 2003b; Siebert and Lynn 1997). However, despite a number of studies on haze formation that focused on proteins and polyphenols, the factors controlling haze formation remain unclear. Proteomics analyses used to identify haze-active proteins have been conducted by several investigators. Robinson et al. (2007) analyzed silica gel eluent proteins by 2DE and identified the chloroform/methanol soluble trypsin inhibitor CMe as a haze-active protein. Based on these results, they determined the gene loci of CMe and detected polymorphisms in the translated amino acid sequence. The 2DE-MS analysis of silica gel eluted proteins suggested that hordeins might be critical factors in haze formation, although only minor haze-active proteins (Jin et al. 2009b; Jin et al. 2011b). CMe, germin E, and protein Z that were absorbed onto silica gel, were identified as major haze-related proteins. Iimure et al. (2009) identified several 2-D protein spots ascribable to four haze-forming gene products absorbed on silica gel. Protein spots included BDAI-1, components of the tetrameric α-amylase inhibitor CMb, and CMe. However, their results suggested that these proteins were nucleation and growth factors of colloidal haze rather than predominant haze-active proteins. It is expected that the outcomes of LC-MS/MS-based strategies opportunely designed for the identification of low molecular weight peptides, combined with data furnished by other analytical methodologies, should allow for further insights into the detailed mechanism of protein-induced haze formation.

21.5 Immunological Aspects of Beer Proteomics and Peptidomics

Proteins that survive the brewing process have structural traits that render them highly resistant to proteases. For this reason, the beer proteins or their domains large enough to be immunogenically active can survive in vivo gastrointestinal digestion. Resistance to gastrointestinal proteolysis is one of the basic characteristics that render allergenic a specific protein domain (Wickham et al. 2009). The modifications that protein chains undergo during brewing steps might also induce the formation of novel antigens and epitopes. Despite the potentially harmful protein content and its large consumption worldwide, allergy to beer occurs at a relatively very low prevalence (Bonadonna et al. 2009). Nevertheless, several case reports have demonstrated that beer can actually trigger immunological responses with the potential of being rather severe. In a study in 1980, urticaria and angioedema, observed in two patients, also exhibiting skin tests positive to malt, were associated with beer consumption (van Ketel 1980). A later study demonstrated that urticaria induced by beer was actually an IgE-mediated immunoreaction (Curioni et al. 1999). The triggers were identified as a barley-derived 10 kDa that was supposed to be LTP1 and an additional 16-kDa component of barley that disappeared during brewing. Beer-induced anaphylaxis in a 21-year-old woman was associated with a series of IgE-binding bands of malt and barley that migrated in the 31–56-kDa region and to a major beer protein of approximately 38 kDa (Figueredo et al. 1999) that was probably a hydrolyzed form of the Z4-serpin. Combining immunodetection, N-terminal sequencing, and skin prick tests, Garcìa-Casado et al. (2001) demonstrated Z4 and ns-LTP1 to be IgE-binding antigens in four beer-allergic individuals. Ns-LTPs, which are almost ubiquitous in vegetables, are the causative agents of severe life-threatening allergies to a variety of foods and have been designated as a family of "pan-allergens." The rigidity of the ns-LTP scaffold is primarily due to the disulfide skeleton significantly conserved across the species and is responsible for the resistance to either gastric pepsinolysis or subsequent digestive steps (Wijesinha-Bettoni et al. 2010). Brewing and digestive conditions only cause a partial disruption of the three-dimensional folding of barley ns-LTP1, that retains its antigenic potential in vivo. It is likely that similar mechanisms might be implicated in the Ig-E binding ability of the chymotrypsin inhibitor Z4. A novel 18-kDa barley-derived beer allergen has recently been described (Hiemori et al. 2008). Although the protein has yet to be definitively identified, the MW and pI values, estimated by 2DE analysis, point to the likelihood of it being the aveninlike protein A.

An even more intricate aspect concerns the celiacogenic potential of beer related to the content of gliadinlike epitopes arising from hordeins. Although many proteomic investigations agree on ruling out the occurrence of major amounts of intact hordeins in beer, the recent peptidomic investigations confirmed that hordein-derived peptides do occur in beer. The panels of the identified peptides encrypt the motifs known to be toxic for individuals with celiac disease. For instance, B-hordein-derived motifs such as PQQPY, shared with the toxic epitope 31–43 of α-gliadin, or the repeated C-hordein-derived motif PQQPF, are considered among the primary triggers of celiac disease (Picariello et al. 2011; Colgrave et al. 2012). These latter motifs belong to both DQ2 and DQ8 restricted epitopes and have been identified as targets of transglutaminase-2 mediated deamidation. A 7-amino acid sequence PQQPYPQ identified in beer contains two overlapping harmful 4-amino acid motifs. Interestingly, this sequence is entirely homologous to the repeated domain of several wheat α/β-gliadins. The QPQPF motif of a B-hordein fragment is common to the toxic 33-mer 56–88 of α2-gliadin (Shan et al. 2002). In addition, the extended QPQPFP sequence found in beer extracts is part of the 25-mer gastrointestinal resistant peptide, a potential trigger of celiac disease (Mamone et al. 2007).

From a quantitative standpoint, preliminary assays based on the R5-monoclonal ELISA suggested that the gliadin content of beer did not exceed the low ppm levels (Iametti et al. 2004). Due to the intrinsic bias of the sandwich ELISA that might underestimate the gluten content, Dostálek et al. (2006) compared the determinations of both sandwich and competitive R5 ELISA in beer. The competitive ELISA provided significantly higher values than those obtained by the sandwich procedures. Some beer samples appeared to contain potentially harmful levels of gluten-like epitopes, however, other brands exhibited values well below the precautionary threshold of 20 ppm indicated for gluten-free foods (Codex Alimentarius Commission 2008). A screening with the R5 antibody-based competitive immunoassays (Guerdrum 2011) determined the level of glutenlike epitopes to be below 20 ppm in 10 out of 28 commercial beer samples.

The determination carried out thus far clearly demonstrates that gluten content can significantly change as a function of the malt and of the brewing process adopted. Furthermore, immunochemical analyses can be affected by a lack of specificity. Therefore, the contribution of reliable and highly specific methodologies, such as those based on the quantification by MRM mass spectrometry, will be indispensable for establishing the drinkability of a specific beer by celiacs. For instance, opportune peptidomic strategies, based on the compared monitoring of internal standards that mimic signature glutenlike peptides, are expected to be addressed in the quantitative determination of gluten in beer. Other authors have proposed the use of either prolyl endopeptidases or tannins during the brewing processes to lower the level of gluten-like epitopes (van Landschott 2011). However, the impact of such treatments on the beer quality is yet to be evaluated. It is also expected that gluten-free beers brewed with cereals other than barley will be studied and assessed for their protein and peptide content. The study of claimed gluten-free beers should not be neglected in light of the need for proper commercial labeling standards. To this purpose it is worth noting that unsuspected sources of wheat gluten have been identified in beers that were claimed to be all-barley malt (Weber et al. 2009). In addition to barley malt, in principle also other ingredients of beer, including other cereals, hops, and yeasts may elicit an allergic reaction (Diel et al. 2009). Because of their specificity, sensitivity, and reliability, the contribution of MS-based and proteomic platforms will undoubtedly be irreplaceable for achieving a full characterization of the beer proteomes and peptidomes and for relating specific components to beer quality traits and human health.

References

Aehle W (2007) Enzymes in industry: production and applications, 3rd edn. Wiley-VCH, Weinheim

Asano K, Shinagawa K, Hashimoto N (1982) Characterization of haze-forming proteins of beer and their roles in chill haze formation. J Am Soc Brew Chem 40:147–154

Baik BK, Ullrich SE (2008) Barley for food: characteristics, improvement, and renewed interest-critical review. J Cereal Sci 30:1–10

Bak-Jensen KS, Laugesen S, Roepstorff P, Svensson B (2004) Two-dimensional gel electrophoresis pattern (pH 6–11) and identification of water-soluble barley seed and malt proteins by mass spectrometry. Proteomics 4:728–742

Bamforth C (2003) Beer – tap into the art of brewing, 2nd edn. Oxford University Press, New York

Bobalova J, Chmelik J (2007) Proteomic identification of technologically modified proteins in malt by combination of protein fractionation using convective interaction media and matrix-assisted laser desorption/ionization time-of-flight mass spectrometry. J Chromatogr A 116:80–85

Bonadonna P, Crivellaro M, Dama A, Senna GE, Mistrello G, Passalacqua G (2009) Beer-induced anaphylaxis due to barley sensitization: two case reports. J Investig Allergol Clin Immunol 9:268–270

Bønsager BC, Finnie C, Roepstorff P, Svensson B (2007) Spatio-temporal changes in germination and radical elongation of barley seeds tracked by proteome analysis of dissected embryo, aleurone layer, and endosperm tissues. Proteomics 7:4528–4540

Brennan CS, Smith DB, Harris N, Shewry PR (1998) The production and characterization of Hor3 null lines of barley provides new information on the relationship of D hordein to malting performance. J Cereal Sci 28:291–299

Briggs DE, Boulton CA, Brookes PA, Sevens R (2004) Brewing science and practice. Woodhead Publishing Limited/CRC press, LLC, Cambridge, UK

Celus I, Brijs K, Delcour JA (2006) The effects of malting and mashing on barley protein extractability. J Cereal Sci 21:203–211

Codex Alimentarius Commission (2008) Joint FAO/WHO Food Standards Programme. Report of the Thirty-first Session International Conference Centre, Geneva (Switzerland) 30 June - 4 July 2008
Colgrave ML, Goswami H, Howitt CA, Tanner GJ (2012) What is in a beer? Proteomic characterization and relative quantification of hordein (gluten) in beer. J Proteome Res 11:386–396
Curioni A, Pressi G, Furegon L, Dal Belin Peruffo A (1995) Major proteins of beer and their precursors in barley: electrophoretic and immunological studies. J Agric Food Chem 43:2620–2626
Curioni A, Santucci B, Cristaudo A, Canistraci C, Pietravalle M, Simonato B, Giannattasio M (1999) Urticaria from beer: an immediate hypersensitivity reaction due to a 10-kDa protein derived from barley. Clin Exp Allergy 29:407–413
Dale CJ, Young TW (1987) Rapid methods for determining the high molecular weight polypeptide components of beer. J Inst Brew 93:465–467
Diel S, Herwald M, Borck H, Diel F (2009) Histamine in beer. In: Preedy VR (ed) Beer in health and disease prevention. Academic, Oxford
Dostálek P, Hochel I, Méndez E, Hernando A, Gabrovská D (2006) Immunochemical determination of gluten in malts and beers. Food Addit Contam 23:1074–1078
Dreyer T, Biedermann K, Ottesen M (1983) Yeast proteinase in beer. Carlsberg Res Commun 48:249–253
Eblinger HM (2009) Handbook of brewing: processes, technology, markets. Wiley-VCH, Weinheim
Ellis HJ, Freedman AR, Ciclitira PJ (1990) Detection and estimation of the barley prolamin content of beer and malt to assess their suitability for patients with coeliac disease. Clin Chim Acta 189:123–130
Enari TM (1986) Proteinases and peptidases of malt and their influence on wort composition and beer quality. Cerevisia 1:19–28
Evans DE, Sheehan MC (2002) Don't be fobbed off, substance of beer foam, a review. J Am Soc Brew Chem 60:47–57
Evans DE, Sheehan MC, Stewart DC (1999) The impact of malt derived proteins on beer foam quality part II: the influence of malt foam-positive proteins and non-starch polysaccharides on beer foam quality. J Inst Brew 105:171–177
Evans DE, Robinson LH, Sheehan MC, Tolhurst RL, Hill A, Skerritt JS, Barr AR (2003) Application of immunological methods to differentiate between foam-positive and haze-active proteins originating from malt. J Am Soc Brew Chem 61:55–62
Fasoli E, Aldini G, Regazzoni L, Kravchuk AV, Citterio A, Righetti PG (2010) Les Maîtres de l'orge: the proteome content of your beer mug. J Proteome Res 9:5262–5269
Figueredo E, Quirce S, del Amo A, Cuesta J, Arrieta I, Lahoz C, Sastre J (1999) Beer-induced anaphylaxis: identification of allergens. Allergy 54:630–634
Finnie C, Svensson B (2009) Barley seed proteomics from spots to structures. J Proteomics 72:315–324
Finnie C, Steenholdt T, Roda Noguera O, Knudsen S, Larsen J, Brinch-Pedersen H, Bach Holm P, Olsen O, Svensson B (2004) Environmental and transgene expression effects on the barley seed proteome. Phytochemistry 65:1619–1627
Fonslow BR, Carvalho PC, Academia K, Freeby S, Xu T, Nakorchevsky A, Paulus A, Yates JR 3rd (2011) Improvements in proteomic metrics of low abundance proteins through proteome equalization using ProteoMiner prior to MudPIT. J Proteome Res 10:3690–3700
Gan ZR (1991) Yeast thioredoxin genes. J Biol Chem 266:1692–1696
García-Casado G, Crespo JF, Rodríguez J, Salcedo G (2001) Isolation and characterization of barley lipid transfer protein and protein Z as beer allergens. J Allergy Clin Immunol 108:647–649
Görg A, Postel W, Weiss W (1992a) Detection of polypeptides and amylase isoenzyme modifications related to malting quality during malting process of barley by two-dimensional electrophoresis and isoelectric focusing with immobilized pH gradients. Electrophoresis 13:759–770
Görg A, Postel W, Baumer M, Weiss W (1992b) Two-dimensional polyacrylamide gel electrophoresis, with immobilized pH gradients in the first dimension, of barley seed proteins: discrimination of cultivars with different malting grades. Electrophoresis 13:192–203

Gorjanović S (2009) A review: biological and technological functions of barley seed pathogenesis-related proteins (PRs). J Inst Brew 115:334–360
Guerdrum LJ (2011) Levels of gliadins in commercial beers. Food Chem 129:5–6
Guido LF, Rodrigues PG, Rodrigues JA, Gonçalves CR, Barros AA (2004) The impact of the physiological condition of the pitching yeast on beer flavor stability: an industrial approach. Food Chem 87:187–193
Hao J, Li Q, Dong J, Yu J, Gu G, Fan W, Chen J (2006) Identification of the major proteins in beer foam by mass spectrometry following sodium dodecyl sulfate-polyacrylamide gel electrophoresis. J Am Soc Brew Chem 64:166–174
Hejgaard J, Kaersgaar P (1983) Purification and properties of the major antigen beer protein of barley origin. J Inst Brew 89:402–410
Hiemori M, Eguchi Y, Kimoto M, Yamasita H, Takahashi K, Takahashi K, Tsuji H (2008) Characterization of new 18-kDa IgE-binding proteins in beer. Biosci Biotechnol Biochem 72:1095–1098
Howard KA, Gayler KR, Eagles HA, Halloran GM (1996) The relationship between D hordein and malting quality in barley. J Cereal Sci 24:47–53
Iametti S, Bonomi F, Ferranti P, Picariello G, Gabrovska D (2004) Characterization of gliadin content in beer by different approach. In: Stern M (ed) Proceedings of the 19th meeting working group on prolamin analysis and toxicity, Praga (Czech Republic), pp 73–78
Iimure T, Takoi K, Kaneko T, Kihara M, Hayashi K, Ito K, Sato K, Takeda K (2008) Novel prediction method of beer foam stability using protein Z, barley dimeric α-amylase inhibitor-1 (BDAI-1) and yeast thioredoxin. J Agric Food Chem 56:8664–8671
Iimure T, Nankaku N, Watanabe-Sugimoto M, Hirota N, Tiansu Z, Kihara M, Hayashi K, Ito K, Sato K (2009) Identification of novel haze-active beer proteins by proteome analysis. J Cereal Sci 49:141–147
Iimure T, Nankaku N, Hirota N, Tiansu Z, Hoki T, Kihara M, Hayashi K, Ito K, Sato K (2010) Construction of a novel beer proteome map and its use in beer quality control. Food Chem 118:566–574
Iimure T, Kihara M, Ichikawa S, Ito T, Takeda K, Sato K (2011) Development of DNA markers associated with beer foam stability for barley breeding. Theor Appl Genet 122:199–210
Iimure T, Nankaku N, Kihara M, Yamada S, Sato K (2012a) Proteome analysis of the wort boiling process. Food Res Int 45:262–271
Iimure T, Kimura T, Araki S, Kihara M, Sato M, Yamada S, Shiryou T, Sato K (2012b) Mutation analysis of barley malt protein Z4 and protein Z7 on beer foam stability. J Agric Food Chem 60:1548–1554
Jégou S, Douliez JP, Mollé D, Boivin P, Marion D (2000) Purification and structural characterization of LTP1 polypeptides from beer. J Agric Food Chem 48:5023–5029
Jégou S, Douliez JP, Mollé D, Boivin P, Marion D (2001) Evidence of the glycation and denaturation of LTP1 during the malting and brewing process. J Agric Food Chem 49:4942–4949
Jin B, Li L, Liu GQ, Li B, Zhu YK, Liao LN (2009a) Structural changes of malt proteins during boiling. Molecules 14:1081–1097
Jin B, Li L, Li B, Liu BG, Liu GQ, Zhu YK (2009b) Proteomics study of silica eluent proteins in beer. J Am Soc Brew Chem 67:183–188
Jin B, Li L, Feng ZC, Li B, Liu GQ, Zhu YK (2011a) Investigation of hordeins during brewing and their influence on beer haze by proteome analysis. J Food Biochem 35:1522–1527
Jin B, Li L, Feng ZC, Li B, Liu GQ, Zhu YK (2011b) Investigation of the relationship of malt protein and beer haze by proteome analysis. J Food Process Pres. doi:10.1111/j.1745-4549.2011.00571.x
Jones BL (2005) Endoproteases of barley and malt. J Cereal Sci 42:139–156
Jones BL, Budde AD (2003) The effect of reducing and oxidizing agents and pH on malt endoproteolytic activities and on malt mashes. J Agric Food Chem 51:7504–7512
Kauffman JA, Clare Mills EN, Brett GM, Fido RJ, Tatham AS, Shewry PR, Onishi A, Proudlove MO, Morgan MRA (1994) Immunological characterisation of barley polypeptides in lager foam. J Sci Food Agric 66:345–355

Kaverva P, Sontag-Strohm T, Lehtonen P (2005) Beer polypeptides and silica gel. J Inst Brew 109:73–79

Kobayashi N, Segawa S, Umemoto S, Kuroda H, Kaneda H, Mitani Y, Watari J, Takashio M (2002) A new method for evaluating foam-damaging effect by free fatty acids. J Am Soc Brew Chem 60:37–41

Koehler SM, Ho TH (1990) Hormonal regulation, processing, and secretion of cysteine proteinases in barley aleurone layers. Plant Cell 2:769–783

Laštovičková M, Mazanec K, Benkovská D, Bobálová J (2010) Utilization of the linear mode of MALDI-TOF mass spectrometry in the study of glycation during the malting process. J Inst Brew 116:245–250

Leiper KA, Stewart GG, McKeown IP (2003a) Beer polypeptides and silica gel. Part II. Polypeptides involved in foam formation. J Inst Brew 109:73–79

Leiper KA, Stewart GG, McKeown IP (2003b) Beer polypeptides and silica gel. Part I. Polypeptides involved in haze formation. J Inst Brew 109:57–72

Mamone G, Ferranti P, Rossi M, Roepstorff P, Fierro O, Malorni A, Addeo F (2007) Identification of a peptide from alpha-gliadin resistant to digestive enzymes: implications for celiac disease. J Chromatogr B Analyt Technol Biomed Life Sci 855:236–241

Mamone G, Picariello G, Addeo F, Ferranti P (2011) Proteomic analysis in allergy and intolerance to wheat products. Expert Rev Proteomics 8:95–115

Marshall T, Williams KM (1987) High resolution two-dimensional electrophoresis of the proteins and macromolecular constituents of beer and wine. Electrophoresis 8:493–495

Molina-Cano JL, Polo JP, Romera E, Araus JL, Zarco J, Swanston JS (2001) Relationships between barley hordeins and malting quality in a mutant of cv. Triumph I. Genotype by environment interaction of hordein content. J Cereal Sci 34:285–294

Moonen JHE, Graveland A, Muts GCJ (1987) The molecular structure of gelprotein from barley, its behaviour in wort-filtration and analysis. J Inst Brew 93:125–130

Okada Y, Iimure T, Takoi K, Kaneko T, Kihara M, Hayashi K, Ito K, Sato K, Takeda K (2008) The influence of barley malt protein modification on beer foam stability and their relationship to the barley dimeric α-amylase inhibitor-I (BDAI-I) as a possible foam-promoting protein. J Agric Food Chem 56:1458–1464

Osborne TB (1895) The proteids of barley. J Am Chem Soc 17:539–567

Osman AM, Coverdale SM, Onley-Watson K, Bell D, Healy P (2003) The gel filtration chromatographic – profiles of proteins and peptides of wort and beer: effects of processing – malting, mashing, kettle boiling, fermentation and filtering. J Inst Brew 109:41–50

Østergaard O, Finnie C, Laugesen S, Roepstorff P, Svennson B (2004) Proteome analysis of barley seeds: identification of major proteins from two-dimensional gels (pI 4–7). Proteomics 4:2437–2447

Pedrajas JR, Kosmidou E, Miranda-Vizuete A, Gustafsson J, Wright APH, Spyrou G (1999) Identification and functional characterization of a novel mitochondrial thioredoxin system in *Saccharomyces cerevisiae*. J Biol Chem 274:6366–6373

Perrocheau L, Rogniaux H, Boivin P, Marion D (2005) Probing heat-stable water-soluble proteins from barley to malt and beer. Proteomics 5:2849–2858

Perrocheau L, Bakan B, Bovin P, Marion D (2006) Stability of barley and malt lipid transfer protein 1 (LTP1) toward heating and reducing agents: relationships with the brewing process. J Agric Food Chem 54:3108–3113

Petry-Podgórska I, Zídková J, Flodrová D, Bobálová J (2010) 2D-HPLC and MALDI-TOF/TOF analysis of barley proteins glycated during brewing. J Chromatogr B Analyt Technol Biomed Life Sci 878:3143–3148

Picariello G, Bonomi F, Iametti S, Rasmussen P, Pepe C, Lilla S, Ferranti P (2011) Proteomic and peptidomic characterization of beer: immunological and technological implications. Food Chem 124:1718–1726

Pressi G, Curioni A, Dal Belin-Peruffo A, Zamorani A (1993) Effectiveness of the electroendosmotic preparative electrophoresis for the purification of all proteins and polypeptides from beer. J Inst Brew 99:63–65
Qi JC, Zhang GP, Zhou MX (2006) Protein and hordein content in barley seeds as affected by nitrogen level and their relationship to beta-amylase activity. J Cereal Sci 43:102–107
Robinson LH, Juttner J, Milligan A, Lahnstein J, Eglinton JK, Evans DE (2007) The identification of a barley haze active protein that influences beer haze stability: cloning and characterisation of the barley SE protein as a barley trypsin inhibitor of the chloroform/methanol type. J Cereal Sci 45:343–352
Schmitt MR, Marinac L (2008) Beta-amylase degradation by serine endoproteinases from green barley malt. J Cereal Sci 47:480–488
Shan L, Molberg Ø, Parrot I, Hausch F, Filiz F, Gray GM, Sollid LM, Khosla C (2002) Structural basis for gluten intolerance in celiac sprue. Science 297:2275–2279
Sheehan MC, Skerritt JH (1997) Identification and characterisation of beer polypeptides derived from barley hordeins. J Inst Brew 103:297–306
Siebert KJ, Knudson EJ (1989) The relationship of beer high molecular weight protein and foam. Tech Q Master Brew Assoc Am 26:139–146
Siebert KJ, Lynn PY (1997) Mechanisms of beer colloidal stabilization. J Am Soc Brew Chem 55:73–78
Siebert KJ, Carrasco A, Lynn PY (1996) Formation of protein–polyphenol haze in beverages. J Agric Food Chem 44:1997–2005
Simpson WJ, Hughes PS (1994) Stabilization of foams by hop-derived bitter acids: chemical interactions in beer foam. Cerevisia Biotechnol 19:39–44
Skerritt JH, Janes PW (1992) Disulphide-bonded gel protein aggregates in barley: quality-related differences in composition and reductive dissociation. J Cereal Sci 16:219–235
Smith DB, Simpson PA (1983) Relationships of barley proteins soluble in sodium dodecyl sulphate to malting quality and varietal identification. J Cereal Sci 1:185–197
Sørensen SB, Ottesen M (1978) Fractionation and characterization of beer proteins. Carlsberg Res Commun 43:133–144
Sørensen SB, Bech LM, Muldbjerg M, Beenfeldt T, Breddam K (1993) Barley lipid transfer protein 1 is involved in beer foam formation. Tech Q Master Brew Assoc Am 30:136–145
van Ketel WG (1980) Immediate type allergy to malt in beer. Contact Dermatitis 6:297–298
van Landschott A (2011) Gluten-free barley malt beers. Cerevisiae 36:93–97
van Nierop SNE, Evans DE, Axcell BC, Cantrell IC, Rautenbach M (2004) Impact of different wort boiling temperatures on the beer foam stabilizing properties of lipid transfer protein 1. J Agric Food Chem 52:3120–3129
Weber D, Cléroux C, Godefroy SB (2009) Emerging analytical methods to determine gluten markers in processed foods-method development in support of standard setting. Anal Bioanal Chem 395:111–117
Weiss W, Postel W, Görg A (1992) Qualitative and quantitative changes in barley seed protein patterns during the malting process analyzed by sodium dodecyl sulfate-polyacrylamide gel electrophoresis with respect to malting quality. Electrophoresis 13:787–797
Wickham M, Faulks R, Mills C (2009) In vitro digestion methods for assessing the effect of food structure on allergen breakdown. Mol Nutr Food Res 53:952–958
Wijesinha-Bettoni R, Alexeev Y, Johnson P, Marsh J, Sancho AI, Abdullah SU, Mackie AR, Shewry PR, Smith LJ, Mills EN (2010) The structural characteristics of nonspecific lipid transfer proteins explain their resistance to gastroduodenal proteolysis. Biochemistry 49:2130–2139
Yasui K, Yokoi S, Shigyo T, Tamaki T, Shinotsuka K (1998) A customer-oriented approach to the development of a visual and statistical foam analysis. J Am Soc Brew Chem 56:152–158
Zhang N, Jones BL (1995) Characterization of germinated barley endoproteolytic enzymes by two-dimensional gel electrophoresis. J Cereal Sci 21:145–153

Chapter 22
Nutritionally Relevant Proteins

Djuro Josić, Spomenka Kovac, and Dajana Gaso-Sokac

22.1 Introduction

Amino acids, peptides, and proteins are essential constituents of food. They are indispensable food components and directly contribute to the flavor of food and are precursors for aroma compounds and colors formed during thermal or enzymatic reactions in production, processing, and storage of food. Proteins contribute to the physical properties of food; they have ability to build or stabilize gels, foams, emulsions, and fibrillar structures.

The most important sources of animal proteins are eggs, meat, and milk. These proteins contain the essential amino acids leucine, isoleucine, lysine, valine, threonine, tryptophan, phenylalanine, and methionine. Most proteins of plant origin do not contain some of these amino acids. Consequently, these proteins do not have full nutritional value. However, essential amino acids are present in the proteins from legumes, oilseeds, and grains. Legumes are added, sometimes after modification to traditional

D. Josić (✉)
Department of Biotechnology, University of Rijeka,
R. Matejcic 2, Rijeka, HR-51000, Croatia

Warren Alpert Medical School, Brown University,
Rhode Island, Providence, RI, USA
e-mail: djuro_josic@brown.edu

S. Kovac • D. Gaso-Sokac
Food Technology, J.J. Strossmayer University of Osijek,
Kuhaceva 20, Osijek 31000, Croatia

Department of Chemistry, J.J. Strossmayer University of Osijek,
Kuhaceva 20, Osijek 31000, Croatia

F. Toldrá and L.M.L. Nollet (eds.), *Proteomics in Foods: Principles and Applications*, Food Microbiology and Food Safety 2, DOI 10.1007/978-1-4614-5626-1_22,

Table 22.1 Digestibility of various proteins in humans (FAO/WHO/UNU 1985)

Protein source	Digestibility (%)	Protein source	Digestibility (%)
Egg	97	Millet	79
Milk, cheese	95	Peas	88
Meat, fish	94	Peanut	94
Maize	85	Soy flour	86
Rice (polished)	88	Soy protein isolate	95
Wheat (whole)	86	Beans	78
Wheat flour, white	96	Corn, cereal	70
Wheat gluten	99	Wheat, cereal	77
Oatmeal	86	Rice cereal	75

foods, such as meat and cereal products, and also used in the production of novel food items such as meat, fish, and milk substitutes. Plant proteins, mostly originating from soybean, are frequently added to processed meat products either for economic reasons or to improve their functional properties (Gaso-Sokac et al. 2011).

Proteins are formed from amino acids through amide linkages. Most proteins are post-translationally modified. The most frequent post-translational modifications are phoshorylation and glycosylation, but other less frequent modifications such as alkylation and sulfatation are also very important for protein function and their nutritive value (Blom et al. 2004).

Proteins differ in their nutritive value. Several factors, such as content of essential amino acids and digestibility, contribute to these differences and to "quality" of a protein. The daily protein requirement therefore depends on the type and composition of proteins in a diet. High-quality proteins are those that contain all the essential amino acids at levels greater than the FAO/WHO/UNU (1985) reference levels, and the digestibility comparable to or better than those of egg white or milk proteins. As a rule, animal proteins have a better nutritive value than proteins of plant origin.

For example, proteins of major cereals and other foods of plant origin are often deficient in at least one of the essential amino acids. The essential amino acids whose concentrations in a protein are below the levels of a reference protein are termed *limiting amino acids*. The nutritional quality of a protein or protein mixture is ideal when it contains all of the essential amino acids in proportions that produce optimum rates of growth and/or optimum maintenance capability.

Digestibility is defined as the proportion of food nitrogen that is absorbed after ingestion. Although the content of essential amino acids is the primary indicator of protein quality, true protein quality also depends on the extent to which these amino acids are utilized in the body. Digestibility of various proteins in humans is listed in Table 22.1 (FAO/WHO/UNU 1985).

Antinutritional factors:

Most plant protein isolates and concentrates contain trypsin and chymotrypsin inhibitors (Kunitz type and Bowman–Birk type) and lectins. These inhibitors impair complete hydrolysis of legume and oilseed proteins by pancreatic proteases. Lectins, which are glycoproteins, bind to intestinal mucosa cells and interfere with absorption of amino acids. Lectins and Kunitz type protease inhibitors are thermolabile, whereas the Bowman–Birk-type inhibitor is stable under normal thermal processing conditions. Thus, heat-treated legume and oilseed proteins are generally more digestible than native protein isolates (despite some residual Bowman–Birk-type inhibitor). Plant proteins also contain other antinutritional factors, such as tannins and phytate. Tannins, which are condensed products of polyphenols, covalently react with ε-amino groups of lysyl residues. This inhibits trypsin-catalyzed cleavage of the lysyl peptide bond.

In last few years proteomics technology has been frequently used in food technology for process validation, optimization, and quality control (Gaso-Sokac et al. 2011). The use of proteomics for characterization of nutritionally important proteins, detection of trace components of protein and peptide origin that are important for human nutrition, and detection of potentially harmful components in human food of both animal and plant origin (Gaso-Sokac et al. 2010) are also very important. The analysis scheme of the food proteome is shown in Fig. 22.1

22.2 Animal Proteins

22.2.1 Egg Proteins

Eggs are a valuable source of protein and are important ingredients in many food products. In food processing whole eggs or egg ingredients (egg white and egg yolk) are often used as coagulating, foaming, and emulsifying agents, while also contributing nutrients and flavor to different foods (Campbell et al. 2003; Kiosseoglou 2003). Individual components of eggs also have potentially useful biological functions, such as antimicrobial activity, protease inhibitory function, and antigenic or immunogenic characteristics (Raikos et al. 2006). Potential allergens in eggs, such as some genetic variants of the ovomucoid from egg white, are also important from a food safety point of view (Rupa and Mine 2008).

Egg white proteins differ markedly from the proteins of egg yolk in biological function and in composition. The technological functions usually assigned to these two distinct portions of the egg are also different. Although the most important functional property of egg white in food products is its ability to form stable foams, the functional significance of egg yolk is largely connected with its capacity to stabilize fat–water emulsions.

The most abundant three proteins in egg white are *ovalbumin, conalbumin* (ovotransferrin), and *ovomucoid* (about 77% of total protein) as listed in Table 22.2. Together with *ovomucin, lysosyme*, and *ovoglobulins* G_2 and G_3, the content of these highly abundant proteins in egg white is over 92% (Belitz et al. 2004). Most egg white proteins are glycosylated, and some of them contain a high amount of carbohydrates. There are some thorough proteomic investigations of both high and low

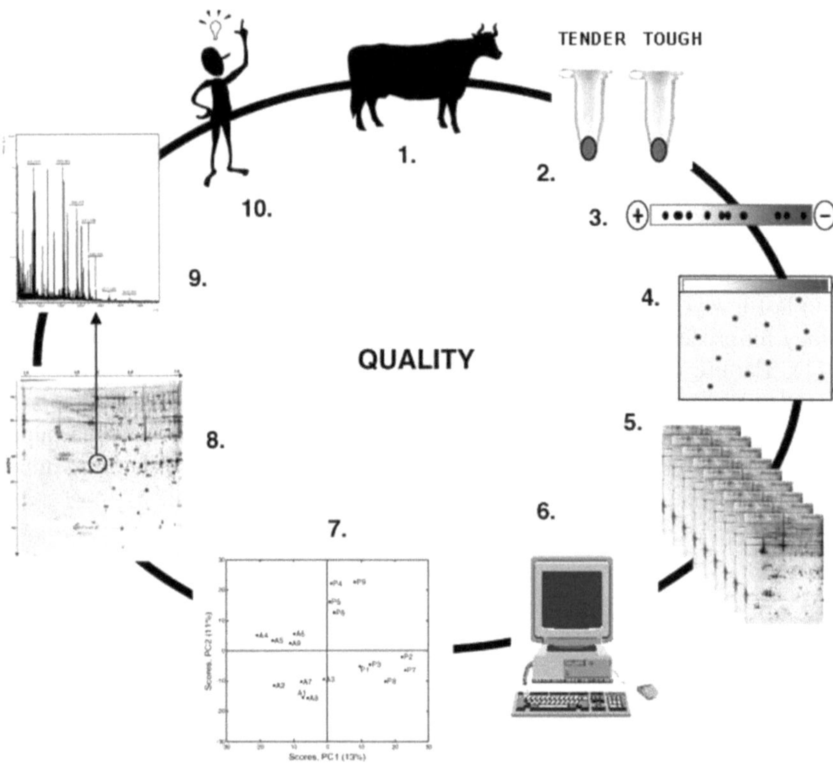

Fig. 22.1 Workflow in the proteome analysis using 2-D electrophoresis or chromatographic separation methods and mass spectrometry. *1*. animal or sample chosen for the analysis; *2*. sample extraction; *3*. isoelectric focusing or first chromatographic step; *4*. SDS-PAGE or next chromatographic step as a second dimension; *5*. alignments and comparisons of the images; *6*. data analysis; *7*. data interpretation and selection of significantly changed proteins; *8*. extraction of significantly changed protein spots; *9*. protein identification by mass spectrometry; *10*. interpretation of the results (Reproduced from Hollung *et al.* (*5*) with permission from Elsevier Ltd, copyright 2007)

abundant egg proteins. Because of their post-translational modifications, some high-abundance proteins, such as ovotransferrin and ovomucoid also show a very high level of polymorphism (Raikos et al. 2006; Guérin-Dubiard et al. 2006). Egg white is a very complex mixture and further proteomic analyses resulted in identification of 148 distinct gene products (D'Ambrosio et al. 2008; Mann and Mann 2008; Farinazzo et al. 2009; D'Alessandro et al. 2010).

Egg yolk is a fat-in-water emulsion with about 50% dry matter; it contains one-third of proteins and two-thirds of lipids. Consequently, the egg yolk contains a high amount of lipoproteins, such as *lipovitellins, lipovitellenins,* and *phosvitin.* Again, most egg yolk proteins are also post-translationally modified. By the use of combinatorial ligand libraries and other high-resolution techniques, more than 250 distinct gene products were identified in egg yolk (Mann and Mann 2008; Farinazzo

Table 22.2 Proteins of egg white (Belitz et al. 2004)

Protein	Percent of the total protein[a]	Denaturation temperature (°C)	Molecular weight (kdal)	Isoelectric point (pH)
Ovalbumin	54	84.5	44.5	4.5
Conalbumin (Ovotransferin)	12	61.5	76	6.1
Ovomucoid	11	70.0	28	4.1
Ovomucin	3.5		$5.5–8.3\times10^6$	4.5–5.0
Lysozyme (Ovoglobulin G_1)	3.4	75.0	14.3	10.7
Ovoglobulin G_2	4	92.5	30–45	5.5
Ovoglobulin G_3	4			5.8
Flavoprotein	0.8		32	4.0
Ovoglycoprotein	1.0		24	3.9
Ovomacroglobulin	0.5		760–900	4.5
Ovoinhibitor	0.1		49	5.1
Avidin	0.05	68.3[b]	9.5	
Cystatin (ficin inhibitor)	0.05		12.7	5.1

[a]Average values are presented
[b]Four times 15.6.kdal + approx. 10% carbohydrate

et al. 2009; D'Alessandro et al. 2010). Some of these proteins play an important role in antimicrobial response and vitamin binding.

The chicken egg proteome is still not complete, but some newly identified proteins are of biopharmaceutical interest as potentially physiologically active substances.

22.2.2 *Proteins of Meat and Meat Products*

In human nutrition, meat is the most valuable source of protein and essential amino acids. The proteins/enzymes of muscle can be categorized based on biological function or chemical properties. Proteins involved in the physical process of contraction are those contained in the sarcomere. These can be divided based on location, such as thick or thin filaments, or on function, such as force generating or regulating proteins (Hollung et al. 2007; Bendixen 2005). An additional, frequently neglected factor that influences meat quality and digestibility is the content and composition of the intramuscular connective tissue (Purslow 2005)

Myosin is the major protein of the thick filaments, which comprises 45% of the myofibrillar proteins. It is an elongated protein molecule about 160 nm in length with a molecular mass of approximately 480,000 D. Myosin contains a total of six polypeptide chains, two heavy chains and four light chains. Myosin heavy chains have "head" and "tail" regions, reflecting the respective globular and rod portions of the molecules. The biological functions of myosin reside in heavy chains. Myosin can be cleaved in the middle region by proteolytic enzymes, such as trypsin, producing two fractions of the protein. One of these is called light meromyosin and the other, which contains the globular head structures of the myosin molecule is called heavy meromyosin. Separated heavy meromyosin retains its ability to interact with actin and its ATPase activity.

Actin is the major protein of the thin filaments and comprises 20% of myofibrilar protein of muscle. Actin is bound to the structure of the muscle much more firmly than is myosin. Its shape can be described as two peanut-shaped domains of equal size lying side by side. Actin monomers, called globular actin or G-actin, are assembled in a double-helical structure called fibrous actin, or F-actin. This constitutes the main portion of the thin filament. G-actin is composed of 374–375 amino acids and has a molecular mass of 42,000–48,000 D.

Tropomyosin, representing 5% of myofibrilar protein, is composed of two alpha-helical polypeptides wound together into a two-strained, coiled-coil supersecondary structure. It resembles the tail or rod portion of the myosin molecule. In skeletal muscle two polypeptides, alpha- and beta-tropomyosin, can combine to form a tropomyosin dimmer. The alpha- and beta-tropomyosin polypeptides have molecular masses of 37,000 and 33,000 D, respectively. They are found in muscle as the alpha-alpha or beta-beta homodimers and the alpha-beta heterodimer. Tropomyosin aggregates end to end and binds to actin filaments along each groove of the actin double helix such that each molecule interacts with seven G-actin monomers.

The content and the relative concentration of different types of *collagen* vary and are dependent on the type of the meat. In conclusion, collagens, as well as different types of proteoglycans are responsible for the so-called "background" feature of the meat and indirectly also for its digestibility. Furthermore, the turnover of the connective tissue, especially the turnover of different types of collagen, and further changes of other main proteins of the meat is controlled by matrix proteases and their specific inhibitors (Belcerzak et al. 2001; Purslow 2002).

In summary, in meat science and for the further exploration of meat proteins, proteomics can be used for: (1) proteome mapping and meat identification, (2) determination of proteome changes due to genetic variations, (3) determination of changes due to the pre-slaughter conditions, (4) determination of post-mortem changes, and (5) study and detection of changes in peptide composition during meat storage and processing (Hollung et al. 2007; Bendixen 2005; Bauchart et al. 2006).

Proteolytic degradation of muscle that occurs post mortem and degradation of proteins during meat processing and aging results in the production of protein fragments (Geesink and Koohmaraie 1999). These polypeptides can be further digested into smaller peptides or even single amino acids (Geesink and Koohmaraie 1999; Mullen et al. 2000). Unfortunately, there are only a few studies dealing with polypeptides and small peptides in aged and cooked meat, but they play a key role for aroma and taste of cooked or dry-cured products (Purslow 2005; Bauchart et al. 2006).

Plant proteins, mostly originating from soybean or other leguminosae, are frequently added to processed meat products either for economic reasons or to improve their functional properties. Leitner et al. (2006) used liquid chromatography-mass spectrometry (LC-MS) and liquid chromatography-tandem mass spectrometry (LC-MS/MS) in order to detect soybean proteins in meat products. In all soybean protein-containing meat samples, the plant protein glycinin G4 subunit A4 was identified, and this protein can be used as a target for a simpler analytical method in order to identify the addition of soybean proteins to meat products, and possible adulteration of meat and meat products.

22.2.3 Seafood and Seafood Proteins

Proteins are also the essential components of seafood which are gaining increasing importance in nutrition, especially in developed countries. The wide variability of proteins that are present in seafood and their highly variable composition offer a large potential to originate a broad variety of different products (Piñeiro et al. 2003). The complexity of seafood also implies the extreme variability of the seafood proteomes, and only some basic questions can be addressed in this short overview. There are some literature data about the use of proteomics for the identification of allergens in seafood (Gaso-Sokac et al. 2010, 2011) and this important point is briefly discussed in this review.

In their review about the use of proteomics as a tool for the investigation of seafood and other marine products, Piñeiro et al. (2003) recommend the use of proteomics for detection of allergens in food of this origin. However, there are still only few studies in this field. Taka et al. (2000) characterized an allergenic parvalbumin from frog by the use of LC-ESI-MS. The main crustacean allergens are proteins tropomyosin and arginine kinase (Lehrer et al. 2003; Ishikawa et al. 2001). Tropomyosin is a myofibrillar protein of 35–38 kDa, and proteins from six species of crustaceans have also been cloned (Motoyama et al. 2007). Arginine kinase from some commercially relevant shrimp species was characterized by the use of proteomic methods (Ortea et al. 2009). Some additional shrimp allergens such as sarcoplasmatic calcium binding protein (SCP) have also been detected (Yu et al. 2003; Shiomi et al. 2008). Interestingly, this protein was previously detected as an allergen in crayfish *Procambarus clarkii* (Gao et al. 2006). The problem of allergens in the seafood was recently discussed (Gaso-Sokac et al. 2010).

22.2.4 Milk Proteins

Because of the immense importance for human nutrition, milk proteins have been studied continuously for more than 100 years. In 1877 Hammarsten distinguished three main proteins in milk: casein, lactalbumin, and lactoglobulin (Belitz et al. 2004). Later it was revealed that the milk protein system is much more complex. Using ultracentrifugation and electrophoresis it was proven that casein consists of tree fractions, namely α-, β- and γ-casein. The most important proteins of milk are listed in Table 22.3. Other protein constituents such as enzymes are present in much lower quantities (they are not listed in Table 22.3). The two important groups of

Table 22.3 Belitz et al. 2004

Fraction	Genetic variants	Portion[a]	Isoionic point	Molecular weight[b](kdal)
Caseins		80		
α_{s1}-Casein	A,B,C,D,E	34	4.92–5.35	23.6[c]
α_{s2}-Casein	A,B,C,D	8		25.2[d]
χ-Casein	A,B,C,E	9	5.77–6.07	19[e]
β-Casein	A^1,A^2,A^3,B,C,D	25	5.20–5.85	24
γ-Casein		4	5.8–6.0	12–21
γ_1-Casein	A^1,A^2,A^3,B			20.5
γ_2-Casein	A^1/A^2,A^3,B			11.8
γ_3-Casein	A^1/A^2/A^3,B			11.6
Whey proteins		20		
β-Lactoglobulin	A,B,C,D,E,F,G	9	5.35–5.41	18.3
α- Lactalbumin	A,B,C	4	4.2–4.5[f]	14.2
Serum albumin	A	1	5.13	66.3
Immunoglobulin		2		
IgG1			5.5–6.8	162
IgG2			7.5–8.3	152
IgA			–	400[g]
IgM			–	950[h]
FSC(s)[i]				80
Proteose-peptone		4	3.3–3.7	

[a]As% of skim milk total protein
[b]Monomers
[c]Variant B
[d]Variant A
[e]Variant A^2
[f]Isoelectric point
[g]Dimer
[h]Pentamer
[i]Free secretory component

proteins, proteins and glycoproteins, are related to the milk fat globule membrane (MFGM, see Fig. 22.2) and whey proteins have recently been the topic of intensive investigation (Gaso-Sokac et al. 2011).

In milk separation by centrifugation, three fractions can be obtained by the following steps: (1) whole milk is centrifuged to obtain the milk fat globule (MFG) and the low fat milk fractions; (2) protein-rich, skimmed milk is ultracentrifuged and separated in two fractions, the whey protein and pellet fraction; and (3) in the last step the casein protein fraction is obtained by washing the pellet from the previous step (Pogacic et al. 2010). The milk proteome and glycoproteome have been topics of extensive investigations since the creation of the proteomics methodology (Gagnaire et al. 2009; Johnson and Lucey 2006). This technology is currently used for: (1) the analysis of high-abundance proteins, (2) the analysis of

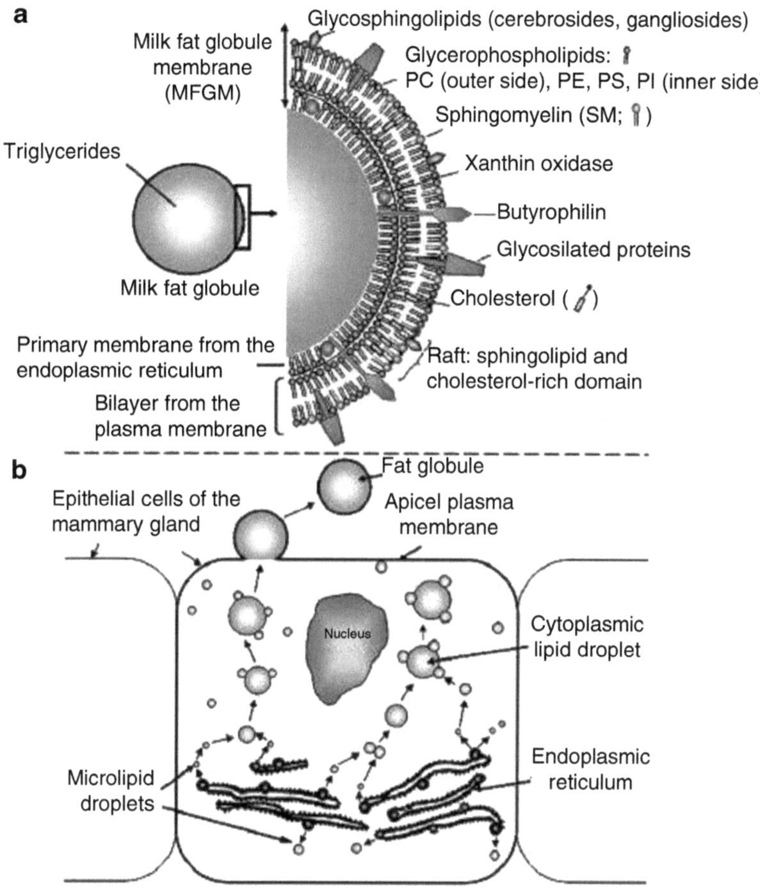

Fig. 22.2 Milk fat globule membrane. Schematic presentation of: (**a**) the structure of the milk fat globule membrane; and (**b**) the pathways for the synthesis and secretion of milk fat globules (Reproduced from C. Lopez *et al.* (*151*) with permission of the American Chemical Society, copyright 2008)

low-abundance proteins, (3) the analysis of proteins in whey, (4) the characterization of milk fat globule and milk fat globule membrane, and (5) the characterization of products containing complex dairy matrices such as cheese and yogurt (Gaso-Sokac et al. 2011).

22.2.4.1 Caseins and Other High-Abundance Proteins in Milk and Milk Products

There are only a few highly abundant proteins that are present in milk from all species. In bovine milk, these proteins are casein (CN), β-lactoglobin (β-LG), α-lactalbumin (α-LA), and bovine serum albumin (BSA) with relative abundances of approximately 80:10:4:1 (Conti et al. 2007; O'Donnell et al. 2004). Similar to other biological fluids, such as blood plasma, despite the fact that it contains a small number of primary proteins, the milk proteome is still extremely complex. A great deal of this complexity is the consequence of post-translational modifications and the presence of numerous genetic variants of this limited list of proteins (Casado et al. 2009).

Caseins are a group of unique milk-specific proteins. These proteins represent about 80% of the total protein in bovine milk. Caseins are a mixture of proteins and subclasses of proteins. The most abundant caseins are α_{S1}-, α_{S2}-, β-, and κ-casein (Fox and Brodkorb 2008). Isolation of κ-casein revolutionized the ideas on the structure of the so-called casein micelle in milk, and a realistic model of its structure became possible only after the characterization of the soluble form of this most abundant milk protein (Waugh et al. 1970). Many technologically and nutritionally important properties of milk, such as its white color, stability to heat or ethanol, coagulation by cheese-making enzymes, and gelation characteristics, are due to the properties of casein micelles. It is for these reasons that properties of casein micelles are still a topic of extensive studies (Johnson and Lucey 2006; Fox and Brodkorb 2008; Waugh et al. 1970; Glantz et al. 2010). It has been known for more than 100 years that the colloidal caseinate particles contain calcium. Therefore, in nutrition, milk is one of the most important sources of calcium. The phosphorylation of caseins and other post-translational modifications of this protein have a vital role in the interaction with calcium phosphate and the organization of the casein micelle (Sørensen et al. 2003). Determination of post-translational modification of major protein components provides the scientific basis for coagulation and cheese-making processes used in dairy production. Further characterization of these proteins is also of fundamental importance for identification of the origin and quality assessment of milk and milk products (Holt 1998; Sørensen et al. 2003; Di Luccia et al. 2009; Matéos et al. 2009; Roncada et al. 2002).

The other high-abundance milk proteins, β-LG, α-LA, and BSA, are major components of the whey (Farrell et al. 2004). The concentration of β-LG in skimmed milk is about 2–4 mg/mL. This protein occurs with high frequency in cows as two genetic variants, variant A and variant B. Because of the different physicochemical characteristics of the two β-LG molecules, the presence of one or the other of these variants significantly affects the properties of the milk. The A variant is expressed at a higher level than the B variant, or the less frequently occurring C variant (Farrell et al. 2004; Ng-Kwai-Hang and Grosclaude 2003). β-lactoglobulin may also be glycosylated, but lactosylation of this protein is a more important chemical modification, which is caused by heating of the milk or whey (Morgan et al. 1998).

Bovine skimmed milk contains α-LA at a concentration of 1.2–1.5 mg/mL. In bovine milk, the mature α-LA is also present in two genetic variants, variants A and B. This protein binds bivalent metals such as zinc and calcium. α-LA is important for normal function of the mammary gland, such as milk secretion and lactose content in milk (Farrell et al. 2004).

22.2.4.2 Low-Abundance Proteins

The low-abundance proteins in milk can be identified by mass spectrometry after proper sample preparation and removal of high-abundance proteins (Manso et al. 2005; Pampel et al. 2007; Al-Ghobashy et al. 2009). Blood plasma proteins, such as BSA, serotransferrin, and lactoferrin, are only observed in colostrum, which may have special physiological importance for children in the early nursing period. Among more than 400 spots in 2-D electrophoresis that were separated from bovine milk, identified proteins include β_2-microglobulin, complement components, α_1-antitrypsin, prealbumin, fructose-biphosphate aldolase A, and casein fragments (Yamada et al. 2002).

The immunoglobulin fraction accounts for about 1% (w/w) of total milk protein, and therefore these proteins can be classified as a kind of "medium-abundance proteins." In milk, IgG, IgA, and IgM have been isolated and characterized. Immunoglobulins in colostrum and milk have a protecting function, especially for the newborn (Farrell et al. 2004).

Lactoferrin is a specific, iron-binding protein that also occurs in milk of most mammalian species. The lactoferrin concentration in milk is relatively low, and varies between 20 and 200 mg/L. This protein increases noticeably in response to inflammation or infection. Consequently, lactoferrin plays an important role in the host defense against infection and inflammation (Ward et al. 2002). Antibacterial and antiviral activity of this protein against both DNA and RNA viruses have been detected (Vogel et al. 2002; van der Strate et al. 2001), and lactoferrin is now being isolated and purified from cheese whey and commercially utilized in the pharmaceutical and food industries (Marshall 2004).

22.2.4.3 Whey Proteins

Whey is considered as a functional milk fraction with a content of proteins and bioactive polypeptides that have a positive effect on the health (Madureira et al. 2007), and whey protein fractions are increasingly incorporated as functional ingredients in food, not only in infant formulas, but also for adults (Panchaud et al. 2005). As ingredients in food, whey proteins can provide antimicrobial activity, immune modulation, improve muscle strength, and may delay and/or ameliorate conditions in different diseases, such as osteoporosis and cardiovascular diseases (Madureira et al. 2007; Marshall 2004).

The identification of low-abundance proteins in whey is challenging due to their wide dynamic concentration range. Namely, in comparison with highly abundant proteins, the concentrations of minor protein components vary by at least a factor of

10^6 (Panchaud et al. 2005). These proteins may play important physiological and eventually therapeutic roles in nutrition and as additives to cosmetic products.

22.2.4.4 Milk Fat Globule and Milk Fat Globule Membrane

Lipid molecules are provided in milk via a unique delivery system, milk fat globules (Argov et al. 2008). In previous research, the lipid fraction in milk was oversimplified as a relatively pure mixture of triacylglycerols (Timmen and Patton 1988), and the unique structure and composition of MFG was overlooked. In MFG, the milk fat globule membrane is the protective coat that surrounds lipid globules. The MFGM prevents flocculation and coalescence of lipid droplets in milk and protects the milk fat against lipolysis (Argov et al. 2008). In order to secrete MFG, portions of the mammary epithelial cell membranes are sacrificed. This process results in a unique structure of MFG (Timmen and Patton 1988). Recent investigations of the MFG and MFGM proteome have provided new insights into mammary function and the mechanism of milk secretion (Reinhardt and Lipolis 2008). Mather (2000) gave an early overview of the proteins found in the MFGM. The eight most abundant MFGM proteins are: mucin1, xantine dehydrogenase/oxidase, periodic acid Schiff III and Schiff 6/7 proteins, CD36, butyrophilin, adipophilin, and fatty-acid binding protein. The MFGM is organized as a trilayer, and its structure is shown in Fig. 22.2 (Lopez et al. 2008).

MFGM proteins also have other important functions, such as defense against pathogens (Sando et al. 2009; Smolenski et al. 2007). One of the major proteins in MFGM, xantine dehydogenase/oxidase, has a direct antibacterial activity, and inhibits bacterial growth through the formation of hydrogen peroxide or the stimulation of lactoperoxidase in milk (Martin et al. 2004). Additional factors with beneficial health properties, such as cholesterol lowering and inhibition of cancer cell growth, were also documented for MFGM (Dewettinck et al. 2008).

22.2.4.5 Peptides

Peptides originating from milk proteins have various health-promoting effects such as: regulation of digestive enzymes and modulation of nutrient absorption (Yamauchi et al. 2003); regulation of the cardiovascular system, for example, anti-hypertensive effects (Pins and Keenan 2006); regulation of the immune system, for example, enhancing immune cell functions and stimulation of the phagocytic activities of macrophages (Meisel and FitzGerald 2003); regulation of the nervous system, for example, by their opoid activity (Yamauchi et al. 2003); and antioxidative and other health-promoting activities (Madureira et al. 2007).

The physiological activity of milk-derived peptides has been the topic of numerous studies during the last 10 years (Madureira et al. 2007). Experimental evidence exists that bioactive peptides can be released from caseins (see above), α-lactalbumin (α-LA), β-lactoglobin (β-LG), lactoferrin, and serum albumin. Some of these bioactive peptides have received special names, such as α- and β-lactorphin, β-lactotensin, serophin, albutensin A, lactoferricin B, lactoferrampin, osteopontin, and many others. Their production and biological properties have been the subject of two comprehensive reviews (Korhonen and Pihlanto 2006; Madureira et al. 2010).

22.2.4.6 Milk Proteins as Allergens

Milk products can also cause allergies. However, proteomics tools have only been sparingly applied in the investigation of allergens in these products. It is well known that changes in the main milk protein casein such as carbonylation (Scaloni et al. 2002) or formation of covalent complexes between casein micelles and β-lactoglobulin (Henry et al. 2002) and modification of other proteins (Gagnaire et al. 2009; Gupta and Lee 2007) during the production process, mainly heating, can cause induction of allergies to milk products, but a thorough proteomic and "allergenomic" investigation has still to be performed.

22.3 Plant Proteins

22.3.1 Seed Proteins

Seeds, mostly cereals, have always played a key role in human nutrition. Cereal product consumption meets close to 50% of the daily requirement for carbohydrates. Additionally, cereals are an important source of vitamin B, minerals, and trace elements. Frequently neglected is the fact that the consumption of cereals also provides about one-third of the requirement for proteins (Belitz et al. 2004).

Wheat and rice are still the most important cereals and the longitudinal section of the wheat grain is shown in Fig. 22.3.

In comparison to other plant tissues, seeds are relatively rich in protein. The biological role of proteins in the seed is still obscure. The process of germination involves intense and many-sided biochemical activity, requiring the rapid biosynthesis of many enzymes. One can assume that seed proteins provide both the machinery and the raw materials for such biosynthesis.

It is also logical to suppose that, as most other constituents of the cotyledons, seed proteins serve as a food reserve for the seedling, providing the young plant with amino acids and nitrogen until the root system and the photosynthetic apparatus are sufficiently developed. Within the cells of the seed cotyledons these "storage proteins" occur in granules with diameters in the range of 2–20 μ, known as *aleurons* or "*protein bodies*".

Interest in seed proteins arose early in the history of protein chemistry. The role of seeds as an important source of proteins is also the topic of many proteomic

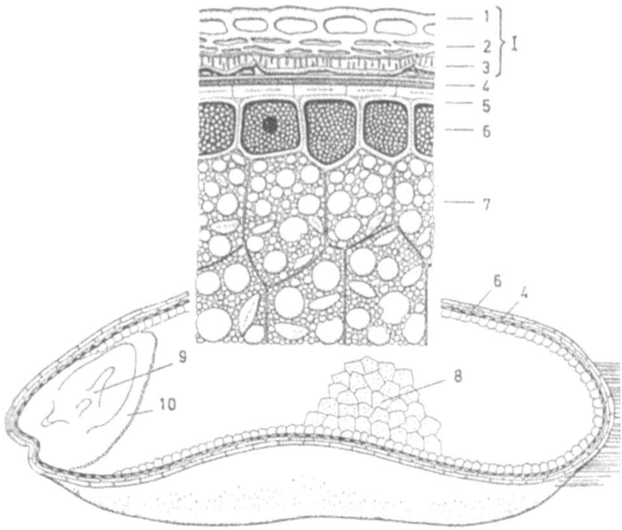

Fig. 22.3 Longitudinal section of a wheat grain. *I* Pericarp, *1* epidermis (epicarp), *2* hypodermis, *3* tube cells, *4* seed coat (testa), *5* nucellar tissue, *6* aleurone layer, *7* outer starchy endosperm cells, *8* inner starchy endosperm cells, *9* germ and *10* scutellum

analyses (Marsolais et al. 2010; Gong et al. 2012). In a broad review, Miernyk and Hajduch (2011) give an excellent overview of seed proteomics.

22.3.1.1 Cereal Proteins

The protein content of cereal grains is in the vicinity of 10% (wheat and barley 13%, rice and maize 9%).

The proteins of different cereal flours vary in their amino acid composition (Table 22.4). Lysine content is low in all cereals. Methionine is also low, particularly in wheat, rye, barley, oats, and corn. Both amino acids are significantly lower in flour than in muscle, egg, or milk proteins. By breeding, attempts are being made to improve the content of all essential amino acids. This approach has been successful in the case of high-lysine barley and several corn cultivars.

Wheat

Wheat is the most important cereal in the western world and wheat seed proteins have been investigated for more than 100 years. In 1907 Osborne separated wheat proteins, on the basis of their solubility, into four fractions. Sequential extraction of a flour sample yielded: water-soluble albumins, salt-soluble (e.g., 0.4 mol/L NaCl)

Table 22.4 Amino acid composition of the total proteins (mole-%) of flours from various cereals (Belitz et al. 2004)

Amino acid	Wheat	Rye	Barley	Oats	Rice	Millet	Corn
Asx	4.2	6.9	4.9	8.1	8.8	7.7	5.9
Thr	3.2	4.0	3.8	3.9	4.1	4.5	3.7
Ser	6.6	6.4	6.0	6.6	6.8	6.6	6.4
Glx	31.1	23.6	24.8	19.5	15.4	17.1	17.7
Pro	12.6	12.2	14.3	6.2	5.2	7.5	10.8
Gly	6.1	7.0	6.0	8.2	7.8	5.7	4.9
Ala	4.3	6.0	5.1	6.7	8.1	11.2	11.2
Cys	1.8	1.6	1.5	2.6	1.6	1.2	1.6
Val	4.9	5.5	6.1	6.2	6.7	6.7	5.0
Met	1.4	1.3	1.6	1.7	2.6	2.9	1.8
Ile	3.8	3.6	3.7	4.0	4.2	3.9	3.6
Leu	6.8	6.6	6.8	7.6	8.1	9.6	14.1
Tyr	2.3	2.2	2.7	2.8	3.8	2.7	3.1
Phe	3.8	3.9	4.3	4.4	4.1	4.0	4.0
His	1.8	1.9	1.8	2.0	2.2	2.1	2.2
Lys	1.8	3.1	2.6	3.3	3.3	2.5	1.4
Arg	2.8	3.7	3.3	5.4	6.4	3.1	2.4
Trp	0.7	0.5	0.7	0.8	0.8	1.0	0.2
Amide group	31.0	24.4	26.1	19.2	15.7	22.8	19.8

globulins, and 70% aqueous ethanol-soluble prolamins. The glutelins remained in the flour residue. They can be separated into two subfractions. For this purpose, all the proteins remaining in the residue are first dissolved in 50% aqueous 1-propanol at 60°C with reduction of the disulfide bonds, for example, with dithioerythritol. The high-molecular (HMW) subunits precipitate out on increasing the propanol concentration to 60%, whereas the low-molecular (LMW) subunits remain in solution. This separation scheme can still be used for fractionation of seed proteins for further proteomic analysis (Miernyk and Hajduch 2011).

Albumins and globulins are derived mostly from cytoplasmic residues and other subcellular fractions that are part of the kernel. Prolamins and glutelins are storage proteins. The most important wheat proteins are glutens. When the flour is mixed with water, gluten proteins form an elastic matrix. This matrix holds carbon dioxide and gives volume to bread during rising. The gluten extracts are also used as additives for many food products. Gluten proteins are composed of monomeric subunits gliadins and polymeric glutenins. Gliadins are subdivided into alpha/beta-, gamma, and omega gliadins. Furthermore, gliadins are subdivided into high molecular weight glutenin subunits (HMW-GS) and low molecular weight glutenin subunits (LMW-GS) that are linked by intermolecular disulfide bonds. Gliadins and glutenin subunits together are called prolamins (see above). Glutens contain large amounts of proline and glutamine. The content of essential amino acids such as arginine, lysine, and histidine is low, and in a complete nutrition, an additional source for these amino acids is necessary (Belitz et al. 2004). Van den Broeck et al. (2008) give an overview of the proteomic analysis of gluten proteins. A negative aspect of gluten proteins is that they can trigger an immune response called celiac disease in those genetically susceptible (Sollid 2002). On the other hand, a lot of effort has been put into analyzing the allelic codes for gluten proteins in order to come to the high yield and highly resistant wheat sorts giving a flour with good baking quality (Yahata et al. 2005; van den Broeck et al. 2008).

Other important proteins in wheat are puroindolins a and b, and some enzymes and enzyme inhibitors, such as amylases, proteolytic enzymes, lipases, and enzymes involved in oxido-reductive processes, as well as amylase and protease inhibitors. Puroindolins are the cysteine-rich proteins that also contain segments in their amino acid sequences (Belitz et al. 2004; Branlard et al. 2003). Puroindolins are the lipid-binding proteins that have influence on the proper texture of dough during the baking process (Branlard et al. 2003). In an early paper, Islam et al. (2003) gave an overview of the wheat proteome and the relationship between chromosome deletion and protein expression.

Rice

Rice is the most important cereal in Asia and in developing countries. This crop feeds one-fourth of the world population, and its genome was sequenced relatively early (Komatsu et al. 2003) and it is also the reason for the early start of proteomic investigation of this important crop (Komatsu et al. 2003). Traditional rice milling involves steeping in hot water and steaming in autoclaves, followed by drying and polishing. This treatment causes removal and destruction of some nutritionally important components, such as proteins and vitamins. It may be the reason that most studies of rice proteomics are focused on the whole plant and plant development (Agrawal and Rakwal 2011), as well as discussion about proteomic changes in transgenic rice (Xue et al. 2012). The proteomic papers dealing with the rice proteins important for nutrition are relatively rare. From the nutritional point of view, together with the proteins from maize and millet, rice proteins are important as a protein source for celiac disease patients (Belitz et al. 2004; Moroni et al. 2010).

22.3.1.2 Celiac Disease

In genetically susceptible individuals, some cereals such as wheat, rye, and barley can cause celiac disease. This disease affects infants as well as adolescents. Recent epidemiological studies indicate that about 1% of the world population suffers from this disease (Moroni et al. 2010). It is associated with a loss of villous structure of the intestinal mucosa, and, depending on severity of the disease, the nutrient absorption function can be impaired (Sollid 2002). After consumption of food containing the above-mentioned glutens, specific peptides from prolamins trigger an immune response that causes damage. This leads to a range of symptoms including altered bowel habits, malnutrition, and weight loss (Rodrigo 2006) Individuals with celiac

Table 22.5 Legumes: protein distribution (%) by *Osborne* fractions

Fraction	Soy-beans	Peanuts	Peas	Mungo beans	Broad beans
Albumin	10	15	21	4	20
Globulin	90	70	66	67	60
Glutelin	0	10	12	29	15

disease are sensitive to the prolamin fractions of wheat, barley, and rye. A simple change of the diet to rice, millet, and maize can eliminate the cause of the disease. Van den Broeck et al. (2008) give the complete proteomic analysis of the gluten proteins involved in celiac disease in different wheat varieties.

22.3.1.3 Legumes

Soybeans are nutritionally the most important legume as a protein source, because of their high content of essential amino acids (see above). The fractionation of legume proteins developed early in the last century by Osborne (1907) using solubility procedures to yield three fractions: albumins, globulins, and glutelins. As shown in Table 22.5, globulins are the predominant fraction in all legumes.

Globulins seem to have a function in seeds as storage proteins. Further simple fractionation of these proteins by ultracentrifugation or chromatography yields separation into two major components present in all legumes: vicilin and legumin. Legumin from soybeans is called glycinin and from peanuts arachin. The modern proteomic analyses of soybean also show that most of the seed proteins (60–80%), for example, in soybeans (Krishnan et al. 2009) as well as in common bean seeds (Marsolais et al. 2010) belong to the above-mentioned group of storage proteins.

The low-abundance proteins in seeds also have important nutritional value as protease inhibitors (e.g., in soybeans), or allergens (mostly in peanuts, see below), and it is important to know the whole proteome of nutritionally important legumes (Krishnan et al. 2009). Many legume proteins, mostly from peanuts are responsible for many allergic reactions. Most allergies in the United States are caused by peanuts and peanut-containing food products (Stevenson et al. 2009). Several proteins detected in peanut seed such as Ara h1-4 are responsible for these reactions. Proteomic analyses show different contents of these allergens in different peanut varieties. Interestingly, these proteins are absent in genetically engineered peanut seeds (Chassaigne et al. 2009; Stevenson et al. 2009). Other potential nutritional risks are lectins that are present in many legume seeds. If not inactivated or degraded during processing, these proteins can cause outbreaks of gastroenteritis, nausea, diarrhea, and other, even more severe, side reactions (Gaso-Sokac et al. 2010; Noah et al. 1980).

Finally, some legumes, especially soybeans, are genetically modified. Such food and food products are already on the market, especially in the United States. Proteomics are widely used for characterization of genetically engineered food, and there is a broad discussion about this topic in nutrition and medicine (Sakata et al. 2009; Stevenson et al. 2009; Gaso-Sokac et al. 2010).

22.3.2 *Fruit and Vegetable Proteomics*

Proteins, such as enzymes and inhibitors, as well as structural proteins play a key role in the molecular physiology of fruit development and ripening, as well as for stability of this food during transport and storage (Palma et al. 2011; Chan et al. 2007). Palma et al. (2011) give a proteomic overview about the proteome change in the ripening process, and the following events where proteins are involved take place during ripening of fruits, such as red pepper: taste alternation, intense metabolism, respiration and emitting volatile components, destruction of chlorophyll and synthesis of new pigments, pectins, and new proteins. Similar changes also occur in other fruits and vegetables during the growth and ripening processes. In this review, proteomic changes in different fruits and vegetables such as tomato, grape, citrus, peach, strawberries, and apple were listed. Chan et al. (2007) demonstrate impressively that proteins play a key role in stability of peach fruits during storage, and describe some enzymatic processes responsible for these changes. Finally, proteins also play an important role for quality of all fruit juice based and other beverages, and there are intensive proteomic studies in this field (Garibaldi and Giuffrida 2010; Colgrave et al. 2012).

References

Agrawal GK, Rakwal R (2011) Rice proteomics: a move toward expanded proteome coverage to comparative and functional proteomics uncovers the mysteries of rice and plant biology. Proteomics 11:1630–1649

Al-Ghobashy MA, Williams MAK, Brophy B, Laible G, Harding DRK (2009) On-line casein micelle disruption for downstream purification of recombinant human myelin basic protein produced in the milk of transgenic cows. J Chromatogr B 877:1667–1677

Argov N, Lemay DG, German JB (2008) Milk fat globule structure and function: nanoscience comes to milk production. Trends Food Sci Technol 19:617–623

Bauchart C, Rémond D, Chambon C, Patureau Mirand P, Savary-Auzeloux I, Reynès C, Morzel M (2006) Small peptides (<5 kDa) found in ready-to-eat beef meat. Meat Sci 74:658–666

Belcerzak D, Querengesser L, Dixson WT, Baracos VE (2001) Coordinate expression of matrix-degrading proteinases and their activators and inhibitors in bovine skeletal muscle. J Anim Sci 79:94–107

Belitz HD, Grosch W, Schieberle P (2004) In: Burghagen M (ed) Food chemistry 3rd revised edn. Springer-Verlag, Berlin/Heidelberg, pp 553–555, 557–559, 673–697

Bendixen E (2005) The use of proteomics in meat science. Meat Sci 71:138–149

Blom N, Sicheritz-Pontén T, Gupta R, Gammeltoft S, Brunak S (2004) Prediction of post-translational glycosylation and phosphorylation of proteins from the amino acid sequence. Proteomics 4:1633–1649

Branlard G, Amiour N, Igrejas G, Gaborit T, Herbette S, Dardevet M, Marion D (2003) Diversity of puroindolines as revealed by two-dimensional electrophoresis. Proteomics 3:168–174

Campbell L, Raikos V, Euston SR (2003) Modification of functional properties of egg-white proteins. Nahrung/Food 47:369–376

Casado B, Affolter M, Kussman M (2009) OMICS-rooted studies of milk proteins, oligosaccharides and lipids. J Proteomics 73:196–208

Chan Z, Qin G, Xu X, Li B, Tian S (2007) Proteome approach to characterize proteins induced by antagonist yeast and salicylic acid in peach fruit. J Proteome Res 6:1677–1688

Chassaigne H, Trégoat V, Nørgaard JV, Maleki SJ, van Hengel AJ (2009) Resolution and identification of major peanut allergens using a combination of fluorescence two-dimensional gel electrophoresis, Western blotting and Q-TOF mass spectrometry. J Proteomics 72:511–526

Colgrave ML, Goswami H, Howitt CA, Tanner GJ (2012) What is in a beer? Proteomic characterization and relative quantification of hordein (gluten) in beer. J Proteome Res 11:386–396

Conti A, Giuffrida MG, Cavaletto M (2007) Proteomics of human milk. In: Thongboonkerd V (ed) Proteomics of human body fluids. Human Press, Totowa, pp 437–451

D'Alessandro A, Righetti PG, Fasoli E, Zolla L (2010) The egg white and yolk interactomes as gleaned from extensive proteomic data. J Proteomics 73:1028–1042

D'Ambrosio C, Arena S, Scaloni A, Guerrier L, Boschetti E, Citterio A, Righetti PG (2008) Exploring the chicken egg white proteome with combinatorial peptide ligand libraries. J Proteome Res 7:3461–3474

Dewettinck K, Rombaut R, Thienpont N, Le TT, Massens K, Van Camp J (2008) Nutritional and technological aspects of milk fat globule membrane material. Int Dairy J 18:436–457

Di Luccia LA, Picariello G, Trani A, Alviti G, Liozzo P, Faccia M, Addeo F (2009) Occurrence of -casein fragments in cold-stored and curdled river buffalo (*Bubalus bubalus L.*) milk. J Dairy Sci 92:1319–1329

FAO/WHO/UNU (1985) Energy and protein requirements. Report of a Joint FAO/WHO/UNU Expert Consultation. World Health Organization Technical Rep. Ser. 724, WHO, Geneva

Farinazzo A, Rastuccia U, Bachi A, Guerrier L, Fortis F, Fasoli E et al (2009) Chicken egg yolk cytoplasmatic proteome, mined via combinatorial ligand libraries. J Chromatogr A 1216:1241–1252

Farrell HM Jr, Jimenez-Flores R, Bleck GT, Brown EM, Butler JE, Craemer LK et al (2004) Nomenclature of the proteins of cow's milk – Sixth revision. J Dairy Sci 87:1641–1674

Fox PF, Brodkorb A (2008) The casein micelle: historical aspects, current concepts and significance. Int Dairy J 18:677–684

Gagnaire V, Jardin J, Jan G, Lortal S (2009) Proteomics of milk and bacteria used in fermented dairy products: from qualitative to quantitative advances. J Dairy Sci 92:811–825

Gao Y, Gillen CM, Wheatly MG (2006) Molecular characterization of the sarcoplasmic calcium-binding protein (SCP) from crayfish Procambarus clarkia. Comp Biochem Physiol B Biochem Mol Biol 144:478–487

Garibaldi M, Giuffrida MG (2010) Heard it through the grapewine: proteomic perspective on grape and wine. J Proteomics 73:1647–1655

Gaso-Sokac D, Kovac S, Josic D (2010) Application of proteomics in food technology and food biotechnology: process development, quality control and product safety. Food Technol Biotechnol 48(3):284–295

Gaso-Sokac D, Kovac S, Josic D (2011) Use of proteomic methodology in optimization of processing and quality control of food of animal origin. Food Technol Biotechnol 49(4):397–412

Geesink GH, Koohmaraie M (1999) Postmortem proteolysis an calpain/calpastatin activity in callipyge and normal lamb biceps femoris during extended postmortem storage. J Anim Sci 77:1490–1501

Glantz M, Devold TG, Vegarud GE, Lindmark Månsson H, Stålhammar H, Paulsson M (2010) Importance of casein micelle size and milk composition for milk gelation. J Dairy Sci 93:1444–1451

Gong CY, Li Q, Yu HT, Wang Z, Wang T (2012) Proteomics insight into the biological safety of transgenic modification of rice as compared with conventional genetic breeding and spontaneous genotypic variation. J Proteome Res 11:3019–3029

Guérin-Dubiard C, Pasco M, Mollé D, Désert C, Croguennec T, Nau F (2006) Proteomic analysis of hen egg white. J Agric Food Chem 54:3901–3910
Gupta P, Lee KH (2007) Genomics and proteomics in process development: opportunities and challenges. Trends Biotechnol 25:324–330
Henry G, Mollé D, Morgan F, Fauquant J, Bouhallab S (2002) Heat-induced covalent complex between casein micelles and -lactoglobulin from goat's milk: identification of an involved disulfide bond. J Agric Food Chem 50:185–191
Hollung K, Veseith E, Jia X, Færgestad EM, Hildrum KI (2007) Application of proteomics to understand the molecular mechanisms behind meat quality. Meat Sci 77:97–104
Holt C (1998) Casein micelle substructure and calcium phosphate interactions studied by Sephacryl column chromatography. J Dairy Sci 81:2994–3003
Ishikawa M, Shiomi K, Suzuki F, Ishida M, Nagashima Y (2001) Identification of tropomyosin as a major allergen in the octopus *Octopus vulgaris* and elucidation of its IgE binding epitopes. Fish Sci 67:934–942
Islam N, Tsujimoto H, Hirano H (2003) Wheat proteomics: relationship between fine chromosome deletion and protein expression. Proteomics 3:307–316
Johnson ME, Lucey JA (2006) Major technological advances and trends in cheese. J Dairy Sci 89:1174–1178
Kiosseoglou V (2003) Egg yolk protein gel and emulsions. Curr Opin Colloid Interface Sci 8:365–370
Komatsu S, Konishi H, Shen S, Yang G (2003) Rice proteomics. Mol Cell Proteomics 2:2–10
Korhonen H, Pihlanto A (2006) Bioactive peptides: production and functionality. Int Dairy J 16:945–960
Krishnan HB, Oehrle NW, Natarajan SS (2009) A rapid and simple procedure for the depletion of abundant storage proteins from legume seeds to advance proteome analysis: a case study using *Glycine max*. Proteomics 9:3174–3188
Lehrer SB, Ayuso R, Reese G (2003) Seafood allergy and allergens: a review. Mar Biotechnol 5:339–348
Leitner A, Castro-Rubio F, Marina ML, Lindner W (2006) Identification of marker proteins for the adulteration of meat products with soybean proteins by multidimensional liquid chromatography–tandem mass spectrometry. J Proteome Res 5:2424–2430
Lopez C, Briard-Bion V, Menard O, Rousseau F, Pradel P, Besle JM (2008) Phospholipid, sphingolipid and fatty acid compositions of the milk fat globule membrane are modified by diet. J Agric Food Chem 56:5226–5236
Madureira AR, Pereira CI, Gomes AMP, Pintado ME, Malcata FX (2007) Bovine whey proteins – overview on their main biological properties. Food Res Int 40:1197–1211
Madureira AD, Tavares T, Gomes AM, Pintado ME, Malcata FX (2010) Physiological properties of bioactive peptides obtained from whey proteins. J Dairy Sci 93:437–455
Mann K, Mann M (2008) The chicken egg yolk plasma and granule proteomes. Proteomics 8:178–191
Manso MA, Léonil J, Jan G, Gagnaire V (2005) Application of proteomics for characterisation of milk and dairy products. Int Dairy J 15:845–855
Marshall K (2004) Therapeutic application of whey protein. Altern Med Rev 9:136–156
Marsolais F, Pajak A, Yin F et al (2010) Proteomic analysis of common bean seed with storage protein deficiency reveals up-regulation of sulfur-rich proteins and starch and raffinose metabolic enzymes, and down-regulation of the secretory pathway. J Proteomics 73:1587–1600
Martin HM, Hancock JT, Salisbury V, Harrison R (2004) Role of xanthine oxidoreductase as an antimicrobial agent. Infect Immun 72:4933–4939
Matéos A, Girardet JM, Mollé D, Dary A, Miclo L, Gaillard JL (2009) Two-dimensional cartography of equine–casein variants achieved by isolation of phosphorylation iso forms and control of the deamidation phenomenon. J Dairy Sci 92:2389–2399
Mather IH (2000) A review and proposed nomenclature for major proteins of the milk-fat globule membrane. J Dairy Sci 83:203–247
Meisel H, FitzGerald RJ (2003) Biofunctional peptides from milk proteins: mineral binding and cytomodulatory effects. Curr Pharm Des 9:1289–1295

Miernyk JA, Hajduch M (2011) Seed proteomics. J Proteomics 74:389–400
Morgan F, Bouhallab S, Mollé D, Henry G, Maubois JL, Léonil J (1998) Lactolation of lactoglobulin monitored by electrospray ionisation mass spectrometry. Int Dairy J 8:95–98
Moroni V, Iametti S, Bonomi F, Arendt EK, Dal Bello F (2010) Solubility of proteins from non-gluten cereals: a comparative study on combination of solubilising agents. Food Chem 121:1225–1230
Motoyama K, Suma Y, Ishizaki S, Nagashima Y, Shiomi K (2007) Molecular cloning of tropomyosins identified as allergens in six species of crustaceans. J Agric Food Chem 55:985–991
Mullen AM, Stoeva S, Laib K, Gruebler G, Voelter W, Troy DJ (2000) Preliminary analysis of amino acids at various locations along the *M. longissimus dorsi* in aged beef. Food Chem 69:461–465
Ng-Kwai-Hang KF, Grosclaude F (2003) Genetic polymorphism of milk proteins. In: Fox PF, McSweeney PLH (eds) Advanced dairy chemistry, vol 1, Proteins, Part B. Kluwer Academic/Plenum, New York, pp 739–816
Noah ND, Bender AE, Readi GB, Gilbert RJ (1980) Food poisoning from raw kidney beans. Brit Med J 281:236–237
O'Donnell R, Holland JW, Deeth HC, Alewood P (2004) Milk proteomics. Int Dairy J 14:1013–1023
Ortea I, Cañas B, Gallardo JM (2009) Mass spectrometry characterization of species-specific peptides from arginine kinase for the identification of commercially relevant shrimp species. J Proteome Res 8:5356–5362
Osborn TB (1907) The proteins of wheat kaernel. Carnegie Institution, Washington, DC
Palma JM, Corpas FJ, del Rio LA (2011) Proteomics as an approach to the understanding of the molecular physiology of fruit development and ripening. J Proteomics 74:1230–1243
Pampel L, Boushaba R, Udell M, Turner M, Titchener-Hooker N (2007) The influence of major components on the direct chromatographic recovery of a protein from transgenic milk. J Chromatogr A 1142:137–147
Panchaud A, Kussmann M, Affolter M (2005) Rapid enrichment of bioactive milk proteins and iterative, consolidated protein identification by multidimensional protein identification technology. Proteomics 5:3836–3846
Piñeiro C, Barros-Velázquez J, Vázquez J, Figueras A, Gallardo JM (2003) Proteomics as a tool for the investigation of seafood and other marine products. J Proteome Res 2:127–135
Pins JJ, Keenan JM (2006) Effects of whey peptides on cardiovascular disease risk factors. J Clin Hypertens 8:775–782
Pogacic T, Kelava N, Zaberlin S, Dolencic-Spehar I, Samarzija D (2010) Methods for culture-independent identification of lactic acid bacteria in dairy products. Food Technol Biotechnol 48:3–10
Purslow PP (2002) The structure and functional significance of variations in the connective tissue within muscle. Comput Biochem Physiol A 133:947–966
Purslow PP (2005) Intramuscular connective tissue and its role in meat quality. Meat Sci 70:435–447
Raikos V, Hansen R, Campbell L, Euston SR (2006) Separation and identification of hen egg protein isoforms using SDS-PAGE and 2-D gel electrophoresis with MALDI-TOF mass spectrometry. Food Chem 99:702–710
Reinhardt TA, Lipolis JD (2008) Developmental changes in the milk fat globule membrane proteome during transition from colostrum to milk. J Dairy Sci 91:2307–2318
Rodrigo L (2006) Celiac disease. World J Gastroenterol 12(41):6585–6593
Roncada P, Gaviraghi A, Liberatori S, Canas B, Bini L, Greppi GF (2002) Identification of caseins in goat milk. Proteomics 2:723–726
Rupa P, Mine Y (2008) Molecular allergology of egg white ovomucoid. ACS Sym Ser 1001:382–399
Sakata K, Ohyanagi H, Nabori H, Nakamura T, Hashiguchi A, Nanjo Y, Mikami Y, Yunokawa H, Komatsu S (2009) Soybean proteome database: a data resource for plant differential omics. J Proteome Res 8:3539–3548
Sando L, Pearson R, Gray C, Parker P, Hawken R, Thompson PC et al (2009) Bovine Muc1 is a highly polymorphic gene encoding an extensively glycosylated mucin that binds bacteria. J Dairy Sci 92:5276–5291

Scaloni A, Perillo V, Franco P, Fedele F, Froio R, Ferrara L, Bergamo P (2002) Characterization of heat-induced lactosylation products in caseins by immunoenzymatic and mass spectrometric methodologies. Biochim Biophys Acta 1598:30–39

Shiomi K, Sato Y, Hamamoto S, Mita H, Shimakura K (2008) Sarcoplasmatic calcium-binding protein: identification as a new allergen of the black tiger shrimp *Penaeus monodon*. Int Arch Allergy Immunol 146:91–98

Smolenski G, Haines S, Kwan FY, Bond J, Farr V, Devis SR et al (2007) Characterisation of host defence proteins in milk using proteomic approach. J Proteome Res 6:207–215

Sollid LM (2002) Coeliac disease: dissecting a complex inflammatory disorder. Nat Rev Immunol 2:647–655

Sørensen ES, Møller L, Vinther M, Petersen TE, Rasmussen LK (2003) The phosphorylation pattern of human s1-casein is markedly different from the ruminant species. Eur J Biochem 270:3651–3655

Stevenson SE, Chu Y, Ozias-Akins P, Thelen JJ (2009) Validation of gel-free, label-free quantitative proteomics approaches: applications for seed allergen profiling. J Proteomics 72:555–566

Taka H, Kaga N, Fujimura T, Mineki R, Imaizumi M, Suzuki Y, Suzuki R, Tanokura M, Shindo N, Murayama K (2000) Rapid determination of parvalbumin amino acid sequence from *Rana catesbeiana* (pI 4.78) by combination of ESI mass spectrometry, protein sequencing, and amino acid analysis. J Biochem 127:723–729

Timmen H, Patton S (1988) Milk fat globules: fatty acid composition, size and in vivo regulation of fat liquidity. Lipids 23:685–689

van den Broeck HC, America AHP, Smulders MJM, Gilissen LJWJ, van der Meer IM (2008) Staining efficiency of specific proteins depends on the staining method: wheat gluten proteins. Proteomics 8:1880–1884

van der Strate BWA, Beljaars L, Molema G, Harmsen MC, Meijer DKF (2001) Antiviral activities of lactoferrin. Antiviral Res 52:225–239

Vogel HJ, Schibli DJ, Jing W, Lohmeier-Vogel EM, Epand RF, Epand RM (2002) Towards a structure-function analysis of bovine lactoferricin and related tryptophan and arginine containing peptides. Biochem Cell Biol 80:49–63

Ward PP, Uribe-Luna S, Conneely OM (2002) Lactoferrin and host defense. Biochem Cell Biol 80:95–102

Waugh DF, Creamer LK, Slattery CW, Dresdner GW (1970) Core polymers of casein micelles. Biochemistry 9:786–795

Xue K, Yang J, Liu B, Xue D (2012) The integrated risk assessment of transgenic rise *Oryza sativa:* a comparative proteomics approach. Food Chem. http://dx.doi.org/10.1016/j.foodchem.2012.04.042

Yahata E, Maruyama-Funatsuki W, Nishio Z, Tabiki T, Takata K, Yamamoto Y, Tanida M, Suriyama H (2005) Wheat cultivar-specific proteins in grain revealed by 2-DE and their application to cultivar identification of flour. Proteomics 5:3942–3953

Yamada M, Murakami K, Wallingfold JC, Yuki Y (2002) Identification of low-abundance proteins of bovine colostral and mature milk using two-dimensional electrophoresis followed by microsequence and mass spectrometry. Electrophoresis 23:1153–1160

Yamauchi R, Usui H, Yunden J, Takenaka Y, Tani F, Yoshikawa M (2003) Characterization of lactotensin, a bioactive peptide derived from bovine lactoglobulin, as a neurotensin agonist. Biosci Biotechnol Biochem 67:940–943

Yu CJ, Lim YF, Chian BL, Chow LP (2003) Proteomics and immunological analysis of a novel shrimp allergen, Pen m 2. J Immunol 170:445–453

Chapter 23
Relevance of Peptides Bioactivity in Foods

Keizo Arihara

23.1 Introduction

Numerous food components having physiological functions have been isolated, characterized, and applied to functional foods (Arihara 2012; Arihara and Ohata 2011; Saarela 2011). Food protein-derived peptides are a group of such functional components (Arihara 2006; Gobbetti et al. 2007; Kannan et al. 2012; Korhonen and Pihlanto 2007). Although the activities of these protein-derived peptides in the sequence of proteins are latent, they are released by proteolytic enzymes. Therefore, food proteins have possible bioactivities beyond a nutritional source of amino acids alone.

Mellander (1950) first reported bioactive peptides generated from food proteins. He suggested that milk casein-derived peptides enhanced bone calcification in infants. Since then, numerous bioactive peptides from food proteins have been studied (Hettiarachchy et al. 2012; Kannan et al. 2012; Owusu-Aspenten 2010). Enzymatic hydrolysis of food proteins such as milk caseins produces various peptides with physiological functions, such as antihypertensive, opioid, immunostimulating, antimicrobial, antithrombotic, hypocholesterolemic, and antioxidative activities.

This chapter includes a brief overview of bioactive peptides generated from food proteins. Also, utilization of modern nutrigenomics techniques for such peptides is discussed here. Nutrigenomics has been rapidly applied to the field of nutrition and health (Affolter et al. 2009; Bidlack and Rodriguez 2012; Thomson-Smith 2010). This powerful tool contains gene, protein, and metabolite profiling (transcriptomics, proteomics, and metabolomics). Although application of this attractive strategy for

K. Arihara (✉)
Department of Animal Science, Kitasato University,
35-1 Higashi-23-Bancho, Towada-shi 034-8628, Japan
e-mail: arihara@vmas.kitasato-u.ac.jp

F. Toldrá and L.M.L. Nollet (eds.), *Proteomics in Foods: Principles and Applications*,
Food Microbiology and Food Safety 2, DOI 10.1007/978-1-4614-5626-1_23,

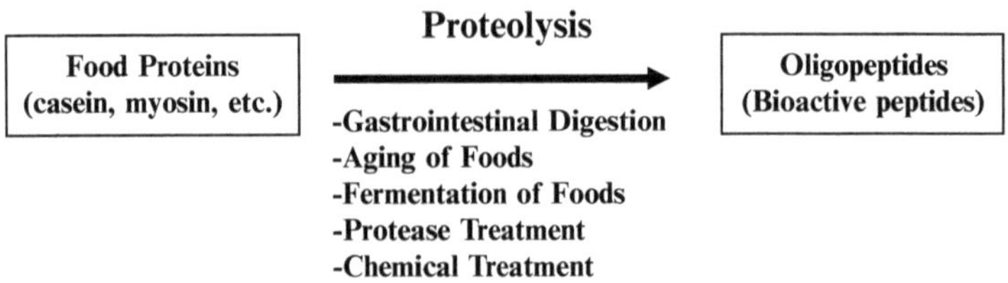

Fig. 23.1 Generation ways of peptides from food proteins

studying bioactive peptides generated from food proteins is still limited, the nutrigenomics approach offers a great possibility for understanding and utilizing bioactive peptides.

23.2 Generation of Peptides from Food Proteins

Most bioactive sequences of food proteins are inactive or incomplete within the parent proteins. Active peptide fragments are released from native proteins via proteolytic digestion. After such peptides are liberated from food proteins, they can act as regulatory (bioactive) compounds. Gastrointestinal proteolysis, aging, fermentation, enzymatic treatment, and other chemical treatments are the principal means for digestion of food proteins to generate bioactivities (Figure 23.1). During gastrointestinal proteolysis, bioactive peptides are liberated from food proteins (Pihlanto and Korhonen 2003). Upon ingestion, pepsin in the stomach digests food proteins into large peptides And once in the small intestine, trypsin and chymotrypsin cleave these peptides into small peptides and amino acids.

In addition to gastrointestinal digestion, there are several ways in which peptides are generated from food proteins. During aging or storage, food proteins are hydrolyzed by endogenous proteases. For example, meat proteins are hydrolyzed by muscle endogenous enzymes, such as calpains and cathepsins (Koohmaraie 1994). Such enzymatic hydrolysis contributes to the improvement of sensory properties (texture, taste, and flavor) of meat (Nishimura et al. 1988). Although there has been no report about the generation of bioactive peptides in meat during post-mortem aging, our preliminary study showed an increase in angiotensin I converting enzyme (ACE) inhibitory and antioxidative activity of beef and pork during storage at 4°C.

Proteolytic events often occur during fermentation of foods and various peptides are generated from food proteins (Hernändez-Ledesma et al. 2004). For example, the proteolysis of raw sausages and dry-cured ham have been studied extensively, because components generated from meat proteins are critical for the development of sensory properties of fermented meat products (Hammes et al. 2003; Toldrá 2004). During fermentation of sausages, the content of peptides and amino acids reaches about 1% dry matter of products (Dainty and Blom 1995). We measured the

ACE inhibitory activities of extracts of several European fermented sausages and found that the activity level of all extracts was higher than those of extracts obtained from nonfermented pork products. Fermented dairy products (e.g., cheese and yogurt) contain bioactive peptides, such as ACE inhibitory peptides, generated by proteolytic activities of starter microorganisms (Bütikofer et al. 2008; Gagnaire et al. 2001; Gobbetti et al. 2000; Gomez-Ruiz et al. 2002; Saito et al. 2000).

Utilization of commercial proteases is the most practical approach for producing bioactive peptides from food proteins (Young and Mine 2009). Inasmuch as protease treatments of food proteins are an efficient method for hydrolyzing proteins, proteolytic enzymes are used for various processes in the food industry. Single proteinases from animal, plant, and microbial sources and combinations of them can be utilized for the digestion of food proteins for releasing bioactive peptides. Also, the production of peptides from food proteins can be achieved by the use of heat, acid, or base hydrolysis.

23.3 Representative Bioactive Peptides from Food Proteins

23.3.1 ACE Inhibitory Peptides

The most extensively studied bioactive peptides generated from food proteins are ACE inhibitory peptides (Aleixandre and Miguel 2012; Meisel et al. 2005). These peptides have attracted particular attention because of their ability to prevent hypertension and have been utilized for pharmaceuticals and physiologically functional foods. These peptides lower blood pressure through inhibition of ACE in the body. ACE is a dipeptidyl carboxypeptidase, which is widely distributed in mammalian bodies, predominantly as a membrane-bound ectoenzyme in vascular endothelial cells. ACE plays a critical role in the regulation of blood pressure in the renin angiotensin system (Figure 23.2). ACE converts the inactive decapeptide angiotensin I into the potent vasoconstricting octapeptide angiotensin II, resulting in increases in blood pressure. ACE also inactivates the antihypertensive vasodilator bradykinin. Therefore, by inhibiting the catalytic action of ACE, the elevation of blood pressure can be suppressed in the body.

ACE inhibitory peptides derived from food proteins were first found in the hydrolysate of gelatin by Oshima et al. (1979). Since then, ACE inhibitory peptides have been identified in the hydrolysates of various proteins from foods such as milk, fish, meat, eggs, soybean, corn, wheat, seaweed, and others (Aleixandre and Miguel 2012; Arihara 2006; Meisel et al. 2005). Some of these peptides have been reported to show antihypertensive effects by oral administration. As the examples of such peptides, two ACE inhibitory peptides (Met-Asn-Pro-Pro-Lys and Ile-Thr-Thr-Asn-Pro) were purified from thermolysin digest of porcine muscle myosin (Arihara et al. 2001). These ACE inhibitory peptides showed antihypertensive effects in spontaneously hypertensive rats (SHR) by oral administration (Nakashima et al. 2002). Also,

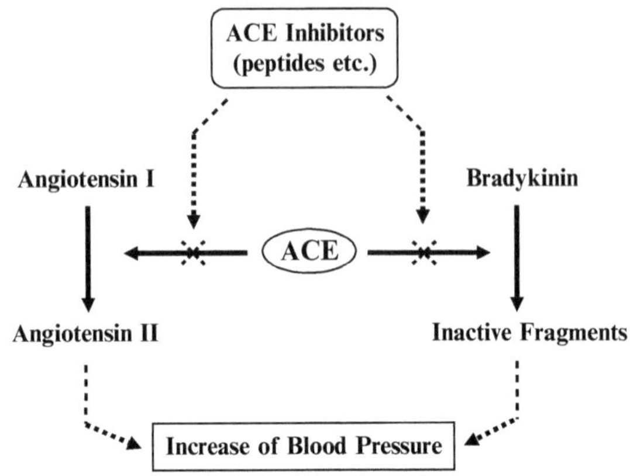

Fig. 23.2 Blood pressure regulation by Angiotensin I-Converting Enzyme (ACE)

six tripeptides that have parts of the sequences of the two peptides were orally administered to SHR. Among the six tripeptides, Met-Asn-Pro caused the most significant decrease in systolic blood pressure. Although many ACE inhibitory peptides have been found in various food protein hydrolysates, their structure–activity relationships have not been fully clarified (Li et al. 2004). However, it has been reported that many ACE inhibitory peptides have tryptophan, phenylalanine, tyrosine, or proline at their C-terminus and these amino acids would be critical for the ACE inhibitory activity of peptides.

23.3.2 *Opioid Peptides*

Opioid peptides are defined as peptides that have an affinity for an opioid receptor as well as opiatelike effects (Pihlanto and Korhonen 2003). These peptides influence the nerve system and gastrointestinal functions. Typical opioid peptides (i.e., endorphins, enkephalin, and prodynorphin) have the same N-terminal sequence, Tyr-Gly-Gly-Phe. Opioid peptides bind to specific receptors of the intestinal target cell. These receptors participate in specific effects, such as emotional behavior, suppression of intestinal motility, and appetite. Brantl et al. (1979) identified the first group of opioid peptides (casomorphins) generated from milk casein. Proteolysis of alfa-lactoalbumin (a milk whey protein) generated a opioid peptide (Tyr-Gly-Leu-Phe) and was named α-lactorphin (Chiba and Yoshikawa 1986). Although administration of α-lactorphin lowered blood pressures in SHR, the antihypertensive mechanism of this peptide is not by ACE inhibition but rather appears to be due to interaction with opioid receptors (Nurminen et al. 2000). Opioid peptides were also identified in hydrolysates of wheat gluten and blood hemoglobin.

23.3.3 Antioxidative Peptides

The intake of antioxidants may decrease the risk of diseases, such as cardiovascular disease and certain types of cancer (Hertog 1996). Endogenous antioxidative peptides (e.g., glutathione, carnosine, and anserine) have been reported to play many physiological roles, such as prevention of diseases related to oxidative stress (Hipkiss and Brownson 2000). In addition to these endogenous nonprotein peptides, several antioxidative peptides have been identified from soybean, milk, eggs, and meat (Kim et al. 2012). For example, Saiga et al. (2003) reported that hydrolysates obtained from porcine myofibrillar proteins by protease treatment exhibited high antioxidant activity. Antioxidative peptides were isolated and sequenced as Asp-Ser-Gly-Val-Thr, Ile-Glu-Ala-Glu-Gly-Glu, Asp-Ala-Gln-Glu-Lys-Leu-Glu, Glu-Glu-Leu-Asp-Asn-Ala-Leu-Asn, and Val-Pro-Ser-Ile-Asp-Asp-Gln-Glu-Glu-Leu-Met. Acidic amino acids, Asp or Glu, were found in all five peptides. Although it has been reported that basic amino acids, such as His and Lys, show strong antioxidative activity (Chen et al. 1995), it was revealed that acidic peptides as well as basic peptides possess antioxidative activity.

23.3.4 Immunomodulating Peptides

Immunomodulatory peptides affect both the immune system and cell proliferation responses (Korhonen and Pihlanto 2007). It has been reported that several hydrolysates of milk caseins stimulate the immune system. Although peptides generated from milk caseins by pancreatin or trypsin inhibited the proliferative responses of murine splenic lymphocytes and Peyer's patch cells, digests generated by pepsin or chymotrypsin did not show such activity (Otani and Hata 1995). Opioid peptides generated from milk proteins also have a modulatory function in the immune system, because opioids can alter the characteristics of cellular components of the immune system (Webster 1998).

23.3.5 Other Bioactive Peptides

In addition to the above-mentioned peptides, it has been found that various bioactive peptides have been found in the hydrolysates of food proteins. Nagaoka et al. (2001) discovered a hypocholesterolemic peptide (Ile-Ile-Ala-Glu-Lys) from enzymatic hydrolysates of β-lactoglobulin. This peptide has a strong effect on serum cholesterol level, and the hypocholesterolemic activity of the peptide was greater than that of the drug β-sitosterol in rats. The peptide is speculated to reduce the micellar solubility of cholesterol and inhibit cholesterol absorption (Nagaoka et al. 2001).

Antimicrobial peptides have been isolated from milk and egg proteins. Antimicrobial peptides generated from milk lactoferrin have been studied most extensively (Tomita et al. 1994). A hen egg ovotransferrin-derived antimicrobial peptide that is active against *Staphylococcus aureus* and *Escherichia coli* has been isolated (Ibrahim et al. 2000).

Several milk protein-derived peptides generated by enzymatic hydrolysis act as mineral trappers and result in enhancement of the absorption efficiency of minerals (Vegarud et al. 2000). Based on the knowledge that the mechanism of blood clotting is similar to that of milk clotting, milk protein-derived antithrombotic peptides have been studied (Fiat et al. 1993). Many studies have shown that hydrolysates of milk proteins exhibited stimulation of the growth of lactic acid bacteria and bifidobacteria (Brody 2000). The hydrolysate of porcine skeletal muscle actomyosin digested by papain also enhanced the growth of *Bifidobacterium* strains (Arihara et al. 2011a). Apart from bioactivities, food protein-derived peptides also contribute to the organoleptic properties of food, such as meat (Arihara 2006). Therefore, generation of peptides from food proteins has the potential to develop novel functional foods with favorable organoleptic properties.

23.4 Conventional Methods for Peptide Identification

After generation of bioactive peptides from food proteins, the next steps are the concentration, purification, and identification of peptides concerning bioactivities. In most hydrolyzed food protein digests, only a few peptides are responsible for objective activities. Also, these bioactive peptides are generated from food proteins in relatively low concentrations. Figure 23.3 shows a typical conventional procedure for the study of bioactive peptides derived from food proteins. After hydrolysis of food proteins, the hydrolysates can be fractionated and the peptides enriched by various methods, such as precipitation with solvents, membrane separation, ultrafiltration, and chromatography. Reversed-phase high-performance liquid chromatography (RP-HPLC) is a standard technique in the purification of peptides (Aguilar 2004). Also, a combination of HPLC and mass spectrometry (LC-MS) is a powerful tool not only for purification of peptides but also for structural identification of peptides (Shen and Noon 2004). Thus, in addition to the Edman degradation method, MS has become a standard tool for sequencing of peptides. LC-MS/MS-based peptide sequencing is also becoming a standard technique. For example, Ghassem et al. (2011) identified ACE inhibitory peptides in fish myofibrillar protein hydrolysates by utilizing HPLC coupled to electrospray ionization–time-of-flight mass spectrometry (ESI-TOF MS/MS). Two peptide sequences were identified as VPAAPPK (IC_{50} = 0.45 μM) at 791.155 *m/z* and NGTWFEPP (IC_{50} = 0.63 μM) at 1085.841 *m/z*, respectively. After purification and identification of peptides, the next step is characterization of peptides in vitro and in vivo. For these studies, synthesized peptides are prepared by a peptide synthesizer in most cases.

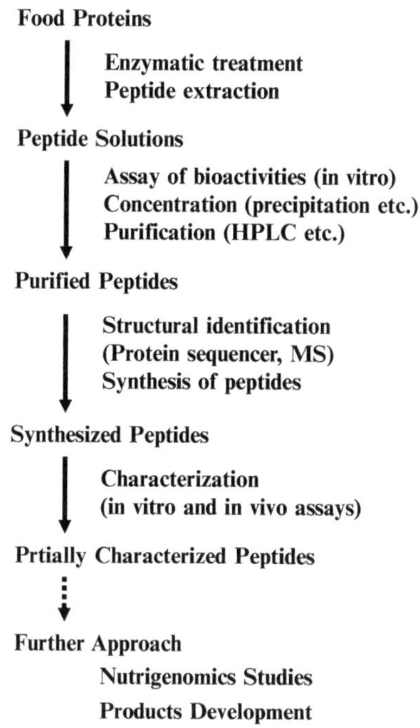

Fig. 23.3 Research strategy for food protein-derived bioactive peptides

Membrane-based separation techniques are utilized in the concentration of bioactive peptides, especially useful for commercial scale preparation (Pouliot et al. 2006). For industrial applications, efficiency of the process is critical, and the purification technique must be a balance between purity and efficiency. The large-scale fractionation of bioactive peptides was reviewed recently (Sato and Hashimoto 2012). Purification methods other than those described above are: chromatographic techniques (e.g., ion-exchange chromatography, hydrophobic interaction chromatography, size-exclusion chromatography, and affinity chromatography), electromembrane filtration, and isoelectric focusing.

23.5 Mechanisms of Bioactivities of Peptides

After oral ingestion of peptides, they are attacked by various enzymes in the stomach and the small intestine. Furthermore, cleaved small peptides can be digested by brush border membrane oligopeptidases and intracellular peptidases (Aito-Inoue et al. 2007). However, some orally ingested peptides can be transported intact into the bloodstream (Pihlanto and Korhonen 2003). After being transported into the bloodstream and once reaching their target sites, peptides induce their bioactivities via several mechanisms. According to the description of Young and Mine (2009), bioactive peptides can be broadly categorized into two groups: peptides that exert their effects by direct physical interaction with another molecule and peptides that interfere with gene expression. As described above, ACE inhibitory peptides lower blood pressure through inhibition of ACE in the body. ACE inhibition involves direct interaction of the peptide with noncatalytic binding sites in the enzyme. However, the inhibitory potencies of ACE inhibitory peptides do not always correlate with their antihypertensive effects. Some peptides with potent ACE inhibitory activities in vitro are inactive with oral administration. Fujita et al. (2000) clarified the discrepancy of ACE inhibitory activity in vitro and antihypertensive effect in vivo. ACE inhibitory activity has been measured according to the method of Cushman and Cheung (1971) in most studies. This in vitro assay is based on the liberation of hippuric acid from Hip-His-Leu catalyzed by ACE. Prior to the assay, a sample solution of peptides is mixed with a solution containing Hip-His-Leu and NaCl and then pre-incubated. After the reaction initiated by the addition of ACE, the hippuric acid liberated by ACE is determined. The concentration of an ACE inhibitory peptide needed to inhibit 50% of ACE activity is defined as the IC50 value. Fujita et al. (2000) pre-incubated ACE peptides from several food proteins with ACE before measurement of ACE inhibitory activity and classified ACE inhibitory peptides into three groups. One group is inhibitor-type peptides, that is, peptides for which IC50 values are not affected by pre-incubation with ACE. Another group is pro-drug-type peptides, that is, peptides that are converted to true inhibitors by an ACE or other proteases. The third group is substrate-type peptides, that is, peptides that are hydrolyzed by an ACE to give peptides with weak activity. Both inhibitor-type and pro-drug-type peptides exert antihypertensive activities with oral administration.

Similarly to ACE inhibitory peptides, antimicrobial and mineral binding peptides directly interfere with other molecules in the body. On the other hand, peptides can produce their bioactivities by altering gene expression (Young and Mine 2009). Some bioactive peptides generated from food proteins can alter gene expression by (1) epigenetic modification of the proteins that attach to the DNA, (2) alteration of the cell's primary signaling ligand to influence transcription factor activity indirectly, and (3) interference with cell signaling and gene expression via the direct binding of peptide ligand to the receptor. Although studies of bioactive peptides classified into these categories are still very limited, the mechanisms of such activities will be revealed with new nutrigenomics and proteomics techniques. Some examples of such studies are described in the following section.

23.6 Nutrigenomics of Bioactive Peptides

Genomics is the study of all DNA sequences in the genome of an organism. Nutrigenomics seeks to discover the interaction between dietary factors and host genes and how genes and their products metabolize these constituents into health-promoting nutrients or disease-causing antinutrients, and bioactive compounds

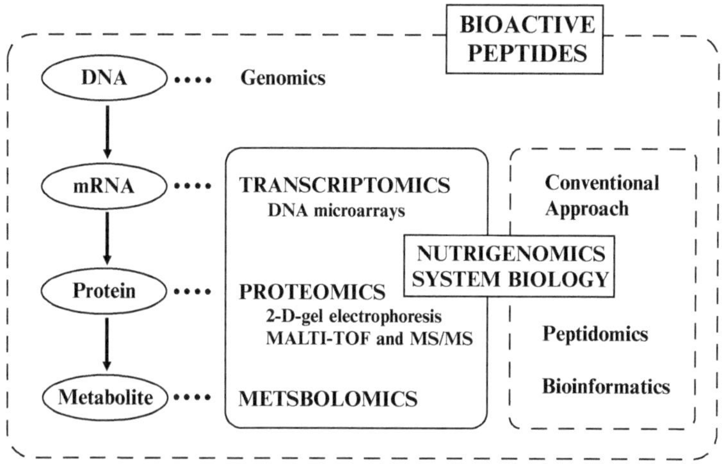

Fig. 23.4 Nutrigenomic approach for food protein-derived bioactive peptides

(Astley and Penn 2009; Kaput 2007; Thomson-Smith 2010). New omics technologies including transcriptomics, proteomics, and metabolomics offer exciting opportunities to address complex issues related to human health (Figure 23.4).

23.6.1 Transcriptomics of Bioactive Peptides

Transcriptomics analysis allows for a genomewide monitoring of expression for the simultaneous assessment of tens of thousands of genes and of their relative expression (Affolter et al. 2009). It measures the relative amounts of mRNAs in a given organism for determining patterns and levels of gene expression. The classical gene analysis approaches, such as Northern blotting and realtime RT-PCR, can only analyze gene expression for a limited number of candidate genes at a time. Microarray technology is a powerful high-throughput genomic tool (Mano et al. 2009). It can be used for profiling and monitoring the expression levels of numerous genes. It can also be used to determine the influence of food nutrients and factors maintaining homeostatic control of gene expression levels. Since transcriptomics is the most successful technology in nutrigenomics approaches, many nutrigenomic studies with microarray technology have investigated the relationship between genes and food intake and bioactive food components (Masotti et al. 2010).

A typical DNA microarray assay includes several procedures (Mano et al. 2009). A microarray is prepared by an arrayed series of many spots of DNA oligonucleotides of specific genes. This array is used to measure the mRNA abundance of a sequence in a sample relative to that in the control. mRNA is extracted from each sample (i.e., animal or cell) treated with food factors. The mRNA is reverse-transcribed to obtain complementary RNA (cRNA) and probes. The labeled mRNA

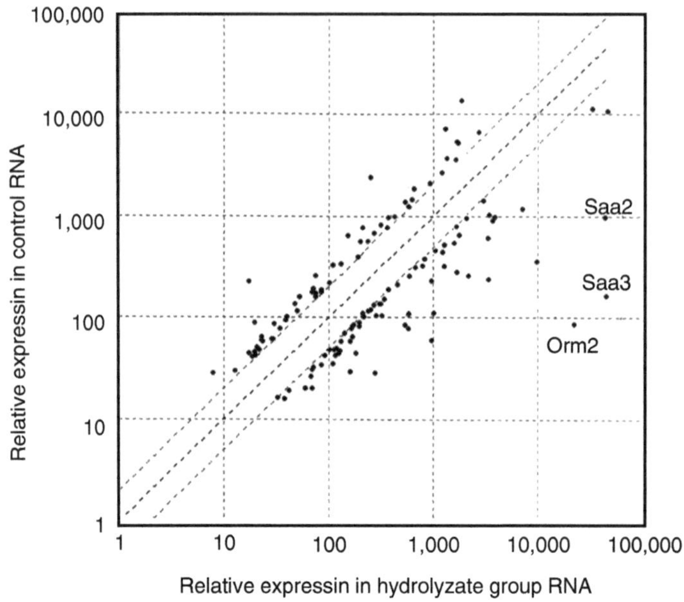

Fig. 23.5 The example result of the DNA microarray assay. This scatter plot represents the relative expression of transcripts in the control RNA sample (Y-axis) versus the relative expression of the transcript in the meat protein hydrolyzate treatment group RNA (X-axis). Each point represents the expression data of an individual transcript (significantly regulated genes)

of the sample functions as probes that hybridize only with the correct target sequence under high-stringency conditions. In microarray analysis, the probes covalently bind to a specific sequence of the bound-labeled probes, measured using microarray scanners. The expression data of tens of thousands of genes are visualized and statistically analyzed using specific bioinformatics tools.

We have studied bioactive peptides in enzymatic hydrolysates of porcine myofibrillar proteins (Arihara 2006). Hydrolysates and identified peptides exhibited ACE inhibitory and antioxidative activities. In addition to bioactivities in vitro, these hydrolysates and peptides have physiological activities in vivo. For example, they showed an antifatigue effect when orally administered to mice and rat in an experiment using a treadmill (Arihara et al. 2011b). Recently, we analyzed gene expression in mice liver by the DNA microarray method. The patterns of gene expression with or without oral administration of meat protein hydrolysates were analyzed. Microarray analysis revealed that the oral administration of the hydrolysate significantly regulated 91 genes (e.g., Saa2, Saa3, and Orm2) of 37,440 genes. Figure 23.5 shows the example result of the DNA microarray assay in our study.

Although studies for bioactive peptides generated from food proteins by DNA microarray are still limited, several examples of such studies are described here. The activity of milk whey proteins and their trypsin-hydrolyzed peptides to suppress the onset of inflammatory reactions was assayed by using the following mouse models:

LPS-induced sepsis and Concanavalin A-induced hepatis (Yamaji and Kume 2008). The patterns of gene expression before and after the onset of inflammation in these models were analyzed by means of DNA microarray methods and compared to those observed in mice fed milk whey proteins or their peptides. The analyses implied that whey protein and their peptides suppressed the onset of inflammation by influencing the gene expression of such factors as involved in signal transduction cascades for the production of pro-inflammatory cytokines.

Asn-Pro-Trp-Asp-Gln (107–111 of milk αs_2-casein) inhibits allergen permeation in Caco-2 cells as an in vitro human intestinal epithelial model. Yasumatsu and Tanabe (2010) have demonstrated that the mechanism underlying this inhibitory activity was examined in Caco-2 cells. Transepithelial resistance value increased in response to the addition of increasing Asn-PropTrp-Asp-Gln concentrations, which suggests that this peptide enhanced epithelial barrier function. Changes in mRNA expression by the addition of peptide were analyzed in Caco-2 cells using the microarray method. From the results of the microarray assay, it was suggested that Asn-Pro-Trp-Asp-Gln upregulated the expression of occludin in particular and enforced the tight junction barrier.

It has been reported that food-derived bioactive peptides express a variety of functions in vivo. Ile-Pro-Pro has been known as an ACE inhibitory and antihypertensive peptide derived from bovine milk protein (Nakamura et al. 1995). Huttunen et al. (2007) studied the in vitro effect of Ile-Pro-Pro on osteoblast proliferation and gene expression. They used UMR-106 osteosarcoma cells, human marrow-derived mesenchymal stem cells (hMSC), and osteoblasts differentiated from hMSC. Treatment with Ile-Pro-Pro increased UMR-106 cell and hMSC proliferation. The gene expression of hMSC-differentiated osteoblasts was analyzed by the DNA microarray method. Microarray analysis revealed that Ile-Pro-Pro upregulated 270 genes and downregulated 100 genes. Realtime PCR confirmed that Ile-Pro-Pro upregulated PTHrP, BMP-5 and CREB-5 and downregulated VDR and caspase-8. These results indicate that Ile-Pro-Pro possesses the potential to increase osteoblast proliferation, differentiation, and signaling.

Nagaoka (2012) studied lactostatin, a hypocholesteromic peptide derived from bovine milk beta-lactoglobulin, by the DNA microarray method. As they expected, the addition of lactostatin to HepG2 (human liver cells) increased the mRNA level of cholesterol 7α-hydroxylase, a key enzyme in cholesterol homeostasis. Also, genes of mitogen activated protein kinase (MAPK) cascades increased.

23.6.2 Proteomics

Although DNA microarray technology is a powerful tool for the study of gene–diet interactions, this technique has some problems or limitations (Zhang et al. 2010). One major problem is the nonreproducibility of gene expression profiles. Different conclusions could be drawn from the same experiment performed at different times or in different laboratories or platforms. Another major issue is the analysis of datasets and their interpretation. Analyses only providing gene lists with significant *p*-values are insufficient to fully understand the underlying biological mechanisms. Also, the changes in mRNA concentration do not necessarily result in differences in the concentration and/or activity of the encoded protein. Therefore, other emerging functional genomics techniques should be considered (Bidlack and Rodriguez 2012; Bagchi et al. 2010; Mine et al. 2009). It is by combining information from nutrigenomics, genomics, proteomics, metabolomics, and appropriate bioinformatics that this will be a viable approach to understanding all aspects and implications of nutrition-modulated beneficial homeostasis. Although a large number of reports regarding gene expression profiling of food-derived bioactive compounds has been studied, the proteomic approach of these compounds has still been limited.

Proteomics is the study of proteins expressed in a cells, tissue, or organism, including all protein isoforms and post-translational modifications (Bidlack and Rodriguez 2012; Bagchi et al. 2010; Mine et al. 2009). Proteomics is the large-scale analysis of a proteome expressed by a genome. A proteome is the entire complement of proteins synthesized in a biological system at a given time and under defined conditions, reflecting the expression of a set of specific genes in the situation pertaining to that time point. Proteomics allows for the high-throughput investigation of numerous proteins simultaneously in cells, tissues, or biological fluids. As an integral part of nutrigenomics, nutritional proteomics examines the effects of food components on protein expression and provides the potential to identify biomarkers sensitive to dietary interventions (Fuchs et al. 2005). The identification of biomarkers that reflect the outcome of peptide utilization will greatly benefit the field.

The numbers of reports on proteomic studies on bioactive peptides generated from food proteins is still very limited. In an attempt to search for novel biomarkers that could monitor the level of stress, we examined the influence of fatigue stress using a treadmill on the differential changes in the blood serum proteome in male SD rats using 2DE followed by peptide mass fingerprinting (PMF) analysis. Of numerous protein spots on 2-D gel maps of rat blood serum, a significantly upregulated protein spot was identified as creatine kinase by MALDI-TOF/MS (Akimoto et al. 2007). Research is now in progress by using this protein as a promising biomarker for antistress activities of a meat protein-derived peptides experiment.

23.6.3 Metabolomics

Metabolomics is a relatively new omics technology in nutritional approaches. It was originally defined as the quantitative measurement of time-related multiparametric metabolic responses of multicellular systems to pathophysiological stimuli or genetic modification (Whitfield and Kirwan 2010). The metabolome consists of the complete set of low molecular weight metabolites produced in a biological system, such as cells and tissues. Nutritional metabolomics has the potential to provide insight into biochemical changes after dietary intervention and to affect food safety issues pertaining to genetically modified food. Although metabolomics has contributed significantly to the omics revolution, a global description of human metabolism is impossible at this point due to limitations in current technologies and diversity among individuals in terms of age, gender, diet, lifestyle, health status, and other internal and external factors. Currently, the extent to which food components in the human diet induce changes in nutritional metabolic profiles is poorly understood. However, with technological advances, the challenges of applying metabolomics in nutrition research of bioactive peptides can be overcome.

23.7 Conclusions and Perspectives

In past decades, much information has been accumulated regarding the bioactive peptides generated from food proteins. In addition to conventional research methods, the nutrigenomics approach promotes the understanding of bioactive peptides. With the advent of new proteomic and genomic techniques, the mechanisms underlying many of the biological properties of food protein-derived peptides will be revealed.

In proteomics, all proteins expressed in a cell or tissue are analyzed to identify the presence or absence of some key proteins. Although proteins are routinely separated by 2-D electrophoresis in proteomics, the physiologically interesting small bioactive peptides are neglected in most proteomic studies. Consequently, the concept of peptidomics was introduced (Boonen 2009; Mine 2009). The purpose of peptidomics is the identification of whole peptidome of a cell, tissue, or organism, because peptides (e.g., hormones, cytokines, and growth factors) play critical roles in many physiological processes. The field of peptidomics is relatively new and has the potential to progress in future with the advent of high-throughput MS-based technologies coupled with bioinformatics and genomic databases.

Bioinformatics involves the integration of computers, software tools, and databases in an effort to address biological questions in the omics, including genomics, proteomics, transcriptomics, metabolomics, and peptidomics. Systems biology involves the integration of genomics, proteomics, and bioinformatics information to create a whole-system view of a biological entity. By combining information from nutrigenomics, genomics, proteomics, metabolomics, peptidomics, and appropriate bioinformatics, it would be possible to understand all aspects and implications of bioactive peptides derived from food proteins.

References

Affolter M, Raymond F, Kussmann M (2009) Omics in nutrition and health research. In: Mine Y, Miyashita K, Shahidi F (eds) Nutrigenomics and proteomics in health and disease. Wiley-Blackwell, Ames, pp 11–29

Aguilar MI (2004) Reversed-phase high-performance liquid chromatography. In: Aguilar MI (ed) HPLC of peptides and proteins. Humana Press, Totowa, pp 9–22
Aito-Inoue M, Lackeyram D, Fan MZ, Sato K, Mine Y (2007) Transport of tripeptide, Gly-Pro-Hyp, across the intestinal brush border membrane of porcine. J Peptide Sci 13:468–474
Akimoto M, Namioka S, Arihara K, Tomita K, Ishikawa S, Itoh M (2007) Novel stress-biomarker and its application. Japan Patent (No. 2007–93597)
Aleixandre A, Miguel M (2012) Food proteins and peptide as bioactive agents. In: Hettiarachchy NS, Sato K, Marshall MR, Kannan A (eds) Bioactive food proteins and peptides, applications in human health. CRC Press, Boca Raton, pp 131–180
Arihara K (2006) Functional properties of bioactive peptides derived from meat proteins. In: Nollet LML, Toldrá F (eds) Advanced technologies for meat processing. CRC Press, Boca Raton, pp 247–273
Arihara K (2012) Meat-based bioactive compounds and functional meat products. Food Sci Technol 26:26–28
Arihara K, Ohata M (2011) Functional meat products. In: Saarela M (ed) Functional foods: concept to products, 2nd edn. Woodhead, Cambridge, pp 512–533
Arihara K, Nakashima Y, Mukai T, Ishikawa S, Itoh M (2001) Peptide inhibitors for angiotensin I-converting enzyme from enzymatic hydrolysates of porcine skeletal muscle proteins. Meat Sci 67:434–437
Arihara K, Ishikawa S, Itoh M (2011a) *Bifidobacterium* growth promoting peptides derived from meat proteins. Japan Patent (No. 4726129)
Arihara K, Tomita K, Ishikawa S, Itoh M, Akimoto M, Sameshima T (2011b) Anti-fatigue peptides derived from meat proteins. Japan Patent (No. 4828890)
Astley S, Penn L (2009) Design of human nutrigenomics studies. Wageningen Academic, Wageningen
Bagchi D, Lau FC, Bagchi M (2010) Genomics, proteomics and metabolomics in nutraceuticals and functional foods. Wiley-Blackwell, Ames
Bidlack WR, Rodriguez RL (2012) Nutritional genomics, the impact of dietary regulation of gene function on human disease. CRC Press, Boca Raton
Boonen K (2009) Peptidomics of the mouse, unraveling the role of bioactive peptides in intercellular signaling. Lambert Academic, Saarbrücken
Brody EP (2000) Biological activities of bovine glycomacropeptide. Br J Nutr 84:S39–S46
Bütikofer U, Meyer J, Sieber R, Walther B, Wechsler D (2008) Occurrence of the angiotensin-converting enzyme-inhibiting tripeptides Val-Pro-Pro and Ile-Pro-Pro in different cheese varieties of Swiss origin. J Dairy Sci 91:29–38
Chen H-M, Muramoto K, Yamauchi F (1995) Antioxidative activity of designed peptides based on the antioxidative peptide isolated from digests of a soybean protein. J Agric Food Chem 44:2619–2623
Chiba H, Yoshikawa M (1986) Biologically functional peptides from food proteins: new opioid peptides from milk proteins. In: Feeney RE, Whitaker JR (eds) Protein tailoring for food and medical users. Marcel Dekker, New York, pp 123–153
Cushman DW, Cheung HS (1971) Spectrophotometric assay and properties of the angiotensin converting enzyme of rabbit lung. Biochem Pharmacol 20:1637–1648
Dainty R, Blom H (1995) Flavor chemistry of fermented sausages. In: Campbell-Platt G, Cook PE (eds) Fermented meats. Blackie Academic & Professional, Glasgow, pp 176–193
Fiat A-M, Migliore-Samour D, Jolles P, Drout L, Collier C, Caen J (1993) Biologically active peptides from milk proteins with emphasis on two examples concerning antithrombotic and immunomodulating activities. J Dairy Sci 76:301–310
Fuchs D, Winkelmann I, Johnson IT, Mariman E, Wenzel U, Daniel H (2005) Proteomics in nutrition research: principles, technologies and applications. Br J Nutr 94:302–314
Fujita H, Yokoyama K, Yoshikawa M (2000) Classification and antihypertensive activity of angiotensin I-converting enzyme inhibitory peptides derived from food proteins. J Food Sci 65:564–569

Gagnaire V, Molle D, Herrouin M, Leonil J (2001) Peptides identified during emmental cheese ripening: origin and proteolytic systems involved. J Agric Food Chem 49:4402–4413

Ghassem M, Arihara K, Babji AS, Said M, Ibrahim S (2011) Purification and identification of ACE inhibitory peptides from Haruan (Channa striatus) myofibrillar protein hydrolysate using HPLC-ESI-TOF MS/MS. Food Chem 129:1770–1777

Gobbetti M, Ferranti P, Smacchi E, Goffredi F, Addeo F (2000) Production of angiotensin-I converting-enzyme-inhibitory peptides in fermented milks started by *Lactobacillus delbrueckii* subsp. *bulgaricus* SS1 and *Lactococcus lactis* subsp. *cremoris* FT4. Appl Environ Microbiol 66:3898–3904

Gobbetti M, Minervini F, Rizzello CG (2007) Bioactive peptides in dairy products. In: Hui YH (ed) Handbook of food products manufacturing: health, meat, milk, poultry, seafood, and vegetables. Wiley, Hoboken, pp 489–517

Gomez-Ruiz JA, Ramos M, Recio I (2002) Angiotensin-converting enzyme-inhibitory peptides in Manchego cheeses manufactured with different starter cultures. Int Dairy J 12:697–706

Hammes WP, Haller D, Gänzle MG (2003) Fermented meat. In: Farnworth ER (ed) Handbook of fermented functional foods. CRC Press, Boca Raton, pp 251–275

Hernändez-Ledesma B, Amigo L, Ramos M, Recio I (2004) Angiotensin converting enzyme inhibitory activity in commercial fermented products. J Agric Food Chem 52:1504–1510

Hertog MGL (1996) Epidemiological evidence on potential health properties of flavonoids. Proc Nutr Soc 55:385–387

Hettiarachchy NS, Sato K, Marshall MR, Kannan A (2012) Bioactive food proteins and peptides, application in human health. CRC Press, Boca Raton

Hipkiss AR, Brownson CA (2000) A possible new role for the anti-aging peptide carnosine. Cell Mol Life Sci 57:747–753

Huttunen MM, Pekkinen M, Ahlstrom MEB, Lamberg-Allardt CJE (2007) Effects of bioactive peptides isoleucine-proline-proline (IPP), valine-proline-proline (VPP) and leucine-lysine-proline (LKP) on gene expression of osteoblasts differentiated from human mesenchymal stem cell. Br J Nutr 98:780–788

Ibrahim HR, Sugimoto Y, Aoki T (2000) Ovotransferrin antimicrobial peptide (OTAP-92) kills bacteria through a membrane damage mechanism. Biochim Biophys Acta 1523:196–205

Kannan A, Hettiarachchy N, Marshall M (2012) Food proteins and peptide as bioactive agents. In: Hettiarachchy NS, Sato K, Marshall MR, Kannan A (eds) Bioactive food proteins and peptides, applications in human health. CRC Press, Boca Raton, pp 1–28

Kaput J (2007) Developing the promise of nutrigenomics through complete science and international collaborations. Forum Nutr 60:209–223

Kim S-K, Wijesekara I, Park EY, Matsumura Y, Nakamura Y, Sato K (2012) Proteins and peptide as antioxidants. In: Hettiarachchy NS, Sato K, Marshall MR, Kannan A (eds) Bioactive food proteins and peptides, applications in human health. CRC Press, Boca Raton, pp 97–115

Koohmaraie M (1994) Muscle proteinases and meat aging. Meat Sci 36:93–104

Korhonen H, Pihlanto A (2007) Bioactive peptides from food proteins. In: Hui YH (ed) Handbook of food products manufacturing: health, meat, milk, poultry, seafood, and vegetables. Wiley, Hoboken, pp 5–38

Li GH, Le GW, Shi YH, Shrestha S (2004) Angiotensin I-converting enzyme inhibitory peptides derived from food proteins and their physiological and pharmacological effects. Nutr Res 24:469–486

Lottspeich F (1979) Novel opioid peptides derived from casein (β-Casomorphins) I. Isolation from bovine casein peptone. Hoppe-Seyler's Z Physiol Chem 360:1211–1216

Mano H, Shimizu J, Wada M (2009) Microarrays: a powerful tool for studying the functions of food and its nutrients. In: Mine Y, Miyashita K, Shahidi F (eds) Nutrigenomics and proteomics in health and disease. Wiley-Blackwell, Ames, pp 341–349

Masotti A, Sacco LD, Bottazzo GF, Alisi A (2010) Microarray technology: a promising tool in nutrigenomics. Crit Rev Food Sci Nutr 50:693–698

Meisel H, Walsh DJ, Murry B, FitzGerald RJ (2005) ACE inhibitory peptides. In: Mine Y, Shahidi F (eds) Nutraceutical proteins and peptides in health and disease. CRC Press, Boca Raton, pp 269–315
Mellander O (1950) The physiological importance of the casein phosphopeptide calcium salts II: Peroral calcium dosage in infants. Acta Soc Med Uppsala 55:247–255
Mine Y (2009) Peptidomics. In: Mine Y, Miyashita K, Shahidi F (eds) Nutrigenomics and proteomics in health and disease. Wiley-Blackwell, Ames, pp 375–386
Mine Y, Miyashita K, Shahidi F (2009) Nutrigenomics and proteomics in health and disease. Wiley-Blackwell, Ames
Nagaoka S (2012) Peptide-lipid interactions and functionalities. In: Hettiarachchy NS, Sato K, Marshall MR, Kannan A (eds) Food proteins and peptides: chemistry, functionality, interactions and commercialization. CRC Press, Boca Raton, pp 263–276
Nagaoka S, Futamura Y, Miwa K, Awano T, Yamauchi K, Kanamaru Y, Kojima T, Kuwata T (2001) Identification of novel hypocholesterolemic peptides derived from bovine β-lactoglobulin. Biochem Biophys Res Commun 281:11–17
Nakamura Y, Yamamoto N, Sakai K, Takano T (1995) Antihypertensive effect of sour milk and peptides isolated from it that are inhibitors to angiotensin I-converting enzyme. J Dairy Sci 78:1253–1257
Nakashima Y, Arihara K, Sasaki A, Ishikawa S, Itoh M (2002) Antihypertensive activities of peptides derived from porcine skeletal muscle myosin in spontaneously hypertensive rats. J Food Sci 67:434–437
Nishimura T, Rhue MR, Okitani A, Kato H (1988) Components contributing to the improvement of meat taste during storage. Agric Biol Chem 52:2323–2330
Nurminen M-L, Sipola M, Kaarto H, Pihlanto-Leppälä A, Piiola K, Korpela R, Tossavainen O, Korhonen H, Vapaatalo H (2000) α-Lactophin lowers blood pressure via radiotelemetry in normotensive and spontaneously hypertensive rats. Life Sci 66:1535–1543
Oshima G, Shimabukuro H, Nagasawa K (1979) Peptide inhibitors of angiotensin I-converting enzyme in digests of gelatin by bacterial collagenase. Biochim Biophys Acta 566:128–137
Otani H, Hata I (1995) Inhibition of proliferative responses of mouse spleen lymphocytes and rabbit Peyer's patch cells by bovine milk caseins and their digests. J Dairy Res 62:339–348
Owusu-Aspenten R (2010) Bioactive peptides, application for improving nutrition and health. CRC Press, Boca Raton
Pihlanto A, Korhonen H (2003) Bioactive peptides and proteins. Adv Food Nutr Res 47:175–276
Pouliot Y, Gauthier SF, Groleau PE (2006) Membrane-based fractionation and purification strategies for bioactive peptides. In: Mine Y, Shahidi F (eds) Nutraceutical proteins and peptides in health and disease. CRC Press, Boca Raton, pp 639–658
Saarela M (2011) Functional foods, concept to product, 2nd edn. Woodhead, Cambridge
Saiga A, Tanabe S, Nishimura T (2003) Antioxidant activity of peptides obtained from porcine myofibrillar proteins by protease treatment. J Agric Food Chem 51:3661–3667
Saito T, Nakamura T, Kitazawa H, Kawai Y, Itoh T (2000) Isolation and structural analysis of antihypertensive peptides that exist naturally in Gouda cheese. J Dairy Sci 83:1434–1440
Sato K, Hashimoto K (2012) Large-scale fractionation of biopeptides. In: Hettiarachchy NS, Sato K, Marshall MR, Kannan A (eds) Food proteins and peptides: chemistry, functionality, interactions and commercialization. CRC Press, Boca Raton, pp 395–408
Shen TL, Noon KR (2004) Liquid chromatography-mass spectrometry and tandem mass spectrometry of peptides and proteins. In: Aguilar MI (ed) HPLC of peptides and proteins. Humana Press, Totowa, pp 111–139
Thomson-Smith LD (2010) Nutrigenomics, nutrition and DNA could go hand in hand. Fastbook, Beau Bassin/ Mauritius
Toldrá F (2004) Dry. In: Jensen WK, Devine C, Dikeman M (eds) Encyclopedia of meat sciences. Elsevier, Oxford, pp 360–365
Tomita M, Takase M, Bellamy WR, Shimamura S (1994) A review, the active peptide of lactoferrin. Acta Paediatr Japan 36:585–591

Vegarud GE, Langsrud T, Svening C (2000) Mineral-binding milk proteins and peptides; occurrence, biochemical and technological characteristics. Br J Nutr 84:91–98

Webster NR (1998) Opioid and immune system. Br J Anesth 81:835–836

Whitfield P, Kirwan J (2010) Metabolomics: an emerging postgenomic tool for nutrition. In: Bagchi D, Lau FC, Bagchi M (eds) Genomics, proteomics and metabolomics in nutraceuticals and functional foods. Wiley-Blackwell, Ames, pp 271–285

Yamaji T, Kume H (2008) Hepatoprotective effects of whey protein and whey peptides on hepatitis. Milk Sci 56:115–118

Yasumatsu H, Tanabe S (2010) The casein peptide Asn-Pro-Trp-Asp-Gln enforces the intestinal tight junction partly by increasing occludin expression in Caco-2 cells. Br J Nutr 104:951–956

Young D, Mine Y (2009) Functional bioactive proteins and peptides in nutrigenomics. In: Mine Y, Miyashita K, Shahidi F (eds) Nutrigenomics and proteomics in health and disease. Wiley-Blackwell, Ames, pp 129–144

Zhang X, Wang W, Xiao K (2010) Novel omics technologies in nutraceutical and functional food research. In: Bagchi D, Lau FC, Bagchi M (eds) Genomics, proteomics and metabolomics in nutraceuticals and functional foods. Wiley-Blackwell, Ames, pp 11–22

Chapter 24
The Role of Proteomics in the Discovery of Marker Proteins of Food Adulteration

Gianfranco Mamone, Gianluca Picariello, Chiara Nitride, Francesco Addeo, and Pasquale Ferranti

Abbreviations

1D	Mono-dimensional
2-D	Two-dimensional
AQUA	Absolute quantification
BLG	Beta-lactoglobulin
CE	Capillary electrophoresis
CID	Collision-induced dissociation
CML	Nε-carboxymethyl-lysine
CZE	Capillary zone electrophoresis
DIGE	Differential in-gel electrophoresis
ELISA	Enzyme-linked immunosorbent assay
ESI	Electrospray ionization
GC	Gas chromatography
GMO	Genetically modified organism

G. Mamone • G. Picariello
Istituto di Scienze dell'Alimentazione (ISA) – CNR,
Via Roma 52 A/C, Avellino 83100, Italy

C. Nitride •
Dipartimento di Scienza degli Alimenti, University of Naples "Federico II",
Parco Gussone, Portici I 80055, Italy

P. Ferranti (✉) • F. Addeo
Istituto di Scienze dell'Alimentazione,
Consiglio and Pasquale Ferranti Nazionale delle Ricerche (CNR),
Via Roma 52 A/C, Avellino 83100, Italy

Dipartimento di Scienza degli Alimenti, University of Naples "Federico II",
Parco Gussone, Portici I 80055, Italy
e-mail: ferranti@unina.it

F. Toldrá and L.M.L. Nollet (eds.), *Proteomics in Foods: Principles and Applications*, Food Microbiology and Food Safety 2, DOI 10.1007/978-1-4614-5626-1_24,

HPLC	High-performance liquid chromatography
ICAT	Isotope-coded affinity tag
IEF	Isoelectric focusing
IEX	Ion exchange
IT	Ion trap
iTRAQ	Isobaric tags for relative and absolute quantitation
LAL	Lysinoalanine
LC	Liquid chromatography
LMW	Low molecular weight
LOD	Limit of detection
LOQ	Limit of quantification
MALDI	Matrix-assisted laser desorption ionization
MM	Mechanically recovered meat
MP	Milk powders
MRM	Multiple reaction monitoring
MS/MS and MS^n	Tandem mass spectrometry
MS	Mass spectrometry
MudPIT	Multidimensional protein identification technology
PAGE	Polyacrylamide gel electrophoresis
PCR	Polymerase chain reaction
PDO	Protected denomination of origin: RP-HPLC: reverse-phase high-performance liquid chromatography
SDS	Sodium dodecyl-sulphate
SILAC	Stable isotope labeling with amino acids in cell culture
SIM	Single ion monitoring
SRM	Selected reaction monitoring
TOF	Time-of-flight
UPLC-FT-ICR-MS	Ultrahigh-pressure liquid chromatography high-resolution Fourier-transform ion cyclotron resonance mass spectrometry

24.1 Introduction

Adulteration in various consumer sectors is now commonly practiced. Raw materials and finished foods are not exempt, and according to several studies, this practice is increasing (French General Directorate for Fair Trading 2001). Adulterant detection and authenticity testing of foods, including dairy and meat products, eggs, cereals, legumes, beverages, and fruit products, are important for value assessment, to check unfair competition and, most important, to ensure consumer protection against fraudulent practices commonly observed in unscrupulous trade.

However, the consequences of food adulteration go very far beyond mere economical factors, and are much more complex. Actually, deceitful adulteration of food products is an objectionable practice also for health reasons, because consumption of products containing undeclared constituents may cause intoxication or problems such as allergy and intolerance in sensitized individuals. For these reasons, the search for methods to exactly define the composition of the various products in order to distinguish them from similar but altered ones is among the most important issues in food and nutrition science. However, the definition of a food at the molecular level is not quite easy. Foods are, in general, complex matrices of very different chemical compound classes, reflecting the complexity of the animal or vegetal organism they derive from, with unique compositional and biochemical characteristics. This means that traditional analytical chemistry methods, traditionally and efficiently applied to discover gross food alterations (one for all, milk dilution by water addition) are inadequate to challenge the complexity of most food matrices at the molecular level.

Just as for any other biological system, the recently born "omic" sciences, intended as the systematic definition of subsets of specific biomolecules, also apply to foods. Thus, for instance, food proteomics include the systematic and complete cartography of the proteins, whereas metabolomics studies the complement of small molecules (low molecular weight arbitrarily assigned below 1–2 kDa) of a foodstuff. Phosphoproteomics, glycomics, and lipidomics concern specific classes of constituents of the systems under analysis. In particular, in consideration of the relevance for biological systems of lipids and sugars, lipidomics and glycomics are emerging as self-standing "omics" from the wider field of metabolomics.

The proteome is defined as the totality of proteins contained in a biological sample in a given state. Thus, in contrast to the static genome (with the exclusion of the epigenetic events), the proteome is highly dynamic, influenced not only by the genome but also by myriad external factors, such as metabolic state, treatments, and various interactions with other systems. Several recent papers have been focused on particular aspects of proteomics and of related "omic" sciences in the field of food analysis, dedicated to food composition and safety (Herrero et al. 2010; Davies 2010), to the search for marker proteins for food authentication, and to identification of food allergens (Picariello et al. 2011).

This chapter critically presents the newest exemplificative achievements of the "omic" sciences, with special emphasis on proteomics and peptidomics, generated by the synergistic efforts of researchers in the fields of analytical chemistry, in particular mass spectrometry, with those working in food technology and biotechnology.

24.2 Methods of Food Adulteration Detection

The impressive improvement of the last decades in food production, processing, and packaging has been paralleled by the escalation of deceitful practices of food adulteration.

The issue of food authenticity has an extremely high relevancy both for human health and for the food market economy and imposes the urgent need to improve the robustness of the available analytical methods for its assessment. Growing consumer awareness of food safety and quality, increased demand for legal regulation and adequate labeling, together with the evolution of the deceptive strategies are fueling the development of up-to-date procedures of food control that have to be developed, standardized, and validated.

Briefly, a food product can be adulterated in the following ways: (1) a food component is replaced wholly or in part with a cheaper ingredient or with forbidden materials; (2) any constituent is wholly or in part subtracted from the food matrices; (3) foodstuffs are incorrectly processed or packed; (4) it has been produced with diseased vegetable or animal organisms; (5) in general, any of the quality parameters fall beyond the prescribed standards. Aside from the undeclared addition of preservatives or potentially harmful ingredients, that have to be assessed by the direct detection of the target molecule(s), the use of undeclared ingredients can be estimated by monitoring the biomolecular asset of the food commodity.

Over the years, an arsenal of analytical methods, many based on morphological/anatomical analysis, organoleptic markers (odor, color, texture), or chemical testing, has been developed to authenticate foods and to check for adulterants. In general, the three basic detection strategies used for demonstrating adulteration in food or agricultural commodity include: (1) demonstrating the presence of a foreign substance or a marker in the commodity, (2) demonstrating that one or more component levels deviate from normal, and (3) demonstrating that a profile is unlikely to occur. Among these, the strategy of detection of adulterans detection by the demonstration of the presence of foreign substances or a marker is considered as the most reliable and simplest (Wilhelmsen 2004, 2006).

For these reasons, in the last years, new emerging technologies are being applied to food authentication analysis. Determination of stable isotope ratio, especially on trace elements, provides a stable isotope signature useful to establish a close link between products and their environment (White et al. 1998). Most recent approaches specifically target proteins or DNA for discovering the signs of raw material derived from foreign organisms in food products. DNA-based methods consist in the polymerase chain reaction (PCR) amplification of DNA fragments arising from extraneous animal or vegetable organisms. DNA-based molecular tools have been proved useful for adulterant detection especially when the adulterants are biological substances (Lum and Hirsch 2006). In particular, it has been successful in species or variety identification as it can identify very subtle changes in the single base pair. In this way a specific DNA sequence can be identified and/or DNA fingerprints can be obtained. However, it is obvious that these methods are complicated when ingredients arising from several species, that often are taxonomically related, occur simultaneously. Furthermore, DNA-based analytical methods are of hardly any usefulness to establish the use of noncompliant processing of the raw material.

Although the detection of DNA markers benefits from having well-defined target analytes and the combined use of database analysis and experimental specificity should lower the chance of obtaining false positive, the techniques relying on the phenotypic expression of specific protein or metabolite markers are less laborious and, in most cases, more reliable.

Monitoring of protein markers generally relies on immunochemical techniques. Commercially available techniques of this type are lateral flow devices or dipsticks that are used for rapid screening and enzyme-linked immunosorbent assays (ELISA) that also provide semiquantitative determinations. Typical limit of detections (LOD) of the tests based on ELISA kits are in the range of 1–5 ppm. Major concerns of the immunochemical methods consist in the fact that the epitopes, that are detected, are usually not well characterized and cross-reactivity with matrix components can result in false positive determinations. The reliability of the detection strongly depends on specificity and stability of the employed antibodies and can be affected by the changes induced on proteins by thermal or other technological treatments. Furthermore, food processing can modify both linear or conformational antigenic sequences, although altering the antibody reactivity. In addition, many protein targets may be underestimated or escape the most commonly utilized sandwich ELISA-based tests.

Chemical/biochemical techniques such as high performance liquid chromatography (HPLC), gas chromatography (GC), capillary electrophoresis (CE), mono-dimensional (1D) or two-dimensional (2-D) polyacrylamide gel electrophoresis (PAGE), capillary electrophoresis, and so on, as well as ELISA have also been proved to be useful in component identification and adulterant detection in foods. However, although they are of considerable value in certain instances of adulterant detection and in routine analyses, these methods are merely descriptive as they compare a profile or a measured value with that expected for a given genuine product and therefore cannot explain the causes of the altered outcome at the molecular level. In other terms, the appearance/disappearance/shift of electrophoretic bands or chromatographic peaks compared to a reference food cannot be considered a definite diagnostic of an occurred food alteration, as the variation of the band/peak could be due to normal food variability (false positive). On the opposite side, an adulterant might be masked by comigration/coelution with a normal food constituent (false negative). In light of this, also conventional electrophoretic and chromatographic techniques alone, routinely used in this kind of analysis, in spite of the tremendous improvements in resolving power and sensitivity due to the technological advances, must be considered inadequate when facing the problem of describing the complex composition of natural or adulterated foods.

Given the limitations of classically used methods, it is clear that confirmatory strategies are also required to provide an unambiguous identification of markers of foreign food components. The proteomic approach can overcome these limitations. The core of proteomic science is the "omics" technologies that rely on well-established analytical platforms, in particular on mass spectrometry (MS) techniques.

24.3 Mass Spectrometry in Food Component Detection and Characterization

The MS-based proteomic/peptidomic approach can represent the ideal solution to such an analytical challenge. Major advantages of electrospray ionization (ESI) and matrix-assisted laser desorption ionization (MALDI) MS techniques are the sensitivity and specificity of the analytical response in the detection of foreign proteins

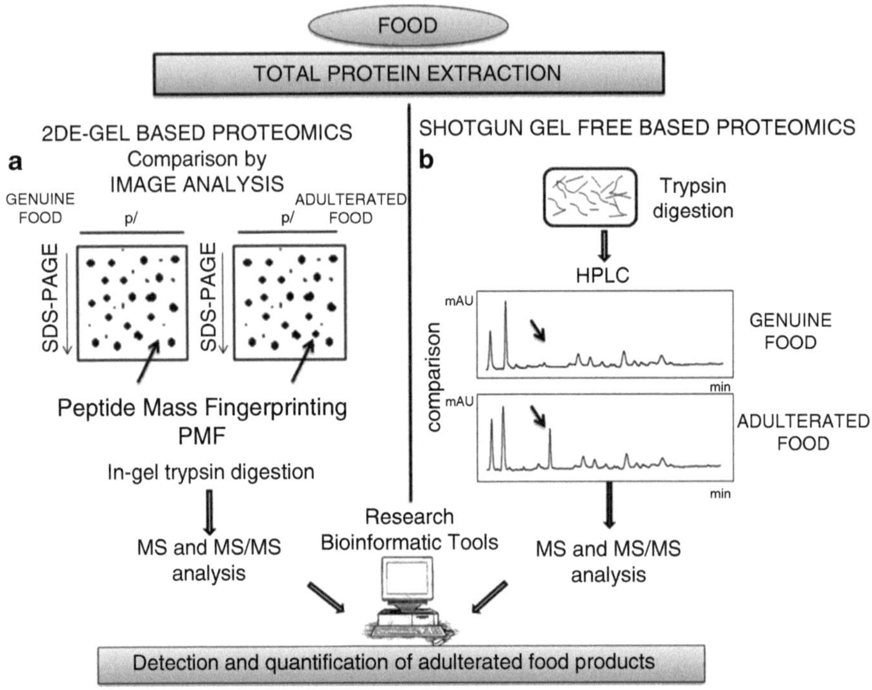

Fig. 24.1 Gel-free and gel-based proteomic approaches for the detection and the quantification of adulterants in food: (**a**) the gel-based approach allows detection of adulterated foods by comparison of protein profiles obtained by 2-D gel-electrophoresis separation followed by peptide mass fingerprinting for target proteins identification; (**b**) the gel-free approach analyzed by LC-MS/MS the total extracted protein digested with trypsin and followed by the identification of target proteins by online research in protein databases

or peptides. It is important to note that the MS response does not depend on the antibody specificity as in ELISA and is only marginally affected by interfering species.

As most foods are highly complex matrices, pre-fractionation of protein is required prior to MS analysis (Mamone et al. 2009); see Fig. 24.1. To date, two-dimensional electrophoresis (2DE) has proven to be a reliable and efficient method to separate in one step, a large number of proteins (Fig. 24.1a) (Gorg et al. 2005). An alternative to electrophoretic methods is liquid chromatography (LC) coupled with MS analysis to identify proteins directly from complex mixtures (Fig. 24.1b).

MS experiments can be addressed either to measure the molecular mass of a protein/derived-peptides (MS1) or to determine additional structural details such as the amino acid sequence and post-translational events, also including process-induced modifications, using tandem MS (MS/MS or MS^n).

The monitoring of intact proteins by the so-called "top-down" approaches is still of limited analytical throughput and efficiency, especially when complex foodstuffs are handled. Direct MALDI analysis of the protein extracts may provide fast and accurate information on food composition and authenticity. Exemplary cases are the analysis of gluten residues in declared gluten-free foods and beverages, exploited by several studies (Hernando et al. 2008; Mamone et al. 2011), and the authenticity assessment of commercial fish species (Mazzeo et al. 2008).

Peptide-centric methodologies ("bottom-up" techniques) that consist in the analysis of proteolitically generated peptides introduce a series of advantages. The bottom-up "discovery-driven" proteomic experiments rely on either 2DE or gel-free procedures and can provide the almost complete protein suite of a system, largely covering the dynamic range of concentrations that usually occur in adulterated foods. In targeted experiments, when the nature of the adulterant is at least in part supposed, it can be defined as a subset of "signature" peptides specific to each foreign protein. These peptides are currently referred to as "proteotypic" peptides. Eligible "proteotypic" peptides are sequences specific to a given protein, generally obtainable through a tryptic digestion, with optimal molecular sizes (10–20 residues long) and high ionizability to be identified and sequenced by mass spectrometry. In addition, the ideal proteotypic peptide should exhibit the following properties: (1) high efficiency to ionize and to fragment; (2) uniqueness for the target protein(s); (3) absence of amino acid residues susceptible to chemical modifications such as Cys (reactive thiol group), Met (oxidable), Asn and Gln (deamidation), N-terminal Gln (pyroglutamate); (4) absence of post-translational modifications and of sequences that are prone to missed proteolytic cleavage (e.g., sequences such as KRXX or XXRK in the case of trypsin).

Once the panel of "proteotypic" peptides is defined, a single LC-ESI-MS/MS analysis of a tryptic digest of the protein fraction extracted from a complex food matrix can provide unambiguous identification of the protein target. The identification relies on three independent and convergent indications: chromatographic retention time, peptide mass, and sequence information (MS/MS experiments). Multiple reaction monitoring (MRM)–MS techniques can achieve sensitivities unparalleled by immunochemical techniques. The use of "proteotypic" peptides as "analytical substitutes" of the parent proteins enhances the potentiality of the MS-based methods, improving sensitivity and enabling the possibility of several confirmations through the monitoring of different peptides and related transitions (multianalyte methods). In addition, by designing appropriate internal standards by peptide synthesis (i.e., synthetic peptides that contain stable modifications, in order to introduce minimal molecular mass shift with respect to the target peptides without affecting ionizability) LC-ESI-MS/MS can provide robust and repeatable quantitative data.

The LC-ESI-MS/MS peptide-centric approach has recently been applied to detection of higher-value milk adulteration with cheaper bovine milk (Cuollo et al. 2010), as detailed in the following section, as well as to monitoring contaminating allergens in processed peanut products (Chassaigne et al. 2007). Also, taking advantage of the recently introduced high-resolution/wide dynamic range LTQ-Orbitrap instrument, coupled with 1D nanoflow RP-HPLC, 158 proteins have been recently identified in hen's egg white by shotgun proteomics, also including several known egg allergens (Mann and Mann 2011). It is almost superfluous to underline how the investigation methodology illustrated by these examples is revealing itself to be invaluable in detecting either intentional or involuntary food contamination by allergenic ingredients.

In the shotgun approach (Fig. 24.1b), the information about the molecular weight of the intact proteins gets lost. However, a careful analysis of MS/MS spectra can turn out to be effective in discriminating foreign proteins. To reduce the complexity of the sample, multidimensional chromatography coupled with mass spectrometry (2-D LC-MS or MudPIT, multidimensional protein identification technology) can be employed. This approach provides a higher separation power than 1D chromatography, in addition to an extended measured dynamic concentration range to analyze complex proteomic samples. Various LC methods in the first dimension prior to RP-HPLC in the second dimension have been proposed. A preliminary ion-exchange (IEX) separation mode has become the most popular implementation (Washburn et al. 2001; Motoyama et al. 2007). At present, multidimensional chromatography-shotgun techniques are largely utilized in proteomics. Nevertheless, up to now only a few articles report on food contaminant identification using this approach. Among the methods related to these, a multiplexed chromatography consisting of sequential HILIC and TiO_2 enrichment and a chemical-based (hydrazide capture) strategy prior to nanoflow RP-HPLC-ESI-LTQ-Orbitrap MS/MS have been addressed to identify glycoproteins of wine potentially cross-reactive with plant allergens (Palmisano et al. 2010). With a slightly different approach, consisting of a preliminary chromatography step at the protein level, followed by RP-HPLC/MS/MS analysis at the peptide level, soybean proteins willingly or accidentally added to meat products could be detected (Leitner et al. 2006). Similarly, protein peanut allergens were identified in breakfast cereals by a 2-D chromatography method consisting of an affinity capture of proteins on immuno-beads followed by tryptic digestion and RP-HPLC-ESI-IT-MS/MS (Careri et al. 2008).

Also capillary electrophoresis mass spectrometry (CE-MS) is a promising methodology in proteomic applications because it is an alternative to 2DE and to chromatographic techniques for complex mixture analysis. It combines high efficiency separation of intact proteins with unique information of MS (Rob et al. 2007). The most recent applications of CE-MS to the analysis of proteins in food, which have been recently reviewed (Garcia-Canas and Cifuentes 2008) include the characterization of protein extracts to provide information about the protein composition with the aim of: (1) determining the origin of the ingredients, the technological process employed to produce a particular food, or the quality of a protein extract; (2) detecting high-quality products adulterated with cheaper products; and (3) monitoring and optimizing technological processes. In addition to these applications, some approaches are worth mentioning, such as those based on CE coupled with MS and developed to predict the amino acid composition of protein hydrolysates (Simó et al. 2005; Tessier et al. 2005), as well as the novel procedure that provides simplified 2-D mapping of CE-MS data in order to simplify the discovery and detection of specific markers (Erny and Cifuentes 2007).

Decisive support to all of the methods applied in proteomic analysis is provided by bioinformatic tools to manage the huge amount of MS/MS datasets. The identifications are usually administered automatically via powerful, commercially accessible software programs such as Mascot and SEQUEST in combination with continually updated public databases such as those maintained by the NCBI and SwissProt (Johnson et al. 2005).

24.4 Proteomic-Based Methods for Food Protein Quantification

Because of its very low limits of detection and quantification (low-ppb range), MS analysis is the basis of modern proteomic methods for quantitative investigation of biological systems. Several conceptually different MS-based proteomic strategies, each with advantages and disadvantages have emerged in the last two decades.

Numerous MS-based methodologies for both relative and absolute quantification are widely used in proteomic studies, but all referable to two main approaches: in the first one, the analyte and its reference standard are the intact proteins; in the second one, the analyte is a peptide arising from protein digestion by proteolytic enzymes.

At the beginning of the "proteomic era," the first established method applied for differentially expressed proteins was based on 2DE fractionation followed by MS analysis (Pietrogrande et al. 2006) (Fig. 24.1a). Classical 2DE employs a two-step separation whereby denatured proteins are first separated according their isoelectric point (isoelectric focusing, IEF) followed by separation as a function of molecular mass (sodium dodecyl-sulphate–polyacrylamide gel electrophoresis, SDS-PAGE). After gel staining, comparison of sample control and case proteomes allows detection of differentially expressed proteins. The gel image analysis requires the use of computer programs that are continuously being developed and improved (i.e., Melanie Software (Amersham Biosciences), PDQuest (Bio-Rad), Progenesis (Nonlinear Dynamics)). The spots of interest are then excised and in-gel digested with trypsin prior to MS (peptide mass fingerprint) or MS/MS analysis in order to determine the protein identity.

2DE represents an excellent tool for mapping differences in protein expression, but certain drawbacks limit its use. First, the gel staining methods are either not sensitive enough or have a limited linearity (Westermeier and Marouga 2005). Second, proteins characterized by extreme pI or MW values can escape detection. Third, the intrinsic gel-to-gel variation of 2DE compromises the quantitative comparison of protein expression levels. (Van den Bergh and Arckens 2004). These drawbacks have been in part circumvented with the introduction of differential in-gel electrophoresis (DIGE) developed by Amersham Biosciences (Ünlü et al. 1997). To identify specific protein components, electrophoresis-based separations, in either the one- or two-dimensional version, coupled with MS analysis may be integrated with immunoblotting detection with antisera raised against specific proteins, as illustrated in the case of milk proteins (Chianese et al. 2010). This approach has been used to characterize milk variants in individual or bulk milks, or to monitor the evolution of dairy products such as yogurt and ripened cheese during maturation (Addeo et al. 1995) by analyzing the modification of milk protein and peptide patterns over time. Today, electrophoretic analysis is still the basis of the official methods in the

EU regulation to assess adulteration of high-value milk with cheaper cow milk (Addeo et al. 2009). Therefore, the implementation of well-established analytical procedures with proteomic platforms for objective molecular marker identification is of greatest promise in this field.

Gel-free MS-based quantitative proteomics is a promising technology capable of bridging the gaps of the 2DE-based analysis (Fig. 24.1b). This approach was made possible by the development of high-resolution mass analyzers and multistage MS instruments (i.e., Q-TOF, orbitrap, Fourier transform ion cyclotron resonance).

In the direct quantification of intact proteins, according to a "top-down" approach, the intensity of analyte multicharged ions is compared with that of internal or external standards. To quantify cow's milk allergens in mixed fruit-juice samples (Kuppannan et al. 2011) and whey drink (Huber et al. 1999), several multiple-charged ions of whey proteins were simultaneously monitored. In these specific cases an external calibration was carried out. With a similar approach, but utilizing beta-lactoglobulin (BLG) from species different from bovine as an internal standard, Czerwenka et al. quantified cow BLG in different milk-derived products (Czerwenka et al. 2007).

Nevertheless, the sensitivity of top-down mass spectrometry methods is limited by the wide distribution of protein charge states. Therefore, the analytical strategies named "shot-gun" proteomics, is being adopted in a number of research laboratories for the quantitative analysis of a number of analytes (Washburn et al. 2001). This approach is built on the direct analysis by high-resolution LC-MS/MS of peptide derived from in-solution digestion of protein complex mixtures.

The use of bottom-up methodologies for quantitative analysis in proteomics is steadily increasing. Quantitative methodologies for large-scale quantification such as stable isotope labeling with amino acids in cell culture (SILAC), isotope-coded affinity tag (ICAT), isobaric tags for relative and absolute quantitation (iTRAQ), or label-free approaches are capable of exploring the dynamics of whole proteomes through the relative quantification among multiplexed samples (Wu et al. 2007; Gingras et al. 2007). However, the above strategies were originally developed to determine differentially expressed proteins. For this reason, these quantitative methods have started to be addressed to the quantification of food adulterants only recently.

The selected reaction monitoring (SRM) MS approaches have emerged as tools for the precise quantitative determination of targeted proteins in complex samples (Picotti et al. 2010). The most common applications of SRM in proteomics rely on the principles of stable isotope dilution (SID) methods (Mayya and Han 2006). Conceptually, SRM and SID are not new techniques, as they were introduced as early as the late 1970s for the quantitative analysis of small molecules by GC-MS. Their integration in proteomic platforms is finding exponentially increasing application so that they can be now considered the gold standard for absolute quantification. Multiple reaction monitoring (MRM) represents an improvement of SRM in terms of selectivity and sensitivity. Both techniques have been broadly adapted to QQQ-based instruments (QQQ, QQQ-IT, QQQ-TOF), in which the first quadrupole (Q1) mass analyzer is set to filter only the preselected peptide precursor ion(s) of choice and other ions are excluded. The second quadrupole (Q2) is a collision cell that fragments the precursor by collision-induced dissociation (CID) and transmits fragment ions to the third quadrupole (Q3). Q3, in turn, transmits to the detector only fragment ion(s) with pre-set *m/z*, whereas all the remaining fragments are excluded (Careri et al. 2007). MS is not intrinsically a quantitative analytical technique, because the ionization efficiency of the protein/peptide is affected by a wide range of physicochemical properties (size, charge, hydrophobicity). For accurate quantification the protein/peptide amount must be referred to that of a suitable standard.

The typical workflow for this methodology is (1) isotopic labeling of protein sample; (2) digestion of combined labeled protein samples to obtain peptide mixtures; (3) chromatographic fractionation of mixed peptide samples; (4) analysis of the separated peptides by LC-ESI- MS/MS; and (5) processing of the MS results to obtain relative protein abundance as well as protein identification by database searching.

The core of quantitative proteomic analysis is the isotopic labeling (i.e., 12C/13C, 14N/15N and 1H/2H) of proteins that permits direct comparison of two proteome states in a single analysis. In this way, the same peptides generated from the hydrolysis of different samples have a similar chemical performance, but a difference in mass detectable by MS techniques. Peptide peak intensities can be used for relative quantification of these peptides.

The purpose of quantification was successfully gained with the introduction of absolute quantification (AQUA) strategy based on the use of specifically designed internal standards. AQUA standard peptides are synthetic copies of proteotypic peptides, which are isotopically labeled to one or more amino acid positions in order to shift molecular mass, without appreciably influencing retention time and ionization properties. Hydrolyzed protein samples are spiked with known amounts of standard AQUA peptides and both native and surrogate peptides are monitored by LC–MS operating in MRM mode. The absolute amount of peptide(s) is determined by the ratio of the ion intensities of AQUA peptide and its native cognate (Gerber et al. 2003; Mallick et al. 2007). Applications to food contamination studies are still scarce but recently increasing in number. By this strategy, allergens in commercial soybean (Abdel Rahman et al. 2011), tomato sauce (Brambilla et al. 2009), and industrial crab production (Abdel Rahman et al. 2010; Barre et al. 2008) have been evaluated.

An alternative to SRM for less complex samples, when an extremely high specificity is not required, could be the single ion monitoring (SIM) which is commonly performed on a single quadrupole. During this process, the MS analysis time is focused only on analytes with a specific mass corresponding to the ion of interest, whereas all others are excluded. SIM analysis often requires a high-resolution separation in order to minimize interference from other species.

The protein extraction from the food matrix is a crucial step; interference by nonprotein compounds, such as tannins, should be carefully prevented. The choice of proteotypic peptides and ion transitions is also decisive and, although predictive software have been developed (Alves et al. 2011; Gallien et al. 2011), they need to be defined experimentally through a prior full-scan LC/MS analysis of the proteolyzed protein extract of the original food matrix (targeted analysis). The possible occurrence of isoforms of protein markers should be considered by selecting distinctive proteotypic peptides if markers are willing to be individually quantified.

A further critical step in large-scale SRM is the control of the conditions and the reproducibility of the HPLC separations. To overcome possible pitfalls in this sense and to assure reproducibility, the efficiency of the stationary phase, the high quality of eluents, and a strict temperature control are mandatory. These precautions also allow us to narrow down the time window used to monitor the transitions of a specific peptide, in order that the LC–MS run can be divided into a higher number of time segments, thus dramatically increasing the number of peptides analyzed during the entire LC–MS run. In this manner sensitivity can be increased at subfemtomole levels enabling the detection and quantification of low ppb of analyte.

A new and promising alternative for quantifying protein abundance changes in shotgun proteomic analyses is the so-called label-free comparative proteomics which avoid any chemical sample alteration (Ono et al. 2006). The label-free approach has been designed to simplify the analytical procedure avoiding use of stable isotopes. Quantification is based on the measurement of signal intensity or of the so-defined mass spectral counting (Liu et al. 2004; Choi et al. 2008), both parameters linked to the protein abundance.

The label-free approach to the analysis of food adulteration is in its early stages and the number of applications is still limited. Using the distinctive peptides as protein markers, a label-free method has been set up for quantifying genetic variants of α_{s1}-casein content in bulk goat milk, which is an important parameter to evaluate the cheese-making aptitude of caprine milk (Picariello et al. 2009a). The targeted monitoring of mass and transitions of selected proteotypic peptides, overcomes the bias due to the presence of a large number of dominant components. These procedures are obviously strictly demanding in terms of chromatographic resolution and mass spectrometry performance, but their relatively low cost for standards and progress in the newest generation of MS instruments and in software development have made them quite attractive at least for semi-quantification.

24.5 Raw Materials and Ingredients: The Proteomic Approach to Quality and Authenticity Assessment

The application of proteomic technologies in food science is eventually aimed at defining the entire and detailed (bio)chemical composition of a food and its modification along the production process to correctly evaluate its nutritional, toxicological, and functional properties.

As a matter of principle, all the macrocomponents of a food may be used to trace its history. Thus, both glycomics and lipidomics strategies have been addressed to this purpose. Lipidomic procedures have been successfully applied to assess, for instance, the authenticity of composition of edible oils and butters (Picariello et al. 2007) as well as to trace the modifications related to heat treatments to reveal the improper or illegal use of frying oils (Picariello et al. 2009b). However, more than any other class of food components, the proteome reflects very closely the (bio)chemical processes a system undergoes. The proteins of a food are, at the same time, a passive source of amino acids but also an active food component giving a food its textural, functional, and sensory characteristics. Proteins also play an active crucial role in inducing food modifications during production, maturation, and storage, by catalysis of processes such as oxidation, proteolysis, or lipolysis. In doing so, proteins are largely responsible for the overall characteristics of many food preparations. Because of the potential to generate a systematic view of protein composition and biological as well as chemical interactions, the application of proteomic analysis in food science is steadily growing.

The aim of these studies is eventually to define a complete database of the protein composition "fingerprint" of a given food. This has also provided important outcomes in the field of food authentication. The European Food Safety Authority has recently established a comprehensive system of authentication and traceability of food and feed in to order to ensure food safety for human consumption at every stage of production. The EC regulation No. 509/2006 (EC Council Regulation 2006) protects consumers through a system of effective and impartial controls that define the safeguard of the Protected Designation of Origin (PDO). From the legislative point of view, quality standards have been established through the requirement of quality labels that specify the chemical composition of each product and in most cases also geographical origin and production methods. This requirement has prompted an increased interest in developing analytical methods to assess quality and authenticity of food products.

The issue of food authentication, particularly urgent in the case of PDO food products, also requires efficient traceability systems, to ensure the correct application of the procedures throughout the productive processes and during distribution. The sensitivity, specificity, and speed in the analytical response candidate MS-based techniques are among the most accurate and versatile strategies to face the challenging tasks in this field. Moreover, proteins encrypt highly distinctive information about the identity and history of a specific product. Thus, it is not surprising that targeted MS-analysis of both proteins and proteolytic peptides have been largely developed for food authentication purposes. Of great relevance are the studies aimed to fingerprint the PDO typical products in terms of their peculiar metabolomic pattern, in order to differentiate them from non-PDO imitation products. One such example is Napoli-type salami, a ripened sausage originated in Campania (South Italy) but now produced in any region of Italy. A consortium of producers in Campania has applied to the EU for the PDO label in order to protect the artisanal product from industrial imitations from other geographical areas. A way to distinguish the PDO product from the industrial resides in the proteomic as well in the metabolomic profile (the overall metabolomic approach is illustrated in Fig. 24.2). In Fig. 24.3 the volatile profile of the two products is compared from where differences emerge and also to find a molecular rationale in the different ingredients and technology employed.

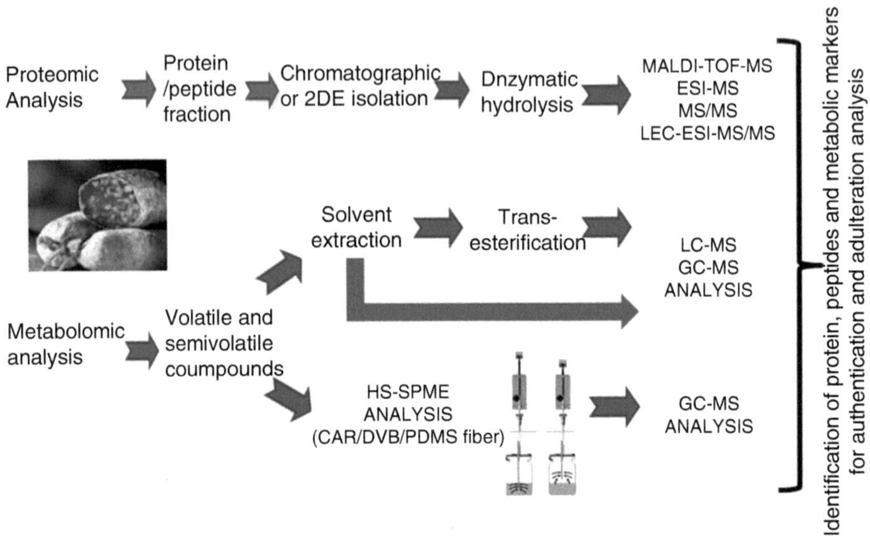

Fig. 24.2 Analytical proteomic and metabolomic approach for fingerprinting of typical products, exemplified by the case of Naples-type salami

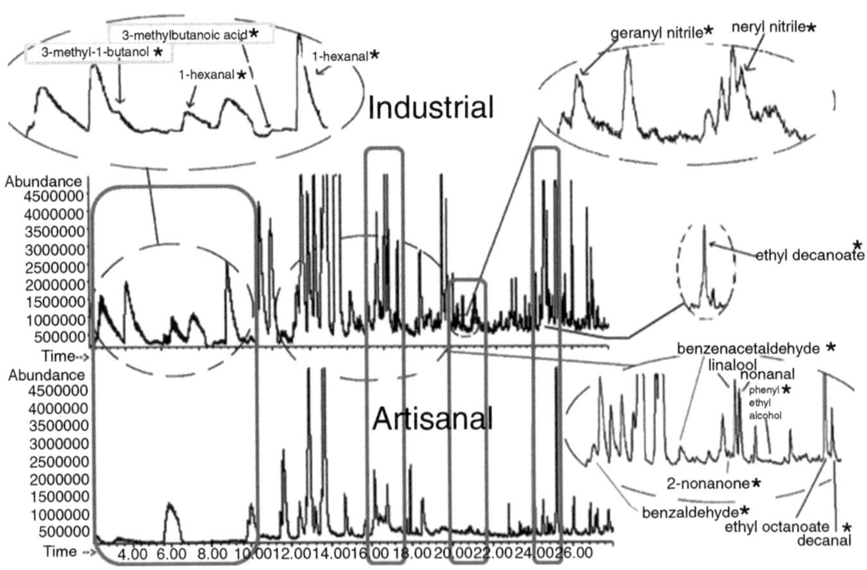

Fig. 24.3 Metabolomic analysis reveals molecular differences and identifies molecular markers to differentiate artisanal products from industrial imitations

The wide-ranging contribution and potentialities of MS to assess the quality of milk-based products has recently been reviewed (Guy and Fenaille 2006). Several MS-based procedures have been developed to authenticate "raw materials" used in manufacturing dairy products. As already mentioned, addition of milk of higher commercial value with cheaper bovine milk is a quite frequent adulteration practice in cheese-making. Taking advantage of the species-specific amino acid substitutions along the homologous protein chains that affect molecular weight of both caseins and whey proteins, it is possible to define fingerprinting profiles of milk from different species either by electrospray or MALDI–TOF–MS.

Angeletti et al. (1998) demonstrated the efficiency of MALDI–TOF–MS in the detection of bovine caseins in water buffalo mozzarella cheese. Analogously, adulteration by bovine milk was detected in ewe cheese (Fanton et al. 1998). On the basis of the identification of a water buffalo-specific β-casein by CE followed by offline MS, an improved strategy to distinguish bovine and buffalo milk by monitoring the products of plasmin hydrolysis has recently been developed (Somma et al. 2008). Selecting the whey proteins as biomarkers, Cozzolino et al. (2002) identified the fraudulent addition of bovine milk to buffalo and ewe dairy products at levels as low as 2%. Similarly, the detection of bovine BLG in goat milk by means of LC/ESI–MS has been reported (Chen et al. 2004).

A CE-MS-based approach was proposed by Muller et al. (2008) for the detection of adulteration of caprine and ovine milk, with lower-value bovine milk. A low-pH electrolyte was selected in order to minimize the adsorption of the proteins onto the inner capillary wall. By analyzing the whey protein fraction from milk, the method allowed the detection of added bovine milk into the nonbovine ones within the concentration range of 5–95%.

As shown by the above example, in CE-MS special attention has to be paid to the selection of MS-compatible volatile separation buffers and the selection of the adequate strategy when capillary coating is required to avoid MS signal suppression (Simpson and Smith 2005). Recently CE-MS of proteins has been studied using both online and offline detection (Haselberg et al. 2007). A comprehensive study of the use of CE volatile separation buffers compatible with ESI-MS detection was carried out in order to obtain as much information as possible from the zein protein fraction from maize (Erny et al. 2007). A coating with ethylpyrrolidine methacrylate- N,N-dimethylacrylamide copolymer was used in order to avoid protein adsorption into the inner capillary wall. It was observed that the use of ammonium gradient during CE separation significantly enhanced the analysis of the zein fraction giving rise to efficiency improvements of up to twentyfold. These findings were next applied in the analysis by CE-ESI- MS of the zein fraction of different maize cultivars using two different mass analyzers, that is, TOF and Ion Trap (Erny et al. 2008). Although both instruments provided good results in terms of sensitivity and repeatability, CE-ESI-TOF/MS allowed the identification of a higher number of proteins.

The molecular profiling performed by MALDI-TOF-MS is emerging as a general tool for the discovery of biomarkers also applied to food authentication. In meat and fish authentication, and in food-borne bacterial identification MS-based techniques have the capability to substitute or complement other available strategies, such as the multiplex PCR assay. In one study (Mazzeo et al. 2008) an innovative method based on MALDI-TOF mass spectrometry has been developed and successfully applied to fish authentication. Highly specific mass spectrometric profiles from 25 different fish species were obtained. Signals generated from proteins with molecular weights of about 11 kDa have been selected as specific biomarkers for unambiguous discrimination. This method is also suitable for verifying commercial product authenticity and to rapidly discriminate species subjected to fraudulent substitutions, such as those belonging to Gadidae and Pleuronectiformes. For example, biomarkers for fillets of sole, European plaice, and Greenland halibut were defined in parvalbumins, differentiated on the basis of the different molecular weight. Proteome analysis has also been applied with the aim of finding new marker proteins suitable for seafood authentication (Pineiro et al. 2003). Using gel-based methods, a polypeptide has been identified as a good candidate for the differentiation of high-value hake species from other, less valuable species of the same fish. Furthermore, marker proteins were suggested to test for shelf life and freshness of seafood as well as for the differentiation between fresh and frozen fish.

24.6 Food Authenticity: Finding Markers of the Technological Process

A relatively new application of proteomic/peptidomic technologies for food authentication is in the discovery of process markers, informative molecular markers indicative of both the use of specific raw ingredients and of specific industrial process. Food production in general requires a series of technological steps involving physical and chemical processes: thermal treatments, spray-drying, cooking, extrusion, gel or dough formation, chemical or enzymatic hydrolysis, cross-linking, and oxidation, just to mention a few, all of which may induce deep and different structural changes in the food constituents. For this reason, although in raw food materials the characterization of protein constituents, still in a relatively "native" state, can be considered quite standardized at present, proteomics of processed foods remains a challenging task and requires properly designed approaches in order to verify conformity of food components and characteristics with those declared and to reveal adulteration.

For example, the skim milk powders (MP) are largely utilized as feed or as ingredients for the manufacture of products ranging from industrial cheeses to fermented meat products, in which they are added as emulsifier, texturizers, and for their high nutritional value. Soybean proteins are also frequently added to a variety of food products not only for economic reasons, as in the case of MP, but also because of their interesting functional properties during the processing of food. The addition of plant proteins to MP is, however, an illegal practice; therefore the detection and identification of adulterations in MP, typically by addition of low levels of cheap plant proteins (legume, cereal proteins) is an important issue to assess genuineness of several widespread industrial food preparation categories.

Because of the complexity of the matrices (both milk and vegetal protein sources), the detection with traditional analytical methods (HPLC, electrophoresis) is impracticable at the very low amounts of adulteration frequently used. Also, the official EU reference capillary zone electrophoresis (CZE) method currently used to prove fraudulent addition of plant proteins to MP has limited reliability and accuracy. For these reasons, the application of proteomic methodologies, based on combined chromatographic and MS methods, opens up new possibilities in the routine quality control of MP. Two LC/ESI–MS/MS methods (Luykx et al. 2007; Cordewener et al. 2009) were successful in detecting peptides originated from the major seed proteins of soy (glycinin, β-conglycinin) and pea (legumin, vicilin) in MP. They share the use of LC/ESI–Q–TOF–MS/MS aimed at targeting specific tryptic peptides from legume proteins. The first one included a pre-fractionation step of the (adulterated) MP samples. The second more general approach is based on an untargeted analysis combined with statistical PCA analysis. This method also allows us to distinguish MP samples from different manufacturers or produced according to different process conditions (milk protein composition, heat treatment).

A further example is that of mechanically recovered meat (MM). MM is industrially obtained by recovering residual raw meat from animal bones or poultry carcasses from which the bulk of the meat has been already removed mechanically. MM has the appearance of finely comminuted meat and is incorporated in a wide range of meat products, as an inexpensive source of meat. Although MM has a similar chemical composition to authentic or hand deboned meat (HDM), it has a lower commercial value. This has led to the exclusion of MM from the EU definition of meat (Directive of the European Parliament N. 101/2001). As a consequence analytical procedures are needed to differentiate it from hand-deboned meat. A very recent study has utilized a proteomic approach to find potential markers for the detection of chicken mechanically recovered meat. Intact proteins were extracted from raw meat and then analyzed with off-gel electrophoresis followed by SDS-PAGE and identification of potential markers by nano-LC-MS/MS. By this analytical workflow it was possible to extract, separate, and identify key proteins from processed meat material and to define chicken MM markers (Surowiec et al. 2011).

As a second example, a proteomic-based method has been developed for the detection of chicken meat within mixed meat preparations (Sentandreu et al. 2010). The procedure is robust and simple, comprising the extraction of myofibrillar proteins, enrichment of target proteins using off-gel isoelectric focusing, in-solution trypsin digestion of myosin light chain 3, and analysis of the generated peptides by LC-MS/MS. Using this approach, it was possible, for example, to detect 0.5% contaminating chicken in pork meat with high confidence. Quantitative detection of chicken meat was done by using AQUA stable isotope peptides synthesized on the basis of the sequence of previously selected species specific peptide biomarkers. Linearity was observed between the amount of the peptide biomarker and the amount of chicken present in the mixture; further independent replication is required now to validate the method. In spite of its simplicity, this approach can be used effectively for the detection of both raw and cooked meat. It is amenable to highly processed foods, which can be particularly problematic as the tertiary protein structure is often affected in processed food precluding immunoassays. In addition, proteomic analysis allowed identification of definitive discriminatory peptide sequences, unlike the DNA/PCR-based methods used presently. Interestingly, the study also envisaged the translation of the technology to routine MS equipment, making the methodology suitable for public inspection laboratories.

Other limiting factors to proteomic analysis of processed foods are: (1) the increased protein complexity (e.g., production of oxidized protein families or of mixtures of hydrolytic fragments) and (2) the interaction of protein with other proteins or with the other molecules within the food matrix (e.g., in the dough network formation or in the case of the condensation products between carbohydrates and proteins in the early stages of the Maillard reaction). However, these reactions not only constitute an analytical complication but, on the other hand, may also be precious markers of the type of process and of the modifications of food quality. For this reason, recent procedures of protein chemistry, sometimes borrowed by classical biochemistry protocols, have been developed or adapted to obtain efficient protein extraction and characterization from processed foods.

In meat products, soy proteins serve as emulsifiers to avoid fat coalescence during heat treatment. However, regulatory authorities in some countries have forbidden soy protein or limited the amount that can be added, and if such threshold levels for soybean protein content in foodstuffs have been established, appropriate analytical techniques are needed for the reliable detection and/or quantitation of soybean proteins (soy protein isolate, SPI) in meat products. Currently, there are only a limited number of methods available for the (semi)quantitative determination of soybean proteins in processed meat products. The AOAC Official Method (AOAC Official Method of Analysis 988.10) based on an ELISA, has been criticized by regulatory authorities (e.g., the U.S. Food and Drug Administration) because of reliability and accuracy. In addition, highly sensitive assays would be desirable to be able to detect low ppm amounts of soybean proteins due to their allergenic potential (Koppelman et al. 2004). Liquid chromatographic assays which have been proposed are only applicable to raw meats. A recent proteomic study applied multidimensional HPLC-MS/MS to identify adulteration with SPI on heat-processed meat products from bovine, pork, chicken, and turkey (Leitner et al. 2006). Pre-fractionation by perfusion chromatography was needed to isolate peaks of interest from meat products containing SPI. After enzymatic digestion using trypsin, the collected fractions were analyzed by nanoflow LC-MS/MS. Several variants and subunits of the major seed proteins, glycinin and conglycinin, were identified in SPI, along with two other proteins. In soybean-protein-containing meat samples, different glycinin A subunits could be identified from the peak discriminating between samples with and without soybean proteins added. Among those, glycinin G4 subunit A4 was consistently found in all samples. Consequently, this protein (subunit) can be used as a target for new analytical techniques in the course of identifying the addition of soybean protein to processed meat products.

Heat treatments preserve or improve organoleptic properties and nutritional value of foodstuffs. Prior to being eaten, packaged, or stored, many foods require thermal processing such as pasteurization, cooking, and roasting. Thermal treatment results in enhanced food safety and quality but at same time may induce some undesirable reactions that result in a loss of nutritional factors or in the formation of mutagenic and carcinogenic molecules (van Boekel et al. 2010; Seal et al. 2008).

Proteins are among the main targets of food thermal damage. Proteins are strictly related to identity and history (in most cases also to geographical origin and to production methods), therefore monitoring of modified protein as a consequence of a heat treatment assesses quality and authenticity of each food product.

The most prominent modifications of food when subjected to heat treatments, concentration, or drying is the nonenzymatic reaction known as the Maillard reaction (Friedman 1996). The early stage of this reaction consists of interaction between amino acid residues and the carbonyl group of the reducing sugar, to give a Schiff base that is subsequently converted to the Amadori product (Chichester 1986). Maillard products are extremely important in food preparation for developing desirable flavor, odor, or color in food such as coffee, bakery products, and roasted meat or for improving the physicochemical properties of milk whey proteins (van Boekel MAJS 2006). On the other hand, an uncontrolled Maillard reaction (in consequence of incorrect or undeclared heat treatment), yields advanced glycation end products resulting in a decreasing of bio-availability of several essential amino acids, (Birlouez-Aragon et al. 2004; Hewedi et al. 1994), decreasing of protein digestibility, and formation of some undesirable compounds (Taylor et al. 2003; Uribarri and Tuttle 2006). Owing to sensitivity, specificity, and speed in the analytical response, MS represents an excellent analytical tool to face the challenge of characterizing Maillard product (Mamone et al. 2009). In the last decades MS has been largely applied for determining Maillard products in complex food matrix.

Malt and beer are an example of food and beverages in which glycated proteins are produced during the malting process. Malt and beer proteins may undergo extensive reactions with glucose generated from starch degradation. The degree of the protein glycation influences the quality and the properties of the beer foam. Glycated forms of lipid transfer protein (LTP) in malt and beer have been identified by MS in combination with of gel-electrophoresis (Jégou et al. 2001) or monolithic chromatographic media (Bobalova and Chmelik 2007). More recently an interesting gel-free MS-based approach (2-D HPLC coupled to MALDI-TOF/TOF), allowed the characterization of barley glycated proteins during brewing, and identification of potential glycation markers for fast and efficient monitoring of the correctness of the malting conversion. (Petry-Podgórska et al. 2010; Zídková et al. 2010)

Because of their high lactose content, milk and dairy products are matrices where the potentiality of MS-based techniques have been better applied to study heat treatment-induced modifications. Based on the 324-Da molecular weight increase, lactosylation protein can be easily detected (Siciliano et al. 2000). Lactosylation of intact milk proteins has been evaluated by LC-ESI-MS analysis within commercial whey-based products (Hau and Bovetto (2001)) and milk (liquid and powdered) samples. (Siciliano et al. 2000; Scaloni et al. 2002). MALDI-TOF-MS has been also applied for characterization of lactosylation of whey proteins in model milks (Meltretter et al. 2007; Catinella et al. 1996), in heat treated milk (Meltretter et al. 2008), and commercially available dairy products (Meltretter et al. 2009).

BLG and α-lactalbumin have been identified as main targets of the Maillard reaction (Siciliano et al. 2000). Whey protein lactosylation accounted for almost 3%, 30%, and 70% of the BLG content in pasteurized, UHT, and dry infant formula samples, respectively. Selective lactosylation upon different thermal processing demonstrated that whey proteins can be exploited as indicators to trace the quality and thermal history of milk (Siciliano et al. 2000; Morgan et al. 1997; Jones et al. 1998; Czerwenka et al. 2006; Losito et al. 2007; Monaci and van Hengel 2007).

MS/MS techniques have been successfully used to localize the main amino acid sites of lactosylated BLG (Leonil et al. 1997; Fogliano et al. 1998), casein (Scaloni et al. 2002), and lysozyme (Yeboah et al. 2004). Identification of minor lactosylated species has required the use of more specific LC-MS/MS technique such as "neutral loss" or "precursor ion scanning" (Molle et al. 1998; Marvin et al. 2002; Fenaille et al. 2004).

In some cases, glycated protein pre-separation alone is insufficient for dealing with Maillard reaction complexity. Therefore, the most abundant casein and whey protein may have to be depleted from the sample. A selective depletion procedure was developed for identification of lactosylated sites in milk and in powdered milk infant formulas (Arena et al. 2010) using a combination of proteomic techniques based on analyte capture by combinatorial peptide ligand libraries, selective trapping of lactosylated peptides by m-aminophenylboronic acid-agarose chromatography, and two different fragmentation MS/MS modes. By this approach 271 nonredundant modification sites in 33 milk proteins have been characterized.

A proteomic enrichment protocol has been recently developed to identify lactosylated casein phosphopeptides in milk samples based on hydroxyapatite chromatography and MALDI-TOF-MS. The amount of modified peptides varied with heat treatment severity. With a 10% limit of detection of UHT milk addition to raw or pasteurized milk, casein phosphopeptides (CPP) were used as markers of thermal treatment in milks heated at various intensities (Pinto et al. 2012). On the other side, proteomic techniques have been scarcely applied to detection of the advanced glycation end products (Henle et al. 1997; Hegele et al. 2008; Hasenkopf et al. 2001) in food samples. Among these products lactulosyl-lysine, Nε-carboxymethyl-lysine (CML), lysine aldehyde, methionine sulfoxide, cyclization of N-terminal glutamic acid, and oxidation of cysteine and tryptophan modification have been screened as the major heat-induced markers of alterations of whey milk proteins by MALDI-TOF-MS analysis (Birlouez-Aragon et al. 2004). A more detailed presentation concerning the distribution of free and protein-bound glycation-induced lysine modifications in raw and processed cow milk has been described (Hegele et al. 2008). More recently a method based on LC-MS/MS using a stable isotope dilution assay has been proposed for quantitative analysis of CML in various dairy products. The method, based on enzymatic digestion and LC–MS/MS analysis, estimates a limit of detection for CML of 8-ng CML per mg protein and the limit of quantification of 27-ng CML per mg protein (Delatour et al. 2009).

Other protein markers of heat treatment, not generated by reaction with lactose, such as lysinoalanine, histidinoalanine, N-terminal pyrrolidone, or partial protein hydrolysis may also occur in milk (Friedman 1999). The most interesting analytically are the heat-susceptible phosphorylated serine and threonine residues of caseins yielding dehydroalanine and methyl-dehydroalanine, respectively. The conjugated carbon–carbon double bond reacting with nucleophilic amino- (Lys), imidazole- (His), or thiol-(Cys) group forms lysinoalanine (LAL), histidinoalanine, and lanthionine residues, respectively. In particular, LAL has been determined in milk and cheese products by LC/ESI/MS in positive selected ion monitoring (Calabrese et al. 2009). The method encompasses acid protein hydrolysis to amino acids prior to LC–MS/MS analysis. This approach has been usefully to demonstrate that LAL is not present in raw milk or some dairy products such as Mozzarella cheese; in contrast, high amounts of LAL are present in calcium caseinate and milk powder, materials frequently used for fraudulent PDO cheeses adulteration. Because of the high specificity, LAL has been proposed as a useful marker to demonstrate either uncorrected heat treatments or the occurrence of heated milk/milk powders that could have been added to fresh milks for cheese adulteration.

The peptide composition dynamically changes in a food as the result of proteolysis, modifications induced by technological treatments, and interactions with other components. They decisively affect both biological as well as functional properties of food products. Thus, food peptidomics, intended as the global characterization of the peptidome of a given food product, embraces a wide research area that aims to establish the origin of the peptidome, its evolution, the impact on the sensorial properties, and its beneficial or adverse effects on human health. Food peptidomics also encompasses the development of analytical strategies to study the food peptidome.

Peptides retain memory of the history a food has undergone. Therefore, peptides are much more than just convenient analytical substitutes of parent proteins in many of the proteomics workflows; peptidome indeed provides matchless information about a given food product, for instance in the case of fermented foods.

The change in peptide composition of cheeses during ripening is a striking example. The peptide profile indicates the origin and evolution of the protein fraction. During the ripening period, the cheese chemical components undergo important chemical, physical, and enzymatic modifications: proteolysis and other reactions, such as lipolysis and lactic and propionic acid fermentation, influence the organoleptic properties of the final product. Proteolysis directly contributes to flavor (release of peptides and amino acids) and off-flavors (bitter hydrophobic peptides), also liberating substrates for other reactions. Thus, for the development of an acceptable cheese flavor, a well-balanced breakdown of the protein (i.e., casein) into small peptides and amino acids is necessary. On the other hand, recently there has been great interest in the protection of typical food products from adulteration, sophistication, and falsification. A large use of MS techniques has been made in the past to trace the proteolytic events that occur in the ripening of Parmigiano-Reggiano, Emmenthal, or Cheddar cheeses (Piraino et al. 2007). The peptidome profile also indicates origin, authenticity, and typicality of PDO cheese. The proteolysis as a function of the rennet typology has been assessed in Fiore Sardo ovine PDO cheese (Pirisi et al. 2007).

A peptidomic approach based on capillary electrophoresis-MS allowed us to identify specific β-casein fragments, referred to as γ_4-casein, in buffalo milk arising from plasmin hydrolysis (Somma et al. 2008). This polypeptide, being unique to water buffalo milk, might be conveniently used as a marker of this species milk.

Although to a minor extent, proteolysis also occurs on frozen foods. This is the case of the γ-caseins produced even during the cold storage of buffalo curd (Di Luccia et al. 2009). Interestingly, peptidomics changes have also been observed in ice-stored fish, in addition to the expected changes due to cooking (Bauchart et al. 2007). The proteolytic peptides specifically generated are molecular targets to distinguish fresh from cold-stored materials, and therefore to assess the fraudulent use of disallowed refrigerated or frozen ingredients.

The biochemical changes occurring during the processing of dry-cured ham have been studied with peptidomics techniques relying on MS-approaches. The intense action of endogenous proteolytic enzymes that are activated in the early post-mortem phase induces extensive degradation of muscle proteins. The recent identification of a large number of small peptides released from enzymes belonging to the glycolytic pathway has confirmed that muscle sarcoplasmic proteins are relevant substrates for proteolysis (Mora et al. 2011a). Generally in real foods, as in this case, the identification of the released peptides is made difficult by the lack of a cleavage specificity. Nevertheless, the multistage MS-based peptide sequencing can furnish information about the activity of early (calpains) as well as the medium- and long-term acting (cathepsins) endogenous proteases. The identification of oligopeptides characterized by the consecutive loss of terminal amino acids has demonstrated an additional intense secondary proteolysis in dry-curing hams that is due to both amino- and carboxy-peptidases (Mora et al. 2009).

The MS-based identification of large- and small-sized fragments of myosin (heavy and light chains) and actin has demonstrated that muscle myofibrillar proteins undergo extensive proteolysis as well, in the maturing of dry-cured ham (Mora et al. 2011b; Di Luccia et al. 2005; Sentandreu et al. 2007).

The proteolytic machinery of lactic acid bacteria concurs to a more extensive proteolysis of fermented sausages (Picariello et al. 2006). However, a few peptidomic investigations have been published related to these products. The peptidomic definition of proteolytic patterns could be helpful in assessing or predicting the suitability of specific strains of lactic acid bacteria as starter cultures for dry fermented sausages. The MS identification of species-specific peptide biomarkers has been proposed as a tool for determining the authenticity of meat-derived products (Sentandreu and Sentandreu 2011).

24.7 Proteomics, Peptidomics, and Metabolomics in Food Maturation, Preservation, and Safety

From the metabolomic perspective, most foods can be considered as complex chemical mixtures consisting of various metabolites and chemical additives in a solid, semi-solid, or liquid matrix. Some foods consist of just a few different compounds (artificial energy drinks, soft drinks, purified vegetable oils) whereas other foods consist of hundreds of compounds (milk, cheese) and still others may have thousands of compounds (fruits, meats, and most prepared foods). These compounds may constitute a fingerprint of the food itself and of its history. In food technology, detection of the metabolomic profile is gaining great importance for several fields of applications. These include tracing the evolution of a product from the starting materials to finished foods; evaluating the effects of either desired or undesired microbial action on food quality; monitoring the production or persistence of functional and bioactive compounds; and, in general, measuring the action of technological treatments (physical, chemical, enzymatic) on the development of rheological and sensory characteristics. Examples in dairy, cereal, wine, or meat products depict the metabolomic fingerprint of these products. Wine metabolome analysis by ultrahigh-pressure liquid chromatography high-resolution Fourier transform ion cyclotron resonance mass spectrometry (UPLC-FT-ICR-MS) in combination with multivariate statistical data processing has been proven able to discriminate the wines according to cultivar, vintage, provenance, maturation, and quality, opening the way to its extensive use in authentication studies (Cuadros-Inostroza et al. 2010; Liger-Belair et al. 2009; Gougeon et al. 2009), at present still based on subjective sensory analysis or analysis of a limited number of compounds. The capability to distinguish among high- and low-quality products on the basis of a comprehensive metabolomic identification of measurable objective biomarkers or molecular pattern related to (sensory) quality is providing new powerful tools for the defense of PDO products from adulteration (the case of salami in Fig. 24.3).

Because of the globalized nature of food production and processing, along with the growing concerns over food safety and food quality, more stringent and precise food monitoring systems are being set up. Proteomic and metabolomic approaches are therefore also being applied to food safety control.

Milk and dairy product quality can be affected by mastitis, which is rather frequent in intensive dairy farming (especially of high-milk yielding cows such as Holstein Friesian). The legal parameter used to evaluate the level of contamination of mastitic milk is the somatic cell count, that is, the mammary gland cell released in the milk by infected animals. However, this parameter is only an indirect indicator of the milk status. Consequently, adulteration of healthy with mastitic milk or forbidden use of mastitic milk is a complex task. A direct evaluation could be provided by identification of specific protein markers, as proposed by studies showing a different degree of milk protein hydrolysis in mastitic milk compared to healthy milk (Hogarth et al. 2004). These proteomic approaches not only provide clues for evaluating technological as well as nutritional losses due to infection, but identification of protein markers may be the basis for the development of analytical methods to evaluate illegal use of poor-quality mastitic milk in bulk milks.

A further issue is the evaluation of proteins specifically expressed in mastitic milks, aimed to evaluate the efficacy of adjunctive therapies for the treatment of inflammation associated with coliform mastitis in order to facilitate new veterinary drug approvals. At present, these studies are accomplished using ELISA tests. However, because a relatively limited number of bovine-specific antibodies are commercially available, reliance on antibodies can be very limiting for biomarker discovery. Proteomic approaches have instead the capability to analyze a large number of protein targets in a single experiment, independent of antibody availability. Proteomic approaches based on LC-MS/MS, recently reviewed (Boehmer et al. 2010), have been used as a means to characterize proteins in various bovine milk fractions, both under normal physiological conditions as well as during clinical mastitis. Temporal expression patterns generated using spectral counts, an LC-MS/MS label-free quantification strategy, corresponded well with ELISA data for acute phase proteins with commercially available antibodies. Combined, the capability to identify low abundance proteins, and the potential to generate temporal expression profiles, indicate the advantages of using proteomics as a screening tool in biomarker discovery analyses to assess biologically relevant proteins modulated during mastitic disease, including previously uncharacterized targets.

The analysis of pathogenic microorganisms is also of particular interest, as the risks associated to their contamination are not limited to their living presence and capacity of infectivity, but they can generally release protein/peptide toxins able to survive for long time even in foods after bacterial cell contamination has been removed, as it happens for many of the microbes that cause food-borne diseases (*L. monocytogenes*, *S. aureus*, *E. coli*, *C. botulinum*, and various *Salmonella* species). These pathogens release protein virulence factors into extracellular medium and to the cell surface that have essential roles in the colonization and insurrection of the host cells, and thus reflect the degree of bacterial pathogenicity. These toxins, being heat-stable and resistant to food processing, can be a danger to consumer health. Bacterial protein toxins have been involved in a number of infectious and food-borne diseases. Chromatographic methods coupled with MS have been carried out to define the toxin contamination levels of milk (Sospedra et al. 2011) and two ripened PDO Italian cheeses, Grana Padano and Pecorino Romano (Schlosser et al. 2007; Ferranti 2005). A procedure combining proteomic approaches with immunochemical, chromatographic, and electrophoretic techniques and MS/MS analyses was developed to monitor production and levels of enterotoxin A and B of *S. aureus* and Shiga-like toxins produced by *E. coli* O157:H7. By producing cheese samples using milk willingly contaminated with bacteria it was possible to monitor 10–100 ppb contamination level, and analysis of market samples allowed to exclude toxin contamination in the two cheese types.

The development of fast and accurate methods for species differentiation of food pathogenic and spoilage bacteria is another important issue to ensure food quality and safety and to detect use of forbidden contaminated materials. MALDI-TOF-MS has been applied to species identification of microorganisms, proving to be a rapid and cost-effective technique and allowing species differentiation due to the highly specific spectral profiles obtained (Bohme et al. 2010). The analysis of cell extracts by MALDI-TOF MS was also applied to create a mass spectral library of the main pathogenic and spoilage bacteria potentially present in seafood and was demonstrated to be a rapid and accurate method for microbial species differentiation, as well as for the classification of unknown strains isolated from seafood.

Among contaminants along the food chain, algal toxins produced by cyanobacteria are another special concern. Cyanobacteria (blue-green algae), present worldwide within water blooms in eutrophic lakes and drinking water reservoirs, producing several different biotoxins. Among these, potent peptide hepatotoxins such as microcystins, cyanopeptolins, and microviridins may be released in the water environment. The occurrence of hepatotoxins-containing algal blooms in freshwaters and subsequently in cultivated fish and seafoods has been implicated in several animal and human poisoning outbreaks worldwide. Also, algae species such as Spirulina are finding large application as ingredients of food, feed, and vitamin integrators. Therefore, methods to monitor the presence of hepatotoxins in water and animal tissues is of growing interest for surveillance agencies, because they can contaminate drinking water reservoirs and the food chain. Proteomic approaches based on MALDI-TOF-MS and Q-TOF/MS were revealed as useful for fast screening and accurate quantitative determination (LOD <0.1 ppb, limit of quantification (LOQ) 1 ppb) in water (Ferranti et al. 2011; Gallo et al. 2009; Ferranti et al. 2008, 2009). The analysis was then extended to the set-up of a method for quantitative toxin determination in food and feed integrators for men and animals and also to detect protein markers to discover the illegal use of unsafe water in fish and seafood aquaculture. It can be foreseen that these methods are being integrated to design sensitive sensors on a microchip surface for automated detection.

24.8 Genetically Modified Foods

A genetically modified organism (GMO) is an organism whose genome has been modified by genetic engineering with the introduction or deletion of specific genes in order to create a new gene set. The constant search for agriculture development led to the introduction of GMOs in food technologies. The main modifications consist in (1) induction of higher resistance to environmental stress, pathogens, and parasitic plants; (2) increase of organoleptic, nutritional, and technological values, as for instance rice containing higher protein and riboflavin A levels or tomato with inhibitors of the enzymes responsible for pectin degradation or soy with greater amounts of lysine; and (3) reduction/inactivation of antinutritional factors. GM food world production is mainly concentrated in the United States, Brazil, Argentina, Canada, and China. Because of the on-going debate on possible hidden adverse effects on humans, GM plant cultivation has expanded quite slowly in Europe, where, to date, only two GM vegetables can be grown: maize Mon810 (1998) and potato Amflora (2010) (http://www.efsa.europa.eu/it).

The gene alteration implies the modification of only a limited group of target compounds, for example, the modification of already existing proteins or the generation of novel proteins. Potential disadvantages of this technique can be the under-expression of the total seed storage protein or the production of unexpected proteins with potential allergic and/or toxic character (Finnie et al. 2004).

Current legislation imposes the monitoring of all the production steps from cultivation to transformation and commercialization, the traceability of raw transgenic materials, and the labeling of the final GM products. These aspects raise the needs for inexpensive and fast techniques for one-step detection of GM contamination in food products.

Furthermore, in several cases the presence of GMOs in food products can be due to an accidental cross-contamination, and the legislation admits the commercialization of products containing levels of GMO-derived materials with an amount <0.5% (Petit et al. 2007).

However, GM raw materials can be also intentionally introduced in the food processing chain following deceptive adulteration, aimed to improve the quality of final products or to reduce costs of production. In all these cases the GM raw materials occur in a much lower amount if compared to the raw material used, and this makes the detection of contamination difficult. Another obstacle to assessment of GM material contamination is the intrinsic high complexity and variability of seed storage proteins and the high influence of soil and climate factors on protein expression.

The GM material assessment is generally carried out by DNA profiling based on PCR methods. However, although the detection of DNA markers benefits from having a well-defined target gene and the combined use of database analysis and experimental specificity reduce false positives, it has to be underlined that the presence of a transgene in a food product does not guarantee the phenotypic expression of the transgene traits.

Therefore, gene products have to consider more reliable target analytes than their gene counterparts. The typical approach for GM protein detection relies on the detection of recombinant proteins by immunochemical analysis and quantification by ELISA. These techniques are easy, inexpensive, and suitable to screen a large number of samples. In general, the set-up of these assays requires identification and characterization of the transgenic protein to raise specific antibodies against the target (Farid 2004). For instance, the lateral flow strip protocol is a useful commercial tool for the detection of RUR, Cry3Bb, and Cry1Ab proteins in corn and of CP4 EPSPS, that is a transgenic protein that confers to several crops, including soybean and maize, the resistance to the herbicide glyphosate. This technique is very fast and sensitive as confirmed by a detection limit of 1 seed out of 1,000. As a relevant limitation, this strategy necessitates that the target protein is in its native and immunoactive form, thus implying that proteins in treated foodstuffs might escape detection. Furthermore, in several processed foods linear epitopes or toxic sequences can be hydrolyzed and released from the parent protein, although retaining their harmful

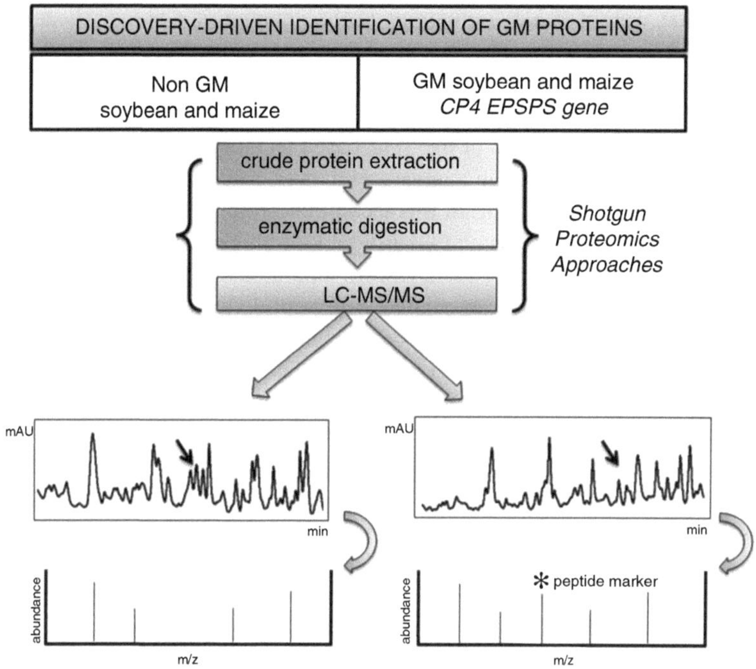

Fig. 24.4 The typical shotgun approach to GM biomarker detection and identification in food matrix. This kind of approach can be applied for the detection of any kind of peptide marker

potential. These sequences tend to be underestimated by the most commonly utilized sandwich ELISA-based tests.

Only in the last years, however, first attempts have been made to apply advanced proteomic techniques for comparison of protein profiles of transgenic material and control wild-type lines in order to detect peculiar modifications of proteins and peptides to be used as molecular markers of GM contamination (a classical analytical scheme in Fig. 24.4). Comparison of 2DE maps of proteins extracted from a wild-type and a hybrid tomato, in which a virus-resistance gene trait, showed, however, that this modification is not detectable by the proteomic approach because the hybridization did not affect protein expression appreciably (Corpillo et al. 2004). Similar results were obtained by Careri et al. (2003) who compared proteins extracted from transgenic potatoes with a control line and by Finnie et al. (2004) who, comparing the protein maps of the Golden Promise barley cultivar with different transgenic barley lines, restricted the potential alteration of the transgenic lines only to the pattern of post-translational modifications.

On the other hand, the combination of 2DE, image analysis, and MS-based approaches has been demonstrated to be very effective for the comparison of GM and non-GM products when the modification interests an encoding gene trait. It has been shown that the introduction of a gene that induces increase of the amount of low molecular weight (LMW) glutenins in the genome of a bread wheat cultivar, caused a positive improvement of dough viscoelastic properties. 2DE-PAGE profile of proteins showed a significant over-expression of the LMW fraction and the down-regulation of the remaining seed storage proteins (Scossa et al. 2008).

One of the main drawbacks of the 2DE-based techniques is the gel-to-gel variation that complicates the image analysis. However, this limit can be overcome by 2DE DIGE analysis. According to this technique proteins from GM and non-GM materials, after being labeled with two different ultrahigh-sensitive fluorescent dyes with different light emission frequency, were separated in a single gel. As described by Islam et al. this technique associated with peptide mass fingerprinting, helped detection of 66 discriminating gene products that were characteristic of transgenic pea lines, properly designed to express α-amylase inhibitor typical of common beans.

Only a few studies published up to now propose MALDI-TOF-MS and LC-MS/MS analysis as additional useful tools for GM food traceability. Among these, Simó et al. (2010) described the comparison of hydrolyzed proteins from GM soybeans versus control lines by a shotgun proteomic approach exploiting CE-ESI-TOF-MS. Remarkably, no detectable changes were identified between the two profiles, thereby demonstrating that the process of GM soy production affects protein expression only to a minor degree. Differently, Fernandez Ocana et al. (2007, 2009) demonstrated that the LC-MS approach may be diagnostic for the detection of the potential target CP4 EPSPS, with a detection limit of 0.9%. These contrasting reports indicate that much extra work in the identification of suitable protein markers is still needed to make the proteomic/peptidomic approach as much versatile as the genomic one. On the other side, these studies are accumulating a huge set of information from which compositional GM material databases are being constructed to direct the upcoming development of proteomic methods.

24.9 Conclusions and Perspectives

Proteomics and the related "omic" sciences are ideally positioned to be used in many areas of food science and nutrition research and to address a multiplicity of aspects, including component analysis, quality/authenticity assessment, consumption monitoring, and study of the physiological impact in nutritional intervention studies. Nevertheless, the number of reports on food adulteration is relatively low when compared with other consolidated techniques of investigation. Despite their steadily increasing use and their enormous potential, marker-assisted proteomic strategies have not yet delivered the expected benefits in the context of the fight against food adulteration. One of the primary limiting factors is related undoubtedly to the instrumentation and maintenance costs. Taking into account the advantages deriving from the short operative time required for analyses, these drawbacks could be at least in part overcome by the establishment of unified monitoring centers and by an adequate policy of resource rationalization. The development of up-to-date databases of food protein components also plods along to balance instrumental advances, actually restraining the informative level of the "omic"-based procedures. A general view displays a scenario where the technological platforms and methodical prerequisites are now available, so that proteomics can massively breach into food science with special focus on quality assessment, whereas the operative procedures are not standardized and the process of data mining still appears to remain a challenge that hampers routine control programs. Thus, it emerges that significant improvements need to be made to optimize and automate the monitoring of target compounds and to render the assessment flowchart less dependent on the skills of the operators. In the view of the evolution toward automation the new trends in sample pre-treatment or pre-fractioning have also to be opportunely comprised. To this purpose, it has to be also underlined that comprehensive maps of validated markers of specific food traits are still missing. The extended high-throughput application of the omic methodologies is already now enlarging the panel of the "probes" to be monitored and is going to provide the means to standardize and validate the procedures for routine analyses and food authentication studies. It can be reasonably expected that in the future descriptive proteomic and omic studies will accelerate the annotation of food components and the definition of the (bio)chemical events induced by food processing, also including the complex pathways of biomolecular interactions that are emerging as key issues of the system biology in general, and of food and nutrition in this special case. As a conclusive stage of the entire process, legislative efforts would be required to progressively introduce innovative regulations, in order to parallel the analytical advances at an applicative level, with the ultimate aim of concretely translating the scientific achievements in terms of human health and wellness.

References

Abdel Rahman MA, Lopata AL, Randel EW, Helleur JR (2010) Absolute quantification method and validation of airborne snow crab allergen tropomyosin using tandem mass spectrometry. Anal Chim Acta 681:49–55

Abdel Rahman MA, Kamath SD, Lopata AL, Robinson JJ, Helleur JR (2011) Biomolecular characterization of allergenic proteins in snow crab (*Chionoecetes opilio*) and de novo sequencing of the second allergen arginine kinase using tandem mass spectrometry. J Proteomics 74:231–241

Addeo F, Garro G, Intorcia N, Pellegrino L, Resmini P, Chianese L (1995) Gel electrophoresis and immunoblotting for the detection of casein proteolysis in cheese. J Dairy Res 62:297–309

Addeo F, Pizzano R, Nicolai MA, Caira S, Chianese L (2009) Fast isoelectric focusing and antipeptide antibodies for detecting bovine casein in adulterated water buffalo milk and derived mozzarella cheese. J Agric Food Chem 57:10063–10066

Alves G, Ogurtsov AY, Yu YK (2011) Assigning statistical significance to proteotypic peptides via database searches. J Proteomics 74:199–211

Angeletti R, Gioacchini AM, Seraglia R, Piro R, Traldi P (1998) The potential of matrix-assisted laser desorption/ionization mass spectrometry in the quality control of water buffalo mozzarella cheese. J Mass Spectrom 33:525–531

AOAC Official Method 998.10, Soy protein in raw and heat processed meat products, Enzyme-Linked Immunosorbent Assay, Official Methods of Analysis, 16th ed., Rev. 1996, 1997, 1998; AOAC International: Gaithersburg, MD.

Arena S, Renzone G, Novi G, Paffetti A, Bernardini G, Santucci A, Scaloni A (2010) Modern proteomic methodologies for the characterization of lactosylation protein targets in milk. Proteomics 10:3414–3434

Barre A, Sordet C, Culerrier R, Rancé F, Didier A, Rougé P (2008) Vicilin allergens of peanut and tree nuts (walnut, hazelnut and cashew nut) share structurally related IgE-binding epitopes. Mol Immunol 45:1231–1240

Bauchart C, Chambon C, Mirand PP, Savary-Auzeloux I, Remond D, Morzel M (2007) Peptides in rainbow trout(*Oncorhynchus mykiss*) muscle subjected to ice storage and cooking. Food Chem 100:1566–1572

Birlouez-Aragon I, Pischetsrieder M, Leclere J, Morales FJ, Hasenkopf K, Kientsch-Engel R, Ducauze CJ, Rutledge D (2004) Assessment of protein glycation markers in infant formulas. Food Chem 87:253–259

Bobalova J, Chmelik J (2007) Proteomic identification of technologically modified proteins in malt by combination of protein fractionation using convective interaction media and matrix-assisted laser desorption/ionization time-of-flight mass spectrometry. J Chromatogr A 1163:80–85

Boehmer JL, DeGrasse JA, McFarland MA, Tall EA, Shefcheck KJ, Ward JL, Bannerman DD (2010) The proteomic advantage: label-free quantification of proteins expressed in bovine milk during experimentally induced coliform mastitis. Vet Immunol Immunopathol 138:252–266

Bohme K, Fernandez-No IC, Barros-Velazquez J, Gallardo JM, Canas B, Calo-Mata P (2010) Comparative analysis of protein extraction methods for the identification of seafood-borne pathogenic and spoilage bacteria by MALDI-TOF mass spectrometry. Anal Method 2:1941–1947

Brambilla F, Resta D, Isak I, Zanotti M, Arnoldi A (2009) A label-free internal standard method for the differential analysis of bioactive lupin proteins using nano HPLC-Chip coupled with Ion Trap mass spectrometry. Proteomics 9:272–286

Calabrese MG, Mamone G, Caira S, Ferranti P, Addeo F (2009) Quantitation of lysinoalanine in dairy products by liquid chromatography–mass spectrometry with selective ion monitoring. Food Chem 116:799–806

Careri M, Elviri L, Mangia A, Zagnoni I, Agrimonti C, Visioli G, Marmiroli N (2003) Analysis of protein profiles of genetically modified potato tubers by matrix-assisted laser desorption/ionization *time-of-flight mass spectrometry*. Rapid Commun Mass Spectrom 17:479–483

Careri M, Costa A, Elviri L, Lagos JB, Mangia A, Terenghi M, Cereti A, Garoffo LP (2007) Use of specific peptide biomarkers for quantitative confirmation of hidden allergenic peanut proteins Ara h 2 and Ara h 3/4 for food control by liquid chromatography–tandem mass spectrometry. Anal Bioanal Chem 389:1901–1907

Careri M, Elviri L, Lagos JB, Mangia A, Speroni F, Terenghi M (2008) Selective and rapid immunomagnetic bead-based sample treatment for the liquid chromatography-electrospray ion-trap mass spectrometry detection of Ara h3/4 peanut protein in foods. J Chromatogr A 1206:89–94

Catinella S, Traldi P, Pinelli C, Dallaturca E, Marsillo R (1996) Matrix-assisted laser desorption/ionisation mass spectrometry in milk science. Rapid Commun Mass Spectrom 10:1629–1637

Chassaigne H, Nørgaard JV, Hengel AJ (2007) Proteomics-based approach to detect and identify major allergens in processed peanuts by capillary LC-Q-TOF (MS/MS). J Agric Food Chem 55:4461–4473

Chen RK, Chang LW, Chung YY, Lee MH, Ling YC (2004) Quantification of cow milk adulteration in goat milk using high-performance liquid chromatography with electrospray ionization mass spectrometry. Rapid Commun Mass Spectrom 18:1167–1171

Chianese L, Calabrese MG, Ferranti P, Mauriello R, Garro G, De Simone C, Quarto M, Addeo F, Cosenza G, Ramunno L (2010) Proteomic characterization of donkey milk caseome. J Chromatogr A 1217:4834–4840

Chichester CO (1986) Advances in food research. In: advances in food and nutrition research, Academic, Boston, pp 77–124

Choi H, Fermin D, Nesvizhskii A (2008) Significance analysis of spectral count data in label-free shotgun proteomics. Mol Cell Proteomics 7:2373–2385
Cordewener JH, Luykx DM, Frankhuizen R, Bremer MG, Hooijerink H, America H (2009) Untargeted LC-Q-TOF mass spectrometry method for the detection of adulterations in skimmed-milk powder. J Sep Sci 32:1216–1223
Corpillo D, Gardini G, Vaira AM, Basso M, Aime S, Accotto GP, Fasano M (2004) Proteomics as a tool to improve investigation of substantial equivalence in genetically modified organisms: the case of a virus-resistant tomato. Proteomics 4:193–200
EC Council Regulation 2006 no. 509/2006 of 20 March 2006 on agricultural products and foodstuff as traditional specialties guaranteed. Off J Eur Union CE L 93, 31/03/06
Cozzolino R, Passalacqua S, Salemi S, Malvagna P, Spina E, Garozzo D (2002) Identification of adulteration in milk by matrix-assisted laser desorption/ionization time-of-flight mass spectrometry. J Mass Spectrom 36:1031–1037
Cuadros-Inostroza A, Giavalisco P, Hummel J, Eckardt A, Willmitzer L, Pena-Cortes H (2010) Discrimination of wine attributes by metabolome analysis. Anal Chem 82:3573–3580
Cuollo M, Caira S, Fierro O, Pinto G, Picariello G, Addeo F (2010) Toward milk speciation through the monitoring of casein proteotypic peptides. Rapid Commun Mass Spectrom 24:1687–1696
Czerwenka C, Maier I, Pittner F, Lindner W (2006) Investigation of the lactosylation of whey proteins by liquid chromatography-mass spectrometry. J Agric Food Chem 54:8874–8882
Czerwenka C, Maier I, Potocnik N, Pittner F, Lindner W (2007) Absolute quantitation of beta-lactoglobulin by protein liquid chromatography-mass spectrometry and its application to different milk products. Anal Chem 79:5165–5172
Davies H (2010) A role for "omics" technologies in food safety assessment. Food Control 21:1601–1610
Delatour T, Hegele J, Parisod V, Richoz J, Maurer S, Steven M, Buetler T (2009) Analysis of advanced glycation endproducts in dairy products by isotope dilution liquid chromatography-electrospray tandem mass spectrometry. The particular case of carboxymethyllysine. J Chromatogr A 1216:2371–2381
Di Luccia A, Picariello G, Cacace G, Scaloni A, Faccia M, Liuzzi V, Alviti G, Musso SS (2005) Proteomic analysis of water soluble and myofibrillar protein changes occurring in dry cured hams. Meat Sci 69:479–491
Di Luccia A, Picariello G, Trani A, Alviti G, Loizzo P, Faccia M, Addeo F (2009) Occurrence of beta-casein fragments in cold-stored and curdled river buffalo (*Bubalus bubalis L.*) milk. J Dairy Sci 92:1319–1329
Directive N 101/2001 of the European Parliament and of the Council on the approximation of the laws of the Member States relating to the labelling, presentation and advertising of foodstuffs. *Official Journal of the European Communities*, L310/19
Erny GL, Cifuentes A (2007) Simplified 2-D CE-MS mapping: analysis of proteolytic digests. Electrophoresis 28:1335–1344
Erny GL, Marina ML, Cifuentes A (2007) Reproducible and efficient separation of aggregatable zein proteins by capillary zone electrophoresis using a volatile background electrolyte. Electrophoresis 28:2988–2997
Erny GL, Leon C, Marina ML, Cifuentes A (2008) Time of flight versus ion trap MS coupled to CE to analyse intact proteins. J Sep Sci 31:1810–1818
Fanton C, Delogu G, Maccioni E, Podda G, Seraglia R, Traldi P (1998) Matrix-assisted laser desorption/ionization mass spectrometry in the dairy industry 2. The protein fingerprint of ewe cheese and its application to detection of adulteration by bovine milk. Rapid Commun Mass Spectrom 12:1569–1573
Farid EA (2004) Testing of genetically modified organisms in foods. In: Farid EA (ed) 1560222735. ISBN-13: 9781560222736. CRC Press
Fenaille F, Morgan F, Parisod V, Tabet JC, Guy PA (2004) Solid-state glycation of beta-lactoglobulin by lactose and galactose: localization of the modified amino acids using mass spectrometric techniques. J Mass Spectrom 39:16–28

Fernandez Ocana M, Fraser P, Patel RKP, Halket JM, Bramley PM (2007) Mass spectrometric detection of CP4 EPSPS in genetically modified soya and maize. Rapid Commun Mass Spectrom 21:319–328

Fernandez Ocana M, Fraser P, Patel RKP, Halket JM, Bramley PM (2009) Evaluation of stable isotope labelling strategies for the quantitation of CP4 EPSPS in genetically modified soy. Anal Chim Acta 634:75–82

Ferranti P (2005) In: Proceedings of the workshop on quality, salubrity and safety of some typical Italian Cheeses, National Research Council/Italian Ministry of University and Research, 5th May, Rome

Ferranti P, Fabbrocino S, Cerulo MG, Bruno M, Serpe L, Gallo P (2008) Characterisation of biotoxins produced by a cyanobacteria bloom in Lake Averno using two LC-MS-based techniques. Food Addit Contam A 25:1530–1537

Ferranti P, Fabbrocino S, Nasi A, Caira S, Bruno M, Serpe L, Gallo P (2009) Liquid chromatography coupled to quadruple time-of-flight tandem mass spectrometry for microcystin analysis in freshwaters: method performances and characterisation of a novel variant of microcystin-RR. Rapid Commun Mass Spectrom 23:1328–1336

Ferranti P, Nasi A, Bruno M, Basile A, Serpe L, Gallo P (2011) A peptidomic approach for monitoring and characterising peptide cyanotoxins produced in Italian lakes by matrix-assisted laser desorption/ionisation and quadrupole time-of-flight mass spectrometry. Rapid Commun Mass Spectrom 25:1173–1183

Finnie C, Steenholdt T, Roda Noguera O, Knudsen S, Larsen J, Brinch-Pedersen H, Bach Holm P, Olsen O, Svensson B (2004) Environmental and transgene expression effects on the barley seed proteome. Phytochemistry 65:1619–1627

Fogliano V, Monti SM, Visconti A, Randazzo G, Facchiano AM, Colonna G, Ritieni A (1998) Identification of a beta-lactoglobulin lactosylation site. Biochim Biophys Acta 1388:295–304

French General Directorate for Fair Trading, Consumer Affairs and Fraud Control (DGCCRF) (2001) Activity reports of DGCCRF; 1997–1999; www.finances.gouv.fr/DGCCRF/activities/index-d.htm

Friedman M (1996) Food browning and its prevention: an overview. J Agric Food Chem 44:631–653

Friedman M (1999) Chemistry, biochemistry, nutrition, and microbiology of lysinoalanine, lanthionine, and histidinoalanine in food and other proteins. J Agric Food Chem 47:1295–1332

Gallien S, Duriez E, Domon B (2011) Selected reaction monitoring applied to proteomics. J Mass Spectrom 46:298–312

Gallo P, Fabbrocino S, Cerulo MG, Ferranti P, Bruno M, Serpe L (2009) Determination of cylindrospermopsin in freshwaters and fish tissue by liquid chromatography coupled to electrospray ion trap mass spectrometry. Rapid Commun Mass Spectrom 23:3279–3284

García-Cañas V, Cifuentes A (2008) Recent advances in the application of capillary electromigration methods for food analysis. Electrophoresis 29:294–309

Gerber SA, Rush J, Stemman O, Kirschner MW, Gygi SP (2003) Absolute quantification of proteins and phosphoproteins from cell lysates by tandem MS. Proc Natl Acad Sci USA 100:6940–6945

Gingras AC, Gstaiger M, Braught B, Aebersold R (2007) Analysis of protein complexes using mass spectrometry. Nat Rev Mol Cell Biol 8:645–654

Gorg A, Weiss W, Dunn MJ (2005) Current two-dimensional electrophoresis technology for proteomics. Proteomics 5:826–827

Gougeon RD, Lucio M, Frommberger M, Peyron D, Chassagne D, Alexandre H, Feuillat F, Voilley A, Cayot P, Gebefugi I, Hertkorn N, Schmitt-Kopplin P (2009) The chemodiversity of wines can reveal a metabologeography expression of cooperage oak wood. Proc Natl Acad Sci USA 106:9174–9179

Guy PA, Fenaille F (2006) Contribution of mass spectrometry to assess quality of milk-based products. Mass Spectrom Rev 25:290–326

Haselberg R, de Jong GJ, Somsen GW (2007) Capillary electrophoresis-mass spectrometry for the analysis of intact proteins. J Chromatogr A 1159:81–109

Hasenkopf K, Ubel B, Bordiehn T, Pischetsrieder M (2001) Determination of the Maillard product oxalic acidmonolysinylamide (OMA) in heated milk products by ELISA. Food 45:206–209

Hau J, Bovetto L (2001) Characterisation of modified whey protein in milk ingredients by liquid chromatography coupled to electrospray ionization mass spectrometry. J Chromatogr A 926:105–112

Hegele J, Buetler T, Delatour T (2008) Comparative LC-MS/MS profiling of free and protein-bound early and advanced glycation-induced lysine modifications in dairy products. Anal Chim Acta 617:85–96

Henle T, Schwarzenbolz U, Klostermeyer H (1997) Detection and quantification of pentosidine in foods. Z Lebensm Unters Forsch A 204:95–98

Hernando A, Mujico JR, Mena MC, Lombardía M, Méndez E (2008) Measurement of wheat gluten and barley hordeins in contaminated oats from Europe, the United States and Canada by Sandwich R5 ELISA. Eur J Gastroenterol Hepatol 20:545–554

Herrero M, Garcıa-Canas V, Simo C, Cifuentes A (2010) Recent advances in the application of capillary electromigration methods for food analysis and foodomics. Electrophoresis 31:205–228

Hewedi MM, Kiesner C, Meissner K, Hartkopf J, Erbersdobler HJ (1994) Effects of UHT heating of milk in an experimental plant on several indicators of heat treatment. J Dairy Res 61:305–309

Hogarth CJ, Fitzpatrick JL, Nolan AM, Young FJ, Pitt A, Eckersall PD (2004) Differential protein composition of bovine whey: a comparison of whey from healthy animals and from those with clinical mastitis. Proteomics 4:2094–2100

Huber CG, Premstaller A, Kleindienst G (1999) Evaluation of volatile eluents and electrolytes for high-performance liquid chromatography-electrospray ionization mass spectrometry and capillary electrophoresis-electrospray ionization mass spectrometry of proteins. II. Capillary electrophoresis. J Chromatogr A 849:175–189

Islam N, Campbell PM, Higgins TJ, Hirano H, Akhurst RJ (2009) Transgenic peas expressing an alpha-amylase inhibitor gene from beans show altered expression and modification of endogenous proteins. Electrophoresis 30:1863–1868

Jégou S, Douliez JP, Mollé D, Boivin P, Marion D (2001) Evidence of the glycation and denaturation of LTP1 during the malting and brewing process. J Agric Food Chem 49:4942–4949

Johnson RS, Davis MT, Taylor JA, Patterson SD (2005) Informatics for protein identification by mass spectrometry. Methods 35:223–236

Jones AD, Tier CM, Wilkins JPG (1998) Analysis of the Maillard reaction products of beta-lactoglobulin and lactose in skimmed milk powder by capillary electrophoresis and electrospray mass spectrometry. J Chromatogr A 822:147–154

Koppelman SJ, Lakemond CM, Vlooswijk R, Hefle SL (2004) Detection of soy proteins in processed foods: literature overview and new experimental work. J AOAC Int 87:1398–1407

Kuppannan K, Albers DR, Schafer BW, Dielman D, Young SA (2011) Quantification and characterization of maize lipid transfer protein, a food allergen, by liquid chromatography with ultraviolet and mass spectrometric detection. Anal Chem 83:516–524

Leitner A, Castro-Rubio F, Marina ML, Lindner W (2006) Identification of marker proteins for the adulteration of meat products with soybean proteins by multidimensional liquid chromatography-tandem mass spectrometry. J Proteome Res 5:2424–2430

Leonil J, Molle D, Fauquant J, Maubois JL, Pearce RJ, Bouhallab S (1997) Characterization by ionization mass spectrometry of lactosyl betalactoglobulin conjugates formed during heat treatment of milk and whey and identification of one lactose-binding site. J Dairy Sci 80:2270–2281

Liger-Belair G, Cilindre C, Gougeon RD, Lucio M, Jeandet P, Gebefugi I, Schmitt-Kopplin P (2009) Unraveling different chemical fingerprints between a champagne wine and its aerosols. Proc Natl Acad Sci USA 106:16545–16549

Liu H, Sadygov RG, Yates JR 3rd (2004) A model for random sampling and estimation of relative protein abundance in shotgun proteomics. Anal Chem 76:4193–4201

Losito I, Carbonara T, Monaci L, Palmisano F (2007) Evaluation of the thermal history of bovine milk from the lactosylation of whey proteins: an investigation by liquid chromatography-electrospray ionization mass spectrometry. Anal Bioanal Chem 389:2065–2074

Lum MR, Hirsch AM (2006) Molecular methods for the authentication of botanicals and detection of potential contaminants and adulterants. Acta Hort 720:59–72

Luykx DM, Cordewener JH, Ferranti P, Frankhuizen R, Bremer MG, Hooijerink H, America H (2007) Identification of plant proteins in adulterated skimmed milk powder by high-performance liquid chromatography—mass spectrometry. J Chromatogr A 1164:189–197

Mallick P, Schirle M, Chen SS, Flory MR, Lee H, Martin D, Ranish J, Raught B, Schmitt R, Werner T, Kuster B, Aebersold R (2007) Computational prediction of proteotypic peptides for quantitative proteomics. Nat Biotechnol 25:125–131

Mamone G, Picariello G, Caira S, Addeo F, Ferranti P (2009) Analysis of food proteins and peptides by mass spectrometry-based techniques. J Chromatogr A 1216:7130–7142

Mamone G, Picariello G, Addeo F, Ferranti P (2011) Proteomic analysis in allergy and intolerance to wheat products. Expert Rev Proteomics 8:95–115

Mann K, Mann M (2011) In-depth analysis of the chicken egg white proteome using an LTQ Orbitrap Velos. Proteome Sci 9:7–11

Marvin LF, Parisod V, Fay LB, Guy PA (2002) Characterization of lactosylated proteins of infant formula powders using two-dimensional gel electrophoresis and nanoelectrospray mass spectrometry. Electrophoresis 23:2505–2512

Mayya V, Han DK (2006) Proteomic applications of protein quantification by isotope-dilution mass spectrometry. Expert Rev Proteomics 3:597–610

Mazzeo MF, Giulio BD, Guerriero G, Ciarcia G, Malorni A, Russo GL, Siciliano RA (2008) Fish authentication by MALDI-TOF mass spectrometry. J Agric Food Chem 56:11071–11076

Meltretter J, Seeber S, Humeny A, Becker CM (2007) Sitespecific formation of Maillard, oxidation, and condensation products from whey proteins during reaction with lactose. J Agric Food Chem 55:6096–6103

Meltretter J, Becker CM, Pischetsrieder M (2008) Identification and site-specific relative quantification of beta-lactoglobulin modifications in heated milk and dairy products. J Agric Food Chem 56:5165–5171

Meltretter J, Birlouez-Aragon I, Becker CM, Pischetsrieder M (2009) Assessment of heat treatment of dairy products by MALDI-TOF-MS. Mol Nutr Food Res 53:1487–1495

Molle D, Morgan F, Bouhallab S, Leonil J (1998) Selective detection of lactolated peptides in hydrolysates by liquid chromatography/electrospray tandem mass spectrometry. Anal Biochem 259:152–161

Monaci L, van Hengel AJ (2007) Effect of heat treatment on the detection of intact bovine beta-lactoglobulins by LC mass spectrometry. J Agric Food Chem 55:2985–2992

Mora L, Sentandreu MA, Fraser PD, Toldrá F, Bramley PM (2009) Oligopeptides arising from the degradation of creatine kinase in Spanish dry-cured ham. J Agric Food Chem 57:8982–8988

Mora L, Sentandreu MA, Toldrá F (2011a) Intense degradation of myosin light chain isoforms in Spanish dry-cured ham. J Agric Food Chem 59:3884–3892

Mora L, Valero ML, Sánchez Del Pino MM, Sentandreu MA, Toldrá F (2011b) Small peptides released from muscle glycolytic enzymes during dry-cured ham processing. J Proteomics 74:442–450

Morgan F, Leonil J, Mollé D, Bouhallab S (1997) Nonenzymic lactosylation of bovine beta-lactoglobulin under mild heat treatment leads to structural heterogeneity of the glycoforms. Biochem Biophys Res Commun 236:413–417

Motoyama A, Xu T, Ruse CI, Wohlschlegel JA, Yates JR III (2007) Anion and cation mixed-bed ion exchange for enhanced multidimensional separations of peptides and phosphopeptides. Anal Chem 79:3623–3634

Muller L, Bartak P, Bednar P, Frysova I, Sevcık J, Lemr K (2008) Capillary electrophoresis-mass spectrometry – a fast and reliable tool for the monitoring of milk adulteration. Electrophoresis 29:2088–2093

Ono M, Shitashige M, Honda K, Isobe T, Kuwabara H, Matsuzuki H, Hirohashi S, Yamada T (2006) Label-free quantitative proteomics using large peptide data sets generated by nano-flow liquid chromatography and mass spectrometry. Mol Cell Proteomics 5:1338–1347

Palmisano G, Antonacci D, Larsen MR (2010) Glycoproteomic profile in wine: a 'sweet' molecular renaissance. J Proteome Res 9:6148–6159

Petit L, Pagny G, Baraige F, Nignol A, Zhang D, Fach P (2007) Characterization of genetically modified maize in weakly contaminated seed batches and identification of the origin of the adventitious contamination. Food Composition and Additive (Report). J AOAC Int 90(4):1098–1106

Petry-Podgórska I, Zídková J, Flodrová D, Bobálová J (2010) 2D-HPLC and MALDI-TOF/TOF analysis of barley proteins glycated during brewing. J Chromatogr B 878:3143–3148

Picariello G, De Martino A, Mamone G, Ferranti P, Addeo F, Faccia M, Spagnamusso S, Di Luccia A (2006) Proteomic study of muscle sarcoplasmic proteins using AUT-PAGE/SDS-PAGE as two-dimensional gel electrophoresis. J Chromatogr B 833:101–108

Picariello G, Sacchi R, Addeo F (2007) One-step characterization of triacylglycerols from animal fat by MALDI-TOF MS. Eur J Lipid Sci Technol 109:511–524

Picariello G, Ferranti P, Caira S, Fierro O, Chianese L, Addeo F (2009a) Fast screening and quantitative evaluation of internally deleted goat alphas1-casein variants by mass spectrometric detection of the signature peptides. Rapid Commun Mass Spectrom 23:775–787

Picariello G, Paduano A, Sacchi R, Addeo F (2009b) Maldi-tof mass spectrometry profiling of polar and nonpolar fractions in heated vegetable oils. J Agric Food Chem 57:5391–5400

Picariello G, Mamone G, Addeo F, Ferranti P (2011) The frontiers of mass spectrometry-based techniques in food allergenomics. J Chromatogr A 1218:7386–7398

Picotti P, Rinner O, Stallmach R, Dautel F, Farrah T, Domon B, Wenschuh H, Aebersold R (2010) High-throughput generation of selected reaction-monitoring assays for proteins and proteomes. Nat Methods 7:43–46

Pietrogrande MC, Marchetti N, Dondi F, Righetti PG (2006) De- coding 2D-PAGE complex maps: relevance to proteomics. J Chromatogr B 833:51–62

Pineiro C, Barros-Velazquez J, Vazquez J, Figueras A, Gallardo JM (2003) Proteomics as a tool for the investigation of seafood and other marine products. J Proteome Res 2:127–135

Pinto G, Caira S, Cuollo M, Fierro O, Nicolai MA, Chianese L, Addeo F (2012) Lactosylated casein phosphopeptides as specific indicators of heated milks. Anal Bioanal Chem 402:1961–1972

Piraino P, Upadhyay VK, Rossano R, Riccio P, Parente E, Kelly AL, McSweeney PLH (2007) Use of mass spectrometry to characterize proteolysis in cheese. Food Chem 101:964–972

Pirisi A, Pinna G, Addis M, Piredda G, Mauriello R, De Pascale S, Caira S, Mamone G, Ferranti P, Addeo F, Chianese L (2007) Relationship between the enzymatic composition of lamb rennet paste and proteolytic, lipolytic pattern and texture of PDO Fiore Sardo ovine cheese. Int Dairy J 17:143–156

Rob H, de Jong GJ, Somsen GW (2007) Capillary electrophoresis-mass spectrometry for the analysis of intact proteins. J Chromatogr A 1159:81–109

Scaloni A, Perillo V, Franco P, Fedele E, Froio R, Ferrara L, Bergamo P (2002) Characterization of heat-induced lactosylation products in caseins by immunoenzymatic and mass spectrometric methodologies. Biochim Biophys Acta 1598:30–39

Schlosser G, Kacer P, Kuzma M, Szilágyi Z, Sorrentino A, Manzo C, Pizzano R, Malorni L, Pocsfalvi G (2007) Coupling immunomagnetic separation on magnetic beads with matrix-assisted laser desorption ionization-time of flight mass spectrometry for detection of staphylococcal enterotoxin B. Appl Environ Microbiol 73:6945–6952

Scossa F, Laudencia-Chingcuanco D, Anderson OD, Vensel WH, Lafiandra D, D'Ovidio R, Masci S (2008) Comparative proteomic and transcriptional profiling of a bread wheat cultivar and its derived transgenic line over expressing a low molecular weight glutenin subunit gene in the endosperm. Proteomics 8:2948–2966

Seal CJ, de Mul A, Eisenbrand G, Haverkort AJ, Franke K, Lalljie SP, Mykkänen H, Reimerdes E, Scholz G, Somoza V, Tuijtelaars S, van Boekel M, van Klaveren J, Wilcockson SJ, Wilms L (2008) Risk-benefit considerations of mitigation measures on acrylamide content of foods–a case study on potatoes, cereals and coffee. Br J Nutr 99:S1–S46

Sentandreu MA, Sentandreu E (2011) Peptide biomarkers as a way to determine meat authenticity. Meat Sci 89:280–285

Sentandreu MA, Armenteros M, Calvete JJ, Ouali A, Aristoy MC, Toldrá F (2007) Proteomic identification of actin-derived oligopeptides in dry-cured ham. J Agric Food Chem 55:3613–3619

Sentandreu MA, Fraser PD, Halket J, Patel R, Bramley PM (2010) A proteomic-based approach for detection of chicken in meat mixes. J Proteome Res 9:3374–3383

Siciliano R, Rega B, Amoresano A, Pucci P (2000) Modern mass spectrometric methodologies in monitoring milk quality. Anal Chem 72:408–415

Simó C, Gonzalez R, Barbas C, Cifuentes A (2005) Combining peptide modeling and capillary electrophoresis-mass spectrometry for characterization of enzymes cleavage patterns: recombinant versus natural bovine pepsin A. Anal Chem 77:7709–7716

Simó C, Domínguez-Vega E, Marina ML, García MC, Dinelli G, Cifuentes A (2010) CE-TOF MS analysis of complex protein hydrolyzates from genetically modified soybeans – a tool for foodomics. Electrophoresis 31:1175–1183

Simpson DC, Smith RD (2005) Combining capillary electrophoresis with mass spectrometry for applications in proteomics. Electrophoresis 26:1291–1305

Somma A, Ferranti P, Addeo F, Mauriello R, Chianese L (2008) Peptidomic approach based on combined capillary isoelectric focusing and mass spectrometry for the characterization of the plasmin primary products from bovine and water buffalo beta-casein. J Chromatogr A 1192:294–300

Sospedra I, Soler C, Mañes J, Soriano JM (2011) Analysis of staphylococcal enterotoxin A in milk by matrix-assisted laser desorption/ionization-time of flight mass spectrometry. Anal Bioanal Chem 400:1525–1531

Surowiec I, Koistinen KM, Fraser PD, Bramley PM (2011) Proteomic approach for the detection of chicken mechanically recovered meat. Meat Sci 89:233–237

Taylor JLS, Demyttenaere JCR, Abbaspour Tehrani K, Olave CA, Regniers L, Verschaeve L, Maes A, Elgorashi EE, van Staden J, De Kimpe N (2003) Genotoxicity of melanoidin fractions derived from a standard glucose/glycine model. J Agric Food Chem 52:318–323

Tessier B, Schweizer M, Fournier F, Framboisier X, Chevalot I, Vanderesse R, Harscoat C, Marc I (2005) Prediction of the amino acid composition of small peptides contained in a plant protein hydrolysate by LC-MS and CE-MS. Food Res Int 38:577–584

Ünlü M, Morgan M, Minden J (1997) Difference gel electrophoresis. A single gel method for detecting changes in protein extracts. Electrophoresis 18:2071–2077

Uribarri J, Tuttle KR (2006) Advanced glycation end products and nephrotoxicity of highprotein diets. Clin J Am Soc Nephrol 1:1293–1299

van Boekel MAJS (2006) Formation of flavour compounds in the Maillard reaction. Biotechnol Adv 24:230–233

van Boekel M, Fogliano V, Pellegrini N, Stanton C, Scholz G, Lalljie S, Somoza V, Knorr D, Jasti PR, Eisenbrand G (2010) A review on the beneficial aspects of food processing. Mol Nutr Food Res 54:1215–1247

Van den Bergh G, Arckens L (2004) Fluorescent two-dimensional difference gel electrophoresis unveils the potential of gel-based proteomics. Curr Opin Biotechnol 15:38–43

Washburn MP, Wolters D, Yates JR III (2001) Large-scale analysis of the yeast proteome by multidimensional protein identification technology. Nat Biotechnol 19:242–247

Westermeier R, Marouga R (2005) Protein detection methods in proteomics research. Biosci Rep 25:19–32

White JW, Winters K, Martin P, Rossmann A (1998) Stable carbon isotope ratio analysis of honey: validation of internal standard procedure for worldwide application. J Assoc Off Anal Chem 81:610–619

Wilhelmsen EC (2004) Food adulteration. Food Sci Technol 138:2031–2056

Wilhelmsen EC (2006) Adulteration determination. In: Meyers RA (ed) Encyclopedia of analytical chemistry: applications, theory and instrumentation. Wiley, United States, p 14344

Wu WW, Wang G, Baek SJ, Shen RF (2007) Comparative study of three proteomic quantitative methods, DIGE, cICAT, and iTRAQ, using 2-D gel- or LC-MALDI TOF/TOF. J Proteome Res 5:651–668

Yeboah FK, Alli I, Yaylayan VA, Yasuo K, Chowdhury SF, Purisima EO (2004) Effect of limited solid-state glycation on the conformation of lysozyme by ESI-MSMS peptide mapping and molecular modeling. Bioconjug Chem 15:27–34

Zídková J, Flodrová D, Bobálová J (2010) 2D-HPLC and MALDI-TOF/TOF analysis of barley proteins glycated during brewing. J Chromatogr B 878:3143–3148

Chapter 25
Evaluation of Genetically Engineered Crops Using Proteomics

Agnès E. Ricroch and Marcel Kuntz

25.1 Introduction

The question addressed in this chapter is whether the use of transgenesis to improve a plant variety (through the expression of a new desired trait) can lead to unintended effects, that is, effects going beyond that of the original genetic modification, and which could have an impact on human and animal health. Such pleiotropic effects could be due to altered expression of untargeted genes leading to metabolic changes, or could be the consequence of an unexpected metabolic effect of a novel gene product.

Genetically-engineered (GE) varieties are assessed for their food and feed safety and nutritional quality in a comparative manner using parental or near isogenic lines as reference (the latter being considered as safe). It is aimed at identifying differences between these comparators and subsequently at evaluating the implications in terms of human and animal health. Current tools to perform such comparative safety assessments are targeted compositional analyses, animal nutrition, and classical toxicology evaluations, as well as agronomic evaluations. A major principle and guiding tool for the food safety assessment of GE crops is the concept of substantial equivalence according to principles outlined in the Organization for Economic Cooperation and Development consensus documents (OECD 2006).

In the last decade, new large-scale profiling methodologies have been developed that allow, in theory, a holistic search for alterations in GE crops. Numerous publications have examined whether the use of transgenesis as a plant breeding tool

A.E. Ricroch
Laboratoire Ecologie, Systématique et Evolution (UMR 8079),
Université Paris-Sud/CNRS/AgroParisTech, Bât. 360 91405, Orsay, France

M. Kuntz (✉)
Laboratoire de Physiologie Cellulaire Végétale (UMR5168),
Université Joseph Fourier/CNRS/CEA/INRA 38054, Grenoble, France
e-mail: kuntz@ujf-grenoble.fr

F. Toldrá and L.M.L. Nollet (eds.), *Proteomics in Foods: Principles and Applications*, Food Microbiology and Food Safety 2, DOI 10.1007/978-1-4614-5626-1_25,

could lead to unexpected changes in the transcriptome or metabolome (see Ricroch et al. 2011). That these two biological levels (transcripts and metabolites) have been most frequently examined is not surprising considering the above-mentioned potential sources of unexpected effects linked to the insertion of a transgene.

However, proteomics have also been used in a number of studies that are summarized here and which yield fairly convergent conclusions. Relevant questions also discussed here are: from a basic research point of view, are proteomics a powerful nontargeted approach to detect unintended effects in GE crops? Do proteomics lead to different conclusions from transcriptomics and metabolomics? Does our current experience with proteomics indicate that current methodologies are robust and sufficiently standardized to be used routinely for reglementary GE crop assessment?

25.2 Proteomic Analysis of GE Varieties of Crop Plants

The main data from the publications discussed below are summarized in Table 25.1, which also briefly lists the methodologies used.

25.2.1 *Grapevine*

Sauvage et al. (2007) used two-dimensional electrophoresis (2DE) to study changes in leaf protein content from GE grapevine plants over- and under-expressing alcohol dehydrogenase (experimental nonmarketed lines). MS identification of peptides indicated that only a few proteins had a different abundance in the GE lines. Interestingly, these proteins were mainly from the chloroplasts and involved in sugar-phosphate metabolism. It should also be noted that a consistency in the range of 53–72% of matching spots was found between the triplicates with a given sample.

25.2.2 *Maize*

MON810 is a transgenic trait providing resistance to certain lepidopteran pests (such as the European corn borer). This trait has been introgressed in a large number of corn varieties.

The grain proteomes of two field-grown MON810/non-GE variety pairs were found to be almost identical, with few spots showing quantitative changes in the 1–1.8-fold range (Coll et al. 2010a). These differences were all variety-specific (not present in both variety pairs). These data confirmed a previous study (Albo et al. 2007) on two different field-grown MON810 varieties. Another previous study (Zolla et al. 2008) found more differences between two MON810 variety pairs. In the latter study, field versus growth chamber growth conditions were also

Table 25.1 2-DE and proteomic comparison of GE varieties with non-GE varieties

Species	References	Varieties or traits/growth	Differences in GE versus control	Methods
Grapevine	Sauvage et al. (2007)	Control, sense, and antisense for alcohol dehydrogenase/greenhouse	Leaves. Changes in abundance in 14 proteins mainly from chloroplast and sugar-phosphate metabolism	2-DE. Gel Image Analysis. MALDI-TOF-MS. LC-MS/MS
Maize	Albo et al. (2007)	Two MON810/non-GE variety pairs/field	Grain proteomes mainly identical	2-DE. MALDI-TOF. LC-MS/MS
	Balsamo et al. (2011)	Four MON810/non-GE pairs (Brazilian varieties)/controlled environmental conditions	Leaves. Quantitative differences in 5 proteins (not consistent in all experiments). 7 exclusive spots for one pair, 5 for the other	2-DE. MALDI-TOF-MS
	Barros et al. (2010)	Two GE lines (MON810 and tolerant to glyphosate) and respective controls/different fields and years	Quantitative differences in 5 grain proteins. Environment exerting greater effects	2-DE
	Batista and Oliveira (2010)	MON810 and non-GE/field/individual or pooled plants	Natural plant-to-plant variability for grain proteins	2-DE. Gel Image Analysis
	Coll et al. (2010a)	Two MON810/non-GE variety pairs/field	Grain proteomes mainly identical, with few spots showing quantitative changes in the 1–1.8 fold range	2-DE. Gel Image Analysis. LC-MS/MS
	Coll et al. (2011)	Two MON810/non-GE variety pairs/real agricultural conditions	Quantitative differences in a particular variety pair not exceeding 1.2% of the grain proteins spots. Approx. 40% non matching proteins between two conventionally varieties	2-DE. Gel Image Analysis. LC-MS/MS
	Zolla et al. (2008)	Two MON810/non-GE variety pairs/field + growth chamber	Grain. Differences in both variety pairs. Growth conditions induced more changes	2-DE. LC-MS/MS

(continued)

Table 25.1 (continued)

Species	References	Varieties or traits/growth	Differences in GE versus control	Methods
Pea	Chen et al. (2009)	Two pea lines producing a bean alpha-amylase inhibitor	Seed. 33 proteins differentially accumulated	2-DE. Gel Image analysis. MALDI-TOF-TOF
	Islam et al. (2009)	Two pea lines producing a bean alpha-amylase inhibitor and non-GE/greenhouse	About 30 seed protein changes in abundance in each GE/non-GE pair (different in both pairs): minor differences for one pair, quantitative and qualitative differences for second	2-DE. Gel Image Analysis. LC-MS/MS
Potato	Lehesranta et al. (2005)	Range of diverse varieties, landraces and GE lines (empty vector or potato sense and antisense). Mal1 gene or antis. S-adenosylmethionine decarboxylase gene/field	Tuber. GE: quantitative differences in a total of 9 proteins out of 730 Qualitative and quantitative differences in 1,077 out of 1,111 when comparing varieties and landraces	2-DE. Gel Image Analysis. LC-MS/MS
Rice	Takahashi et al. (2005)	Expression of a homolog of maize HC-toxin reductase, conferring tolerance to several stresses/cell cultures	Proteome of cultured cells: higher levels for 5 spots out of 668	2-DE. Image gel analysis
Soybean	Brandao et al. (2010)	Glyphosate-tolerant and non-GE	Differences (at least 90% variation) for 10 seed proteins)	2-DE. Gel Image Analysis. MALDI-TOF-MS.
Tomato	Corpillo et al. (2004)	Hybrid variety expressing the nucleoprotein gene of TSW virus (TSWV-*N*)+*nptII* for selection and non-GE/growth chamber	Seedlings. No major changes	2-DE. MALDI-TOF-MS

Tomato and tobacco	Di Carli et al. (2009)	*Lycopersicon esculentum* (cv. MicroTom) and *Nicotiana benthamiana* producing recombinant antibodies against cucumber mosaic virus and tomato spotted wilt virus, resp.	Leaves. Quantitative differences (average less than 2.4) in 10 proteins out of 2,000 spots	2-DE. Gel Image Analysis. MALDI-TOF-MS. LC-MS/MS
Wheat	Scossa et al. (2008)	Bread wheat overexpressing a GE low MW glutenin subunit gene and non-GE	Seeds. Differential accumulation of several classes of endosperm proteins, paralleled by corresponding changes in transcript levels	

compared and more changes detected than in the case of the genetic modification. Concerning the differences, these authors speculated that genome rearrangement occurred, but other explanations seem equally likely such as the use of not fully isogenic comparator lines. It is, however, difficult to explain the discrepancy between the results by Zolla et al. (2008) and by Coll et al. (2010a) inasmuch as the latter team also used one of the two pairs used by the former.

More recently, another study on two variety pairs of MON810 and the comparable non-GE counterpart grown in real agricultural conditions has been published (Coll et al. 2011). A very small number of quantitative differential spots was found in a particular variety pair, not exceeding 1.2% of the proteins analyzed by 2DE. The differences between two conventional varieties were much greater (approx. 40% nonmatching proteins). It should be mentioned that the same team found similar results when using a transcriptomic approach in leaf (Coll et al. 2008, 2010b).

Barros et al. (2010) performed a transcriptomic, proteomic, and metabolomic comparison of kernels of two GE maize lines (MON810 and one line tolerant to the herbicide glyphosate) with the respective control lines. When plants were grown in the same location over three seasons, the authors found more differences in gene expression, protein distribution, and metabolite content between seasons than differences linked to the genetic modification. That environment exerts a greater effect was also shown by the distinct profiles observed when plants were grown for one season in three different locations.

The occurrence of natural plant-to-plant variability (not linked to a genetic modification) was investigated by Batista and Oliveira (2010). They compared 2DE protein patterns from MON810 and comparator lines obtained either from individual plants or from pooled plants. For individual samples, five different ears of five different maize plants were harvested. They observed, for some spots, a high quantitative variability between individual samples from one line and this variability was masked when plants were pooled. For other spots, variability existed between individual samples and also between pools.

Leaf proteome was compared in two pairs of GE (MON810) and non-GE isogenic Brazilian varieties grown under controlled environmental conditions (Balsamo et al. 2011). Leaf proteomic profiles of the four GE varieties were similar to their non-GE counterparts. Observed quantitative differences (in five spots) were not reproducible in all six 2DE performed. Reproducible qualitative differences were as follows: in the first pair, 1 exclusive spot in the GE line and 6 in the non-GE line; in the second pair, 1 exclusive spot in the GE line and 4 in the non-GE line. Thus, 12 exclusive proteins were observed in total; all of these leaf proteins were variety-specific. Previous studies also found similar maize leaf transcriptome patterns in GE and their non-GE counterparts (Coll et al. 2008, 2009).

25.2.3 Pea

Islam et al. (2009) performed a proteomic study on two pea lines producing a bean alpha-amylase inhibitor (AI1). About 30 seed protein spots were found to differ in abundance between each GE/comparator pair, but they were generally different between pairs, although the GE lines produced AI1 at similar levels. These differences were minor for one pair, but strikingly quantitatively and qualitatively different (appearance and disappearances of 36 protein spots) for the second pair. According to the authors, differences of a "similar magnitude" exist between different pea cultivars.

Chen et al. (2009) found that 33 proteins differentially accumulated in AI1-expressing lines compared with the parental line. They concluded that three of these differences were associated with the production of AI1 and the remaining 30 differences were due to the transformation events. Seed storage proteins, which are common food allergens, were among the protein exhibiting changes in their amounts. This illustrates the interest of using 2DE protein analysis and proteomics to detect new allergens in food (see below).

25.2.4 Potato

Important differences, both qualitative and quantitative, were found in the tuber proteome of field-grown varieties and landraces, and such differences were limited (quantitative) between the pairs constituted by experimental GE lines modified in cell wall structure or in ethylene/polyamine metabolism and their comparators (Lehesranta et al. 2005). It should be noted that the same and related lines, plus lines expressing a sense and antisense fructokinase gene were also studied using metabolomics (Defernez et al. 2004) or targeted composition analysis (Shepherd et al. 2006), with similar conclusions.

25.2.5 Rice

Takahashi et al. (2005) used in vitro cultured cells from experimental GE rice over-expressing the *YK1* gene, the homologue of maize HC-toxin reductase (HCTR) in rice. Out of a total of 668 polypeptides visualized by 2DE, 5 were increased in cells over-expressing *YK1* with respect to the control line, which included stress-related proteins such as osmotin-like protein and osr40c1 (an abscisic acid-responsive protein normally associated with salt tolerance).

25.2.6 Soybean

Using 2DE, Brandao et al. (2010) compared soybean GE (tolerant to the herbicide glyphosate) and non-GE comparator seeds. They found differences (greater than 90% variation in protein spot area and/or intensity) for ten proteins, six of them with differences in area or intensity and four of them in volume and intensity. Two proteins could not be identified. The other eight proteins identified by MS were seed storage proteins of glycinin and β-conglycinin types (responsible for the main nutritional, physicochemical, and physiological properties of soybeans), actin, sucrose-binding protein (involved in the storage of nutrients and sugar binding) and allergen Gly mBd 28 k (less than twofold increase).

25.2.7 *Tomato*

Corpillo et al. (2004) considered a virus-resistant GE tomato line to be "substantially equivalent" to its non-GE counterpart from both proteomic and agronomic points of view, because no reproducible differences could be found in peptide abundance in the GE line versus the control line.

Di Carli et al. (2009) used two transgenic plant models, a dwarf tomato line and tobacco, producing recombinant antibodies against two plant viruses. In each model, around 10 proteins out of around 2,000 spots were found to accumulate differentially in the transgenic lines. The variation ratio was less than 2.4 on average. Most of the differentially accumulated proteins were related to photosynthesis or plant defense.

25.2.8 *Wheat*

Scossa et al. (2008) performed, in parallel, a transcriptional and proteomic comparison of seeds from a GE bread wheat line (over-producing a low molecular weight glutenin subunit) with respect to the nontransformed line. Differential accumulation of several classes of endosperm proteins was observed, and paralleled by corresponding changes in transcript levels. The authors interpreted these observations as a compensatory mechanism of the strong over-expression of the transgenic glutenin gene (a consequence of the diversion of amino acid reserves).

25.3 Proteomic Analysis of *Arabidopsis thaliana* Relevant to GE Plant Safety

A. thaliana is a basic research model plant that can also provide information about the potential impact of transgenesis (Wienkoop et al. 2010).

A line containing the selectable bar marker gene (encoding phosphinotricin acetyl transferase) was used in several studies. Ren et al. (2009a) concluded that differences in metabolic profiles (major contributors were altered levels of alanine and threonine) were due to the bar gene. However, 2DE analysis on 12 bar-containing lines showed little consistent differences in the 4–14 protein spots that changed quantitatively in these lines (Ren et al. 2009b).

Natural variability can also be important in *A. thaliana* ecotypes. After growing various ecotypes under controlled growth conditions, and using 2DE to analyze seed proteins, Ruebelt et al. (2006) found that nearly half the spots were present or absent depending on the ecotype. In addition, 95% of the spots consistently found in all ecotypes were found to vary quantitatively. The seed proteome of 12 transgenic lines (using three different genes and three different promoters) were also compared to their parental line and to 12 ecotype lines, with no evidence for unintended changes.

25.4 Allergen Detection

Evaluation of their allergenicity potential is part of the reglementary safety assessment before marketing. 2DE of protein extracts may provide a complementary approach as shown by the following examples.

Batista et al. (2007) performed 2DE of flour protein extracts from a glyphosate-tolerant soybean and its non-GE comparator. Blots were probed with sera from soybean-sensitive individuals. MS was also used to identify IgE-binding proteins. Allergen production apparently remained unaltered as a consequence of the genetic modification.

Nakamura et al. (2010) also used 2DE-based techniques to compare an experimental GE potato line (producing an Arabidopsis transcription factor) with a control. The patterns of IgE-binding proteins were almost identical, with, however, several quantitative differences in these proteins (identified by MS/MS).

Satoh et al. (2011) used a similar approach in the case of GE rice. Salt-soluble proteins were probed with patient sera and antigen-specific animal sera. No differences were found between GE or non-GE lines.

25.5 Discussion

These proteomic studies are heterogeneous and have to be considered as exploratory. Considering all sources of difficulties in data interpretation (such as plant to plant variation, the possibility that comparator lines are not necessarily fully isogenic, experimental errors, etc.), care has to be taken as to their biological signification. These studies form merely a compilation of data; no normalized and validated approaches are available for routine assessment of GE plants.

Nevertheless, some lessons can be learned from these studies. Some differences are found in the proteome of GE varieties compared to control varieties. However, these differences can be generally considered as quantitatively minor and, when data are available, comparison of various conventional varieties consistently shows more differences. This is due, on one hand, to the biodiversity of existing varieties and, on the other, to the fact that GE lines have been selected from the laboratory to the field by phenotypic comparison with a close comparator. In addition, for marketed varieties, the transgenic trait has usually been introgressed into elite lines (these crosses contribute to the elimination of unintended genomic modifications). In some studies, environmental factors (such as field location or year of sampling) have consistently been shown to exert a greater impact than transgenesis.

As shown by Ricroch et al. (2011), metabolomics are the prevalent "omic" approach to assess GE crops, followed by transcriptomics. Few laboratories have used different omics comparatively. Therefore, an exhaustive comparative assessment of these techniques is not yet possible. Nevertheless, when data are available, there is no indication that proteomics will arrive at different general conclusions regarding the extent of unintended impact of transgenesis.

Perhaps the most useful role for 2DE protein analysis would be for allergenicity predictions. Proteomic and mass spectrometry methods can be used for qualitative and quantitative estimation of the allergen levels, including new ones.

None of these proteomic assessments has raised new safety concerns about marketed GE varieties. This is not surprising considering the experience acquired after 16 years of GE crop marketing. However, considering the highly polarized opinions on GE crops and a certain distrust in data provided by seed companies, it is important that new approaches such as transcriptomics, proteomics, and metabolomics have been used by public research laboratories.

References

Albo AG, Mila S, Digilio G, Motto M, Aime S, Corpillo D (2007) Proteomic analysis of a genetically modified maize flour carrying Cry1Ab gene and comparison to the corresponding wild-type. Maydica 52:443–455

Balsamo GM, Cangahuala-Inocente GC, Bertoldo JB, Terenzi H, Aris AC (2011) Proteomic analysis of four Brazilian MON810 maize varieties and their four non-genetically-modified isogenic varieties. J Agric Food Chem 59:11553–11559

Barros ELS, Anttonen MJ, van Dijk JP, Röhlig RM, Kok EJ, Engel KH (2010) Comparison of two GM maize varieties with a near-isogenic non-GM variety using transcriptomics, proteomics and metabolomics. Plant Biotechnol J 8:436–451

Batista R, Oliveira M (2010) Plant natural variability may affect safety assessment data. Regul Toxicol Pharmacol 58:8–12

Batista R, Martins I, Jeno P, Ricardo CP, Oliveira MM (2007) A proteomic study to identify soya allergens – the human response to transgenic versus non-transgenic soya samples. Int Arch Allergy Immunol 144:29–38

Brandao AR, Barbosa HS, Arruda MAZ (2010) Image analysis of two dimensional gel electrophoresis for comparative proteomics of transgenic and non-transgenic soybean seeds. J Proteomics 73:1433–1440

Chen HC, Bodulovic G, Hall PJ, Moore A, Higgins TJV, Djordjevic MA, Rolfe BG (2009) Unintended changes in protein expression revealed by proteomic analysis of seeds from transgenic pea expressing a bean alpha-amylase inhibitor gene. Proteomics 9:4406–4415

Coll A, Nadal A, Palaudelmás M, Messeguer J, Mele E, Puigdoménech P, Pla M (2008) Lack of repeatable differential expression patterns between MON810 and comparable commercial varieties of maize. Plant Mol Biol 68:105–117

Coll A, Collado ANR, Capellades G, Messeguer J, Melé E, Palaudelmàs M, Pla M (2009) Gene expression profiles of MON810 and comparable non-GM maize varieties cultured in the field are more similar than are those of conventional lines. Transgenic Res 18:801–808

Coll A, Nadal A, Rossignol M, Puigdoménech P, Pla M (2010a) Proteomic analysis of MON810 and comparable non-GM maize varieties grown in agricultural fields. Transgenic Res 20:939–949
Coll A, Nadal A, Collado R, Capellades G, Kubista M, Messeguer J, Pla M (2010b) Natural variation explains most transcriptomic changes among maize plants of MON810 and comparable non-GM varieties subjected to two N-fertilization farming practices. Plant Mol Biol 73:349–362
Coll A, Nadal A, Rossignol M, Puigdomenech P, Pla M (2011) Proteomic analysis of MON810 and comparable non-GM maize varieties grown in agricultural fields. Transgenic Res 20:939–949
Corpillo D, Gardini G, Vaira AM, Basso M, Aime S, Accotto GP, Fasano M (2004) Proteomics as a tool to improve investigation of substantial equivalence in genetically modified organisms: the case of a virus-resistant tomato. Proteomics 4:193–200
Defernez M, Gunning YM, Parr AJ, Shepherd LVT, Davies HV, Colquhoun IJ (2004) NMR and HPLC-UV profiling of potatoes with genetic modifications to metabolic pathways. J Agric Food Chem 52:6075–6085
Di Carli M, Villani ME, Renzone G, Nardi L, Pasquo A, Franconi R, Scaloni A, Benvenuto E, Desiderio A (2009) Leaf proteome analysis of transgenic plants expressing antiviral antibodies. J Proteome Res 8:838–848
Islam N, Campbell PM, Higgins TJV, Hirano H, Akhurst RJ (2009) Transgenic peas expressing an alpha-amylase inhibitor gene from beans show altered expression and modification of endogenous proteins. Electrophoresis 30:1863–1868
Lehesranta Satu J, Davies HV, Shepherd LVT, Nunan N, McNicol JW, Auriola S, Koistinen KM, Suomalainen S, Kokko HI, Kärenlampi SO (2005) Comparison of tuber proteomes of potato varieties, landraces, and genetically modified lines. Plant Physiol 138:1690–1699
Nakamura R, Satoh R, Shimazaki T, Kasuga M, Yamaguchi-Shinozaki K, Kikuchi A, Watanabe KN, Teshima R (2010) Immunoproteomic and two-dimensional difference gel electrophoresis analysis of arabidopsis dehydration response element-binding protein 1A (DREB1A)-transgenic potato. Biol Pharma Bull 33:1418–1425
OECD (2006) An introduction to the food/feed safety consensus documents of the task force. In: Series on the safety of novel foods and feeds, No 14, 7–9. Paris: Organization for Economic Cooperation and Development
Ren Y, Wang T, Peng Y, Xia B (2009a) Distinguishing transgenic from non-transgenic *Arabidopsis* plants by 1H NMR-based metabolic fingerprinting. J Genet Genomics 36:621–628
Ren Y, Jun L, Wang H, Li L, Peng Y, Qu LJ (2009b) A comparative proteomics approach detect unintended effects in transgenic *Arabidopsis*. J Genet Genomics 36:629–639
Ricroch A, Bergé J-B, Kuntz M (2011) Evaluation of genetically engineered crops using transcriptomic, proteomic and metabolomic profiling techniques. Plant Physiol 55:1752–1761
Ruebelt MC, Lipp M, Reynolds TL, Astwood JD, Engel K-H, Jany K-D (2006) Application of two-dimensional gel electrophoresis to interrogate alterations in the proteome of genetically modified crops. 2. Assessing natural variability. J Agric Food Chem 54:2162–2168
Satoh R, Nakamura R, Komatsu A, Oshima M, Teshima R (2011) Proteomic analysis of known and candidate rice allergens between non-transgenic and transgenic plants. Regul Toxicol Pharmacol 59:437–444
Sauvage F-X, Pradal M, Chatelet P, Tesniere C (2007) Proteome changes in leaves from grapevine (*Vitis vinifera* L.) transformed for alcohol dehydrogenase activity. J Agric Food Chem 55:2597–2603
Scossa F, Laudencia-Chingcuanco D, Anderson OD, Vensel WH, Lafiandra D, D'Ovidio R, Masci S (2008) Comparative proteomic and transcriptional profiling of a bread wheat cultivar and its derived transgenic line overexpressing a low molecular weight glutenin subunit gene in the endosperm. Proteomics 8:2948–2966
Shepherd LVT, McNicol JW, Razzo R, Taylor MA, Davies HV (2006) Assessing the potential for unintended effects in genetically modified potatoes perturbed in metabolic and developmental processes. Targeted analysis of key nutrients and anti-nutrients. Transgenic Res 15:409–425

Takahashi H, Hotta Y, Hayashi M, Kawai-Yamada M, Komatsu S, Uchimiya H (2005) High throughput metabolome and proteome analysis of transgenic rice plants (*Oryza sativa* L.). Plant Biotechnol 22:47–50

Wienkoop S, Baginsky S, Weckwerth W (2010) Arabidopsis thaliana as a model organism for plant proteome research. J Proteomics 73:2239–2248

Zolla L, Rinalducci S, Antonioli P, Righetti PG (2008) Proteomics as a complementary tool for identifying unintended side effects occurring in transgenic maize seeds as a result of genetic modifications. J Proteome Res 7:1850–1861

Chapter 26
Microbial Proteomics for Food Safety

Jasminka Giacometti and Djuro Josić

26.1 Introduction

For millennia, microorganisms have been indispensable tools in food technology and biotechnology (Josic and Kovac 2008). On the other hand, more than 250 known pathogens, mostly microbial ones and their toxins are known to cause food-borne illness. In the United States alone, about 48 million illnesses, 128,000 hospitalizations, and 3,000 deaths caused by food poisoning have been registered (CDC 2011a, b). Although not or only partially documented, this number is much higher in the Third World countries. Climatic factors influence the growth and survival of pathogens in food, as well as transmission pathways (Kovats 2011; Risks 2009). New risks are being encountered because of changing characteristics of the relevant microorganisms, changing production methodologies, changes in the environment and the ecology, and an increase of the global trade of foodstuffs. In addition, demands on food safety increase steadily. Due to the nature of both food pathogens and our food chain, measures to ensure food safety have to be implemented on a global scale, necessitating also a global approach.

Outbreaks of food-borne pathogens appear to be on the rise again in some industrialized countries, with a shift from traditional problems with food from mostly animal origin such as meat, eggs, milk, and milk products to fresh food of

J. Giacometti
Department of Biotechnology, University of Rijeka,
R. Matejčić 2 HR-51000, Rijeka, Croatia
e-mail: jgiacometti@biotech.uniri.hr

D. Josić (✉)
Department of Biotechnology, University of Rijeka,
R. Matejčić 2 HR-51000, Rijeka, Croatia

Warren Alpert Medical School, Brown University,
Rhode Island, Providence, RI, USA
e-mail: djosic@biotech.uniri.hr; djuro_josic@brown.edu

F. Toldrá and L.M.L. Nollet (eds.), *Proteomics in Foods: Principles and Applications*, Food Microbiology and Food Safety 2, DOI 10.1007/978-1-4614-5626-1_26,

plant origin (fruits and vegetables), shellfish, and dry products and ingredients (Havelaar et al. 2010). In the last years numerous incidents have occurred in contamination of fresh and processed food with food-borne pathogens (Austin et al. 1998). Despite numerous reported cases of food contamination with pathogen agents, throughout the developed countries generally were reported decreasing numbers of cases of food poisoning. It is due to advanced control measures regarding general hygiene, food safety, and improved methods of food processing and storage (Hartung and Koeter 2008).

Proteomic techniques offer considerable opportunities to assess animal and plant health and production and to monitor quality and safety of food of animal and plant origin (Gaso-Sokac et al. 2010). Proteome analysis of pathogens in infected food provides reliable information about pathogen activities during infection, outbreak of disease, and healing period. Importantly, functionally relevant proteins are preferentially identified and *in vivo* proteome data can thus help to guide further functional analysis efficiently (Scott and Cordwell 2009). Technical limitations in the past concern quantification, identification of post-translational modifications, and poor coverage of secreted and low-abundance proteins. New methods in genomics, proteomics, peptidomics, glycomics, and metabolomics together with already established genomic methods are emerging to overcome most of these limitations and challenges. This will further improve *in vivo* investigations and enhance the value of these techniques as an important resource to investigate and combat infectious diseases (Bumann 2010). Around 3,705 microbial genomes have been completely sequenced, including 1,750 bacteria, 94 archaea, and 119 eukaryotes (http://www.microbesonline.org/). The sequenced microbial genomes provide large amounts of data that can be used for fast and high-throughput screening technologies (Cravatt et al. 2008; O'Flaherty and Klaenhammer 2011).

This chapter focuses on the role of proteomics in assuring food safety. Pathogenic bacteria *Escherichia coli*, *Salmonella*, and *Campylobacter* are those of greatest current concern in fresh products, and *Listeria monocytogenes* contamination is of primary concern in processed products (Tompkin 2002). The pathogenic fungi *Aspergillus, Fusarium, Penicillium, Alternaria*, and *Claviceps* are the eukaryotic food pathogens with the highest toxigenic potential (Capriotti et al. 2011). Viruses and prions have recently been discussed as a further concern for safety, especially in food of animal origin (Gaso-Sokac et al. 2010) although this topic is beyond the scope of this review.

26.2 Factors Affecting the Survival and Growth of Pathogens

Various fundamental changes in an organism's cellular physiology are caused by stress exposure. Stress responses are of particular importance to microorganisms, because their habitats are subjected to continuous changes of temperature, osmotic pressure, and availability of nutrients (Capozzi et al. 2009). Pathogen survival and growth on food-produce is influenced by a number of interdependent factors such as storage temperature, product type/combinations, minimal processing operations (e.g., slicing, shredding, washing, and decontamination treatments), mild processing methods (e.g., high pressure, pulsed electric field processing, cold plasma, advanced heating), package atmosphere, and competition from the natural microflora present on food-produce (Capozzi et al. 2009). Antimicrobial washing by using different disinfectants and adding natural antimicrobial agents is another important factor that diminishes populations of microorganisms on fresh produce (Adams et al. 1989; Allende et al. 2006; Aureli et al. 1992; Burrows et al. 1999; Park et al. 2001).

The latest consumer preference for less processed food containing lower amounts of preservatives significantly changed the current situation regarding food safety (Allende et al. 2006; Elmnasser et al. 2007). Especially the so-called "ready to eat" minimally processed vegetables that are popular among consumers. Such foods retain much of the indigenous microflora, and also some potential bacterial pathogens, such as *Aeromonas* spp. and *Yersinia* spp. (Xantopoulos et al. 2010). It is also documented that the serious outbreak of food poisoning in Northern Europe in years 2010/2011, caused by a novel strain of *E. coli* appeared after consuming such minimally processed food (Fenugreek seeds with high probability for EHEC O104; German Federal Institute of Risk Assessment 2011).

26.3 Proteomics in Microbial Stress Adaptation

The adaptive stress response of bacteria is a crucial mode of cellular protection towards environmental and food-relevant stresses. Cellular biomarkers are quantitatively correlated to adaptive stress and can also predict the impact of changing environments on bacterial ability to resistance and survival. Cross-protection phenomena preservation strategy is another problem to ensure microbial safety and stability, as well as the sensory and nutritional quality of minimally processed food (Konieczny and Kijowski 2005). The right combination of hurdles may be effective in controlling growth bacteria and pathogens, and food spoilage and minimizing organoleptic changes in food. Long and short time exposures to mild stress conditions such as heat, acid, salt, and oxidative agents induce (cross-) protection towards otherwise lethal stress conditions (Abee and Wouters 1999; Skandamis et al. 2008) and might even affect the virulence of pathogens (Cotter and Hill 2003; Gahan and Hill 1999). The ability of microorganisms to develop resistance upon activation of adaptive stress responses is also beneficial for various industrial applications (Serrazanetti et al. 2009), including the development of reliable starter cultures (van de Guchte et al. 2002) and selection of stabile probiotic strains (Culligan et al. 2009). Den Besten et al. (2010) identified several cellular biomarkers for bacterial stress adaptive behavior at the transcriptome, proteome, and metabolome levels. In *Bacillus cereus* as a model microorganism, potential candidate biomarkers to stress response were identified, such as transcriptional regulator σ^B, catalases involved in H_2O_2-scavenging, chaperones, and ATP-dependent Clp proteases involved in protein repair and maintenance. In an overview, Abee et al. (2011) integrated three different research strategies that have led to the identification of biomarkers important for robustness in *Bacillus cereus*, and based on two-component systems (TCS) using comparative genomics, a meta-transcriptome data analysis of the RsbKY-controlled σ^B regulon and protease transcript levels and catalase in prediction of the robustness level of stress-adapted *B. cereus* cells towards lethal stresses.

The investigations using a wide variety of environmental changes were further performed with *B. cereus*, and were extended to the other bacteria including *Bacillus subtilis*, *Listeria monocytogenes*, *Escherichia coli* (Petersohn et al. 2001; Allen et al. 2008; van der Veen et al. 2007), and *Chronobacter turicensis* (Carranza et al. 2009), as well as with different yeast specii (Causton et al. 2001; Gasch et al. 2000). This quantitative approach leads to prediction of microbial performance using cellular biomarkers for the early detection of food pathogens and to the control of their adaptive behavior that results in enhanced resistance (Callahan et al. 2006; Dupuis et al. 2008; Levin 2009).

26.3.1 Cold- and Heat-Stress Adaptation

Cold stress impairs cellular replication, transcription, and translation processes (Schmid et al. 2009). After cold shock, mesophilic bacteria have to cope with several recognized problems such as too low membrane fluidity, too high superhelical density of the DNA for opening of the double helix, decrease in enzyme activities and adjusted protein levels, adaptation of ribosome function at low temperatures, and initiation of translation of secondary structures in RNA. Upon these conditions, appropriate signal transduction processes are activated and subsequently lead to the mobilization of necessary stress protection measures through modifications in gene expression and protein function activities (Graumann and Marahiel 1999). Insights into gene expression changes mobilized during stress adaptation responses have recently been gained through transcriptome and proteome stress analysis in bacteria. The transcriptional regulator σ^B functions as a central regulator of general stress responses in Gram-positive bacteria including the genera *Bacillus*, *Staphylococcus*, and *Listeria.* This sigma factor is absent in various lactic acid bacteria (van de Guchte et al. 2002; van Schaik and Abee 2005a), suggesting that these bacteria have developed different stress regulatory networks.

In *B. cereus*, heat shock also had a strong σ^B-activating effect, but other stresses (such as ethanol shock, osmotic shock, and acid stress) were also found to lead to the activation of σ^B (van Schaik et al. 2004). A mechanistic understanding of the σ^B-activation processes and assessment of its regulons could provide tools for pathogen control and inactivation both in the food industry and clinical settings.

Some investigations studied the cold-shock response and cold-shock proteins using *E. coli* and *B. subtilis* as model systems. Using a combined transcriptomic and proteomic approach, Budde et al. (2006) identified the members of the chill-stress stimulon of *B. subtilis* on a genome-wide scale. A total of 580 genes, representing approximately 14% of the protein-coding capacity of *B. subtilis*, displayed temperature-dependent alterations: 279 genes were induced and 301 genes were repressed (Budde et al. 2006). The complementation of the transcriptional profiling by a proteomic approach allowed the identification of a substantial number of proteins up- or downregulated in low-temperature-grown *B. subtilis* cells whose structural genes were not identified as differentially expressed by this transcriptional analysis. The rather large size of this group of proteins indicates that post-transcriptional regulatory phenomena make major contributions to the adaptation of *B. subtilis* to growth at low temperature. Accumulation of stress proteins including disaggregation chaperones (DnaK and ClpB) and components of the RNA degradosome, enolase, and polynucleotide phosphorylase (PNP) was increased at higher temperatures in *E. coli*. At 42°C, over-expression of the inclusion body-associated proteins A and B (ibpAB) and increased alpha-glucosidase activity were noted in the study of Lethanh et al. (2005). These authors suggest that the ratio of IbpA/IbpB has heat-shock proteins a major function in cell protection during stressful situations. Better understanding of the cold-shock response of food-borne pathogens is necessary to prevent food spoilage and poisoning. The enterohemorrhagic *E. coli* O157 (EHEC) strain (B-1) showed significantly higher resistance to cold stress compared to non-pathogenic *E. coli* K-12 DH5α. Using comparative proteomic analysis of hypothermally adapted *E. coli* O157 wild-type and rpoS mutant strains, Vidovic et al. (2011, 2012) identified proteins that were differentially expressed upon cold temperature, for example, RpoS, a 37.8-kDa protein that regulates the expression of proteins involved in homeoviscous adaptation during cold shock, as well as various proteins involved in central metabolic pathways of this food-borne pathogen.

Bacterial thermosensing functions probably stem from the membrane and are compatible with the proposed membrane location of peptidyl glutamyl peptide hydrolase (PgpH). Liu et al. (2006) suggest that the PgpH multidrug resistance protein is a potential mediator of cold signaling in *L. monocytogenes*. Altered expression of numerous gene transcripts and proteins was observed in *L.monocytogenes* cells adapted to cold exposure (Cacace et al. 2010; Schmid et al. 2009; Chan et al. 2008). Cacace et al. (2010) analyzed proteomic changes underlying the cold adaptation response into key cellular events of *L. monocytogenes* that allow the survival and multiplication even at refrigeration temperatures. Adaptation processes related to proteins implicated in metabolic pathways for energy production, such as glycolysis and Pta-AckA pathway, were present at a higher level in the cells grown at 4°C. Increased cold-shock protein (Csp) synthesis and activity at low temperatures provides DNA and RNA chaperone functions, which are needed in cold-exposed *L. monocytogenes* cells to help resolve structural hurdles of nucleic acids (Schmid et al. 2009). These proteins are required for efficient cold as well as osmotic stress tolerance of this bacterium. Schmid et al. (2009) showed that their functional importance differs, depending on the environmental stress conditions: CspA > CspD > CspB in response to cold stress versus CspD > CspA/CspB in response to NaCl salt osmotic stress. In addition, CspA homologues are involved in diverse phenomena such as response to freezing conditions, stationary phase, osmotic stress, starvation, antibiotic biosynthesis, resistance to antimicrobial peptides, inhibition of replication, heat resistance of the spores, UV sensitivity, and so on (Derzelle et al. 2003; Katzif et al. 2003; Leblanc et al. 2003; Mangoli et al. 2001; Movahedi and Waites 2002; Yamanaka and Inouye 1997).

In *Cronobacter turicensis 3032* (Carranza et al. 2009) 14 housekeeping factors were identified under various stress conditions. These proteins belong to membrane-bound zinc metalloproteases, cytoplasmic proteases, complex-forming ATPase, periplasmic serine proteases, and peptidases. In a following study Carranza et al.'s (2010) quantitative proteomics approach was used to investigate the adaptive resilience of this bacterium grown at different temperatures. The presented study suggests that the opportunistic pathogen compensates suboptimal growth temperatures by multifaceted alterations in its protein expression profile and increases the production of various potential virulence factors when exposed to elevated temperature.

Inasmuch as temperature treatment for food occurs at approximately 12°C, it is expected that *Staphylococcus aureus* adjusts its metabolism to colder temperatures before growing in this matrix. Sánchez et al. (2010) performed a proteomic comparison following two molecular strategies, *in vivo*, for the process of acclimatisation of *S.aureus* to cold temperatures. The identification of two different protein patterns associated with cold acclimation shows differentiation between patterns in glycolytic proteins (pattern A) and general stress and regulatory proteins (pattern B). It should be mentioned that the protein cell-binding factor (PEB1), fibronectin binding protein (CadF), and cytolethal distending toxin B (CDT) have been known to be virulence factors of *Campylobacter jejuni* (van Vliet and Ketley 2001; Konkel et al. 2001).

Fifteen and twenty differentially expressed proteins were defined for *C. jejuni* cultured at 37°C and 42°C on agar and in broth, respectively. Subsets of these proteins are optimally expressed at 37°C and may contribute to the host adaptation and/or the pathogenicity in the human intestinal tract (Zhang et al. 2009). Hwang et al. (2011) studied survivability of an rpoN missing mutant compared with the wild-type *C. jejuni* under various stress conditions. The rpoN mutant was more sensitive to osmotic stress (0.8%, w/v, NaCl), more susceptible to acid stress (pH 5), more resistant to hydrogen peroxide, and had little resistance to alkaline pH, heat, cold, and antimicrobials. The authors suggest that RpoN plays an important role in *C. jejuni's* defense against various stresses which this bacterial pathogen may encounter during transmission to and infection of humans.

Better cryotolerance is exhibited by cells that are cold-shocked prior to freezing. Therefore, food-spoilage bacteria can be sensitized to damage caused by cold temperatures through direct freezing of the food (Willimsky et al. 1992; Azizoglu et al. 2009; Grzadkowska and Griffiths 2001). Similarly, cryotolerance of economically important microorganisms such as lactic acid bacteria starter cultures can be improved, so that economic losses resulting from reduced starter culture's activity/viability due to harsh temperature changes in the fermentation processes are minimized (Kim and Dunn 1997; Bâati et al. 2000; Streit et al. 2008).

Under heat stress, upregulated genes are coding proteins involved in processes of immediate pertinence, including energy production through carbohydrate and lipid metabolism, metabolite transport, respiration, redox balance, and ROS detoxification, cell wall modification, and DNA damage repair as well as protein chaperones that are used for refolding and degradation. Görner et al. (2002) show that phosphorylation of the nuclear localization signal of Msn2p transcription factors by protein kinase A (PKA) controls the cytoplasmic relocalization of the protein specifically in response to glucose repletion. On the other hand, the relocalization of Msn2p in response to other conditions is independent of this domain and may be primarily controlled by alterations in nuclear export. Transcription factors Msn2 and Msn4 are involved in heat-stress response. The mRNA levels of translation-related genes tend to decrease, until the heat-shock factor Hsf1 does not change (Fonseca et al. 2012). Hsf1 and Msn2/Msn4 transcription factors in *Saccharomyces cerevisiae* play important roles in cellular homeostasis by activating gene expression in response to multiple stresses including heat shock, oxidative stress, and nutrient starvation. Yak1 kinase plays a central role in mediating PKA-dependent inhibition of Hsf1 and Msn2/Msn4 (Lee et al. 2008). This knowledge will help us to understand the remarkable ability of yeast and other organisms to survive stressful variations in their environment.

The mold *Aspergillus fumigatus* is able to survive at temperatures higher than 55°C (around 70°C). This heat resistance of *A. fumigatus* differs from other *Aspergilli* and may contribute to its virulence by increased level of heat-shock proteins (Hsp) (Bhabhra and Askew 2005). During the heat-shock response, the proteins Hsp30/Hsp42 and Hsp90 showed the highest increase in abundance (Albrecht et al. 2010).

In order to survive stress caused by processing, food-borne pathogens have developed distinct regulatory mechanisms as an adaptation to new environmental conditions (see Table 26.1). For example, a diverse range of post-translational modifications after cold- and heat-stress factors can further regulate the activity and abundance of specific proteins in bacterial, yeast, and fungal cells, thereby controlling key processes that contribute to the virulence of these organisms.

26.3.2 Osmotic Stress Adaptation

Salt (NaCl) is one of the most commonly employed agents for food preservation, allowing considerable increase in storage time by reducing water activity. Osmotic stress of pathogens is linked to survival and growth at high salt levels and low water activity. Cellular stress adaptation can contribute to a better understanding of strategies and possible cross-protection mechanisms on the basis of transcriptomic and phenotypic characteristics (den Besten et al. 2009). The transcriptome profiles for two salt conditions were compared in order to investigate the overlap and to identify specific responses of mildly (2.5% NaCl,w/v) and severely (5% NaCl, w/v) salt-stressed cells. The whole-genome expression analyses of mildly and severely salt-stressed *Bacillus cereus* cells revealed an overlap in the transcriptome response.

The physiological response of *Bacillus subtilis* to changing osmolality has been analyzed in substantial detail at the level of single genes or proteins (Bremer 2002). It is known that many membrane proteins, especially transporters of compatible solutes and ions, play a crucial role in adaptation of *B. subtilis* to salt stress. Transcriptomic and proteomic approaches can provide a view of the dynamic

Table 26.1 Overview of some gene and proteins associated with heat- and cold stress in major foodborne pathogens

Stress adaptation gene/protein system	Functional description	*Listeria monocytogenes*	*Bacillus cereus*	*Campylobacter jejuni*	*E. coli*
Chaperons					
ClpB, ClpC, ClpE, ClpP	ATP dependent Clp chaperone and protease		Periago et al. (2002)	Thies et al. (1999)	Etchegaray and Inouye (1999)
groEL, GrpE, GroES	Chaperone Hsp60, 60 kDa chaperonin (cpn60)			Parkhill et al. (2000)	Hasan and Shimizu (2008)
	HSP-70 cofactor			Andersen et al. (2005)	
	10 kDa chaperonin (cpn10)			Stintzi (2003)	
dnaK/Hsp70	Chaperone Hsp70, co-chaperone with DnaJ			Parkhill et al. (2000)	Hasan and Shimizu (2008)
Hfq	RNA chaperone	Christiansen et al. (2004)			
Metabolism					
Carbohydrate metabolism					
pgm	Phosphoglycerate mutase 2	Cacace et al. (2010)			
ldhA	L-lactate dehydrogenase A chain				Hasan and Shimizu (2008)
PflA	Pyruvate formate-lyase activating enzyme				
Pgk	Phosphoglycerate kinase		Browne and Dowds (2001)		
Tpi	Triosephosphate isomerase				
Pfk	Phosphofructokinase				
Hydrogen peroxide, superoxide metabolic process					
Catalase	Catalase	Azizoglu and Kathariou (2010)			
SOD	Superoxide dismutase		Periago et al. (2002)		
Nucleotide metabolic process					
drm	Phosphopentomutase	Cacace et al. (2010)			

Guanosine tetraphosphate metabolic process				
PgpH	Bifunctional (p)ppGpp synthase/hydrolase SpoT	Liu et al. (2006)		
Lipid metabolic process				
RpoH				Park (2002)
Pyridoxal phosphate biosynthetic process				
lmo2101	Pyridoxine synthesis protein Pdx1	Dussurget et al. (2005)		
Deoxyribonucleotide catabolic process				
dra	Deoxyribose-phosphate aldolase	Dussurget et al. (2005)		
Protein folding				
PpiB	Peptidyl-prolyl cis-trans isomerase B		Periago et al. (2002)	
Phosphoenolpyruvate-dependent sugar phosphotransferase system				
lmo0783	Mannose-specific PTS component IIB	Dussurget et al. (2005)		
Transcription				
lisRK	Two-component regulatory system	Cotter et al. (1999)		
RsbV	Anti-antisigma factor protein	Chaturongakul and Boor (2004)	van Schaik et al. (2005c)	
SigB	Truncated RNA polymerase sigma factor SigB	Chaturongakul et al. (2011)	van Schaik et al. (2004)	
rsbY	Activator of SigB (σ^B)		van Schaik et al. (2005a)	

(continued)

Table 26.1 (continued)

Stress adaptation gene/protein system	Functional description	*Listeria monocytogenes*	*Bacillus cereus*	*Campylobacter jejuni*	*E. coli*
SigC	RNA polymerase sigma factor C	Chaturongakul et al. (2011)			
SigH	RNA polymerase sporulation specific sigma factor SigH				
SigL/RpoN	RNA polymerase, sigma-54 (Sigma L) subunit				
CtsR	Transcriptional regulator CtsR				
HrcA	Heat-inducible transcription repressor HrcA				
Csps	Cold shock domain proteins, CspA, CspB, CspG, CspI	Schmid et al. (2009)	Mayr et al. (1996)		Goldstein et al. (1990) Lee et al. (1994) Nakashima et al. (1996)
Transport					
OppA	Oligopeptide-binding protein oppA	Bergholz et al. (2010)			
opuA	Osmoprotectant ABC transporters				
opuB/C	Osmoprotectant ABC transporters				
opuD/betL	Glycine betaine transporter				

Redox homeostasis and antioxidant activity					
TrxA	Thioredoxin		Periago et al. (2002)		
Fri	DNA binding protein of starved cells (Dps)	Dussurget et al. (2005)			
Stress response					
OtsA	Alpha,alpha-trehalose-phosphate synthase				Kandror et al. (2002)
Hfq	RNA chaperone	Christiansen et al. (2004)			
HtrA	Serine protease	Stack et al. (2005)			
htpG	Chaperone protein				Hasan and Shimizu (2008)
ibpB	Small heat shock protein				
OsmC	Putative osmotically inducible peroxidase	Dussurget et al. (2005)			
HspR	Heat shock regulator protein HspR			Andersen et al. (2005)	
HSP16.4	Heat shock protein		Periago et al. (2002)		
ATP binding and nucleotide-binding					
RsbT	RsbT protein	Chaturongakul and Boor (2004)			
RsbW	Serine-protein kinase rsbW	Dussurget et al. (2005)	van Schaik et al. (2004)		
OpuC, Gbu	Glycine betaine-binding ABC transporter protein (osmolyte transporter)	Wemekamp-Kamphuis et al. (2004a)			

(continued)

Table 26.1 (continued)

Stress adaptation gene/protein system	Functional description	*Listeria monocytogenes*	*Bacillus cereus*	*Campylobacter jejuni*	*E. coli*
Others					
YflT	Putative uncharacterized protein		Periago et al. (2002)		
YbbT	Heat-shock protein				
YloH	DNA-directed RNA polymerase subunit omega				
MreB	Rod shape-determining protein				
lmo0796	YceI-like family protein	Dussurget et al. (2005)			
lmo1830	Conserved hypothetical dehydrogenase				

changes occurring in salt-stressed *B. subtilis* cultures because these studies provide an unbiased view of cells copying with high salinity. Furthermore, salt stress induced the transcription of osmoprotectant transporters and can also provide protection against other stresses, such as heat (Holtmann and Bremer 2004) and chilling (Wemekamp-Kamphuis et al. 2004a).

Hahne et al. (2010) applied whole genome microarray technology and metabolic labeling by use ^{14}N- and ^{15}N-labeled media for quantitative proteomic analysis. Their results indicate a well-coordinated induction of gene expression subsequent to an osmotic upshift, that involves large parts of the SigB, SigW, SigM, and SigX regulons (genes that regulated by the same regulatory protein). About 500 osmotic upregulation genes of *B. subtilis* genes were detected and they are a rich basis for further in-depth investigation of the physiological and genetic responses of *B. subtilis* to hyperosmotic stress.

Listeria monocytogenes can survive a variety of environmental stresses, such as high salt levels, as well as a range of temperatures and also it is able to tolerate acidic conditions. Consequently this high degree of bacteria resistance is a reason for difficulties in controlling this pathogen in food (Soni et al. 2011a). Another problem is the *L.monocytogenes* resistance related to treatment during food processing and preservation. Accumulation of glycine betaine and carnitine occurs via glycine betaine transporters encoded by the betL gene and the gbu operon (Ko and Smith 1999; Sleator et al. 1999) and carnitine transporter encoded by the opuC operon (Angelidis et al. 2002). Increased uptake of glycine betaine and carnitine osmolytes via betL, gbu, and opuC encoded transporters is crucial under hyperosmotic conditions and prevents intracellular water loss by counteracting external osmolarity and keeping the macromolecular structure of the cells intact (Gardan et al. 2003; Soni et al. 2011a, b). In order to achieve a better understanding of the impact of salt stress during food processing and preservation, Duché et al. (2002a, b) studied the effect of NaCl (3.0– 6% NaCl, w/v, concentration) on protein expression of *L. monocytogenes*. Identification of salt-shock proteins and salt-acclimation proteins was performed by 2-D electrophoresis followed by mass spectrometry (MS) or N-terminal sequence analysis and database searching (Duché et al. 2002a, b). Salt presence stimulates the synthesis of glyceraldehyde-3-phosphate dehydrogenase (Gap), pyruvate dehydrogenase A (PdhA), pyruvate dehydrogenase D (PdhD), and phosphoglycerate mutase (Pgm). These proteins may represent vital enzymes of catabolic metabolism. Other salt-stress proteins identified in *L. monocytogenes* also appeared to have a broad spectrum of functions. Duché et al. (2002a) revealed that one protein, GbuA, over-expressed in the presence of salt, is directly connected to the salt-stress response. This response is also connected with the general stress response, as two general stress proteins, Gsp (heat-shock protein, DnaK, and general stress protein, Ctc), were induced after salt stress.

These transporter systems expressed under hyperosmotic stress conditions were also found as the ones expressed under cold stress (Mendum and Smith 2002; Schmid et al. 2009; Wemekamp-Kamphuis et al. 2004a), suggesting that some of the mechanisms counteracting osmotic and cold stress may be common in *L. monocytogenes*. These observations also reveal a complex process in understanding the mechanisms of actions of both salt-shock proteins (Ssp) and stress acclimation proteins (Sap) (Duché et al. 2002a).

The overview of some genomics and proteomics investigations and identified genes and proteins associated with salt stress in two most studied bacteria *L. monocytogenes* and *B. cereus* is shown in Table 26.2.

26.3.3 High Hydrostatic Pressure (HHP) Stress Adaptation

High-pressure processing (HHP) is a nonthermal process used in inactivation and elimination of pathogenic and food spoilage microorganisms (Considine et al. 2008). The use of HHP technology is one of the recently developed methods for food preservation (Hugas et al. 2002).

Pressure treatment induced various changes in microbial cells, including alterations in the cell membrane, cell morphology, effects on proteins, including enzymes, and effects on genetic mechanisms of microrganisms. However, the mechanisms of microbial inactivation by HHP are still not fully understood (Pagán and Mackey 2000). Martínez-Gomariz et al. (2009) analyzed changes in the proteome of *Bacillus cereus* during the HHP treatment and found qualitative differences in some identified differently expressed proteins.

Microorganisms vary in their response to pressure (Patterson 2005). Generally, Gram-positive bacteria are more resistant to pressure than Gram-negatives and cocci. Certain bacteria and strains, such as *Escherichia coli* O157:H7 and three strains of *Listeria monocytogenes* can be exceptionally pressure resistant. HHP treatment ranges from 200 to 800 MPa for a period from 5 to 30 min depending on the food matrices and microorganisms (Patterson 2005). Little is known about the pressure sensitivity of toxic mold growth. Their inactivation was often combined with shock or other processes such as post heat shock (Tribst et al. 2009).

E.coli shows that its physiological tolerance limits are well below pressure levels applied during food HPP, which typically ranges from 300 to 800 MPa. It is assumed that many cellular processes are involved in cell maintenance and repair (Manãs and Mackey 2004; Bozoglu et al. 2004).

Vanlint et al. (2012) examined and compared the intrinsic potential for HHP resistance development among strains of *Escherichia coli, Shigella flexneri, Salmonella typhimurium, Salmonella enteritidis, Yersinia enterocolitica, Aeromonas hydrophila, Pseudomonas aeruginosa*, and *Listeria innocua* using a selective enrichment approach.

It was found that HHP resistance could only be detected in some of the *E. coli* strains, indicating that a specific genetic predisposition might be required for resistance development. HHP resistance proved to be a very stable trait that was maintained for >80 generations in the absence of HHP exposure. The authors suggest that, at the mechanistic level, HHP resistance not necessarily linked to derepression of the heat-shock genes, and not related to the phenomenon of persistence.

Table 26.2 Overview of some gene and proteins associated with salt stress in *Listeria monocytogenes* and *Bacillus cereus*

Stress adaptation gene/protein system	Functional description	*Listeria monocytogenes*	*Bacillus cereus*
Chaperons			
dnaK/Hsp70	Chaperone Hsp70, co-chaperone with DnaJ	Duché et al. (2002a)	
ClpB, ClpC, ClpE, ClpP	ATP dependent Clp chaperone and protease		den Besten et al. (2009)
Metabolism			
Carbohydrate metabolism			
odhA	L-lactate dehydrogenase A chain		den Besten et al. (2009)
PdhA	PdhA protein	Duché et al. (2002a)	
Hydrogen peroxide; superoxide metabolic process			
katA			den Besten et al. (2009)
katB			
PerR		Olsen et al. (2005)	
Primary metabolic process			
YvyD		Duché et al. (2002a)	
Organic acid metabolic process			
Ack A	Acetate kinase	Duché et al. (2002a)	
Amino-acids synthesis			
CysK	Cysteine synthase A	Duché et al. (2002a)	
Protein biosynthesis			
TufA	Protein translation elongation factors		den Besten et al. (2009)
EF-Tu	Elongation factor Tu	Duché et al. (2002a)	
Guanosine tetraphosphate metabolic process			
PgpH	Bifunctional (p) ppGpp synthase/hydrolase SpoT	Liu et al. (2006)	
Translation			
rpsJ	30S ribosomal protein S10		den Besten et al. (2009)
S6	30S ribosomal protein S6	Duché et al. (2002a)	
Transcription			
lisRK	Two-component regulatory system	Sleator and Hill (2005)	
RsbV	Anti-antisigma factor protein	Raengpradub et al. (2008)	
SigB	Truncated RNA polymerase sigma factor SigB	den Besten et al. (2009)	
rsbY	Activator of SigB (σ^B)		

(continued)

Table 26.2 (continued)

Stress adaptation gene/ protein system	Functional description	*Listeria monocytogenes*	*Bacillus cereus*
Csps	Cold shock domain proteins, CspA, CspB, CspG, CspI	Schmid et al. (2009)	
CarD-like transcriptional regulators	Transcriptional regulator		den Besten et al. (2009)
CcpA	Catbolite control protein A	Duché et al. (2002a)	
Transport			
OppA	Oligopeptide-binding protein oppA	Bergholz et al. (2010)	
opuA	Osmoprotectant ABC transporters		den Besten et al. (2009)
opuB/C	Osmoprotectant ABC transporters		
opuD/betL	Glycine betaine transporter		
opuE	Osmoregulated proline transporter		
AppA	Oligopeptide ABC transporter	Duché et al. (2002a)	
Redox homeostasis and antioxidant activity			
AhpC	Alkyl hydroperoxide reductase		den Besten et al. (2009)
RpoB	RNA polymerase		
PdhD	Dihydrolipoyl dehydrogenase	Duché et al. (2002a)	
Fri	DNA binding protein of starved cells (Dps)	Olsen et al. (2005)	
Stress response			
Hfq	RNA chaperone	Raengpradub et al. (2008)	
Ctc	General stress protein CTC	Duché et al. (2002a)	
HtrA	Serine protease	Stack et al. (2005)	
ATP binding and nucteotide-binding			
OpuC, Gbu	Glycine betaine-binding ABC transporter protein (osmolyte transporter)	Wemekamp-Kamphuis et al. (2004a)	
Others			
Gap		Duché et al. (2002a)	
GuaB			
PbhD			

The survival of *L. monocytogenes* when exposed to HHP may be influenced by the pre-induction of stress response regulators such as SigB, which can activate several protective genes under stressful conditions (Karatzas et al. 2005). Cold-shock proteins have been observed previously to be strongly upregulated by HHP treatments and low temperature in *L. monocytogenes* (Wemekamp-Kamphuis et al. 2004b).

C. jejuni had a specific response to HHP treatment which could not be anticipated from the responses of other bacterial species to this stress. Examining the response of *C. jejuni* to HHP shock, only proteins specific to oxidative stress were induced (Bièche et al. 2012). This could explain why *C. jejuni* is rather more sensitive to this treatment than other Gram-negative bacteria (Solomon and Hoover 2004).

26.3.4 Other Stress Factors

Antimicrobial chemicals are widely applied to clean and disinfect food-contacting surfaces. Oxidative stress is commonly induced by chemicals. However, the cellular response of bacteria to various disinfectants is unclear. Using a genome-wide comparative transcriptional approach, Ceragioli et al. (2010) analyze the general and specific transcriptome responses of *B. cereus* ATCC 14579 upon exposure to various disinfectant treatments, such as benzalkonium chloride (BC), sodium hypochlorite (SH), hydrogen peroxide (HP), and peracetic acid (PAA). In this study, exposure to BC resulted in the upregulation of genes involved in fatty acid metabolism. Exposure of *B. cereus* cells to BC and SH treatments also resulted in upregulation of quaternary ammonium compounds (QACs) resistance membrane-bounded proteins that are able to enhance bacterial tolerance towards other antibacterial compounds. Exposure of *B. cereus* ATCC 14579 to HP and PAA resulted in increased levels of genes that activated the DNA damage repair system, including the SOS response. These results may affect the recontamination capacity of *B. cereus* and, in this way, food quality and safety.

The central role of RsbY in regulating the activity of σ^B shows that in *B. cereus* the pathway leading to the activation of σ^B is markedly different from that in other Gram-positive bacteria (van Schaik et al. 2005b).

Catalase production and activation is the important line of defense of bacteria against H_2O_2. Most bacteria produce one or more catalases, which are a by-product of aerobic growth of bacteria. *B. cereus* has two catalases: a thermolabile catalase present in vegetative cells and a thermostable catalase present in spores. van Schaik et al. (2005c) investigated the hyperresistance to H_2O_2 of the σ^B deletion mutant of *B. cereus* ATCC 14579. It was found that the upregulated transcription of the katA gene coded the main vegetative cell catalase of this bacterium played an important role in the hyperresistance to H_2O_2 of this deletion of mutant.

Abee et al. (2011) clearly display a pattern of organization of the samples using three of the transcriptome samples and clustered transcriptoma data including stressors as detergents (benzalkonium chloride, hydrogen peroxide, peracetic acid, sodium hypochlorite), ethanol, heat, various acids (HCl, lactic acid, acetic acid), and salt stress at different time intervals and concentrations.

QACs such as benzalkonium chloride (BC) and cetrimide are frequently used for disinfection and preservation. Several studies have linked intrinsic/natural resistance of Gram-negative bacteria to tenside-based disinfectants, such as QACs, to the low permeability of the outer membrane, or broad-spectrum efflux systems

(Denyer and Maillard 2002; Nikaido 2001). Acquisition of resistance in *Escherichia coli* has been related mainly to changes in the composition of lipopolysaccharide (LPS) and fatty acids in the membrane (Ishikawa et al. 2002; Sakagami et al. 1989). In contrast to antibiotics, which have very specific targets within the bacterial cell, the biocides affect multiple cellular components.

Hypochlorite-responsive transcription factor, YjiE, was identified as regulator of a large number of genes upon hypochlorite stress. YjiE confers hypochlorite resistance to *E. coli* cells (Gebendorfer et al. 2012). Adapted strains, such as *E. coli* K-12 showed increased tolerance to BC including responses normally related to the multiple antibiotic resistance (Mar) regulon and protection against oxidative stress (Bore et al. 2007).

σ^B controls a large regulon in *Staphylococcus aureus*, whose activation requires RsbU, a protein phosphatase 2C (PP2C)-type phosphatase. Pané-Farré et al. (2009) investigated the role of RsbU (a stress signaling protein phosphatase) in controlling σ^B activity in *Staphylococcus aureus* following alkaline stress. Contrary to the case in *B. subtilis*, the induced expression of RsbU was sufficient to stimulate σ^B-dependent transcription in *S. aureus*. The obtained results suggest that the activity of the *S. aureus* RsbU and hence σ^B may be subjected to different regulation in comparison to that in *B. subtilis*.

Prolonged exposure to a high concentration of propionate (PA) during food processing systems and within the gut of infected hosts have a stress effect on the proteome of *Salmonella enteritidis*. Calhoun et al. (2010) noted a significant difference in the proteomes of PA adapted and unadapted *S. enteritidis* and affirmed the contribution of Dps and CpxR in PA-induced acid resistance.

A DNA-binding protein, designated as SEp22 is related to the pathogenicity of *Salmonella enteridis* from poultry farms. Amano (2011) analyzed diverse roles of this protein that are expressed during starvation of bacterial cells, and demonstrates its importance in the stress responses in infection and survival, as well as regulation of bacterial growth.

In conclusion, new strategies and technologies have been developed in order to minimize the number of cases of food contamination with pathogen agents. However, there are also new risks, caused by the adaptation of food-borne pathogens and development of new survival strategies. As a part of a survival strategy to defend against the host response and treatment by microbicidal agents, many food-borne pathogens, bacteria as well as fungi, are able to form biofilms. Studies of biofilm-forming pathogenic bacteria and fungi are presented in the next chapter.

26.4 Biofilms

Many pathogenic and nonpathogenic microorganisms such as bacteria and fungi are able to form biofilms (Sauer 2003). Formation of bacterial biofilm is the process by which bacteria attach to abiotic surfaces, the surfaces of other unicellular organisms, the epithelia of multicellular organisms, and interfaces such as that between air and water (Absalon et al. 2011). The type of surface adhesion enables bacteria to adapt themselves favorably in their environment and to survive. Surface-attached bacteria, which may form monolayer and multilayer biofilms (Karatan and Watnick 2009), use distinct structures for adhesion in biofilms, and develop distinct transcriptional profiles within these two structures (Moorthy and Watnick 2005). Extracellular polymeric substances (EPS) are comprised of biological polymers such as exopolysaccharide, protein, and DNA (Flemming and Wingender 2010); mediate bacterial aggregation and surface attachment; as well as possibly also serving as a reservoir for extracellular degradative enzymes and the nutrients released by their function.

EPS substances (or matrix) is important both in the formation and structure of the biofilm and also protects the cells by preventing access of the antimicrobial and xenobiotics to the cells in the biofilm and confers protection against environmental stresses such as UV radiation, pH change, osmotic shock, and dessication (Hall-Stoodley and Stoodley 2005; de Carvalho and da Fonseca 2007).

The propensity to form a biofilm is guided by numerous environmental signals such as mechanical signals, nutritional and metabolic pathways, inorganic molecules, osmolarity, host-derived signals, antimicrobials, and quorum sensing (QS) (Karatan and Watnick 2009).

Quorum sensing (QS) regulated gene expression also contributes to biofilm tolerance (Kjelleberg and Givskov 2007). In addition to QS molecules, a diversity of other signals triggers biofilm formation. These include secondary metabolites such as antibiotics, pigments, and siderophores (López et al. 2010). A study of these signals in their natural setting is critical to an understanding of their role in adaptation of the bacterium to its environment.

The nutritional status of the environment affects the propensity of a bacterium to form a multilayer biofilm (Karatan and Watnick 2009). In response to nutrient limitation, some bacteria such as *Salmonella enterica serovar Typhimurium* join a multilayer biofilm (Gerstel and Romling 2001). In these organisms, the stationary-phase sigma factor, RpoS, activates many of the genes needed for biofilm formation (Gerstel and Romling 2003).

On the other hand, nutrient-rich environments promote biofilm formation in *V. cholerae*, where RpoS participates in repression of genes required for biofilm formation (Yildiz et al. 2004). It follows that formation of biofilm depends on the environment and bacterium that inhabit it.

The phenomenon of resistance within biofilms can be explained by several mechanisms, including delayed penetration of the antimicrobial agents into the biofilm extracellular matrix, slowing of growth rate of organisms within the biofilm, or other changes that include interaction of the organisms with a surface. Better knowledge of biofilm formation and the mechanism of its degradation will provide important information for successful development of sanitation plans (Lewis and Klibanov 2005). The ultimate composite is a biofilm that is resistant to cleaners and sanitizers and is extremely difficult to remove (Bridier et al. 2011; Krysinski et al. 1992; Zottola 1994).

A significant number of studies reported on the persistence of some food-borne pathogens on food contact surfaces and biofilms, affecting the quality and safety of the food products (Simões et al. 2010). If microorganisms from food contact surfaces are not completely removed, they can contribute to biofilm formation and also increase the potential of cross-contamination (Hinton and Mead 1991). These new control strategies are constantly emerging with main incidence in the use of biosolutions (enzymes, phages, interspecies interactions, and antimicrobial molecules from microbial origin) (Sillankorva et al. 2004; Chaignon et al. 2007; Dean et al. 2011).

In the food industry, the occurrence of biofilms in food-processing environments can cause post-processing contamination leading to lower shelf-life of products and transmission of diseases (Poulsen 1999). Various structures such as flagella, fimbriae, outer membrane proteins, curli, and EPS are involved in biofilm formation. Flagella play an important role in the early stages of bacterial attachment by overcoming the repulsive forces associated with the substratum (Giaouris and Nychas 2006).

Some microorganisms, such as the sporogen bacterium *Bacillus cereus* (Oosthuizen et al. 2002), the Gram-positive bacterium *Listeria monocytogenes* (Trémoulet et al. 2002a), and some pathogenic *E. coli strains* (Trémoulet et al. 2002b), form biofilm on the surface of stainless containers and other abiotic surfaces. *B. cereus*, a dairy-associated toxigenic bacterium, readily forms biofilm on various surfaces and was used to gain a better understanding of biofilm development by Gram-positive aerobic rods. The uniquely secreted PGW-biofilm proteins include both enzymes and signal molecules. The amounts of several proteins in attached and planktonic cells grown in the presence of glass wool (PGW) cells are significantly different from the amounts in planktonic cells.

It was found that increased levels of sporulation protein YhbH in both attached and PGW cells were observed within 2 h, indicating that this protein may play an important role in regulation of the biofilm phenotype of *B. cereus*. Authors conclude that *B. cereus* DL5 readily adapt to an attached mode of growth, with dense biofilm structures developing within 18 h after inoculation when glass wool was used as a surface (Oosthuizen et al. 2002).

The presence of microorganisms on food contact surfaces due to inadequate cleaning and disinfection of food-processing environments as well as the ability of pathogens to adhere and form biofilms, is one of the most common causes of food spoilage and transmission of food-borne diseases. *Listeria monocytogenes* and *Salmonella enterica* are two of the most common food-borne pathogens responsible for numerous disease outbreaks worldwide every year (CDC 2011a, b). The biofilm-forming bacteria can resist very aggressive cleaning procedures, and can cause serious contaminations in the pharmaceutical and food industries. Better knowledge of biofilm formation and the mechanism of its degradation can be useful to prevent contamination by these bacteria (Rodrigues et al. 2011).

Mangalappalli-Illathu et al. (2008) examined the adaptive response and survival of planktonic and biofilm phenotypes of *Salmonella enterica* serovar *Enteritidis* adapted to benzalkonium chloride (BC). Their results suggest that enhanced biofilm-specific upregulation of 17 unique proteins, along with the increased expression of CspA, TrxA, Tsf, YjgF, and a probable peroxidase, phenotype-specific alterations in cell surface roughness, and a shift in fatty acid composition were affected to enhance survival to the BC-adapted biofilm cell population relative to their BC-adapted planktonic cell counterparts.

Biofilm formation can be important in the colonization of the food-borne pathogen *Listeria monocytogenes* in the food-processing environment (Hefford et al. 2005). After growth at 37°C, 568 formed adherent multicellular layers were examined on a variety of test surfaces with multiple transfers of the test surface into fresh medium. The proteomes from biofilm- and planktonic-grown cells from the same cultures were compared. Nineteen proteins were successfully identified from the 2-D gels using a combination of MALDI-TOF and MS/MS. Highly expressed proteins in biofilm-grown cells were involved in stress response, envelope and protein synthesis, biosynthesis, energy generation, and regulatory functions.

van der Veen and Abee (2010) investigated the role of sigB in *L. monocytogenes* static and continuous-flow biofilm formation and its function in the resistance of biofilm cells to the disinfectants benzalkonium chloride and peracetic acid.

Quantitative real-time PCR and promoter reporter studies showed that sigB is activated in static and continuous-flow biofilms. Biofilm formation studies using an in-frame sigB deletion mutant and complementation mutant showed that the presence of SigB is required to obtain wild-type levels of both static and continuous-flow biofilms. Finally, disinfection treatments of planktonically grown cells and cells dispersed from static and continuous-flow biofilms showed that SigB is involved in the resistance of both planktonic cells and biofilms to the disinfectants benzalkonium chloride and peracetic acid.

Escherichia coli 0157:H7 biofilms were studied in order to identify some of the proteins involved in the biofilm phenotype (Trémoulet et al. 2002b). Using proteomic analysis of sessile or planktonic bacteria of the same age and comparison of two-dimensional gels showed clear differences between protein patterns of sessile and planktonic cells. Seventeen proteins were found in biofilms, of which 14 increased and 3 decreased in biofilms. Ten proteins were identified by MALDI-TOF-MS and classified into four categories according to their function: (1) general metabolism proteins (malate dehydrogenase, thiamine–phosphate pyrophosphorylase), (2) sugar and amino acid transporters (d-ribose-binding periplasmic protein, d-galactose-binding protein, YBEJ), (3) regulator proteins (DNA starvation protein and H-NS), and (4) three proteins with unknown function. The results of this study showed that *E. coli* O157:H7 modified the expression of several proteins involved in biofilm growth mode.

Camphylobacter jejuni 11168 can form biofilms on a variety of surfaces (Kalmokoff et al. 2006). Proteomic analyses of planktonic and biofilm-grown cells showed differences in protein expression profiles between these two growth modes. Proteins in biofilms involved in the motility complex, such as the flagellins (FlaA, FlaB), the filament cap(FliD), the basal body (FlgG, FlgG2), and the chemotactic protein (CheA), exhibited higher levels of expression than those found in the stationary-phase planktonic cells. In addition, Kalmokoff et al. (2006) showed the increased expression of proteins involved in the general (GroEL, GroES) and oxidative (Tpx, Ahp) stress responses, two known adhesins (Peb1, FlaC), and proteins involved in biosynthesis, energy generation, and catabolic functions. Inactivation of genes that affect the flagellar filament (fliA, flaA, flaB, flaG) or the expression of the cell adhesin (flaC) also resulted in a delay in pellicle formation. These results indicate that the flagellar motility complex plays a crucial role in the initial attachment of *C. jejuni* 11168 to solid surfaces during biofilm formation as well as in the cell-to-cell interactions required for pellicle formation, and suggests a role for the flagellar apparatus in the biofilm phenotype.

Proteomic analysis of the biofilm matrix of a Gram-negative bacterium, *Vibrio cholerae*, was performed in order to explain the function of biofilm matrix proteins (Absalon et al. 2011). The prepared protein mixtures were analyzed by MS/MS. Using *in silico* strategy to predict the subcellular localization of identified proteins, Absalon et al. (2011) found that 10 of 74 extracytoplasmic proteins were secreted and were candidate biofilm matrix-associated proteins. NusA, a transcription elongation factor, was identified in the proteomic analysis, predicted to be secreted. Other identified secreted proteins included bacterial appendages such as the mannose-sensitive hemagglutinin type IV pilus (MshA) and the flagellum as well as RbmA and RbmC, co-regulated with the VPS synthesis genes. Three proteins not previously associated with biofilms were also identified, such as a hemolysin (HlyA), a chitinase, and the hemagglutinin/protease (HAP). Of these proteins 17 were located in the outer membrane (OM), and 26 were located in the periplasm. The location of 18 proteins could not be predicted with certainty. It was found that the RbmA protein, which is evenly distributed in the biofilm, strengthens intercellular interactions, whereas communally synthesized Bap1 concentrated at the biofilm–surface interface and stabilized the association of the multilayer biofilm with the surface. This study provided evidence for spatial and functional differentiation of proteins in the biofilm matrix and suggested bacterial cooperation in stabilization of multilayer biofilm surface association and recruitment of new members.

Candida albicans is the most commonly isolated fungus associated with biofilms, drug-resistant microbial communities formed on medical devices. Lattif et al. (2008) used two-dimensional difference-in-gel electrophoresis (DIGE)-based proteomics to identify differentially expressed proteins in C. albicans biofilms grown to early and mature phases, compared to planktonic cells. This study provided a proteomic dataset subjected to pathway mapping to reveal phase-specific pathways that are differentially expressed in biofilm cell walls and extracellular matrix (EPS). There were 107 proteins differentially expressed in EPS, and 44 were differentially expressed in cell walls during biofilm formation, compared to planktonic controls. Furthermore, 95% (102/107) and 68% (30/44) of these differentially expressed proteins were upregulated in EPS and cell walls of biofilms and were mapped to cellular pathways. Taking into account the results obtained from this study, the authors demonstrated that EPS and cell walls of *C. albicans* biofilms increase levels of specific proteins within pathways in a phase-dependent manner. They suggested that these pathways, especially glycolysis/gluconeogenesis, play critical roles in fungal biofilm formation and maintenance.

Villena et al. (2009) investigated an initial profiling of the intracellular proteome of *Aspergillus niger* ATCC 10864 biofilm cultures developed on polyester cloth. Using 2D-PAGE and MS-TOF analysis, the proteome of conventionally grown free-living submerged cultures was compared to *A. niger* ATCC 10864. Proteomic maps showed different expression patterns in both culture systems with differentially expressed proteins in each case, where in biofilm cultures were over-expressed and differentially expressed 19% and 32% of the selected protein spots. These presented results shows significant differences between the proteomes of *A. niger* biofilm and free-living mycelia. These results indicate that cell adhesion is the most important stimulus responsible for biofilm development which is the basis of surface adhesion fermentation (Villena et al. 2009).

Proteomics have been helpful to gain some insight to increased microorganism resistance in biofilms. Despite the remarkable improvements in proteomics technologies in the past few years, the number of these studies is still insufficient.

For this reason, more aggressive, innovative, high-throughput, global-level proteomic studies are required, particularly for biofilms from *in vivo* clinical sources as well as in food safety.

The use of proteomics with the simultaneous application of genomics and other multidimensional technologies may provide a complete view of these problems.

Acknowledgments This work was supported by the Croatian Ministry of Science, Projects No. 335-0000000-0221. None of the authors has conflicts of interest with respect to this work.

References

Abee T, Wouters JA (1999) Microbial stress response in minimal processing. Int J Food Microbiol 50:65–91

Abee T, Wels M, de Been M, den Besten H (2011) From transcriptional landscapes to the identification of biomarkers for robustness. Microb Cell Fact 10(Suppl 1):S9

Absalon C, van Dellen K, Watnick PI (2011) A communal bacterial adhesin anchors biofilm and bystander cells to surfaces. PLoS Pathog 7(8):e1002210

Adams MR, Hartley AD, Cox LJ (1989) Factors affecting the efficacy of washing procedures used in the production of prepared salads. Food Microbiol 6:69–77

Albrecht D, Guthke R, Brakhage AA, Kniemeyer O (2010) Integrative analysis of the heat shock response in *Aspergillus fumigatus*. BMC Genomics 11:32

Allen KJ, Lepp D, McKellar RC, Griffiths MW (2008) Examination of stress and virulence gene expression in *Escherichia coli O157:H7* using targeted microarray analysis. Foodborne Pathog Dis 5:437–447

Allende A, Tomás-Barberán FA, Gil MI (2006) Minimal processing for healthy traditional foods. Trends Food Sci Technol 17(9):513–519

Amano F (2011) SEp22, *Salmonella* Dps, a key molecule bearing both pathogenicity and resistance to environmental stresses in *Salmonella*. J Health Sci 57(6):458–471

Angelidis AS, Smith LT, Hoffman LM, Smith GM (2002) Identification of OpuC as a chill-activated and osmotically activated carnitine transporter in *Listeria monocytogenes*. Appl Environ Microbiol 68:2644–2650

Andersen MT, Brøndsted L, Pearson BM, Mulholland F, Parker M, Pin C, Wells JM, Ingmeret H (2005) Diverse roles for HspR in *Campylobacter jejuni* revealed by the proteome, transcriptome and phenotypic characterization of an hspR mutant. Microbiology 151(Pt 3):905–915

Aureli P, Costantini A, Zolea S (1992) Antimicrobial activity of some plant essential oils against *Listeria monocytogenes*. J Food Prot 55:344–348

Austin JW, Dodds KL, Blanchfield B, Farber JM (1998) Growth and toxin production by *Clostridium botulinum* on inoculated fresh-cut packaged vegetables. J Food Prot 61:324–328

Azizoglu RO, Osborne J, Wilson S, Kathariou S (2009) Role of growth temperature in freeze-thaw tolerance of *Listeria spp.*. Appl Environ Microbiol 75(16):5315–5320

Azizoglu RO, Kathariou S (2010) Temperature-dependent requirement for catalase in aerobic growth of *Listeria monocytogenes* F2365. Appl Environ Microbiol 76(21):6998–7003

Bâati L, Fabre-Gea C, Auriol D, Blanc PJ (2000) Study of the cryotolerance of *Lactobacillus acidophilus*: effect of culture and freezing conditions on the viability and cellular protein levels. Int J Food Microbiol 59(3):241–247

Bergholz TM, den Bakker HC, Fortes ED, Boor KJ, Wiedmann M (2010) Salt stress phenotypes in *Listeria monocytogenes* vary by genetic lineage and temperature. Foodborne Pathog Dis 7(12):1537–1549

Bhabhra R, Askew DS (2005) Thermotolerance and virulence of *Aspergillus fumigatus*: role of the fungal nucleolus. Med Mycol 43(Suppl 1):S87–S93

Bièche C, de Lamballerie M, Chevret D, Federighi M, Tresse O (2012) Dynamic proteome changes in *Campylobacter jejuni 81–176* after high pressure shock and subsequent recovery. J Proteomics 75:1144–1156

Bore E, Hébraud M, Chafsey I, Chambon C, Skjaeret C, Moen B, Møretrø T, Langsrud Ø, Rudi K, Langsrud S (2007) Adapted tolerance to benzalkonium chloride in *Escherichia coli* K-12 studied by transcriptome and proteome analyses. Microbiology 153:935–946

Bozoglu F, Alpas H, Kaletunç G (2004) Injury recovery of foodborne pathogens in high hydrostatic pressure treated milk during storage. FEMS Immunol Med Microbiol 40:243–247

Bremer E (2002) Adaptation to changing osmolarity. In: Sonenshein AL, Hoch JA, Losick R (eds) *Bacillus subtilis* and its closest relatives: from genes to cells. ASM Press, Washington, DC, pp 385–391

Bridier A, Briandet R, Thomas V, Dubois-Brissonnet F (2011) Resistance of bacterial biofilms to disinfectants: a review. Biofouling 27(9):1017–1032

Browne N, Dowds BC (2001) Heat and salt stress in the food pathogen *Bacillus cereus*. J Appl Microbiol 91(6):1085–1094

Budde I, Steil L, Scharf C, Völker U, Bremer E (2006) Adaptation of *Bacillus subtilis* to growth at low temperature: a combined transcriptomic and proteomic appraisal. Microbiology 152(3):831–853

Bumann D (2010) Pathogen proteomes during infection: a basis for infection research and novel control strategies. J Proteomics 73(11):2267–2276

Burrows JP, Weber M, Buchwitz M, Rozanov V, Ladstter-Weienmayer A, Richter A, de Beek R, Hoogen R, Bramstedt K, Eichmann K-U, Eisinger M, Perner D (1999) The global ozone monitoring experiment (GOME): mission concept and first scientific results. J Atmos Sci 56:151–175

Cacace G, Mazzeo MF, Sorrentino A, Spada V, Malorni A, Siciliano R (2010) Proteomics for the elucidation of cold adaptation mechanisms in *Listeria monocytogenes*. J Proteomics 73(10):2021–2030

Calhoun LN, Liyanage R, Lay JO, Kwon YM (2010) Proteomic analysis of *Salmonella enterica* serovar Enteritidis following propionate adaptation. BMC Microbiol 10:249

Callahan JH, Shefcheck KJ, Williams TL, Musser SM (2006) Detection, confirmation, and quantification of staphylococcal enterotoxin B in food matrixes using liquid chromatography-mass spectrometry. Anal Chem 78:1789–1800

Capozzi V, Fiocco D, Amodio ML, Gallone A, Spano G (2009) Bacterial stressors in minimally processed food. Int J Mol Sci 10(7):3076–3105

Capriotti AL, Caruso G, Cavaliere C, Foglia P, Samperi R, Lagana A, Moro PA (2011) Multiclass mycotoxin analysis in food, environmental and biological matrices with chromatography// mass spectrometry. Mass Spectrom Rev. doi:10.1002/mas

Carranza P, Hartmann I, Lehner A, Stephan R, Gehrig P, Grossmann J, Barkow-Oesterreicher S, Roschitzki B, Eberl L, Riedel K (2009) Proteomic profiling of *Cronobacter turicensis* 3032, a food-borne opportunistic pathogen. Proteomics 9(13):3564–3579

Carranza P, Grunau A, Schneider T, Hartmann I, Lehner A, Stephan R, Gehrig P, Grossmann J, Groebel K, Hoelzle LE, Eberl L, Riedel K (2010) A gel-free quantitative proteomics approach to investigate temperature adaptation of the food-borne pathogen *Cronobacter turicensis 3032*. Proteomics 10(18):3248–3261

Causton HC, Ren B, Koh SS, Harbison CT, Kanin E, Jennings EG, Lee TI, True HL, Lander ES, Young RA (2001) Remodeling of yeast genome expression in response to environmental changes. Mol Biol Cell 12(2):323–337

Centers for Disease Control and Prevention (CDC) (2011a). Estimates of foodborne illness in the United States. http://www.cdc.gov/foodborneburden Published October 19, 2011. Accessed 10 Nov 2011

Centers for Disease Control and Prevention (CDC) (2011b) Vital signs: incidence and trends of infection with pathogens transmitted commonly through food--foodborne diseases active surveillance network, 10 U.S. sites, 1996–2010. Morb Mortal Wkly Rep 60:749–755. http://www.medscape.com/viewarticle/744862 Accessed 27 Oct 2011

Ceragioli M, Mols M, Moezelaar R, Ghelardi E, Senesi S, Abee T (2010) Comparative transcriptomic and phenotypic analysis of the responses of *Bacillus cereus* to various disinfectant treatments. Appl Environ Microbiol 76(10):3352–3360

Chaignon P, Sadovskaya I, Ragunah C, Ramasubbu N, Kaplan JB, Jabbouri S (2007) Susceptibility of *Staphylococcal* biofilms to enzymatic treatments depends on their chemical composition. Appl Microbiol Biotechnol 75:125–132

Chan Y-C, Hu Y, Chaturongakul S, Files KD, Bowen BM, Boor KJ, Wiedmann M (2008) Contributions of two-component regulatory systems, alternative sigma factors, and negative regulators to *Listeria monocytogenes* cold adaptation and cold growth. J Food Prot 1:420–425

Chaturongakul S, Boor KJ (2004) RsbT and RsbV contribute to sigmaB-dependent survival under environmental, energy, and intracellular stress conditions in *Listeria monocytogenes*. Appl Environ Microbiol 70:5349–5356

Chaturongakul S, Raengpradub S, Palmer ME, Bergholz TM, Orsi RH, Hu Y, Ollinger J, Wiedmann M, Boor KJ (2011) Transcriptomic and phenotypic analyses identify coregulated, overlapping regulons among PrfA, CtsR, HrcA, and the alternative sigma factors sigmaB, sigmaC, sigmaH, and sigmaL in *Listeria monocytogenes*. Appl Environ Microbiol 77(1):187–200

Christiansen JK, Larsen MH, Ingmer H, Sogaard-Andersen L, Kallipolitis BH (2004) The RNA-binding protein Hfq of *Listeria monocytogenes*: role in stress tolerance and virulence. J Bacteriol 186:3355–3362

Considine KM, Kelly AL, Fitzgerald GF, Hill C, Sleator RD (2008) High-pressure processing-effects on microbial food safety and food quality. FEMS Microbiol Lett 281(1):1–9

Cotter PD, Hill C (2003) Surviving the acid test: responses of Gram-positive bacteria to low pH. Microbiol Mol Biol Rev 67:429–453

Cotter PD, Emerson N, Gahan CG, Hill C (1999) Identification and disruption of lisRK, a genetic locus encoding a two-component signal transduction system involved in stress tolerance and virulence in *Listeria monocytogenes*. J Bacteriol 181(21):6840–6843

Cravatt BF, Wright AT, Kozarich JW (2008) Activity-based protein profiling: from enzyme chemistry to proteomic chemistry. Annu Rev Biochem 77:383–414

Culligan E, Hill C, Sleator RD (2009) Probiotics and gastrointestinal disease: successes, problems and future prospects. Gut Pathog 1:19

de Carvalho CCCR, da Fonseca MMR (2007) Preventing biofilm formation: promoting cell separation with terpenes. FEMS Microbiol Ecol 61(3):406–413

den Besten HMW, Mols M, Moezelaar R, Zwietering MH, Abee T (2009) Phenotypic and transcriptomic analyses of mildly and severely salt-stressed *Bacillus cereus ATCC 14579* cells. Appl Environ Microbiol 75(12):4111–4119

den Besten HMW, Arvind A, Gaballo HMS, Moezelaar R, Zwietering MH, Abee T (2010) Short- and long-term biomarkers for bacterial robustness: a framework for quantifying correlations between cellular indicators and adaptive behavior. PloS one 5(10):e13746. doi:10.1371/journal.pone.0013746

Dean SN, Bishop BM, van Hoek ML (2011) Natural and synthetic cathelicidin peptides with anti-microbial and anti-biofilm activity against *Staphylococcus aureus*. BMC Microbiol 11(1):114
Denyer SP, Maillard J-Y (2002) Cellular impermeability and uptake of biocides and antibiotics in gram-negative bacteria. J Appl Microbiol 92:35S–45S
Derzelle S, Hallet B, Ferain T, Delcour J, Hols P (2003) Improved adaptation to cold-shock, stationary phase, and freezing stresses in *Lactobacillus plantarum* overproducing cold-shock proteins. Appl Environ Microbiol 69:4285–4290
Duché O, Trémoulet F, Glaser P, Duché O, Trémoulet F, Glaser P, Labadie J (2002a) Salt Stress Proteins Induced in *Listeria monocytogenes* Salt Stress Proteins Induced in *Listeria monocytogenes*. Appl Environ Microbiol 68(4):1491–1498
Duché O, Trémoulet F, Namane A, Labadie J (2002b) A proteomic analysis of the salt stress response of *Listeria monocytogenes*. FEMS Microbiol Lett 215(2):183–188
Dupuis A, Hennekinne JA, Garin J, Brun V (2008) Protein standard absolute quantification (PSAQ) for improved investigation of staphylococcal food poisoning outbreaks. Proteomics 8:4633–4636
Dussurget O, Dumas E, Archambaud C, Chafsey I, Chambon C, Hébraud M, Cossart P (2005) *Listeria monocytogenes* ferritin protects against multiple stresses and is required for virulence. FEMS Microbiol Lett 250(2):253–261
Elmnasser N, Guillou S, Leroi F, Orange N, Bakhrouf A, Federighi M (2007) Pulsed-light system as a novel food decontamination technology: a review. Can J Microbiol 53:813–821
Etchegaray JP, Inouye M (1999) CspA, CspB and CspG major cold shock proteins of are induced at low temperatures under conditions that completely block protein synthesis. J Bacteriol 181(6):1827–1830
Fenugreek seeds with high probability for EHEC O104: H4 responsible outbreak, Federal Institute for Risk Assessment. June 30, 2011
Flemming HC, Wingender J (2010) The biofilm matrix. Nat Rev Microbiol 8:623–633
Fonseca LL, Chen P-W, Voit EO (2012) Canonical modeling of the multi-scale regulation of the heat stress response in yeast. Metabolites 2(1):221–241
Gahan CGM, Hill C (1999) The relationship between acid stress responses and virulence in *Salmonella typhimurium* and *Listeria monocytogenes*. Int J Food Microbiol 50:93–100
Gardan R, Duché O, Leroy-Sétrin S, the European Listeria Genome Consortium, Labadie J (2003) Role of ctc from *Listeria monocytogenes* in Osmotolerance. Appl Environ Microbiol 69(1):154–161
Gasch AP, Spellman PT, Kao CM, Carmel-Harel O, Eisen MB, Storz G, Botstein D, Brown PO (2000) Genomic expression programs in the response of yeast cells to environmental changes. Mol Biol Cell 11:4241–4257
Gaso-Sokac D, Kovac S, Josic D (2010) Application of proteomics in food technology and food biotechnology: process development, quality control and product safety. Food Technol Biotechnol 48(3):284–295
Gebendorfer KM, Drazic A, Le Y, Gundlach J, Bepperling A, Kastenmüller A, Ganzinger KA, Braun N, Franzmann TM, Winter J (2012) Identification of a hypochlorite-specific transcription factor from *Escherichia coli*. J Biol Chem 287:6892–6903
Gerstel U, Romling U (2001) Oxygen tension and nutrient starvation are major signals that regulate agfD promoter activity and expression of the multicellular morphotype in *Salmonella typhimurium*. Environ Microbiol 3:638–648
Gerstel U, Romling U (2003) The csgD promoter, a control unit for biofilm formation in *Salmonella typhimurium*. Res Microbiol 154:659–667
Giaouris ED, Nychas G-JE (2006) The adherence of *Salmonella Enteritidis* PT4 to stainless steel: the importance of the air–liquid interface and nutrient availability. Food Microbiol 23(8):747–752
Goldstein J, Pollitt NS, Inouye M (1990) Major cold shock protein of *Escherichia coli*. PNAS 87:283–287

Görner W, Durchschlag E, Wolf J, Brown EL, Ammerer G, Ruis H, Schüller C (2002) Acute glucose starvation activates the nuclear localization signal of a stress-specific yeast transcription factor. EMBO J 21(1–2):135–144

Graumann PL, Marahiel M (1999) Cold shock response in *Bacillus subtilis*. J Mol Microbiol Biotechnol 1(2):203–209

Grzadkowska D, Griffiths MW (2001) Cryotolerance of *Escherichia coli O157: H7* in laboratory media and food. J Food Sci 66(8):1169–1173

Hahne H, Mäder U, Otto A, Bonn F, Steil L, Bremer E, Hecker M, Becher D (2010) A comprehensive proteomics and transcriptomics analysis of *Bacillus subtilis* salt stress adaptation. J Bacteriol 192(3):870–882

Hall-Stoodley L, Stoodley P (2005) Biofilm formation and dispersal and the transmission of human pathogens. Trends Microbiol 13(1):7–10

Hartung T, Koeter H (2008) Food for thought ... on food safety testing. ALTEX 25(4):259–264

Hasan CM, Shimizu K (2008) Effect of temperature up-shift on fermentation and metabolic characteristics in view of gene expressions in *Escherichia coli*. Microb Cell Fact 7:35

Havelaar AH, Brul S, de Jong A, de Jonge R, Zwietering MH, Ter Kuile BH (2010) Future challenges to microbial food safety. Int J Food Microbiol 139(Suppl):S79–94

Hefford MA, D'Aoust S, Cyr TD, Austin JW, Sanders G, Kheradpir E, Kalmokoff ML (2005) Proteomic and microscopic analysis of biofilms formed by *Listeria monocytogenes* 568. Can J Microbiol 51(3):197–208

Hinton M, Mead GC (1991) Opinion, *Salmonella* control in poultry: the need for the satisfactory evaluation of probiotics for this purpose. Lett Appl Microbiol 13:49–50

Holtmann G, Bremer E (2004) Thermoprotection of *Bacillus subtilis* by exogenously provided glycine betaine and structurally related compatible solutes: involvement of Opu transporters. J Bacteriol 186:1683–1693

Hugas M, Garriga M, Monfort JM (2002) New mild technologies in meat processing: high pressure as a model technology. Meat Sci 62(3):359–371

Hwang S, Jeon B, Yun J, Ryu S (2011) Roles of RpoN in the resistance of *Campylobacter jejuni* under various stress conditions. BMC Microbiol 11(1):207

Ishikawa S, Matsumura Y, Yoshizako F, Tsuchido T (2002) Characterization of a cationic surfactant-resistant mutant isolated spontaneously from *Escherichia coli*. J Appl Microbiol 92:261–268

Josic D, Kovac S (2008) Application of proteomics in biotechnology–microbial proteomics. Biotechnol J 3(4):496–509

Kalmokoff M, Lanthier P, Tremblay T-L, Foss M, Lau PC, Sanders G, Austin J, Kelly J, Szymanski CM (2006) Proteomic analysis of *Campylobacter jejuni* 11168 biofilms reveals a role for the motility complex in biofilm formation. J Bacteriol 188(12):4312–4320

Kandror O, DeLeon A, Goldberg AL (2002) Trehalose synthesis is induced upon exposure of *Escherichia coli* to cold and is essential for viability at low temperatures. PNAS 99:9727–9732

Karatan E, Watnick P (2009) Signals, regulatory networks, and materials that build and break bacterial biofilms. Microbiol Mol Biol Rev 73:310–347

Karatzas KAG, Valdramidis VP, Wells-Bennik MHJ (2005) Contingency locus in ctsR of *Listeria monocytogenes* Scott A: a strategy for occurrence of abundant piezotolerant isolates within clonal populations. Appl Environ Microbiol 71:8390–8396

Katzif S, Danavall D, Bowers S, Balthazar JT, Shafer WM (2003) The major cold shock gene, cspA, is involved in the susceptibility of *Staphylococcus aureus* to an antimicrobial peptide of human cathepsin G. Infect Immun 71:4304–4312

Kim WS, Dunn NW (1997) Identification of a cold shock gene in lactic acid bacteria and the effect of cold shock on cryotolerance. Curr Microbiol 35(1):59–63

Kjelleberg S, Givskov M (2007) The biofilm mode of life. In: Kjelleberg S, Givskov M (eds) The biofilm mode of life: mechanisms and adaptations. Horizon Scientific Press, Norfolk, pp 5–22

Ko R, Smith LT (1999) Identification of an ATP-driven, osmoregulated glycine betaine transport system in *Listeria monocytogenes*. Appl Environ Microbiol 65:4040–4048

Konieczny P, Kijowski J (2005) Animal origin food preservation and its safety issues. Pol J Food Nutr Sci 14:21–29
Konkel ME, Monteville MR, Rivera-Amill V, Joens LA (2001) The pathogenesis of *Campylobacter jejuni*-mediated enteritis. Curr Issues Intest Microbiol 2:55–71
Kovats S (2011) Climate change, crop yields, and undernutrition, with Sari Kovats by Ashley Ahearn. Environ Health Perspect 119(12):1–49
Krysinski EP, Brown LJ, Marchisello TJ (1992) Effect of cleaners and sanitizers on *Listeria monocytogenes* attached to product contact surfaces. J Food Prot 55:246–251
Lattif AA, Chandra J, Chang J, Liu S, Zhou G, Chance MR, Ghannoum MA, Mukherjee PK (2008) Proteomics and pathway mapping analyses reveal phase-dependent over-expression of proteins associated with carbohydrate metabolic pathways in *Candida albicans* biofilms. Open Proteomic J 1(1):5–26
Leblanc L, Leboeuf C, Leroi F, Hartke A, Auffray Y (2003) Comparison between NaCl tolerance response and acclimation to cold temperature in *Shewanella putrefaciens*. Curr Microbiol 46(3):157–162
Lee SJ, Xie A, Jiang W, Etchegaray J-P, Jones PG, Inouye M (1994) Family of the major cold-shock protein, CspA (CS7.4), of *Escherichia coli*, whose members show a high sequence similarity with the eukaryotic Y-box binding proteins. Mol Microbiol 11:833–839
Lee P, Cho BR, Joo HS, Hahn JS (2008) Yeast Yak1 kinase, a bridge between PKA and stress-responsive transcription factors, Hsf1 and Msn2/Msn4. Mol Microbiol 70(4):882–895
Lethanh H, Neubauer P, Hoffmann F (2005) The small heat-shock proteins IbpA and IbpB reduce the stress load of recombinant *Escherichia coli* and delay degradation of inclusion bodies. Microb Cell Fact 4(1):6
Levin RE (2009) The use of molecular methods for detecting and discriminating *Salmonella* associated with foods: a review. Food Biotechnol 23:313–367
Lewis K, Klibanov AM (2005) Surpassing nature: rational design of sterile-surface materials. Trends Biotechnol 23:343–348
Liu S, Bayles DO, Mason TM, Wilkinson BJ (2006) A cold-sensitive *Listeria monocytogenes* mutant has a transposon insertion in a gene encoding a putative membrane protein and shows altered (p)ppGpp levels. Appl Environ Microbiol 72(6):3955–3959
López D, Vlamakis H, Kolter R (2010) Biofilms. Cold Spring Harb Perspect Biol 2(7):a000398
Manãs P, Mackey BM (2004) Morphological and physiological changes induced by high hydrostatic pressure in exponential and stationary-phase cells of *Escherichia coli*: relationship with cell death. Appl Environ Microbiol 70:1545–1554
Mangalappalli-Illathu AK, Vidović S, Korber DR (2008) Differential adaptive response and survival of *Salmonella enterica* serovar enteritidis planktonic and biofilm cells exposed to benzalkonium chloride. Antimicrob Agents Chemother 52(10):3669–3680
Mangoli S, Sanzgiri VR, Mahajan SK (2001) A common regulator of cold and radiation response in *Escherichia coli*. J Environ Pathol Toxicol Oncol 20:23–26
Martínez-Gomariz M, Hernáez ML, Gutiérrez D, Ximénez-Embún P, Préstamo G (2009) Proteomic analysis by two-dimensional differential gel electrophoresis (2-D DIGE) of a high-pressure effect in *Bacillus cereus*. J Agric Food Chem 57(9):3543–3549
Mendum ML, Smith LT (2002) Characterization of glycine betaine porter I from *Listeria monocytogenes* and its roles in salt and chill tolerance. Appl Environ Microbiol 68:813–819
Moorthy S, Watnick PI (2005) Identification of novel stage-specific genetic requirements through whole genome transcription profiling of *Vibrio cholerae* biofilm development. Mol Microbiol 57:1623–1635
Mayr B, Kaplan T, Lechner S, Scherer S (1996) Identification and purification of a family of dimeric major cold shock protein homologs from the psychrotrophic *Bacillus cereus* WSBC 10201. J Bacteriol 178:2916–2925
Movahedi S, Waites W (2002) Cold shock response in sporulating *Bacillus subtilis* and its effect on spore heat resistance. J Bacteriol 184:5275–5281
Nakashima K, Kanamaru K, Mizuno T, Horikoshi KJ (1996) A novel member of the cspA family of genes that is induced by cold shock in *Escherichia coli*. Bacteriology 178:2994–2997

Nikaido H (2001) Preventing drug access to targets: cell surface permeability barriers and active efflux in bacteria. Semin Cell Dev Biol 12(3):215–223
O'Flaherty S, Klaenhammer TR (2011) The impact of omic technologies on the study of food microbes. Annu Rev Food Sci Technol 2:353–271
Olsen KN, Larsen MH, Gahan CG, Kallipolitis B, Wolf XA, Rea R, Hill C, Ingmer H (2005) The Dps-like protein Fri of *Listeria monocytogenes* promotes stress tolerance and intracellular multiplication in macrophage-like cells. Microbiology 151(Pt 3):925–933
Oosthuizen MC, Steyn B, Theron J, Cosette P, Lindsay D, Holy AV, Brözel VS (2002) Proteomic analysis reveals differential protein expression by *Bacillus cereus* during biofilm formation. Appl Environ Microbiol 68(6):2770–2780
Pagán R, Mackey B (2000) Relationship between membrane damage and cell death in pressure-treated *Escherichia coli* cells: differences between exponential and stationary phase cells and variation among strains. Appl Environ Microbiol 66:2829–2834
Pané-Farré J, Jonas B, Hardwick SW, Gronau K, Lewis RJ, Hecker M, Engelmann S (2009) Role of RsbU in controlling SigB activity in *Staphylococcus aureus* following alkaline stress. J Bacteriol 191(8):2561–2573
Park C-M, Hung Y-C, Doyle MP, Ezeike GOI, Kim C (2001) Pathogen reduction and quality of lettuce treated with electrolysed oxidizing and acidified chlorinated water. J Food Sci 66:1368–1372
Park SF (2002) The physiology of *Campylobacter species* and its relevance to their role as food-borne pathogens. Int J Food Microbiol 74:177–188
Parkhill J, Wren BW, Mungall K, Ketley JM, Churcher C, Basham D, Chillingworth T, Davies RM, Feltwell T, Holroyd S, Jagels K, Karlyshev AV, Moule S, Pallen MJ, Penn CW, Quail MA, Rajandream MA, Rutherford KM, van Vliet AHM, Whitehead S, Barrell BG (2000) The genome sequence of the food-borne pathogen *Campylobacter jejuni* reveals hypervariable sequences. Nature 403:665–668
Patterson MF (2005) A review: microbiology of pressure-treated foods. J Appl Microbiol 98:1400–1409
Periago PM, van Schaik W, Abee T, Wouters JA (2002) Identification of proteins involved in the heat stress response of *Bacillus cereus* ATCC 14579. Appl Environ Microbiol 68:3486–3495
Poulsen LV (1999) Microbial biofilm in food processing. LWT- Food Sci Technol 32(6):321–326
Petersohn A, Brigulla M, Haas S, Hoheisel JD, Völker U, Heckeret M (2001) Global analysis of the general stress response of *Bacillus subtilis*. J Bacteriol 183:5617–5631
Risks G (2009) First meeting of ECDC expert group on climate change executive summary 1 introduction. Prevention and Control (October):7–8
Raengpradub S, Wiedmann M, Boor KJ (2008) Comparative analysis of the sigma B-dependent stress responses in *Listeria monocytogenes* and *Listeria innocua* strains exposed to selected stress conditions. Appl Environ Microbiol 74(1):158–171
Rodrigues D, Cerca N, Teixeira P, Oliveira R, Ceri H, Azeredo J (2011) *Listeria monocytogenes* and *Salmonella enterica* enteritidis biofilms susceptibility to different disinfectants and stress-response and virulence gene expression of surviving cells. MDR 17(2):181–189
Sakagami Y, Yokoyama H, Nishimura H, Ose Y, Tashima T (1989) Mechanism of resistance to benzalkonium chloride by *Pseudomonas aeruginosa*. Appl Environ Microbiol 55(8):2036–2040
Sánchez B, Cabo ML, Margolles A, Herrera JJR (2010) A proteomic approach to cold acclimation of *Staphylococcus aureus CECT 976* grown at room and human body temperatures. Int J Food Microbiol 144(1):160–168
Sauer K (2003) The genomics and proteomics of biofilm formation. Genome Biol 4(6):219
Schmid B, Klumpp J, Raimann E, Loessner MJ, Stephan R, Tasara T (2009) Role of cold shock proteins in growth of *Listeria monocytogenes* under cold and osmotic stress conditions. Appl Environ Microbiol 75(6):1621–1627
Scott NE, Cordwell SJ (2009) *Campylobacter proteomics*: guidelines, challenges and future perspectives. Expert Rev Proteomics 6:61–74
Serrazanetti DI, Guerzoni ME, Corsetti A, Vogel R (2009) Metabolic impact and potential exploitation of the stress reactions in lactobacilli. Food Microbiol 26:700–711

Sillankorva S, Oliveira DR, Vieira MJ, Sutherland IW, Azeredo J (2004) Bacteriophage Φ S1 infection of *Pseudomonas fluorescens* planktonic cells versus biofilms. Biofouling 20:133–138

Simões M, Simões LC, Vieira MJ (2010) A review of current and emergent biofilm control strategies. LWT- Food Sci Technol 43(4):573–583

Skandamis PN, Yoon Y, Stopforth JD, Kendall PA, Sofos JN (2008) Heat and acid tolerance of *Listeria monocytogenes* after exposure to single and multiple sublethal stresses. Food Microbiol 25:294–303

Sleator RD, Gahan CG, Abee T, Hill C (1999) Identification and disruption of BetL, a secondary glycine betaine transport system linked to the salt tolerance of *Listeria monocytogenes* LO28. Appl Environ Microbiol 65:2078–2083

Soni KA, Nannapaneni R, Tasara T (2011a) An overview of stress response proteomes in *Listeria monocytogenes*: review. Agric Food Anal Bacteriol 1(1):66–85

Soni KA, Nannapaneni R, Tasara T (2011b) The contribution of transcriptomic and proteomic analysis in elucidating stress adaptation responses of *Listeria monocytogenes*. Foodborne Pathog Dis 8(8):843–852

Solomon E, Hoover D (2004) Inactivation of *Campylobacter jejuni* by high hydrostatic pressure. Lett Appl Microbiol 38:505–509

Stack HM, Sleator RD, Bowers M, Hill C, Gahan CGM (2005) Role for HtrA in stress induction and virulence potential in *Listeria monocytogenes*. Appl Environ Microbiol 71(8):4241–4247

Sleator RD, Hill C (2005) A novel role for the LisRK two-component regulatory system in *Listerial* osmotolerance. Clin Microbiol Infect 11:599–601

Stintzi A (2003) Gene expression profile of *Campylobacter jejuni* in response to growth temperature variation. J Bacteriol 185(6):2009–2016

Streit F, Delettre J, Corrieu G, Béal C (2008) Acid adaptation of *Lactobacillus delbrueckii subsp. bulgaricus* induces physiological responses at membrane and cytosolic levels that improves cryotolerance. J Appl Microbiol 105(4):1071–1080

Thies FL, Karch H, Hartung HP, Giegerich G (1999) The ClpB protein from *Campylobacter jejuni*: molecular characterization of the encoding gene and antigenicity of the recombinant protein. Gene 230(1):61–67

Tompkin RB (2002) Control of *Listeria monocytogenes* in the food-processing environment. J Food Prot 65(4):709–725

Trémoulet F, Duché O, Namane A, Martinie B, Labadie JC (2002a) Comparison of protein patterns of *Listeria monocytogenes* grown in biofilm or in planktonic mode by proteomic analysis. FEMS Microbiol Lett 210(1):25–31

Trémoulet F, Duché O, Namane A, Martinie B, Labadie J-C (2002b) A proteomic study of *Escherichia coli* O157:H7 NCTC 12900 cultivated in biofilm or in planktonic growth mode. FEMS Microbiol Lett 215(1):7–14

Tribst AA, Franchi MA, Cristianini M, de Massaguer PR (2009) Inactivation of *Aspergillus niger* in mango nectar by high-pressure homogenization combined with heat shock. J Food Sci 74(9):M509–514

van de Guchte M, Serror P, Chervaux C, Smokvina T, Ehrlich SD, Maguin E (2002) Stress responses in lactic acid bacteria. Antonie Van Leeuwenhoek 82:187–216

van Schaik W, Tempelaars MH, Wouters JA, De Vos WM, Abee T (2004) The alternative sigma factor σ^B of *Bacillus cereus*: response to stress and role in heat adaptation. J Bacteriol 186:316–325

van Schaik W, Abee T (2005a) The role of σ^B in the stress response of Gram-positive bacteria: targets for food preservation and safety. Curr Opin Biotechnol 16:218–224

van Schaik WV, Tempelaars MH, Zwietering MH, Vos WMD, Abee T (2005b) Analysis of the role of RsbV, RsbW, and RsbY in regulating σB activity in *Bacillus cereus*. J Bacteriol 187(16):5846–5851

van Schaik W, Zwietering MH, Vos WMD, Schaik WV, Abee T (2005c) Deletion of the sigB gene in *Bacillus cereus* ATCC 14579 leads to hydrogen peroxide hyperresistance. Appl Environ Microbiol 71(10):6427–6430 der Veen S, Hain T, Wouters JA, Hossain H, de Vos WM, Abee T, Chakraborty T, Wells-Bennik MH (2007) The heat-shock response of *Listeria monocytogenes* comprises genes involved in heat shock, cell division, cell wall synthesis, and the SOS response. Microbiology 153:3593–3607

van der Veen S, Abee T (2010) Importance of SigB for *Listeria monocytogenes* static and continuous-flow biofilm formation and disinfectant resistance. Appl Environ Microbiol 76(23):7854–7860

Vanlint D, Rutten N, Michiels CW, Aertsen A (2012) Emergence and stability of high pressure resistance in different foodborne pathogens. Appl Environ Microbiol AEM.00030-12; published ahead of print 17 February 2012. doi:10.1128/AEM.00030-12

van Vliet AH, Ketley JM (2001) Pathogenesis of enteric *Campylobacter* infection. Symp Ser Soc Appl Microbiol 30:45S–56S

Vidovic S, Mangalappalli-Illathu AK, Korber DR (2011) Prolonged cold stress response of *Escherichia coli O157* and the role of rpoS. Int J Food Microbiol 146(2):163–169

Vidovic S, Mangalappalli-Illathu AK, Xiong H, Korber DR (2012) Heat acclimation and the role of RpoS in prolonged heat shock of *Escherichia coli O157*. Food Microbiol 30(2):457–464

Villena GK, Venkatesh L, Yamazaki A, Tsuyumu S (2009) Initial intracellular proteome profile of *Aspergillus niger* biofilms. Rev Peru Biol 16(1):101–108

Wemekamp-Kamphuis HH, Sleator RD, Wouters JA, Hill C, Abee T (2004a) Molecular and physiological analysis of the role of osmolyte transporters BetL, Gbu, and OpuC in growth of *Listeria monocytogenes* at low temperatures. Appl Environ Microbiol 70:2912–2918

Wemekamp-Kamphuis HH, Wouters JA, de Leeuw PPLA, Hain T, Chakraborty T, Abee T (2004b) Identification of sigma factor sigmaB-controlled genes and their impact on acid stress, high hydrostatic pressure, and freeze survival in *Listeria monocytogenes* EGD-e. Appl Environ Microbiol 70:3457–3466

Willimsky G, Bang H, Fischer G, Marahiel MA (1992) Characterization of cspB, a *Bacillus subtilis* inducible cold shock gene affecting cell viability at low temperatures. J Bacteriol 174:6326–6335

Xantopoulos V, Tzanetakis N, Litopoulou-Tzasnetaki E (2010) Occurrence and characterization of *Aeromonas hydrophila* and *Yersinia enterocolitica* in minimally processed fresh vegetable salads. Food Control 21:393–398

Yamanaka K, Inouye M (1997) Growth-phase dependent expression of cspD, encoding a member of the CspA family in *Escherichia coli*. J Bacteriol 179:5126–5130

Yildiz FH, Liu XS, Heydorn A, Schoolnik GK (2004) Molecular analysis of rugosity in a *Vibrio cholerae* O1 El Tor phase variant. Mol Microbiol 53:497–515

Zhang M-J, Xiao D, Zhao F, Gu Y-X, Meng F-L, He L-H, Ma G-Y, Zhang J-Z (2009) Comparative proteomic analysis of *Campylobacter jejuni* cultured at 37°C and 42°C. Jpn J Infect Dis 62(5):356–361

Zottola EA (1994) Scientific status summary "microbial attachment and biofilm formation: a new problem for the food industry?". Food Technol 48:107–114

Chapter 27
Prion Biomarkers

Rhiannon L.C.H. Huzarewich, Philippe Simon, and Stephanie A. Booth

27.1 Introduction

Diseases caused by infectious prions, otherwise called transmissible spongiform encephalopathies (TSEs) make up a rare group of invariably fatal neurodegenerative diseases that affect both humans and animals (Prusiner 1998). The most notable being bovine spongiform encephalopathy (BSE) or mad cow disease in cattle that has had major agricultural and economic impacts worldwide, especially in Western Europe and North America. The prion is a unique infectious agent composed entirely, or almost entirely, of an abnormal conformer (PrP^{Sc}) of the normal host prion protein (PrP^{C}) that accumulates during disease progression. The major public health concern is that BSE prions can be transmitted to humans by ingestion of contaminated beef to cause the new variant Creutzfeldt–Jakob disease (vCJD). Recently, human-to-human transmission of vCJD, has been reported through blood transfusion (Llewelyn et al. 2004). Of concern recent evidence suggests that such

R.L.C.H. Huzarewich
Viral Diseases, National Microbiology Laboratory, Public Health Agency of Canada,
Winnipeg, MB, R3E 3R2, Canada
e-mail: Rhiannon.Huzarewich@phac-aspc.gc.ca

P. Simon
Special Pathogens, National Microbiology Laboratory, Public Health Agency of Canada,
Winnipeg, MB, R3E 3R2, Canada

Department of Medical Microbiology, University of Manitoba,
Winnipeg, MB, R3E 0W3, Canada
e-mail: Philippe.Simon@phac-aspc.gc.ca

S.A. Booth (✉)
Molecular Pathobiology, National Microbiology Laboratory, Public Health Agency of Canada,
Winnipeg, MB, R3E 3R2, Canada

Department of Medical Microbiology, University of Manitoba,
Winnipeg, MB, R3E 0W3, Canada
e-mail: Stephanie.Booth@phac-aspc.gc.ca

F. Toldrá and L.M.L. Nollet (eds.), *Proteomics in Foods: Principles and Applications*, Food Microbiology and Food Safety 2, DOI 10.1007/978-1-4614-5626-1_27,

human-adapted prions are much more readily transmitted from human to human via this route than via the ingestion of BSE prions from contaminated meat products (Bishop et al. 2006). Other food animals affected by TSEs are sheep and goats (Scrapie) and deer and elk (Chronic Wasting Disease, CWD). Although scrapie has been endemic for hundreds of years in many parts of the world its transmission to humans has never been reported. CWD has only recently been recognized, however, the incidence has increased markedly within North America and although it has not been linked to CJD either epidemiologically, or by laboratory confirmation, there is still a concern over the possibility for cross-species transmission (Belay et al. 2004; Raymond et al. 2000).

The prion protein, PrP^{C}, is a ubiquitously expressed, 35–36-kDa glycoprotein linked to the cell membrane, in lipid rafts, by a glycosylphosphatidylinoisitol (GPI) anchor. Its normal structure is predominantly alpha-helical whereas the disease associated conformer, PrP^{Sc}, is enriched in beta-sheets and is protease resistant. The molecular process that drives this conformational change is still poorly understood but specific post-translational modifications do not appear to play a role. PrP^{Sc} has characteristic biochemical and biophysical properties, most notably resistance to treatments such as denaturants, hydrolysis, radiation, and high temperatures that would normally inactivate infectious agents. These include autoclaving at 121°C, 0.5% bleach, hydrogen peroxide, or ultraviolet (UV) irradiation. Decontamination of prions requires on autoclaving at 134°C for extended periods of an hour or more, chemical treatment with 2M NaOH for an hour or mixing with bleach for a final concentration of 2% available chlorine for one hour prior to disposal. In practical terms this means that prions that are contaminating food-stuffs cannot be inactivated by usual means.

On a positive note methodologies to detect PrP^{Sc} are both sensitive and completely specific for post-mortem diagnosis of infected animals by analysis of brain tissue. PrP^{C} is most highly expressed in the brain and this tissue therefore provides the substrate pool and site for the majority of prion replication and PrP^{Sc} accumulation. PrP^{Sc} is resistant to the enzyme proteinase K (PK) and can be selectively detected by Western Blot or ELISA after PK digestion of brain tissue. It is this PK resistance of PrP^{Sc} that is exploited by all commercially available tests used for TSE's. Live animal surveillance, however, would require methodologies suitable for testing readily accessible tissues such as blood or urine. However, PrP^{C}, and hence PrP^{Sc}, are present in extremely small amounts in body fluids as well as in peripheral tissues such as muscle; this precludes direct testing. Recently novel amplification techniques have been developed that can considerably increase the sensitivity of PrP^{Sc} detection. These are based on the ability of PrP^{Sc} to recruit and catalyze the conversion of a pool of normal PrP^{C}, thus "amplifying" the original infectious seed. Amplification can be increased by breaking down the resulting aggregated seeds of PrP^{Sc} to smaller units which in turn act as seeds for further replication until levels of PrP^{Sc} are detectable by Western blot or ELISA. This methodology is almost analogous to that of the polymerase chain reaction for nucleic acids and the term *protein misfolding cyclic amplification* (PMCA) has been coined to describe it (Barria et al. 2012). These developments may provide the sensitivity necessary for a blood or food screening test useful for some of the transmissible TSEs. One caveat, however, is the recently reported existence of PK-sensitive, pathological isoforms of PrP that may have a significant role in the pathogenesis of some prion diseases (Gambetti et al. 2008). These novel isoforms of PrP^{Sc} have unique biochemical properties including increased sensitivity to PK digestion, and may be generated in sporadic or acquired disease. Given this nuance and the tiny amounts of PrP^{Sc} present in readily sampled tissues, the development of alternative diagnostic tests that do not rely on PK digestion would be useful.

The search for nonprion biomarkers of TSE disease has had a long history, primarily in regard to diagnosis of human cases of Creutzfeldt–Jakob disease, rather than animal TSEs. A biomarker can be any discriminative feature that can be measured objectively and used as an indicator of biological processes such as normal health, pathogenic processes, or pharmacological responses to a therapeutic treatment. Biomarkers include physical traits such as temperature or blood pressure, imaging of pathological features such as amyloid deposition or ventricular volume changes in the brain, and the presence of biological molecules in tissues and body fluids such as blood or urine. Proteomics are particularly suited to finding protein markers that may be indicative of disease and these can subsequently be harnessed in a manner appropriate for high-throughput testing, such as ELISA.

The detection of nonprion protein markers by Western blotting or ELISA of cerebrospinal fluid (CSF) from patients is routinely used in the diagnosis of CJD. One of these proteins is a 30-kDa polypeptide that belongs to the 14–3–3 family of proteins, which is a normal neuronal protein that is released into the CSF after neuronal insult. Its detection, in conjunction with clinical indicators of prion disease, is widely used as diagnostic evidence of sCJD (sporadic CJD), iCJD (iatrogenic SJD), and the genetic form of CJD (Blennow et al. 2005). It is, however, much less sensitive for diagnosis of vCJD (variant CJD), GSS (Gerstmann-Sträussler-Scheinker syndrome), and FFI (fatal familial insomnia) (Green et al. 2001). Other proteins are also increased in the CSF of CJD patients including Tau and phospho-Tau, S-100ß, and neuron-specific enolase (NSE). Levels of the tau protein are raised in patients affected by all forms of CJD including vCJD (Sanchez-Juan et al. 2006). Given the utility of these CSF biomarkers in confirmatory diagnostics of CJD cases it is reasonable to assume that the use of the latest sensitive proteomic tools will uncover new biomarkers for human TSEs. In animals, however, CSF is not a convenient tissue for testing. Any biomarker to have comprehensive value for testing for prions within human food would ideally be detectable in animal muscle, blood, or urine, and be present even at preclinical stages of infection.

This review focuses on efforts to date to harness proteomic technologies to identify potential diagnostic biomarkers in livestock and animal products as a means of protecting the food supply.

27.2 Detection of PrP^{Sc}

PrP^{Sc} itself is the biomarker most intimately associated with disease and its detection is the cornerstone of all approved confirmatory and screening-based diagnostics currently in use. These tests all rely on the acquisition of PK resistance when the normal host prion protein is refolded into the disease related form PrP^{Sc}. Although this form is biophysically distinct, both PrP^{C} and PrP^{Sc} are covalently identical and therefore cannot be directly identified by methods such as mass spectroscopy. New methods have been developed in the last 5 years that depend on the ability of TSE-associated forms of prion to seed the conversion of their normal counterpart. This PrP^{C} substrate can conveniently be provided as bacterially expressed recombinant PrP. The resulting protease-resistant PrP can be readily detected either by Western blot, or by binding to an amyloid indicator such as the small molecule dye thioflavin T (ThT), which has the advantage of being more readily adapted to a high-throughput assay. A number of methodologies show significant potential as ultrasensitive prion detection methods including variations of the protein misfolding cyclic amplification (PMCA) (Saa et al. 2006; Chen et al. 2010; Atarashi et al. 2007), the amyloid seeding assay (ASA) (Colby et al. 2007) and quaking-induced conversion (QuiC) (Colby et al. 2007; Atarashi et al. 2008; Wilham et al. 2010; Orru et al. 2012). Amplifications of over one billion fold have been reported, enabling the detection of close to a single infectious unit ($\leq 10^{-15}$ μg) of prion.

The real-time quaking induced conversion assay (RT-QuIC) developed by Atarashi et al. includes a high-throughput multiwall plate format and a readout based on ThT amyloid (Atarashi et al. 2011a). One drawback of the ThT detection method is that over time negative control samples can attain positive signals due to the spontaneous formation of amyloid fibrils independent of PrP^{Sc}. In Atarashi's method conditions were selected to minimize this reaction and the method proved highly successful for the detection of sporadic CJD (sCJD) in samples of brain homogenate containing only 1 fg of CJD prion. Subsequently RT-QUIC was used to examine more than 200 CSF samples from patients with sCJD and without sCJD but with other neurodegenerative conditions with over 80% sensitivity and 100% specificity for sCJD, figures higher than obtained by the standard 14–3–3 methods currently in use (Atarashi et al. 2011b).

Further to this a recent study reports the development of eQuIC which appears to substantially improve on this method, increasing sensitivity by several orders of magnitude (Orru et al. 2011). In eQuIC an initial immunoprecipitation step is used to separate prion seeds from blood, plasma, or tissue material that may be inhibitory to the conversion process prior to initiating the reaction. In addition, fresh substrate is introduced during the RT-QuIC step, a method that has previously been shown to enhance conversion (Rubenstein et al. 2010). This method has only been tested with scrapie-infected hamster blood to date and it will be interesting to determine whether eQuIC will increase the sensitivity of diagnosis in sCJD in comparison to Atarashi RT-QuIC study. The accuracy of such tests, if translated to practical screening, could be applicable to pre-mortem detection, the testing of some foodstuffs, and pre-clinical screening; thus precluding the need for nonprion predictive biomarkers, except perhaps in the case of newly described prion etiologies involving proteinase K sensitive PrP (Gambetti et al. 2008).

27.3 Nonprion Protein Biomarkers

Mass spectrometry-based proteomics technology is a burgeoning field for the detection of biomarkers predictive of numerous diseases including neurodegenerative diseases. Given the public health concerns regarding prions and the need for pre-clinical test development it is perhaps surprisingly under-represented in the literature regarding TSE pathobiology. The main reason for this is undoubtedly the requirement for strict biosafety conditions when analyzing prion-contaminated samples, which precludes the use of expensive mass spectrometry equipment primarily located in core facilities. The majority of published studies are therefore not performed on state-of-the-art equipment but using methodologies that are affordable and practical in a containment laboratory.

Cerebrospinal fluid (CSF) as an anatomical component to the CNS is accordingly considered as perhaps the most sensitive window for evaluating alterations in the CNS proteome during disease. Given the close association with brain tissue as the location for the majority of prion replication there may also be a greater potential for CSF to contain biomarkers specific to prion replication in comparison to some other peripheral tissues. CSF constituents are secreted by both active and passive transport systems from the choroid plexuses in brain ventricles or via drainage of the interstitial liquid of the nervous tissues. Proteins and polypeptides that pass through the blood–brain barrier are found here as well as secreted biologically activated peptides. Therefore alteration in the CSF protein components may reflect the pathological and physiological status of the CNS and be a reservoir for diagnostic markers. It should be noted also that the CSF also contains salts and hormones and other cellular metabolites as a method of active homeostasis from the brain and in addition to proteomic strategies the metabolomic profiling of these CSF constituents is undoubtedly exceedingly promising as a new field for diagnostic biomarker discovery.

CSF protein profiling has been performed in numerous CNS degenerative diseases over the past few years, including a small number of studies on prion diseases, mainly human in origin. As mentioned, the presence of the 14–3–3 protein in CSF is part of the diagnostic criteria for the ante-mortem diagnosis of sCJD, however, it is not 100% specific and can give false-positive results in several neurodegenerative conditions, as well as conspicuous false-negatives in vCJD patients (Green et al. 2001). This has prompted further studies, mainly employing two-dimensional electrophoresis (2DE), to be performed in an attempt to improve the specificity of diagnosis of human prion diseases. In 2002 Choe et al. reported the identification of seven CSF proteins that can distinguish vCJD from sCJD using a 2DE approach, one of these being apolipoprotein E, a protein well known to be implicated in neurodegeneration (Choe et al. 2002). A similar approach used by Sanchez et al. in 2003 identified a potential CJD marker, fatty acid binding protein (FABP), that was increased in eight affected patients (Guillaume et al. 2003). The authors also validated a significant increase in one isoform of the FABP protein in plasma samples specifically in a small group of patients affected with CJD over a control group of patients with other neurodegenerative conditions. These data illustrate the potential usefulness of CSF analysis for translation to a specific blood test.

A subsequent study by the same group used a different technology, SELDI-TOF (surface-enhanced laser desorption/ionization-TOF) spectrometry and identified a protein of 13.4 kDa that was more abundant in the CSF of a small group of CJD patients (Sanchez et al. 2004). Further analysis using cationic exchange chromatography, sodium dodecyl sulfate-polyacrylamide gel electrophoresis (SDS-PAGE), and liquid chromatography-tandem mass spectrometry (LC-MS/MS) revealed this protein to be cystatin C. Immunoblotting further confirmed the significant increase in cystatin C abundance in all eight CJD-affected patients. However, alteration in the abundance of cystatin C has also been identified in studies of other degenerative conditions including Alzheimer's and therefore is likely not to be specific to prion diseases. It is also of interest to note from this paper that the authors found that cystatin C undergoes proteolysis on long-term storage at −20°C which may affect its use as a pre-clinical marker. Given that in prion diseases the numbers of cases are few and far between, it is therefore often the case that clinical material available for proteomic profiling has often been stored for lengthy periods, and freeze-thawed, which should be an important consideration in study design.

In 2006 a second group also used 2DE to profile CSF from six patients in comparison with the same number of healthy controls (Piubelli et al. 2006). In total, 28 polypeptide chains were differentially modulated in the sCJD samples, of which 10 were upregulated, and the remaining 18 being downregulated. Of these, 13 were identified upon digestion and MALDI-TOF, MS analysis. Cystatin C was again one of the proteins determined to be increased in CJD CSF along with 14–3–3 proteins, transferrin, ubiquitin, gelsolin, complement factor/3a, Apoliprotein J, and some unidentified polypeptides. Use of a panel of such proteins may provide increased specificity for diagnosis. Interestingly many, if not all, of the genes mentioned above have been identified as differentially expressed at the gene level in brain tissue of rodents infected with prions. A study by Brown et al. also links the levels of differentially expressed genes to protein levels in the CSF, an observation that suggests that candidate gene panels identified from animal genomic studies could be used for prediction of disease-associated CSF biomarkers (Brown et al. 2004). This type of strategy is advantageous in prion diseases given the fact that prion-infected tissues can generally not be disseminated to mass spectrometry core facilities and validation of gene expression studies by Western blotting is practical for many laboratories.

One caveat of many of the published studies is the use of CSF from healthy subjects as control material for the proteomic comparisons. To determine prion-specific biomarkers for diagnosis it is absolutely necessary to be able to discriminate with a range of neurodegenerative conditions that are undoubtedly much more common in the population. This needs to be addressed in future studies. A recent study published in 2008 by Brechlin et al. does indeed employ a pool of control subjects suffering from a number of neurodegenerative conditions as well as including a larger number of CJD patients than previously reported (Brechlin et al. 2008). The resulting data confirmed the increase in several known surrogate markers of sCJD in the 36 affected patients including 14–3–3 protein, neuron-specific enolase, and lactate dehydrogenase. They also identified a protein with an approximate molecular mass of 85 kDa as a marker for sCJD with high diagnostic specificity and sensitivity

In a study employing the more sophisticated technique of MALDI-TOF (matrix-assisted laser desorption/ionization-TOF) protein profiling, Qualtieri et al. report the detection of 61 peptides differentially expressed in 15 patients affected by sporadic and genetic forms of CJD versus 10 healthy control subjects plus 8 non-CJD relapsing-remitting multiple sclerosis patients (Qualtieri et al. 2010). Among the 61 signals, the three signals that had increased levels and the highest statistical significance were further characterized and the identities of two were determined as thymosin β4 N-acetylated and free ubiquitin (increased abundance of ubiquitin had previously been associated with CJD, (Manaka et al. 1992)). Ubiquitin was also identified in a recently reported study of CJD CSF biomarker analysis using the SELDI-TOF methodology by Steinaker et al. (2010). The validity of thymosin β4 levels as a marker for CJD was further tested by the Qualtieri group in a panel of CSFs from three CJD patients, five healthy subjects, and six non-CJD relapsing-remitting multiple sclerosis samples and determined congruent with the initial analysis (Urso et al. 2010). In a subsequent study reported very recently in 2012, these authors further tested the contribution of CSF thymosin β4 levels to the clinical differentiation of CJD on 21 patients with neuropathologically confirmed CJD; 15 patients with frontotemporal dementia; 18 patients with probable Alzheimer disease; and 9 patients with a rapid-onset progressive dementia (Le Pera et al. 2012). The control group consisted of 25 noncognitively impaired individuals without CJD or dementia. The thymosin β4 cerebrospinal fluid levels were markedly increased in CJD samples compared with frontotemporal cases and patients with Alzheimer disease. A lower significance was observed versus the group with rapid-onset progressive dementia thus by employing a cutoff value of thymosin β4 use of this protein as a CSF biomarker for CJD was estimated at 100% sensitivity with 98.5% specificity.

A novel methodology profiling only proteins or peptides between 1 and 10 KD was published in early 2012 by Cao Chen et al. (2012). CSF from 40 probable sporadic CJD patients, 32 non-CJD patients with dementia, and 17 non-CJD patients without dementia were enriched by magnetic bead-based weak cation exchange chromatography, and trypsin digested and identified by reverse phase-HPLC-QTOF MS/MS. In total, 42, 53, and 47 signals of proteins were identified in the CSF of probable sCJD, non-CJD with dementia and non-CJD without dementia, respectively. Nine CSF proteins were found to be specifically observed in the probable sCJD group and included cystatin C, gelsolin, secreted phosphoprotein, and dikkopf-related protein 3. No further validation of any of the nine potential markers was, however, done in this study but the analysis of cleaved fragments of cellular proteins in CSF, perhaps created by disease-related protease activity, may be a promising strategy to determine biomarkers.

Animal models of prion disease have been little used as a basis for CSF proteomic screens, presumably as this would be a rather impractical tissue for any kind of live-animal screening protocol. However, as the data from Sanchez et al. showed, CSF biomarkers may sometimes be translated to blood as found for the fatty acid binding protein, although this type of cross-validation hasn't been either explored, or reported further, in the literature (Guillaume et al. 2003). Recently, a proof of principle study to identify pre-clinical protein signatures from the CSF of prion-infected animals was found to have reasonable predicative accuracy (Herbst et al. 2009. In this study the combination of MALDI-FTMS, in addition to machine learning for the classification of mass spectral features, was able to identify pre-clinical protein signatures from the CSF of prion-infected animals with reasonable predicative accuracy. CSF was isolated from 21 infected and 22 control hamsters at a time-point when approximately 80% of the expected incubation period had been completed. Peptide profiles were identified and the peaks compared using IonSpec peak picking software; a number of peptide peaks exhibiting differential abundances were identified. A linear support vector machine (SVM) and tenfold leave-one-out cross-validation was used to evaluate the predictive accuracy of the peptide peaks showing the greatest differences in abundance between infected and noninfected hamsters. The predictive accuracy was determined to be 72%; a true positive rate of 73% and false positive rate of 27% using a tenfold leave-one-out cross-validation. However, the identity of these peptides was not resolved in this study. Although specificity was fairly low, as mentioned by Herbst et al., this type of approach combined with a prefractionation step to improve the accuracy of biomarker detection in the range of low-abundance proteins could well result in identification of a panel of markers with diagnostic potential. In this case the small size of hamsters and small volumes of CSF precludes this approach, so larger animal models would be optimal for future studies.

For preclinical testing of animals due to enter the human food chain, blood or urine provide much more accessible tissues for screens than CSF. Blood or plasma is a notoriously difficult tissue for biomarker analysis due to the limited dynamic range of detection and sample complexity. Very few proteomic studies to detect potential biomarkers in food animal species that have the potential to be infected by prions, scrapie in sheep and BSE in cattle, have been performed. Samples from such animals, especially naturally BSE-infected cattle for validation are extremely scarce. Recently Batxelli-Molina et al. reported the determination of proteins differentially expressed at early and late time-points in sheep plasma using SELDI-TOF MS technology (Batxelli-Mollina et al. 2010). A protein signature pattern, which discriminated between early phase scrapie samples and control sera, was identified using a small number of low molecular weight proteins. These biomarkers showed a sensitivity of 87% and specificity of 90% for all the studied sheep in the early stage of the disease. One of these potential biomarkers was identified as a major fragment of the transthyretin monomer. This was validated in a SELDI-TOF MS kinetic study of sera from Syrian hamsters infected by scrapie, by Western blot analysis and ELISA quantitation. Diagnostic signatures of BSE have also been identified in serum of infected cattle by multivariate analysis of infrared spectra in a study by Martin et al., at a sensitivity of 85% and a specificity of 90%. This supports the hypothesis that infection with prion agents leads to specific changes in the molecular content of serum, however, similar proteomic data have not yet been reported (Martin et al. 2004).

One approach to tackling the obstacle of complexity within blood or plasma is to target a specific fraction of proteins that intuitively may be involved in the disease process. In one such study published in 2001 glycoproteins were specifically targeted for analysis (Wei et al. 2011). Glycosylation is altered in cells infected with prions that may indicate perturbations in the general glycosylation machinery of cells during prion replication (Rudd et al. 1999). This may in turn lead to glycoprotein profile changes within the cell resulting in a targeted pool of disease-specific biomarkers. Wei et al. used lectin affinity chromatography and multidimensional separations to enrich and isolate low abundance glycoproteins. The relative quantitation of a panel of proteins was obtained by a combination of isotopic labeling and validated by spectral counting. Overall 708 proteins were identified, 53 of which showed more than twofold increase in concentration whereas 58 exhibited more than twofold decrease. Western blot validation was performed on serum amyloid protein, suggesting its potential use as a biomarker for prion disease. Given that a number of potential biomarkers identified in previous studies were also picked out as were a significant number of novel proteins, this method shows promise for determining new biomarkers that are potentially disease specific.

Alterations in the pattern of urinary protein excretion have been shown for a multitude of conditions, even those where pathology is restricted to remote organs. Recently, Simon et al. found BSE-induced biomarkers in the urine of infected cattle over the time course of the disease using a combination of 2-D-DIGE and mass spectrometry analysis (Simon et al. 2008). Their results suggest that the differences observed over time were due specifically to changes in the protein concentration in the blood plasma from the diseased state, rather than a general response to renal injury. They identified four classifier proteins that were of significance. Two of these proteins, immunoglobulin Gamma-2 chain C region and clusterin had significantly increased abundance over time. Clusterin discrimination of infected versus noninfected animals was achieved with 100% accuracy over the course of their study. The over-abundance of immunoglobulins has been previously reported in the urine of scrapie-infected hamsters (Serban et al. 2004). However, the study size in these experiments was very small and limited to a single experimental sample group so further validation is required. A second group has also identified urine biomarkers of prion disease using a targeted approach. A list of secreted proteins that were identified as differentially expressed in prion-infected brain tissue from a rodent model was used as the starting point for further validation. One protein in particular, urinary alpha1-antichymotrypsin, was found to be dramatically increased in the urine of patients suffering from sporadic Creutzfeldt–Jakob disease and a number of other animal models of prion disease (Miele et al. 2008).

Table 27.1 Potential biomarkers of prion diseases identified by *Mass spectrometry (MS)* and 2-D-gel electrophoresis

Marker	Fluid	Disease	Identification method	Reference
Alpha1-antichymotrypsin	Urine	CJD	2-DE	Miele et al. (2008)
Cathelicidin antimicrobial peptide (Bos taurus)	Urine	BSE	2-DE	Simon et al. (2008)
Clusterin				
Ig Gamma-2 chain C region (Bos taurus)				
Uroguanylin				
Gelsolin	CSF		2-DE	Brechlin et al. (2008)
Fibrinogen [beta]				
Hp2-[alpha] haptoglobin				
Serotransferrin				
L-lactate dehydrogenase B chain				
Heart-type fatty acid binding protein (H-FABP)	Plasma and CSF	CJD		Guillaume et al. (2003)
Apolipoprotein J	CSF	CJD	2-DE	Piubelli et al. (2006)
Transferrin				
Ubiquitin				Steinaker et al. (2010)
Transthyretin	Serum and CSF	CJD	2-DE	Brechlin et al. (2008)
Thymosin-[beta] 4	CSF	CJD	MALDI-TOF	Urso et al. (2010)
Cystatin C	CSF	CJD	SELDI-TOF 2-DE	Sanchez et al. (2004), Brechlin et al. (2008)

In summary, we provide Table 27.1 as an overview of those nonprion biomarkers that have been identified to date in human and animal biofluids using proteomic methodologies.

27.4 Detection of CNS as Prion Risk Material in Meat Products

Following confirmation of the presence of BSE in domestic cattle populations many countries introduced a requirement that brain and spinal cord, some lymphoid tissue, and part of the intestine have to be removed from cattle destined for the human food chain at slaughter. This ban on so-called specified risk material (SRM) is currently the most important measure to protect against any risk of human ingestion of prion-contaminated meat products, in particular BSE as found in cattle. Development of an analytical method for the detection of SRM contamination in meat products themselves would be beneficial. The detection of CNS tissue is the major focus of this work as these tissues contain the highest titer of prions. In addition, to develop a comprehensive test for all SRMs, which by definition contain a complex mix of tissues restricted to animals of certain ages and species, would likely be impossible. A proteomics approach to the detection of biomarkers to identify these tissues is one option, although this hasn't been explored as yet. A number of protein markers have already been tested in this respect, however. These include neuron-specific enolase, myelin basic protein, synaptophysin, and glial acidic protein (GFAP) as well as several brain-specific fatty acids. Recently a sensitive immunopolymerase chain reaction assay for sensitive detection of GFAP protein has been developed and used to perform a marker survey on 687 sausage samples (Weigel et al. 2010). At least 11 of these samples tested positive by the GFAP test for the presence of contaminating CNS material. However, translation of this type of methodology to widespread testing is beset by practical difficulties due to meat processing. For example heat treatment, which severely compromises the detectability of CNS protein markers. To date, an analytical approach using GC/MS is the only procedure that appears to perform quantitatively and at present may be the best methodology for CNS contaminant detection in heat-treated meat products (Lucker et al. 2010). Lucker's data determined that the undeclared addition of CNS tissue in the preparation of meat products is fairly rare, however, the development of further sensitive tests may be warranted.

27.5 Conclusion

Considerable advances have been made in regard to the detection of prions in biological materials in recent years. Most significant are the rapid strides in increasing the sensitivity of methodologies based on PrP^{Sc} seeding, such as PMCA using recombinant prion substrates, and the QuIC, RT-QuIC, and eQuIC assays. These provide practical, sensitive, and quantitative methods to detect prion seeding in multiple sample types including tissues such as muscle, and dilute fluids, even those containing inhibitors such as blood plasma. Signal amplifications of a billion fold promise highly specific detection of infectious prions with sensitivities pushing closer to 100%. Less progress has been made in the development of predictive tests based on identification of nonprion biomarkers despite significant advances in the sensitivity of proteomics and prefractionation methods. This is due to a number of reasons; not least the requirement for prion samples to be treated according to specific biosafety provisos based upon the inability to decontaminate infectious material by conventional chemical and physical procedures. This precludes their transfer to core facilities for analysis by state-of-the-art expensive mass spectrometers. In addition there are few samples from clinical cases available for validatory studies and experimental studies on food animals such as cattle and sheep are particularly expensive and lengthy. Studies in rodents are therefore useful, but it is unclear whether markers detected in such studies can be translated to larger animals.

The complexity and etiological heterogeneity of TSEs along with their unique molecular mechanisms pose huge challenges to understanding their biology and for the identification of common ante-mortem biomarkers. TSEs transmitted by digestion, such as BSE, characteristically involve initial replication of the agent in the periphery prior to neuroinvasion and progression of the disease. The implication of this diversity in the distribution of infectivity within the host during pre-clinical stages of disease, and between the different forms and strains of prions, is that diagnostic biomarkers will unlikely be inclusive for all TSEs. The maintenance of the ban on specified risk material from foods destined for human consumption is therefore. Strategies for the detection of biomarkers to indicate the contamination of foods by SRMs may well present one of the most beneficial applications of proteomics to the prion food safety issue. An interesting extension of this issue is the recent finding using proteomic profiling that prion protein copurifies with urine-derived gonadotrophins that are extracted from urine for the treatment of infertility. These data suggest that if the urine donor was a pre-clinical TSE patient then infectious prions could conceivably be concentrated in the urine preparations and be a source of iatrogenic transmission (Vab Dorsselaer et al. 2011). This identifies a use for proteomics in yet another context in relation to prion disease, as a proactive approach to identify and mitigate potential risks for prion transmission.

TSE biomarker studies remain an active area of research and the development of direct detection methods for PrPSc is proceeding at a rapid rate. It is, however, clearly evident that a significant amount of further work needs to be conducted before TSE preclinical diagnostic testing of blood and urine samples as well as post-mortem testing of foods becomes a reality. Primary objectives for such studies will be to determine the disease stage at which prions can be detected in a particular tissue, and the specificity of this detection. The combination of a number of different types of test will undoubtedly be required to ensure the maintenance of a prion-free food supply.

References

Atarashi R, Moore RA, Sim VL, Hughson AG, Dorward DW, Onwubiko HA, Priola SA, Caughey B (2007) Ultrasensitive detection of scrapie prion protein using seeded conversion of recombinant prion protein. Nat Methods 4:645–650

Atarashi R, Wilham JM, Christensen L, Hughson AG, Moore RA, Johnson LM, Onwubiko HA, Priola SA, Caughey B (2008) Simplified ultrasensitive prion detection by recombinant PrP conversion with shaking. Nat Methods 5:211–212

Atarashi R, Sano K, Satoh K, Nishida N (2011a) Real-time quaking-induced conversion: a highly sensitive assay for prion detection. Prion 5:150–153

Atarashi R, Satoh K, Sano K, Fuse T, Yamaguchi N, Ishibashi D, Matsubara T, Nakagaki T, Yamanaka H, Shirabe S, Yamada M, Mizusawa H, Kitamoto T, Klug G, McGlade A, Collins SJ, Nishida N (2011b) Ultrasensitive human prion detection in cerebrospinal fluid by real-time quaking-induced conversion. Nat Med 17:175–178

Barria MA, Gonzalez-Romero D, Soto C (2012) Cyclic amplification of prion protein misfolding. Methods Mol Biol 849:199–212

Batxelli-Mollina I, Salvetat N, Andreoletti O, Guerrier L, Vicat G, Molina F, Mourton-Gilles C (2010) Ovine serum biomarkers of early and late phase scrapie. BMC Vet Res 6:49

Belay ED, Maddox RA, Williams ES et al (2004) Chronic wasting disease and potential transmission to humans. Emerg Infect Dis 10:977–984

Bishop MT, Hart P, Aitchison L et al (2006) Predicting susceptibility and incubation time of human-to-human transmission of vCJD. Lancet Neurol 5:393–398

Blennow K, Johansson A, Zetterberg H (2005) Diagnostic value of 14-3-3beta immunoblot and T-tau/P-tau ratio in clinically suspected Creutzfeldt-Jakob disease. Int J Mol Med 16:1147–1149

Brechlin P, Jahn O, Steinacker P, Cepek L, Kratzin H, Lehnert S, Jesse S, Mollenhauer B, Kretzschmar HA, Wiltfang J, Otto M (2008) Cerebrospinal fluid-optimized two-dimensional difference gel electrophoresis (2-D DIGE) facilitates the differential diagnosis of Creutzfeldt-Jakob disease. Proteomics 8:4357–4366

Brown AR, Webb J, Rebus S, Williams A, Fazakerley JK (2004) Identification of up-regulated genes by array analysis in scrapie-infected mouse brains. Neuropathol Appl Neurobiol 30:555–567

Chen B, Morales R, Barria MA, Soto C (2010) Estimating prion concentration in fluids and tissues by quantitative PMCA. Nat Methods 7:519–520

Chen C, Xiao D, Zhou W, Zhang YC, Shi Q, Tian C, Zhang J, Zhou CX, Zhang JZ, Dong XP (2012) Comparative peptidome analyses of the profiles of the peptides ranging from 1–10 KD in CSF samples pooled from probable sporadic CJD and non-CJD patients. Prion 6(1):46–51 [epub ahead of print]

Choe LH, Green A, Knight RS, Thompson EJ, Lee KH (2002) Apolipoprotein E and other cerebrospinal fluid proteins differentiate ante mortem variant Creutzfeldt-Jakob disease from ante mortem sporadic Creutzfeldt-Jakob disease. Electrophoresis 23:2242–2246

Colby DW, Zhang Q, Wang S, Groth D, Legname G, Riesner S, Prusiner SB (2007) Prion detection by an amyloid seeding assay. Proc Natl Acad Sci U S A 105:1774

Gambetti P, Dong Z, Yuan J et al (2008) A novel human disease with abnormal prion protein sensitive to protease. Ann Neurol 63:697–708

Green AJ, Thompson EJ, Stewart GE et al (2001) Use of 14-3-3 and other brain-specific proteins in CSF in the diagnosis of variant Creutzfeldt-Jakob disease. J Neurol Neurosurg Psychiatry 70:744–748

Guillaume E, Zimmermann C, Burkhard PR, Hochstrasser DF, Sanchez JC (2003) A potential cerebrospinal fluid and plasmatic marker for the diagnosis of Creutzfeldt-Jakob disease. Proteomics 3:1495–1499

Herbst A, McIlwain S, Schmidt JJ et al (2009) Prion disease diagnosis by proteomic profiling. J Proteome Res 8:1030–1036

Le Pera M, Urso E, Sprovieri T, Bossio S, Aguglia U, Cupidi C, Ferraro T, Gambardella A, Qualtieri A, Quattrone A (2012) Contribution of cereprospinal fluid thymosin β4 levels to the clinical differentiation of Ceutzfeldt-Jakob disease. Arch Neurol 69(7):868–872

Llewelyn CA, Hewitt PE, Knight RS et al (2004) Possible transmission of variant Creutzfeldt-Jakob disease by blood transfusion. Lancet 363:417–421

Lucker E, Biedermann W, Alter T, Hensel A (2010) GC/MS detection of central nervous tissue as specified BSE risk material in meat products and meat and bone meals: thermal stability of markers in comparison with immunochemistry and RT-PCR. Anal Bioanal Chem 398:963–972

Manaka H, Kato T, Kurita K, Katagiri T, Shikama Y, Kujirai K, Kawanami T, Suzuki Y, Nihei K, Sasaki H et al (1992) Marked increase in cerebrospinal fluid ubiquitin in Creutzfeldt-Jakob disease. Neurosci Lett 139:47–49

Martin TC, Moecks J, Belooussov A, Cawthraw S, Dolenko B, Eiden M, Von Frese J, Kohler W, Schmitt J, Somorjai R, Udelhoven T, Verzakov S, Petrich W (2004) Classification of signatures of Bovine Spongiform Encephalopathy in serum using infrared spectroscopy. Analyst 129:897–901

Miele G, Seeger H, Marino D, Eberhard R, Heikenwalder M, Stoeck K, Basagni M, Knight R, Green A, Chianini F, Wüthrich RP, Hock C, Zerr I, Aguzzi A (2008) Urinary alpha1-antichymotrypsin: a biomarker of prion infection. PLoS One 3(12):e3870

Orru CD, Wilham JM, Raymond LD, Kuhn F, Schroeder B, Raeber AJ, Caughey B (2011) Prion disease blood test using immunoprecipitation and improved quaking–induced conversion. MBio 2:e00078–11

Orru CD, Wilham JM, Vascellari S, Hughson AG, Caughey B (2012) New generation QuIC assays for prion seeding activity. Prion 6(2):147–152 [epub ahead of print]

Piubelli C, Fiorini M, Zanusso G et al (2006) Searching for markers of Creutzfeldt-Jakob disease in cerebrospinal fluid by two-dimensional mapping. Proteomics 6(Suppl 1):S256–S261

Prusiner SB (1998) Prions. Proc Natl Acad Sci USA 95:13363–13383

Qualtieri A, Urso E, Le Pera M, Sprovieri T, Bossio S, Gambardella A, Quattrone A (2010) Proteomic profiling of cerebrospinal fluid in Creutzfeldt-Jakob disease. Expert Rev Proteomics 7:907–917

Raymond GJ, Bossers A, Raymond LD et al (2000) Evidence of a molecular barrier limiting susceptibility of humans, cattle and sheep to chronic wasting disease. EMBO J 19:4425–4430

Rubenstein R, Chang B, Gray P, Piltch M, Bulgin MS, Sorensen-Melson S, Miller MW (2010) A novel method for preclinical detection of PrPSc in blood. J Gen Virol 91:1883–1892

Rudd PM, Endo T, Colominas C, Groth D, Wheeler SF, Harvey DJ, Wormald MR, Serban H, Prusiner SB, Kobata A, Dwek RA (1999) Glycosylation differences between the normal and pathogenenic prion protein isoforms. Proc Natl Acad Sci U S A 96:13044–13049

Saa P, Castilla J, Soto C (2006) Ultra-efficient replication of infectious prions by automated protein misfolding cyclic amplification. J Biol Chem 281:35245–35252

Sanchez JC, Guillaume E, Lescuyer P, Allard L, Carrette O, Scherl A, Burgess J, Corthals GL, Burkhard PR, Hochstrasser DF (2004) Cystatin C as a potential cerebrospinal fluid marker for the diagnosis of Creutzfeldt-Jakob disease. Proteomics 4:2229–2233

Sanchez-Juan P, Green A, Ladogana A et al (2006) CSF tests in the differential diagnosis of Creutzfeldt-Jakob disease. Neurology 67:637–643

Serban A, Legname G, Hansen K, Kovaleva N, Prusiner SB (2004) Immunoglobulins in urine of hamsters with scrapie. J Biol Chem 279:48817–48820

Simon SLR, Lamoureux L, Plews M et al (2008) The identification of disease-induced biomarkers in the urine of BSE infected cattle. Proteome Sci 6:23

Steinaker P, Rist W, Swiatek-de-Lange M, Lehnert S, Jesse S, Pabst A, Tumani H, von Arnim CA, Mitrova E, Kretzschmar HA, Lenter M, Wiltfang J, Otto M (2010) Ubiquitin as a potential cerebrospinal fluid marker of Creutzfeldt-Jakob disease. Proteomics 10:81–89

Urso E, Le Pera M, Bossio S, Sprovieri T, Qualtieri A (2010) Quantification of thymosin beta(4) in human cerebrospinal fluid using matrix-assisted laser desorption/ionization time-of-flight mass spectrometry. Anal Biochem 402:13–19

Vab Dorsselaer A, Carapito C, Delalande F, Schaeffer-Reiss C, Thierse D, Diemer H, McNair DS, Krewski D, Cashman NR (2011) Detection of prion protein in urine-derived injectable fertility products by a targeted proteomic approach. PLoS One 6:e17815

Wei X, Herbst A, Ma D, Aiken J, Li L (2011) A quantitative proteomic approach to prion disease biomarker research: delving into the glycoproteome. J Proteome Res 10:2687–2702

Weigel I, Schulze G, Pischetsrieder M (2010) Immunochemical detection of tissue from the central nervous system via myelin proteolipid protein: adaptation for food inspection and development of recombinant bivalent Fab mini-antibodies. J Agric Food Chem 58:6587–6593

Wilham JM, Orru CD, Bessen RA, Atarashi R, Sano K, Race B, Meade-White KD, Taubner LM, Timmes A, Caughey B (2010) Rapid end-point quantitation of prion seeding activity with sensitivity comparable to bioassays. PLoS Pathog 6:e1001217

Chapter 28
Proteomics of Filamentous Fungi

Mark W.J. van Passel, Peter J. Schaap, and Leo H. de Graaff

28.1 Introduction

In food production fungi are Jekyll and Hyde. Fungal spoilage is a major cause of food spoilage and as such they represent a health hazard. In contrast fungi have been used for many centuries to produce and preserve food and food ingredients in a way that still today is considered to be safe. The application of fungi and of products derived from fungi in the food industry is nowadays very broad; fungal enzymes are ubiquitous, used in starch processing, in the bakery and brewery, and to produce beverages including wines and in food fermentation. The major classes of fungal food enzymes are proteases, pectinases, amylases, xylanases, lipases, and cellulases. The over-representation of fungal polysaccharide-modifying enzymes as food enzymes is a reflection of the high proportion of plant-derived material in our food products. The application of these enzymes has been reviewed elsewhere (de Vries and Visser 2001; van den Brink and de Vries 2011).

In this chapter we concentrate on the application of the use of proteomics tools in the analysis of proteins that are important in food and food processes. We have made one important restriction by limiting ourselves to filamentous fungi. We do realize in this way we have excluded a very important group of fungi, that is, yeasts, but we think this is beneficial for the clarity and structure of this chapter.

Traditionally many enzymes that are used in food and food production are extracellular enzymes that are secreted by fungi into their environment. Before the development of sophisticated proteomics methods, these culture broths were screened

---

M.W.J. van Passel • P.J. Schaap • L.H. de Graaff (✉)
Systems and Synthetic Biology, Wageningen University,
Dreijenplein 10 6703 HB, Wageningen, The Netherlands
e-mail: Leo.deGraaff@wur.nl

F. Toldrá and L.M.L. Nollet (eds.), *Proteomics in Foods: Principles and Applications*, Food Microbiology and Food Safety 2, DOI 10.1007/978-1-4614-5626-1_28,

in application tests to discover the desired enzyme activities. Upon the development of molecular biology of filamentous fungi, genes encoding enzymes used by the food industry were cloned and a more refined testing and screening could be performed under conditions free of "contaminating" activities. In this aspect the work of Jaap Visser and his coworkers on the cloning of genes encoding pectin lyases and polygalacturonases has been pioneering in this field. On the basis of biochemical data it was expected for both the pectin lyases and the polygalacturonases to find two genes encoding pectin lyase I and II and polygalacturonase I and II, respectively. Screening of an *Aspergillus niger* gene library showed the two protein bands were actually a family of proteins encoded by *pelA*, *pelB*, *pelC*, and *pelD* (Gysler et al. 1990; Harmsen et al. 1990; Kusters-van Someren et al. 1991) and the *pgaI*, *pgaII* and *pgaC* (Bussink et al. 1992). In addition to the endo-acting enzymes *A. niger* produces many exo-polygalacturonases (Martens-Uzunova et al. 2006). This work actually showed the genomic complexity of the encoding genes involved and preluded a decade of search for genes encoding novel enzymes.

With the publication of the *A. niger* genome sequence by Pel et al. (2007) the search for novel enzymes entered the genomics era. Although many new carbohydrate-modifying enzymes could be recognized on the basis of their resemblance to previously characterized genes and many of these new enzymes could be assigned to a particular class of carbohydrate-active enzymes (CAZY) (Cantarel et al. 2009), the elucidation of the specific characteristics of these new enzymes still needs biochemical characterization. In this aspect the availability of many genome sequences of filamentous fungi that are of importance for food industry has its most important impact on the correlation of genes and proteins in proteomics analysis. By using functional annotation and bioinformatics tools a theoretical proteome is inferred from the genome sequence, which is used to screen the peptides against that result from the proteomic analysis.

The increase in importance of proteomics in recent years can be deduced from the increase in the number of publications. As of 2008 there is a strong increase, as illustrated in Fig. 28.1 for Aspergilli, which is paralleled by an increase in the number of citations in the same period. This increase is the result of different developments, for example, the increased number of genome sequences that have become available, the technological developments with respect to the analysis of protein mixtures and the improved bioinformatics tools. The still relatively low number of publications is a reflection of the degree of difficulty in generating the protein samples and interpretation of the results from MS analysis in a biological context.

28.2 Proteomics Methodology

With the introduction of next generation sequencing technologies it is quite economical to sequence the genome of a microorganism. Sequencing and genome assembly is usually followed by structural and functional genome annotation, a two-step process that tries to identify genetic elements on the genome, mostly gene,

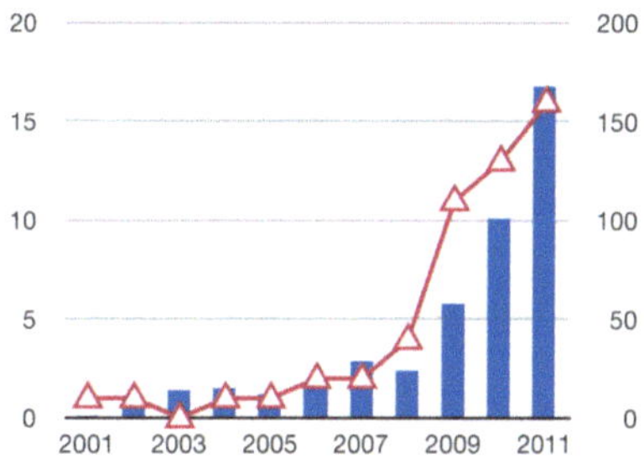

Fig. 28.1 Illustration of the increased importance of proteomics in research on secreted proteins by filamentous fungi. Publication in the Web of Science database were selected by using the keywords "secretome" or "secretomes" or "proteome" or "proteomics" or "proteomic" AND aspergillus

gene structure, and ORF prediction followed by the attachment of functional information to these elements. Automatic annotation tools try to perform both steps computationally resulting in a theoretical proteome. Experimental validation of predicted genes via proteogenomics is currently the best option to identify protein-coding genes (Ansong et al. 2008). Proteogenomics is the parallel analysis of the genome and the proteome.

In such a bottom-up approach, proteins are proteolytic digested with a protease with a strict specificity, for instance, by trypsin prior to analysis by mass spectrometry. With trypsin, cleavage occurs exclusively at the C-terminal to arginine or lysine. Subsequently the masses of proteolytic peptides or their tandem mass spectra are compared with theoretical peptide masses or tandem mass spectra of potentially matching theoretical peptide sequences that conform to the rules of tryptic specificity and are derived from the inferred proteome. Many search engines such as Sequest, Mascot, X!Tandem, OMSSA, and Myrimath have been developed to match the spectra generated by tandem mass spectrometry (MS/MS) with peptides from a target protein sequence database (Eng et al. 1994; Perkins et al. 1999; Craig and Beavis 2004; Geer et al. 2004; Tabb et al. 2007). In this way peptides correlating with a measured mass or with a theoretical spectrum correlating with an experimental tandem mass spectrum can be automatically identified. The identified peptides receive a correlation confidence score and are combined into lists of identified proteins. Multiple correlating peptide identifications scoring above a certain confidence threshold can subsequently be used for protein identification (Aebersold and Mann 2003; Chait 2006). In a technique known as shotgun proteomics this setup is applied to complex or crude protein extracts that are digested directly, followed by separation of the peptides by liquid chromatography coupled with mass spectrometry (Wolters et al. 2001). Because on average every tenth amino acid residue is either an arginine or a lysine, the resulting peptide mixtures tend to be very complex. Sample complexity can be effectively reduced through pre-fractionation of the protein extracts by 1D SDS-PAGE followed by gel slicing and in-gel digestion. Peptides from each slice are subsequently analyzed by MS/MS (Fig. 28.2).

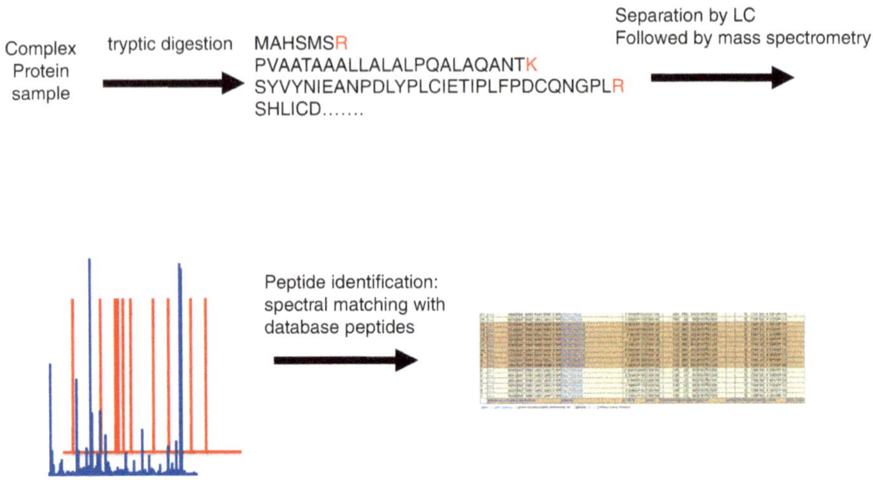

Fig. 28.2 Shotgun proteomics workflow. Sample proteins are directly digested into peptides with a protease of strict specificity such as trypsin and then separated by liquid chromatography directly coupled to a mass spectrometer. Next peptides are sequentially subjected to tandem mass spectrometry analysis resulting in an experimental MS/MS spectrum (*blue*), which is compared with a theoretical MS/MS spectrum of candidate peptides of similar mass (*red*). Possible peptide matches are reported with an associated confidence score

Due to the highly complex nature of peptide samples, automatically assigned peptide spectral matches (PSM) are associated with varying degrees of confidence. This confidence level is the measure for the similarity between the experimental MS/MS spectra of the actual peptide and theoretical spectra of candidate peptides contained in the target database. In proteomics a standard metric for estimation of confidence is the false discovery rate (FDR) (Elias and Gygi 2007). The FDR is the likelihood that a positive identification is a false positive. For example, a 1% FDR indicates that out of 100 PSMs above a set confidence threshold level, one is expected to be a false positive.

FDRs for PSMs are usually derived with a target–decoy database search strategy. In this strategy the search is performed simultaneously against a target protein peptide database and against a decoy peptide database derived from compositionally similar but unrelated protein sequences. The decoy database sequences are usually generated by reversing or randomly reshuffling the target protein database sequences but can also be taken from a pool of unrelated sequences in other organisms.

Proteins are indirectly identified through their peptide matches. Various techniques are available to estimate the FDR for protein identification by considering the decoy matches as an estimate for the amount of false positive PSMs (Nesvizhskii et al. 2003; Ma et al. 2009).

In principle MS-based proteomics can correct genome annotation errors by discovering unannotated novel genes and reading frames and can validate gene functions, and hypothetical open reading frames. In addition MS-based proteomics can facilitate the discovery of important post-translational modifications (Ansong et al. 2008; Armengaud 2009).

The computational strategy of automatic peptide spectrum matching of (shotgun) proteomics data, however, depends on a high-quality theoretical proteome. In other words one needs to know very accurately which peptides one can expect. Although by and large correct, gene-model predictions of filamentous fungi suffer from incorrect exon-identification and incorrect start and termination sites, leading to identification failures despite the acquisition of good-quality tandem mass spectra. Other reasons for not obtaining peptide to spectrum matches can be due to alternative splicing, missed genes or genetic variation, and single nucleotide polymorphisms leading to coding SNP protein isoforms between the investigated and the sequenced strains.

Alternative computational workflows have been developed to overcome the computational bias imposed by the use of a computationally predicted protein sequence database. A simple alternative approach would be a naive six-frame translation of the genome and basic filtering of the obtained hypothetical proteins for the presence of trypsinated peptides of sufficient size. Although this approach is not dependent on the quality of the functional genome annotation, this analysis method may, however, suffer from a high false discovery rate and thus results in low proteomics identification efficiency inasmuch as most of the generated hypothetical peptides used for matching will be nonexistent (Nesvizhskii 2010).

Because mRNA expression is a clear prerequisite for protein expression research groups have started to use mRNA, EST, and also RNA-Seq data to build customized sample-specific protein databases for searches (Desgagne-Penix et al. 2010; Lundberg et al. 2010; Adamidi et al. 2011). Despite their obvious potential, transcriptome-derived customized databases have seen limited use in improving protein identification in fungal shotgun proteomics studies.

28.2.1 Post-Translational Modifications

A second way of not getting a peptide to spectrum match is due to covalent post-translational modification (PTM) of proteins at one or more sites. Two categories exist: post-translational modification due to post-harvest sample treatment such carbamidomethylation of cysteines and naturally occurring enzyme-catalyzed modifications on the side chains or amino acid backbones. The five most common types of naturally occurring enzyme-catalyzed covalent additions to proteins are phosphorylation, acylation, alkylation, glycosylation, and oxidation.

Modifications due to sample treatment are usually treated as fixed modifications and different masses are used for the modified residue(s). For example, selecting carboxymethyl (C) as a fixed modification means that all calculations will use 161 Da as the new mass of cysteine. Naturally occurring enzyme-catalyzed covalent additions to proteins with a known difference in mass are usually treated as variable modificationa. In other words, both peptides with the unmodified and the modified form of the particular amino acid will be put to the test. However, even a single variable modification with a fixed mass difference with the unmodified amino acid will generate many possible additional peptides to be tested, thereby significantly increasing the search space and hence the false discovery rate.

Two types of covalent glycosylation are quite common for secreted proteins in eukaryotes. N-glycosylation of proteins is the most common and is typically more complex in structure and more diverse in architecture than O-glycosylation. The N-linked glycosylation process occurs in the lumen of the endoplasmic reticulum. A pre-assembled core glycan unit usually consisting of three glucose, nine mannose, and two N-acetylglucosamine molecules is covalently attached to the side chain of selected asparagine residues. N-Linked glycosylation usually occurs at Asn residues in the Asn-X-Ser/Thr core sequence motif where X denotes any amino acid except proline in the polypeptide chain of the target protein. Further modification occurs in the Golgi apparatus. Three main subtypes are generated: high-mannose, hybrid, and complex N-glycan. Most fungal secreted N-glycans are of the high-mannose subtype. The precise composition and architecture of the complex N-glycans and therefore the mass difference cannot be predicted and thus a "variable modification" search strategy cannot be directly applied for N-glycosylation. N-Glycosylation, however, can be effectively removed with PNGase F (Koutsioulis et al. 2008). Important for further analysis is that PNGase F digestion also deaminates the asparagine residue to an aspartic acid residue.

O-Mannosylation is also commonly found in fungal glycoproteins. O-Mannosylation has been studied extensively in the unicellular yeast S. cerevisiae. In S. cerevisiae, the oligosaccharides are attached to the protein by a mannosyl residue which is transferred to Ser/Thr residues in the endoplasmic reticulum by protein O-D-mannosyltransferase. The oligosaccharide is then linearly elongated in the Golgi by up to seven mannose residues by mannosyltransferases that utilize guanosine diphosphate mannose (GDP-mannose) as the mannosyl donor for the elongation process (Lussier et al. 1999). Again the precise composition and architecture in filamentous fungi cannot be predicted.

28.3 Comparative Genomics

28.3.1 Comparing Protein Repertoires: The XlnR Regulon

The advent of high-throughput whole genome sequencing techniques has enabled comprehensive gene repertoire evaluations between organisms, simplifying phenotypic characterizations. Comparative analyses have revealed major differences between prokaryotes and eukaryotes in interspecific gene content variations. For example, whereas prokaryotes experience horizontal gene transfer as a dominant force in genome evolution (Dagan and Martin 2007; van Passel et al. 2008), eukaryotes seem to be generally less prone to gene acquisitions (Andersson 2005), and specifically yeasts and filamentous fungi are thought to favor gene or even whole genome duplications, followed by diversification and differential gene loss events (Ochman et al. 2005; Fedorova et al. 2008).

In Aspergilli, the gene repertoires of 11 strains have been elucidated (Galagan et al. 2005; Machida et al. 2005, Pel et al. 2007, Andersen et al. 2011). Interest lies mostly with the citric acid producing *A. niger*, which is also highly relevant in the food industry for its food enzyme production, and the pathogenic *Aspergillus fumigatus*, both of which have multiple fully sequenced representatives. *Aspergillus oryzae* is a species widely recognized for its use in traditional oriental food fermentations (e.g., soy sauce, sake, and vinegar production) (Machida et al. 2005).

Next to gene and genome duplication events, some studies suggest that species-specific genes are still acquired by horizontal gene transfer (Khaldi and Wolfe 2008; Andersen et al. 2011), whereas the analysis of core and lineage-specific elements in *A. fumigatus, Aspergillus clavatus*, and *Neosartorya fischeri* revealed that the most-encountered gene novelty is thought to have originated by duplication, representing potential "gene factories," hotspots of genetic innovation (Fedorova et al. 2008). This teaches us that pursuing the comprehensive reconstructing of Aspergillus genome evolution is particularly precarious due to the range of molecular mechanisms that shape genomes over large time-scales, such as whole genome duplications, massive gene loss, and horizontal gene transfer.

Here we focus on a specific suite of genes encoding xylan backbone-degrading enzymes, their associated accessory enzymes, cellulose-degrading enzymes, and enzymes involved in the pentose catabolic pathway, which are cooperatively involved in the breakdown of plant cell wall material (van Peij et al. 1998b, Stricker et al. 2008; Mach-Aigner et al. 2012). These enzymes coordinately degrade the heterogeneous plant cell wall polysaccharide xylan and cellulose, one of the most abundant natural biopolymers. The XlnR regulon in *A. niger* has been intensively studied, with respect to the range of genes encoding the variety of enzymes (de Vries et al. 1997; Gielkens et al. 1999; Hasper et al. 2000, 2002) and the dedicated regulatory switch XlnR (van Peij et al. 1998a, b), as well as regulation on the level of catabolite repression (de Vries et al. 1999; Mach-Aigner et al. 2012). Some of the genes encoding xylanolytic enzymes have in fact first been discovered in Aspergilli other than *A. niger*, and several comparative analyses have been performed to assess the presence of homologous xylanolytic regulons in a limited number of sequenced Aspergilli (van den Brink and de Vries 2011). With the availability of the full genomes of 11 *Aspergillus* species, we set out to uncover the variations in xylanolytic capacity in all sequenced Aspergilli. The dynamics in gene content of glycosyl hydrolases could identify targets for strain improvement, and further catalogue the array of enzymatic activities that these fungi have at their disposal.

In brief, the proteomes were obtained from NCBI, and the proteins encoded by the *A. niger* XlnR regulon described previously (van Peij et al. 1998b; Mach-Aigner et al. 2012) were used as queries in a reciprocal best blast hit (E-value 10^{-10}) analysis for orthologues. For the detection of putative related auxiliary enzymes, the proteins encoded by the predicted xylanolytic regulons of each strain were used as an internal query with its respective proteome, using cutoff values of 60% amino acid similarity and 80% alignment length. Protein descriptions were taken from NCBI.

First, we find that the xylose-activated regulator XlnR is conserved among all sequenced isolates (Table 28.1), and displays a phylogenetic pattern very similar to that of the entire Aspergillus genus based (Wang et al. 2009), suggesting the regulon is functionally conserved between all isolates over evolutionary time. Next, with respect to the set of 22 proteins described by Mach-Aigner et al. (Mach-Aigner et al. 2012), substantial variations in orthologue distributions are observed, which are in agreement with the relationship between the strains. For example, the 22 orthologues are fully conserved between the two *A. niger* and *A. kawachii* isolates, which are closely related. In contrast, *A. fumigatus* A1163 has lost the a-L-arabinofuranosidase B gene since it split from *A. fumigatus* A293. Overall, the conservation of this xylanolytic regulon varies from 22/22 (in *A. niger* CBS and *A. kawachii*) to 17/22 in *Aspergillus clavatus*.

When using these orthologous proteins as an internal query for additional putative xylan- or cellulose-degrading enzymes, we find between two (*A. clavatus* and *A. nidulans*) and seven (*N. fischeri*) auxiliary enzymes (Table 28.2). Most of these genes fall in the category of "xylan-backbone modifying enzymes" (such as XlnB and XlnC) or the "cellulose-degrading enzymes" (such as CbhA and EglB), with frequent multiple extra copies of genes encoding these enzymes. This could point to a degree of variation in activity on, for example, substrate composition or accessibility. Genes belonging to the categories xylan-accessory enzymes and D-xylose metabolic enzymes seem to be more conserved among these isolates.

These glycosyl hydrolase repertoire variations emphasize the dynamical nature of adaptive processes that may have occurred, perhaps even during the domestication of the important food strains. These protein sets could also be checked for signatures of positive selection based on synonymous and nonsynonymous substitution frequencies between closely related strains. Such analyses could aid in customizing more functional enzymes. More interestingly, these variations in the observed xylanolytic enzymes seem to be primarily due to the endo-acting enzymes (such as the endo xylanases (XlnB) and endo glucanases (EglB)). This suggests that there have been selective duplications in these strains to specialize in different extracellular substrate compositions or conditions. For applications in the food industry, this means that there may be an even larger variety of relevant enzymes encoded in other related Aspergillus strains, potentially expanding the catalytic potential both in specificities as well as activities.

28.4 Proteomics Analysis

Most of the proteomics work on filamentous fungi has concentrated on the analysis of secretomes since these protein mixtures have relatively low complexity. One of the first publications in this field is a proteomics analysis of the secretome of *A. oryzae* after growth on wheat bran (Oda et al. 2006). In this work protein spots on 2DE gels were analyzed and 29 proteins were identified from 85 protein spots from a solid-state fermentation and 110 spots from a submerse fermentation. The proteins

Table 28.1 Distribution of predicted orthologs (with accession numbers) of the genes of the *A. niger* XlnR regulon in 11 sequenced Aspergillus species

Category	Enzyme	*A. flavus*	*A. oryzae*	*A. terreus*	*A. niger ATCC1015*	*A. niger CBS*	*A. kawachii*	*A. fumigatus Af293*	*A. fumigatus A1163*	*N. fischeri*	*A. clavatus*	*A. nidulans*
Xylan regulon regulator	Transcriptional regulator (XlnR)	EED54287	BAE60472	EAU37394	EHA24805	XP_001397110	GAA90473	EAL93859	EDP55066	EAW19127	EAW14420	EAA61796
Xylan backbone-modifying enzymes	Endoxylanase B (XlnB)	EED48943	BAE62665	EAU31723	EHA26077	XP_001388522	GAA92017	EAL86316	EDP53627	EAW16741	EAW12347	EAA66432
	Endoxylanase C (XlnC)	EED44928	BAE65852	EAU31038	EHA20975	XP_001389996	GAA92552	EAL89839	EDP50325	EAW25166	EAW10405	EAA64983
	β-Xylosidase (XlnD)	EED57542	BAE55977	EAU34175	EHA26824	XP_001389416	GAA82061	EAL91022	EDP56909	EAW16964	EAW07155	EAA64470
Xylan accessory enzymes	α-L-Arabinofuranosidase B (AbfB)	EED53056	BAE58593	EAU32201	EHA24585	XP_001396769	GAA90571	EAL89298	*Absent*	EAW16707	EAW11011	EAA64278
	α-Galactosidase B (AglB)	EED55259	BAE58368	*Absent*	EHA23366	XP_001400244	GAA84472	EAL86091	EDP51064	EAW24423	EAW14901	EAA61810
	α-Glucuronidase (AguA)	EED51045	BAE59621	EAU33846	EHA27966	XP_001401203	GAA90668	EAL91181	EDP52172	EAW17337	EAW13280	EAA66353
	Acetyl xylan esterase (AxeA)	EED52855	BAE65196	EAU30034	EHA25274	XP_001395572	GAA86514	EAL85420	EDP48681	EAW20289	EAW06435	EAA66943
	Arabinoxylan arabinofuranohydrolase (AxhA)	EED52086	BAE62533	EAU29520	EHA20976	XP_001389998	GAA92551	EAL87191	EDP53757	EAW23755	EAW12189	EAA62979
	Esterase A (EstA)	EED51751	BAE62240	EAU34085	EHA27365	XP_001396326	GAA86871	EAL87843	EDP49915	EAW23619	*Absent*	EAA64563
	Ferulic acid esterase (FaeA)	EED49792	BAE56750	EAU31039	EHA28107	XP_001393337	GAA93172	*Absent*	*Absent*	*Absent*	*Absent*	EAA62870
	β-Galactosidase (LacA)	EED54458	BAE60622	EAU39262	EHA26981	XP_001389622	GAA86664	EAL90749	EDP56654	EAW22442	EAW07417	EAA65398
Cellulose-degrading enzymes	β-Glucosidase (BglA)	EED47060	BAE54829	EAU36321	EHA19005	XP_001398816	GAA83366	EAL88289	EDP55914	EAW23194	EAW08156	EAA59363
	Cellobiohydrolase A (CbhA)	EED54935	BAE61042	EAU35529	EHA24304	XP_001392008	GAA85531	EAL88562	EDP49274	EAW15926	EAW11196	EAA62357

(continued)

Table 28.1 (continued)

Category	Enzyme	*A. flavus*	*A. oryzae*	*A. terreus*	*A. niger ATCC1015*	*A. niger CBS*	*A. kawachii*	*A. fumigatus Af293*	*A. fumigatus A1163*	*N. fischeri*	*A. clavatus*	*A. nidulans*
	Cellobiohydrolase B (CbhB)	*Absent*	*Absent*	*Absent*	EHA26941	XP_001389576	GAA87307	*Absent*	*Absent*	*Absent*	*Absent*	EAA66593
	Endoglucanase A (EglA)	EED51074	BAE59597	EAU34588	EHA27741	XP_001400902	GAA87033	EAL86857	EDP48456	EAW20396	EAW12022	*Absent*
	Endoglucanase B (EglB)	EED58088	BAE56461	*Absent*	EHA24278	XP_001391969	GAA85499	EAL88602	EDP49318	EAW15969	*Absent*	EAA62395
	Endoglucanase C (EglC)	*Absent*	*Absent*	EAU35155	EHA26208	XP_001388626	GAA91782	EAL85019	EDP49062	EAW19879	EAW13711	EAA60055
D-Xylose metabolism enzymes	L-Arabitol dehydrogenase (LadA)	EED57627	BAE56054	EAU34245	EHA26884	XP_001389509	GAA87367	EAL90955	EDP56840	EAW22238	EAW07224	EAA65971
	Transaldolase B (TalB)	EED46167	BAE63755	EAU33023	EHA23875	XP_001391399	GAA86166	*Absent*	*Absent*	EAW17822	*Absent*	EAA66811
	Xylitol dehydrogenase (XdhA)	EED46119	BAE64390	EAU38639	EHA26001	XP_001395093	GAA91097	EAL84639	EDP48196	EAW20990	EAW10868	EAA61897
	D-Xylulose kinase (XkiA)	EED46161	BAE63761	EAU33016	EHA23873	XP_001391397	GAA86164	EAL91618	EDP51726	EAW17810	EAW12861	EAA60583
	Xylose reductase (XyrA)	EED55688	BAE57985	EAU37586	EHA26354	XP_001388804	GAA88683	EAL88193	EDP55820	EAW23286	EAW08252	EAA66522
Total in genome		**20**	**20**	**19**	**22**	**22**	**22**	**19**	**18**	**20**	**17**	**21**

Table 28.2 Xylanolytic, cellulolytic and accessory enzyme protein repertoires derived from 11 sequenced Aspergillus genomes. Absent means no homologs have been detected, and *highlighted in grey boxes* are enzymes that are encoded by multiple genes

Category	Enzyme	EC no.	GH family no.	*A. flavus*	*A. oryzae*	*A. terreus*	*A. niger* ATCC1015	*A. niger* CBS	*A. kawachii*	*A. fumigatus* Af293	*A. fumigatus* A1163	*N. fischeri*	*A. clavatus*	*A. nidulans*
Xylan regulon regulator	Transcriptional regulator (XlnR)			1	1	1	1	1	1	1	1	1	1	1
Xylan backbone-modifying enzymes	Endoxylanase B (XlnB)	EC 3.2.1.8	11	2	2	2	2	2	2	3	3	3	2	2
	Endoxylanase C (XlnC)	EC 3.2.1.8	10	2	3	2	1	1	1	1	1	1	1	1
	b-Xylosidase (XlnD)	EC 3.2.1.37	3	1	1	1	1	1	1	1	1	1	1	1
Xylan accessory enzymes	a-L-Arabinofuranosidase B (AbfB)	EC 3.2.1.55	54	1	1	1	1	1	1	1	absent	1	1	1
	a-Galactosidase B (AglB)	EC 3.2.1.22	27	1	1	absent	1	1	1	1	1	1	1	1
	a-Glucuronidase (AguA)	EC 3.2.1.139	67	1	1	1	1	1	1	1	1	1	1	1
	Acetyl xylan esterase (AxeA)	EC 3.2.1.72		1	1	1	1	1	1	1	1	1	1	1
	Arabinoxylan arabinofuranohydrolase (AxhA)	EC 3.2.1.55	62	1	1	2	1	1	1	1	1	1	1	2
	Esterase A (EstA)	-		1	1	1	1	1	1	1	1	1	absent	1
	Ferulic acid esterase (FaeA)	EC 3.1.1.73		1	1	1	1	1	1	absent	absent	absent	absent	1
	b-Galactosidase (LacA)	EC 3.2.1.23	35	1	1	1	1	1	1	1	1	1	1	1
Cellulose-degrading enzymes	b-Glucosidase (BglA)	EC 3.2.1.21	3	1	1	1	1	1	1	1	1	1	1	1
	Cellobiohydrolase A (CbhA)	EC 3.2.1.91	7	2	2	2	1	1	1	2	2	3	2	1
	Cellobiohydrolase B (CbhB)	EC 3.2.1.91	7	absent	absent	absent	1	1	1	absent	absent	absent	absent	1
	Endoglucanase A (EglA)	EC 3.2.1.4	2	1	1	1	1	1	1	1	1	1	1	absent
	Endoglucanase B (EglB)	EC 3.2.1.4	5	2	2	absent	3	3	3	3	3	4	absent	1
	Endoglucanase C (EglC)	EC 3.2.1.4	74	absent	absent	1	1	1	1	1	1	1	1	1
D-Xylose metabolism enzymes	L-Arabitol dehydrogenase (LadA)	EC 1.1.1.12		1	1	1	1	1	1	1	1	1	1	1
	Transaldolase B (TalB)	EC 2.2.1.2		1	1	1	1	1	1	absent	absent	1	absent	1
	Xylitol dehydrogenase (XdhA)	EC 1.1.1.9		1	2	1	1	1	1	1	1	1	1	1
	D-Xylulose kinase (XkiA)	EC 2.7.1.17		1	1	1	1	1	1	1	1	1	1	1
	Xylose reductase (XyrA)	EC 1.1.1.21		1	1	1	1	1	1	1	1	1	1	1
Total in genome				24	26	23	25	25	25	24	23	27	19	23

identified are relevant enzymes for the food industry such as α-amylase, glucoamylase, β-glucosidase, xylanase, and protease including alanyl dipeptidyl peptidase.

A number of studies have analyzed the *A. niger* secretome under conditions of polysaccharidase production. These studies have in common a combined analytical and computational approach exploiting the genome sequence to select gene models that encode proteins that have a signal peptide (Tsang et al. 2009; Braaksma et al. 2010; de Oliveira et al. 2011). In their study Tsang et al. could identify 222 secreted proteins from the analyzed culture broths. Two different methods' predictive tools were used to predict 691 secreted proteins encoded by the *A. niger* ATTC1015 genome; 23% of these 691 proteins were identified as secreted proteins. These predictions depend to a high degree on the correctness of the gene models in the annotated genome. Braaksma et al. 2010 used the genome sequences available for *A. niger* ssp. and closely related Aspergilli such as a.o. *A. oryzae,* to cross-validate and improve the quality of the predictions of signal sequences. With these improved methods they have analyzed the secretome of *A. niger* under conditions of pectinase production and of carbon starvation and were able to identify 209 secreted proteins. Moreover, they were able to identify 40 additional proteins in the datasets of Tsang et al. (2009) and Lu et al. (2010), indicative of the superiority of this approach.

Secretome analysis as described above has not only validated the expression and secretion of proteins that are of great importance in the food industry, but also has contributed to the expression and secretion of novel enzymes under specific growth conditions. An iTRAQ-based (quantitative) analysis revealed that many proteins relevant for food and food processing could be detected in the secretome of *A. niger* when grown on 50 mM D-glucose for 144 h (Adav et al. 2010). Although not specifically induced, many polysaccharidases and peptidases were found using these conditions, probably resulting from derepression due to carbon starvation. In addition to many characterized enzymes, such as, for example, pectinases, α-L-arabinofuranosidase (Flipphi et al. 1993a, b, c), and α-L-rhamnosidase (Manzanares et al. 1997), the authors were able to show that in many genes that were annotated as encoding hypothetical enzymes, the corresponding proteins indeed were found in the fermentation broth.

In a more specific approach, Lu et al. (2010), Ferreira de Oliveira et al. (2010), and de Oliveira et al. (2011) studied the induction of enzymes by D-xylose and D-maltose in *A. niger.* D-xylose induces the hemicellulase encoding genes of the XlnR regulon (van Peij et al. 1998a, b). D-maltose induces the amylolytic enzymes of the AmyR regulon, whereas D-xylose acts as a repressor. Whereas Lu et al. (2010) analyzed selected spots from 2DE, Ferreira de Oliveira et al. (2010) used a shotgun approach for the analysis of the secretome. Despite these technical differences in the approach of the analysis of the secretome; the outcome of both studies was essentially the same. In D-xylose-induced conditions the enzyme mixture consists mostly of the xylanolytic enzymes of the XlnR regulon, and when induced with D-maltose glucoamylase is dominantly present in the secretome, representing over 50% of the extracellular proteome. Xylanolytic enzymes were completely absent under D-maltose induction conditions. In addition to these polysaccharidases enzymes such as sulfhydryl oxidase (Sox) and catalase (Cat) were detected.

Both the studies of Lu et al. (2010) and Ferreira de Oliveira et al. (2010) have analyzed the changes of the intracellular proteome upon changes in the induction regime. Lu et al. (2010) studied the proteins in a cleared total cell free extract. Their results show a major difference in the intracellular proteome in mycelium from shake flask and bioreactor grown cultures. The differences between these two growth conditions directly relate to the availability of oxygen. Bioreactor grown mycelium is rich in glycolytic enzymes and enzymes involved in protection against oxygen stress, while these are present in far lower amounts or are even absent in shake flask grown mycelium.

In addition to these metabolic enzymes, bioreactor-grown mycelium contained many chaperones and foldases. In the study of Ferreira de Oliveira et al. (2010) specifically, the proteins involved in secretion of the extracellular enzymes were studied by a proteomic analysis of enriched microsomal fractions from D-xylose and D-maltose induced conditions. In this analysis more than 1,000 proteins were identified of which 25% were predicted to be involved in the secretion process. This study confirms that D-maltose and D-xylose each induce the secretion of specific extracellular enzymes. Moreover, their data show that induction by D-xylose and D-maltose also results in the association of the 20S core of the proteasome with secretory organelles. This suggests that the recruitment of the proteasome may be a general feature of the shift to a secretion state of the cell.

The differences in the intracellular proteome of shake flask-grown mycelium versus bioreactor-grown mycelium as was shown by Lu et al. (2010) underpins the importance of controlled growth conditions in the study of these basic processes. Essentially this study shows the intracellular proteome largely depends on the amount of oxygen available during growth. Secretion-related proteins such as chaperones and foldases are present in larger amounts in bioreactor cultures, resulting in higher amounts of extracellular proteins.

28.5 Conclusions and Future Prospects

We have given an overview here of the recent developments of proteomics analysis applied to fungi that are of major importance in the production of food enzymes. Most of the studies have been performed using *A. niger* reflecting the dominance of this organism in food biotechnology.

The proteomics analysis of complex protein mixtures goes through a process of rapid development; analytical techniques become more sophisticated and easier to apply. This combined with the availability of much more and also more reliable genome data and more advanced bioinformatics tools, will lead to a strong increase in the application of proteomics analysis as an aid in strain improvement and in understanding fermentation phenotypes. As we discussed, proteome data will help to improve gene models, which contributes to the availability of more reliable genome data. This combined with more user-friendly bioinformatics tools, contributes to a wider application of proteomics studies.

The availability of many more genome sequences, also from fungi, allows comparative genomic and proteomic studies. The comparative studies contribute to a better understanding of the recruitment or loss of genetic information by fungi as part of their adaption process or specialization. Also these studies provide insight to the generation of variation of enzymes in organisms and between organisms. This not only contributes to a better understanding of the biological processes, but also enlarges the resources that can be applied in food biotechnology.

The results of these recent studies have shown that the proteomics analysis of extracellular proteins of fungi, that are mostly enzymes involved in food processes, provide valuable additional information in addition to the data derived from genome sequences. The conditional expression of specific enzymes can be studied in this way and gives leads for the function of particular proteins. However, it is very important to note that specific characterization of the protein is needed to determine its activity and mode of action. The characterization of these novel enzymes is still a challenging task to perform. Developments in the field of synthetic biology, easing the process of high-level expression of proteins, combined with high-throughput analysis of the protein activity will contribute to ease the characterization.

References

Adamidi C, Wang Y, Gruen D et al (2011) *De novo* assembly and validation of planaria transcriptome by massive parallel sequencing and shotgun proteomics. Genome Res 21:1193–1200

Adav SS, Li AA, Manavalan A, Punt P, Sze SK (2010) Quantitative iTRAQ secretome analysis of *Aspergillus niger* reveals novel hydrolytic enzymes. J Proteome Res 9:3932–3940

Aebersold R, Mann M (2003) Mass spectrometry-based proteomics. Nature 422:198–207

Andersen MR, Salazar MP, Schaap PJ et al (2011) Comparative genomics of citric-acid-producing *Aspergillus niger* ATCC 1015 versus enzyme-producing CBS 513.88. Genome Res 21: 885–897

Andersson JO (2005) Lateral gene transfer in eukaryotes. Cell Mol Life Sci 62:1182–1197

Ansong C, Purvine SO, Adkins JN, Lipton MS, Smith RD (2008) Proteogenomics: needs and roles to be filled by proteomics in genome annotation. Brief Funct Genomic Proteomic 7:50–62

Armengaud J (2009) A perfect genome annotation is within reach with the proteomics and genomics alliance. Curr Opin Microbiol 12:292–300

Braaksma M, Martens-Uzunova ES, Punt PJ, Schaap PJ (2010) An inventory of the *Aspergillus niger* secretome by combining in silico predictions with shotgun proteomics data. BMC Genomics 11:584

Bussink HJ, Buxton FP, Fraaye BA, de Graaff LH, Visser J (1992) The polygalacturonases of *Aspergillus niger* are encoded by a family of diverged genes. Eur J Biochem/FEBS 208:83–90

Cantarel BL, Coutinho PM, Rancurel C, Bernard T, Lombard V, Henrissat B (2009) The Carbohydrate-Active EnZymes database (CAZy): an expert resource for glycogenomics. Nucleic Acids Res 37:D233–D238

Chait BT (2006) Mass spectrometry: bottom-up or top-down? Science 314:65–66

Craig R, Beavis RC (2004) TANDEM: matching proteins with tandem mass spectra. Bioinformatics 20:1466–1467

Dagan T, Martin W (2007) Ancestral genome sizes specify the minimum rate of lateral gene transfer during prokaryote evolution. Proc Natl Acad Sci USA 104:870–875

de Oliveira JM, van Passel MW, Schaap PJ, de Graaff LH (2011) Proteomic analysis of the secretory response of *Aspergillus niger* to D-maltose and D-xylose. PLoS One 6:e20865
de Vries RP, Visser J (2001) Aspergillus enzymes involved in degradation of plant cell wall polysaccharides. Microbiol Mol Biol Rev: MMBR 65:497–522, table of contents
de Vries RP, Michelsen B, Poulsen CH et al (1997) The *faeA* genes from *Aspergillus niger* and *Aspergillus tubingensis* encode ferulic acid esterases involved in degradation of complex cell wall polysaccharides. Appl Environ Microbiol 63:4638–4644
de Vries RP, Visser J, de Graaff LH (1999) CreA modulates the XlnR-induced expression on xylose of *Aspergillus niger* genes involved in xylan degradation. Res Microbiol 150:281–285
Desgagne-Penix I, Khan M, Schriemer D, Cram D, Nowak J, Facchini P (2010) Integration of deep transcriptome and proteome analyses reveals the components of alkaloid metabolism in opium poppy cell cultures. BMC Plant Biol 10:252
Elias JE, Gygi SP (2007) Target-decoy search strategy for increased confidence in large-scale protein identifications by mass spectrometry. Nat Methods 4:207–214
Eng JK, McCormack AL, Yates Iii JR (1994) An approach to correlate tandem mass spectral data of peptides with amino acid sequences in a protein database. J Am Soc Mass Spectrom 5: 976–989
Fedorova ND, Khaldi N, Joardar VS et al (2008) Genomic islands in the pathogenic filamentous fungus *Aspergillus fumigatus*. PLoS Genet 4:e1000046
Ferreira de Oliveira JM, van Passel MW, Schaap PJ, de Graaff LH (2010) Shotgun proteomics of *Aspergillus niger* microsomes upon D-xylose induction. Appl Environ Microbiol 76:4421–4429
Flipphi MJ, Visser J, van der Veen P, de Graaff LH (1993a) Cloning of the *Aspergillus niger* gene encoding alpha-L-arabinofuranosidase A. Appl Microbiol Biotechnol 39:335–340
Flipphi MJ, van Heuvel M, van der Veen P, Visser J, de Graaff LH (1993b) Cloning and characterization of the *abfB* gene coding for the major alpha-L-arabinofuranosidase (ABF B) of *Aspergillus niger*. Curr Genet 24:525–532
Flipphi MJ, Panneman H, van der Veen P, Visser J, de Graaff LH (1993c) Molecular cloning, expression and structure of the endo-1,5-alpha-L-arabinase gene of *Aspergillus niger*. Appl Microbiol Biotechnol 40:318–326
Galagan JE, Calvo SE, Cuomo C et al (2005) Sequencing of *Aspergillus nidulans* and comparative analysis with *A. fumigatus* and *A. oryzae*. Nature 438:1105–1115
Geer LY, Markey SP, Kowalak JA et al (2004) Open mass spectrometry search algorithm. J Proteome Res 3:958–964
Gielkens MM, Dekkers E, Visser J, de Graaff LH (1999) Two cellobiohydrolase-encoding genes from *Aspergillus niger* require D-xylose and the xylanolytic transcriptional activator XlnR for their expression. Appl Environ Microbiol 65:4340–4345
Gysler C, Harmsen JA, Kester HC, Visser J, Heim J (1990) Isolation and structure of the pectin lyase D-encoding gene from *Aspergillus niger*. Gene 89:101–108
Harmsen JA, Kusters-van Someren MA, Visser J (1990) Cloning and expression of a second *Aspergillus niger* pectin lyase gene (*pelA*): indications of a pectin lyase gene family in *A. niger*. Curr Genet 18:161–166
Hasper AA, Visser J, de Graaff LH (2000) The *Aspergillus niger* transcriptional activator XlnR, which is involved in the degradation of the polysaccharides xylan and cellulose, also regulates D-xylose reductase gene expression. Mol Microbiol 36:193–200
Hasper AA, Dekkers E, van Mil M, van de Vondervoort PJ, de Graaff LH (2002) EglC, a new endoglucanase from *Aspergillus niger* with major activity towards xyloglucan. Appl Environ Microbiol 68:1556–1560
Khaldi N, Wolfe KH (2008) Elusive origins of the extra genes in *Aspergillus oryzae*. PLoS One 3:e3036
Koutsioulis D, Landry D, Guthrie EP (2008) Novel endo-Œ±–N-acetylgalactosaminidases with broader substrate specificity. Glycobiology 18:799–805
Kusters-van Someren MA, Harmsen JA, Kester HC, Visser J (1991) Structure of the Aspergillus niger pelA gene and its expression in *Aspergillus niger* and *Aspergillus nidulans*. Curr Genet 20:293–299

Lu X, Sun J, Nimtz M, Wissing J, Zeng AP, Rinas U (2010) The intra- and extracellular proteome of *Aspergillus niger* growing on defined medium with xylose or maltose as carbon substrate. Microb Cell Fact 9:23

Lundberg E, Fagerberg L, Klevebring D et al (2010) Defining the transcriptome and proteome in three functionally different human cell lines. Mol Syst Biol 6:450

Lussier M, Sdicu A-M, Bussey H (1999) The KTR and MNN1 mannosyltransferase families of *Saccharomyces cerevisiae*. Biochimica et Biophysica Acta (BBA) – General Subjects 1426:323–334

Ma Z-Q, Dasari S, Chambers MC et al (2009) IDPicker 2.0: improved protein assembly with high discrimination peptide identification filtering. J Proteome Res 8:3872–3881

Mach-Aigner AR, Omony J, Jovanovic B, van Boxtel AJ, de Graaff LH (2012) d-Xylose concentration-dependent hydrolase expression profiles and the function of CreA and XlnR in *Aspergillus niger*. Appl Environ Microbiol 78:3145–3155

Machida M, Asai K, Sano M et al (2005) Genome sequencing and analysis of *Aspergillus oryzae*. Nature 438:1157–1161

Manzanares P, de Graaff LH, Visser J (1997) Purification and characterization of an alpha-L-rhamnosidase from *Aspergillus niger*. FEMS Microbiol Lett 157:279–283

Martens-Uzunova ES, Zandleven JS, Benen JA et al (2006) A new group of exo-acting family 28 glycoside hydrolases of *Aspergillus niger* that are involved in pectin degradation. Biochem J 400:43–52

Nesvizhskii AI (2010) A survey of computational methods and error rate estimation procedures for peptide and protein identification in shotgun proteomics. J Proteomics 73:2092–2123

Nesvizhskii AI, Keller A, Kolker E, Aebersold R (2003) A statistical model for identifying proteins by tandem mass spectrometry. Anal Chem 75:4646–4658

Ochman H, Daubin V, Lerat E (2005) A bunch of fun-guys: the whole-genome view of yeast evolution. Trends Genet 21:1–3

Oda K, Kakizono D, Yamada O, Iefuji H, Akita O, Iwashita K (2006) Proteomic analysis of extracellular proteins from *Aspergillus oryzae* grown under submerged and solid-state culture conditions. Appl Environ Microbiol 72:3448–3457

Pel HJ, de Winde JH, Archer DB et al (2007) Genome sequencing and analysis of the versatile cell factory *Aspergillus niger* CBS 513.88. Nat Biotechnol 25:221–231

Perkins DN, Pappin DJC, Creasy DM, Cottrell JS (1999) Probability-based protein identification by searching sequence databases using mass spectrometry data. Electrophoresis 20:3551–3567

Stricker AR, Mach RL, de Graaff LH (2008) Regulation of transcription of cellulases- and hemicellulases-encoding genes in *Aspergillus niger* and *Hypocrea jecorina* (*Trichoderma reesei*). Appl Microbiol Biotechnol 78:211–220

Tabb DL, Fernando CG, Chambers MC (2007) MyriMatch: Äâ highly accurate tandem mass spectral peptide identification by multivariate hypergeometric analysis. J Proteome Res 6:654–661

Tsang A, Butler G, Powlowski J, Panisko EA, Baker SE (2009) Analytical and computational approaches to define the *Aspergillus niger* secretome. Fungal Genet Biol: FG and B 46(Suppl 1):S153–S160

van den Brink J, de Vries RP (2011) Fungal enzyme sets for plant polysaccharide degradation. Appl Microbiol Biotechnol 91:1477–1492

van Passel MW, Marri PR, Ochman H (2008) The emergence and fate of horizontally acquired genes in *Escherichia coli*. PLoS Comput Biol 4:e1000059

van Peij NN, Visser J, de Graaff LH (1998a) Isolation and analysis of *xlnR*, encoding a transcriptional activator co-ordinating xylanolytic expression in Aspergillus niger. Mol Microbiol 27:131–142

van Peij NN, Gielkens MM, de Vries RP, Visser J, de Graaff LH (1998b) The transcriptional activator XlnR regulates both xylanolytic and endoglucanase gene expression in *Aspergillus niger*. Appl Environ Microbiol 64:3615–3619

Wang H, Xu Z, Gao L, Hao B (2009) A fungal phylogeny based on 82 complete genomes using the composition vector method. BMC Evol Biol 9:195

Wolters DA, Washburn MP, Yates JR (2001) An automated multidimensional protein identification technology for shotgun proteomics. Anal Chem 73:5683–5690

Index

F. Toldrá and L.M.L. Nollet (eds.), *Proteomics in Foods: Principles and Applications*, Food Microbiology and Food Safety 2, DOI 10.1007/978-1-4614-5626-1,

C

N

W

X

Y

MIX
Papier aus verantwortungsvollen Quellen
Paper from responsible sources
FSC® C105338
www.fsc.org

If you have any concerns about our products,
you can contact us on
ProductSafety@springernature.com

In case Publisher is established outside the EU,
the EU authorized representative is:
Springer Nature Customer Service Center GmbH
Europaplatz 3, 69115 Heidelberg, Germany

Printed by Libri Plureos GmbH
in Hamburg, Germany